PHYSICS OF ENGINEERING MATERIALS

Daniel D. Pollock

Department of Mechanical and Aerospace Engineering
State University of New York at Buffalo

PRENTICE HALL

Englewood Cliffs, New Jersey 07632

Library of Congress Cataloging-in-Publication Data

Pollock, Daniel D.
   Physics of engineering materials/Daniel D. Pollock.
     p. cm.
   Includes bibliographies and index.
   ISBN 0-13-669326-1
     1. Solids. 2. Materials. 3. Semiconductors. 4. Quantum theory.
5. Thermoelectricity. I. Title.
QC176.P65   1990
530.4'1--dc19

88-36995
CIP

Editorial/production supervision
  and interior design: GERTRUDE SZYFERBLATT
Cover design: LUNDGREN GRAPHICS, LTD.
Manufacturing buyer: MARY ANN GLORIANDE

Printed in the United States of America

10 9 8 7 6 5 4 3 2 1

ISBN   0-13-669326-1

PRENTICE-HALL INTERNATIONAL (UK) LIMITED, *London*
PRENTICE-HALL OF AUSTRALIA PTY. LIMITED, *Sydney*
PRENTICE-HALL CANADA INC., *Toronto*
PRENTICE-HALL HISPANOAMERICANA, S.A., *Mexico*
PRENTICE-HALL OF INDIA PRIVATE LIMITED, *New Delhi*
PRENTICE-HALL OF JAPAN, INC., *Tokyo*
SIMON & SCHUSTER ASIA PTE., *Singapore*
EDITORA PRENTICE-HALL DO BRASIL, LTDA., *Rio de Janeiro*

CONTENTS

10 MAGNETISM   524

PREFACE

The fantastic growth in discoveries of new materials, new phenomena, and new theoretical understandings makes it imperative that engineers be prepared to comprehend and utilize them. This book is intended to introduce the basic concepts underlying the physical properties of materials in ways that minimize the necessity for extensive backgrounds in physics and mathematics. Maximum access thus is provided to the understanding of those properties of materials of greatest importance to engineering students and to practicing engineers.

The introductory chapters are designed to acquaint the reader with quantum mechanics in a logical way. This is accomplished by demonstrating the limitations of classical mechanics, at a level familiar to most students and engineers, and the consequent need for better explanations. Quantum mechanics, unfamiliar to most engineers, is then derived and developed in uncomplicated ways. This is done only to the extent required for the understanding of the topics that follow; it also constitutes a sound background for more advanced study.

The presentations of the various topics have been made as reader-friendly as possible. The physics and mathematics are those at the least complicated levels consistent with clear explanations; the results are neither oversimplified nor distorted. All the "algebra" is shown, and no portions of the presentations are treated as being obvious to the reader.

The physical units used are those currently employed by engineers involved with the topic under discussion. The use of a single system of units would be counterproductive. Means are provided for the conversion of units to other systems.

The assumption is made that the reader will have the equivalent of first-year, college-level calculus, physics, chemistry, and either physical metallurgy or materials science. Information beyond these levels is provided and explained where needed. In some cases, alternate explanations are given. This has been done to offer the reader different ways of examining a topic, to assist those with different backgrounds, and to minimize the necessity for recourse to other sources. This also facilitates self-study. Bibliographies are provided for those desiring additional information or for explanations beyond the scope of this book.

Introductory material is provided to guide the reader into a topic under discussion. Fundamental physical relationships, selected from concepts familiar to those with engineering backgrounds, form the bases for the discussions that follow. Brief reviews of related information, presented elsewhere in the book, also are provided. These are intended to emphasize interrelationships and provide continuity of ideas.

The first four chapters constitute the foundation upon which the topics that follow are built; they demonstrate some of the limitations of the classical theory of solids, show the necessity for statistical treatments, explain the required statistics, and develop modern theories of solids. The six chapters that follow are unusual in that they are structured as units that are based on broad topics, rather than as conventional chapters. This has been done to emphasize unifying concepts and interrelationships. Many of the subsections within these topical groupings are given in sufficient detail to justify their inclusion as chapters in more conventionally structured books. Chapter 7 is a good example. The various ways of examining thermoelectricity also illustrate the versatility of quantum mechanics. This is intended to emphasize some of the different approaches that may be used to explain phenomena or solve problems. Various selections from these six chapters or their subsections may be studied depending on the needs or interests of the reader or on the emphasis desired for a given course. Sufficient material is available to tailor courses for engineering students of different disciplines.

Practicing engineers will find this book useful in acquiring information on topics of interest. As a text for college-level courses, this book provides elementary, uncomplicated, theoretical presentations and their applications to real materials that are based on over forty years of wide academic and engineering experience. Readers with some familiarity with this field will find this book easy to read.

ACKNOWLEDGMENTS

I wish to express my deep appreciation to Professor C. W. Curtis, of Lehigh University, for the insights and practical approaches to solid-state phenomena that he provided early in my career.

My debts for the editorial assistance of Dr. T. Charalampopoulos, of Louisiana State University, and Mr. Bill Thomas are acknowledged with gratitude.

The assistance of Bonnie Boskat in typing the manuscript is noted with deep appreciation.

Above all, homage must be paid to *She-Who-Must-Be-Obeyed* for her cooperation and forbearance during the preparation of this book.

Daniel D. Pollock
Williamsville, New York

1

THE NECESSITY FOR
THE QUANTUM THEORY

The purpose of this introductory material is to show the necessity for the development of the quantum theory. The historical approach is used to describe the inability of the classical theory to explain selected physical phenomena. The demonstrations of the needs for modern physics in the solutions of these difficulties provide means for the introduction and explanation of some of the more fundamental concepts. This is done to the extent required here for introductory purposes.

1.1 THE DRUDE–LORENTZ (CLASSICAL) THEORY OF METALS

The first attempt to show that the properties of metals could be explained by their valence electrons was made by Drude (1900). Drude's ideas were modified later (1916) by Lorentz. Their combined ideas now are known as the Drude–Lorentz theory.

Valence electrons were considered to behave within a metal in a way corresponding to gas molecules in a container, obeying the same laws. This implies that the electrons are free, have a continuous energy distribution, and can be described by the classical, Maxwell–Boltzmann statistics.

When a voltage (potential difference) is applied to a metallic conductor, *all* the valence electrons were considered to be affected by it, in accordance with Lentz's law, and, in so doing, create an electric current.

Such uninhibited freedom of valence electrons converts the metal atoms into ions because their original electrical neutrality no longer exists; they now have unbalanced positive charges. The positive metal ions, oscillating about fixed lattice positions, are considered to be immersed in an "electron gas" in which the negative charges on the highly mobile electrons that constitute this gas exactly balance the positive charges on the relatively immobile ions. Therefore, it is necessary to consider only the responses of the valence electrons when this theory is applied to electrical conduction in metals.

The voltage, V, is treated in terms of the electric field, $\bar{E}_x$, that it creates: $\bar{E}_x = V/\text{length}$, where the length is the distance over which the voltage exists within the conductor. The effect of the field is to provide an additional small velocity component to each valence electron that is parallel to the field. This increment is called the drift velocity, $\mathbf{v}_d$. This velocity component is in addition to the velocity of each electron in the absence of a voltage at a given temperature. The latter velocity is the thermal velocity, $\mathbf{u}$.

A valence electron under the influence of $\bar{E}_x$ will acquire a $\mathbf{v}_d$ until it "collides" with an ion in the lattice. The energy equivalent of $\mathbf{v}_d$ is entirely given up to that ion. (The energy equivalent of $\mathbf{v}_d$ is $E_d = \frac{1}{2}m\mathbf{v}_d^2$, where m is the mass of the electron.) This causes the ion to increase the amplitude of its oscillation. At the instant that an electron transfers this energy to an ion, $\mathbf{u}$ remains unchanged, but $\mathbf{v}_d = 0$. The electron then continues to move in the lattice, in a random direction, because its thermal velocity is considered to be unaffected by the field. The electron simultaneously acquires another drift velocity as this takes place; it travels through the lattice until another "collision" occurs. Electron "collisions" of this kind are called lattice scatterings. This process is repeated as long as the voltage is maintained. *All* the valence electrons were considered to engage in this process. Thus, the electric current consists of the effects of the sum of all the drift velocities of *all* the electrons for the time that the voltage is present. The energies, E_d, transferred by all the electrons to the ions can provide a means for explaining the increase in the temperature of a conductor as the current continues to flow.

The Drude–Lorentz analysis of this behavior starts with the effect of the electric field on the acceleration, a, of the electron. This is shown by beginning with Newton's second law:

$$F = ma = m\frac{d\mathbf{v}}{dt}$$

or

$$a = \frac{F}{m} = \frac{d\mathbf{v}}{dt}$$

where F is the force acting on an electron, m is its mass, and $\mathbf{v}$ is its velocity at a time, t. The force may be expressed as $\bar{E}_x e$, where e is the charge on an electron. Thus, by substituting this quantity for F in the preceding equation,

$$\frac{d\mathbf{v}}{dt} = \frac{F}{m} = \frac{\bar{E}_x e}{m}$$

The integration of this equation with respect to time gives the drift velocity as

$$\mathbf{v} = \frac{\bar{E}_x e}{m} t$$

But the drift velocity of the electron is zero at the instant of collision and increases with time to $\mathbf{v}$ at the instant prior to its subsequent collision. It is convenient to approximate the average of these two velocities as being the drift velocity, $\mathbf{v}_d$. Thus, the drift velocity of a valence electron may be approximated by its average, or

$$\mathbf{v}_d \simeq \frac{\mathbf{v}}{2} \simeq \frac{\bar{E}_x e}{2m} t$$

The use of an average drift velocity implies that the electron will travel an average distance, or mean free path, L, between collisions. Thus, using the vectorial sum of the thermal and drift velocities,

$$\mathbf{u} + \mathbf{v}_d \simeq \frac{L}{t}$$

This equation is rearranged to obtain an expression for the time between collisions in order to simplify the preceding equation for $\mathbf{v}_d$. This gives the time between collisions as

$$t \simeq \frac{L}{\mathbf{u} + \mathbf{v}_d}$$

This may be simplified further by recognizing that the thermal velocity, $\mathbf{u}$, is very much greater than the drift velocity, $\mathbf{v}_d$. Thus, the approximation may be made that

$$t \simeq \frac{L}{\mathbf{u}}$$

And, upon substitution of this expression for t, the equation for $\mathbf{v}_d$ becomes

$$\mathbf{v}_d \simeq \frac{\bar{E}_x e}{2m} \cdot \frac{L}{\mathbf{u}}$$

The equation for $\mathbf{v}_d$ now is in a form that can be applied to the electrical properties of interest here. The current density, j, is given by

$$j = ne\mathbf{v}_d$$

in which n is the number of valence electrons per unit volume of the conductor. The substitution of the expression for $\mathbf{v}_d$ into the equation for j gives

$$j = ne\mathbf{v}_d \simeq ne \, \frac{\bar{E}_x eL}{2m\mathbf{u}} = \frac{ne^2 \bar{E}_x L}{2m\mathbf{u}}$$

This expression for j now is substituted into the equation for Ohm's law to obtain

$$j = \frac{\bar{E}_x}{\rho} \simeq \frac{ne^2 \bar{E}_x L}{2m\mathbf{u}}$$

where ρ is the electrical resistivity. This relationship is solved for ρ, and noting that $\bar{E}_x$ vanishes, it is found that

$$\rho \simeq \frac{2m\mathbf{u}}{ne^2 L} = \frac{1}{\sigma}$$

where the electrical conductivity, σ, is shown as the reciprocal of the resistivity.

The temperature dependence of ρ is approximated by equating two classical expressions for energy:

$$E = \frac{m\mathbf{u}^2}{2} = \frac{3k_B T}{2}$$

where k_B is Boltzmann's constant and T is the absolute temperature. This relationship is solved to give the thermal velocity as

$$\mathbf{u} \simeq \left(\frac{3k_b T}{m}\right)^{1/2}$$

The expression for the thermal velocity is substituted into the equation for ρ to obtain the Drude–Lorentz equation for the electrical resistivity of a metal as

$$\rho \simeq \frac{2m}{ne^2 L} \left(\frac{3k_B T}{m}\right)^{1/2}$$

Thus, according to the Drude–Lorentz theory, the electrical resistivity is expected to vary as $T^{1/2}$. It is of interest to note that if the statistical average velocity had been used instead of $\mathbf{v}_d = \mathbf{v}/2$, a factor of $3\pi/8$ would replace the factor $\frac{1}{2}$ in this equation. However, this discrepancy is small considering the other approximations and the incorrect temperature dependency of ρ that is obtained.

The experimentally determined temperature dependencies of the normal metals at and above room temperature are virtually linear functions of temperature; ρ varies directly as T. The Drude–Lorentz analysis, thus, gives an erroneous $T^{1/2}$ dependency. A tantalizing fact regarding this is that the equation derived here does provide reasonable values for ρ near room temperature, despite its incorrect prediction of the temperature variation.

This theory also attempts to explain the heat capacity of a metal (the heat required to raise 1 gram-atomic weight of the metal 1 K) by treating the oscillating ions of a metal, in addition to the free electrons, as though both species were gas molecules. Using the theory of ideal gases, the average energy of a single ion is

$$\bar{U}_i = \text{KE} + \text{PE} = \tfrac{3}{2}k_B T + \tfrac{3}{2}k_B T = \tfrac{6}{2}k_B T \quad \text{(cal/ion)}$$

The energy of a gram-atomic weight of the metal is obtained by multiplying both sides of the equation for $\bar{U}_i$ by Avogadro's number, N_A, to obtain

$$\bar{U}_I = N_A \bar{U}_i = \tfrac{6}{2} N_A k_B T \quad \text{(cal/mol)}$$

The product $N_A k_B = R$, the ideal gas constant, so that the molar energy of the ions is

$$\bar{U}_I = \tfrac{6}{2} RT \quad \text{(cal/mol)}$$

The value of $R = 1.987$ (cal/mol)/K $\simeq 2$ (cal/mol)/K. The substitution of this quantity into the equation for $\bar{U}_I$ gives that portion of the internal energy of the solid residing in its oscillating atoms as

$$\bar{U}_I \simeq \tfrac{6}{2} \cdot 2T \cong 6T \quad \text{(cal/mol)}$$

The heat capacity is defined as $d\bar{U}/dT$, so the heat capacity of the ions is

$$C_{v,I} = \frac{dU_I}{dT} \simeq \frac{d}{dT}(6T)$$

Then the indicated operation gives the heat capacity of the ions as

$$C_{v,I} \simeq 6 \quad \text{(cal/mol)/K}$$

The energy content in the free valence electrons is treated in a similar way to that shown for the ions. Starting with the average energy of a valence electron, considered as an ideal gas molecule, its energy is given by

$$U_e = \tfrac{3}{2}k_B T \quad \text{(cal/electron)}$$

And the average energy of a mole of such electrons is, where Z is the valence of the metal,

$$\bar{U}_E = N_A \bar{U}_e = \tfrac{3}{2} Z N_A k_B T = \tfrac{3}{2} ZRT \simeq 3ZT \quad \text{(cal/mol)}$$

and the molar heat capacity component of the electrons is

$$C_{v,E} = 3Z$$

The heat capacity of the metal is given by the sum of the contributions of the ions and electrons, or

$$C_v = C_{v,I} + C_{v,E} = 6 + 3Z \quad \text{(cal/mol)/deg}$$

It had been shown by Dulong and Petit (1818) that the heat capacity of many metallic elements was approximately 6 (cal/mol)/deg. However, when the Drude–Lorentz model is used to calculate the heat capacity of the given metal, it yields a result that is at least 50% too large. (The valence, Z, may be 1, 2, or 3 for a normal metal.) This error results from the assumption that *all* the valence electrons are involved in these physical processes. Attempts to vary the number of electrons so that the heat capacity calculations give reasonable results always cause the calculated values of the electrical resistivity to be too large. This model does not give consistent results; it, evidently, contains an inherent error. Its inconsistency has been termed the Drude–Lorentz dilemma. It was not to be resolved until quantum mechanics were employed to explain and eliminate these problems.

1.2 BLACK-BODY RADIATION

Planck's radical approach to the explanation of black-body radiation marked the beginning of the quantum theory. It clearly demonstrated the limitations of the classical theory in dealing with electromagnetic radiation and provided a new approach for the solution of other phenomena that the prior theories could not explain.

Solids and liquids give off light, visible electromagnetic radiation, upon suitable heating. The energy of this radiation varies with the temperature and its spectrum is continuous. Visible radiation becomes apparent when substances are heated to above about 700°C (Fig. 1-1). Invisible radiation, at lower energies in the infrared range, is produced at lower temperatures. These emanations result from thermally activated oscillations of the atoms composing these substances.

An ideal black body is one that absorbs all incident radiation. It has zero reflectivity. All the radiation emitted by such a body has its origin within it. Thus, black-body radiation is a measure of the energy of the radiating substance. The thermally induced spectra are virtually the same for all substances at a given temperature; these spectra approach that of the black body. Such spectra are continuous because the atoms are not independent oscillators, but interact with each other. (The light spectra of gases, such as neon, are discontinuous because the atoms show little interaction.)

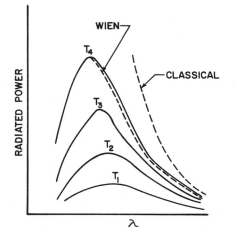

Figure 1-1 Radiated power as a function of wavelength for temperatures $T_1 < T_2 < T_3 < T_4$. (Modified from F. K. Richtmyer, E. H. Kennard, and T. Lauritsen, *Introduction to Modern Physics*, 5th ed., McGraw-Hill Book Co., New York, 1955.)

The basis for understanding black-body radiation resides in the way in which the radiated energy is related to its wavelength, or to its frequency, and to the temperature of the substance from which it emanates. The approach of classical physics to this problem may be represented by the Rayleigh–Jeans equation:

$$E(\lambda) = \frac{8\pi k_B T}{\lambda^4} \tag{1-1}$$

Here $E(\lambda)$ is the energy per unit volume of the heated substance that is radiated in the range of wavelengths between λ and $\lambda + d\lambda$, k_B is Boltzmann's constant, and T is the absolute temperature. It will be seen (Fig. 1-1) that this equation approaches agreement with experimental data only at very long wavelengths. It is apparent that classical physics is unable to explain the observed behavior at any of the shorter wavelengths.

Wien showed that the wavelength at which the maximum energy radiation occurred could be described by

$$\lambda_{\max} = \frac{\text{constant}}{T} \tag{1-2}$$

This inverse relationship, known as the *Wien displacement law*, is essentially that which had long been utilized by metal workers. Increases in heating cause metals to change their colors over a range from a barely visible dull red to a condition known as "white-hot". This is one of the bases for modern optical pyrometry.

By means of classical thermodynamics, Wien was able to show that black-body radiation as a function of wavelength and temperature could be expressed by an equation of the form of

$$E(\lambda) = \lambda^{-5} f(\lambda, T) \tag{1-3}$$

His application of classical electromagnetic theory to a solid composed of molecular-sized oscillators permitted the reexpression of Eq. (1-3) as

$$E(\lambda) = \frac{C_1}{\lambda^5} \cdot \frac{1}{e^{C_2/\lambda T}} \tag{1-4}$$

where C_1 and C_2 are undefined constants. This equation provides excellent agreement with the observed data from short wavelengths up to those in the neighborhood of the maximum radiated energy. At longer wavelengths, the radiated energies as calculated by means of Eq. (1-4) become smaller than the experimentally observed values. Thus, Eq. (1-4) fails to represent the observed phenomena over the entire spectrum.

One example of other empirical attempts to express these phenomena is given by the more general equation, in which a, b, c, and d are adjustable parameters:

$$E(\lambda) = a T^{5-b} \lambda^{-b} \exp[-c/(\lambda T)^d]$$

It will be observed that this equation reduces to Eq. (1-1) for the case where $b = 4$ and $c = 0$. The case in which $b = 5$ and $d = 1$ is that given by Eq. (1-4). Thus, this equation can be made to agree with the Rayleigh–Jeans or the Wien equations by the selection of arbitrary values for the parameters a, b, c, and d. As such, it does provide some unification of the prior work, but it is incapable of adding to the understanding of any mechanisms involved in black-body radiation.

1.3 THE PLANCK THEORY OF RADIATION

Planck's work (1901) was undertaken in an effort to eliminate the deficiencies of the classical Rayleigh and Wien approaches. He assumed that the radiating substance was composed of electric dipoles that acted as simple harmonic oscillators. Classical physics is based on the concept that such oscillators may absorb and emit energy in a continuous way; no restrictions are placed on their energies. Planck's departure from this idea constitutes one of the bases of modern physics. His assumptions are as follows:

1. The energy of an oscillator must be discrete. It cannot have a continuous energy spectrum. This is expressed as

$$E = nh\nu \qquad (1\text{-}5)$$

   Here n is an integer, h is a constant of proportionality, equal to 6.6256×10^{-34} J·s, now known as Planck's constant, and ν is the frequency of the oscillation. The integral values of n cause the energy to be in multiples of $h\nu$ and this results in a discrete energy spectrum. This discontinuous behavior represents an important fundamental innovation in contrast to the classical theory, which permits an oscillator take on any energy in conformance with a continuous spectrum.

2. Therefore, any changes in energy must be discontinuous. Such changes in energy may result either from the emission or the absorption of discrete amounts, or quanta, of energy.

These assumptions were taken into account to obtain the average energy of an oscillator. The product of the number of oscillators per unit volume and the average energy of an oscillator is given by

$$E(\lambda) = \frac{8\pi}{\lambda^4} \cdot \frac{h\nu}{e^{h\nu/k_B T} - 1} \qquad (1\text{-}6)$$

When use is made of the relationship $\nu = c/\lambda$, where c is the speed of light, Eq. (1-6) may be reexpressed as

$$E(\lambda) = \frac{8\pi ch}{\lambda^5} \cdot \frac{1}{e^{ch/\lambda k_B T} - 1} \qquad (1\text{-}7)$$

Equation (1-7) is *Planck's radiation law.* It provides excellent agreement with the observed data over the entire spectrum of wavelengths. It also is in agreement with the fundamental thermodynamics as given in a general way by Eq. (1-3).

For small values of λ, $\exp(ch/\lambda k_B T) \gg 1$, so Eq. (1-7) may be closely approximated by

$$E(\lambda) \simeq \frac{8\pi ch}{\lambda^5} \cdot \frac{1}{e^{ch/\lambda k_B T}} \qquad (1\text{-}8)$$

The similarity between Eqs. (1-8) and (1-4) will be noted. The constants of the Wien equation, Eq. (1-4), now are defined by Eq. (1-8) as

$$C_1 = 8\pi ch \qquad (1\text{-}8a)$$

and

$$C_2 = \frac{ch}{k_B} \qquad (1\text{-}8b)$$

At the other extreme, that is for long λ, the exponential term in the denominator of Eq. (1-7) may be approximated from the first two terms of the series

$$e^x \simeq 1 + x + \frac{x^2}{2!} + \frac{x^3}{3!} + \cdots$$

where $x = ch/\lambda k_B T$. Thus, the denominator of the second factor of Eq. (1-7) may be written as

$$1 + \frac{ch}{\lambda k_B T} - 1 = \frac{ch}{\lambda k_B T} \qquad (1\text{-}9)$$

The substitution of Eq. (1-9) into Eq. (1-7) results in

$$E(\lambda) \simeq \frac{8\pi ch}{\lambda^5} \cdot \frac{\lambda k_B T}{ch} = \frac{8\pi k_B T}{\lambda^4} \qquad (1\text{-}10)$$

Equation (1-10) is the same as the classical equation, Eq. (1-1). Thus, Planck not only was able to eliminate the difficulties encountered by his predecessors, but he was also able to include their findings as limiting cases of his theory.

Classical physics is incapable of providing an analysis that would give the results obtained by Planck [Eq. (1-6) or (1-7)]. The limitations of the classical approach are most clearly represented by Wien's findings [Eq. (1-4)] as being the best description that may be obtained from this theory. In contrast, Planck's radiation law conforms very closely to the observed data over the entire spectrum. This was made possible only by his radical new concept: the energies of oscillators must be discrete and vary with their frequencies; they cannot vary continuously as decreed by classical thought. Planck's "quantum leap of imagination" was to have far-reaching consequences and became one of the foundations of the new physics.

1.4 EINSTEIN'S EXPLANATION OF THE PHOTOELECTRIC EFFECT

It had been shown by Hertz (1887) that electrodes that were irradiated by ultraviolet light from nearby electric arcs could be made to arc at lower voltages than they would in the absence of such irradiation. It is now known that such irradiation can cause electrons to be ejected from the surface of a material. When a suitable voltage (potential difference) exists between a pair of electrodes, the radiation-generated electrons emitted by one electrode will flow to the other electrode. Lower voltages are required to strike an arc between the electrodes when such an externally induced electron flow takes place between them. The behavior observed by Hertz is a manifestation of the photoelectric effect.

This phenomenon can be explained most simply by assuming that the electrode that is emitting the electrons is a metal. Use now may be made of the "electrons-in-a-box" model, which is shown in Fig. 1-2. The potential barriers at the surfaces of the metal are such that the volume of the metal constitutes a lower energy condition in which the electrons move freely. The electrons, thus, do not normally leave the metal unless they have their energies increased at least by an amount equal to W. W is known as the *work function* of a metal; it is a constant for a given metal with a chemically clean metal surface. Each metal has its own work function. Suitable incident radiant energy, as described, can impart sufficient energy to the electrons to induce them to leave the metal. When this behavior is accomplished by thermal energy, the mechanism is known as *thermionic emission*.

On the basis of the classical theory, the electric field associated with certain electromagnetic radiation should be sufficient to cause an electron to be ejected from the metal surface, if the field is strong enough. The force, F, exerted by this electric field, $\bar{E}_r$, on an electron is $F = \bar{E}_r e$. Simply restated, if the kinetic energy increment that the radiation imparts to the electron is equal to or greater than W, the electron will be excited enough to overcome the potential barrier and to be ejected. It was assumed in classical physics that the kinetic energy imparted to the electron should be a function of the intensity of the radiation. Therefore, it was considered that the greater the intensity of the incident radiation, the greater would be the kinetic energy

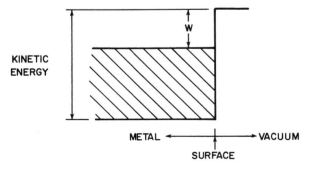

Figure 1-2 Work function at the surface of a metal. Note that the symbol ϕ is widely used to denote the work function.

of the emitted electrons. And, since W is the minimum excitation energy, a lower intensity limit would exist below which no electron emission could take place. In summary, the kinetic energy of the emitted electrons was considered to be only a function of the intensity of the incident radiation. Einstein showed that this was erroneous and that the kinetic energy of the emitted electrons is a function of the *frequency* of the incident radiation.

Without describing the experimental details, it is sufficient to note that a set of emitting and collecting electrodes was sealed in an evacuated glass tube. The external electrical circuit was such that it permitted the application of accelerating and retarding voltages between the two electrodes. The retarding voltage, V, could be adjusted so that

$$eV = \mathrm{KE} \geqslant W$$

where e is the charge on an electron and KE is its kinetic energy as a result of irradiation. In other words, the retarding voltage could be adjusted such that the electrical energy between the electrodes would just balance the kinetic energy of the irradiated electrons and no current would flow between the electrodes; such a retarding voltage is denoted by V_0. In this case, $eV_0 \equiv W$. The measurement of V_0 provided a direct means to measure the kinetic energy imparted to the electrons by the radiation.

When "monochromatic" light (essentially radiation with a single frequency) was directed on the emitter electrode at various intensities, the current (or number of electrons) between the electrodes increased as the intensity increased. Of greater significance was the fact that the retarding voltage remained constant; it was not a function of the intensity. This meant that, contrary to the classical theory, the intensity of the incident radiation had no effect on the energy of the electrons. The classical theory was thus shown to be incorrect.

Very different behavior was demonstrated when incident light of constant intensity, but of different frequencies, was employed. The retarding voltage was found to increase as the frequency of the incident light was increased. This showed that the kinetic energy of the emitted electrons was a function of the frequency of the radiation, in agreement with Planck's findings.

A plot of the retarding voltage as a function of frequency of the radiation for a given metal emitter gives a linear relationship. Similar plots for other metals show that all are linear and all have identical slopes. The only difference between them is that their voltage intercepts are not the same. In other words, the plots for the various metals are parallel to one another, but each shows a different voltage intercept. The voltage intercept is a measure of the work function. (The work function is the product of the voltage intercept and the charge on an electron.)

Einstein (1905) explained this behavior by making assumptions similar to those of Planck. Einstein quantized the radiation in a way corresponding to that which Planck had used to quantize the oscillators. The assumptions used to explain this behavior are as follows:

1. The incident radiation is made up of discrete pulses called *photons*.
2. These photons are absorbed or given off in discrete amounts.
3. Each photon has an energy that is an integral multiple of hv.
4. The photons behave like waves of corresponding frequency.

This fourth assumption is famous and is very important because it assumes, for the first time, a relationship between wavelike and particlelike behavior that now is known as *duality*. It anticipates the more formal work of de Broglie and its experimental verification by Davisson and Germer by about 20 years.

Consider what happens to an electron with energy W (Fig. 1-2) when it reacts with and absorbs energy from an incident photon. The increment of energy absorbed by it will be hv. It will require a minimum energy of W in order to leave the metal surface. Thus, the net kinetic energy of the emerging electron will be

$$KE = eV = hv - W$$

This equation is in agreement with the experimentally observed linear behavior. The differentiation of this equation with respect to v gives, since W is a constant for a given metal and vanishes,

$$\frac{dV}{dv} = \frac{h}{e}$$

This finding predicts a constant slope that is independent of the emitting metal. Thus, in agreement with the experimental data, the slope is the same for any metal; the plots of voltage versus frequency are parallel straight lines. The product of the voltage intercept, V_0, of any of these curves and the charge on the electron gives the work function of that metal:

$$V_0 e = W \quad \text{(electron volts)}$$

This theoretical approach constitutes another important milestone in the understanding of physical properties. It explains the photoelectric effect in terms of the frequency of quantized radiation. This phenomenon could not have been explained by a classical theory that is based on the intensity of continuous radiation rather than on its dependence on the frequency of discrete particles.

1.5 THE BOHR ATOM

Much experimental work had been done involving the spectra of the elements. The spectrum of each element had been shown to consist of a series of discrete diffraction lines, and each element shows different sets of spectral lines. (This fact, along with the intensity, or density, of the spectral lines, makes possible the chemical analyses of unknown materials.) Classical mechanics, based on the assumption of continuous energies and spectra, could not be used to explain the observed behavior. Some early efforts to express the observed spectral lines by means of empirical equations showed that these could be described only by use of squared integers and their differences. The attempts to understand such discontinuous behavior led to the presently accepted model of the atom.

The difficulties inherent in the description of electron behavior by means of classical mechanics may be illustrated simply as follows. The rate of radiation given off by an accelerated electron, treated as a point charge, is given by

$$I = \frac{2}{3}\frac{e^2 a^2}{c^3} \quad \text{(erg/s)}$$

where a is its acceleration. Thus, the energy of the electron will decrease in a continuous way as long as the electron is accelerated. Its acceleration may be expressed as $a = v^2/r$, in which v is its velocity and r is the radial distance of its orbit around the nucleus. This expression is substituted for the acceleration to give

$$I = \frac{2e^2(v^2/r)^2}{3c^3} = \frac{2e^2 v^4}{3c^3 r^2}$$

Greater insight may be obtained by rearranging these factors, then by multiplying the numerator and denominator by 2, and finally by noting that the kinetic energy is $E = e^2/2r$. The resulting expression for the radiation rate is

$$I = \frac{2}{3c^3} \cdot \frac{e^2}{2r} \cdot \frac{2v^4}{r} = \frac{2}{3c^3} \cdot E \cdot \frac{2v^4}{r}$$

This may be rewritten again in the form

$$\frac{I}{E} = \frac{2}{3c^3} \cdot \frac{2v^4}{r} = \frac{4}{3} \frac{v}{r} \left(\frac{v}{c}\right)^3$$

The inversion of this equation gives the time, τ, required for the radiated energy to consume all the kinetic energy of the electron as

$$\frac{E}{I} = \frac{3}{4} \frac{r}{v} \left(\frac{c}{v}\right)^3 = \tau \quad \text{(s)}$$

The substitution of $r \cong 10^{-8}$ cm, $v \cong 10^8$ cm/s, and $c \cong 10^{10}$ cm/s gives an approximation of $\tau \cong 10^{-10}$ s; this is the classical lifetime of an accelerated electron in motion around a nucleus. The time for such an electron to complete one orbit around the nucleus is given by $2\pi r/v \cong 10^{-16}$ s. These results may be used to calculate the number of orbits made by an electron before all its energy is radiated away and it crashes into the nucleus: $\tau/(2\pi r/v) \cong 10^6$ orbits. Thus, according to classical mechanics, and contrary to experiment, any atoms containing accelerated electrons should have continuous spectra, be unstable, and exist for about 10^{-10} s. It is apparent that such findings are incorrect and cannot be used to explain the observed spectra. This is the reason why classical mechanics had to be abandoned and another means had to be found that would provide a realistic description of the previously noted discontinuous spectral behavior. In so doing, this gave a new model of the electron structure of atoms.

The Rutherford model of the atom was considered to consist of a positive nucleus surrounded by numbers of negative electrons so that the atom was electrically neutral. The distribution of the electrons was not defined. The experimentally observed spectral lines could be explained only if the electrons behaved in a consistent, discrete, systematic manner. This forced Bohr to assume (at first) that the electrons moved in fixed, discrete, nonradiating, *circular* orbits around a nucleus. It follows from this that electrons in such arrays could take on only specific and discontinuous momenta and energies. The difference between Bohr's model of the atom and that of classical physics, in which the electrons were permitted to have any energies or momenta, constituted a revolutionary new approach to the study of atomic structure. Bohr postulated that the spectral lines were a result of the release of radiant energy by an electron after having been excited to a higher-energy orbit and dropping down to one of lower energy. Since the orbit energies were postulated to be discrete, such released radiant energy also had to be discrete, thus accounting for the experimentally observed, discontinuous spectral behavior. And, at the same time, the prediction that an electron should crash into the nucleus was eliminated.

Bohr's findings for the hydrogen atom may be obtained by equating the electrostatic (coulombic) force between an electron and the nucleus to the mechanical centripetal force of the electron. This leads to an equation for the square of the velocity, $\mathbf{v}$, of an electron, as follows:

$$\frac{e^2 Z}{r^2} = \frac{m\mathbf{v}^2}{r}$$

so that

$$\mathbf{v}^2 = \frac{e^2 Z}{mr} \quad \text{or} \quad m\mathbf{v}^2 = \frac{e^2 Z}{r}$$

in which Z is the atomic number and m is the electron mass. Its kinetic energy is obtained directly from this, noting that the mass vanishes, as

$$\mathrm{KE} = \frac{1}{2} m\mathbf{v}^2 = \frac{1}{2} m \cdot \frac{e^2 Z}{mr} = \frac{e^2 Z}{2r}$$

The potential energy of the electron, defined as zero when it is infinitely separated from the nucleus, is given by

$$\mathrm{PE} = -\frac{e^2 Z}{r}$$

It will be noted that the magnitude of the potential energy is twice that of the kinetic energy; the magnitude of the potential energy must be greater than that of the kinetic energy in order that the electron maintain its association with the nucleus. The total energy of the electron, E_T, the algebraic sum of these energies, is

$$E_T = -\frac{e^2 Z}{r} + \frac{e^2 Z}{2r} = -\frac{e^2 Z}{2r} = -\text{KE}$$

The kinetic energy may be written in terms of angular frequency, ω, or its linear frequency, v, where $\omega = 2\pi v$, as

$$\text{KE} = \tfrac{1}{2}mr^2\omega^2 = \tfrac{1}{2}mr^2(2\pi v)^2$$

The change in energy is obtained, by differentiating this equation, as

$$\Delta E = 4\pi^2 mr^2 v\, \Delta v$$

However, as postulated by Bohr, such energy changes must be discrete so that, when the Planck expression for energy is used,

$$\Delta E = 4\pi^2 mr^2 v\, \Delta v = nhv$$

And, noting that v vanishes, it is found that

$$\frac{nh}{2\pi} = 2\pi mr^2\, \Delta v$$

The change in the angular momentum, $\Delta p(\theta)$, is obtained from the differentiation of the equation for angular momentum as given by

$$p(\theta) = mr^2\omega$$

to obtain

$$\Delta p(\theta) = mr^2\, \Delta\omega = mr^2 \cdot 2\pi\, \Delta v = \frac{nh}{2\pi}$$

when the preceding equation is used. The expression for $p(\theta)$ may be rewritten, using $\omega = \mathbf{v}/r$, to obtain an equation for the radius, r, of an electron about the nucleus:

$$p(\theta) = mr^2\,\frac{\mathbf{v}}{r} = mr\mathbf{v}$$

so that

$$r = \frac{p(\theta)}{m\mathbf{v}}$$

Recourse again is made to the postulation that $p(\theta)$ must be discrete. Thus, the radius must be expressed as

$$r = \frac{p(\theta)}{m\mathbf{v}} = \frac{nh}{2\pi m\mathbf{v}}$$

Or, in terms of momentum, this equation may be written as

$$m\mathbf{v} = \frac{nh}{2\pi r}$$

Squaring both sides of this equation gives

$$m^2v^2 = \frac{n^2h^2}{4\pi^2r^2}$$

Another expression for m^2v^2 may be obtained from that previously derived from the square of the velocity by multiplying both sides of that equation by m:

$$m^2v^2 = \frac{me^2Z}{r}$$

Both expressions for m^2v^2 now may be equated and solved to obtain the Bohr radius as

$$\frac{n^2h^2}{4\pi r^2} = \frac{me^2Z}{r}$$

$$r = \frac{n^2h^2}{4\pi me^2Z}$$

The Bohr radius for hydrogen may be obtained from this equation by setting the variables n and Z both equal to unity. The substitutions for the other factors in this equation, all of which are constants, give the hydrogen radius as 0.528×10^{-8} cm. Thus, according to the Bohr theory, the radius of an electron in a circular orbit about a nucleus may be given by

$$r = 0.528 \frac{n^2}{Z} \times 10^{-8} \quad \text{(cm)}$$

The equation for the Bohr radius also permits the kinetic energy of an electron to be described in greater detail. Thus,

$$\text{KE} = \frac{e^2Z}{2r} = \frac{e^2Z}{2} \cdot \frac{4\pi^2me^2Z}{n^2h^2} = 2\pi^2m\left(\frac{e^2Z}{nh}\right)^2$$

This, in turn, permits the calculation of the radiant energy given off by an excited electron when it drops to a lower energy level. This energy must be the difference between that of the higher and the lower energy levels. Therefore, the energy, and consequently the frequency, of the emitted radiation may be obtained from

$$\Delta E = hv = E_{T_2} - E_{T_1} = \text{KE}_1 - \text{KE}_2$$

So, using the previously derived expression for KE and factoring, it is found that

$$\Delta E = hv = 2\pi^2m\frac{e^4Z^2}{h^2}\left(\frac{1}{n_1^2} - \frac{1}{n_2^2}\right)$$

Thus, the energy levels responsible for the spectrum of the hydrogen atom are given by

$$\Delta E = hv = \frac{2\pi^2me^4}{h^2}\left(\frac{1}{n_1^2} - \frac{1}{n_2^2}\right)$$

and their corresponding frequencies by

$$v = \frac{2\pi^2 me^4}{h^3} \left(\frac{1}{n_1^2} - \frac{1}{n_2^2} \right)$$

These frequencies correspond to the observed spectral emissions of hydrogen, within experimental error.

The velocity of a Bohr electron may be obtained from its kinetic energy, noting that the mass vanishes:

$$\text{KE} = \frac{1}{2}mv^2 = \frac{2\pi^2 me^4 Z^2}{n^2 h^2}$$

$$v^2 = \frac{4\pi^2 e^4 Z^2}{n^2 h^2}$$

$$v = \frac{2\pi e^2 Z}{nh}$$

This result may be used to obtain another important property of electrons. The velocity of an electron relative to the speed of light, c, is, using the equation for its velocity and n and Z equal to unity,

$$\frac{v}{c} = \frac{2\pi e^2 Z}{nhc} \cong 0.0073 \cong \frac{1}{137} \cong \alpha$$

This ratio of $v/c = \alpha$ for hydrogen is known as the *fine-structure constant*. Here α is shown to be about 0.7% of the speed of light. The small value of α makes it unnecessary to treat electrons by means of relativistic mechanics, v/c being very much less than unity. This would not have been the case if $v/c \rightarrow 1$.

In addition, the fine-structure constant plays an important role in the electron structures of atoms in the modern theory, beyond the capability of description by the Bohr model. As will be discussed later (Sections 2.5 and 3.10), electron excitations in magnetic fields result in small differences in the energies of electrons with a given principal quantum number, n. Such excitations are represented by the splitting of the spectral lines into multiple lines. Modern theory now accounts for these small energy changes by means of the third quantum number, m_ℓ. The very small value of α permits the observation of the electrons as having been in such split states. If α were large, the electrons could not remain in any of these excited states long enough for their return to their original states to be manifested as the observable radiant energy emissions responsible for the split spectral lines; their times of occupancy would be much too short. Thus, the magnitude of α is responsible for the nonrelativistic description of electrons as being in discrete energy levels, or energy states. This permits the characteristic energy levels of hydrogen to be given in terms of the previously derived relationships as

$$E_n = E_T = -\text{KE} = -\frac{e^2 Z}{2r} = -\frac{e^2 Z}{2} \cdot \frac{4\pi m e^2}{n^2 h^2} = -\frac{2\pi m e^4}{n^2 h^2} = -\frac{13.6}{n^2} \quad \text{(eV)} \qquad (1\text{-}11)$$

where E_n is the energy of the orbit or state n and n is an integer. The condition for which $n = 1$ is the lowest possible orbit, or energy, known as the *ground state* of the electron. Excited, or energy-emitting, states are present when electrons occupy states for all other orbits for which n is greater than unity. Equation (1-11) provides an excellent description for the spectra of the hydrogen atom.

When applied to explain the spectra of elements with larger atomic numbers, Bohr's model was capable only of a qualitative description of their electron structures. Bohr and others attempted to extend this limited degree of applicability by considering the electrons to be in orbits with varying degrees of ellipticity. This, too, failed because, as indicated in subsequent sections, the classical mechanics employed by them to describe electron properties simply are not applicable.

1.6 INTERRELATIONSHIPS BETWEEN WAVES AND PARTICLES

The wavelike and particlelike behavior of electrons will be considered further. Einstein assumed this duality for radiation (Section 1.4). Experiments involving electrons and diffraction effects will be described to illustrate this behavior. These will show the need for the statistical treatment of electrons.

Classical physics (geometric and physical optics) and the findings of Einstein unmistakably showed that light possessed both wave and particle properties. If light is corpuscular in nature, how can diffraction phenomena be explained? This question could be resolved if it were assumed that the waves and particles were associated in some way. It had already been assumed by Planck, Einstein, and others that oscillating particles, photons, wave quanta, and the like, had energies associated with their frequencies: $E = h\nu$. The nature of this wave–particle relationship and its consequences will be considered in this section.

1.6.1 Electrons Treated as Waves or Particles

The optical behaviors of electrons will be examined, in ways similar to their application in electron microscopes, to explain those of their properties of significance here. Their reactions to electric fields constitute evidence that they may be treated as negatively charged particles. They travel in linear paths in electric fields, and they change directions of motion and velocities in response to changes in the field strength. They can be made to reflect and to refract. Thus, beams of electrons can be considered to act as waves of matter that obey the laws of optics.

An electron treated as a particle in motion in a given electric field will have a velocity vector of $\mathbf{v}_1$, which makes an angle of θ_1 with respect to the field. An increase in the field strength will increase its velocity from $\mathbf{v}_1$ to $\mathbf{v}_2$, leaving that component of its velocity perpendicular to the field, $\mathbf{v}_y$, unchanged, but the angle now is changed to $\theta_2 < \theta_1$. This set of conditions may be expressed by equating the unchanged perpendicular components of the velocities for both of these cases in the following way:

$$\mathbf{v}_y = \mathbf{v}_1 \sin \theta_1 = \mathbf{v}_2 \sin \theta_2$$

Thus, the relationship between the angles and the velocities of the electron in the two different electric fields is

$$\frac{\sin \theta_1}{\sin \theta_2} = \frac{v_2}{v_1} \tag{1-12}$$

Such a change in velocity, occurring when the electron passes from one electric field to another, may be considered as a refraction of the electron. Now, if the electron is treated as a bundle of matter waves, it may have its behavior described optically by Snell's law as

$$\frac{\sin \theta_1}{\sin \theta_2} = \frac{v_1}{v_2} \tag{1-13}$$

Equations (1-12) and (1-13) appear to give contradictory results. This seeming contradiction may be resolved by considering the ways in which waves and particles may be related.

An electron as a particle may be simulated by means of a group of waves, as shown in Fig. 1-3. Such a group of waves is known as a *wave packet*. Packets of this kind may be obtained as a result of interference effects between at least two waves of slightly different wavelength. The waves from which the packet is formed will have a velocity **v**, known as the *phase velocity*. The packet, or bundle, representing a particle will have a *group velocity*, v_g. The relationship between these two velocities is given by

$$v = \frac{c^2}{v_g} \tag{1-14}$$

where c is the speed of light. It is seen that the velocity of a particle is inversely proportional to that when it is treated as a packet of waves. This removes the apparent inconsistency shown by Eqs. (1-12) and (1-13).

An important restriction must be placed on Eq. (1-14). This limitation is that the group and the phase velocities must be equal. Thus,

$$v_g = v \tag{1-15}$$

When Eq. (1-15) is not obeyed, the packet spreads out and quickly disappears in a manner similar to ripples on the surface of a pond. Thus, particles may be treated as

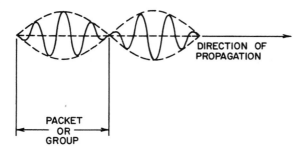

DIRECTION OF
PROPAGATION

PACKET
OR
GROUP

Figure 1-3 Schematic representation of wave packets.

waves if and only if Eq. (1-15) is obeyed. This may be extended to include the corpuscular properties of radiation, as well as those of matter.

The ability to treat electrons either as particles or waves is of great importance. Under certain conditions, as will be shown later, it is not possible to describe them precisely either as waves or particles. This dual behavior makes it necessary to treat electrons either as waves or as particles, depending on which approach provides the better explanation of the phenomenon being considered.

Some useful relationships may be obtained from the phase velocity of an electron when it is treated as a wave or as a particle. Using the *wave number*, defined as $k = 1/\lambda$, this gives

$$\mathbf{v} = \mathbf{v}_g = \lambda v = \frac{v}{k}$$

This may be rewritten as

$$\mathbf{v}k = \mathbf{v}_g k = v$$

Upon differentiation, this gives

$$\mathbf{v}\, dk = \mathbf{v}_g\, dk = dv$$

or

$$\mathbf{v} = \mathbf{v}_g = \frac{dv}{dk}$$

The linear frequency is given by $v = \omega/2\pi$, where ω is the angular frequency. This gives $dv = d\omega/2\pi$, in differential form, so that

$$\mathbf{v} = \mathbf{v}_g = \frac{dv}{dk} = \frac{d\omega}{2\pi\, dk}$$

The *wave vector* $\bar{\mathbf{k}} = 2\pi k$, so, upon differentiation, $d\bar{\mathbf{k}} = 2\pi dk$, and

$$\mathbf{v} = \mathbf{v}_g = \frac{dv}{dk} = \frac{d\omega}{d\bar{\mathbf{k}}}$$

It is helpful to obtain an approximation of the "size" of electrons for better understanding those cases in which they are treated as particles. The electrostatic energy of an electron, treated as a spherical particle, of "radius" r and charge e is

$$E = \frac{e^2}{r}$$

Equating this electrostatic energy to its relativistic energy gives

$$E = \frac{e^2}{r} = m_0 c^2$$

where m_0 is the rest mass of the electron and c is the velocity of light. Thus, the "radius" may be expressed as

$$r = \frac{e^2}{m_0 c^2}$$

The substitution of $e \cong 4.8 \times 10^{-10}$ esu, $m_0 \cong 9.1 \times 10^{-28}$ g, and $c \cong 3 \times 10^{10}$ cm/s gives $r \cong 2.8 \times 10^{-13}$ cm. Here it is implicit that the electrostatic energy of an electron arises from its mass. This is to be expected, because the electrostatic and relativistic energies were equated to obtain this result. Despite the fact that this conclusion remains to be verified, the magnitude of r appears to be reasonable.

1.6.2 The de Broglie Theory of Duality

Dual behavior was first considered by Einstein (Section 1.4). The wave–particle duality was directly related by de Broglie (1924). His findings may be most simply shown by considering the energy of corpuscular electromagnetic radiation in terms of the Planck–Einstein approach. This is equated to the relativistic energy to give

$$E = h\nu = mc^2 \tag{1-16}$$

Dividing both sides of Eq. (1-16) by c, recalling that $c/\nu = \lambda$, and noting that the momentum is given by $p = mc$, and then inverting gives

$$\frac{c}{h\nu} = \frac{\lambda}{h} = \frac{1}{mc} = \frac{1}{\mathbf{p}}$$

Then, de Broglie's relationship is found by solving for λ:

$$\lambda = \frac{h}{mc} = \frac{h}{mv} = \frac{h}{\mathbf{p}} \tag{1-17}$$

where $\mathbf{p}$ is the momentum of the corpuscle, or photon. It will be noted that other particles with nonzero rest masses, such as electrons, neutrons, and alpha particles, usually have velocities very much less than c, the speed of light. The treatment of the velocities of such particles in terms of their group velocities, rather than by the simplified approach given here, constituted the basis for de Broglie's original derivation of Eq. (1-17). This assumption of the dual nature of such particles permits the extension of Eq. (1-17), as derived previously, to cases where $\mathbf{p} = mv$ rather than $\mathbf{p} = mc$, $\mathbf{v} \ll c$.

A wavelength is most commonly associated with a wave and a momentum is normally ascribed to a particle. The de Broglie equation relates the wavelike properties of a particle with its particlelike properties. The usefulness of this relationship also is enhanced by its simplicity. Its importance here is that electrons may be treated either as particles or as waves, depending on the conditions being considered.

The same is true for photons. Further insight may be obtained by first treating the photon as a particle, then as a wave and, finally, by comparing the equations thus obtained. Starting by treating the photon as a particle, its momentum is

$$\mathbf{p} = mc$$

Both sides of this equation are squared and multiplied by c^2 to give

$$c^2\mathbf{p}^2 = m^2c^4$$

The relativistic energy is $E = mc^2$ and its square is $E^2 = m^2c^4$ so that

$$E^2 = c^2\mathbf{p}^2$$

Thus,

$$E = c\mathbf{p}$$

Now, treating the photon as a wave,

$$E^2 = h^2v^2$$

Substituting c/λ for v and then using Eq. (1-17) results in

$$E^2 = \frac{h^2c^2}{\lambda^2} = c^2\mathbf{p}^2$$

So, again,

$$E = c\mathbf{p}$$

Thus, in each case, the same results are obtained for the energy of the photon. The addition of the two expressions for the energy of a photon gives

$$2E = mc^2 + c\mathbf{p}$$

However, as a particle, the momentum of a photon may be expressed as $\mathbf{p} = m_0v \ll mc$, so it may be approximated that its energy is

$$E \cong \frac{mc^2}{2} = \frac{m^2c^2}{2m} = \frac{\mathbf{p}^2}{2m}$$

Therefore, it would be expected that a photon, a particle of zero rest mass, could have momentum. The experimental proof that this, indeed, was the case was provided by Compton (Section 1.6.4.).

1.6.3 The Davisson–Germer Proof of Duality

The de Broglie relationship was verified by Davisson and Germer (1927) while experimenting with accelerated electrons. This was accomplished by varying the voltage between an electron source and a nickel target in a vacuum chamber. Some of the energy of the incident electrons, converted to heat, was sufficient to recrystallize grains in the impingement zone at the surface of the nickel specimen. The grains grew as the process continued and, as these experimenters showed, these crystal lattices acted as diffraction gratings for the electrons.

The intensities of the scattered electrons were scanned and plotted on polar coordinates. The intensities of the scattered electrons for each given voltage were represented by the length of the radius vector at each angle of measurement. The shapes of these curves showed a dramatic change as the scans were made with increasing voltages. A peak, or spur, appeared that attained its greatest extent at 50° and 54 V. If electrons may be treated as waves, they should satisfy the Bragg diffraction equation. This is given by

$$n\lambda = 2d \sin \theta \qquad (1\text{-}18)$$

where the order of the diffraction is $n = 1$ and d is the interplanar spacing of the nickel crystals. Using $d = 2.15 \,\text{Å}$ for the nickel specimen and $\theta = 50°$, it was found that $\lambda = 1.65 \,\text{Å}$.

If electrons are particles with wavelike properties, their wavelengths should be determined by the de Broglie equation. When the voltage is converted to statvolts, it is found that Eq. (1-17) gives $\lambda = 1.67 \,\text{Å}$, in excellent agreement with the wavelength determined from the calculation based on wave properties.

Similar experiments were performed on about 20 other crystals. In each case the results obtained by use of the Bragg equation agreed with those obtained from the de Broglie equation, within the limits of experimental error. This good agreement verified the de Broglie equation and left no doubt as to the wave–particle duality of electrons.

1.6.4 Interaction of Radiation with Particles

Another finding concerning the interaction of radiation with matter had a very important part in the development of quantum mechanics. For example, the reflection or diffraction of a beam of electromagnetic radiation by matter would be expected to change its intensity, but changes in its wavelength would not have been anticipated ordinarily. The fact that such changes do take place is of great significance.

Compton (1923) explained the mechanisms that are involved when essentially monochromatic x-rays are scattered by a solid and the resultant beams show an additional, different wavelength, or frequency, instead of just the original one. This is shown schematically in Fig. 1-4. The peak at λ_0 is that of the original beam of x-rays

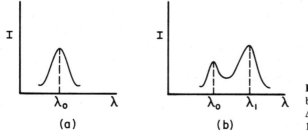

Figure 1-4 Intensities of x-rays (a) before and (b) after scattering. $\Delta\lambda = \lambda_1 - \lambda_0 = h/mc(1 - \cos\theta)$. See Fig. 1-5.

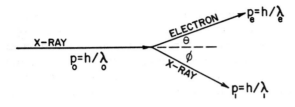

Figure 1-5 Schematic diagram of the momenta vectors involved in the Compton effect.

prior to any reaction with the solid.   Compton showed that the additional peak at the longer wavelength represents the result of an elastic interaction between the x-rays and the relatively free electrons in the solid.   This was explained by considering the change in the momentum of a photon of the x-rays when it impinges on a valence electron and imparts a momentum, p_e, to it.   This is shown in Fig. 1-5.

When the conservation of energy and momentum are taken into account, the shift of wavelength can be explained.   This shift depends on the scattering angle, but is independent of the wavelength of the incident radiation and of the material doing the scattering.   This phenomenon clearly shows that electromagnetic radiation treated as a particle, a photon, can impart momentum to a particle of matter, in this case to the valence electron.

Since some of the incident radiation is unchanged by this process, another reaction must take place.   Here the incident photons interact elastically with those electrons that are much more strongly bound to ions of the solid and are in the form of the completed electron shells that surround the nuclei and constitute the ion cores. The great difference in the masses involved between a photon and an ion core is responsible for the unchanged momentum of the scattered photon.   Thus, its energy, frequency, and wavelength are unchanged.   This accounts for the presence of the diminished intensity peak, which appears at exactly the same wavelength as the original, unscattered beam.

The changes in the properties of a photon (electromagnetic radiation) and an electron (a particle of matter) after mutual interaction are important considerations in understanding the Heisenberg uncertainty principle.

1.6.5 Influence of Space and Time on Particles

The fact that electrons and photons show dual behavior makes it possible to explore some of their properties by means of diffraction phenomena, where the classical theory cannot explain the observed results.   This permits their examination by means of the wave theory.   As a result, it will become apparent that the degree of error with which some of the parameters can be known becomes increasingly greater as the attempt is made to know others more precisely.   In other words, the more exactly one of these is known, the greater is the uncertainty, or imprecision, with which it is possible to know the other one.

Consider the diffraction phenomenon caused by the passage of a beam of light, made up of an uninterrupted stream of waves or of photons, through a slit (Fig. 1-6).

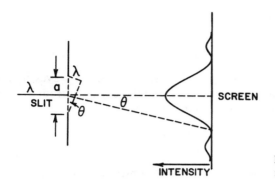

Figure 1-6 Diagram of diffraction of monochromatic light by a slit.

This diffraction behavior can be explained only by wave theory. The intensity maxima and minima result from constructive and destructive wave interference, respectively. The angle between the central maximum of the intensity and the first intensity minimum is θ. The condition for this behavior is given by

$$\sin \theta = \frac{\lambda}{a} \simeq \theta \tag{1-19}$$

for small values of θ.   Equation (1-19) is valid only for a suitable range of slit openings, a, for a given wavelength, λ. This relation of wavelength to slit size constitutes a space limitation for the following reasons. Where the slit opening is very much larger than the wavelength, no interference, and hence essentially no diffraction, will occur. A ray, or beam or light, will be virtually unaffected by the opening; its optical behavior may be predicted by classical means. The beam may be considered as being corpuscular in nature. But, when the slit opening is decreased to a critical size with respect to λ, diffraction takes place and classical geometric optics can no longer be used. The beam now must be considered to be composed of waves. Further reduction in the slit size causes a more pronounced diffraction pattern, rather than the isolation of a ray. Equation (1-19) constitutes a space limitation because the size of the slit opening determines whether particlelike or wavelike behavior will be observed. This also holds for electrons, since matter waves behave similarly.

What happens when the number of waves passing through the slit is limited? Now suppose that a shutter is placed behind the slit and that a detector that is sensitive to small changes in wavelength, λ, or frequency, v, is placed so that it can scan the screen. What happens when the shutter is opened for a very short time, Δt? The short grouplet of waves that passes through will have been changed by this. This may be shown as described next.

The phase velocity of the incident waves is given by $\mathbf{v} = \lambda v$. Upon differentiation and rearrangement, this becomes, since $\mathbf{v}$ is a constant and vanishes,

$$\left| \frac{dv}{v} \right| = \left| \frac{d\lambda}{\lambda} \right|$$

or, neglecting the negative sign,

$$\frac{\Delta v}{v} = \frac{\Delta \lambda}{\lambda}$$

The quantity $\Delta v/v$ is the reciprocal of the number of waves in the grouplet. Another expression for the number of waves in this very short train is given by $\mathbf{v}\,\Delta t/\lambda$. Both of these expressions may be equated to give

$$\frac{\Delta v}{v} = \frac{\lambda}{\mathbf{v}\,\Delta t} \qquad\qquad (1\text{-}20)$$

The equation for the phase velocity, $\mathbf{v} = \lambda v$, is substituted in Eq. (1-20) to obtain

$$\frac{\Delta v}{v} = \frac{\lambda}{\lambda v\,\Delta t}$$

Or, since λ and v vanish,

$$\Delta v = \frac{1}{\Delta t} \qquad\qquad (1\text{-}21)$$

Thus, the frequency of the waves comprising the short train is no longer necessarily a constant but can vary over a range of frequencies, Δv, depending on Δt. This range of variation of frequency increases as the open time of the shutter (exposure time) decreases. Thus, the attempt to examine the small group of waves more closely by the use of shorter exposure times results in greater frequency range spreads, which cause the frequency to become increasingly uncertain. In contrast to this, relatively long exposure times have virtually no effect on the frequency. The situation in which decreased exposure time affects the frequency thus introduces a time limitation.

An uninterrupted stream of photons was used to describe the space limitation. The attempted use of a limited number of waves introduced a time limitation. What happens when just one photon at a time passes through a pair of slits? This is shown schematically in Fig. 1-7.

The intensity of the light in this optical system can be decreased so that there is a high probability of finding just one photon between the slits and the film at any given time. If this very low intensity exposure is permitted to continue for a sufficiently long time, so that the sum of such single photons approximates the number that would reach the film in a given shorter time at more usual intensities of exposure, the diffraction pattern is found to be the same as that caused by plane waves of light passing through *two* slits. This experiment was performed by G. I. Taylor.

If a photon is a particle, it should be able to pass only through one slit at a time, not both. The resultant diffraction pattern then would be quite different from that actually obtained. The observed pattern can be caused only by the passage of a photon through both slits simultaneously. How could the photon, if it is a particle, have passed through both slits at the same time? It would thus appear that the nature of a photon must be different from that given by the concepts presented earlier for

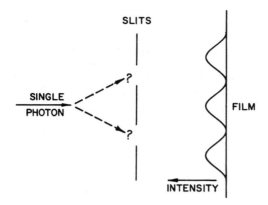

Figure 1-7 Schematic illustration of the G. I. Taylor experiment.

either particles or waves. It turns out that it is not possible to know the exact nature of a photon or of any other subatomic particle when an attempt is made to examine it too closely, as was done in this experiment. It also may be concluded that the same is true for an electron as well.

1.6.6 The Uncertainty Principle

It has been shown that photons and electrons, as particles, also may be described in terms of waves and that this approach has definite limitations. The closer the attempt to study the behavior of a photon, or a similar particle, the greater will be the difficulty in ascertaining its properties. Heisenberg summarized this in his uncertainty, or indeterminacy, principle. This concept was derived by means of idealized "thought" experiments.

If it is desired to "see" an electron in such an experiment, photons must be reflected from the electron and enter the microscope so that the electron may be observed. Both the photons and the electron interact in the same way as described to explain the Compton effect (Section 1.6.4). A photon "reflected" from the electron will undergo a change in momentum, and it also will be diffracted if and when it enters the microscope. Knowledge of the position of the electron will depend on the diffraction pattern of the photon (Fig. 1-8) and the degree of fineness with which the microscope can resolve an object. It is most probable that a photon will fall within the central intensity maximum of the diffraction pattern, since this region should contain the largest number of photons. However, the sharpness of the image of the electron, which is projected on the screen, will be affected by the several small maxima that are concentric about the central maximum; the outline of the image will be fuzzy. This occurs because there are much smaller, but equally valid, probabilities that the photon will fall in any of the smaller intensity maxima that are concentric about the central maximum.

A photon reflected from the electron, having undergone a momentum change from $\mathbf{p}$ to $\mathbf{p}'$, can enter the objective lens of the microscope at any angle between $\pm\phi$

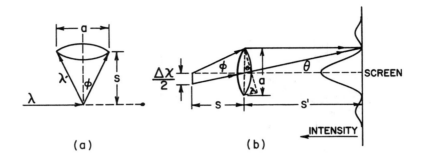

(a)   (b)

Figure 1-8 Factors involved in the Heisenberg "thought" experiment. (a) The photon interacts with the electron and enters the microscope. (b) Optical system combining Figs. 1-8(a) and 1-6.

(Fig. 1-9). The maximum change of its momentum in the x direction is

$$\Delta \mathbf{p}_x = 2\mathbf{p}' \sin \phi$$

and, for small angles, it may be approximated that

$$\Delta \mathbf{p}_x \simeq 2\mathbf{p}'\phi$$

This may be rearranged to obtain an expression for the small angle, ϕ, as

$$\phi \simeq \frac{\Delta \mathbf{p}_x}{2\mathbf{p}'}$$

The degree with which any microscope can form a sharply defined image is limited by the factors in the relationship

$$\Delta x = \frac{\lambda}{2n \sin \phi}$$

Here Δx is the fineness with which an object may be resolved (resolving power), and n

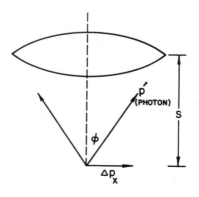

Figure 1-9 Momentum diagram of the photon in the Heisenberg analysis.

is the index of refraction of the medium in which the light travels in passing from the object being viewed to the objective lens of the microscope. The degree of resolution is indicated as being centered on the optical axis, but, for simplicity, only half of this, $\Delta x/2$, is shown in Fig. 1-8(b). The medium in which the photon travels is a vacuum, so $n = 1$. Thus, where ϕ is small and the wavelength of the photon has been changed from λ to λ' by the Compton effect,

$$\Delta x = \frac{\lambda'}{2\phi}$$

This permits another expression for ϕ to be obtained by rearrangement:

$$\phi = \frac{\lambda'}{2\Delta x}$$

The two expressions for ϕ are equated to give

$$\frac{\Delta \mathbf{p}_x}{2\mathbf{p}'} = \frac{\lambda'}{2\Delta x}$$

The coefficient $\frac{1}{2}$ vanishes, and this equation may be rewritten as

$$\Delta \mathbf{p}_x \Delta x = \lambda' \mathbf{p}'$$

The de Broglie relationship, Eq. (1-16), is used to reexpress λ' as $h/\mathbf{p}'$. This substitution for λ' gives

$$\Delta \mathbf{p}_x \Delta x = \frac{h}{\mathbf{p}'} \mathbf{p}' = h$$

More precisely, this equation should be written as

$$\Delta \mathbf{p}_x \Delta x \geqslant h \tag{1-22}$$

Figure 1-8 also may be used to obtain Eq. (1-22) by another means. This approach involves the use of $\tan \theta$ and $\tan \phi$ in conjunction with a suitable expression for $\Delta \mathbf{p}_x$, when the space limitation, Eq. (1-19), is taken into consideration.

Since momentum and energy are closely related, any situation that changes the momentum of a photon would be expected to affect its energy, since $E = \mathbf{p}^2/2m$. The energy uncertainty can be shown in an elementary way starting with Eq. (1-5),

$$E = h\nu$$

and differentiating it to obtain

$$\Delta E = h\,\Delta \nu \tag{1-23}$$

The time limitation, Eq. (1-21), is now used to substitute for $\Delta \nu$ as $1/\Delta t$ in Eq. (1-23) to find

$$\Delta E = h\,\frac{1}{\Delta t}$$

Or, more precisely,

$$\Delta E\, \Delta t \geqslant h \qquad (1\text{-}24)$$

Thus, the Heisenberg uncertainty principle places limits on the degree to which the product of the conjugate parameters in Eqs. (1-22) and (1-24) can be known. The accuracies to which the *individual* measurements of x and $\mathbf{p}$ or of E and t may be known are not theoretically limited in themselves. The only limitations to the accuracies with which these *single* factors may be known are those inherent in the methods of their measurement. However, no two members of a pair of these conjugates can be exactly known simultaneously when being considered as pairs. The greater the accuracy with which one conjugate is known, the lesser must be the accuracy with which the other is known. For example, if a particle has $\Delta\mathbf{p}_x = 0$, then it follows that its position cannot be known because $\Delta x = \infty$; complete uncertainty exists with respect to its position. Since the conjugates $\Delta\mathbf{p}_x$ and Δx cannot equal zero, the particle cannot be stopped. Thus, *a particle obeying these rules must always be in motion*. A logical consequence of this is that such a particle may never have zero energy. This constitutes another way of considering the relationship between Eqs. (1-22) and (1-24). Thus, particles obeying quantum mechanics may never be at rest or have zero energy, as permitted by classical mechanics, especially at 0 K. This result is the basis for the concept of the *zero-point energy* of particles.

The Heisenberg relationships also impose restrictions on wave packets when they are treated as particles. This may be seen by starting with Eq. (1-22) reexpressed as

$$\frac{\Delta\mathbf{p}_x\, \Delta x}{h} \geqslant 1$$

Then, using the de Broglie equation and the wave number, $k_x = 1/\lambda$, an expression is found for $\Delta\mathbf{p}_x$ as follows:

$$\mathbf{p}_x = \frac{h}{\lambda} = hk_x$$

Upon differentiation,

$$\Delta\mathbf{p}_x = h\Delta k_x$$

This is substituted into the reexpressed Heisenberg equation to obtain

$$\frac{\Delta\mathbf{p}_x\Delta x}{h} = \frac{h\Delta k_x\Delta x}{h} = \Delta x\,\Delta k_x \geqslant 1$$

Or, using the wave vector, $\bar{\mathbf{k}}_x = 2\pi k_x$, this equation is given as

$$\Delta x\, \frac{\Delta\bar{\mathbf{k}}_x}{2\pi} \geqslant 1$$

and

$$\Delta x\, \Delta\bar{\mathbf{k}}_x \geqslant 2\pi$$

Here Δx is the uncertainty in the location (position) of the wave packet and $\mathbf{k}_x$ is that of the wave vector of the waves of which the packet is composed.   See Section 1.6.1.

If the product of pairs or conjugates was permitted to equal zero, it would correspond to permitting either one or both members of a pair to be known exactly (to have zero uncertainty).   Here h would have to equal zero.   This would effectively transform quantum mechanics into classical mechanics.

Classical mechanics assumes that position, momentum, energy, and time can be known exactly.   It contains no analog of Heisenberg's principle.   This is not the case in quantum mechanics; these conjugate parameters cannot simultaneously be known exactly.   Indeed, if one of them is known exactly, complete uncertainty must exist with regard to its conjugate.   Since the precision allowed in classical mechanics is not permitted in quantum mechanics, the best that can be done is to make use of the most likely values.   It is for this reason that quantum mechanics must be based on a statistical approach that provides the most probable values.

1.6.7 Bases for Quantum Mechanics

The situations that create difficulties for classical mechanics include the following basic concepts: electromagnetic radiation may be treated as photons; photon energy is exchanged in discrete amounts; photons can have momentum; matter particles can show dual behavior and may be treated as particles or in terms of wave properties; and limits affect the degree of precision with which these things can be known by classical means.   Quantum mechanics, therefore, must take into account this particular set of characteristics.   As has been shown in the preceding sections, this must be done in ways that describe their most probable properties.

A diffraction pattern will be used to provide an insight into the statistical approach to quantum mechanics.   While an incident beam of photons is used in the following explanation, it must be remembered that identical results may be obtained for any quantized, indistinguishable particle, such as an electron, proton, or neutron, that shows dual behavior.   See Sections 1.6.2 and 1.6.3.

A circular aperture with the same diameter, a, as the small slit shown in Fig. 1-6 is used here for simplicity.   A beam of "monochromatic" light passing through this hole will result in a circular diffraction pattern.   This diffraction pattern and its intensity as a function of θ are shown in Fig. 1-10(a) and (b) as taken along any diameter of the pattern.

Upon passing through the aperture, a photon may be diffracted in any direction in space.   This becomes increasingly apparent as the ratio of a/λ becomes very small. See Section 1.6.5.   An approximation of the number of photons passing through the aperture and being diffracted per unit area per unit time, or the photon flux, f, is made by reference to the hemisphere indicated in Fig. 1-10(c).   The radius of the hemisphere is $a/2$, so its area is $\pi a^2/2$.   The radius of an element of surface area of this hemisphere is $r = (a/2) \cos \theta$, for $\theta < \pm \pi/2$, so the element of hemispheric surface area is $\pi r\, d\theta = \pi(a/2) \cos \theta\, d\theta$.   Let N be the number of photons that cross the surface of the hemisphere in unit time.   Then dN will be the number of photons that cross

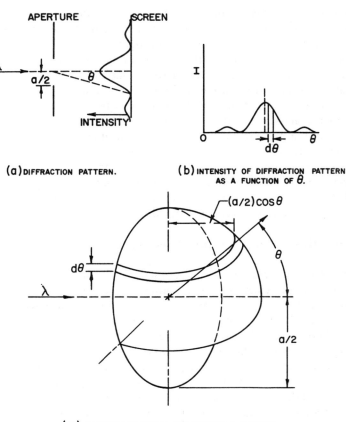

(a) DIFFRACTION PATTERN.

(b) INTENSITY OF DIFFRACTION PATTERN AS A FUNCTION OF θ.

(c) APERTURE OF RADIUS $a/2$ SHOWING A PORTION OF THE REFERENCE SPHERE

Figure 1-10 Factors involved in the analysis of photon flux.

the element of surface area in the same unit time. The ratio of dN/N will be given by the ratio of their respective areas:

$$\frac{dN}{N} = \frac{\pi(a/2)\cos\theta\, d\theta}{(\pi/2)a^2} = \frac{\cos\theta\, d\theta}{a}$$

since $\pi a/2$ vanishes, and may be expressed more conveniently as

$$dN = \frac{N\cos\theta\, d\theta}{a}$$

The flux across a plane perpendicular to the optical axis and parallel to the planes of the aperture and the screen is given by

$$df = dN\sin\theta$$

or, upon substitution of the preceding expression for dN, this equation for df becomes

$$df = \frac{N \sin \theta \cos \theta \, d\theta}{a}$$

The intensity of the pattern is greatest in the region of the central maximum. So, for the small angle between the optical axis and the first intensity minimum, it may be approximated that $\sin \theta \simeq \theta$ and $\cos \theta \simeq 1$. These approximations result in

$$df \simeq \frac{N\theta \, d\theta}{a}$$

Use now is made of the space limitation, Eq. (1-19), to substitute for the angle θ as λ/a:

$$df \simeq \frac{N}{a} \cdot \frac{\lambda}{a} \, d\theta = \frac{N\lambda \, d\theta}{a^2}$$

The implicit coefficient of unity in this equation should be represented by an integer, α, which determines all the minima in the diffraction pattern. This, more generally, is given by

$$df \simeq \frac{\alpha\lambda N \, d\theta}{a^2}$$

The fraction N/a^2 is the number of photons per unit time per unit area or the intensity, I, of the photons. Therefore,

$$df \simeq \alpha\lambda I \, d\theta$$

and the probability of finding a photon anywhere on the screen will vary as df. Since prime interest is attached to the greatest probability of finding a photon from the monochromatic beam, and the number of photons is greatest where the intensity is highest in the central maximum, the optimum likelihood of finding a photon will lie within the central maximum, and the flux may be expressed as

$$df \simeq \text{constant} \cdot I \, d\theta$$

for this case, since α and λ now are constants.

A means now is sought to translate these findings into a more general and useful form. This may be done by means of a function, ψ, which may be regarded as representing the equation for the *amplitude* of a de Broglie wave and which is called a *wave function*. This wave function is meaningless by itself, especially since most wave functions are complex; that is, they involve $(-1)^{1/2}$ which is denoted by i. In classical optics, the intensity of a wave is proportional to the square of its amplitude. It was postulated by Born that the *intensity* of a wave could be given by $\psi^*\psi$, where ψ^* is the complex conjugate of ψ, in a way analogous to the classical case. Thus, if $\psi = a + ib$, $\psi^* = a - ib$, then $\psi^*\psi = a^2 + b^2$; the product of the complex or imaginary conjugate amplitudes gives a real intensity that does have meaning; it is known as the *probability density*.

This concept may be applied to the probability of finding a photon. On the basis of the foregoing, the intensity of the diffraction pattern may be expressed as

$$I \propto \psi^*\psi$$

and, multiplying both sides of this variation by $d\theta$,

$$I\,d\theta \propto \psi^*\psi\,d\theta$$

Or, in terms of the expression previously obtained for df, upon substituting the probability density for the intensity, I, df is given by

$$df = \text{constant} \cdot I\,d\theta = C^2\psi^*\psi\,d\theta$$

This gives the probability of finding a photon in the element $d\theta$ in terms of the amplitude of a de Broglie wave. This frequently is written as

$$df = \text{constant} \cdot I\,d\theta = C^2|\psi|^2\,d\theta$$

In both of these expressions, C is a constant.

These equations were derived on the basis of the most probable angular range in which a photon would be found. But this represents only a small fraction of the entire range in which it is possible to find a photon. The probability that a photon will be found anywhere on the screen is

$$B^2 \int_{-\pi/2}^{\pi/2} \psi^*\psi\,d\theta$$

where B^2 is an arbitrary constant selected such that

$$B^2 \int_{-\pi/2}^{\pi/2} \psi^*\psi\,d\theta = 1$$

This is called *normalization*. The probability of finding a photon in the most probable range (that is, within the central maximum of the intensity) with respect to the entire range is given by

$$\frac{C^2\psi^*\psi\,d\theta}{B^2 \int_{-\pi/2}^{\pi/2} \psi^*\psi\,d\theta} = C^2\psi^*\psi\,d\theta$$

since the denominator is equal to unity. This is called the *probability density*. This concept may be applied to any range of interest.

Since Heisenberg's uncertainty principle states that the position of the photon cannot be known precisely, the only alternative is to find its most likely location. In the preceding attempt to do this, it must be emphasized that this is only a probability and that the photon may be found anywhere on the screen, in angular positions ranging from $\pi/2$ to $-\pi/2$. The fact that the intensity is greatest in the central maximum was used here to focus on the maximum probability of finding a photon and to simplify the approximation. This now takes on greater significance because it means that the probability densities obtained from properly selected wave functions should be greatest in that region.

The ideas about wave functions outlined here will be used in the development of Schrödinger's equation in Chapter 3 and elsewhere.

1.7 BIBLIOGRAPHY

DAY, M. C., and SELBIN, J., *Theoretical Inorganic Chemistry*, 2nd ed., Van Nostrand Reinhold, New York, 1969.

HALLIDAY, D., and RESNICK, R., *Physics*, Part II, Wiley, New York, 1966.

POLLOCK, D. D., *Physical Properties of Materials for Engineers*, Vol. I, CRC Press, Boca Raton, Fla., 1982.

RICHMYER, F. K., KENNARD, E. H., and LAURETSON, T., *Introduction to Modern Physics*, 5th ed., McGraw-Hill, New York, 1955.

SPROULL, R. L., *Modern Physics*, Wiley, New York, 1956.

1.8 PROBLEMS

1.1. Given that the work function of a metal is 4.5 eV. Approximate the maximum wavelength (in units of angstroms) required of a photon to eject an electron from the metal.

1.2. Calculate the minimum energy required to promote the valence electron of a hydrogen atom to (a) the first and (b) the second excited states.

1.3. Calculate the wavelengths of the energies emitted by the electrons in Problem 1.2 when they return to the ground state.

1.4. Given that a free electron has an energy of 1 eV. Calculate its frequency, wavelength (cm), and momentum when treated as a wave of zero mass.

1.5. Use the data given in Section 1.6.3. to verify the findings of Davisson and Germer, treating the electrons as particles and waves.

1.6. (a) Calculate the uncertainty in the velocity of an electron $(9.1 \times 10^{-28} \, g)$ when its uncertainty in position is $\sim 10^{-7} \, cm$.

   (b) Do the same for a particle with a mass of 10 g and $\Delta x \cong 1 \, cm$.

   (c) Explain why these results conform to the uncertainty principle.

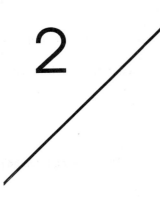

2 / STATISTICAL BACKGROUND

All solids are composed of vast, but definite numbers of particles. In general, their macroscopic properties may be described by two different, generic, statistical approaches. One of these, thermodynamic statistics, is based on continuous energy spectra of any number of countable particles and provides the average properties of a solid. The second, quantum mechanics statistics, is based on discrete energy spectra and provides the most probable properties of fixed numbers of quantized indistinguishable particles. The statistics used for quantized particles must take spin into consideration in addition to the other factors discussed in Sections 1.6.6 and 1.6.7.

The examinations of these statistical approaches and their capabilities provide good bases for the understandings of the physical properties of materials.

2.1 THE BOLTZMANN EQUATION

The thermodynamic properties of a solid (or liquid or gas) are considered to be those averaged over time for a given set of conditions (that is, a given *state*). The solid, for thermodynamic purposes, is known as a *system*. The properties of such a system cannot be known at any given instant for two reasons. First, its initial state is unknown and, second, fluctuations in its properties take place continuously.

This inability to describe the time dependencies of the properties of the solid is overcome by means of an imaginary mathematical construct that is known as a *canonical ensemble.* This is arrived at in the following way. The solid is treated as a *macroscopic* system surrounded by a thermal reservoir. This reservoir is made to be sufficiently large so that it may be regarded as being unaffected by any changes induced in it by the system. This system–reservoir configuration may be considered to be duplicated as many times as may be required for mathematical or theoretical purposes. The number of such imaginary system–reservoir duplications is not necessarily related to the number of particles of which the system is composed; the numbers of such "clones" may be very much greater than the actual number of particles in a system. Such a vast number of isolated clones is called a *canonical ensemble.* Each clone has exactly the same state parameters (pressure, volume, temperature, and constitution) as every other one. In addition the total energy, E_T, of each system–reservoir pair is constant and independent of the system.

The difficulties noted previously for the determination of the thermodynamic properties may be eliminated if it is assumed that the average of the thermodynamic properties of all the clones of an ensemble at a given instant is identical to the time average of the corresponding properties of the system. This assumption simplifies calculations because the time factor need no longer be taken into account; such calculations are said to be time-independent.

Now consider a single, isolated member of an ensemble with a constant total energy, E_T, and whose system is in a given energy state, E_i. The energy of the reservoir, then, is given by $E_T - E_i \gg E_i$. Each particle in the system may be described as being in a state that is defined both by its spatial coordinates and by an energy factor, such as its velocity or its momentum, also in Cartesian coordinates, and actually is a six-dimensional space known as *state space*. The system, then, is depicted by the most probable way in which its particles can possess the energies (occupy states) with coordinates $(x, y, z, \mathbf{p}_x, \mathbf{p}_y, \mathbf{p}_z)$ in the available energy range. A very simple derivation based on this approach is given here.

The probability, P_i, that the system will be in the state E_i will vary directly as the maximum number of equally likely states, n_i, that are available to its particles to possess the given E_i, with respect to all the possible states, N, in the energy range $E_T - E_i$. Thus,

$$P_i \propto \frac{n_i}{N} \tag{2-1}$$

This implies that the particles comprising the system are distinguishable and that they can be counted.

This probability also may be expressed in terms of the number of possible ways, W, that the particles can occupy states in the range $E_T - E_i$:

$$P_i \propto W(E_T - E_i) \tag{2-2}$$

Equations (2-1) and (2-2) are combined to give

$$p_i \propto \frac{n_i}{N} \propto W(E_T - E_i) \tag{2-3}$$

Here n_i/N varies with the given energy difference. Such a variation is a characteristic property of logarithmic functions. Thus, Eq. (2-3) may be rewritten as

$$\ln P_i \propto \ln \frac{n_i}{N} \propto \ln[W(E_T - E_i)] \qquad (2\text{-}4)$$

Attention now is directed at the logarithmic energy term. This is more easily dealt with in the form of a polynomial. Such a conversion may be obtained by starting with the general expression for Taylor's expansion:

$$f(x) = f(a) - f'(a)\frac{x-a}{1!} + f''(a)\frac{(x-a)^2}{2!} \cdots \qquad (2\text{-}5)$$

In the case under consideration $a = E_T$ and $x = E_i$, keeping in mind that $E_i \ll E_T$. Thus, E_i may be regarded as being negligibly small with respect to E_T. This results in the approximation that $(E_T - E_i) \cong E_T$ and is the equivalent of saying that the relative value of a in the Taylor expansion is such that it may be set equal to zero. This is convenient since it results in a special case of the Taylor expansion that is known as McLaurin's theorem and is the equivalent of evaluating the expansion about E_T:

$$f(x) = f(0) + f'(0)\frac{x}{1!} + f''(0)\frac{x^2}{2!} \cdots \qquad (2\text{-}6)$$

Now, using just the first two terms of this series, since the higher powers of E_i are small and may be neglected, the factors needed for Eq. (2-6) that are obtained from Eq. (2-4) are

$$f(0) = \ln W(E_T) \qquad (2\text{-}7a)$$

$$f'(0) = -\left(\frac{\partial \ln W}{\partial E}\right)_{E=E_T} \cdot E_i \qquad (2\text{-}7b)$$

It will be noted that Eq. (2-7b) is negative because the energy difference as originally stated in Eq. (2-4) is opposite in sign to that given in Eq. (2-5). The substitution of Eqs. (2-7) into Eq. (2-6) in the form of Eq. (2-4) results in

$$\ln P_i \propto \ln W(E_T) - \left(\frac{\partial \ln W}{\partial E}\right)_{E=E_T} \cdot E_i \qquad (2\text{-}8)$$

Since E_T is constant and independent of E_i, $\ln W(E_T)$ is a constant. The coefficient of E_i, in Eq. (2-8), is evaluated in terms of E_T so that $(\partial \ln W/\partial E)_{E=E_T}$ may be represented by the parameter β:

$$\frac{\partial \ln W}{\partial E} = \beta \qquad (2\text{-}9)$$

These considerations applied to Eq. (2-8) give

$$\ln P_i \propto -\beta E_i \qquad (2\text{-}10a)$$

or

$$P_i \propto e^{-\beta E_i} \qquad (2\text{-}10b)$$

Equation (2-10b) is essentially the Boltzmann equation for the probability of finding a system in a state E_i.

Equations (2-10a, b) were obtained from one state of the system. The probability of finding the ensemble in this state E_i is obtained by summing over all the systems in the canonical assembly so that

$$P = \sum P_i = C \sum_{i=1}^{\infty} e^{-\beta E_i} = 1 \qquad (2\text{-}11)$$

Here the coefficient C forces the probability to equal unity. This has the effect of ensuring that all n_i in N are considered [Eq. (2-1)]; therefore, there is a real probability of finding the E_i in the range $E_T - E_i$. This technique of normalization is the same as that employed in Section 1.6.7.

The parameter β may be defined more exactly by starting with the first law of thermodynamics:

$$dE = dQ - p\,dV \qquad (2\text{-}12)$$

where E is the internal energy of the system, Q is the externally applied energy, p is the pressure, and V is the volume of the system. The second law defines the entropy as

$$dS = \frac{dQ}{T}; \qquad dQ = T\,dS \qquad (2\text{-}13)$$

The substitution of Eq. (2-13) into Eq. (2-12) gives the Maxwell relation

$$dE = T\,dS - p\,dV \qquad (2\text{-}14)$$

For virtually all solids (and many liquids), the work done by the system against pressure is essentially constant. Such constants are very small for solids, so it may be approximated that $p\,dV \cong 0$ with a high degree of accuracy. This approximation results in

$$dE \cong T\,dS \qquad (2\text{-}15)$$

Or

$$\left(\frac{\partial S}{\partial E}\right)_V = \frac{1}{T} \qquad (2\text{-}16)$$

This expression of the combined first and second laws may be made more useful for present needs by means of Boltzmann's definition of entropy:

$$S = k_B \ln W \qquad (2\text{-}17)$$

in which k_B is Boltzmann's constant. This permits the reexpression of Eq. (2-16) as

$$\left(\frac{\partial S}{\partial E}\right)_V = \frac{\partial (k_B \ln W)}{\partial E} = \frac{1}{T} \qquad (2\text{-}18a)$$

or as

$$\frac{\partial \ln W}{\partial E} = \frac{1}{k_B T} \qquad (2\text{-}18b)$$

Equation (2-18b) is exactly the same as Eq. (2-9), so

$$\beta = \frac{1}{k_B T} \tag{2-19}$$

Then, using Eq. (2-10b),

$$P_i \propto e^{-E_i/k_B T} \tag{2-20}$$

Equation (2-20) gives the probability of a system being in state E_i. More generally, the probability that a single particle will occupy state E_i, sometimes called the occupation number, is given by

$$P_i = \frac{1}{N} = f_B = \frac{1}{e^{E_i/k_B T}} \tag{2-21}$$

This probability for a single event, denoted here by f_B, is known as the Boltzmann distribution function.

Now, the average probability of finding the number of occupied states, $\langle n \rangle$, may be determined by dividing the probable number of states with energy E_i by the average probable number of available states as shown by

$$\langle n \rangle = \frac{\sum_{i=1}^{\infty} n_i e^{-E_i/k_B T}}{\sum_{i=1}^{\infty} e^{-E_i/k_B T}} \tag{2-22}$$

The average of any physical property, A, may be obtained by the means indicated by Eq. (2-22). This is illustrated in a general way by

$$\langle A \rangle = \frac{\sum_{i=1}^{\infty} A_i e^{-E(A)_i/k_B T}}{\sum_{i=1}^{\infty} e^{-E(A)_i/k_B T}} \tag{2-23}$$

It will be noted that no restrictions were imposed on E_i or $E_T - E_i$ other than the assumption that an equally likely number of ways were available for the accommodation of E_i within the energy range $E_T - E_i$. Implicit in this is the idea that such an energy range is a continuous one. Thus, as given up to this point, the concepts presented here are applicable only to the macroscopic classical approach in which an identifiable particle may occupy any energy state in a continuous spectrum.

2.2 THE MAXWELL–BOLTZMANN EQUATION

Again consider a system consisting of distinguishable, or countable, particles. From this, the total number of particles in the system is given by

$$N = \sum_i n_i \tag{2-24}$$

Then the energy of the system is

$$E = \sum_i n_i E_i \tag{2-24a}$$

As in the Boltzmann statistics, no limitations are placed on N or E_i. Equation (2-24) is differentiated to obtain

$$\partial E = \sum_i E_i \, \partial n_i + \sum_i n_i \, \partial E \qquad (2\text{-}25)$$

But, according to Eq. (2-12), ∂E may change only if the volume changes. This is taken into account, under isothermal conditions, by rewriting Eq. (2-25) in the form

$$\partial E = \sum_i E_i \, \partial n_i + \sum_i n_i \frac{\partial E}{\partial V} \, \partial V \qquad (2\text{-}26)$$

But $p = -\partial E/\partial V$, so

$$\partial E = \sum_i E_i \, \partial n_i - p \, \partial V \qquad (2\text{-}27)$$

And, by comparison with Eq. (2-12),

$$\partial Q = \sum_i E_i \, \partial n_i \qquad (2\text{-}28)$$

At equilibrium, when the system does no work against its surroundings, $p \, \partial V = 0$ and

$$\partial E = \partial Q = \sum_i E_i \, \partial n_i \qquad (2\text{-}29)$$

Now, if extremely small changes occur very slowly in the system at equilibrium, it may be stated that

$$\partial \ln W = 0 \qquad (2\text{-}30\text{a})$$

$$\partial \ln N = \sum_i \partial n_i = 0 \qquad (2\text{-}30\text{b})$$

and

$$\partial E = \sum_i E_i \, \partial n_i = 0 \qquad (2\text{-}30\text{c})$$

Since Eqs. (2-30) all equal zero, they may be related, using the undetermined coefficients α and β, in the form

$$\partial \ln W - \alpha \sum_i \partial n_i - \beta \sum_i E_i \, \partial n_i = 0 \qquad (2\text{-}31)$$

The number of equally likely ways that E_i can be accommodated in the energy range is given by W. This factor is expressed as the number of combinations of the N states taken n_i at a time, or

$$W = \frac{\sum_i N!}{\sum_i (N - n_i)! \sum_i n_i!} \qquad (2\text{-}32)$$

Equation (2-32) is reexpressed in logarithmic form as

$$\ln W = \sum_i [\ln N! - \ln n_i! - \ln(N - n_i)!] \tag{2-33}$$

Stirling's approximation for sufficiently large N,

$$\ln a! \cong a \ln a - a$$

is now used to simplify Eq. (2-23) by transforming it into

$$\ln W = \sum_i [N \ln N - N - n_i \ln n_i + n_i - (N - n_i) \ln (N - n_i) + (N - n_i)] \tag{2-34}$$

Upon collection of terms, Eq. (2-34) reduces to

$$\ln W = \sum_i [N \ln N - n_i \ln n_i - (N - n_i) \ln (N - n_i)] \tag{2-35}$$

However large, N is a constant for a given system; thus, upon differentiation the term $N \ln N$ vanishes and Eq. (2-35) becomes

$$\partial \ln W = \sum_i \left\{ \left[-n_i \cdot \frac{1}{n_i} - \ln n_i \right] \partial n_i - \left[(N - n_i) \frac{-1}{N - n_i} - \ln (N - n_i) \right] \partial n_i \right\}$$

Upon simplification this reduces to

$$\partial \ln W = \sum_i [\ln (N - n_i) - \ln n_i] \partial n_i \tag{2-36a}$$

or

$$\partial \ln W = \sum_i \ln \left(\frac{N - n_i}{n_i} \right) \partial n_i \tag{2-36b}$$

Equation (2-36b) is substituted into Eq. (2-31) to obtain

$$\sum_i \left[\ln \left(\frac{N - n_i}{n_i} \right) - \alpha - \beta E_i \right] \partial n_i = 0 \tag{2-37}$$

Equation (2-37) is valid only if

$$\ln \left(\frac{N - n_i}{n_i} \right) - \alpha - \beta E_i = 0 \tag{2-38a}$$

Or

$$\ln \left(\frac{N - n_i}{n_i} \right) = \alpha + \beta E_i \tag{2-38b}$$

Equation (2-38b) may be rewritten as

$$\ln \left(\frac{N}{n_i} - 1 \right) = \alpha + \beta E_i \tag{2-39}$$

Now, since no limitations have been placed on N, and since $N \gg n_i$, then $N/n_i \gg 1$ and it may be approximated that

$$\ln \left(\frac{N}{n_i} - 1 \right) \cong \ln \left(\frac{N}{n_i} \right) \cong \alpha + \beta E_i \tag{2-40}$$

The probability, P_i, will vary as n_i/N. This may be obtained from Eq. (2-40) as

$$\ln P_i \propto \ln \left(\frac{n_i}{N} \right) = -\alpha - \beta E_i \tag{2-41}$$

Or

$$P_i \propto \frac{n_i}{N} \propto \frac{1}{e^{\alpha + \beta E_i}} \tag{2-42}$$

This is the general form of the Maxwell–Boltzmann equation. The probability for the occurrence of a *single* event is

$$P_i = \frac{1}{N} = f_{\text{MB}} = \frac{1}{e^{\alpha + \beta E_i}} \tag{2-43}$$

where f_{MB} is the Maxwell–Boltzmann distribution function. Here $\beta = 1/k_B T$ [Eq. (2-19)]. The parameter α is, as yet, undefined.

The meaning of α may be obtained by starting with the Maxwell expression for the first and second laws of thermodynamics,

$$dE = T \, dS - p \, dV \tag{2-14}$$

Equation (2-14) is incomplete as it stands because the energy of a system varies with the number of particles of which it is composed. Each particle entering or leaving the system changes the energy of the system. Therefore, Eq. (2-14) must be modified to include this influence as

$$\partial E = T \, \partial S - p \, \partial V + \mu \, \partial N \tag{2-44}$$

in which μ is the chemical potential. Thus, μ is the energy associated with the presence or the absence of a particle. Greater insight into the meaning of this concept may be obtained by considering an isolated system composed of at least two phases, α and β, at thermal equilibrium. The *total number* of particles of which the system is composed will be constant and, in addition $\partial S = 0$ and $\partial V = 0$. These conditions reduce Eq. (2-44) to

$$\partial E = \mu \, \partial N$$

or

$$\mu = \frac{\partial E}{\partial N} \tag{2-45}$$

If ∂N particles of a given species diffuse from the α phase to the β phase, the energy of the α phase will be changed by an absolute amount

$$\partial E_\alpha = \mu_\alpha \, \partial N \tag{2-46a}$$

and that of the β phase will be changed by

$$\partial E_\beta = \mu_\beta \, \partial N \tag{2-46b}$$

Then, since $\partial E_\alpha = \partial E_\beta$ at equilibrium, Eqs. (2-46a) and (2-46b) may be equated to obtain

$$\mu_\alpha = \mu_\beta \tag{2-46c}$$

And μ_i, the chemical potential, must be the energy associated with a single particle in a phase or as it transfers to another phase. This concept is important in the theory of alloy phases.

The definition as given by Eq. (2-45) permits the parameter α to be expressed more specifically by starting with the logarithmic form of Eq. (2-43):

$$\ln \frac{1}{N} = -(\alpha + \beta E_i) \tag{2-47}$$

the derivative of Eq. (2-47) is taken with respect to N to obtain

$$\frac{\partial \ln(1/N)}{\partial N} = -\frac{\partial \alpha}{\partial N} - \beta \frac{\partial E_i}{\partial N} \tag{2-48}$$

And, when maximized, Eq. (2-48) is set equal to zero to get

$$\frac{\partial \alpha}{\partial N} = -\beta \frac{\partial E_i}{\partial N} \tag{2-49}$$

Use now is made of Eq. (2-45) so that

$$\frac{\partial \alpha}{\partial N} = -\beta \mu \tag{2-50}$$

Equation (2-50) may now be reexpressed as

$$\partial \alpha = -\beta \mu \, \partial N \tag{2-51}$$

To evaluate α, it must be remembered that μ is the energy associated with just one particle. Thus, upon integration,

$$\alpha = -\beta \mu \tag{2-52}$$

The definition of α is completed by referring to Eq. (2-19). This gives

$$\alpha = -\frac{\mu}{k_B T} \tag{2-53}$$

Thus, Eq. (2-43) now may be expressed in its usual form as

$$f_{\text{MB}} = \frac{1}{e^{(E_i - \mu)/k_B T}} \tag{2-54}$$

Equation (2-54) is the classical Maxwell–Boltzmann distribution function (Fig. 2-1).

The proper use of Eq. (2-54) requires that the bases on which it was derived be understood.   No limitations were placed on E_i [Eq. (2-24)], so it may represent a continuous spectrum of energy for countable particles.   In addition no upper limit was placed on N.   A lower limit on N is implicit in the reexpression of Eq. (2-33) as Eq. (2-34).   The magnitude of N required for this transformation to be valid is negligibly small compared to the magnitude of N encountered in virtually all thermodynamic systems; it, therefore, may be neglected as a lower limit.   Thus, N usually will be very large and, using the condition to arrive at Eq. (2-40), may be expressed as

$$\frac{n_i}{N} \ll 1 \tag{2-55}$$

This means that there is a very much larger number of energy states available for occupancy by the n_i particles than the actual number of particles being considered. This is the same as saying that the occupation number is very low.   It follows from this that the probability of multiple occupancy of a given state is negligibly small and may be disregarded.   Another consequence of Eq. (2-55) is that it is highly probable that the n_i particles will be randomly distributed in the N states rather than in a few states or in closely bunched states.   Since interactions do not take place between particles under these conditions, the macroscopic properties of a system composed of such particles are independent of the specific nature of the particles.   This also is the case for the Boltzmann equation, Eq. (2-21).   The criterion for this behavior is given by Eq. (2-55) and such systems are called *nondegenerate* systems; each state is singly occupied.   As an example, the properties of ideal gases may be determined by using

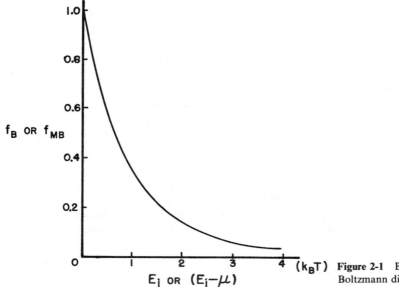

Figure 2-1 Boltzmann or the Maxwell–Boltzmann distribution functions.

the Boltzmann or the Maxwell–Boltzmann statistics. The same is true for real gases at high temperatures, at low pressures, or when both of these conditions prevail simultaneously; such qualifications ensure that any van der Waals, London, or other electrostatic interactions between particles may be considered to be absent or negligible.

At much lower energies than those just considered, the particles of the system will show a trend toward the occupancy of the states at lower energies. Under these conditions, where interactions between the real gas particles are present, Eqs. (2-21) and (2-55) are no longer accurate.

In summary, the classical Boltzmann and Maxwell–Boltzmann statistics represent nondegenerate systems because the noninteracting, identifiable (countable) particles have an unlimited number of states available for occupancy within a continuous energy range.

2.3 BASES FOR THE STATISTICAL TREATMENT OF QUANTUM PARTICLES

The radically different behaviors of particles that do not obey classical mechanics have been indicated in Chapter 1. Any statistics treating such particles must, of necessity, take such differences into account. These will be considered by starting with a short examination of the capabilities of classical statistics.

2.3.1 Capabilities of Classical Statistics

The classical Boltzmann and Maxwell–Boltzmann distribution functions [Eqs. (2-21) and (2-54), respectively] are based on the idea that all the particles in a system are identifiable and, therefore, countable. There is no duality or uncertainty with respect to the properties of particles, such as those noted in Sections 1.6.2 and 1.6.6. The number of states is unlimited, and their energies are unrestricted and are considered to lie in a continuous spectrum. As a result, particles may singly occupy any state or coordinates $(x, y, z, \mathbf{p}_x, \mathbf{p}_y, \mathbf{p}_z)$ in state space. See Section 2.1. These statistics cannot be applied to quantum particles because the specific nature of these particles must be taken into account. This is not true of classical thermodynamic statistics (Sections 2.1 and 2.2). This may be shown by attempting to set up conditions that include the nature of a particle by ensuring that only single-state occupancy prevails.

Under certain conditions, Eq. (2-21) can be controlled so that there is a very high probability of single occupancy of states. For this to be the case, the condition that

$$N_v \leqslant 1/V_s$$

must be obeyed. Here N_v is the number of particles per unit volume of state space, and V_s is the maximum volume of state space available to each particle. (It must be recognized that V_s is not related to particle volume.) The mean interparticle distance, λ, must be given by

$$\lambda \ll \frac{1}{N_v}$$

so that

$$\lambda \ll V_s^{1/3} \quad \text{and} \quad \lambda^3 \ll V_s$$

Now, using Eq. (2-21), the classical probability of single occupancy of state E_i is

$$n_{v_i} = N_v \lambda^3 \exp\left(\frac{-E_i}{k_B T}\right)$$

or, since $N_v \leqslant 1/V_s$,

$$n_{v_i} = \frac{\lambda^3}{V_s} \exp\left(\frac{-E_i}{k_B T}\right)$$

So, if λ^3 is considered as being the probable assignable volume of state space per particle, V_i, then

$$n_{v_i} = \frac{V_i}{V_s} \exp\left(\frac{-E_i}{k_B T}\right)$$

In effect, state space may be manipulated so that a high probability of single occupancy of a state exists by decreasing the arbitrary position and momentum unit coordinates of state space. However, a lower limit to this occurs as a result of Eq. (1-22) if these are too small. Also, a limit arising from Eq. (1-17) becomes apparent before the effects of Eq. (1-22) are manifested. This may be shown in the following way, starting with

$$\lambda = \frac{h}{p} = \frac{h}{(2mE_i)^{1/2}}$$

This minimizes the volume of a state as

$$V_{i(\text{min})} = \lambda_{\text{min}}^3 = \frac{h^3}{(2mE_i)_{\text{max}}^{3/2}}$$

So

$$n_{v_i} = \frac{h^3/(2mE_i)_{\text{max}}^{3/2}}{V_s} \exp\left(\frac{-E_i}{k_B T}\right)$$

The important point here is than when the de Broglie wavelength, in terms of particle energy, is small compared to the interparticle distance, classical statistics can be considered as representing highly probable single occupancy of states by particles. However, manipulation of state space, such as shown here, cannot ensure that the nature of a particle will be taken into account by single occupancy of state space. This is particularly true for indistinguishable particles where Eq. (2-55) is not obeyed and n_i/N may approach unity. Here, the principle of exclusion must be invoked to secure particle individuality. See Sections 2.3.2, 2.5, and 3.9.

An example of high probability of single occupancy of a state is given by real gases either at high temperatures or at low pressures or densities or where both conditions are present; these are conditions under which the macroscopic properties of the ensemble are independent of the nature of the particles. In the Maxwell–Boltzmann function, Eq. (2-54), for the case in which $E_i = \mu = \langle E \rangle$, the average energy of the ensemble, $f_{MB} = 1$. But, as E_i becomes increasingly larger than $\mu = \langle E \rangle$, the probability of occupancy of state E_i by a particle becomes decreasingly probable, and single occupancy is therefore more likely. Where E_i is very much larger than $\mu = \langle E \rangle$, Eq. (2-54) may be considered to approach Eq. (2-21). This is sometimes called the "*Boltzmann tail.*"

2.3.2 Limitations Imposed on Quantum Particles

Particles obeying quantum mechanics are permitted much less "freedom" than classical particles (Chapter 1). While quantum particles can have energies that approach a continuous spectrum, known as a *quasicontinuum*, their energies must always be considered as being discrete. They also are indistinguishable because their properties change upon attempts at close observation. In addition the Heisenberg uncertainty principle does not allow the precise location of the coordinates of particles $(x, y, z, \mathbf{p}_x, \mathbf{p}_y, \mathbf{p}_z)$ in state space [Eq. (1-22)]. A further constraint on these particles is given by the Pauli exclusion principle. In its simplest form, this principle states that no two particles can have the same coordinates in state space at a given time.

The state space for quantum particles is more complicated than that for classical particles. Here, an additional dimension, or degree of freedom known as *spin*, must be considered in addition to the six coordinates required for a classical particle. This is discussed more fully in Sections 2.5 and 3.9.

The inclusion of spin results in two kinds of statistical treatments. Particles with integral spin (integral values of intrinsic angular momentum) are treated statistically by the Bose–Einstein method (Sections 2.4 and 2.5). Particles with half-integral spin are treated using the Fermi–Dirac statistics (Section 2.6). When spin is explicitly considered, the Fermi–Dirac approach allows a state either to be empty or to be singly filled by a single particle. Where spin is implicitly considered, two particles with otherwise identical coordinates in state space may occupy a state, provided that their spins are opposite to each other. No other ways of filling a state are permitted.

2.4 THE BOSE–EINSTEIN EQUATION

Unlike the classical particle treated in previous sections, those considered here may occupy only discrete states. Any meaningful statistical treatment of quantum particles must take this into account and, in effect, be restricted by the limitations described in Sections 1.6.7 and 2.3.2. In addition because of the uncertainty principle and the Compton effect (Sections 1.6.6 and 1.6.4), any attempt to count such particles will change their properties. Thus, the quantum particles with integral spin (Section 2.5) to be treated statistically here must be considered to be indistinguishable. This means that any combinations of particles with a given energy, such as AB, AC, or BC, now are not countable as such, as is the case for the Maxwell–Boltzmann statistics; they can no longer be treated as different ways of filling a given state. The paired particles now must be counted as a single way of treating multiple pairs of indistinguishable particles. Thus, many such particle pairs may have the same energy. Such systems are said to be *degenerate*. Another important difference is that the energy spectrum is discontinuous and the states now are discrete, quantized energy levels because of Eq. (1-5). Where large numbers of particles are involved, these states are very closely packed and, while still discontinuous, approach a quasicontinuum.

With these constraints in mind, the particles now can be treated in a way similar to ideal gas molecules and without restrictions on the order or the number in which these particles occupy states. These concepts are important because they permit the derivation of the Bose–Einstein distribution function in an uncomplicated way. The energy of the ith particle is $E_i = n_i h v_i$. Now, using Eq. (2-54) and the averaging method indicated by Eq. (2-22), the probable relative number of particles is

$$\frac{n}{N} = \frac{\sum_{n=0}^{\infty} n_i e^{-(n_i h v_i + \mu)/k_B T}}{\sum_{n=0}^{\infty} e^{-(n_i h v_i + \mu)/k_B T}} \tag{2-56}$$

The chemical potential, μ, is constant at equilibrium, so the treatment of Eq. (2-56) may be facilitated by letting

$$n_i \varepsilon_i = -\frac{n_i h v_i + \mu}{k_B T} \tag{2-57}$$

This reduces Eq. (2-56) to

$$\frac{n}{N} = \frac{\sum_{n=0}^{\infty} n_i e^{n_i \varepsilon_i}}{\sum_{n=0}^{\infty} e^{n_i \varepsilon_i}} \tag{2-58}$$

The summations now are readily written as

$$\frac{n}{N} = \frac{0 + e^{\varepsilon} + 2e^{2\varepsilon} + \cdots + n e^{n\varepsilon} + \cdots}{1 + e^{\varepsilon} + e^{2\varepsilon} + \cdots + e^{n\varepsilon} + \cdots} \tag{2-59}$$

Each term in the numerator of Eq. (2-59) is the derivative of the corresponding term in its denominator. This may be expressed as

$$\frac{n}{N} = \frac{d}{d\varepsilon} \ln \left(1 + e^{\varepsilon} + e^{2\varepsilon} + \cdots + e^{n\varepsilon} + \cdots \right) \tag{2-60}$$

The series in Eq. (2-60) is given by

$$1 + e^{\varepsilon} + e^{2\varepsilon} + \cdots + e^{n\varepsilon} + \cdots = \frac{1}{1 - e^{\varepsilon}} \tag{2-61}$$

so the use of Eq. (2-61) in Eq. (2-60) gives

$$\frac{n}{N} = \frac{d}{d\varepsilon} \ln \left(\frac{1}{1 - e^{\varepsilon}}\right) \tag{2-62}$$

And, when the indicated derivative is taken,

$$\frac{n}{N} = \frac{e^\varepsilon}{1 - e^\varepsilon} \tag{2-63}$$

Then the numerator and denominator of Eq. (2-63) are divided by e^ε to obtain

$$\frac{n}{N} = \frac{1}{e^{-\varepsilon} - 1} \tag{2-64}$$

The substitution of Eq. (2-57) for ε, where $n_i = 1$, gives

$$\frac{1}{N} = f_{\mathrm{BE}} = \frac{1}{e^{(h\nu_i + \mu)/k_B T} - 1} \tag{2-65}$$

Here f_{BE} is the Bose–Einstein distribution (Fig. 2-2).

It is instructive to apply Eq. (2-65) to the photons emanating from a black body (Sections 1.2 and 1.3). At thermal equilibrium, the number of such photons is constant. So, according to Eq. (2-45),

$$\mu = \frac{\partial E}{\partial N} = 0 \tag{2-66}$$

And, if the energy of a photon is $E = h\nu$, its *average* energy is given by

$$\langle E \rangle = E f_{\mathrm{BE}} = \frac{h\nu}{e^{(h\nu + 0)/k_B T} - 1} = \frac{h\nu}{e^{h\nu/k_B T} - 1} \tag{2-67}$$

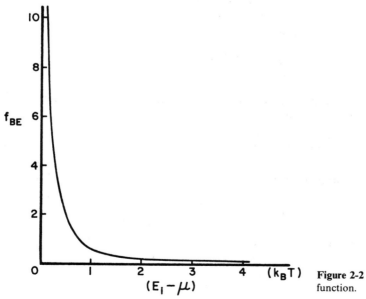

Figure 2-2 Bose–Einstein distribution function.

This is the same average photon energy used by Planck in Eq. (1-6). Any number of photons may be absorbed or emitted, so their numbers are not necessarily conserved; this depends on the photon states. The case used here, that is, $\mu = 0$, is representative of degenerate states. In those instances in which the particles are nondegenerate, the chemical potential has a relatively large, negative value [Section 2.6 and Eq. (2-80a)].

A comparison of Eqs. (2-54) and (2-65) shows that $f_{MB} < f_{BE}$ for the case of indistinguishable, degenerate particles. This arises because indistinguishable particles would be expected to tend toward higher state occupancy (higher occupation numbers) than distinguishable particles (Figs. 2-1 and 2-2).

It also is of interest to note that, for high temperatures, $T \geqslant 300\,\mathrm{K}$, the exponent in Eq. (2-67) may be expressed as the first two terms of the series expansion for e^x in the same way in which Eq. (1-10) was obtained from Eq. (1-9). Thus, Eq. (2-67) may be rewritten as

$$\langle E \rangle \cong \frac{h\nu}{1 + (h\nu/k_B T) - 1} = h\nu \cdot \frac{k_B T}{h\nu} = k_B T \tag{2-68}$$

This gives the average energy of a quantized particle as being the same as that for a classical particle at sufficiently high temperatures. This agreement is known as the *Bohr correspondence principle*. This states that, in the appropriate limit, quantum mechanics must agree with classical mechanics. This constitutes a test for quantum mechanics.

Equation (2-65) also conforms to this test. Where Eq. (2-66) is obeyed and $\exp(h\nu_i/k_B T) \gg 1$, f_{BE} is identical to f_B [Eq. (2-21)].

2.5 INFLUENCE OF SPIN

The concept of "spin" must be taken into account in any statistics dealing with particles that obey quantum mechanics. Spin has no analog in classical mechanics. For example, it is difficult to picture a particle with a radius of about $10^{-13}\,\mathrm{cm}$ (Section 1.6.1), virtually a point mass, as being unsymmetrical and behaving as having an intrinsic angular momentum; this now is usually designated as spin. It turns out that, for reasons not yet fully understood, all indistinguishable, quantum particles behave as though they belong to one of two spin types. One of these classes of particles shows integral spin in which the intrinsic angular momentum of the particles is given by $n(h/2\pi)$, where $n = 0, 1, 2, \ldots$, a unit derived in Section 1.5. (Where $n = 0$, the particle is spherically symmetrical.) The second type of particle shows half-integral spin with units of $(n/2)(h/2\pi)$, where n is an integer; these may be both positive and negative spins. It will be noted that the difference in angular momentum of two such particles of opposite spin and a given n is just $h/2\pi$, one unit of angular momentum.

Spin is taken into account in the following way. As noted in Section 2.1, a particle is basically defined by its coordinates in state space. The wave function ψ, discussed in Section 1.6.7, actually is a function of state space and time. Thus, ψ can represent a particle. And, in a way analogous to that described in Section 1.5, ψ most simply may be considered to be a time-independent function only of its radius vector in state space, or $\psi = \psi(r)$, in the absence of spin. See Sections 1.6.7 and 2.1. This should not be construed to be a return to the Bohr "planetary" or "orbital" model for a particle, such as an electron, since this approach was previously shown to be incorrect.

In the case of an electron, the application of an external magnetic field can cause it to behave as though it had an intrinsic angular momentum or spin either parallel or antiparallel to the field. These configurations are considered to be an additional degree of freedom; this in effect is two additional possible coordinates in state space, represented by $\psi_{\pm}(\sigma)$. Thus, the total wave function for the electron must be a function of both state space and spin as represented by

$$\Psi_1 = \psi_1(r)\psi_+(\sigma) \tag{2-69a}$$

for parallel spin and by

$$\Psi_2 = \psi_1(r)\psi_-(\sigma) \tag{2-69b}$$

for antiparallel spin.

Therefore, as a result of the inclusion of spin, Ψ_1 and Ψ_2 are slightly different amplitudes of de Broglie waves. As such, they represent very similar, but significantly different states. And, in addition changes in spin, or spin reversals, must be considered to change the wave functions, and consequently the probability densities, even though such changes usually correspond to negligibly small energy changes. As an example, two electrons with a given $\psi(r)$ and opposite $\psi(\sigma)$ may usually be considered as being at the same energy level. Only one such pair of indistinguishable particles can occupy this state at any given time. This is known as the Pauli exclusion principle and is discussed in Sections 2.3.2 and 3.9.

In cases in which permutations of the position and spin coordinates leave the total wave functions unchanged, such wave functions are said to be symmetrical. Particles such as phonons and photons, with zero or integral spin, behave in this way. Since such changes involving these particles do not change the total wave functions, they obey the Bose–Einstein statistics and are called *bosons*.

As shown previously, particles with half-integral spin have their total wave functions changed when their state space (position) and spin coordinates are permuted. Such wave functions are said to be antisymmetrical. They, therefore, must be treated statistically in ways that take this into account. Such particles obey the Fermi–Dirac statistics and are called *fermions*.

2.6 THE FERMI–DIRAC EQUATION

The inclusion of spin in state space [as in Eqs. (2-69a) and (2-69b)], in addition to the constraints reviewed in Section 2.3, is required for any description of the properties of indistinguishable particles with half-integral spin. This may be accomplished most simply by again referring to

$$\Psi_1 = \psi_1(r)\psi_+(\sigma) \tag{2-69a}$$

and

$$\Psi_2 = \psi_2(r)\psi_-(\sigma) \tag{2-69c}$$

When two particles have identical spatial coordinates in state space, so that $\psi_1(r) = \psi_2(r)$, but have opposite spins, they may be considered as being in the same energy state. This has the effect of implicitly including spin in the spatial coordinates. No other particle or pair of particles may simultaneously occupy this state (Section 3.9). When spin is explicitly considered, only one particle can occupy a state. However, other particles with total wave functions with different $\psi_i(r)$ and $\psi_\pm(\sigma)$ may have the same energies. These particles are said to be in degenerate states because, even though they have different total wave functions, they have the same energies. Their discrete energy spectrum can consist of large numbers of closely packed states that approach a quasicontinuum.

Let a number of degenerate states, N_s, be selected for a given energy, E_i, in a quasicontinuum. These will be occupied by a number of indistinguishable particles, n_s, that is smaller than N_s, but which may approach N_s. Thus, n_s of the available N_s states will be filled, and $N_s - n_s$ will be empty. The first particle can be accommodated in N_s ways. Then, one of the states having been occupied, the second particle can occupy any one of the remaining $N_s - 1$ states. When two states have been allocated, the third particle has a chance of being found in $N_s - 2$ states. As this process continues, it can be seen that the states may be filled in any one of $N_s - n_s + 1$ ways. But some of these ways are indistinguishable (Section 2.3) and must not be counted. This is accomplished by eliminating any permutations and counting only the number of combinations of indistinguishable particles in distinguishable states, W_s. This is the same as counting the number of N_s things taken n_s at a time as given by

$$W_s = \frac{N_s!}{(N_s - n_s)!n_s!} \tag{2-70}$$

Then the probability that the n_s particles will be in the N_s states is

$$P_s = P^{n_s}W_s = P^{n_s}\frac{N_s!}{(N_s - n_s)!n_s!} \tag{2-71}$$

The coefficient in Eq. (2-71) arises because the probability of filling any single state is P, and n_s equally likely particles are involved. The resulting probability for any specific distribution of the n_s particles is P^{n_s}.

Equation (2-71) gives the probability of finding a particle in a degenerate state of a given energy within the quasicontinuum. This may be extended to include all the particles with energies in the entire range under consideration by using the same approach given previously for one such energy. This is obtained from the product of all such P_s. The probability of finding particles in all the energy states in the quasicontinuum, then, is

$$P = P^N W_{s_1} \cdot W_{s_2} \cdot W_{s_3} \cdots = P^N W \qquad (2\text{-}72)$$

Here N is the total number of particles and W is the product of all the W_{s_i} for each energy level; this is given by

$$W = \prod_s \frac{N_s!}{(N_s - n_s)! n_s!} \qquad (2\text{-}73)$$

The most probable situation is obtained by maximizing Eq. (2-72). Here P^N is a constant, so the maximum probability may be obtained from Eq. (2-73). This equation now is treated in a way corresponding to the way in which Eq. (2-32) was used to obtain Eq. (2-39). This result is

$$\ln\left(\frac{N_s}{n_s} - 1\right) = \alpha + \beta E_i \qquad (2\text{-}74)$$

In the classical case, cited previously, no limitations are placed on N, so N/n_i is considered to be very large. Consequently, $\ln(N/n_i - 1) \cong \ln(N/n_i)$. In the quantum case under consideration here, while N_s may be large, n_s may be of comparable magnitude, and N_s/n_s may not be large with respect to unity. Another way of expressing this, in contrast to Eq. (2-55), is

$$0 \ll \frac{n_s}{N_s} < 1 \qquad (2\text{-}75a)$$

And, in some cases, n_s/N_s is close to unity and may be treated as

$$\frac{n_s}{N_s} \to 1 \qquad (2\text{-}75b)$$

Therefore, the simplification used to obtain Eq. (2-40) is not valid for the quantum case. So Eq. (2-74) must be expressed in the exponential form, on the basis of Eq. (2-75a), as

$$\frac{N_s}{n_s} - 1 = \exp(\alpha + \beta E_i) \qquad (2\text{-}76a)$$

or as

$$\frac{N_s}{n_s} = \exp(\alpha + \beta E_i) + 1 \qquad (2\text{-}76b)$$

Then the probability of finding a single particle in a state E_i is obtained from the reciprocal of Eq. (2-76b) as

$$P_i = \frac{1}{N} = f_{\text{FD}} = \frac{1}{\exp(\alpha + \beta E_i) + 1} = f = \frac{1}{e^{(E_i - \mu)/k_B T} + 1} \qquad (2\text{-}77)$$

Here α and β have the same meanings given by Eqs. (2-19) and (2-53), respectively. Equation (2-77) is the Fermi–Dirac distribution function. And, because of the requirement of single-state occupancy, f_{FD} cannot exceed unity. Since this function will be used more frequently than the other distribution functions, the subscript FD will not normally be used to designate it; it will be indicated simply by f, as indicated in Eq. (2-77).

The primary use of Eq. (2-77) in this book will be to treat electrons in a solid as a degenerate gas. The condition given by Eq. (2-75a) ensures this. However, in such cases as virtually pure (intrinsic) or highly dilute (lightly doped) semiconductors, the electrons are in nondegenerate states. That is, in a way corresponding to Eq. (2-55),

$$\frac{n_s}{N_s} \ll 1 \tag{2-55a}$$

When this is the case,

$$f \ll 1 \tag{2-78}$$

so that the exponential term in the denominator of Eq. (2-77) must be

$$\exp\left(\frac{E_i - \mu}{k_B T}\right) \gg 1 \tag{2-79}$$

Thus, for the condition in which E_i is small with respect to the chemical potential,

$$\exp\left(\frac{-\mu}{k_B T}\right) \gg 1 \tag{2-80}$$

and

$$-\mu \gg k_B T \tag{2-80a}$$

so the chemical potential is negative and is much greater than the absolute value of the classical energy of an electron as represented by $k_B T$. Under these conditions, Eq. (2-77) may be reduced to f_{MB} [Eq. (2-54)] or to f_M [Eq. (2-21)].

2.7 CHEMICAL POTENTIAL AND THE FERMI ENERGY

The Fermi–Dirac function will be used in this book primarily to determine the probability of occupancy of states by valence electrons in solids. A metal is considered to constitute a *potential well* for the valence electrons in a way analogous to that described in Chapter 1. It will be apparent that the electron gas under consideration here is considerably different from that of the Drude–Lorentz model (Section 1.1).

It is instructive to approach the study of valence electron behavior by starting with the case in which the metal is at 0 K. The degenerate states available for electron occupancy form a quasicontinuum in the potential well. These states are successively filled by electrons, starting with the lowest state in the well and continuing until all the electrons are accommodated. The last state that is occupied by an electron at 0 K is analogous to an electron requiring an energy increment W to leave the well shown in Fig. 1-2. This highest occupied level is tentatively defined, for present purposes, as the Fermi energy and is denoted by E_F. All states up to and including E_F are filled; all states above E_F are empty. Thus, a distribution function, a function that provides the probability of the occupancy of a state, must have a value of unity for all states equal to or lower in energy than E_F and a value of zero for all states with energies higher than E_F; it must drop discontinuously, as a step function from 1 to 0 at $E = E_F$.

These constraints actually are the boundary conditions for Eq. (2-77). Or, stated mathematically for 0 K,

$$f = f(E, 0) = \begin{vmatrix} 1 \text{ for } E < E_F \\ 0 \text{ for } E > E_F \end{vmatrix} \tag{2-81}$$

In order that Eq. (2-77) conform to these conditions, it must follow that

$$\mu = E_F \tag{2-82}$$

This is shown by applying the conditions given by Eq. (2-81) to Eq. (2-77). Where $E < \mu = E_F$, the denominator becomes $e^{-\infty} + 1 = 0 + 1$, and $f(E, 0) = 1$. Where $E > \mu = E_F$, the denominator becomes $e^{+\infty} + 1 = \infty$, and $f(E, 0) = 0$. This discontinuous behavior at $T = 0$ K conforms to the required boundary conditions of Eq. (2-81).

Now consider the case in which $E = \mu = E_F$ for any temperature $T > 0$ K. Here the denominator of Eq. (2-77) becomes $e^0 + 1 = 2$, so $f = f(E_F, T) = \frac{1}{2}$. This is the complete definition of the Fermi energy: *the Fermi energy is that energy at which the probability of finding an electron in a state is exactly 0.5.* Now Eq. (2-77) may be used for any $T \neq 0$ K.

As is shown in Section 4.4 on the Fermi–Sommerfeld theory, the value of E_F for most metals is quite high, with an average value of about 4 to 6 eV. This confirms Eq. (2-80a), since

$$|\mu| = E_F \gg k_B T \tag{2-83}$$

This shows that a valence electron with $E \cong E_F$ is relatively high in energy. Thus, the addition of normal energies, of the order of $k_B T$, will have only minor effects on the inherent energies of such electrons. As a result, it would be expected that normal energies would have only very small effects on E_F. This effect is given, without derivation, by

$$E_F(T) = E_F = \mu = E_F(0) \left[1 - \frac{\pi^2}{12} \left(\frac{k_B T}{E_F(0)} \right)^2 \right] \tag{2-84}$$

in which $E_F(T)$ and $E_F(0)$ are the Fermi energies at T and 0 K, respectively. The factor $k_B T$ is very small compared to $E_F(0)$, so in most cases the second term in the brackets of Eq. (2-84) is negligible with respect to unity. Therefore, with few exceptions, the effect of temperature may be neglected, and it may be considered that

$$E_F(T) = E_F = \mu \cong E_F(0) \qquad (2\text{-}85)$$

with a very high degree of accuracy. In other words, for most stituations, μ and E_F may be considered to be invariant with temperature. The equivalence of μ (the chemical potential of the electron gas) and E_F may be used as a bridge between the thermodynamics and the quantum mechanics. This relationship is very helpful in the theory of alloy phases. And Eq. (2-77) most frequently is written in the form

$$f = \frac{1}{e^{(E-E_F)/k_BT} + 1} \qquad (2\text{-}86)$$

Now consider the effects of energy variations above and below E_F on Eq. (2-86). These are shown schematically in Fig. 2-3. Keeping in mind that k_BT usually is considered to represent normal energies, let $E - E_F = -2k_BT$. Then $f = 0.88$. In other words, a state with this energy has an 88% probability of being filled. In addition states with energies at deeper levels below E_F will have greater probabilities of occupancy. And, since the electrons must occupy discrete states in the degenerate quasicontinuum, it is very unlikely that such electrons in lower-lying states can be promoted by normal energies to occupy states with such high probabilities of prior occupancy simply because no suitable states are available for their occupancy. States with energies in the neighborhood of E_F have a probability of occupancy of about 0.5, or 50%. Where states have $E - E_F = +2k_BT$, $f = 0.12$; these have a low 12% probability of occupancy and, therefore, have a high probability of availability for any electrons that may be promoted to them from lower states. States with $E - E_F > 2k_BT$ have an even greater degree of availability for electrons, but require abnormal energies. In general, energies of the order of $\pm k_BT$ are considered to be normal energies. Thus, under normal conditions, only those valence electrons within the electron gas that have an energy range of less than $E_F \pm 2k_BT$ can be excited to higher states and enter into physical processes. This energy range imposes a strong limitation on the number of available electrons. An approximation of this may be obtained from the ratio of $\pm k_BT/E_F$. At 300 K, $k_BT \cong 0.026$ eV, and, for a normal metal, $E_F \cong 5$ eV; these give a ratio of $0.05/5 \cong 1\%$. Thus, a fraction of the order of 1% or less of the total number of valence electrons is responsible for most of the physical properties of materials. Electrons in the energy range of $E_F \pm 2k_BT$ may be approximated as being free electrons. These are designated as being "nearly free" electrons.

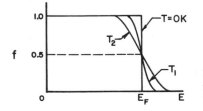

Figure 2-3   Influence of temperature on the Fermi–Dirac function; $T_2 > T_1 \gg 0$ K.

For those cases in which $(E - E_F)/k_B T \gg 1$, that is, where E is significantly greater than E_F, at a given moderate temperature, as is the situation for many intrinsic (virtually pure) or highly dilute (lightly doped) semiconductors,

$$f \cong \frac{1}{\exp(E - E_F)/k_B T} \tag{2-87}$$

This is the same as Eq. (2-65) for $\mu = E_F$ and for the conditions given previously. Here both Eqs. (2-87) and (2-65) may be reduced to f_{MB} [Eq. (2-54)]. These reductions are the equivalent of treating the indistinguishable particles of the electron gas as though they behaved classically.   See Section 2.3.1.

2.8 SUMMARY OF STATISTICS

Particle Type	Particle Property	Energy Spectrum	Ensemble Property	Applicable Statistics	Typical Applications
Classical	Distinguishable (countable)	Continuous	Nondegenerate Independent of particle	Boltzmann Maxwell–Boltzmann	Ideal gases
Classical	Distinguishable (countable)	Continuous	Nondegenerate Independent of particle	Boltzmann Maxwell–Boltzmann	Real gases at high temperature and/or low pressure
Quantum	Indistinguishable (noncountable) Zero or integral spin	Discrete	Degenerate Dependent on particle	Bose–Einstein	Phonons, photons, Cooper pairs, magnons, excitons
Quantum	Indistinguishable (noncountable) Half-integral spin	Discrete	Degenerate Dependent on particle	Fermi–Dirac	Electrons, protons, neutrons
Quantum	Indistinguishable (noncountable) Half-integral spin	Discrete	Nondegenerate Dependent on particle	Fermi–Dirac	Conduction electrons in intrinsic or lightly doped semiconductors

2.9 BIBLIOGRAPHY

ATKINS, P. W., *Physical Chemistry*, W. H. Freeman, San Francisco, 1978.

DEKKER, A. J., *Solid State Physics*, Prentice-Hall, Englewood Cliffs, N.J., 1957.

LEVY, R. A., *Principles of Solid State Physics*, Academic Press, New York, 1968.

HURD, C. M., *Electrons in Metals*, Wiley-Interscience, New York, 1975.

PAULING, L., *General Chemistry*, 3rd ed., W. H. Freeman, San Francisco, 1970.

POLLOCK, D. D., *Physical Properties of Materials for Engineers*, Vol. I, CRC Press, Boca Raton, Fla., 1982.

———, *Thermoelectricity: Theory, Thermometry, Tool*, STP 852, American Society for Testing and Materials, Philadelphia, 1985.

2.10 PROBLEMS

2.1. Use the Boltzmann equation to estimate the probability of escape of an electron from a hydrogen atom when heated to (a) 1500°C and (b) 3000°C.

2.2. Use the Maxwell–Boltzmann equation setting $(E_i - \mu) = \Delta E$, to determine the probabilities requested in Problem 2.1. Explain any similarities or differences in your results.

2.3. Graph $\Delta E = nk_B T$ for $E > E_F$. What does this tell about the probable density of electrons with $E \gg E_F$?

2.4. Calculate the difference in the energies of two electrons with the same quantum numbers, but with opposite spin. Does this conform to the approximation used in Section 2.6?

3

WAVE FUNCTIONS, ATOMS, AND BONDS

The probable properties of large numbers of quantum particles are determined by the statistical methods outlined in Chapter 2. It is reasonable to assume that different techniques are required to describe numbers of quantum particles that, by comparison, are vanishingly small. This problem may be reduced to the determination of the properties of a single particle. How is this accomplished? The general approach to the treatment of these elusive particles is sketched in Section 1.6.7. These basic concepts provide the means to solve the problem.

The approach here will be to start with some elementary ideas of classical mechanics and to convert these concepts to their equivalents in quantum mechanics. This will be done by changing the total energy of an electron as expressed in classical terms into their quantum mechanics equivalents and arriving at the Schrödinger equation. The solutions to the time-independent form of this equation will provide expressions for Ψ that will result in naturally discrete energy values for the particles being considered.

The solutions to the Schrödinger equation obtained here will be applied to describe the behaviors of electrons. More complex applications of Schrödinger's equation will be outlined for the hydrogen and helium atoms. These will show the basis for the Pauli exclusion principle and explain why the quantum numbers that are the same as those obtained for the hydrogen atom may be employed for atoms with many more electrons. The solutions for electrons will be used to describe the electron configurations of atoms and bonding mechanisms and, in subsequent chapters, to develop some of the fundamental ideas of the modern theories of solids.

3.1 CLASSICAL MECHANICS

A brief review of some of the elementary classical ideas is presented as an introduction to quantum mechanics. Consider a mechanical system in which a particle of mass m is acted upon by a force. For simplicity, the particle will be considered to move in the x direction. The force, then, is given by $F(x)$. This force function is defined as

$$F(x) = -\frac{d}{dx} V(x) \tag{3-1}$$

where $V(x)$ is a potential energy function.

The following Newtonian relationships are given, on this basis, as a review. The familiar relationship between force, mass, m, and acceleration, a, is given by

$$F(x) = m \frac{d^2x}{dt^2} = ma \tag{3-2}$$

The momentum, $\mathbf{p}$, is given in terms of velocity, $\mathbf{v}$, by

$$\mathbf{p} = m \frac{dx}{dt} = m\mathbf{v} \tag{3-3a}$$

or

$$\frac{dx}{dt} = \frac{\mathbf{p}}{m} \tag{3-3b}$$

Differentiation of Eq. (3-3a) with respect to time gives

$$\frac{d\mathbf{p}}{dt} = m \frac{d^2x}{dt^2} = m \frac{d\mathbf{v}}{dt} = ma = F(x) = -\frac{dV(x)}{dx} \tag{3-4}$$

The kinetic energy, KE, of the particle is given by

$$KE = \frac{1}{2} m\mathbf{v}^2 = \frac{\mathbf{p}^2}{2m} \tag{3-5}$$

It should be noted that this is a function of motion, whereas the potential energy, PE, is one of position:

$$PE = V(x) \tag{3-6}$$

The total energy of the system, E, is

$$E = KE + PE = \frac{1}{2} m\mathbf{v}^2 + V(x) = \frac{\mathbf{p}^2}{2m} + V(x) \tag{3-7}$$

A function may be defined as the total energy of the system in terms of the variables x and $\mathbf{p}$. This is called the Hamiltonian function of the system and is written for the case under consideration here as

$$H(x, \mathbf{p}) = H = \frac{\mathbf{p}^2}{2m} + V(x) \equiv E \tag{3-8}$$

where E is the total energy of the system. For simplicity, $H(x, \mathbf{p})$ will be written as H. Differentiating Eq. (3-8) with respect to momentum and noting that $V(x)$ vanishes gives

$$\frac{\partial H}{\partial \mathbf{p}} = \frac{\mathbf{p}}{m} \tag{3-9}$$

Since both Eqs. (3-3b) and (3-9) equal $\mathbf{p}/m$, they may be equated:

$$\frac{\mathbf{p}}{m} = \frac{dx}{dt} = \frac{\partial H}{\partial p} \tag{3-10}$$

The differentiation of the Hamiltonian function [Eq. (3-8)] with respect to position, x, recognizing that the momentum term vanishes, results in

$$\frac{\partial H}{\partial x} = \frac{dV(x)}{dx} \tag{3-11}$$

so that, by use of Eq. (3-4),

$$-\frac{dV(x)}{dx} = \frac{d\mathbf{p}}{dt} = -\frac{\partial H}{\partial x} \tag{3-12}$$

Equations (3-10) and (3-12) constitute the Hamiltonian equivalents of the classical expressions given by Eqs. (3-3b) and (3-4). In this way the momentum and force functions may be replaced by the appropriate forms of the Hamiltonian functions that define the variables x and $\mathbf{p}$.

The Hamiltonian function for the total energy of a system will be used as a basis for obtaining expressions from which Schrödinger's equation can be obtained. The solutions to this equation will form the basis for understanding the topics that follow.

3.2 THE SCHRÖDINGER WAVE EQUATION

The variables, position and momentum, in the Hamiltonian functions can be determined sufficiently well for classical mechanics applications. This is not the situation for the quantum mechanics case. The Heisenberg uncertainty principle (Section 1.6.6) places limits on the accuracy to which the position–momentum and energy–time conjugates may be known. Therefore, as previously indicated, the approach is based on the postulates that the most probable values can be determined from the wave function and that the Hamiltonian for the system can be expressed as in Eq. (3-8). This equation is transformed into the wave equation by the replacement of the variables by means of differential operators.

Let the momentum of the particle be represented by the operator such that it becomes, where $i = \sqrt{-1}$,

$$\mathbf{p} \rightarrow \frac{h}{2\pi i} \frac{\partial}{\partial x} \tag{3-13}$$

and the total energy be represented by the operator

$$E \rightarrow -\frac{h}{2\pi i}\frac{\partial}{\partial t} \qquad (3\text{-}14)$$

It will be observed that each of these relationships involves the conjugates treated in the Heisenberg equations. The position of the particle, x, will remain untransformed by an operator since it is the only variable in $V(x)$ and its conjugate is not involved in $V(x)$. This being the case, x may be known to any desired degree of accuracy. Therefore, x is not affected by a transformation and

$$x \rightarrow x \qquad (3\text{-}15)$$

Now consider the Hamiltonian function, for one dimension, that was given previously:

$$H \equiv H(x, \mathbf{p}) = \frac{\mathbf{p}^2}{2m} + V(x) \equiv E \qquad (3\text{-}8)$$

It will be recalled from Section 1.6.6 that, in dealing with these variables, an approach based on probability must be employed. The wave function, Ψ (previously defined as a function of space and time), was employed in such a way that $\Psi^*\Psi$ was proportional to the probability of finding the particle. (This can be applied more generally. The function Ψ may be made to represent any observable quantity.) Let Ψ be a function of position, x, and time, t. In the present case, momentum, position, time, and energy are the factors involved. The quantum mechanics representation is obtained, for each of these variables, by performing the appropriate operations on Ψ as given by Eqs. (3-13), (3-14), and (3-15). Starting with Eq. (3-8),

$$H(x, \mathbf{p}) = E \qquad (3\text{-}8a)$$

and allowing $H(x, \mathbf{p})$ and E to operate on $\Psi(x, t)$ changes Eq. (3-8a) to

$$H(x, \mathbf{p})\Psi(x, t) = E\Psi(x, t) \qquad (3\text{-}8b)$$

or, more simply,

$$H\Psi = E\Psi \qquad (3\text{-}8c)$$

The wave function is often expressed simply as Ψ, for convenience. The variables x and t are implicit in this notation in the same way that x and $\mathbf{p}$ are included in H in Eq. (3-8).

The operation to determine $\mathbf{p}^2/2m$ in Eq. (3-8) requires that the operator, Eq. (3-13), be used twice because the momentum is squared. This is done as follows:

$$\frac{1}{2m}\cdot\frac{h}{2\pi i}\frac{\partial}{\partial x}\cdot\frac{h}{2\pi i}\frac{\partial}{\partial x}\cdot\Psi = -\frac{1}{2m}\frac{h^2}{4\pi^2}\frac{\partial^2\Psi}{\partial x^2} = -\frac{h^2}{8\pi^2 m}\frac{\partial^2\Psi}{\partial x^2} \qquad (3\text{-}16)$$

The operation to determine $V(x)$ is simply

$$V(x)\Psi \qquad (3\text{-}17)$$

since x remains unchanged. Similarly, E is transformed by Eq. (3-14) as

$$E \to \frac{-h}{2\pi i} \frac{\partial \Psi}{\partial t} \tag{3-18}$$

These operations transform Eq. (3-8) into

$$-\frac{h^2}{8\pi^2 m} \frac{\partial^2 \Psi}{\partial x^2} + V(x)\Psi = -\frac{h}{2\pi i} \frac{\partial \Psi}{\partial t} \tag{3-19a}$$

Equation (3-19a) is the quantum mechanic equivalent of Eq. (3-7). Or, when multiplied through by -1, Eq. (3-19a) may be rewritten as

$$\frac{h^2}{8\pi^2 m} \frac{\partial^2 \Psi}{\partial x^2} - V(x)\Psi = \frac{h}{2\pi i} \frac{\partial \Psi}{\partial t} \tag{3-19b}$$

Since this is only for the one dimension, x, Ψ is, as previously defined, a function of x and t only, rather than of x, y, z, and t. Equation (3-19b) is a one-dimensional form of the Schrödinger equation for a single particle. Equation (3-19a) also may be written as

$$-\frac{h^2}{8\pi^2 m} \frac{\partial^2 \Psi}{\partial x^2} + V(x)\Psi = E\Psi \tag{3-20}$$

where the operation of E on Ψ is indicated, but not performed. Equations (3-19a), (3-19b), and (3-20) are the quantum mechanics equivalents of Eq. (3-7).

The properties of the factors of these equations are of interest: The mass relates to a particle; Ψ, being an amplitude, relates to a wave; the potential energy function represents an external influence; and the energy, according to the Heisenberg uncertainty principle, cannot be known exactly. Thus, duality, uncertainty, and external factors are implicit in Eq. (3-20).

Equation (3-8b) also may be written as

$$H\Psi = E\Psi = -\frac{h}{2\pi i} \frac{\partial \Psi}{\partial t} \tag{3-21}$$

The problem becomes one of finding suitable expressions for Ψ that constitute meaningful solutions for Schrödinger's equation. This can be done only when certain criteria are satisfied (Section 3.3).

However, Eq. (3-20) will serve to provide a simple, but very useful, illustration of the use of the Schrödinger equation. Consider the case of a free electron. All its energy can be considered to be kinetic; hence the potential energy factor, $V(x)$, will be zero. The equation now becomes, using the form of Eq. (3-20),

$$-\frac{h^2}{8\pi^2 m} \cdot \frac{\partial^2 \Psi}{\partial x^2} = E\Psi \tag{3-22}$$

When the total kinetic energy of the free electron is constant, a solution to this equation can be obtained by selecting the wave function

$$\Psi = e^{i\mathbf{k} \cdot x} \tag{3-23}$$

in which $\bar{\mathbf{k}}$ is the wave vector of the electron, this is equal to $2\pi/\lambda$ (λ being the wavelength of the electron), and x denotes its position.   Upon the successive differentiation of Ψ, the following are obtained:

$$\frac{\partial \Psi}{\partial x} = i\bar{\mathbf{k}}e^{i\bar{\mathbf{k}}\cdot x}$$

and, using Eq. (3-23), the second derivative is

$$\frac{\partial^2 \Psi}{\partial x^2} = -\bar{\mathbf{k}}^2 e^{i\bar{\mathbf{k}}\cdot x} = -\bar{\mathbf{k}}^2 \Psi \tag{3-24}$$

When Eq. (3-24) is substituted into Eq. (3-22), it becomes

$$-\frac{h^2}{8\pi^2 m}(-\bar{\mathbf{k}}^2 \Psi) = E\Psi$$

It is seen that Ψ [Eq. (3-23)] is a solution to the Schrödinger equation when

$$E = \frac{h^2 \bar{\mathbf{k}}^2}{8\pi^2 m} \tag{3-25}$$

This elementary relationship has important applications.   It will be used to describe "nearly free" electron behavior in metals and will serve as a basis for the development of electron band structure and its implications.   See Section 2.7 and Chapter 4.

The solution given by Eq. (3-25) can be verified in a simple way.   This is done by starting with the deBroglie equation [Eq. (1-17)]:

$$p = mv = \frac{h}{\lambda} \tag{1-17}$$

When this equation is squared and rearranged, it is found that

$$v^2 = \frac{h^2}{m^2 \lambda^2}$$

This may be substituted into the classical expression for kinetic energy to get

$$E = \frac{1}{2}m\mathbf{v}^2 = \frac{h^2}{2m\lambda^2}$$

Use is now made of the expression for the wave vector ($\bar{\mathbf{k}} = 2\pi/\lambda$) to substitute for λ^2. When this is done,

$$E = \frac{h^2}{2m} \cdot \frac{\bar{\mathbf{k}}^2}{4\pi^2} = \frac{h^2 \bar{\mathbf{k}}^2}{8\pi^2 m} \tag{3-26}$$

Thus, Eq. (3-26) verifies Eq. (3-25), since identical results for the energy of a free electron were obtained by both classical and quantum means.   Verifications such as the one given here are not normally this easy; the case presented here is the simplest possible case. It also provides the agreement required by the Bohr correspondence principle (Section 2.4).

The general form of Schrödinger's equation is given by

$$-\frac{h^2}{8\pi^2 m}\left[\frac{\partial^2 \Psi}{\partial x^2} + \frac{\partial^2 \Psi}{\partial y^2} + \frac{\partial^2 \Psi}{\partial z^2}\right] + V(x, y, z)\Psi = E\Psi \tag{3-27}$$

The wave function or amplitude, Ψ, now is a function of x, y, z, and t instead of just x and t, as was given for the one-dimensional case.

Many applications of the Schrödinger equation deal only with the influence of the potential energy. As previously noted, this factor depends only on position; it is not a function of time. Thus, if the time factor were separated out of Eq. (3-27), it would be simpler to use. The one-dimensional expressions, Eqs. (3-20) and (3-21), will be used to show this.

This problem is approached by using other wave functions to separate the variables. Each of these is a function of a single variable such that

$$\Psi(x, t) = \Psi_1(x)\phi(t) \tag{3-28}$$

Equation (3-28) is substituted into Eq. (3-21) and rearranged to obtain

$$\Phi(t)H(x, \mathbf{p})\Psi_1(x) = -\frac{h}{2\pi i}\,\Psi_1(x) \cdot \frac{\partial \phi(t)}{\partial t}$$

Both sides of this equation are divided by $\Psi_1(x)$ so that this function vanishes from the right side of the equation:

$$\phi(t)\,\frac{H(x, \mathbf{p})\Psi_1(x)}{\Psi_1(x)} = -\frac{h}{2\pi i}\,\frac{\partial \phi(t)}{\partial(t)}$$

The equation is next divided by $\phi(t)$ to eliminate this function from the left side of the equation and rearranged for clarity as

$$\frac{H(x, \mathbf{p})\Psi_1(x)}{\Psi_1(x)} = -\frac{h}{2\pi i} \cdot \frac{\partial \phi(t)}{\partial t} \cdot \frac{1}{\phi(t)} \tag{3-29}$$

The left side of Eq. (3-29) will be a function only of x after the indicated operation is performed; the right side will be a function only of t. Since both sides of this equation are functions of two different variables and are equal to each other, they must equal a certain class of constants, E, real numbers, known as *eigenvalues*. This class of parameters will be discussed in Section 3.5.

Attention may now be centered on the time-dependent portion of Eq. (3-29), This is equated to an eigenvalue to give

$$-\frac{h}{2\pi i} \cdot \frac{1}{\phi(t)} \cdot \frac{\partial \phi(t)}{\partial t} = E$$

This may be rearranged as

$$\frac{\partial \phi(t)}{\phi(t)} = -\frac{2\pi i}{h} E \, \partial t \tag{3-30}$$

After integration, this becomes

$$\ln \phi(t) = -\frac{2\pi i}{h} t + \ln[\phi(t_0)]$$

Or, in exponential form,

$$\phi(t) = \exp\left(-\frac{2\pi i}{h} Et\right) \tag{3-31}$$

This is the *time-dependent* form of Schrödinger's equation. Here the constant of integration, $\ln[\phi(t_0)]$, is arbitrarily equated to zero. This will not affect its use since $\phi(t)$ is infrequently used alone, but usually with $\Psi_1(x)$ in the form $\Psi(x, t) = \Psi_1(x)\phi(t)$; in any event, where energy is conserved, it is not necessary to obtain an exact time dependence. The general solution provided by Eq. (3-31) is of assistance in dealing with the energy equivalents of Hamiltonians. See Eqs. (3-72a) and (3-72b).

The time-independent form of the Schrödinger equation is obtained from the left portion of Eq. (3-29). For the reasons given previously, it too is equated to same eigenvalue previously used for the determination of the time-dependent equation. Thus,

$$\frac{H(x, \mathbf{p})\Psi_1(x)}{\Psi_1(x)} = E$$

Upon rearrangement, this becomes

$$H(x, \mathbf{p})\Psi_1(x) = E\Psi_1(x) \tag{3-32}$$

Equation (3-32) is of the same form as Eq. (3-8c), but with the time factor eliminated. When the original Hamiltonian function is used [Eq. (3-8)], the operations shown by Eqs. (3-16) and (3-17) transform Eq. (3-32) into

$$-\frac{h^2}{8\pi^2 m} \frac{\partial^2 \Psi_1(x)}{\partial x^2} + V(x)\Psi_1(x) = E\Psi_1(x) \tag{3-33}$$

This is of exactly the same form as Eq. (3-20), only now it is a function only of the single variable *position*. The coefficient of the second derivative term may be replaced by unity if Eq. (3-33) is multiplied through by $-8\pi^2 m/h^2$:

$$\frac{\partial^2 \Psi_1(x)}{\partial x^2} - \frac{8\pi^2 m}{h^2} V(x)\Psi_1(x) = -\frac{8\pi^2 m}{h^2} E\Psi_1(x)$$

Or this is frequently rearranged and used in the form

$$\frac{\partial^2 \Psi_1(x)}{\partial x^2} + \frac{8\pi^2 m}{h^2} [E - V(x)]\Psi_1(x) = 0 \tag{3-34}$$

These are *time-independent* forms of Schrödinger's equation for one dimension.

The separation of Eqs. (3-20) and (3-21) into two equations, where each is a function of a single variable, Eq. (3-31) and Eq. (3-34), has reduced the difficulty and increased the utility of their application. The principal area of interest here will be the effect of a potential field on an electron as a function of its position in the potential field. Thus, Eq. (3-34) provides this. It is the form of the Schrödinger equation that will be of most use in this book.

3.3 CRITERIA FOR SOLUTIONS OF SCHRÖDINGER'S EQUATION

It will be recalled [Eq. (3-27)] that Ψ was defined as a wave function whose independent variables are space and time. The probability of finding the particle in a given range (Section 1.6.7) was shown to be proportional to $\Psi^*\Psi\, d\theta$. If the variables are changed so that the probable location of the photon on the two-dimensional screen is of interest, rather than its angular range, the probability of finding the photon will be proportional to $\Psi^*\Psi\, dx\, dy$. Or, for one dimension, the probability of finding the particle will vary as $\Psi^*\Psi\, dx$. This one-dimensional case will be used in the following discussion for simplicity. The three-dimensional case is treated in exactly the same way.

No restrictions were expressly stated for the wave function in the illustration given previously for the case of the free electron, the result of which is expressed by Eq. (3-25). However, not all the solutions to Schrödinger's equation correspond to natural phenomena. It is, therefore, necessary to impose conditions on Ψ and on its derivatives to ensure that meaningful solutions are obtained.

First, the conditions must be such that the particle is certain to be at some point within the system. To make sure that this is so, the condition

$$\int_{-\infty}^{\infty} \Psi^*\Psi\, dx = 1$$

must be satisfied.

Second, both Ψ and $d\Psi/dx$ must be continuous. If this were not the case, a singularity (discontinuity) could exist. This could be the equivalent of the creation or destruction of particles. In the case of an electron, or any other particle, such behavior would violate the laws of conservation.

Also consider what would happen if Ψ provided more than one solution. This would lead to more than one probability of finding the particle under a given set of conditions. Such a situation would be confusing rather than helpful. So, a third condition must be imposed that Ψ be single valued. In addition to these constraints, Ψ must never be identical to zero. That is, there must always be a real probability of finding the particle in the system, however small that probability may be.

These limitations on the wave functions will not only ensure that meaningful solutions to Schrödinger's equation will be obtained, but they also will be of assistance in simplifying the solutions to this equation.

3.4 ONE-DIMENSIONAL WELLS

The positive charge on an ion in a metallic lattice exerts an omnidirectional electrostatic attraction for a valence electron. In effect, the positive ion acts as a three-dimensional potential *well* for the electron. The actual description of an approximation of this case will be outlined in Section 3.8. As a start, approximate, one-dimensional models will be used to illustrate the way in which the Schrödinger equation is used and to begin to describe electron behavior. These will provide a basis for approximating the influence of a three-dimensional potential well on an electron (Section 3.8).

The significant differences between the findings of quantum and classical mechanics will then be apparent. In addition, a basis for understanding the properties of the general classes of solids will have been established.

3.5 ELECTRON IN A WELL OF INFINITE POTENTIAL

The simplest case is given by the solution of the Schrödinger equation for an electron in a one-dimensional well whose sides effectively constitute infinitely high potential barriers. This is shown schematically in Fig. 3-1.

According to classical physics, a particle (electron) in such a well, with constant energy $E < V_0$, should move with a constant velocity, since $E = \frac{1}{2}mv^2$. And it should never leave the well because of the high barriers. The probability of finding such a particle anywhere within the well is the same, since its velocity is constant. The probability of finding this particle outside the well is zero. It will be seen later that these statements are very different from some of the findings of quantum mechanics.

The well itself has been designated as region II; portions external to the well have been designated as regions I and III. The factors operative in each of the regions are given as follows:

$$\text{Region I:}\quad x < 0;\ E < V;\ V = V_0$$

$$\text{Region II:}\quad 0 < x < L;\ E > V;\ V = 0$$

$$\text{Region III:}\quad x > L;\ E < V;\ V = V_0$$

These factors will be used to obtain a solution to Schrödinger's equation. Here, only the influence of the potential well on the behavior of the electron must be considered.

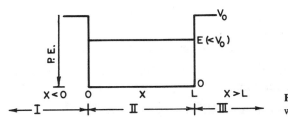

Figure 3-1 One-dimensional potential well.

Since this is the case, the time-independent form of the Schrödinger equation [Eq. (3-34)] will be used. This can be rewritten, for convenience, for this particular set of conditions, by reexpressing the coefficient of Ψ in the following way:

$$\frac{\partial^2 \Psi}{\partial x^2} + \frac{4\pi^2}{h^2}[2m(E - V)]\Psi = 0 \tag{3-35}$$

The wave function Ψ in Eq. (3-35) is identical to Ψ_1 used in Eq. (3-34). The subscript has been dropped for convenience. Both are independent of time.

Solutions to Eq. (3-35) may be more readily understood by expressing it simply in its general form as

$$\psi'' + R\psi = 0$$

where $R = 4\pi^2/h^2[2m(E - V)]$. Let the wave function be $\psi = e^{ax}$. Then, $\psi' = ae^{ax}$ and $\psi'' = a^2 e^{ax}$. The substitution of these quantities into the general equation results in

$$a^2 e^{ax} + Re^{ax} = 0$$

The exponential factors vanish and, for the case where $V > E$, R is negative:

$$a^2 - R = 0$$

One pair of equations that leads to solutions of Eq. (3-35) may be based on this equation by factoring:

$$a^2 - R = (a + R^{1/2})(a - R^{1/2}) = 0$$

so that the coefficients of the exponent of the wave function are given by

$$a = \pm R^{1/2}$$

Another pair of equations that gives solutions of Eq. (3-35) is based on the case where $E > V$, and R is positive. This results in

$$a^2 = -R$$

so that a second set of coefficients of the exponent of the wave function is given by

$$a = \pm iR^{1/2}$$

The solutions for the coefficients of the exponent of the wave function for the first case, $V > E$, gives the wave functions

$$\psi_1 = c_1 \exp(R^{1/2}x)$$

and

$$\psi_2 = c_2 \exp(-R^{1/2}x)$$

where c_1 and c_2 are arbitrary constants. The sum of these two equations results in a solution to the general equation:

$$\psi = c_1 \exp(R^{1/2}x) + c_2 \exp(-R^{1/2}x)$$

The second set of coefficients for the exponent of the wave function for the second case, $E > V$, gives another solution to the general equation:

$$\psi = c_3 \exp(iR^{1/2}x) + c_4 \exp(-iR^{1/2}x)$$

where c_3 and c_4 are arbitrary constants. It will be recognized that

$$\exp(iR^{1/2}x) = \cos(R^{1/2}x) + i\,\sin(R^{1/2}x)$$

and that

$$\exp(-iR^{1/2}x) = \cos(R^{1/2}x) - i\,\sin(R^{1/2}x)$$

These periodic, trigonometric expressions are used to give the wave function:

$$\psi = c_3[\cos(R^{1/2}x) + i\,\sin(R^{1/2}x)] + c_4[\cos(R^{1/2}x) - i\,\sin(R^{1/2}x)]$$

The sine and cosine terms are collected and

$$\psi = i(c_3 - c_4)\,\sin(R^{1/2}x) + (c_3 + c_4)\,\cos(R^{1/2}x)$$

Attention now is directed at the arbitrary constants. Since these are arbitrary, their difference, their sum and the product $i(c_3 - c_4)$, must be arbitrary, too. Therefore, this wave function may be written in terms of other arbitrary constants, C and D, as

$$\psi = C\,\sin(R^{1/2}x) + D\,\cos(R^{1/2}x)$$

This is the periodic form of the wave equation that is used later in Eq. (3-37).

When these wave functions are used in solving Eq. (3-35), $R^{1/2} = 2\pi/h[2m(V - E)]^{1/2}$ in Eqs. (3-36) and (3-38), where $V > E$. However, since $E > V$ and $V = 0$, $R^{1/2} = 2\pi/h(2mE)^{1/2}$ in Eq. (3-37).

It will be recalled that E represents discrete energy levels, or eigenvalues, and V is the external potential acting on the electron. The solutions to Eq. (3-35) just derived now can be applied to the one-dimensional well by setting Ψ equal to the following appropriate wave functions, as explained previously. In region I, $V > E$, $x \leqslant 0$:

$$\Psi_{\text{I}} = A \exp\left\{\frac{2\pi}{h}[2m(V - E)]^{1/2}x\right\} + B \exp\left\{-\frac{2\pi}{h}[2m(V - E)]^{1/2}x\right\} \qquad (3\text{-}36)$$

In region II, $V = 0$, $0 \leqslant x \leqslant L$:

$$\Psi_{\text{II}} = C \sin\left[\frac{2\pi}{h}(2mE)^{1/2}x\right] + D \cos\left[\frac{2\pi}{h}(2mE)^{1/2}x\right] \qquad (3\text{-}37)$$

Finally, in region III, $V > E$, $x \geqslant L$:

$$\Psi_{\text{III}} = F \exp\left\{\frac{2\pi}{h}[2m(V - E)]^{1/2}x\right\} + G \exp\left\{-\frac{2\pi}{h}[2m(V - E)]^{1/2}x\right\} \qquad (3\text{-}38)$$

Now the criteria for meaningful solutions (Section 3.3) must be applied to these expressions for Ψ. These will simplify the solutions to the wave functions. First, the constraint that Ψ must be finite will be invoked:

As $x \to -\infty$, B must $\to 0$, so that Ψ_I does not become infinite

As $x \to +\infty$, F must $\to 0$, so that Ψ_{III} does not become infinite

In addition, Ψ and $\partial\Psi/\partial x$ must be continuous. Thus, at the boundaries of the well, that is, in region II,

At $x = 0$	At $x = L$
$\Psi_I = \Psi_{II}$	$\Psi_{II} = \Psi_{III}$
$\dfrac{\partial\Psi_I}{\partial x} = \dfrac{\partial\Psi_{II}}{\partial x}$	$\dfrac{\partial\Psi_{II}}{\partial x} = \dfrac{\partial\Psi_{III}}{\partial x}$

These constraints are applied first at $x = 0$. Taking the partial derivative of Ψ_{II} with respect to x results in

$$\frac{\partial\Psi_{II}}{\partial x} = C \frac{2\pi}{h} (2mE)^{1/2} \cos\left[\frac{2\pi}{h}(2mE)^{1/2}x\right] - D\frac{2\pi}{h}(2mE)^{1/2}\sin\left[\frac{2\pi}{h}(2mE)^{1/2}x\right]$$

At $x = 0$, $\cos 0 = 1$ and $\sin 0 = 0$, so this derivative reduces to

$$\frac{\partial\Psi_{II}}{\partial x} = C\frac{2\pi}{h}(2mE)^{1/2}$$

Now the partial derivative of Ψ_I is obtained with respect to x, recalling that $B \to 0$, so

$$\frac{\partial\Psi_I}{\partial x} = A\frac{2\pi}{h}[2m(V-E)]^{1/2}\exp\left\{\frac{2\pi}{h}[2m(V-E)]^{1/2}x\right\}$$

At $x = 0$, the exponential factor equals unity, so

$$\frac{\partial\Psi_I}{\partial x} = A\frac{2\pi}{h}[2m(V-E)]^{1/2}$$

And, since continuity must be preserved, $\partial\Psi_I/\partial x = \partial\Psi_{II}/\partial x$ at $x = 0$. In order that Ψ_I not be infinite, A must approach zero since V was postulated as being infinitely deep. Thus, Ψ_I must approach zero at the well boundary at $x = 0$. This condition results in a negligibly small probability of finding the particle in region I because $\Psi_I^*\Psi_I \to 0$ very rapidly for $x < 0$.

In a similar way, in region III, Ψ_{III} also will approach zero at $x = L$, since G also must approach zero because of the magnitude of the potential barrier, V. The criteria for meaningful solutions (Section 3.3) require that the wave functions be continuous. Thus, at $x = 0$ and $x = L$, Ψ_{II} must approach zero to preserve the requisite continuity, since both Ψ_I and Ψ_{III} do so. To meet this condition at $x = 0$, the coefficient D of Eq. (3-37) must equal zero, since $\cos 0 = 1$. Thus, at $x = 0$, the wave function within the well reduces to

$$\Psi_{II} = C\sin\left[\frac{2\pi}{h}(2mE)^{1/2}x\right]_{x=0} \to 0$$

In order that Ψ_{II} equal zero at $x = L$,

$$C \sin \left[\frac{2\pi}{h} (2mE)^{1/2} x \right]_{x=L} \to 0$$

But C does not equal zero; so, for this condition to be the case, it must follow that

$$\frac{2\pi}{h} (2mE)^{1/2} L \to n\pi$$

where n is an integer. Solving for E, the energy of the electron is given by

$$E = \frac{n^2 h^2}{8mL^2} \tag{3-39}$$

Now it is seen that the eigenvalues, E, can only take on certain allowed, *integral* values that are determined by the integer n. This limitation arose from the restriction placed on Ψ_{II}; it was contained within the well. Since Ψ_{II} is made to approach zero at the boundaries of the well, and recalling that it is a periodic function of position, x, this means that it can only have an *integral* number of half-wavelength in the potential well over the interval between 0 and L. This number is determined by the value of the *integer, n*. This means that the energies (eigenvalues) will vary discretely with n. The case $n = 0$ is not allowed because it is the equivalent of $\Psi = 0$, a condition in violation of the criteria given in Section 3.3.

It will be recalled that when the variables of Eq. (3-29) were separated into two equations, Eqs. (3-30) and (3-32), it was noted that only certain values of E would satisfy the conditions. Members of this class of real numbers are natural solutions to Schrödinger's equation and represent discrete energy states. These numbers are the eigenvalues. The Ψ functions that correspond to these eigenvalues are known as *eigenfunctions*.

The eigenvalues, thus, arise directly and naturally from solutions of the Schrödinger equation. This is in contrast to the Bohr approach, wherein the angular momenta of electrons had to be postulated to be in integral multiples of $h/2\pi$ in order to ensure discrete energies (Section 1.5). Thus, the application of the Schrödinger equation, based on probability rather than on classical mechanics, gives results that correspond to the observed behavior independently of such forced postulations.

3.6 COMPARISON OF QUANTUM AND CLASSICAL MECHANICS

The eigenvalues are discrete since n is an integer [Eq. (3-39)]. This results from the boundary conditions imposed on Ψ_{II}. Consequently, the lowest allowed energy is given by $n = 1$. It must be emphasized that the case $n = 0$ is forbidden because it is the equivalent of allowing the wave function to equal zero (Section 3.3). However, average energies, which can be calculated only from discrete values, may have any nonzero value (Sections 2.4 and 2.6). The constraints placed on Ψ_{II} also affect the

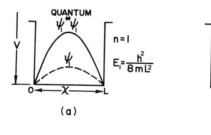

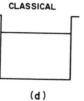

(a)

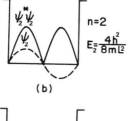

(b)

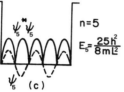

(c)

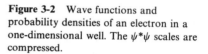

Figure 3-2 Wave functions and probability densities of an electron in a one-dimensional well. The $\psi^*\psi$ scales are compressed.

probability of finding an electron in the well; this varies periodically across the well because of the nature of Eq. (3-37).

Classical mechanics, in contrast, is based on a continuous energy spectrum. And since no constraints are placed on it, the energy of the particle can equal zero. Another result of this is that the probability of finding the electron is the same at any position across the well (Section 3.5).

The probabilities of finding the electron for both quantum and classical treatments are shown in Fig. 3-2. In Fig. 3-2(a), where $n = 1$, just one-half of a wavelength of Ψ is contained in the well. The probability density of finding the electron, $\Psi^*\Psi$, varies and goes through a maximum between $x = 0$ and $x = L$. In Fig. 3-2(b), where one complete wavelength of Ψ is contained within the well, the probability density varies in a more complicated way. The probability of finding an electron goes through zero in Fig. 3-2(b). How can this be? This question is not allowed, since, according to the Heisenberg uncertainty principle, when the exact position of the electron is known, none of its other conjugate properties can be known. For $n = 5$, in Fig. 3-2(c), the average probability of finding the electron becomes less variable. It thus can be seen that, as n becomes sufficiently large, the probability of finding the electron in the well approaches the classical case, as shown in Fig. 3-2(d), as a limit. Here the probability is constant across the well, in agreement with the Bohr correspondence principle (Section 2.4).

3.7 ELECTRON IN A DEEP POTENTIAL WELL

The behavior of an electron in a deep, but not an infinite, one-dimensional well will now be considered. It was found in the previous section that Eq. (3-36) becomes

$$\Psi_I = A \exp\left\{\frac{2\pi}{h} [2m(V - E)]^{1/2}x\right\} \qquad x \leqslant 0$$

since it was shown that the coefficient B must approach zero. The wave function in the well was given as

$$\Psi_{II} = C \sin\left[\frac{2\pi}{h} (2mE)^{1/2}x\right] + D \cos\left[\frac{2\pi}{h} (2mE)^{1/2}x\right], \qquad 0 \leqslant x \leqslant L \qquad (3\text{-}37)$$

Evaluating these functions at $x = 0$, it is seen that $\Psi_I = A \exp[0]$, so

$$\Psi_I = A \neq 0$$

and, since $\sin 0 = 0$ and $\cos 0 = 1$,

$$\Psi_{II} = D \neq 0$$

Now Ψ_I and Ψ_{II} are nonzero and may have significant values in their respective ranges, in contrast to the case of an infinite well (Section 3.5). The criteria for meaningful solutions are invoked again. Since continuity must exist across the boundary, $\Psi_I = \Psi_{II}$ and so $A = D$; the values of A and D now may be appreciable. These parameters increase inversely as the well depth. Also, the derivatives must be continuous at the wall. So, at $x = 0$, the partial derivatives of Ψ_I and Ψ_{II} are

$$\frac{\partial \Psi_I}{\partial x} = A \frac{2\pi}{h} [2m(V - E)]^{1/2}$$

because the exponential factor equals unity, and

$$\frac{\partial \Psi_{II}}{\partial x} = C \frac{2\pi}{h} (2mE)^{1/2}$$

again considering that $\cos 0 = 1$ and $\sin 0 = 0$ at $x = 0$. And, since continuity must exist, these derivatives must also be equal. Thus, at $x = 0$,

$$A \frac{2\pi}{h} [2m(V - E)]^{1/2} = C \frac{2\pi}{h} (2mE)^{1/2}$$

and, rearranging, after noting that both $2\pi/h$ and $(2m)^{1/2}$ vanish,

$$A = \frac{C(2mE)^{1/2}}{[2m(V - E)]^{1/2}} = C\left[\frac{E}{V - E}\right]^{1/2}$$

It was just shown that $A = D$, so a more useful equation may be obtained from this:

$$D = C\left[\frac{E}{V - E}\right]^{1/2}$$

This will be used shortly.

Continuity also must exist at $x = L$; that is, $\Psi_{II} = \Psi_{III} > 0$; thus,

$$C \sin \left[\frac{2\pi}{h} (2mE)^{1/2} L \right] + D \cos \left[\frac{2\pi}{h} (2mE)^{1/2} L \right] = G \exp \left\{ - \frac{2\pi}{h} [2m(V - E)]^{1/2} L \right\}$$

For convenience, let the parameters

$$\beta = \frac{2\pi}{h} (2mE)^{1/2}$$

and

$$\alpha = \frac{2\pi}{h} [2m(V - E)]^{1/2}$$

Another helpful relation, for later use, may be obtained from the ratio of these two parameters, again noting that $2\pi/h$ and $(2m)^{1/2}$ vanish:

$$\frac{\beta}{\alpha} = \left[\frac{E}{V - E} \right]^{1/2}$$

Now, the equation representing $\Psi_{II} = \Psi_{III}$, using the preceding parameters, may be written more simply as

$$C \sin \beta L + D \cos \beta L = Ge^{-\alpha L}$$

The derivatives of these functions with respect to L also must be equal. Thus, according to the criteria for meaningful solutions, and noting that dL vanishes,

$$\beta C \cos \beta L - \beta D \sin \beta L = -\alpha Ge^{-\alpha L}$$

This expression now is divided by the original equation to obtain

$$\frac{\beta C \cos \beta L - \beta D \sin \beta L}{C \sin \beta L + D \cos \beta L} = - \frac{\alpha Ge^{-\alpha L}}{Ge^{-\alpha L}}$$

Noting that $Ge^{-\alpha L}$ vanishes, factoring β, and dividing both sides of this equation by β gives

$$\frac{C \cos \beta L - D \sin \beta L}{C \sin \beta L + D \cos \beta L} = - \frac{\alpha}{\beta}$$

Multiplying by (-1) and rearranging results in

$$\frac{D \sin \beta L - C \cos \beta L}{D \cos \beta L + C \sin \beta L} = \frac{\alpha}{\beta}$$

Using the relationship previously found between D and C and substituting this for D, the following is obtained:

$$\frac{C[E/(V - E)]^{1/2} \sin \beta L - C \cos \beta L}{C[E/(V - E)]^{1/2} \cos \beta L + C \sin \beta L} = \frac{\alpha}{\beta}$$

The factor C vanishes from the fraction. Then, making use of the relationship previously found for β/α,

$$\frac{(\beta/\alpha) \sin \beta L - \cos \beta L}{(\beta/\alpha) \cos \beta L + \sin \beta L} = \frac{\alpha}{\beta}$$

Clearing α and rearranging the terms results in

$$\beta^2 \sin \beta L - \alpha\beta \cos \beta L = \alpha\beta \cos \beta L + \alpha^2 \sin \beta L$$

Collecting terms,

$$(\beta^2 - \alpha^2) \sin \beta L = 2\alpha\beta \cos \beta L$$

or

$$\tan \beta L = \frac{\sin \beta L}{\cos \beta L} = \frac{2\alpha\beta}{\beta^2 - \alpha^2}$$

Finally, substituting the values of the parameters of α and β and simplifying gives the relationship being sought:

$$\tan \left[\frac{2\pi}{h} (2mE)^{1/2} L \right] = \frac{2[E(V - E)]^{1/2}}{2E - V}$$

This expression cannot be solved directly for E. It can be solved graphically. However, an approximate solution will be given here. Consider the right side of this equation and recall that $V \gg E$; then, dividing numerator and denominator by V, so that each E is compared directly with V,

$$\frac{2[E(V - E)]^{1/2}}{2E - V} = \frac{2\left[\dfrac{E}{V} \left(1 - \dfrac{E}{V} \right) \right]^{1/2}}{2\dfrac{E}{V} - 1} \cong \frac{2\dfrac{E}{V} [(1 - 0)]^{1/2}}{2 \times 0 - 1} \cong -2\left[\frac{E}{V} \right]^{1/2} \cong 0$$

Thus, it may be approximated that

$$\tan \left[\frac{2\pi}{h} L(2mE)^{1/2} \right] \cong 0$$

And, for this to be so, the following must be true:

$$\frac{2\pi}{h} L(2mE)^{1/2} = n\pi$$

Squaring both sides of this equation and noting that π vanishes gives

$$\frac{4L^2}{h^2} \cdot 2mE = n^2$$

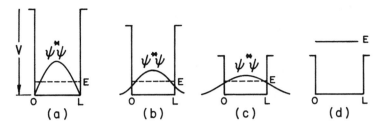

Figure 3-3 Probability of finding an electron in a one-dimensional well: (a), (b), and (c) decreasing well depth; (d) a free electron.

Then, solving for E, the eigenvalue is found to be

$$E = \frac{n^2 h^2}{8mL^2} \tag{3-40}$$

This is the same result as that obtained for a much higher barrier; see Eq. (3-39). Discrete energy levels are obtained again.

Now consider the effect of further decreasing the depth, V, of the potential well. The same equations for the three regions are still used. However, as V decreases, the values of Ψ_I, Ψ_{II}, and Ψ_{III} increase at $x = 0$ and $x = L$. Thus, Ψ_I and Ψ_{III} extend increasingly exponentially into regions I and III as V diminishes. The closer that V approaches E (the shallower the depth of the well), the greater are the values of Ψ_I and Ψ_{III} at $x = 0$ and $x = L$, respectively, and they drop off less rapidly in their respective regions. At the point where $E > V$, the particle is no longer affected by the potential well. This removes all the restrictions on E, and the classical case of the unrestricted, or free, particle prevails. These effects are shown schematically in Fig. 3-3.

It also should be noted that when $x = L$ is small, of atomic dimensions, the energy levels are relatively widely separated. However, as L gets larger and normal crystal dimensions are approached, the energy levels become closer and closer together until a quasicontinuum is approached, in agreement with the Bohr correspondence principle. This is shown schematically in Fig. 3-4.

The case where many electrons are in a well of crystal dimensions, such as that shown in Fig. 3-4(b), has many important applications that are discussed in later chapters. Some of these include the Fermi–Dirac statistics (Section 2.6), the Sommerfeld theory, and electron transport processes (Chapters 5 and 6).

It also should be noted that when $x = L$ is small, of atomic dimensions, the energy levels are relatively widely separated. However, as L gets larger and normal crystal dimensions are approached, the energy levels become closer and closer together until a quasicontinuum is approached, in agreement with the Bohr correspondence principle. This is shown schematically in Fig. 3-4.

The case where many electrons are in a well of crystal dimensions, such as that shown in Fig. 3-4(b), has many important applications that are discussed in later chapters. Some of these include the Fermi–Dirac statistics (Section 2.6), the Sommerfeld theory, and electron transport processes (Chapters 5 and 6).

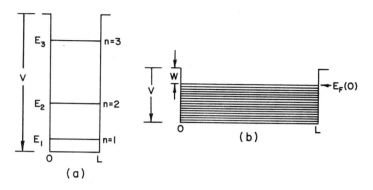

Figure 3-4   Effect of well width, *L*, on energy levels. (a) For atomic distances the states are widely separated. These separations decrease with increasing well width until at macroscopic distances, (b) the states approach a quasicontinuum.

3.8 THREE-DIMENSIONAL POTENTIAL WELL

Up to this point, solutions to Schrödinger's equation have been found only for one-dimensional wells. Consideration now will be given to the behavior of an electron in a three-dimensional well with infinitely high walls, as shown schematically in Fig. 3-5. The variation of the potential along the *x-y*, *x-z*, and *y-z* planes is the same as for the one-dimensional case (Section 3.5):

$$V(x, y, z) = 0 \text{ in the well}$$

$$V(x, y, z) = \infty \text{ outside the well}$$

Since $V(x, y, z)$ is zero within the well, the three-dimensional form of Schrödinger's time-independent equation, Eq. (3-34), becomes

$$\frac{\partial^2 \Psi}{\partial x^2} + \frac{\partial^2 \Psi}{\partial y^2} + \frac{\partial^2 \Psi}{\partial z^2} + \frac{8\pi^2 m}{h^2} E\Psi = 0 \qquad (3\text{-}41)$$

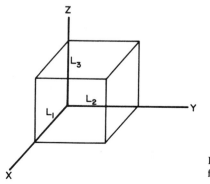

Figure 3-5   Three-dimensional well used for Schrödinger's equation.

where Ψ is a function only of x, y, and z, the position of the particle in the well. It is helpful to separate these variables, so solutions will be sought of the form

$$\Psi = \Psi(x, y, z) = X(x)Y(y)Z(z) \tag{3-42}$$

in which each of the individual, separable expressions is a function of a single variable. Such separable solutions do not always exist; they depend on the form of the potential energy. The second derivatives of Eq. (3-42) are found to be

$$\frac{\partial^2 \Psi(x, y, z)}{\partial x^2} = Y(y)Z(z)\frac{\partial^2 X(x)}{\partial x^2} \tag{3-43a}$$

$$\frac{\partial^2 \Psi(x, y, z)}{\partial y^2} = X(x)Z(z)\frac{\partial^2 Y(z)}{\partial y^2} \tag{3-43b}$$

$$\frac{\partial^2 \Psi(x, y, z)}{\partial z^2} = X(x)Y(y)\frac{\partial^2 Z(z)}{\partial z^2} \tag{3-43c}$$

By substituting Eqs. (3-43a), (3-43b), and (3-43c), Eq. (3-41) becomes

$$Y(y)Z(z)\frac{\partial^2 X(x)}{\partial x^2} + X(x)Z(z)\frac{\partial^2 Y(y)}{\partial y^2} + X(x)Y(y)\frac{\partial^2 Z(z)}{\partial z^2} + \frac{8\pi^2 m}{h^2} EX(x)Y(y)Z(z) = 0$$

$$\tag{3-44}$$

When Eq. (3-44) is divided through by $X(x)Y(y)Z(z)$, it is found that

$$\frac{1}{X(x)}\frac{\partial^2 X(x)}{\partial x^2} + \frac{1}{Y(y)}\frac{\partial^2 Y(y)}{\partial y^2} + \frac{1}{Z(z)}\frac{\partial^2 Z(z)}{\partial z^2} + \frac{8\pi^2 m}{h^2} E = 0 \tag{3-45}$$

Each of three terms in Eq. (3-45) is a function of a single, different variable. The sum of these three terms is a constant. This can only be true if each of the terms is equal to a constant, or eigenvalue. Therefore, each of these terms may be expressed as

$$\frac{1}{X(x)}\frac{\partial^2 X(x)}{\partial x^2} = -\frac{8\pi^2 m}{h^2} E_x \tag{3-46a}$$

$$\frac{1}{Y(x)}\frac{\partial^2 Y(y)}{\partial y^2} = -\frac{8\pi^2 m}{h^2} E_y \tag{3-46b}$$

and

$$\frac{1}{Z(z)}\frac{\partial^2 Z(z)}{\partial z^2} = -\frac{8\pi^2 m}{h^2} E_z \tag{3-46c}$$

where the eigenvalues E_x, E_y, and E_z are constants and

$$E_x + E_y + E_z = E \tag{3-46d}$$

These equations are treated individually as shown by using Eq. (3-46a) as an example.

Upon rearrangement, they take the form

$$\frac{\partial^2 X(x)}{\partial x^2} + \frac{8\pi^2 m}{h^2} E_x X(x) = 0$$

These equations are of the same form as that previously obtained for the one-dimensional well (Section 3.5). The application of the same criteria and boundary conditions as were used previously gives the following results:

$$E_x = \frac{n_x^2 h^2}{8mL_x^2} \qquad (3\text{-}47\text{a})$$

$$E_y = \frac{n_y^2 h^2}{8mL_y^2} \qquad (3\text{-}47\text{b})$$

$$E_z = \frac{n_z^2 h^2}{8mL_z^2} \qquad (3\text{-}47\text{c})$$

in which n_x, n_y, and n_z are integers. Now, making use of Eq. (3-46d), it is found that

$$E = E_x + E_y + E_z = \frac{h^2}{8m}\left[\frac{n_x^2}{L_x^2} + \frac{n_y^2}{L_y^2} + \frac{n_z^2}{L_z^2}\right] \qquad (3\text{-}48)$$

If the well is a cube, $L_x = L_y = L_z = L$, then

$$E = \frac{h^2}{8mL^2}\left[n_x^2 + n_y^2 + n_z^2\right] \qquad (3\text{-}49)$$

The three integers n_x, n_y, and n_z are the three quantum numbers required to specify each state in rectangular coordinates in state space (Section 2.1). In effect, Eq. (3-49) quantizes the volume of the well. Again, none of these integers may be equated to zero, since this would be the same as setting a wave function equal to zero; it will be recalled (Section 3.3) that this violates the criteria for meaningful wave functions. Equations (3-48) and (3-49) give the probable energy of an electron in terms of its most likely position within the well. A portion of this energy has been associated with each coordinate in state space, the total energy being the sum of these.

It must be noted that this description is incomplete. The presence of just three quantum numbers in Eq. (3-49) has important implications. This arises because it actually represents solutions to wave functions, $\psi(r)$, where $r = (n_x^2 + n_y^2 + n_z^2)^{1/2}$, that do not include either $\psi_+(\sigma)$ or $\psi_-(\sigma)$ of Eqs. (2-69a) and (2-69b). In other words, the influence of spin is not included in this equation. As noted in Section 2.5, the inclusion of spin has small but significant effects on a given total wave function. However, these may be considered to have negligible effects on the eigenvalues of two electrons with a given $\psi(r)$ but with opposite values of $\psi(\sigma)$. Thus, the eigenvalues provided by each set of three quantum numbers in Eq. (3-49) may be considered to represent two implicit electron states of opposite spin. This is the same as saying that two indistinguishable particles of opposite spins may occupy the same state in state space at the same time; it does not violate the principle of exclusion because of the requirement of different spins (Section 3.9). Another way of accounting for spin is to consider that Ψ [Eq. (3-41)] is a double-valued function whose eigenvalues for opposite spins usually may be regarded as being virtually the same. Frequent use is made of Eq. (3-49) in explaining the properties of solids on this basis.

3.9 THE PAULI EXCLUSION PRINCIPLE

The preceding derivations and illustrations have considered only one electron in potential wells. It might be thought that as more electrons are added that all these would be found at some common, lowest energy level. If this were so, all the elements would have very similar properties. However, this is known not to be the case. The work begun by Mendeleev on the Periodic Table shows that the elements have widely differing chemical behavior that varies uniformly and periodically with their atomic number. Experiments on light spectra also show that electrons must occupy various different energy levels. It also has been shown experimentally that the discrete x-ray spectrum of an element is characteristic of that element. Since such spectra involve the inner electrons, which normally do not enter into bonding or chemical reactions, these too must be in different energy states.

In considering the presence of more than one electron in a potential well, such as that constituted by a positively charged nucleus, it is natural to think in terms of the repulsions of the similarly charged electrons and their attractions to the nucleus. The behavior of each electron would affect the behavior of all the others. In a sense, this mutual interaction could be thought of as the sole basis for differences in atomic behaviors. However, such a model based only on undifferentiated coulombic attractions and repulsions could not account for the phenomena noted.

It will be recalled that the three-body problem (sun, earth, and moon or a nucleus and two electrons) cannot be solved exactly. The complications arising in the consideration of atoms with atomic numbers of the order of 100 can be appreciated readily. This difficulty may be avoided. As has been shown in this chapter, electron behaviors can be described readily in terms of energies and quantum numbers and not in terms of their mutual electrostatic behaviors. A way of doing this is by the use of the integers such as are used in Eq. (3-48). If additional electrons are successively added about a nucleus or to a well, the energy of the system and the quantum numbers of the electrons must change accordingly.

Based on the behavior of spectra, Pauli (1925) stated the exclusion principle that describes this behavior. This postulates that no two electrons can be in the same quantum state at the same time. In terms of Eq. (3-49), this means that no two electrons can have the same integers n_x, n_y, and n_z in the same order when spin is implicitly included (Section 2.5). This description is incomplete since a fourth quantum number, spin, must also be considered. This is more fully discussed, in terms of atomic structure, in Section 3.10.

When the Pauli exclusion principle is applied to the electrons about a nucleus, it results in electron configurations that are in agreement with the behaviors previously noted: the periodic chemical behavior and light and x-ray spectra, to mention only a few phenomena, are explained readily.

However, electrons with different quantum numbers can have the same energies. As examples, electrons with quantum numbers $(5, 1, 1)$, $(1, 5, 1)$, $(1, 1, 5)$, and $(3, 3, 3)$ all have different quantum numbers and wave functions, but all have the same energy: $27h^2/8mL^2$, according to Eq. (3-49). These are known as degenerate states. A nondegenerate electron state is one for which only one wave function exists corresponding to the given energy. The state $(1, 1, 1)$ is nondegenerate; no other state can have $E = 3h^2/8mL^2$. However, each state can accommodate two electrons, provided that their spins are opposite. See Sections 2.5, and 3.10.

In a well of small crystal dimensions, such as is shown in Fig. 3-4(b), the order of 10^{20} or more valence electrons may be involved. With such a large number of electrons, a small crystal will contain many electrons in degenerate states. The large number of electron energy levels packed into a relatively small range of energy results in the energy levels being very close together but still remaining discrete. This condition, as previously noted in Sections 2.6 and 2.7 and shown schematically in Fig. 3-5, is called a quasicontinuum and it approaches the classical idea of a continuous range of energies as a limit. This too conforms to the Bohr correspondence principle. See Figs. 4-5 and 4-6.

3.10 APPLICATION TO ATOMIC STRUCTURE

Previous sections have considered the Schrödinger model of the behavior of individual electrons as affected by potential wells. The wave mechanics approach is extended in this section to provide an outline of the interactions of electrons with nuclei.

The simplest atom, hydrogen, is used as the basis for understanding the electron configurations of atoms. It consists of a positive nucleus, a single proton, with one electron in "orbit" around it. Instead of a cubic well, as in Section 3.8, the potential is spherically symmetrical about the nucleus. Since this potential is a function only of the distance, r, between the charged particles, the spherical potential well is described by

$$V(r) = -\frac{e^2}{4\pi\varepsilon_0 r} \tag{3-50}$$

where $4\pi\varepsilon_0$ is the dielectric constant of vacuum and the origin is taken at the nucleus, which is considered to be a point mass. This system is shown in Fig. 3-6.

Here

$$x = r \sin\theta \cos\phi$$
$$y = r \sin\theta \sin\phi \tag{3-51a}$$
$$z = r \cos\theta$$

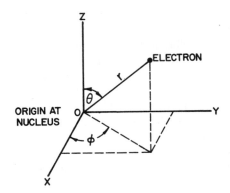

Figure 3-6 Spherical coordinates for an electron in a hydrogen atom.

and

$$r^2 = x^2 + y^2 + z^2 \tag{3-51b}$$

The wave equation is now a function of the variables, r, θ, and ϕ, and corresponds to Eq. (3-34):

$$\frac{1}{r^2} \frac{\partial}{\partial r} \left[r^2 \frac{\partial \Psi}{\partial r} \right] + \frac{1}{r^2 \sin \theta} \frac{\partial}{\partial \theta} \left[\sin \theta \frac{\partial \Psi}{\partial \theta} \right] + \frac{1}{r^2 \sin^2 \theta} \frac{\partial^2 \Psi}{\partial \phi^2} + \frac{8 \pi^2 m}{h^2} [E - V(r)]\Psi = 0$$

$$\tag{3-52}$$

Solutions are sought which are of the form

$$\Psi(r, \theta, \phi) = R(r)\Theta(\theta)\Phi(\phi) \tag{3-53}$$

in a way analogous to the treatment of Eq. (3-28).

Separations are performed and three ordinary, linear differential equations are obtained; their solutions provide three quantum numbers. The notations now used for these were introduced by Goudsmit. The first of these is the principal quantum number, n. The second is the angular momentum, or azimuthal, quantum number, l. The third is m_l, the magnetic quantum number. The quantum numbers n and l are related to $R(r)$; l and m_l are related to $\Theta(\theta)$; and m_l is related to $\Phi(\phi)$.

The principal quantum number, n, is a positive integer. In terms of the Bohr model, it specifies the major axis of an elliptical "orbit". It denotes the probable size of the "orbit" and the energy shell. As n increases, the probability of finding the electron moves away from the nucleus. The letters K, L, M, N, and so on, corresponding to $n = 1, 2, 3, 4$, and so on, respectively, are commonly used to designate electron levels, or *shells*, in x-ray and spectroscopic work. The gross energy level of an electron is determined by n. The total possible number of electrons with a given value of n is $2n^2$.

The second quantum number, l, is a measure of the angular momentum of the orbital motion of the electron. In terms of the Bohr representation, l is a measure of the minor axes of the "orbits". It can assume integral values from zero to $n - 1$, each

TABLE 3-1   DESIGNATION OF ELECTRON STATES

n	Maximum l	Maximum Number of States	Designation of States
1	0	2	s
2	1	6	p
3	2	10	d
4	3	14	f

of which denotes small differences in energy. For a given value of l, the maximum number of states is given by $2(2l + 1)$. These relationships are given in Table 3-1.

As the numerical value of l increases for a given n, the probability of finding the electron at slightly greater average distances from the nucleus increases. As such, l represents small energy variations in the sublevels within an energy level given by a particular n. The energy differences resulting from electron transitions in l states within a given n level are smaller compared to electron transitions between levels of different principal quantum numbers.

Where the wave functions are such that l is relatively large, the electron transitions from a level n to one at $n - 1$ results in a more circular "orbit". When l decreases, the "orbits" become more elliptical for increasing n. At $l = 0$ (zero angular momentum), a linear wave function exists; here the probability of finding the electron passes through the nucleus. This state is spherically symmetrical because all spatial orientations of the linear function are equally probable. Some probability densities are shown in Fig. 3-7. These give the most probable distances of an electron from the nucleus. It must be remembered that these are *probabilities* and as such are inexact. As a result of this inability to define them more precisely, electrons in a given state, at best, must be considered to be most likely within a probability "cloud" about the nucleus.

The third quantum number m_l also explains experimentally observed spectral behavior and is operative only in the presence of a magnetic field. Using the Bohr model, it determines the spatial orientations of the planes of the "orbits" with respect to the magnetic field; m_l is also called the magnetic quantum number because of this. In the presence of a magnetic field, m_l can vary in integral steps from $-l$ to l, including

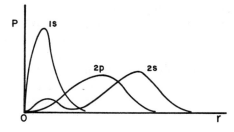

Figure 3-7 Probability densities, $P = 4\pi r^2 \psi^* \psi \, dr$, for a lithium atom.

zero. Here, different values of m_l become apparent from small variations in the energy of an atom depending on its orientation in the magnetic field. This explained early observations of the appearance of multiple spectral lines under these conditions. Such multiple lines represent wavelengths that are very close to the wavelength of the single line that appears in the absence of the magnetic field. This is one of the bases for the association of a magnetic moment with an electron. Such spectral line splitting is known as the *Zeeman effect*. In the absence of a magnetic field, m_l is unaffected and degenerate states result, in agreement with the single spectral line observed under these conditions. This is the reason that electrons usually are identified by their first two quantum numbers: 2*p*, 3*d*, 4*s*, and so on.

As noted in Section 3.9, three quantum numbers are insufficient to describe electron behavior. Solutions to Eq. (3-53) provide only three quantum numbers; electron spin is not included in these solutions. Pauli found that an additional quantum number was needed to explain the observed spectral behavior of atoms in nonhomogeneous magnetic fields. In this case, spectral lines of alkali atoms with complete inner shells and one outer *s* electron ($l = 0$) were expected to be unaffected by the field. It was found, however, that some spectral lines of such atoms split into two lines. This behavior was explained originally by Uhlenbeck and Goudsmit on the basis that these electrons could be considered to be spinning about their own axes while in orbit about the nucleus. The splitting of the spectral lines meant that two spin orientations must exist. In one of these orientations, an electron can be thought to spin about its own axis in the same direction as the external magnetic field; this is called *parallel spin*. The other spin direction can be considered as an intrinsic electron rotation in a direction opposite to this magnetic field, or *antiparallel spin*. This condition is a very slightly lower energy state than that of parallel spin; the energy difference usually is considered to be negligible. For two such spin orientations, the spin quantum number m_s is either $\pm\frac{1}{2}$. The spin quantum number, as described here, satisfied Pauli's requirement for a fourth quantum number and explained the spectra. In actuality, spin cannot be described (Section 2.5). The properties of electrons are such that they can be explained when they are considered to behave as though they have intrinsic spins about their own axes.

Thus, the concept of spin also arose from the necessity to explain observed spectral data. It was shown later by Dirac, in considering relativistic effects, that electrons must possess both spins and angular momenta. This is the basis for the statement given in Section 3.8 that Ψ in Eq. (3-41), and in the present case [Eq. (3-53)], may usually be considered as a double-valued function whose eigenvalues for opposite spins generally may be regarded as being virtually the same.

Reference has been made to the Bohr atom to help explain the quantum numbers obtained from Schrödinger's equation. The properties of an electron in an "orbit" about a nucleus are easy to picture using this model. However, when wave functions are used, probability distributions are obtained for the properties of an electron "orbiting" about the nucleus. It will be recognized from prior discussions that it is not possible to know these properties in the same sense as in the Bohr atom.

It thus is necessary to expect on an intuitive basis that some type of electron "orbital" behavior must take place that includes mass, momentum, magnetic moment, charge, and spin. This must be the case because the experimentally observed phenomena can be explained only by the inclusion of these factors.

The solution outlined previously for hydrogen is that for a two-body problem: one electron and a nucleus (proton). Where many electrons are involved such solutions become much more difficult primarily because $V(r)$ [Eq. (3-50)] can become very complicated. This complexity arises from the necessity to include the combined effects of the coulombic attraction between the nucleus and each electron, as well as all the electron–electron interactions and their individual spins in the Hamiltonians.

The inclusion of spin increases the number and complexity of the wave functions. The helium atom is the simplest case, since only two electrons are involved. Linear combinations of symmetric wave functions are assumed, each corresponding to a given eigenvalue. Thus, for a helium atom

$$E_1 = H_{11} + H_{12} \qquad\qquad (3\text{-}54\text{a})$$

and, for the equally probable case in which the electrons are interchanged,

$$E_1 = H_{11} - H_{12} \qquad\qquad (3\text{-}54\text{b})$$

Here, the first subscript denotes the eigenvalue and the second denotes the electron. Where equations of the type represented by Eq. (3-54b) are used, the wave functions are termed antisymmetric. This change in sign arises from the interchange of the two electrons and is a consequence of the changes in their coordinates in state space. The total wave function must include such changes when this occurs. See Section 2.5. This is the quantum mechanic basis for the Pauli exclusion principle, since it prevents electrons from having the same quantum numbers including spin. The concepts of symmetric and antisymmetric wave functions are used later in this chapter to explain bonding and to derive the band theory of solids in Chapter 4.

The solution of Schrödinger's equation for the hydrogen atom, with one electron in a spherical potential, was used to obtain three quantum numbers to which a fourth, spin, was added. The theoretical basis for the Pauli exclusion principle, governing allowed values of the quantum numbers, was described using helium, a two-electron atom. A justification for the extension of these ideas to atoms composed of many more electrons must be provided.

The results of one method, known as the central field approximation, verify their extension to atoms with more complex electron configurations. In this method, the potential function sums both the individual coulombic attractions between the nucleus and all the electrons, neglecting any nuclear effects, and includes the individual electron–electron interaction potentials, neglecting magnetic effects arising from electron motion. Exact solutions to Schrödinger's equation can be obtained only for spherical potential fields. A spherical approximation can be made since those energy levels that are completely filled, assuming the Pauli exclusion principle, are very nearly spherical. This provides the basis for considering the electrons to be

TABLE 3-2   ALLOWED COMBINATIONS OF QUANTUM NUMBERS, UP TO $n = 4$

n	l	Designation	m_l	m_s	Number	Total
1	0	$1s$	0	$\pm\frac{1}{2}$	2	2
2	0	$2s$	0	$\pm\frac{1}{2}$	2	8
2	1	$2p$	$-1, 0, 1$	$\pm\frac{1}{2}$	6	
3	0	$3s$	0	$\pm\frac{1}{2}$	2	
3	1	$3p$	$-1, 0, 1$	$\pm\frac{1}{2}$	6	18
3	2	$3d$	$-2, -1, 0, 1, 2$	$\pm\frac{1}{2}$	10	
4	0	$4s$	0	$\pm\frac{1}{2}$	2	
4	1	$4p$	$-1, 0, 1$	$\pm\frac{1}{2}$	6	
4	2	$4d$	$-2, -1, 0, 1, 2$	$\pm\frac{1}{2}$	10	32
4	3	$4f$	$-3, -2, -1, 0, 1, 2, 3$	$\pm\frac{1}{2}$	14	

in a spherically symmetrical potential field in a way analogous to that of the hydrogen atom. Electrons closer to the nucleus are acted on by a much greater attractive potential than outer electrons since the latter are shielded, or screened, from the nucleus by the intervening electrons, and thus are subject to a smaller attractive potential. A given outer electron is thus considered as being influenced by a smaller net nuclear charge than that acting on an inner electron. In this way, the model approximates a symmetry of potential like that of a hydrogen atom. This simplification gives a potential function that approximates the resultant of the nuclear attraction, the screening effects for each electron, and the electron–electron interaction potentials. An expression for each electron in the atom is contained in the Hamiltonian. Each of these equations is similar to that for hydrogen, so hydrogenlike wave functions can be used. The ionization energies of electrons calculated in this way are approximate. However, these are in sufficient agreement with experiment to ensure that the hydrogenlike wave functions and their corresponding quantum numbers are realistic and are applicable to atoms with many electrons. Therefore, the quantum numbers characterized earlier for the hydrogen atom may be applied to atoms with complex electron configurations when taken in conjunction with the Pauli exclusion principle. The basis for this application is shown in Table 3-2.

3.11 ELECTRON ENERGY STATES

The permissible values for the four quantum numbers must be used in accordance with the Pauli exclusion principle for the reasons explained in the previous section. This acts as a guide to the way in which the electron energy states are filled, as shown in Table 3-2. Consider, for example, a given set of $2p$ electrons. Here $n = 2$ and $l = 1$. Then, m_l can take the values of $-1, 0$, and 1 and $m_s = \pm\frac{1}{2}$. From these factors it follows that only six states are possible:

State	$2p$	$2p^2$	$2p^3$	$2p^4$	$2p^5$	$2p^6$
m_l	-1	0	1	-1	0	1
m_s	$-\frac{1}{2}$	$-\frac{1}{2}$	$-\frac{1}{2}$	$\frac{1}{2}$	$\frac{1}{2}$	$\frac{1}{2}$

This is in agreement with the prior statement that the maximum number of states for any value of l is given by $2(2l + 1)$. In a similar way, the s level ($l = 0$) can accommodate two electrons, the d level ($l = 2$) can contain up to 10 states, and the f level ($l = 3$) can accept 14 electrons at most.

Thus, for a given n and l, only specified numbers of electrons can be accommodated. If this were not the case, both multiple occupancy of a given state and the case in which all electrons might be at one lowest level might exist. Under such conditions, the observed chemical, spectral, and other physical phenomena noted in Section 3.9 could not have been explained.

The application of the Pauli exclusion principle to electrons as they are successively added to nuclei with increasing numbers of positive charges leads to the organization of atoms in the same way as they are given in the Periodic Table (see Table 3-3). The isolated atoms are electrically neutral.

The valence electrons, those in outer, incompletely filled shells, are those that normally enter into chemical reactions and determine the types of compounds thus formed. Many such compounds have stoichiometries that are governed by the valences as given for the ground state.

The nearly free electrons in the lattice of a metal or alloy may be approximated to constitute a degenerate "electron gas" similar to that of the classical theory (Sections 1.1 and 2.7). These outer electrons were given up by the atoms when the lattice was formed. Thus, a metal atom occupying a lattice site is *not* electrically neutral when examined in isolation. *It is for this reason that metal atoms are considered to be ions in the subsequent discussions.*

For most purposes of this book, the inner electrons, occupying completed shells of an atom, are considered to be tightly bound to the nucleus; these do not ordinarily enter into physical processes. As such, they are designated as constituting an ion core. However, these bound electrons must be taken into consideration to explain certain magnetic properties. The valence electrons of metals are much more loosely bound; consequently, these may participate in many physical processes. Therefore, the nearly free valence electrons are of major interest in considering the physical properties of solids (Section 2.7).

Normal metallic elements (those with completed inner shells and with three or less valence electrons) in the crystalline state frequently do not show the integral valences that would be expected from their ground-state electron configurations. One reason for nonintegral valences arises from the concept that a fraction of these electrons may be closely associated with the atoms or involved primarily in the bonding of the atoms of the crystal; when either of these is the case, the electrons are not normally available for engaging in physical processes. As a result of this behavior, the number of available, or uninvolved, nearly free electrons per atom results in a nonintegral valence for the atom.

TABLE 3-3   GROUND STATE ELECTRON CONFIGURATIONS OF ATOMS

Z^a	Element	Outer Configuration	Z^a	Element	Outer Configuration	Z^a	Element	Outer Configuration
1	H	$1s$	33	As	$4s^2\,4p^3$	66	Dy	$4f^9\,5d\,6s^2$
2	He	$1s^2$	34	Se	$4s^2\,4p^4$	67	Ho	$4f^{10}\,5d\,6s^2$
			35	Br	$4s^2\,4p^5$	68	Er	$4f^{11}\,5d\,6s^2$
3	Li	$2s$	36	Kr	$4s^2\,4p^6$	69	Tm	$4f^{12}\,5d\,6s^2$
4	Be	$2s^2$				70	Yb	$4f^{13}\,5d\,6s^2$
5	B	$2s^2\,2p$	37	Rb	$5s$	71	Lu	$4f^{14}\,5d\,6s^2$
6	C	$2s^2\,2p^2$	38	Sr	$5s^2$	72	Hf	$5d^2\,6s^2$
7	N	$2s^2\,2p^3$	39	Y	$4d\,5s^2$	73	Ta	$5d^3\,6s^2$
8	O	$2s^2\,2p^4$	40	Zr	$4d^2\,5s^2$	74	W	$5d^4\,6s^2$
9	F	$2s^2\,2p^5$	41	Nb	$4d^4\,5s$	75	Re	$5d^5\,6s^2$
10	Ne	$2s^2\,2p^6$	42	Mo	$4d^5\,5s$	76	Os	$5d^6\,6s^2$
			43	Tc	$4d^6\,5s$	77	Ir	$5d^9$
11	Na	$3s$	44	Ru	$4d^7\,5s$	78	Pt	$5d^9\,6s$
12	Mg	$3s^2$	45	Rh	$4d^8\,5s$			
13	Al	$3s^2\,3p$	46	Pd	$4d^{10}$	79	Au	$5d^{10}\,6s$
14	Si	$3s^2\,3p^2$				80	Hg	$6s^2$
15	P	$3s^2\,3p^3$	47	Ag	$4d^{10}\,5s$	81	Tl	$6s^2\,6p$
16	S	$3s^2\,3p^4$	48	Cd	$5s^2$	82	Pb	$6s^2\,6p^2$
17	Cl	$3s^2\,3p^5$	49	In	$5s^2\,5p$	83	Bi	$6s^2\,6p^3$
18	Ar	$3s^2\,3p^6$	50	Sn	$5s^2\,5p^2$	84	Po	$6s^2\,6p^4$
			51	Sb	$5s^2\,5p^3$	85	At	$6s^2\,6p^5$
19	K	$4s$	52	Te	$5s^2\,5p^4$	86	Rn	$6s^2\,6p^6$
20	Ca	$4s^2$	53	I	$5s^2\,5p^5$	87	Fr	$7s$
21	Sc	$3d\,4s^2$	54	Xe	$5s^2\,5p^6$	88	Ra	$7s^2$
22	Ti	$3d^2\,4s^2$				89	Ac	$6d\,7s^2$
23	V	$3d^3\,4s^2$	55	Cs	$6s$	90	Th	$6d^2\,7s^2$
24	Cr	$3d^5\,4s$	56	Ba	$6s^2$	91	Pa	$6d^3\,7s^2$
25	Mn	$3d^5\,4s^2$	57	La	$5d\,6s^2$	92	U	$6d^4\,7s^2$
26	Fe	$3d^6\,4s^2$	58	Ce	$4f^2\,6s^2$	93	Np	$5f^5\,7s^2$
27	Co	$3d^7\,4s^2$	59	Pr	$4f^3\,6s^2$	94	Pu	$5f^5\,6d\,7s^2$
28	Ni	$3d^8\,4s^2$	60	Nd	$4f^4\,6s^2$	95	Am	$5f^6\,6d\,7s^2$
			61	Pm	$4f^5\,6s^2$	96	Cm	$5f^7\,6d\,7s^2$
29	Cu	$3d^{10}4s$	62	Sm	$4f^6\,6s^2$	97	Bk	$5f^8\,6d\,7s^2$
30	Zn	$4s^2$	63	Eu	$4f^7\,6s^2$	98	Cf	$5f^9\,6d\,7s^2$
31	Ga	$4s^2\,4p$	64	Gd	$4f^7\,5d\,6s^2$	99	Es	$5f^{10}\,6d\,7s^2$
32	Ge	$4s^2\,4p^2$	65	Tb	$4f^8\,5d\,6s^2$	100	Fm	$5f^{11}\,6d\,7s^2$

^{a}Atomic number.

An additional reason for nonintegral valences results from the fact that some of the outer electrons can resonate to other unfilled levels in some of the atoms composing the solids. This is particularly true of transition elements and alloys involving transition elements. Nickel provides a good example of this (Table 3-3). In the ground state, elemental Ni has a $3d^8\,4s^2$ array. In the crystalline state, the configuration of Ni is considered to have a $3d^{9.4}\,4s^{0.6}$ hybridized configuration as a result of this effect.   In the case of alloys of normal metal elements with transition elements, the valence electrons from the normal elements tend to occupy the vacant *d* states of the transition element.   Electron mechanisms of this kind are based on what is known as the rigid-band model.   Strong evidence for this is shown by the thermoelectric and magnetic properties of these alloys.

Elements with valences of four or greater usually form covalent bonds in solids. The electrons form these bonds by being mutually shared in pairs of opposite spin by adjacent atoms.   This causes the electrons to be highly directional and very immobile. The resulting bonding energies are about four times the strength of a metallic bond. The result, in most cases, is the virtually complete inhibition of electron freedom. Elements and compounds formed by this type of bonding are, consequently, insulators or, in a relatively small number of cases, semiconductors (Section 3.12.3).

3.12 SOLID-STATE BONDING

One means for the classification of the bonding of the elements to form crystalline solids was indicated in the previous section.   This is based on the number of valence electrons of an atom.   Atoms of elements with three or fewer valence electrons, the normal metals, bond metallically; those with four or more valence electrons tend to bond covalently.   Two other bond types, ionic and van der Waals bonds, are also of importance in bonding and crystal formation.   All these bond types will be described more completely here.

3.12.1 Ionic Bonding

Ionic bonding results from the strong electrostatic attraction between oppositely charged particles.   The ionizations of the originally electrically neutral atoms involved results from the *transfer* of valence electrons between the atom species involved.   One of these gives up its valence electrons and becomes a positively charged ion.   The other accepts the transferred valence electrons and becomes a negatively charged ion.   The resulting coulombic attraction between the charged particles binds the two ions together.   Atoms most likely to participate in this mechanism may be identified by their positions in the Periodic Table relative to those of the noble gases.   Atoms with positions just preceding the noble gases tend to accept electrons; those just following the noble gases tend to donate or to transfer their electrons.   The selection of noble gas atoms as bench marks for atoms that are involved in this bond type will be apparent.

The almost universal example used to illustrate this bonding is that of rock salt, NaCl.   The ground-state, outer, electron configuration of Na is $2p^6 3s$.   This may be considered to be the equivalent of the highly stable, low-energy electron array of neon with one additional valence electron.   The $2p^6$ neon structure very effectively shields the outer $3s$ valence electron from the nuclear charge and, as a consequence, causes it to be relatively weakly associated with its nucleus.   This is the equivalent of considering the electron to be in a very shallow potential well analogous to that shown in Fig. 3-3(c).

The $3p^5$ array of atomic chlorine lacks one electron to complete its $3p$ levels that would enable it to form the $3p^6$ configuration of argon. The very stable $3p^6$ argon structure is much lower in energy than that of $3p^5$. The strong tendency for the chlorine atom (or any other halogen) to lower its energy by completing its $3p$ levels accounts for its high chemical reactivity. It is on this basis that the $3p^5$ structure may be considered to constitute a very deep well for an additional electron.

When Na and Cl atoms are brought together, the $\Psi^*\Psi$ function of the loosely bound $3p$ valence electron from the Na atom [Fig. 3-3(c)] penetrates (tunnels) the deep well associated with the Cl atom and the electron is absorbed by it. The acquisition of this electron completes the $3p$ level of Cl to $3p^6$, changes its charge from that of electrical neutrality to a highly symmetric, electrically negative electron cloud similar to that of electrically neutral argon, and causes its $\Psi^*\Psi$ to be analogous to that of Fig. 3-3(a). The Na atom, having transferred its $3s$ electron to the Cl, is now electrically positive with a highly symmetric electron cloud resembling that of Ne; this also results in its $\Psi^*\Psi$ being analogous to that of Fig. 3-3(a). The transfer of the valence electron from the Na atom to the Cl atom has oppositely charged both species and has caused both ions to have highly spherical, noble-gas-like electron configurations. This is one of the major reasons why classical electrostatics may be used to describe their bonding mechanism.

Some typical ionic lattices are shown in Fig. 3-8. The energy of attraction between two such isolated particles is given by Coulomb's law as

$$U_A = \frac{(Z_1 e)(-Z_2 e)}{4\pi\varepsilon_0 r} \tag{3-55}$$

Here Z_1 and Z_2 are the respective numbers of charges on the positive and negative ions, e is the charge on an electron, $4\pi\varepsilon_0$ is the dielectric constant of vacuum, and r is the distance between the ions. Equation (3-55) may be simplified by changing its

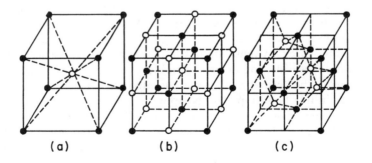

| (a) | (b) | (c) |

Figure 3-8   Some common ionic lattice types: (a) CsCl, (b) Nacl, and (c) ZnS.

units by means of using the relative dielectric constant of vacuum: $\varepsilon_r = 1$. This gives

$$U_A = -\frac{Z_1 Z_2 e^2}{r} \tag{3-56}$$

The minimization of the energy of attraction (maximization of the negative U_A) between the ions requires that they be as close together as possible. As the ions are brought close to each other, the $\Psi^*\Psi$ functions start to interact. This is the equivalent of saying that the electron "clouds" just abut and start to effectively overlap each other. As the "clouds" start their mutual penetration, the Pauli exclusion principle begins to be manifested. Electrons with the same quantum numbers will be present in each "cloud". This is not permitted because of the requirement of exclusion. The clouds resist further penetration because of this, especially since both of the ions have filled, noble-gas, p^6 electron arrays. Further penetration of the clouds can occur only if some of the electrons are promoted to higher states in accordance with the requirement of exclusion. However, abnormally high energies are required for this to occur. While this approach, pioneered by P. W. Bridgeman, is employed to make such synthetic materials as diamonds, sapphires, and rubies, the requisite abnormal pressures and temperatures far exceed those normally available. The "clouds", therefore, resist any additional penetration and behave as though they constituted mutually repelling hard spheres. Strictly speaking, the "clouds" may only be treated as hard spheres when averaged over time, in the absence of such external factors as magnetic fields. Instantaneous, asymmetric charge distributions are always present and can be shown to account for some properties of solids.

The mutual repulsion between ions that arises from the Pauli exclusion principle has several important implications. Of greatest interest in this section is the fact that it, along with the nearly spherical $\Psi^*\Psi$ functions, enables the use of classical electrostatics to explain ionic bonding. Also of great significance is that both of these factors permit the use of the hard-sphere models commonly used to describe crystal structures. The combined effects of these factors can also account for the observed variations in the so-called atomic "sizes" or "diameters." These may be explained by small variations in the degrees of overlap or penetrations of the $\Psi^*\Psi$ functions of different species of atoms or ions in various compounds or in different solid solutions. *Atomic "size" is not a constant.* It is for this reason that atomic "size" as used in this book refers to the distance of closest approach of atoms in elemental crystals at a given temperature.

It was shown by Born that the mutual repulsion energy between a pair of ions in a crystal is given by

$$U_R = \frac{b}{r^n} \tag{3-57}$$

where b is a constant for a given ion pair [Eq. (3-64c)] and n is known as the Born exponent or the repulsion exponent [Table 3-4]. The total energy of bonding of an

TABLE 3-4   REPULSION EXPONENT

Noble Gas Ion Core	Outer Core Configuration	Repulsion Exponent, n
He	$1s^2$	5
Ne	$2s^2 2p^6$	7
Ar	$3s^2 3p^6$	9
Kr	$3d^{10} 4s^2 4p^6$	10
Xe	$4d^{10} 5s^2 5p^6$	12

isolated ion pair is obtained from the sum of Eqs. (3-56) and (3-57):

$$U(r) = -\frac{Z_1 Z_2 e^2}{r} + \frac{b}{r^n} \tag{3-58}$$

The strong electrostatic attractions between ions in ionic crystals are both omnidirectional and long-range in nature. This causes each ion of a given sign to be attracted to every other ion of opposite sign in the crystal. Since these attractive energies vary inversely with the distances between the ions, the more closely situated neighbors make larger contributions to this than do the more widely separated pairs. The summation of all these contributions is a function of the type of lattice of the solid. This may be expressed by considering the electrostatic potential, ϕ, at a reference ion in a lattice, that results from the contributions of all other ions. When surface effects are neglected, this is given by

$$\phi = -\frac{Ze\alpha_m}{r} \tag{3-59}$$

where α_m is the *Madelung constant*, and, summing over all the ions,

$$\alpha_m \propto \sum_j \frac{\pm 1}{r_{ij}} \tag{3-60}$$

The plus and minus signs in Eq. (3-60) refer to the attractions between both a positive and a negative reference ion and all other ions in the crystal. The value of α_m will, therefore, vary with the lattice type (Table 3-5). Thus, the combined effects of all the attractive energies on a pair of ions in a lattice is given by

$$U_A = -\frac{Z_1 Z_2 e^2 \alpha_m}{r} \tag{3-61}$$

This permits the total energy of an ion pair in a lattice to be given by

$$U_T(r) = U_A + U_R = -\frac{Z_1 Z_2 e^2 \alpha_m}{r} + \frac{b}{r^n} \tag{3-62}$$

At equilibrium, $U_T(r)$ will be a minimum and the variation of $U_T(r)$ with respect to r

**TABLE 3-5   SOME TYPICAL VALUES FOR THE
MADELUNG CONSTANT**

Compound	Crystal Lattice	α_m
NaCl	FCC	1.74756
CsCl	CsCl	1.76267
CaF_2	Cubic	2.51939
$CdCl_2$	Hexagonal	2.244
MgF_2	Tetragonal	2.381
ZnS (wurtzite)	Hexagonal	1.64132
TiO_2 (rutile)	Tetragonal	2.408
βSiO_2	Hexagonal	2.2197
Al_2O_3 (corundum)	Rhombohedral	4.1719

Abstracted from *Handbook of Chemistry and Physics*,
60th ed., p. F-242, CRC Press, Boca Raton, Fla., 1980.

will be zero so that, where n now is a constant for a given ionic solid and $r = r_0$ is the equilibrium spacing,

$$\frac{dU_T(r)}{dr} = \frac{\alpha_m Z_1 Z_2 e^2}{r_0^2} - \frac{nb}{r_0^{n+1}} = 0 \tag{3-63}$$

Then

$$nbr_0^{-(n+1)} = \frac{\alpha_m Z_1 Z_2 e^2}{r_0^2} \tag{3-64}$$

Equation (3-64) is multiplied through by r_0 to give

$$nbr_0^{-n} = \frac{\alpha_m Z_1 Z_2 e^2}{r_0} \tag{3-64a}$$

The repulsion term in Eq. (3-62) is then found to be

$$\frac{b}{r_0^n} = \frac{\alpha_m Z_1 Z_2 e^2}{nr_0} \tag{3-64b}$$

And the parameter b is obtained as

$$b = \frac{\alpha_m Z_1 Z_2 e^2}{nr_0^{1-n}} \tag{3-64c}$$

The Born exponent, n, based on this approach, can be shown to vary inversely with the compressibilities of these solids.

The inclusion of Eq. (3-64b) in Eq. (3-62) results in

$$U_T(r) = -\frac{Z_1 Z_2 e^2 \alpha_m}{r_0} + \frac{Z_1 Z_2 e^2 \alpha_m}{nr_0} = -\frac{Z_1 Z_2 e^2 \alpha_m}{r_0}\left(1 - \frac{1}{n}\right) \tag{3-65}$$

It is emphasized that the interionic distance r_0 in Eqs. (3-63) through (3-65) is the distance of closest approach of the ions in a crystal lattice. It is the shortest center-to-center distance of two oppositely charged ions of differing "sizes" treated as hard spheres. This is the sum of the "radii" of both such spheres. It is *not* the ionic radius. The "radii" of given ions will vary depending on factors that include lattice type and the nature of other ions in the lattices with the given ion; these variations result from small interactions between the $\Psi^*\Psi$ electron "clouds".

Reasonable approximations of the repulsion exponent may be made based on the noble gas configurations of the ions being considered. If one of these has a neonlike configuration and the other has an argonlike structure, then, using Table 3-4, $n \cong (7 + 9)/2 \cong 8$. Average errors of less than about 20% may be anticipated for such approximations. These effects are small in Eq. (3-65).

The molar bonding energy is obtained from Eq. (3-65) by multiplying both sides by Avogadro's number:

$$U_{\text{mol}} = N_A U_T(r) = -\frac{N_A Z_1 Z_2 e^2 \alpha_m}{r_0}\left(1 - \frac{1}{n}\right) \qquad (3\text{-}66)$$

Calculations based on this relationship usually provide molar bonding energies that are slightly less than those determined experimentally (Table 3-6). The $\Psi^*\Psi$ "clouds" that constitute the ions are not perfectly hard spheres; some mutual penetration appears to take place. Some of the electrons in the interpenetrating "clouds" may be shared by both ions. This is the equivalent of adding a small component of covalent bonding to the predominating ionic bonding. This effect is illustrated by noting the differences between Figs. 3-9(a) and 3-9(b). The bonding energy of oppositely charged ions, when treated as perfectly hard spheres, is indicated in Fig. 3-9(a). The effect of a small covalent bonding component, resulting from $\Psi^*\Psi$ overlap, is shown in Fig. 3-9(b). It is considered to be highly probable that perfectly pure ionic bonding

TABLE 3-6   BONDING ENERGIES OF IONIC CRYSTALS WITH NaCl LATTICE

Compound	Bonding Energy (kJ/mol)		Compound	Bonding Energy (kJ/mol)	
	Experimental	Calculated		Experimental	Calculated
LiF	−1014.5	−1014.1	KF	−794.7	−791.8
LiCl	−832.8	−807.7	KCl	−694.2	−676.6
LiBr	−794.7	−757.8	KBr	−663.6	−646.8
LiI	−744.0	−695.5	KI	−627.6	−605.0
NaF	−897.7	−901.0	RbF	−759.5	−755.3
NaCl	−764.5	−747.8	RbCl	−667.0	−650.7
NaBr	−726.9	−708.4	RbBr	−638.9	−620.9
NaI	−683.3	−655.7	RbI	−606.7	−584.5

Note for conversions: 4.187 J = 1 cal; 23.05 kcal/mol = 1 eV.

Based upon C. Kittel, *Introduction to Solid State Physics*, 3rd ed., p. 98, John Wiley & Sons, Inc., New York, 1967.

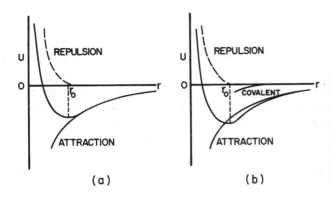

Figure 3-9 Bonding energy as a function of distance between two oppositely charged ions: (a) electrostatic attraction and repulsion; (b) mixed ionic and covalent bonding. (Modified from N. F. Mott and R. W. Gurney, *Electronic Processes in Ionic Crystals*, 2nd ed., p. 9, Colver Publications, New York, 1948.)

normally does not exist and that some small degrees of covalency are very apt to be present. This could account for the small differences between the experimental and calculated values given in Table 3-6.

The minima in the curves of Fig. 3-9 determine the equilibrium interionic spacings, r_0 [Eq. (3-63)]; these are the primary determinants of the crystal lattices formed by various paired ionic species. The inflection points can be considered to be proportional to the minimum quantity of energy required to destroy their crystalline array; this is the equivalent of melting. This occurs where the second derivative of Eq. (3-62), is equal to zero. It is the equivalent of saying that the inherent bonding forces between the ions are minimum. Any externally added energies, or their derivative forces, equal to or larger than the inherent ones are thus expected to disrupt the crystalline array; and r_0 no longer is a constant. The crystal becomes a liquid.

It also is of interest to note that the slope of Eq. (3-63) evaluated at r_0 is proportional to Young's modulus, E_y, of the crystal. This is another way of saying that

$$\left.\frac{d^2 U(r)}{dr_0^2}\right|_{r=r_0} \cong CE_y \qquad (3\text{-}67)$$

in which C is a constant.

Based on the foregoing, the properties of ionically bonded solids may be derived on the basis that the electrons involved are in very deep wells analogous to that of Fig. 3-3(a). Consequently, their bonding energies are expected to be high. The $\Psi^*\Psi$ functions are very highly localized about the ions, and the probability of finding them anywhere else is vanishingly small. Thus, the electrons may be considered as being virtually immobile. As a result, their electrical and thermal conductivities are anticipated to be very low, and ionic solids would be expected to be insulators. The deep wells and the corresponding high bond strengths are manifested by deep and narrow energy versus distance curves. These lead to the prediction of high modulii of elasticity, high melting points, and, as will be shown later, low coefficients of thermal expansion.

3.12.2 The Hydrogen Molecule

The hydrogen molecule represents the least complicated case for the illustration of covalent bonding. The wave mechanics that describe this bonding mechanism also may be used to derive the band theory of solids. The method employed here will be to bring two widely separated hydrogen atoms together and to examine the ways in which their increasing proximity affects their wave functions and eigenvalues.

Each hydrogen atom consists of a positive nucleus that acts as an *omnidirectional* potential well for the single electron in "orbit" around it. The three-dimensional $\psi^*\psi$ function for the electron is analogous to that shown schematically in Fig. 3-3(c) for one dimension. This function (here three dimensional) has appreciable values at short distances away from the nucleus, but it becomes negligibly small at slightly longer distances. The $\psi^*\psi$ functions of each atom may be superimposed. Thus, when the two atoms have a large internuclear distance, r_{12}, of about 40 to 50 atom "diameters", any interaction between either of the electrons associated with each of the atoms is negligible. And where the ground-state energy of each of the remote atoms is E_∞, the energy of this two-atom system is $2E_\infty$. At separations of $r_{12} \leqslant 5$ "diameters," the superimposed $\psi^*\psi$ functions of each atom overlap significantly and their combined value is appreciable. This means that there is now a probability that the electrons can resonate and transfer or "exchange" from one nucleus to the other. As the separation between the nuclei continues to be decreased, the overlapping increases the combined $\psi^*\psi$ value to a degree such that it no longer is possible to consider the electrons as being associated with their original nuclei; the electrons resonate between both nuclei. This constitutes a new, lower-energy state in which the electrons must be considered as being simultaneously associated with both nuclei; in effect, the electrons are shared between them. This is the basic mechanism of covalent bonding.

The increase in $\psi^*\psi$ may be considered to result from the constructive interference of the individual ψ functions. Their superposition, in effect, results in new ψ and $\psi^*\psi$ functions for the two-atom system. These reflect the relatively high electron probability density *between* the two nuclei, and they drop off exponentially at all other positions around the two nuclei. The new ψ function for the two-atom system, resulting from their very close proximity, is expected to have its own eigenvalues. The energy of the system no longer is $2E_\infty$. As is shown next, the bonding results in energies lower (more negative) than $2E_\infty$, a requisite for any bonding to take place.

A simple analysis will be made based on the qualitative description just given. Consider two isolated hydrogen atoms. The Hamiltonian for each such atom and its respective electron is

$$H_{11} = H_{22} = E_\infty \tag{3-68}$$

where E_∞ is the ground-state eigenvalue of each of the remote atoms. This is the equivalent of $E_\infty = E_1$ in Eq. (1-11) and equals $-13.6\,\text{eV}$. Equation (3-68) is obeyed, as the two atoms are brought more closely together, as long as their respective wave functions, ψ_1 and ψ_2, do not overlap. When overlapping starts, interactions occur between the two atoms and these are taken into account by an additional term in the Hamiltonians of the atoms:

$$H_{12} = H_{21} = -E_I \tag{3-69}$$

Thus, the total energy of atom 1 is

$$E = H_{11} + H_{12} = H_{11} - E_I = E_\infty - E_I \tag{3-70a}$$

And that for atom 2 is

$$E = H_{22} + H_{21} = H_{22} - E_I = E_\infty - E_I \tag{3-70b}$$

Here E_∞ is the eigenvalue of the isolated H atom and E_I is the interaction energy between the two atoms. The major components of E_I are discussed next.

The energy equivalents of the Hamiltonians given by Eqs. (3-70a) and (3-70b) are made to operate on their respective ψ functions using Eq. (3-14). These give for the hydrogen molecule, where $E_M = 2E_\infty$,

$$-\frac{h}{2\pi i} \frac{\partial \psi_1}{\partial t} = E_M \psi_1 - E_I \psi_2 \tag{3-71a}$$

and

$$-\frac{h}{2\pi i} \frac{\partial \psi_2}{\partial t} = E_M \psi_2 - E_I \psi_1 \tag{3-71b}$$

Now, making use of the general solution to Schrödinger's equation previously obtained as Eq. (3-31), let

$$\psi_1 = A_1 \exp\left(-\frac{2\pi i}{h} Et\right) \tag{3-72a}$$

and

$$\psi_2 = A_2 \exp\left(-\frac{2\pi i}{h} Et\right) \tag{3-72b}$$

The substitution of Eqs. (3-72a) and (3-72b) into Eq. (3-71a) results in

$$-\frac{h}{2\pi i} A_1 \left(-\frac{2\pi i}{h} E\right) \exp\left(-\frac{2\pi i}{h} Et\right) = E_M A_1 \exp\left(-\frac{2\pi i}{h} Et\right) - E_I A_2 \exp\left(\frac{2\pi i}{h} Et\right)$$

Both the coefficient $-h/2\pi i$ and the exponentials vanish, so Eq. (3-71a) reduces to

$$A_1 E = E_M A_1 - E_I A_2 \tag{3-73a}$$

Similarly, Eq. (3-71b) is found to be expressed as

$$A_2 E = E_M A_2 - E_I A_1 \tag{3-73b}$$

These equations are rewritten, for clarity, as

$$(E_M - E)A_1 - E_I A_2 = 0 \tag{3-74a}$$

and

$$-E_I A_1 + (E_M - E)A_2 = 0 \tag{3-74b}$$

Equations (3-74a) and (3-74b) may be solved if the determinant

$$\begin{vmatrix} E_M - E & -E_I \\ -E_I & E_M - E \end{vmatrix} = 0$$

This is the equivalent of saying that if $H = E$ then $H - E$ must equal zero. This being the case, it is found that the determinant reduces to

$$(E_M - E)^2 = E_I^2$$

The result is *two solutions* for the energy:

$$E = E_M \pm E_I \tag{3-75}$$

This apparently simple result is very important because it permits the effects of the symmetric and antisymmetric wave functions to be visualized with relative ease. These are analogous to those shown in Fig. 3-11. Under the conditions discussed next, these cause the eigenvalue to split into two values: bonding "orbitals" and antibonding "orbitals". Such splitting also may be used to derive the band theory of solids.

It, therefore, is important to examine the term E_I more closely. For the antisymmetric case, it can be shown that

$$E_I = E_{IA} = \frac{B - C}{1 - S_{12}} \tag{3-76}$$

And, for the symmetric case,

$$E_I = E_{IS} = \frac{B + C}{1 + S_{12}} \tag{3-77}$$

The parameter B includes all the energies of interaction between both nuclei, between both electron "clouds", and those between each of the "clouds" and the other nucleus. The parameter C represents the large changes of the energy of the system that result from the changed probability densities of the electrons. The parameter S_{12} denotes $\psi_1 \psi_2$ and is a measure of the overlap of the wave functions. It has values in the range $0 \leqslant S_{12} \leqslant 1$. Thus, for the symmetric case,

$$E_S = 2E_\infty + E_{IS} = 2E_\infty + \frac{B + C}{1 + S_{12}} \tag{3-78}$$

And, for the antisymmetric case,

$$E_A = 2E_\infty + E_{IA} = 2E_\infty + \frac{B - C}{1 - S_{12}} \tag{3-79}$$

At relatively large separations of the H atoms, virtually no interaction occurs between them; thus, $E_{IS} = E_{IA} \cong 0$ and $E_S = E_A = 2E_\infty$. However, as the atoms are brought sufficiently close together, so that their wave functions begin to overlap, the effects of the symmetric and antisymmetric wave functions begin to become manifested and the energy splits into the two levels indicated by Eq. (3-75).

Consider the symmetric case, E_S [Eq. (3-78)]. As overlap takes place, the parameters B and C are negative, but $|C| \gg |B|$. This value of C is greatest *between* the nuclei where the electrons are shared; it is called the *exchange interaction energy*. Here the denominator of the fraction is greater than unity. Thus the second term in Eq. (3-78) is negative. The value of E_S becomes increasingly more negative with decreasing separation in a way analogous to that shown in Figs. 3-9(a) and 3-11. However, a distance is reached at which the Pauli repulsion energy begins to make itself manifested, a result of the interpenetration of the "clouds", as described in Section 3.12.1. This is similar to that indicated by E_R in Fig. 3-11 for metallic bonding. In the case of hydrogen, where the shielding of the nucleus by an electron is minimal, the nuclei, particles that obey the exclusion principle, also contribute to this repulsion. (It should be noted that, for atoms with large atomic numbers, where the nuclei are shielded much more effectively, the nuclear contribution to the repulsion may be neglected.) Thus, as the interpenetration of the "clouds" becomes appreciable, the parameter C changes its sign, becomes positive, and increases exponentially as the separation between the atoms diminishes because of its repulsion component. This causes E_S to become less negative and to pass through a minimum. Such energy minima correspond to bonding, and E_S is more negative than $2E_\infty$. Further decreases in the distance between the nuclei cause very large, positive increases in the parameter C, and E_S then becomes large and positive. The effect described here is analogous to that shown by the repulsion energy in Eq. (3-57) and in Figs. 3-9(a) and 3-11 for metals. The minimum in E_S is highly localized between the ions; it also determines the distance between the ions. The sharing of the electrons responsible for this is reflected by the high degree of directionality, as well as by the strength of these bonds, E_b.

Attention now is directed at the antisymmetric case [Eq. (3-79)]. Here, again, B and C are negative, but $|C|$ is only slightly larger than $|B|$. The denominator of the fraction is positive and is less than unity. Now, when the energy split occurs, the second term in Eq. (3-79) is positive and small. Thus, as the separation between the nuclei diminishes, E_A very slowly becomes more positive, so E_A is less negative than $2E_\infty$. Then, when the "cloud" interpenetration becomes appreciable, the repulsion component of C increases exponentially, as previously noted. The net result is that E_A then becomes very large and positive. The behavior described for the antisymmetric case has no classical analog. It has been known as the antibonding "orbital". It is analogous to that shown in Fig. 3-11 for the case of metallic bonding.

The single covalent bond between two hydrogen atoms, consisting of two electrons of opposite spin that are shared between the atoms, is considered to be representative of the single covalent bonding that occurs between atoms of larger atomic numbers. In the cases of greatest interest here, each atom is considered to form one such single bond with each of its nearest neighbors in its lattices.

3.12.3 Covalent Bonding: General

Covalent bonding (also known as homopolar or valence bonding) takes place between neutral atoms whose outer $(s + p)$ electrons total four or more. The electrons of a given atom of this kind are *shared in pairs* of opposite spin with *each* of its nearest neighbors in a crystal. Each such electron pair is like that of hydrogen. Here the crystal structure is largely determined by the $(s + p)$ sum and results from the paired sharing of electrons with its neighbors. With the exception of hydrogen, this tendency is summarized by the $(8 - N)$ rule. The number eight is the sum of the maximum allowed $s + p$ electrons. The parameter N is the actual number of $(s + p)$ electrons associated with a given atom. The difference between these factors, given by $(8 - N)$, represents the number of electrons required by that atom to acquire a filled, noble gas, p^6 electron configuration. One such atom contributes one of its $(s + p)$ electrons along with a similar electron of opposite spin from one of its nearest neighbors to form a mutual bond. The result is the formation of shared bonds consisting of paired electrons between an atom and each of its nearest neighbors. The number of such shared electron bonds is determined by $(8 - N)$. This limits the number of nearest neighbors and, thus, controls the crystal structure; it is the equivalent of saying that the coordination number, CN, or number of nearest neighbors of a given atom, and consequently the lattice type of that atom, is largely determined by the value of $(8 - N)$. As is shown next, the basis for this resides in the interactions of the ψ functions of the s and p electrons and the resulting $\psi^*\psi$ functions.

The bonding by the shared pairs of electrons usually occurs without the necessity to excite electrons to higher energy states. In general, this results from the condition that atoms engaging in this bond type have up to four unoccupied states in their outer shells (Table 3-7). This is most readily visualized by considering the covalent bonding of elements other than hydrogen. The outer levels of the atoms

TABLE 3-7   ELEMENTAL SINGLE COVALENT BOND ENERGIES[a]

Element	Outer Electron Configuration	Bond Energy (kJ/mol)	Element	Outer Electron Configuration	Bond Energy (kJ/mol)
H	$1s$	463	O	$2s^2 2p^4$	139
C	$2s^2 2p^2$	348	S	$3s^2 3p^4$	213
Si	$3s^2 3p^2$	177	Se	$4s^2 4p^4$	184
Ge	$4s^2 4p^2$	157	Te	$5s^2 5p^4$	155
Sn	$5s^2 5p^2$	143			
N	$2s^2 2p^3$	161			
P	$3s^2 3p^3$	215	F	$2s^2 2p^5$	153
As	$4s^2 4p^3$	134	Cl	$3s^2 3p^5$	243
Sb	$5s^2 5p^3$	126	Br	$4s^2 4p^5$	193
Bi	$6s^2 6p^3$	105	I	$5s^2 5p^5$	151

[a]This source also provides data for some compounds.
Based on L. Pauling, *The Chemical Bond*, p. 60, Cornell University Press, Ithaca, N.Y., 1967.

have the same quantum numbers, except for spin; their values of n and l are the same, and those for m_l are degenerate and equal to zero. Thus, if the spins of each of the two electrons that constitute a single bond are antiparallel (opposite to each other), the atoms can exchange and mutually share electrons, a process known as exchange interaction [Eq. (3-78)]. The wave functions of these electrons are symmetric [Section 2.5], and bonding occurs in a way in accordance with Pauli exclusion; little or no added energy is required as indicated by Eq. (3-78).

The creation of covalent bonds by the mutual sharing of pairs of electrons according to the $(8 - N)$ rule has the effect of completing the p^6 states of each atom. The net result is that each atom participating in this sharing of electrons has its energy lowered by acquiring a noble-gas-like electron configuration. However, unlike the electron "clouds" associated with the noble gases, the probability densities of these outer electrons are *not* spherical. These now have strong directionalities and, thus, have relatively high probability densities *between* the two bonded atoms. This results from the symmetric wave functions of the electrons forming each of the paired bonds. If the members of an electron pair had parallel spins, their wave functions would be antisymmetric and repulsion rather than bonding would occur [Eq. (3-79)].

The high values of the $\psi^*\psi$ functions between the participating atoms cause covalent bonds to be highly directional. This has strong effects on their crystal structures. Carbon provides a good example of the effects of the electron configurations, the corresponding variations in the $\psi^*\psi$ functions, and the lattice types. The ground state of carbon is $2s^2 2p^2$. Its first excited state is $2s\, 2p^3$, a state of slightly higher energy than that of the ground state. In this case the four outer electrons are unpaired. These are called *hybridized* sp^3 states. Atoms with this configuration can form four single bonds of paired electrons. This results from the interactions of the ψ functions of the single s and the three p electrons that give four equivalent $\psi^*\psi$ functions. These are oriented between the centroid and the corners of a regular tetrahedron and constitute deep wells for the electron pairs. In so doing, as is shown by Eq. (3-78) and Fig. 3-10, their energies are lowered considerably below that of the ground state. The high directionality of the bonding is shown by the resulting array of the participating atoms.

Carbon, with a valence of four, based on the $(8 - N)$ rule, requires that each atom have four nearest neighbors. As previously noted, the $\psi^*\psi$ functions cause the four nearest neighbors to be situated on the apexes of an equilateral tetrahedron with the given atom at its centroid. The sp^3 configuration also is known as tetrahedral hybridization because of this [Fig. 3-10(a)].

The resulting crystal lattice, built up from these tetrahedra, is known as the diamond cubic lattice type [Fig. 3-10(b)]. It is a very inefficient utilization of the available lattice space. The *atomic packing factor* (APF) of this type is 0.34. (The APF equals the volume of the atoms in a unit cell divided by the volume of the unit cell of the lattice.) The lattice types of normal metals, with very much lower bonding energies than that of carbon, show much more efficient use of lattice space. The APF of the body-centered cubic (BCC) is 0.68 and that of the face-centered cubic (FCC) or hexagonal close-packed (HCP) lattices is 0.74 and is equal to that of the maximum

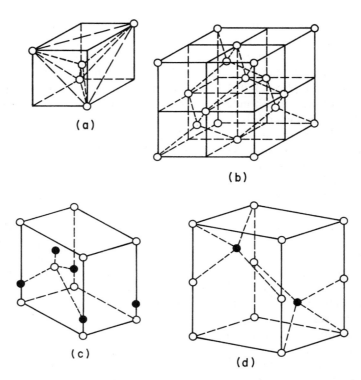

Figure 3-10 Tetrahedral bonding: (a) Tetrahedral bonding arising from the sp^3 configuration is shown by the broken lines; (b) Diamond cubic lattice; (c) Wurtzite unit cell; (d) NiAs unit cell. Note that (c) and (d) show the unit cells of hexagonal structure cells.

packing of hard spheres.   The optimum filling of space by atoms, known as the *space principle*, is considered to constitute an important condition for stable crystal lattices. The high degree of bond directionality and the very strong bonding energy associated with the hybridized sp^3 array, of about 700 kJ/mol for diamond, accounts for the tetragonal geometry of carbon and its apparent exception to the space principle. Silicon, germanium, and tin (below about 13°C) also have diamond cubic lattices for reasons similar to those given for carbon.   The respective approximate bonding energies of silicon and germanium are 400 and 350 kJ/mol.   These elements are semiconductors.

Another important hybridized configuration of carbon is that of $2s2p^22p$, where a 2s state has been raised to the 2p level.   This may be rewritten, in an oversimplified way, as $2s2p_x^22p_z$.   When an electron from a second carbon atom is brought sufficiently close to a given atom, its ψ function interferes constructively with those ψ functions of the 2s and the $2p_x^2$ electrons of the reference atom.   The result of this is three coplanar maxima for the $\psi^*\psi$ functions that form an equilateral triangle and constitute deep wells for the electron pairs. Hence the name trigonal hybridization is given to the sp^2 array. Destructive interference takes place between the $2s$ and the $2p_z$ functions so that the value of the resulting $\psi^*\psi \to 0$, and this $2p_z$ electron may be considered to be unaffected by a well; it also is known as a π electron.

The net result of this electron reaction is that the carbon atoms form very stable triangular arrays that fit together in planes that are composed of equilateral hexagons, each hexagon being composed of six of the triangles. The $\psi^*\psi \to 0$ condition for the $2p_z$ (or π) electron causes it to be excluded from the planes and only very weakly associated with the planes. This behavior has important and useful consequences. One of these is that the bonding energy between planes is very low. This enables the stable planes to glide across each other when very low shear stresses are applied; such behavior accounts for the lubrication properties of graphite. Another result is that interplanar π electrons are relatively highly mobile and can enter into such physical processes as thermal and electrical conduction. Those engaged in the hybridized bonding within the planes are strongly bound and cannot contribute to conduction. The unusual combination of structural stability and conductivity, at relatively low cost, is responsible for the wide use of graphite electrodes and their common use at temperatures up to about $3700°C$.

With the exception of the hybridized trigonal bond, the properties of most covalently bonded solids may be considered to be derived on the basis that the highly localized, paired, bonding electrons are in deep wells analogous to the one given in Fig. 3-3(b). These deep wells are formed by each of the atoms that contribute to an electron pair. The $\psi^*\psi$ functions of the electrons in such newly constituted wells are considered to have their maxima concentrated at their centers; these have little or no tails outside the wells. This can account for the high bonding energies, the high strengths, and the very low electrical and thermal conductivities and low thermal expansions expected of insulators. Those cases in which the $\psi^*\psi$ tails show small, but significant, extensions beyond the wells can account for the semiconducting properties of elements such as silicon and germanium.

It should be noted that, when different species of atoms are involved in covalent bonding, their electronegativities may influence their bonding. Electronegativity is a construct introduced by Pauling (1932) as a means of describing the predisposition of an atom to absorb electrons; its units are the square root of bond strength. Thus, an atom with a higher electronegativity will be more reactive chemically than one with a lower electronegativity. When the electronegativities of atoms engaged in covalent bonding are similar, this factor has little influence on the bonding. However, when the *difference* between the electronegativities of the two species is equal to or more than about 0.2 units, small amounts of ionic bonding may take place along with the covalent bonding. Such mixed bonding is analogous to that shown in Fig. 3-9(b). Larger electronegativity differences induce higher degrees of ionicity in the bonding. As was noted for ionic bonding in Section 3.12.1, it is probable that perfectly pure covalent bonding normally does not exist in compounds because no two atoms have identical electronegativities; small degrees of ionicity probably are present.

3.12.4 Metallic Bonding

The valence electrons of metals are weakly bound to their atoms. In their ground states, metals may be considered to consist of ion cores that are surrounded by their valence electrons. In this way the ion cores may be pictured as omnidirectional potential wells for their valence electrons in a manner analogous to that shown in Fig. 3-3(c). These wells are relatively shallow, compared to those involved in ionic or in covalent bonding, and the $\psi^*\psi$ functions are considerably larger both in their magnitudes and in their extents outside the wells.

As the two widely separated wells are brought more closely together, the behaviors of their $\psi^*\psi$ functions and their eigenvalues closely resemble those described in Sections 3.12.2 and 3.12.3. One significant difference between the atoms involved in metallic bonding and that just noted is that the number of valence electrons is equal to three or less. The paired sharing of electrons of opposite spin by nearest neighbors for the mutual completion of their outermost p levels is not possible under this condition; covalent bonding cannot take place because the number of valence electrons is too small.

The bonding of metallic atoms cannot be ionic in character because, even if some electron transfer takes place as in some alloys of transition metals, all the ion cores are positively charged. No oppositely charged ions are present to provide the strong coulombic attractions.

The positively charged metal ions in a solid are immersed in an "electron gas" of the kind proposed originally by Drude (Section 1.1) that now is treated as a degenerate gas, as described in Sections 2.7 and 3.11. If the covalent and ionic mechanisms are not involved in the bonding of metals, it appears that the valence electrons must be responsible for this. The question then arises regarding the ways in which the highly fluid "electron gas" is enabled to bind the metallic ions. The model for any interactions between the positive metallic ions must, therefore, be basically different from those previously described for the other bond types.

The bonding mechanism of metals may most readily be understood by considering the ways in which the wave functions and eigenvalues of monovalent metals are changed by the bonding. Consider two such ions of the same element. Their valence electrons are loosely bound so that they may be considered as being ions with a charge of $+1e$ with one valence electron associated with each. This reduces the problem to one similar to that of hydrogen given in Section 3.12.2.

The picture may be further simplified by considering the most common metallic lattices. A monovalent atom in the body-centered cubic lattice has eight nearest neighbors. Thus, an atom, such as sodium, may be considered as simultaneously sharing its valence electron with each of these neighbors. This may be thought of as the instantaneous sharing of $\frac{1}{8}$ of the electron from each neighbor by each of the atoms. In the case of the body-centered cubic lattice, with 12 nearest neighbors, the bonding consists of $\frac{1}{12}$ of an electron per atom from each nearest neighbor. Where the hexagonal, close-packed lattices are formed, the number of nearest neighbors also is 12, but such atoms usually have a valence of 2. Thus, the bonding consists of $\frac{2}{12}$ or $\frac{1}{6}$

of an electron per bond contributed by each atom.   The point of importance here is that the charge density of electrons around the ions in metallic lattices is extremely dilute and more uniform compared with that in covalent bonding.   Thus, a model for metallic bonding may be developed by considering the simplest case of the interactions between two ions and one electron; this gives $\frac{1}{2}$ of an electron per bond contributed by each ion.

On the basis of this configuration, let the Hamiltonian for each ion, when the electron is associated with it, be, in a way similar to that for Eq. (3-69),

$$H_{11} = H_{22} = E_1 = E_\infty \tag{3-80}$$

where E_1 is the ground state given by Eq. (3-68).   This results from the equal probability of finding the electron with each ion.   When the wave functions overlap, and electron resonance, or exchange, takes place between the ions, the interactions between them are given by

$$H_{12} = H_{21} = E_I \tag{3-81}$$

Then the total energies for the ions are given, in a way analogous to Eqs. (3-70a) and (3-70b), as

$$E = E_1 - E_I \tag{3-82}$$

Now, using the same approach as was given to obtain Eqs. (3-73a) and (3-73b) from Eqs. (3-70a) and (3-70b), it is found that

$$A_1 E = E_1 A_1 - E_I A_2 \tag{3-83a}$$

and

$$A_2 E = E_1 A_2 - E_I A_1 \tag{3-83b}$$

These equations are treated in the same way as that used to obtain Eqs. (3-78) and (3-79), and the results are given by

$$E_S = E_1 + \frac{B + C}{1 + S_{12}} \tag{3-84}$$

for the symmetric case and

$$E_A = E_1 + \frac{B - C}{1 - S_{12}} \tag{3-85}$$

for the antisymmetric case.   Here the parameters B, C, and S_{12} have essentially the same meanings as those previously given.

Where relatively large distances separate the ions, the electron has an equal probability of association with each of the ions.   And, since virtually no interactions occur between them, the fractions in Eqs. (3-84) and (3-85) may be approximated as being equal to zero; and $E_S = E_A = E_1 = E_\infty$ at relatively large separations.   This condition prevails, as the ions are brought more closely together, as long as their $\psi^*\psi$ functions do not overlap.

The splitting of the energy levels, indicated by Eqs. (3-84) and (3-85), starts when overlap begins. As previously described, the parameters B and C are negative and $|C| > |B|$, but the value of $|C|$ is not as large as that for covalent bonding. Thus, for the symmetric case, where the denominator of Eq. (3-84) is greater than unity, the fraction is increasingly more negative as the ions are brought more closely together; this results in correspondingly smaller values for E_S. This continues until the interpenetration starts to become appreciable. As previously described, C changes sign and becomes positive; it increases exponentially; this component of C is shown by E_R in Fig. 3-11. This causes E_S to go through a minimum at r_0, after which it rapidly increases. Here the minimum is not as deep as that for covalent bonding. This minimum energy, indicated by E_b, is responsible for the metallic bond and determines the equilibrium separation between the metal ions and the strength of the bond. The smaller contribution of $|C|$, described previously, however, is large enough to bond the metal ions, but it is not large enough to keep the electrons localized, as was the case in covalent bonding; the bond strength is less than that of the covalent bond. The electron, therefore, is very mobile. Thus, in a crystal lattice, the electrons are not normally closely associated with their atoms, as is the case in covalent bonding. They are, relatively, very much freer and move randomly (in the absence of external influences) throughout the lattice. This accounts for their treatment as a degenerate electron gas. In turn, it also explains their high thermal and electrical conductivities and their ductilities. The random behavior of the electrons results in their transient associations with countless ions in very rapid succession. These very rapid successive electron–ion associations make possible the ductile behaviors of metals without fracture. In contrast, the covalent bond is both highly localized and strongly directional. So, even though the bonding energy is greater than that of a metallic bond, the bonds cannot allow appreciable deformation; such bonds result in fracture without permitting deformation. Thus, many covalently bonded materials are considered to be brittle because they show only elastic behavior up to fracture.

The antisymmetric case for metals, Eq. (3-85) and Fig. 3-11, is very much like that described for covalent bonding (Section 3.12.2). The same relationship given previously for B and C holds for this case. In addition, the denominator of Eq. (3-85) is positive and less than unity. Thus, when the "clouds" begin to overlap, and the

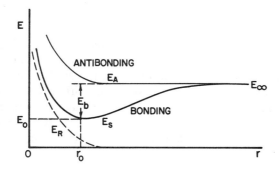

Figure 3-11 Energies acting upon two ions as a function of separation, r. $E_b = E_A - E_s - E_R$. E_R is the repulsion component of E_I, Eq. (3-81). E_b is the bonding energy.

energy splits, E_A very slowly becomes more positive and greater than $E_1 = E_\infty$ with decreasing separation. When significant overlap occurs, C increases exponentially; this component of C is indicated by E_R in Fig. 3-11. No energy minimum is present under these conditions; hence the name antibonding "orbital". As indicated for the analogous situation in covalent bonding, this case has no classical analog.

The extremely low electron density around each ion has led in the past to the consideration of the metallic bond as being an extremely dilute case of covalent bonding. Such a model leads to readily apparent incorrect descriptions of electrical and magnetic properties among others. This concept appears to have been abandoned.

The metallic bond also has been considered to be similar to the ionic bond because the bonding may be pictured to be electrostatic. This approach is incorrect in that the energies involved are very much smaller and the electrons are in constant motion. So, even though the randomly moving electron gas of negatively charged particles may be considered to cement the ions in a lattice, it leads to inaccurate results for metals.

Thus, the bonding and the resultant properties of metals are derived from the degenerate gas of nearly free valence electrons. The metallic ions may be pictured as constituting relatively shallow potential wells for the valence electrons. These have $\psi^*\psi$ functions with appreciable magnitudes and extents outside the wells. Their relatively high probabilities outside the wells or away from the ion cores is responsible for their gaslike behavior. This behavior readily explains the high thermal and electrical conductivities of metals and their ductility. It also explains the relatively low bonding energies of metals of about $\frac{1}{4}$ to 1 eV.

3.12.5 Molecular Bonding

The ideal gas law, $pV = RT$, in which p is the pressure, V the volume, R the gas constant, and T the temperature, is valid only for noninteracting gas particles. Corrections must be made to the ideal gas law when interactions take place between the particles in real gases. These are taken into account in the van der Waals equation:

$$\left(p + \frac{a}{V^2}\right)(V - b) = RT \tag{3-86}$$

The terms a/V^2 and b, respectively, include attraction and repulsion interactions between the particles. These purely electrostatic forces account for the changes in state of the noble gases, for covalently bonded gases of elements, for gaseous compounds, and for solids of covalently bonded molecules.

Covalently bonded compounds are characterized by atoms with completed outer levels. These have no electrons available to engage in any transfer or sharing processes involved in any further bonding.

The electrons associated with any atom or any molecule are in constant motion around their nuclei. Their most probable locations and properties have been described in Section 2.6 and earlier in this chapter. Their charge densities are nonuniform at virtually any given instant. The methods cited here simplify this situation by considering the average charge densities of the nuclei and of the electrons to be constant. In actuality, these charge densities are in a constant state of fluctuation. Such fluctuations give rise to the molecular or van der Waals binding forces.

Molecular solids and liquids are bonded by such forces. These result from the constantly varying charge distributions around the atoms and molecules. The asymmetries of the charges cause these particles to act as constantly changing electric dipoles. The formation of a dipole on one particle induces an electric field in adjacent particles. These respond by forming a group of oppositely oriented dipoles that are electrostatically attracted to the first dipole. The second group of dipoles simultaneously induces an electric field in the first particle that can improve the dipolar alignment. The net result of these electrostatic interactions may be expressed by means of an equation analogous to Eq. (3-62). In the present situation, the attraction and repulsion energies vary as r^{-6} and r^{-12}, respectively. The high exponential variation for the attraction energy is in agreement with the anticipated low bonding strengths afforded by this mechanism. The observed low melting and boiling points of organic compounds confirm this. Typical bond strengths of these compounds are approximated to lie in the range of about -10^{-14} to -10^{-12} erg.

The energies involved in molecular bonding may be approximated in a general way by

$$U(r) \cong -\frac{C_m}{r^6} + \frac{C_R}{r^n} \tag{3-87}$$

where C_m and C_R are parameters that depend on the molecular species involved in the bonding. They are analogous to the attraction and repulsion terms in Eq. (3-86) and may be determined by deviations from ideal gas behavior. The case described for $n = 12$ is known as the Lennard–Jones potential. This case has been expressed as

$$U(r) \cong 4E_b \left[\left(\frac{\sigma}{r} \right)^{1/2} - \left(\frac{\sigma}{r} \right)^6 \right] \tag{3-88}$$

where E_b is the energy at r_0 (the separation between the molecules), and as such is the bonding energy; the parameter $\sigma = r_0/2^{1/6}$.

3.13 BIBLIOGRAPHY

ATKINS, P. W., *Physical Chemistry*, W. H. Freeman, San Francisco, 1975.

DAY, M. C., JR., and SELBIN, J., *Theoretical Inorganic Chemistry*, 2nd ed., Van Nostrand Reinhold, New York, 1969.

DEKKER, A. J., *Solid State Physics*, Prentice-Hall, Englewood Cliffs, N.J., 1959.

HURD, C. M., *Electrons in Metals*, Wiley-Interscience, New York, 1975.

POLLOCK, D. D., *Physical Properties of Materials for Engineers*, Vols. I and III, CRC Press, Boca Raton, Fla., 1982.

POLLOCK, D. D., *Thermoelectricity: Theory, Thermometry, Tool*, ASTM STP 852, American Society for Testing and Materials, Philadelphia, 1985.

RICHTMYER, R. K., KENNARD, E. H., and LAURENTSON, T., *Introduction to Modern Physics*, 5th ed., McGraw-Hill, New York, 1955.

SOLYMAR, L., and WALSH, D., *Lectures on the Electrical Properties of Materials*, 2nd ed., Oxford University Press, New York, 1975.

SPROULL, R. L., *Modern Physics*, Wiley, New York, 1956.

STRINGER, J., *An Introduction to the Electron Theory of Solids*, Pergamon Press, New York, 1967.

3.14 PROBLEMS

3.1. Make plots of Ψ_I and Ψ_{III} for the condition $0 \ll V \ll V_0 = \infty$, where $x \to 0$ and $+x \to L$. How do these results conform to the behaviors noted for these functions in the solution of Ψ_{II} for a one-dimensional well in Section 3.5?

3.2. Evaluate the coefficient C of Ψ_{II} at $x = L$ for a one-dimensional well.

3.3. Explain the conditions, if any, under which Eq. (3-49) may be used to describe the energy of a particle with four quantum numbers. Does Problem 2.4 shed any light on this problem?

3.4. Explain the meaning of the Pauli exclusion principle in terms of the occupancy of cells in four-dimensional state space.

3.5. Use the rules for the allowed combinations of quantum number to identify all the electrons in isolated sodium and chlorine atoms.

3.6. Calculate the bonding energy of an atom of rock salt. Compare the result with the data given in Table 3-6 and explain any differences. Discuss the reason for the high bonding energies of these materials.

3.7. Calculate the approximate value of Young's modulus for NaCl.

3.8. Graph Eqs. (3-76) and (3-77) using constant, arbitrary values for E_∞ and for $S_{12} < 1$, while separately varying the parameters B and C. (Recall that $C \gg B$.) Include the ranges in which C changes its sign. Use the resulting four graphs to explain the influences of symmetric and antisymmetric behaviors in covalent bonding.

3.9. Why may ionized monovalent metallic ions in a lattice be treated in a way similar to hydrogen atoms? What is the major difference? Can this provide insight into the treatment of metallic valence electrons as an electron gas?

3.10. Explain the reasons for the relatively weak molecular bonding. How does this affect such properties as melting point, boiling point, and coefficient of thermal expansion?

4 / THEORIES OF SOLIDS

The Drude–Lorentz theory of metals is discussed in Section 1.1. Its findings are based on the concept that *all* the valence electrons contribute to the properties of metals and that they may be treated in the same way as molecules of an ideal gas. Its results may be summarized as follows. It provides an incorrect temperature dependence of the electrical resistivity and gives a reasonable approximation of the Weidemann–Franz ratio and an excessively large value for the heat capacity. The disparity between the heat capacity and the electrical resistivity reveals an inherent inconsistency in this theory.

These difficulties arise primarily from the basic assumptions that all the valence electrons are involved and that they obey classical statistics (Sections 2.1, 2.2, and 2.3.1). As indicated in Section 2.3.2, the conditions for the treatment of fermions impose considerably greater constraints on electrons than are placed on classical particles. When these limitations are taken into account, as described in Sections 2.5 and 2.6, the problems of the Drude–Lorentz theory are removed. This results from the treatment of the valence electrons as constituting a degenerate gas in a quasicontinuum of energy. In addition, it was shown in Section 2.7 that only a very small fraction of these electrons are able to participate in most physical processes under normal conditions [Eq. (2-87)]. Just those electrons within an energy range of about $E_F \pm 2k_B T$ are available; these constitute 1%, or less, of the total number of available electrons. These significant differences eliminate the difficulties of the earlier theory and provide the bases for modern theories of solids.

112

4.1 THE SOMMERFELD EQUATION FOR THE DENSITY OF ELECTRON STATES

The Sommerfeld theory marks the real beginning of the application of quantum mechanics as a means for understanding the physical properties of solids. Metals are treated as homogeneous isotropic solids. The electron representation is similar to that employed by the Drude–Lorentz model. The free electrons, treated as noninteracting particles, are considered to be in a well of constant internal potential (Fig. 1-2). Electron–electron and electron–ion interactions are not taken into account. The significant difference here is that the electrons retained in the well obey quantum mechanics instead of classical mechanics. This gives further insight into the reasons for the boundary conditions imposed on the solutions to Schrödinger's equation (Chapter 3).

It will be recalled that the presence of a large number of electrons in a potential well whose dimensions are of crystal sizes would result in a quasicontinuum of energy levels (Fig. 3-4). Here the separation between electron states may be illustrated by noting that the coefficient of Eq. (3-49) is $\sim 3.7 \times 10^{-15}\,\mathrm{eV}$ $(0.6 \times 10^{-33}\,\mathrm{J})$ for $L = 10^{-2}\,\mathrm{m}$. The Pauli exclusion principle dictates the way in which these energy states are filled. Neglecting spin, only two electrons can occupy each state. If a crystal of a monovalent element of the order of one molecular weight is being considered, about 6×10^{23} electrons must be taken into account. Each state described by Eq. (3-49) can accept two electrons of opposite spin. Thus, starting with the lowest energy level, about 3×10^{23} states must be filled consecutively until all electrons occupy states. This means that only a small fraction of the total number of the valence electrons can occupy energy states in the neighborhood of those with potential energies close to W in Fig. 1-2.

The quantum mechanic rules that state that electrons can change their energy only by discrete amounts hold here. Thus, those valence electrons with potential energies considerably greater than W (those occupying states well below W) cannot enter into a physical process unless abnormally large amounts of energy are supplied. Vacant states are not available for occupation by such low-lying electrons when normal amounts of energy are applied; all these states are already occupied. Only those valence electrons occupying states with energies close to E_F are available for participation in physical processes (Section 2.7) when energies of the usually encountered magnitudes are applied, because vacant states just above these are available for them to occupy; this is the classical equivalent of those electrons in the range $E_F \pm 2k_B T$. Recall that the Drude–Lorentz theory included all the valence electrons as contributing to physical processes. This difference between the classical and the modern theories is very important.

The question then becomes one of the determination of the ways in which the electrons enter into physical processes such as thermal and electrical conductivity. This may be resolved into two other questions:

1. How many energy states are available for electron occupation? Or, in other words, what is the density of states?

2. How are the electrons distributed in these states and how do they enter into physical processes?

Attention will now be directed to the first of these questions, the density of electron states, by beginning with Eq. (3-49) for a three-dimensional well. This equation can be reexpressed as

$$E = \frac{h^2}{8m}\left[\frac{n_x^2}{L_x^2} + \frac{n_y^2}{L_y^2} + \frac{n_z^2}{L_z^2}\right] = \frac{1}{2m}\left[\left(\frac{n_x h}{2L_x}\right)^2 + \left(\frac{n_y h}{2L_y}\right)^2 + \left(\frac{n_z h}{2L_z}\right)^2\right] \quad (3\text{-}49)$$

where each state implicitly includes two spins (Sections 3.8 and 3.10). The energy also may be expressed in terms of momentum as

$$E = \frac{\mathbf{p}^2}{2m} = \frac{\mathbf{p}_x^2 + \mathbf{p}_y^2 + \mathbf{p}_z^2}{2m} \quad (4\text{-}1)$$

Equations (3-49) and (4-1) are equated to give

$$\mathbf{p}_x^2 + \mathbf{p}_y^2 + \mathbf{p}_z^2 = \left(\frac{n_x h}{2L_x}\right)^2 + \left(\frac{n_y h}{2L_y}\right)^2 + \left(\frac{n_z h}{2L_z}\right)^2$$

so that

$$\mathbf{p}_i = \pm \frac{n_i h}{2L_i} \quad (4\text{-}2)$$

The $\pm$ sign refers to the fact that the electron can be moving in either the positive or negative directions. Equation (4-2) quantizes momentum space. Now, if the well is considered to be a cube of atomic dimensions, and using the de Broglie equation where $\lambda = 2L$, the unit of momentum space is $h/2L$. This permits two electrons of opposite spin to occupy a given state (Section 2.5). Consider the first octant of momentum space (shown in two dimensions) in Fig. 4-1(a). Here each point represents an allowed momentum state. Each momentum state occupies an area of

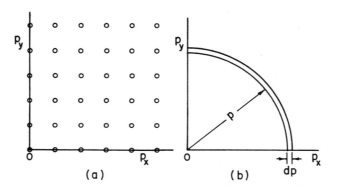

Figure 4-1 (a) Momentum space showing quantized states in two dimensions, (b) Diagram for the calculation of the density momentum states.

$h^2/4L^2$ in the figure.   Three-dimensional momentum space must be employed in order to satisfy Eq. (4-1).   Here the volume occupied by each state is $h^3/8L^3$.   Only the first octant is considered because the other octants give redundant values of momentum. This results from the character of Eq. (4-1) in which all of the momentum coordinates are squared.

The function for the *density of momentum states* is determined from the volume between $\mathbf{p}$ and $\mathbf{p} + d\mathbf{p}$ and the volume of each state as expressed by

$$N(\mathbf{p})\,d\mathbf{p} = \frac{\text{volume between } \mathbf{p} \text{ and } \mathbf{p} + d\mathbf{p}}{\text{volume of one momentum state}} = \frac{(4\pi\mathbf{p}^2\,d\mathbf{p})/8}{h^3/8L^3}$$

This reduces to the number of states between $\mathbf{p}$ and $\mathbf{p} + d\mathbf{p}$, or the density of momentum states, as given by

$$N(\mathbf{p})d\mathbf{p} = \frac{4\pi V \cdot \mathbf{p}^2\,d\mathbf{p}}{h^3} \tag{4-3}$$

since $L^3 = V$, the volume of one momentum state.   It is more convenient to describe the density of states in terms of energy, E, rather than in terms of momentum.   This conversion may be accomplished by means of the classical relationships:

$$\mathbf{p} = [2mE]^{1/2} \quad \text{and} \quad d\mathbf{p} = m[2mE]^{-1/2}\,dE \tag{4-4}$$

in which m is the electron mass.   These are substituted into Eq. (4-3) to give

$$N(E)\,dE = \frac{4\pi V}{h^3} \cdot 2mE \cdot m[2mE]^{-1/2}\,dE$$

Upon rearranging and combining the energy terms, this becomes

$$N(E)\,dE = \frac{2\pi V}{h^3}\,(2m)^2(2m)^{-1/2}E^{1/2}\,dE$$

or, when the terms containing the electron mass are combined, the *density of electron states*, or the number of states available between E and $E + dE$, is obtained as

$$N(E)\,dE = \frac{2\pi V}{h^3}\,(2m)^{3/2}E^{1/2}\,dE = dN \tag{4-5}$$

Here m actually is the *effective mass* of the electron; it can be significantly different from that of a free electron.   This property, usually designated by m^*, is discussed more fully later in this chapter and in Chapter 6.

Basically, Eq. (4-5) is the answer to the question regarding the number of states available for occupation by electrons; this equation is the Sommerfeld function for the density of electron states, where each state implicitly includes two spins.   This is so because Eq. (4-5) was derived from Eq. (3-49).

The second question posed earlier regarding the distribution of the electrons and how they enter into physical processes may now be considered.   The probability of occupancy of an electron in an available state given by Eq. (4-5) is determined by Eq.

(2-87). This can be expressed as

$$N(E)_T = f \cdot N(E, 0) \tag{4-6}$$

where $N(E)_T$ is the density of states for a $T > 0$ K, f is the Fermi–Dirac distribution function [Eq. (2-87)], and $N(E, 0)$ is the density of states for $T = 0$ K [Eq. (4-5)].

The Sommerfeld equation, by itself, constitutes only an indefinite description of available states. At most, it can provide the way in which states are filled at 0 K. The application of Eq. (2-87) to Eq. (4-5) removes this inadequacy; their combination gives the probability of occupancy of an available state at any $T \geqslant 0$ K (Fig. 2-3). In so doing, its greatest utility resides in the prediction of the probable number of electrons, described by the density-of-states curve, that are capable of excitation and, therefore, of involvement in physical processes. Approximations given in Section 2.7 show that only those electrons in the energy range of about $E_F \pm 2k_B T$ are available for such purposes. Thus, the Fermi energy may be regarded as a reference energy, or as a bench mark, in dealing with any available valence electrons.

4.2 THE FERMI LEVEL

It will be recalled (Fig. 3-4) that when the crystal size becomes large the energy levels form a quasicontinuum. In addition, according to the Pauli exclusion principle, only two electrons can occupy each level if spin is implicitly included. Under these conditions the right side of Eq. (4-5) need not be multiplied by a factor of 2, since two spins were originally included in this application of Eq. (3-49). The curve for the density of states given by Eq. (4-5) is shown schematically in Fig. 4-2 at 0 K.

As previously noted, the situation shown in Fig. 4-2 is obtained when the energy states are filled successively with valence electrons, starting with the lowest state, until all the electrons are accommodated. The electron in the highest energy state to be filled at 0 K is analogous to the electron at the highest energy level in the well shown in Fig. 1-2. Such an electron would require the energy increment equal to or greater than W in order to leave the well. This energy state, that is, the highest level to be occupied by an electron at 0 K, was tentatively defined as the Fermi level, E_F. As has been shown in Section 2.7, this "definition" is incomplete. It is, nevertheless, sufficient for the present purposes. See Section 2.7 and Eq. (2-87).

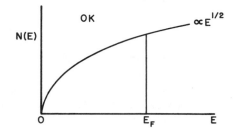

Figure 4-2 Sketch of the Sommerfeld theory for the density of electron states at 0 K.

The spherical model, illustrated in two dimensions in Fig. 4-1(b), can be used to determine the number of occupied energy states in a somewhat different way. Here, the radius vector, **p**, is the momentum vector from the origin to the outermost, filled, momentum state in three dimensions. The unit of momentum space is obtained from the de Broglie relationship, $\mathbf{p} = h/\lambda$, now as given by $\mathbf{p} = h/L$. This explicitly permits only one electron to occupy a given state. Using the spherical model, the number of occupied states is obtained by calculating the number of momentum states in the spherical volume occupied in momentum space:

$$N = 2 \cdot \frac{4}{3} \pi \mathbf{p}^3 \div \frac{h^3}{L^3} = \frac{8\pi \mathbf{p}^3 V}{3h^3} \tag{4-7}$$

Here the factor 2 is used explicitly to account for electrons of both spins. The relationships given by Eq. (4-4) are used again in Eq. (4-7) to obtain

$$N = \frac{8\pi(2m)^{3/2} E^{3/2} V}{3h^3} \tag{4-8}$$

The differentiation of Eq. (4-8) with respect to E gives Eq. (4-5) when spin is taken into account. Conversely, Eq. (4-5) may be integrated to give Eq. (4-8) for the conditions noted.

Either of the preceding equations for N can be solved to obtain an expression for E_F. The rearrangement of Eq. (4-8) gives

$$E_F^{3/2} = \frac{1}{8} \cdot \frac{3N}{\pi V} \left(\frac{1}{2m}\right)^{3/2} h^3$$

Then, the Fermi energy is found to be

$$E_F = \frac{1}{4} \left(\frac{3N}{\pi V}\right)^{2/3} \left(\frac{1}{2m}\right) h^2 = \frac{h^2}{8m} \left(\frac{3N}{\pi V}\right)^{2/3} \tag{4-9}$$

If the factor V is taken as the volume of an ion in the crystal lattice, then N/V is the electron:ion ratio. Thus, E_F may be considered to be a function of the electron:ion ratio. This concept is useful in the discussion of the role of the Fermi level on the physical properties of alloys and on the formation of alloy phases. See Section 4.7.

It is interesting and instructive to obtain an approximation of the average energy of an electron. This can be done by means of the method given in Section 2.1, where the average energy is given by using Eq. (4-5) in the way indicated by Eq. (2-22). This results in

$$\langle E \rangle = \frac{\displaystyle\int_0^{E_F} EN(E)\,dE}{\displaystyle\int_0^{E_F} N(E)\,dE} = \frac{\displaystyle\int_0^{E_F} E^{3/2}\,dE}{\displaystyle\int_0^{E_F} E^{1/2}\,dE}$$

since the constant factors in Eq. (4-5) vanish. Thus, the average energy of all of the valence electrons in the distribution is found, after integration, to be

$$\langle E \rangle = \frac{\frac{2}{5}E_F^{5/2}}{\frac{2}{3}E_F^{3/2}} = \frac{3}{5}E_F \tag{4-10}$$

A valence electron with $E = \frac{3}{5}E_F$ cannot enter into a physical process when normal energies are applied. It is too far below E_F. Such an electron can respond only to discrete energies. Those states that it would be required to occupy are already filled; it has no place to go unless abnormal energies are applied.

Since only those electrons with $E \cong E_F$ are of interest here, since they are the only ones that can enter into physical processes, it is helpful to approximate the value of E_F to obtain an idea of their energies. Such an approximation can be obtained readily from Eq. (4-8) for a normal, monovalent metal with a typical atomic "diameter" of about 3 Å. Using $h \cong 6.6 \times 10^{-27}$ erg s, $m \cong 9.1 \times 10^{-28}$ g, and $V \cong (3 \times 10^{-8})^3$ cm^3 results in $E_F \cong 6.5 \times 10^{-12}$ erg or, using 1 eV $\cong 1.6 \times 10^{-12}$ erg, $E_F \cong 4$ eV.

The magnitude of E_F may be more readily appreciated by considering the properties of an electron with an energy equal to E_F in terms of those of a classical electron. This is accomplished by means of the equation for the energy of a classical electron in the form $E_F = k_B T_F$, where T_F is the *Fermi temperature*. Using $E_F \cong 4$ eV and $k_B \cong 8.6 \times 10^{-5}$ eV/K gives $T_F \cong 4.7 \times 10^4$ K. In essence, this means that a classical electron would have to be at about 50,000 K to possess the energy equivalent of an electron at E_F. The rough approximations of E_F and T_F given here lie close to the averages of the more accurately calculated values given in Table 4-1; another unit for expressing E_F also is given.

The addition of normal energies, of the order of $k_B T$, to electrons in the energy ranges close to those shown for E_F thus represents very small increments when compared to E_F. This confirms the behavior described earlier by Eq. (2-85), so E_F may

TABLE 4-1 CALCULATED FERMI ENERGIES AND TEMPERATURES FOR FREE ELECTRONS

Element	E_F (eV)	T_F (K $\times 10^{-4}$)
Li	4.7	5.5
Na	3.1	3.7
K	2.1	2.4
Rb	1.8	2.1
Cs	1.5	1.8
Cu	7.0	8.2
Ag	5.5	6.4
Au	5.5	6.4

1 eV = 23,050 cal/mol.
After C. Kittel, *Introduction to Solid State Physics*, p. 208, John Wiley, & Sons, Inc., New York, 1966.

be considered to be invariant with temperature for most cases. One important exception to this occurs when alloying additions to a transition element cause its *d* band to approach maximum filling.

Additional important relationships may be obtained from the data given here. Using $E_F \cong 6.5 \times 10^{-12}$ erg and the electron rest mass $\cong 9.1 \times 10^{-28}$ g, the velocity of an electron at or near E_F is $\cong 1.2 \times 10^8$ cm/s. This is approximately 0.4% of the velocity of light. This leads to another important conclusion that is obtained from the relativistic equation for mass: $m = m_0/[1 - (v/c)^2]^{1/2}$. The ratio of electron velocity to that of light gives $(v/c)^2 \cong 2 \times 10^{-5}$, so the denominator of the relativistic equation is virtually equal to unity. Thus, in agreement with the fine-structure constant (Section 1.5), relativistic effects need not be considered and $m \cong m_0$. The electron rest mass may be used in all relationships except those in which the effective mass, m^*, must be used.

4.3 QUASIELECTRONS

The concept of quasielectrons may be most readily understood by starting with all electron states filled up to E_F at 0 K, as described in Section 4.1. Now, if an external energy $\Delta E = k_B T$ is added to the system, consider one of the electrons that has been promoted to a state at or near $E_F + \Delta E$. The energy state now occupied by the electron is only very slightly above E_F because, as shown in Section 2.7, $\Delta E/E_F \cong 1\%$. This excitation creates a hole (an absence of an electron) in the state that it originally occupied. If it is assumed that all the excited electrons behave like ideal gas particles and do not interact, the result of the excitations is electron–hole pairs, each component of each pair occupying its assigned state relative to E_F; the electrons are in appropriate states above E_F and the holes are in the formerly occupied states below E_F. And, since no interactions occur, the sum of these excitations is just equal to ΔE. These pairs, or quasiparticles, exist for the length of time (lifetime) that the external energy is present. Quasiparticles of this kind are considered to be zero-order excitations because of the absence of interactions between them. As such, they effectively are excited electrons with properties essentially the same as those of nearly free electrons.

If interactions are considered to take place between the many quasiparticles at different states within the energy range of $E_F + \Delta E$, their lifetimes would be expected to decay to zero in shorter times than that noted for noninteracting particles, depending on the interaction. The interactions must be such that the decay of an excited state takes place so that the electron resides in successively lower levels of excitation before returning to a level at or near E_F. The lifetimes of the particle while in each of the intermediate states in which it resides during decay must be sufficiently long so that each of these may be treated as being a noninteracting excitation. This permits the decay to be treated as a succession of events. The minimum lifetime for a transition between two such intermediate states is given by $h/\Delta E_{max}$ [Eq. (1-24)]; this is too short a time for this process to be observed. Therefore, the energy increments involved in these excitations must be much less than the ΔE_{max} for this mechanism, but still must be large enough so that the succession of decreasingly excited states may be clearly defined and have lifetimes greater than the minimum. The lifetime may be approximated as diminishing approximately as $(\Delta E)^2$.

The changes in energies during the excitation and decay of the quasiparticles have corresponding effects on their velocities and their effective masses. However, since $\Delta E \ll E_F$, these changes as considered here are very small. Thus, except for the relatively slow decay, systems of interacting and noninteracting quasiparticles may be considered as being virtually the same.

Each excited electron component of a quasiparticle, or quasielectron, is surrounded by a volume of approximately 1 Å in diameter that is lower in electron density and thus has an effective charge equal to and opposite to that of the electron; this is called a *correlation hole*. This, in turn, is encompassed by a volume of higher electron density; it consists of unexcited electrons that are repelled from the central quasielectron and move around it in a way similar to that of a liquid flowing around a moving object. These quasiparticles, therefore, mutually repel one another, and they behave as though they were electrically neutral unless their separations are less than about 1 Å. This strong tendency of avoidance, termed *correlation*, between quasiparticles also exists between quasiparticles and electrons where it is manifested as the previously described correlation holes. The correlation holes are maintained and are maximized by continual, reversible, mutual interactions between excited and unexcited electrons. This may be pictured as the mutual, reversible, reciprocal sharing of the excitation energy of one quasiparticle with many surrounding electrons during which the energy is conserved.

In reality, quasiparticles are constructs that are known as *virtual particles*. Their properties and behaviors are based on the Heisenberg uncertainty discussed previously. Deviations from energy conservation may take place if they occur for times shorter than the minimum time required by Eq. (1-24); this is allowed simply because such deviations cannot be observed. The changes in mass and velocity, noted previously, result from the association of the properties of the virtual particle with those of the original particle. In other words, the original, or precursor, particle carries the surrounding virtual particles that are created by the excitation as that much "excess baggage." In the case of major interest here, the properties of the quasielectron are affected by those of the hole.

The fundamental object behind the use of the construct of quasielectrons, introduced by Landau, is to permit the application of just one individual excitation resulting in only one excited particle as being a representative, elementary excitation in those cases in which the role of one particle would be completely submerged and "smeared out" among large numbers of reciprocally interacting particles. However, the small energies, $\Delta E \cong k_B T$, that are of major interest in this book are comparatively weak. Any resultant electron excitations, then, are relatively very small. The net result is that any corresponding quasielectron interactions are very weak and their lifetimes are relatively long. These circumstances simplify most situations that will be considered in subsequent treatments of excited electrons because they make possible their treatment as individual particles with little loss of accuracy.

4.4 THE FERMI–SOMMERFELD THEORY OF METALS

The Sommerfeld theory provides a means for computing the density of states [Eq. (4-5)]. The Fermi–Dirac function [Eq. (2-87)] shows how the electrons are distributed in those states, as well as which of them can enter into physical processes. These equations along with Eq. (4-6) can now be used to show their combined effects graphically, as in Fig. 4-3.

It is apparent from the discussion in Sections 2.7 and 4.1 and in Fig. 4-3(c) that only a relatively few electrons in states near E_F are excited, or promoted, by the thermal energy to states at or above E_F. These electrons are indicated by the highly exaggerated area of the curve of Fig. 4-3(c) for energies at and above E_F.

Other ways of showing this are given in Fig. 4-4(a) in terms of momentum and in Fig. 4-4(b) in terms of a well. In Fig. 4-4(a), given for momentum space, the ripples are the momenta equivalents of less than $\pm 2k_B T$, or $\pm (2mk_B T)^{1/2}$. The well picture [Fig. 4-4(b)] shows little "waves" on the Fermi "sea." In both cases the variations are within the order of $\pm 2k_B T$. The ratios of the wavy areas to the total areas are intended to give a rough visual approximation of the small numbers of electrons that are made available by thermal activation.

Another useful way of representing electron states is shown in Fig. 4-4(c) for nearly free electrons. Here the states occupied up to the Fermi level are shown as a spherical surface in wave-vector space; this is known as the Fermi surface. This is a three-dimensional plot of Eq. (3-26). Since energy and momentum are related, $E = \mathbf{p}^2/2m$, Figs. 4-4(a) and (c) are different ways of expressing the same behavior. Use is made of this correspondence later in the calculation of the electrical conductivity of metals. The radius vector $\mathbf{p}$ shown in this figure is the momentum of all electrons occupying states at the Fermi surface.

The relative simplicity of the Fermi–Sommerfeld theory limits its application primarily to isotropic metals and alloys. More complex representations, based on

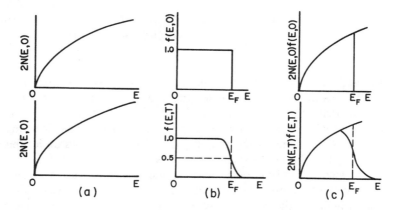

Figure 4-3   (a) Sommerfeld distribution of electron states, (b) Fermi-Dirac function at 0 and TK, (c) products of (a) and (b) at 0 and TK.

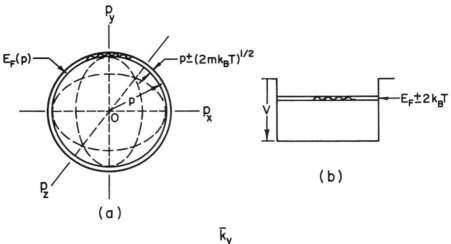

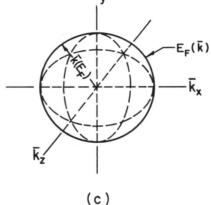

Figure 4-4   The effect of temperature on the Fermi surface, (a) in terms of electron momentum, (b) in terms of the energy of an electron in a potential well, (c) the Fermi surface in wave-vector space.

this approach, can be used to illustrate the basis for the anisotropic properties of other crystals, as well as those of metals and alloys. The influences of the bounding surfaces of the Brillouin zones (Section 4.6) have important effects on the properties of electrons when the Fermi surface approaches them. The electrons may no longer be considered as being nearly free; Eq. (3-26) does not hold under these conditions, and the shape of the Fermi surface becomes more complicated.

Despite these complications, a model that makes use of a spherical Fermi surface can be used to provide uncomplicated and accurate descriptions of the electron component of the heat capacity of metals and of their electrical and thermal conductivities.

4.5 BAND THEORY OF SOLIDS

The foregoing theories have considered the electrons within the metal to be in a constant internal potential as shown in Figs. 1-2 and 3-4. However, the potential within a metallic lattice is not constant. Each ion within the lattice constitutes a potential well (Section 3.12.4). This is shown schematically in Fig. 4-5(b). This periodicity of the potentials gives rise to the band model of solids. It should be noted that this model considers the electrons as being noninteracting and neglects electron–electron interactions. This problem has been eliminated by the concept that quasiparticles are involved rather than electrons. Electrons are considered here for the reasons discussed in Section 4.3.

This theory is developed by considering the results of bringing two widely separated wells, of the type shown in Fig. 3-1, increasingly closer to each other. This is the same as bringing together two widely separated ions. The initial situation can be approximated by a simple model consisting of two widely separated hydrogen ions, as in Fig. 4-6.

The analysis given here closely parallels those presented for the derivations of Eqs. (3-75), (3-84), and (3-85) in Sections 3.12.2 and 3.12.4. The results obtained from the band theory are more general and include covalent and metallic bonding as special cases. This generality provides a basis for the classification of solids and for the prediction of their inherent properties.

Here the ions are considered to be motionless, since, by comparison, their thermal oscillations are small compared to the motions of the electrons. If an electron is introduced into this two-well system, the resulting quantum mechanic effects are as indicated in Fig. 4-7(a). The energy in each well is given by Eq. (3-40) for $n = 1$, or $E_1 = h^2/8mL^2$. The wave functions are designated by two subscripts; the

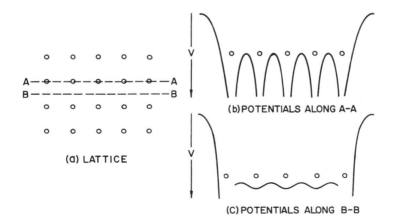

(a) LATTICE

(b) POTENTIALS ALONG A–A

(c) POTENTIALS ALONG B–B

Figure 4-5   Schematic diagrams of the effect of interplanar position upon the periodic potentials in a lattice.

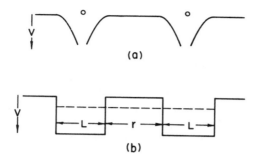

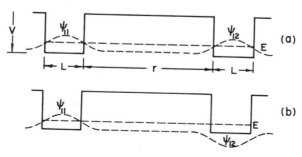

(a)

(b)

Figure 4-6 Basis for the use of the square-well approximation of the periodic potential in a lattice.

first describes the state and the second, the well. Here ψ_{11} and ψ_{12} are identical within the wells. They quickly drop off and become very small between the wells. The wave function for the two-well system is

$$\psi = \psi_{11} + \psi_{12} \tag{4-11}$$

And, since ψ_{11} and ψ_{12} are identical,

$$\psi_{11}^{*}\psi_{11} \equiv \psi_{12}^{*}\psi_{12} \tag{4-12}$$

It is, therefore, equally likely that the electron will be found in either well. The probability densities become vanishingly small between the wells. As a result of these conditions, there is a negligible probability of finding the electron between the wells and an equal probability of finding it in either well, Fig. 4-7(a).

As shown in Fig. 4-7(b), another solution to Schrödinger's time-independent equation also exists. ψ_{12} reverses its sign when the electron changes its spatial coordinates in passing from one well to the other, and it is said to be antisymmetric. See Sections 2.5, 3.12.2, and 3.12.4. However, the shapes of ψ_{11} and ψ_{12} are identical, and Eqs. (4-11) and (4-12) are valid. The probability density of finding the electron remains the same as in the symmetric case because the product of $\psi_{12}^{*}\psi_{12}$ is positive. In addition, the energies remain unchanged. So, for the case in which the ions are very far from each other, Eq. (3-40) may be used:

$$E_{\infty} = E_1 = \frac{h^2}{8mL^2} = E_S = E_A, \qquad n = 1 \tag{4-13}$$

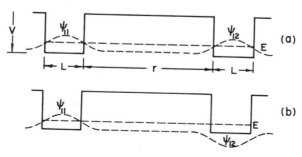

Figure 4-7 Schematic diagrams of (a) the symmetric and (b) the antisymmetric wave functions in two potential wells. [Modified from Sproull, R. L., *Modern Physics*, p. 205, John Wiley and Sons, Inc., NY, 1956.]

where E_S is the energy in the symmetric case and E_A that of the antisymmetric case.

Interesting and significant changes take place in the wave functions as the wells are brought closer together until they finally abut. An attempt to describe these changes with decreasing separation, r, of the well is shown in Fig. 4-8. As the wells are brought increasingly closer together, the probability of finding the electron between the wells, in the symmetric case, becomes increasingly greater. At the position where the wells just touch each other, the greatest probability of finding the electron is at the common wall. In the antisymmetric case, the probability of finding the electron is zero at the common wall when the wells just come into contact with each other.

The symmetric case, thus, results in a wave function similar to that in Fig. 3-2(a), but in which the well length now is $2L$. When Eq. (3-40) is used to calculate the energy of the electron for this case and $n = 1$,

$$E_S = \frac{n^2 h^2}{8ML^2} \rightarrow \frac{1 h^2}{8m} \cdot \frac{1}{4L^2} \tag{4-14}$$

Thus, when the wells abut, the symmetric case represents an energy that is one-quarter that of the widely separated wells, as is given by Eq. (4-13).

In the antisymmetric case, the resulting wave function is similar to that shown in Fig. 3-2(b). Here the well length also is $2L$, but now n must equal 2 to agree with ψ_2 as shown in this figure. The factors characteristic of this case are substituted into Eq. (3-40) to obtain

$$E_A = \frac{n^2 h^2}{8mL^2} \rightarrow \frac{4 h^2}{8m} \cdot \frac{1}{4L^2} = \frac{h^2}{8mL^2} \tag{4-15}$$

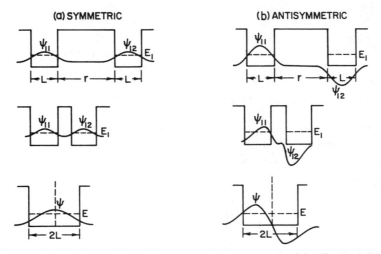

Figure 4-8   The effects of diminishing separation between potential wells upon wave functions. [Based on Sproull, R. L., *Modern Physics*, p. 206, John Wiley and Sons, Inc., NY, 1956.]

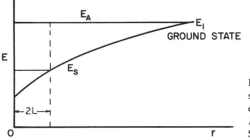

Figure 4-9   Effects of potential well
separation on the energy levels of an
electron. (Modified from Sproull, R. L.,
Modern Physics, p. 207, John Wiley &
Sons, Inc., New York, 1956.)

From this it is seen that the energy of the electron in the antisymmetric case remains unchanged (to a good first approximation) as the wells are brought closer together [Eq. (4-13)].

This elementary analysis neglects repulsion energy between the wells. This increases when the wave functions overlap as the distance between them decreases. As discussed later, the repulsion arises from the requirement of exclusion. Thus, in Fig. 4-9, the curve for E_A should have a small negative slope and the magnitude of E_A at $2L$ should be slightly greater than given by this approximation. The small increase in E_A with decreasing distance between the wells is discussed in greater detail in Sections 3.12.2 and 3.12.4.

The effects of well separation, r, on the energy of an electron may be shown as in Fig. 4-9. This indicates that at any finite separation of ions the ground state (E_1 in this case) splits into two levels. Furthermore, this splitting becomes greater as the distance between them diminishes. This behavior conforms to those indicated by Eqs. (3-75), (3-84), and (3-85). It also can be shown that the splitting, or separation, of the levels occurs at larger values of r with greater energy ranges between E_S and E_A for the higher energy levels (above the ground state). Also see Fig. 4-13.

The physical significance of the two cases is shown in a simplified way in Fig. 4-10. In the symmetric case, a low-energy, stable configuration is one in which the greatest probability density of finding the electron is between the two wells or ions [Fig. 4-10(a)]. The stability could be considered to arise from the $+$, $-$, $+$ order similar to that which gives rise to coulombic attraction. In reality, the electron, not the electrostatic attraction, constitutes the cement that bonds the ions. In the antisymmetric case, the greatest probability of finding an electron [Fig. 4-10(b)] could

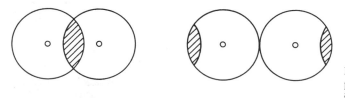

(a)   (b)

Figure 4-10   Schematic diagrams of the
probability of finding electrons from two
ions (a) with symmetric and (b) with
antisymmetric wave functions.

be considered to result in the formation of two oppositely charged dipoles. This picture leads to repulsion and is a higher-energy, lower-stability configuration than the symmetric case. Actually, no electron is present to bond the ions, and the repulsion component, E_R, of E_I predominates in this case.

The effect of the repulsion energy on E_A has been noted previously. Its influence on E_S is of greater consequence. When the two wells are brought together, the probability "clouds" surrounding each of them begin to interact. See Fig. 3-7. Here the Pauli exclusion principle starts to become operative. As the clouds appreciably penetrate one another, electrons with the same quantum numbers can approach each other. This condition is forbidden by the principle of exclusion. Further mutual penetration of the probability clouds can take place only if the quantum states of some of the electrons are changed. This usually is the equivalent of promoting some to higher energy states. If a suitable amount of energy is not added to the system to accomplish this, the electron clouds will resist further penetration. This is the basis for the origin of repulsion energies in metallic crystals. This phenomenon also is the reason why atoms or ions may be treated as hard spheres in descriptions of the geometries of crystal structures. These mechanisms are treated in greater detail in Sections 2.5 and 3.9. The particulars of some of their applications are discussed in Sections 3.12.1, 3.12.2, and 3.12.4.

An approximation of the repulsion energy between the two wells has been made based on the properties of ionic crystals (Section 3.12.1), where the repulsion energy is taken as varying as

$$E_R \propto \frac{b}{r^s}, \qquad 5 < s < 12 \qquad (3\text{-}57)$$

b is a constant, r is the distance between ions, and s is the Born exponent or repulsion exponent (Table 3-4). Without attempting to specify s for this application, except to note that it must be large for the mechanism just described, the E_R component of E_I is negligible at large distances because of the value of s. However, E_R becomes very large at small distances where r is less than unit distance. This must also be included, in terms of the repulsion energy involved, in the model obtained thus far, as is shown in Fig. 4-11. When E_S is added to the repulsion component, E_R, a minimum occurs in

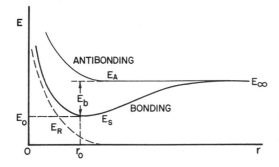

Figure 4-11 Energies acting on two electrons as a function of separation, r. $E_b = E_A - E_S - E_R$. E_R is the repulsion component of E_I, Eq. (3-81). E_b is the bonding energy.

E_S at r_0; this determines the approximate equilibrium distance between the wells or the approximate interionic distance at 0 K.

A more realistic, but seemingly less frequently used, expression for E_R is given by

$$E_R \propto e^{-r/\alpha} \tag{4-16}$$

where α is a constant that depends on the degree of shielding of the nuclei. For those ions of low atomic numbers, whose nuclei are minimally shielded by surrounding electrons, the component of E_R arising from nuclear interactions is relatively large. In such cases, α may be approximated by the Bohr radius (Section 1.5). For those ions of higher atomic number, where the electron shielding of the nuclei is more effective, E_R results primarily from the previously noted electron component of repulsion. In this case, empirical values usually are used for α. This probably accounts for continued use of Eq. (3-57), despite its large extrapolation from ionic bonding. On the other hand, Eq. (4-16) more accurately describes the strong exponential increase of E_R with decreasing r than does Eq. (3-57).

Thus far, only the effects of bringing two wells together have been examined. The effects of assembling larger numbers of wells are shown in Fig. 4-12. As has been shown, two wells cause two levels. Three wells cause three levels. N wells result in N levels. If N is sufficiently large, as in a real crystal, a quasicontinuum of levels results [Sections 3.7–3.9 and 4.1; also see Fig. 3-4(b)]. The band widths increase very slowly as N increases; the width of the outer band increases slightly more than that of the next inner band. These small band-width variations continue until a sufficient number of ions form a *stable* crystal. The unit of the factor L of Eq. (3-40) becomes constant when this takes place, no matter how small that stable crystal may be (neglecting surface effects). Further growth (or increasing N) of that crystal has no effect on the band width because the unit of L now is the lattice parameter. All the ions occupy stable lattice positions in stable geometric arrays. The lattice parameter of a stable crystal is not a function of its size as governed by N. Under equilibrium conditions, in the absence of solid-state transformations, it is a function only of temperature.

The inner, completed levels are very tightly bound to their nuclei and are virtually unaffected by the presence of the other ions in the lattice. The outermost bands, whether they overlap or whether they are completely filled or not, provide a basis for the prediction of the properties of solids and for their classification. A band model for several classes of solids is shown schematically in Fig. 4-13.

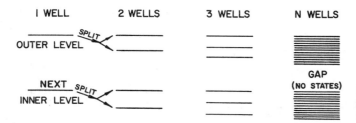

Figure 4-12 Effects of increasing numbers of potential wells on the number of states within a band.

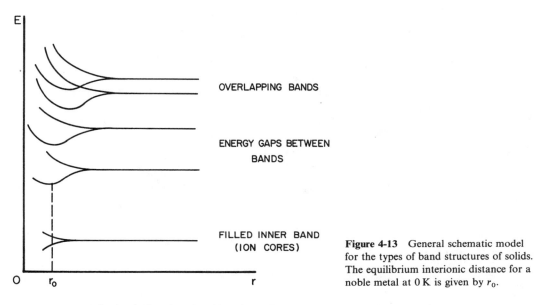

OVERLAPPING BANDS

ENERGY GAPS BETWEEN
BANDS

FILLED INNER BAND
(ION CORES)

Figure 4-13 General schematic model for the types of band structures of solids. The equilibrium interionic distance for a noble metal at 0 K is given by r_0.

The basis for the classification of solids and the prediction of their properties by the band model depends on the extent to which the outermost band is filled and whether or not it overlaps other bands. The width of any gaps, or energy intervals, between outer bands also plays a significant part in these predictions.

Consider first normal, monovalent metals [Fig. 4-14(a)]. By this is meant those metals whose atoms have completed inner bands and only half-filled outer, or valence, bands. Since the valence bands are incompletely filled, the application of thermal or electrical energy can excite electrons near E_F to unoccupied levels. The electrons have some place to go and can enter into a physical process. Such elements are good thermal and electrical conductors. See Sections 2.7 and 3.12.4.

Now consider a situation in which the valence bands are *completely filled* and the next highest band is completely vacant [Fig. 4-14(b)]. This is the situation for most ionically and covalently bonded solids, as described in Sections 3.12.1–3.12.3. Much depends on the extent of the energy gap between them. If the gap is very large ($\gg k_B T$) the element will be an insulator. This results from the fact that ordinary energies cannot cause the valence electrons to participate because they are unable to absorb the energies and jump across the gap. They are immobilized in the filled valence band where no empty states are available for them to occupy and so engage in a physical process. The gap represents a range of energies that the electrons are forbidden to occupy. See Section 4.6. Neither thermal nor electrical conduction is possible by the electrons under these conditions. Such substances are insulators. Generally, these materials demonstrate *very low ionic* thermal and electrical conductivities.

However, if the energy gap between the outermost completely filled state in the valence band and the lowest state in the next outer, completely vacant band

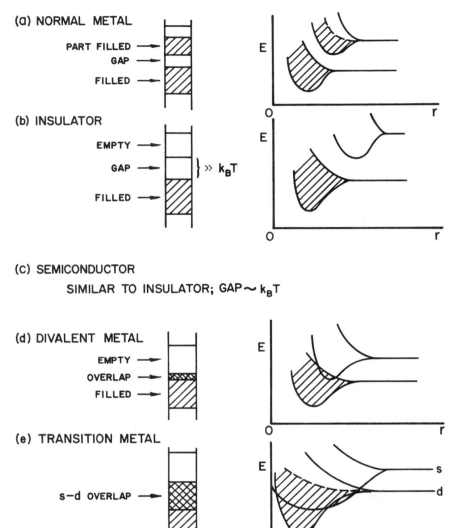

Figure 4-14   Use of the band theory as a means for the classification of solids.

(conduction band) is of the order of $k_B T$, then conduction can occur.   Normal energies are sufficient to permit electrons from the top of the filled valence band to jump across the energy gap and occupy states in the bottom of the unfilled conduction band.   The more electrons that enter into this process, the better the conduction becomes.   The resulting holes (absences of electrons) in the formerly filled valence band also can enter into the conduction process.   Very pure silicon, germanium, and carbon (diamond) show this behavior. Elements of this type are known as intrinsic semiconductors (Section 3.12.3).

Some of the divalent elements, such as the alkaline earths [Fig. 4-14(d)], have completed, outer *s* valence levels. On this basis, it might be thought that such elements should be either insulators or semiconductors. However, their *s* levels overlap empty *p* levels. Such elements show conduction behavior similar to that of the normal monovalent metals. This is to be expected since the band overlap enables the *s* electrons to absorb energy and provides them with easy access to the unoccupied *p* states.

The transition elements [Fig. 4-14(e)], atoms with incompletely filled outer *d* and *s* bands in the solid state, also show reasonably good metallic behaviors [Section 3.11]. Here the holes (absences of electrons) also may enter into the conduction process. The transition elements are discussed more fully in later sections devoted to heat capacity, thermoelectricity, and magnetism.

4.5.1 Applications of the Fermi–Sommerfeld Theory

The band concept may be applied to the Fermi–Sommerfeld theory by considering each band in terms of the individual function for its density of states. This is shown in Fig. 4-15. Here the electron behavior is the same as that shown in Fig. 4-14, but is in terms of the density of states and energy.

In the case of a monovalent normal metal, the outer valence band is only half-filled. Electrons near E_F can accept additional energy by occupying some of the many available vacant states and readily enter into an electron-transport process. The thermal and electrical conductivities of elements such as these would be expected to be good.

The electron band configuration of an insulator, because it includes a gap much larger than $k_B T$, prevents the electrons from participating in conduction processes. Under normal conditions, the available energy is insufficient to excite outer electrons in the completely filled valence band across the large, forbidden energy gap into empty states in the conduction band. Any thermal or electrical conduction, then, must be performed by the ions. This, almost without exception, is extremely small compared to that accomplished by electrons. Where the energy gap is relatively small, of the order of $k_B T$, conduction by electrons becomes possible. Normally available energies can promote electrons in states at the top of the valence band across the energy gap to unoccupied states at the bottom of the unfilled conduction band. As the added external energy, or temperature, increases, more electrons are promoted. The increased number of activated electrons, as well as the holes that they leave behind in the valence band, increase the conductivity and semiconducting behavior is observed. Another way of stating this is that the electrical resistivity decreases with increasing temperature. Such materials, therefore, have negative temperature coefficients of electrical resistivity.

Divalent metals, such as the alkaline earths, show metallic behavior because of the overlap of the outer bands. This permits the valence electrons to occupy states of

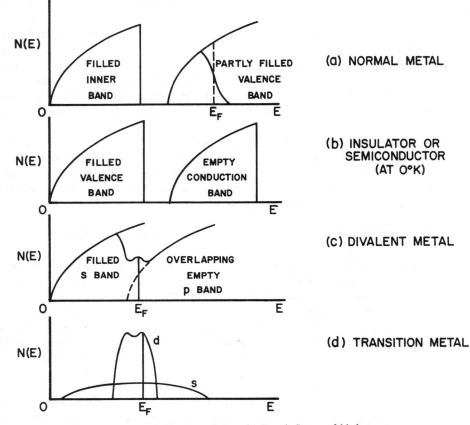

Figure 4-15 The band theory applied to the Fermi–Sommerfeld theory to describe the band structures of classes of elements. Note that (a), (b), and (c) are oversimplified because the variations in the densities of states near the tops of the bands are neglected.

higher energies and to enter into conduction processes easily without the necessity for jumping across a gap.   The properties of the trivalent normal metals are explained on the same basis as for the divalent metals except that the overlapping outer band is partly filled.

4.6 BRILLOUIN ZONE THEORY OF SOLIDS

With the exception of the band theory, the preceding theories consider the electrons to be in a constant internal potential in the solid.   However, each ion core, as previously shown in Fig. 4-5, constitutes a potential well.   In real crystals, the ions and, consequently, the potential wells are arranged in periodic, geometric arrays.   Thus, the potential within a crystal is periodic. This periodicity of potential varies with the direction in the crystal because the distances between the ions in crystals vary with direction. The electrons in the crystal are affected by the periodic potential. Therefore, virtually all the physical properties, including the electron-transport properties of anisotropic, real crystals are expected to vary with the crystallographic direction. This is known as *anisotropic behavior*. In neglecting the periodic potential and its variation with lattice direction, the prior theories can deal only with ideal, isotropic materials. The properties of such materials are the same in all crystallographic directions. Such approaches are best applied to polycrystalline materials; these consist of extremely large numbers, usually of the order of $10^3/mm^2$, of very small, virtually randomly oriented grains that average out to approach the anisotropic effects. However, it is possible to account for the anisotropic properties of single crystals when the influence of the directional variation of the periodic internal potential on the nearly free electrons is taken into account. This is the basis of the Brillouin theory.

In the case of the free electron,

$$E = \frac{h^2 \bar{k}^2}{8\pi^2 m} \tag{3-25}$$

gives its $E - \bar{k}$ relationship. No restrictions were placed on $\bar{k}$, so the energy of a free electron is a continuous, parabolic function of the wave vector. As will be seen, this must be modified for the case of real crystals.

In the case of real crystals, the variation in the magnitude of the periodic potential varies with the distance between parallel planes of ions. The variation in the potential as measured along the centers of a plane of ions is much greater than that measured halfway between two planes [Fig. 4-5]. The problem of dealing with this behavior can be simplified if most of the nearly free electrons are considered to be equidistant from the ionic planes. When this is the case, the variation in the periodic potential from crest to trough is small, and the consequent periodic variation in the density of the electrons also is small. This means that there is a nearly equal probability of finding an electron on a potential crest as in a trough. This nearly equal crest–trough probability allows the approximation that the electrons are practically unaffected by the small variation in the periodic potential. This is the same as setting $V(x) \cong 0$ as in Eq. (3-22) and permits the approximation that the electrons are "nearly free," so Eq. (3-25) is still applicable. See Section 2.7.

However, at certain critical values of $\bar{k} = \bar{k}_c$, the critical wave vector, the nearly free electron model breaks down. Here, even for small potential differences between the crests and troughs, the electrons undergo an elastic reversal in direction and behave like standing waves for the critical values of $\bar{k}$. This arises from the periodicity of the lattice.

Consider electron diffraction similar to that described in Section 1.6.3. An electron injected into the lattice "sees" the same periodic potential as a nearly free electron present prior to any injection and is scattered. At certain energies and angles, however, the injected electrons undergo constructive interference and are diffracted from the lattice planes, and Bragg's law is obeyed. Apparently, the nearly free electrons already within the lattice, as well as those that are injected and scattered, do not satisfy the Bragg relationship. The conditions for Bragg reflection, therefore, must constitute boundary conditions for the properties of the nearly free electrons.

Consider an incident, plane wave of the nearly free electrons, $\psi_1 = e^{i\bar{k}x}$, to be moving in the x direction perpendicular to a family of lattice planes. This wave will propagate across the lattice until it approaches the conditions for Bragg reflection. It then begins to deviate from the "nearly free" electron behavior, and at $\bar{k}_c$ the plane wave is reflected back on itself. This reversal of propagation is described by $\psi_2 = e^{-i\bar{k}x}$.

The conditions under which Bragg's law is obeyed now must be examined. The Bragg equation gives the conditions for diffraction as

$$n\lambda = 2a \sin \theta \qquad (4\text{-}17)$$

where n is an integer, λ is the wavelength, a is the interplanar spacing, and θ is the angle of incidence. From this equation and the definition of $\bar{k}$ ($\bar{k} = 2\pi/\lambda$), Eq. (4-17) may be written in reciprocal form as

$$\frac{\bar{k}}{2\pi} = \frac{1}{\lambda} = \frac{1}{2a \sin \theta} \qquad (4\text{-}18)$$

Constructive interference occurs when twice the interplanar distance is an integral multiple of the wavelength. In the present case, the electrons are moving perpendicular to the planes, so $\theta = \pi/2$; and, for this critical condition,

$$\bar{k}_c = \pm \frac{n\pi}{a} \qquad (4\text{-}19)$$

The condition for Bragg reflection thus represents wave vectors that are forbidden to the nearly free valence electrons within the crystal. From Eq. (3-25), it can be seen that the disallowed values of $\bar{k}$ result in forbidden energy ranges, or energy gaps; the "nearly free" valence electrons cannot occupy states in these gaps. This condition is responsible for the breakdown of the applicability of Eq. (3-25), which was indicated previously. The inverse relationship of $\bar{k}$ to the interplanar distance a in Eq. (4-19) is the basis for the concept of *reciprocal space* when three dimensions are used.

The wave function for the nearly free electrons undergoing reflections at $\bar{k}_c$ must include both the incident and the reflected waves. Thus, the total wave function is

$$\Psi = \psi_1 \pm \psi_2 = B(e^{i\bar{k}x} \pm e^{-i\bar{k}x}) \qquad (4\text{-}20)$$

in which the constant B is that employed for normalization as in Section 1.6.7. It is convenient to reexpress the exponential terms in Eq. (4-20) by their trigonometric equivalents. This is done by recalling that

$$e^{i\bar{k}x} = \cos \bar{k}x + i \sin \bar{k}x$$

and

$$e^{-i\bar{k}x} = \cos \bar{k}x - i \sin \bar{k}x$$

Thus,

$$\Psi = \psi_1 \pm \psi_2 = 2B(\cos \bar{k}x + i \sin \bar{k}x) \tag{4-21}$$

Equation (4-21) provides the information necessary to calculate the potential energy, V, that constitutes the gap. The periodic potential $V(x)$ varies with the position in the one-dimensional lattice as shown by Fig. 4-5. The probable position of an electron is given by the probability density, denoted here by $|\Psi|^2$. Thus, in a way analogous to Eq. (1-17),

$$V = \frac{1}{L} \int |\Psi|^2 V(x) \, dx \tag{4-22}$$

Now, by substituting Eq. (4-21), Eq. (4-22) becomes

$$V = \frac{4B^2}{L} \int (\cos^2 \bar{k}x + \sin^2 \bar{k}x) V(x) \, dx \tag{4-23}$$

It will be recalled that

$$\sin^2 \bar{k}x = \tfrac{1}{2} - \tfrac{1}{2} \cos 2\bar{k}x$$

and

$$\cos^2 \bar{k}x = \tfrac{1}{2} + \tfrac{1}{2} \cos 2\bar{k}x$$

So that, if $B = 2^{-1/2}$,

$$V = \frac{2}{L} \int \left(\frac{1}{2} + \frac{1}{2} \cos 2\bar{k}x \right) V(x) \, dx + \frac{2}{L} \int \left(\frac{1}{2} - \frac{1}{2} \cos 2\bar{k}x \right) V(x) \, dx$$

or

$$V = \frac{1}{L} \int (1 + \cos 2\bar{k}x) V(x) \, dx + \frac{1}{L} \int (1 - \cos 2\bar{k}x) V(x) \, dx \tag{4-24}$$

$V(x)$ is a periodic, trigonometric function of x, so it will vanish upon integration as the first term of each of the integrals in Eq. (4-24). Thus, Eq. (4-24) is reduced to

$$V = \pm \frac{1}{L} \int \cos 2\bar{k}x \cdot V(x) \, dx \tag{4-25}$$

The function Ψ contains an integral number of periods, n, in the crystal length L [Eq. (3-39) and Fig. 3-2]. It, therefore, is necessary to integrate Eq. (4-25) over just one of these periods as represented by the interplanar distance, a. So Eq. (4-25) is reexpressed as

$$V = \pm \frac{1}{a} \int_0^a \cos 2\bar{k}x \cdot V(x)\, dx \qquad (4\text{-}26)$$

Upon integration of Eq. (4-26), since $\bar{k} = \bar{k}_c = \pm n\pi/a$ and $n = 1$, it is found that

$$V = \pm V_n \qquad (4\text{-}27)$$

Equation (4-27) represents the potential energy component of the total energy acting on an electron when its wave vector is at or near the boundary conditions. The kinetic energies of ψ_1 and ψ_2 are the same, as shown by Eq. (3-25), for $\bar{k} < \bar{k}_c$. Thus, *at or close to the boundary*, the total energy of an electron is given by

$$E = \text{KE} + \text{PE} = \frac{h^2\bar{k}^2}{8\pi^2 m} \pm V_n \qquad (4\text{-}28)$$

As long as $\bar{k} < \bar{k}_c$, the electrons show the "nearly free" behavior as described by Eq. (3-25). When $\bar{k}$ is at or near $\bar{k}_c$, electrons depart from "nearly free" behavior as shown by Eq. (4-28). The electrons are forbidden to occupy any states in the energy range of $\pm V_n$.

So, at each critical value of $\bar{k}$, the electron energy divides into two values. The magnitude of the energy separation, $\pm V_n$, or gap, between these two energy values is a function of the magnitude of the variation of the periodic lattice potential, $V(x)$. Since both the magnitude of this potential variation and the periodicity of the lattice potential vary with the direction taken in the lattice, it is expected that these energy gaps will occur at different values of $\bar{k}$ and E for different lattice directions. This is the basis for understanding the anisotropic properties of crystals. The surfaces formed in $\bar{k}$ space by the planes perpendicular to the various values of $\bar{k}_c$ enclose a volume that constitutes the Brillouin zone for the given lattice. The one-dimensional case is illustrated in Fig. 4-16. The energy gap at the first discontinuity, $\bar{k}_x = \pi/a$, is $\pm V_n = \pm V_1 = 2V_1$. Similarly, the gap at the second discontinuity, $\bar{k}_x = 2\pi/a$ is $\pm V_2 = 2V_2$. The subscripts refer to the first and second zones, respectively. Equation (4-28) accounts for $V_2 > V_1$, as shown in Fig. 4-16(b).

The illustration in Fig. 4-16(a) for the case of the free electron [Eq. (3-25)] shows continuous parabolic behavior, since no restrictions were placed on $\bar{k}$. The energy discontinuities at the critical values, $\bar{k}_c$ [Eq. (4-28)], result from the limitations placed on $\bar{k}$ that arise from the periodicity of the lattice [Fig. 4-16(b)]. The projection parallel to the $\bar{k}_x$ axis gives the linear Brillouin zones for the x direction. This is the Brillouin zone for a one-dimensional lattice [Fig. 4-16(c)]. The projection parallel to the E axis gives the band model [Fig. 4.16(d)]. The energy gaps and the extent to which the bands are filled, as well as any overlap, play exactly the same roles as described in Section 4.5.

Figure 4-16(b) is sometimes called the *extended zone* model. Here, $\bar{k}_x$ is defined over the range between $\pm \infty$. The same information can be furnished by a "reduced

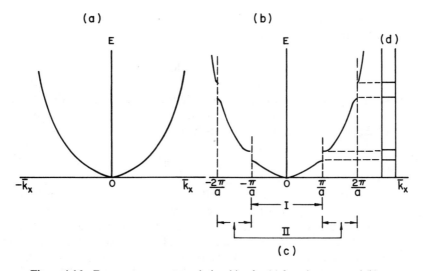

Figure 4-16   Energy–wave vector relationships for (a) free electrons and (b) electrons in a periodic potential. (c) Relationship to a one-dimensional Brillouin zone. (d) Relationship to band theory.

zone" picture in which $\bar{\mathbf{k}}_x$ is defined between its critical values of $\pm\pi/a$. As shown in Fig. 4-17, the second and any higher zones are folded over back into the first zone.

The simplest case for the illustration of the three-dimensional character of the Brillouin zone may be obtained by equating Eqs. (3-25) and (3-49),

$$E = \frac{h^2\bar{\mathbf{k}}^2}{8\pi^2 m} = \frac{h^2}{8m}\left[\frac{n_x^2}{a^2} + \frac{n_y^2}{a^2} + \frac{n_z^2}{a^2}\right] \qquad (4\text{-}29)$$

in which the interplanar distance, a, again is substituted for the well dimension, L. From this it can be seen that

$$\frac{\bar{\mathbf{k}}}{\pi} = \pm\frac{1}{a}(n_x^2 + n_y^2 + n_z^2)^{1/2} \qquad (4\text{-}30)$$

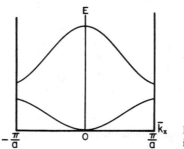

Figure 4-17   First two zones of Fig. 4-16 in the reduced-zone representation.

and for a given crystallographic direction $n_i = (n_x^2 + n_y^2 + n_z^2)^{1/2}$, the boundary of the Brillouin zone is given by

$$\bar{\mathbf{k}}_c = \pm \frac{n_i \pi}{a} \qquad (4\text{-}31)$$

so that Eqs. (4-31) and (4-19) are identical for any given direction.

The critical condition for the three-dimensional case can be expressed by the use of Eq. (4-31) in Eq. (4-30). This gives

$$\bar{\mathbf{k}}_c = \pm \frac{\pi}{a} (n_{xc}^2 + n_{yc}^2 + n_{zc}^2)^{1/2} \qquad (4\text{-}32)$$

Here the integers n_i, n_{xc}, n_{yc}, and n_{zc} represent the critical or limiting values. These integers define planes in $\bar{\mathbf{k}}$ space (wave vector space) that outline and enclose all the allowed values of $\bar{\mathbf{k}}$. Suitable sets of such planes, which depend on the unit cell of the crystal, enclose a polygon in $\bar{\mathbf{k}}$ space. The smallest volume enclosed by such planes in $\bar{\mathbf{k}}$ space, which obeys Eq. (4-31), is known as a *Brillouin zone*. Brillouin zones of the most common metallic lattices are shown in Fig. 4-18. The Jones zone may be used to explain the properties of divalent, HCP metals.

Brillouin zones also may be constructed by passing planes perpendicular to the midpoints of vectors joining nearest neighbors in the unit cell in reciprocal space.

Electrons within a given Brillouin zone may be treated as being nearly free when the Fermi surface does not approach the zone boundaries. This is the case where the wave vector at the Fermi surface is less than $\bar{\mathbf{k}}_c$. Here the Fermi surface is spherical, as is shown in Fig. 4-4(c). In this case, the energy of an electron at the Fermi surface will lie on the parabolic portion of the E versus $\bar{\mathbf{k}}$ relationship given by Eq. (3-25) and

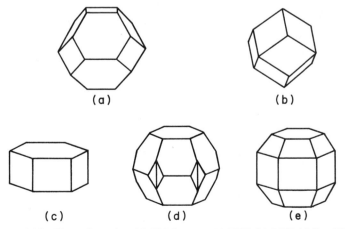

Figure 4-18 Three-dimensional Brillouin zones (a) FCC, (b) BCC, (c) first HPC zone, (d) second HCP zone, and (e) Jones zone for the HCP lattice.

shown in Fig. 4-16(a). This approximation does not hold for electrons with wave vectors, or energies, at or close to the zone boundaries. Here a considerable increase in E occurs near $\bar{\mathbf{k}}_c$ for a relatively small change in $\bar{\mathbf{k}}$ [Fig. 4-16(b)]; this portion of the curve is no longer parabolic, and the electrons show different behaviors from those of the nearly free case.

An insight into these differences in electron behavior may be obtained from an examination of the implications of Eq. (3-25). The second derivative of this equation is a constant that is given by

$$\frac{d^2E}{d\bar{\mathbf{k}}^2} = \frac{h^2}{4\pi^2 m} = \frac{\hbar^2}{m}, \qquad \hbar = \frac{h}{2\pi} \tag{4-33}$$

Equation (4-33) is rearranged to give the effective mass of the electron as

$$m^* = \hbar^2 \left(\frac{d^2E}{d\bar{\mathbf{k}}^2}\right)^{-1} \tag{4-34}$$

As long as the E, $\bar{\mathbf{k}}$ coordinates of the electron lie on the parabolic portion of Eq. (3-25), the second derivative is constant and $m^* \cong m$, the mass of a free electron. However, when $\bar{\mathbf{k}} \rightarrow \bar{\mathbf{k}}_c$, the E versus $\bar{k}$ curve is no longer parabolic and the second derivative is no longer constant. This causes significant changes in the effective electron mass; it becomes considerably different from that of a free electron. In the case of normal metals with partially filled bands or with overlapping bands, the ratio of $m^*/m \cong 1$. This good approximation is used frequently for purposes of simplification. It should be understood, however, that all equations involving the use of the mass of the electron should use m^* instead of m when greater precision is desired. The properties of electrons close to the top of the valence band are described in greater detail in Chapter 6.

4.7 PHYSICAL SIGNIFICANCE

As in band theory, the ways in which the electrons are accommodated by the Brillouin zones determine the properties of the solid. This also permits the classification of solids and the prediction of their properties, as will be shown here. The Brillouin zone clearly shows the effect of crystalline anisotropy on the energies of the electrons, and consequently on physical properties.

The "nearly free" electrons can be treated by Eqs. (3-25) and (3-49), as indicated by Eq. (4-29):

$$E = \frac{h^2 \bar{\mathbf{k}}^2}{8\pi^2 m} = \frac{h^2}{8m}\left[\frac{n_x^2}{a^2} + \frac{n_y^2}{a^2} + \frac{n_z^2}{a^2}\right] \tag{4-29}$$

Then, for given energies, isoenergy contours can be drawn in $\bar{\mathbf{k}}$ space. These will be nearly spherical surfaces for a cubic lattice, since E varies as $\bar{\mathbf{k}}^2$. A breakdown of this behavior occurs when the electron energies approach critical values of $\bar{\mathbf{k}} = \bar{\mathbf{k}}_c$ [Eq.

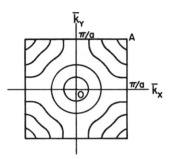

Figure 4-19 Two-dimensional Brillouin zone showing isoenergy contours.

(4-19) and Fig. 4-16(b)]. Here the isoenergy surface contours and that of E_F are no longer spherical. This is shown in Fig. 4-19 in two dimensions.

As the Fermi surface approaches the zone wall, the values of $\bar{k}$ become increasingly greater than that predicted by the "nearly free" electron theory. This continues until the discontinuity at the zone wall, $\bar{k}_c$, is encountered. When the entire zone is filled, it accommodates a maximum electron:ion ratio of 2.

It can be seen that the energy–wave vector relationships along the [100], or $\bar{k}_x$, direction will be different from that along the [110], or OA, direction in Fig. 4-19.

For a given direction in $\bar{k}$ space, where the isoenergy contours are spherical, the electrons are considered to be nearly free; the Sommerfeld distribution function [Eq. (4-5)] is followed and $N(E)$ varies as $E^{1/2}$. However, as the wave vector approaches $\bar{k}_c$, the density of states becomes greater than that predicted by the Sommerfeld parabola and the Fermi surface is no longer spherical. States begin to fill in other directions. Then, fewer and fewer states become available as the higher energy levels in the corners of the zone are filled, and $N(E)$ decreases. At the point where the zone is completely filled, $N(E)$ equals zero. This relationship is illustrated in Fig. 4-20 for a two-dimensional cubic lattice.

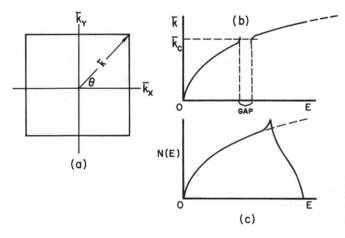

Figure 4-20 Representations of relationships between filled bands and zones: (a) Brillouin zone, (b) band theory, and (c) density of states.

The directional properties of the E versus $\bar{k}$ relationship are important because they permit the prediction of properties as a function of crystalline direction within the zone.   What appears to be a large energy gap in one direction may be spanned in another.   This is shown in Fig. 4-21.   If this zone appeared to be filled in the $\bar{k}_x$ direction, the solid could be mistaken for a semiconductor or an insulator, if only the one direction had been considered.   The overlap of the states in the [110] direction with the second zone indicates that no effective gap exists; such a solid will demonstrate metallic behavior.

The solid under consideration could have had an effective energy gap if the curve for the [110] direction had shown a discontinuity at $4 < E < 5$ (arbitrary units).   The energy gap between the first and second zones would then have consisted of about one such energy unit.   If this gap had been of the order of $k_B T$, the solid would have been

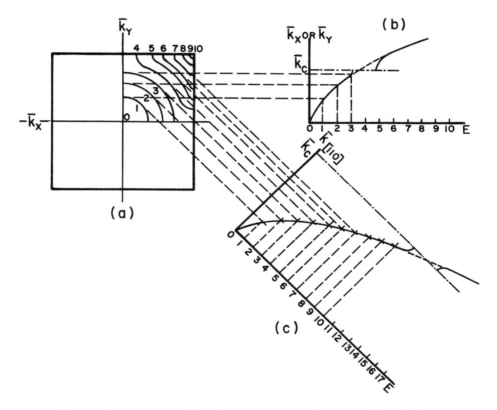

Figure 4-21   Boundaries of the effects of wave-vector directions within a Brillouin zone on electron energies: (a) isoenergy contours within a zone; (b) variation of $\bar{k}$ and E in the [100], $\bar{k}_x$, direction; (c) variation of $\bar{k}$ and E in the [100], $\bar{k}_x = \bar{k}_y$, direction. (Method after W. Hume-Rothery, *Atomic Theory for Students of Metallurgy*, p. 204, Institute of Metals, London, 1952.)

classified as a semiconductor. Had it been much larger than $k_B T$, it would have been classed as an insulator.

The normal metals, such as copper, silver, or gold, each have one electron per atom; their zones thus are half-filled. At this electron concentration, the isoenergy contours may be approximated to be spherical in $\bar{k}$ space; that is, their curves of density of states are considered to lie on the parabolic portion of the Sommerfeld curve and to show nearly free behavior. This is an approximation, since, in actuality, small necks protrude from the sphere to the hexagonal, octahedral planes of the first Brillouin zone of these FCC elements (Fig. 4-22).

The Hume-Rothery theory of terminal solid solutions has been rationalized on the basis of the relationship of the spherical approximation of the Fermi surface [Fig. 4-4(c)] to the zone walls. The radius of this sphere is a function of the electron:atom, or e:a, ratio [Eq. (4-9)]. That concentration of an alloying element whose e:a ratio causes the Fermi surface to become tangent to the zone boundaries is considered to define the solubility limit. This approach gives reasonable results for FCC solvents and avoids the complications introduced by the necks on the Fermi surface. This approach also has been applied to other phases appearing in alloys with FCC metals.

The variation of electron energy with the wave vector, as in Fig. 4-19, has important consequences best visualized by means of the Brillouin zone. Since E varies with $\bar{k}$, $N(E)$ also will be affected by this variation. The resulting anisotropy in $N(E)$ will give corresponding variations in $n(E_F)$. [It will be recalled that $n(E_F)$ is the number of electrons per unit volume having $E \cong E_F$.] Since these are the electrons that enter into physical processes, variations in their number strongly affect those physical properties that take $n(E_F)$ into account. See Chapter 6.

The E versus $\bar{k}$ variations within a given Brillouin zone also help to explain the properties of alloys and alloy phases. This is especially true when the Brillouin zone is more than half-filled. For example, if there are almost as many valence electrons as there are available states within a zone, it is to be expected that $n(E_F)$ will be quite small, especially in the absence of overlapping zones. In such a case, physical properties involving this factor would be significantly different from the situation in

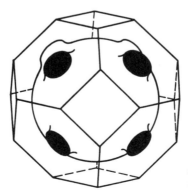

Figure 4-22 Brillouin zone of a noble metal showing the inscribed Fermi surface.

which $n(E_F)$ was comparatively large. This also is important in the anisotropy of electrical and mechanical properties of materials with HCP lattices [Figs. 4-18(d) and (e)].

Another consequence of the variation of electron energy with the wave vector within the zone results in a corresponding anisotropy in the effective mass of the electron. This follows from its behavior, previously noted in Section 4.6, as a function of $(d^2E/d\bar{\mathbf{k}}^2)^{-1}$. This, too, has a profound influence on the variation of many physical properties as a function of crystal direction.

The anisotropy of physical properties must be taken into account in applying engineering materials to obtain their optimum utilization. This is particularly true of wrought (textured) titanium and other HCP alloys. In such cases, the elastic constants, coefficients of thermal expansion, and thermal conductivities vary considerably. Other materials, such as those used for hard and soft magnets and for semiconductor devices, are processed to have specific crystal orientations to achieve the most desirable properties.

4.8 BIBLIOGRAPHY

HUME-ROTHERY, W., *Atomic Theory for Students of Metallurgy*, Institute of Metals, London, 1952.

HURD, C. M., *Electrons in Metals: An Introduction to Modern Topics*, Wiley-Interscience, New York, 1975.

KITTEL, C., *Introduction to Solid State Physics*, 5th ed., Wiley, New York, 1976.

LEVY, R. A., *Principles of Solid State Physics*, Academic Press, New York, 1968.

POLLOCK, D. D., *Physical Properties of Materials for Engineers*, Vol. I, CRC Press, Boca Raton, Fla., 1982.

————, *Electrical Conduction in Solids: An Introduction*, American Society of Metals, Metals Park, Ohio, 1985.

RICHTMYER, F. K., KENNARD, E. H., and LAURITSEN, T., *Introduction to Modern Physics*, McGraw-Hill, New York, 1955.

SEITZ, F., *The Physics of Metals*, McGraw-Hill, New York, 1943.

SOLYMER, L., and WALSH, D., *Lectures on the Electrical Properties of Materials*, 2nd ed., Oxford University Press, New York, 1979.

SPROULL, R. L., *Modern Physics*, Wiley, New York, 1956.

STRINGER, J., *An Introduction to the Electron Theory of Solids*, Pergamon Press, Elmsford, N.Y., 1967.

4.9 PROBLEMS

4.1. Why is it that only those electrons with $E \geqslant E_F$ can enter into a physical process?

4.2. What is the importance of E_F in treating electrons in solids?

4.3. Calculate the average velocity of an electron with $E \cong E_F$ at 0 K. What does this mean in contrast to classical mechanics?

4.4. (a) Approximate the minimum lifetime of a quasielectron for $\Delta E_{max} \sim 3.7 \times 10^{-15}$ eV. Why is this a minimum time?

   (b) Compare this with the lifetime of an electron in a normal metal at room temperature.

4.5. How do the graphs of Problem 3.8 help to explain the band theory?

4.6. Plot a curve for E_R versus r of Eq. (4-16) for α equal to the Bohr radius. Use this to make an approximation of the average "size" of an atom.

4.7. Discuss the effect of the magnitude of the energy gap between outer bands on the properties of a solid.

4.8. Show how the Brillouin zones of BCC and FCC lattices may be constructed directly from {110} and {111} planes, respectively. Why is this so?

5 / THERMAL PROPERTIES OF SOLIDS

Atomic structure, bonding, and the classification of solids having been considered, the thermal properties of solids are introduced at this point because the thermal vibrations of the ions comprising a crystal affect other physical properties. As a means of simplifying the treatment of lattice vibrations, at first, only the ions of pure, elemental solids will be considered. The additional contributions of the electrons to thermal and other physical properties are discussed later.

The simplest way of treating ions in this manner is to consider the atoms in the crystal to be covalently bonded (Sections 3.12.3, 4.5.1, and 4.7). Under this condition, the valence electrons are bonded very strongly to the atoms, forming ions. Crystals bonded in this way show minimal electrical and thermal conductivities; they are classed as insulators because of the extremely low number and mobility of any valence electrons that might be available. In other words, these materials are essentially nonconductors. Intrinsic semiconductors also may be included for those temperature ranges in which the electron population of the conduction band is negligibly small.

The thermally induced vibrations of the ions are responsible for such physical properties as heat capacity, temperature coefficient of thermal expansion, and thermal conductivity. All these are important physical properties of materials.

While not included in this book, the understanding of thermally induced lattice vibrations provides a basis for a better comprehension of many solid-state reactions that include diffusion, nucleation and growth, allotropic changes, and order–disorder.

5.1 HEAT CAPACITY

On the basis of the foregoing introduction, this section will consider only elemental solids. The heat capacity* of a substance at constant volume is defined as

$$C_V = \left(\frac{\partial U}{\partial T}\right)_V \tag{5-1}$$

where U is the internal energy and T is the absolute temperature. The heat capacity at constant pressure is given by

$$C_P = \left(\frac{\partial H}{\partial T}\right)_P = \left(\frac{\partial (U + PV)}{\partial T}\right)_P \tag{5-2}$$

in which H is the enthalpy, P the pressure, and V the volume. Performing the indicated differentiation,

$$C_P = \left(\frac{\partial U}{\partial T}\right)_P + P \left(\frac{\partial U}{\partial T}\right)_P + V \left(\frac{\partial P}{\partial T}\right)_P$$

and noting that $\partial P/\partial T = 0$ at constant pressure,

$$C_P = \left(\frac{\partial U}{\partial T}\right)_P + P \left(\frac{\partial U}{\partial T}\right)_P \tag{5-3}$$

The quantity ∂U can be expressed as

$$\partial U = \left(\frac{\partial U}{\partial V}\right)_T \partial V + \left(\frac{\partial U}{\partial T}\right)_V \partial T$$

and when divided by ∂T, this becomes

$$\frac{\partial U}{\partial T} = \left(\frac{\partial U}{\partial V}\right)_T \left(\frac{\partial V}{\partial T}\right)_P + \left(\frac{\partial U}{\partial T}\right)_V \tag{5-4}$$

Thus, C_P [Eq. (5-3)] becomes

$$C_P = \left(\frac{\partial U}{\partial V}\right)_T \left(\frac{\partial V}{\partial T}\right)_P + \left(\frac{\partial U}{\partial T}\right)_V + P \left(\frac{\partial V}{\partial T}\right)_P \tag{5-5}$$

When this is used with Eq. (5-1),

$$C_P - C_V = \left(\frac{\partial U}{\partial V}\right)_T \left(\frac{\partial V}{\partial T}\right)_P + \left(\frac{\partial U}{\partial T}\right)_V + P \left(\frac{\partial V}{\partial T}\right)_P - \left(\frac{\partial U}{\partial T}\right)_V$$

$$C_P - C_V = \left(\frac{\partial U}{\partial V}\right)_T \left(\frac{\partial V}{\partial T}\right)_P + P \left(\frac{\partial V}{\partial T}\right)_P = \left[\left(\frac{\partial U}{\partial V}\right)_T + P\right] \left(\frac{\partial V}{\partial T}\right)_P \tag{5-6}$$

---

*The *heat capacity* is defined as the quantity of heat required to raise the temperature of *one mol* of a substance 1 K. The *specific heat* is the quantity of heat required to raise *one gram* of a substance 1 K.

For most solids, especially metals at normal pressure, $(\partial V/\partial T)_P$ is relatively small, $\Delta V/\Delta T$ being of the order of 10^{-4} to 10^{-5} cm^3/°C. So for many purposes it can be assumed that this quantity is negligible and that

$$C_P \cong C_V \qquad (5\text{-}7)$$

This approximation will be employed for solids where applicable and convenient. It gives a maximum error of less than 0.5 cal/mol K.

Dulong and Petit (1818) showed that the heat capacities of many substances were related to their atomic weights, the product of their specific heats and atomic weights being approximately constant, about 6 (cal/mol)/deg. This is only an approximation. For many elements, the heat capacity lies between 5 and 7 cal/mol-deg at 0°C and has an average value of 6.2. The heat capacities are not constant but increase about 0.04%/°C for temperatures above 0°C. This is shown in Fig. 5-1. The approximate nature of the Dulong and Petit rule is apparent from the figure. Nevertheless, it is a very helpful rule of thumb in practical applications. It also is apparent from the figure that the curves for the elements shown all have the same type of sigmoidal-shaped curve.

The curves of each of the solids shown in Fig. 5-1 can be made to fall on a single, "universal" curve. This is done by means of a parameter that is specific for each

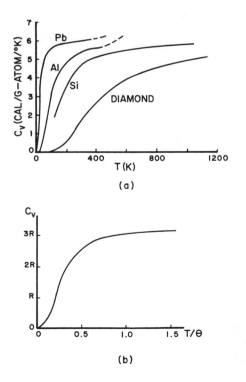

Figure 5-1 (a) The heat capacities of some elements. (From F. K. Richtmyer, E. H. Kennard, and T. Lauritsen, *Introduction to Modern Physics*, 5th ed., p. 410, McGraw-Hill, New York, 1955.) (b) The "universal" curve. (From F. Seitz, *Modern Theory of Solids*, p. 109, McGraw-Hill, New York, 1940.)

element, the characteristic temperature, θ, which is used to normalize the curves. When the data for C_V are plotted versus T/θ, the data fall very close to a common curve. It should be noted that many exceptions to this curve exist. It is really only valid for elemental, nonmolecular solids. It is not valid where allotropic phase or solid-state transformations such as magnetic changes take place. It is also sensitive to the presence of order–disorder reactions in alloys. Peaks and discontinuities in heat capacity curves have been employed to detect such changes and reactions in the solid state.

In the very low temperature range, approximately to about $\theta_D/20$ for many materials, C_V varies as T^3. The high-temperature portion of the "universal" curve and the rule of Dulong and Petit can be explained by showing that the average rate of change of energy of an atom is approximately constant at elevated temperatures. This is done by dividing the average energy of a mol of a solid by Avogadro's number: 6 cal/mol·K $\div$ 6.02×10^{23} atoms/mol $\cong 1 \times 10^{-23}$ (cal/atom)/K, an approximation in good agreement with experimental results. As is shown in subsequent treatments, low-temperature behavior requires more complex explanations.

It should be noted that some elements with low atomic weights have heat capacities less than the predicted values. Among these are Li, 5.0; Be, 5.2; B, 4.6; C, 3.2; and Li, 5.0 (cal/mol K). Others, such as Na and K, have values of 6.7 and 7.1 cal/mol K, respectively.

Kopp's rule may be used to approximate the heat capacity of a solid as the sum of the heat capacities of its constituent atoms. The accuracy of this estimation improves with increasing temperature for temperatures higher than about 500°C.

5.1.1 Classical Theory

The rule of Dulong and Petit can be explained by means of classical physics (Section 1.1). The classical theory assumed that all the internal energy of a solid could be considered to reside in the ion cores of a solid. A solid was thought of as being made up of an assembly of noninteracting ion cores that behaved as simple harmonic oscillators, vibrating about an equilibrium position, in thermal equilibrium at a given temperature. This neglects any contribution of the valence electrons to the internal energy of the solid. The thermal equilibrium of the ion cores is treated, as in the case of ideal gases, as though the energy distribution is continuous and makes use of the Maxwell–Boltzmann statistics. The ions were considered to have three degrees of freedom that correspond to their energies of translation parallel to the three Cartesian coordinates. This approach, as shown in Section 1.1, gives the heat capacity as

$$C_{V,i} \cong 6 \text{ (cal/mol)/deg} \tag{5-8}$$

The result given by Eq. (5-8) is in agreement with the findings of Dulong and Petit. However, it fails to provide a temperature dependence for the heat capacity. This arises because the solid is treated as an assembly of independent oscillators. The very close proximities of the ions in the three-dimensional geometric arrays that constitute crystalline solids cause this assumption to be a great oversimplification.

The oscillations of a given ion may readily be visualized as affecting those of its neighbors. The neighboring ions' oscillations, in turn, then influence those of their neighbors', and so on. And, if the internal energy of a solid resides primarily in the ions, their amplitudes of oscillation must be expected to vary with temperature. Thus, any attempt to provide a realistic description of heat capacity must take such factors into consideration.

5.1.2 The Einstein Model

Einstein (1906) developed a model that overcame the failure of the classical approach to describe the heat capacity of a solid as a function of temperature. Instead of using the classical expression for energy, he used the then new concept of describing energy in terms of the frequency of the oscillating ion. Based on the Planck idea of discrete energies, the average energy of one such oscillator is found by obtaining the average of all the quantized energies of all the ions in the solid, as is shown by the derivation of Eq. (2-67). The energy of a single quantized oscillator, $E = hv$, is called a phonon (Section 5.1.7). The average energy of such an oscillator is

$$\langle E \rangle = \frac{hv}{e^{hv/k_B T} - 1} \tag{2-67}$$

This expression is quite different from that given by classical mechanics: $E = k_B T$. It has been noted that Eq. (2-67) equals the classical case at high temperatures [Eq. (2-68)].

To apply this average energy, Einstein based his ideas on the classical picture of a solid and extended these ideas. As in the classical case, it was assumed that the internal energy of a solid was associated only with the ions. The energy of the electrons was not taken into account. The solid thus was treated as an assembly of *independent* simple-harmonic oscillators in thermal equilibrium. That is, each ion of the solid independently oscillated with the same given frequency, v_0. As noted for the classical case, this is too great an oversimplification, because each oscillating ion at least would be expected to affect the oscillations of its nearest neighbors, so their frequencies could not possibly be all the same. The inclusion of the quantum mechanic approach, represented by Eq. (2-67), did constitute major progress.

If each ion of the solid has three translational degrees of freedom and N_A ions constitute a mol, then, by Eq. (2-67), the internal energy, U, of a mol of the solid is given by the sum of the average energies of all the ions in the solid:

$$U = \frac{3N_A hv_0}{\exp(hv_0/k_B T) - 1} \tag{5-9}$$

Neglecting zero point energy [Eq. (1-24)] and differentiating,

$$C_V = \left(\frac{\partial U}{\partial T}\right)_V = \frac{3N_A(hv_0)(\exp hv_0/k_B T)(hv_0/k_B T^2)}{(\exp(hv_0/k_B T) - 1)^2} \tag{5-10}$$

The numerator is multiplied and divided by k_B to obtain the Einstein equation for the molar heat capacity as

$$C_V = \frac{3N_A k_B \exp(h v_0/k_B T)(h v_0/k_B T)^2}{(\exp(h v_0/k_B T) - 1)^2} \tag{5-11}$$

Equation (5-11) is evaluated at elevated temperatures by expanding the exponential terms as the first two terms of a series, and noting that the squared factors vanish, to obtain

$$C_V \cong \frac{3N_A k_B(1 + h v_0/k_B T)(h v_0/k_B T)^2}{(1 + h v_0/k_B T - 1)^2} = 3N_A k_B \left(1 + \frac{h v_0}{k_B T}\right) \tag{5-12}$$

At elevated temperatures, C_V approaches the $3N_A k_B$ limit, the classical value, because the fraction $h v_0/k_B T$ is negligibly small. Thus, the Einstein model agrees with the findings of Dulong and Petit at high temperatures.

How does Einstein's model behave at lower temperatures? This may be determined by starting with Eq. (5-11) and expanding the denominator to find

$$C_V = \frac{3N_A k_B \exp(h v_0/k_B T)(h v_0/k_B T)^2}{\exp(2h v_0/k_B T) - 2\exp(h v_0/k_B T) + 1}$$

Then, by dividing both numerator and denominator by $\exp(h v_0/k_B T)$, an expression for the ready evaluation of C_V at low temperatures is obtained as

$$C_V = \frac{3N_A k_B(h v_0/k_B T)^2}{\exp(h v_0/k_B T) - 2 + \exp(-h v_0/k_B T)} \tag{5-13}$$

As the temperature decreases and approaches zero, the negative exponential term in the denominator will decrease rapidly and approach zero, while the positive one correspondingly will become very large. The denominator will become large at a much faster rate than the squared fraction in the numerator. This will cause C_V to diminish and to approach zero in an exponential fashion.

Thus, the Einstein model fits the observed conditions at the limits. Figure 5-2 shows how this model agrees with the experimental data over the range of intermediate temperatures. It is apparent that the Einstein relationship gives a poor

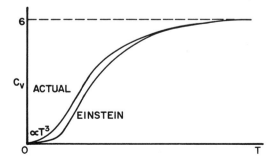

Figure 5-2 Comparison of the Einstein theory of heat capacity with experimental behavior.

representation of C_V in the lower range of temperatures. This arises because of its prediction of exponential rather than the experimentally observed T^3 behavior; the differences between the predicted and actual values decrease as the temperature increases. This degree of agreement is surprisingly good considering the simplicity of the model.

The major difficulty with this model is that it was assumed that all the ions were independent and oscillated with one given frequency, v_0. It is emphasized again that this condition cannot exist because each oscillating ion and all of its neighbors must mutually interact, and so on, throughout the solid. Thus, each ion of the lattice affects and is affected by every other ion of the lattice.

It should be noted, however, that the Einstein model is useful for approximating lattice vibrational effects. This usually is done at temperatures above the range in which the exponential effects are pronounced. Such approximations usually are made close to or above the Debye temperature [see Eqs. (5-101), (5-112), and (5-113), Section 5.1.5, and Table 5.1] to minimize any error and to simplify the calculations. An illustration of this usage is given later in the analysis of Matthiessen's rule.

This model continues to have other important applications. One of these is to calculate the heat capacities of gases when information is incomplete or unavailable. Ideal behavior is assumed and the internal energy is considered to consist of three components. Two of these, the translational and rotational modes, are specified by $\frac{1}{2}RT$ per mol for each degree of freedom. The vibrational modes are included by appropriate summations of Eq. (2-67) over all applicable modes. The heat capacity is determined by the derivative of the sum of the operative components with respect to temperature. Tables are available for evaluating the Einstein summations. This treatment gives reasonable and useful approximations at or above room temperature.

5.1.3 Elastic Vibrations in Solids

The thermal energy of elemental solids resides in the oscillations of their ions about their equilibrium positions. Since the oscillations of one particle ultimately affect the oscillations of all the other particles, this results in highly complicated motions for each particle. Such complex behavior may be approximated by considering these oscillations in totality rather than individually. The basis for this is that the interactions between particles are transferred very rapidly and that their resultant may be described by an elastic wave. The wave integrates the motions of all the particles and thus may be employed to represent the net effects of the particles involved in the wave. This approach has the additional advantage of providing a spectrum of lattice vibrations.

5.1.3.1 Normal vibrational modes.
Vibrational modes are most clearly described by picturing the ions on a one-dimensional lattice. This is the same as uniformly positioned identical particles on a string. Consider two such particles [Fig. 5-3(a)]. In the first case, the two vibrating particles are in synchronization; their wavelength is $6d$, where d is the interionic distance. The vibrations of the particles are

exactly 180° out of phase in the second case and their wavelength is $3d$. The third case shows the particles at rest. The fourth case again shows the particles 180° out of phase; this is redundant. Similar cases are shown in Fig. 5-3(b) for three particles. This behavior may be extended to include any number of oscillating particles.

Vibrations of this kind are called *normal modes* of vibration and are like standing waves because the ends are fixed. It is possible to have any number of particles. Any complex particle vibrations can be described by the addition, or superposition, of the independent vibrational motions of particles in normal modes. These normal modes are preserved independently of other modes of motion. This permits their addition.

It is seen that, in the plane of the paper, one particle on the string would have one normal mode, one wavelength, and one frequency. Two particles would have two normal modes, two wavelengths, and two frequencies. Three particles would have three normal modes, three wavelengths, and three frequencies. N particles would have N normal modes, N wavelengths, and N frequencies. The other two vibrational planes, perpendicular to the one described here, must be considered also. Thus, for N particles with oscillations polarized in three planes, $3N$ normal vibrational modes exist.

It also may be seen [Fig. 5-3] that the smallest possible wavelength is equal to two particle spacings:

$$\lambda_{min} = 2d \qquad (5\text{-}14)$$

This determines the maximum wave vector as

$$\bar{k}_{max} = \frac{2\pi}{\lambda_{min}} = \frac{2\pi}{2d} = \frac{\pi}{d} \qquad (5\text{-}15)$$

In turn, the maximum frequency of oscillation is found to be

$$\omega_{max} = 2\pi v_{max} = \frac{2\pi c_s}{\lambda_{min}} = \bar{k}_{max} \cdot c_s \qquad (5\text{-}16)$$

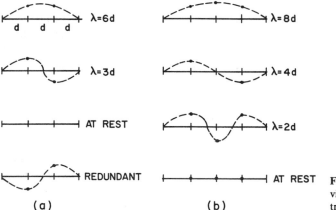

(a)   (b)   **Figure 5-3** Sketches of normal vibrational modes of vibrating particles treated as being on strings.

where c_s is the velocity of sound, or the velocity of the wave, in the one-dimensional lattice.

What happens when two species of particles of different masses are separated from each other at a constant distance, a, along the string? The oscillations of each of the species of particles are virtually the same over the spectrum of sound waves in the solid.   In this case, the vibrational modes of the string composed of two species of particles are exactly the same as those described for the string of identical particles; these are known as *acoustical modes* of vibration.   Particle responses of this kind are important in the understanding of such thermal properties as heat capacity, thermal expansion, and thermal conductivity.

A second case occurs when the waves that represent the oscillating ions have frequencies in the spectrum of electromagnetic radiation.   Here the oscillations of each type of particle are out of phase with each other. Their vibrational modes behave as though each species belonged to an independent, but interpenetrating string of particles.   Such particle behavior is called *optical modes* of vibration and is of importance in explaining the response of solids to such electromagnetic influences as radio waves and infrared and ultraviolet radiation.

The behavior of each vibrational mode differs widely from the other.   The acoustical mode has its minimum frequency at $\bar{k} = 0$ and its maximum frequency at $\bar{k} = \pi/d$. See Eqs. (5-51) and (5-52).   Here d is the distance between particles of the same species; using the interspecies spacing a, $d = 2a$. See Eq. (5-53). This is one reason for considering that each species belonged to an independent string.   In contrast, the optical mode has its minimum frequency at $\bar{k} = \pi/d = \pi/2a$ and its maximum frequency at $\bar{k} = 0$.   See Fig. 5-6.

The vibrational frequencies of the normal modes are dependent on the volume of the solid.   [This is the reason why $C_P > C_V$, Eq. (5-6).]   This relationship may be expressed as $v \propto V^{-\gamma}$, where V is the volume and γ is the Grüneisen constant.   This constant is independent of temperature and ranges between 1.3 and 3, depending on the solid, for metals and halide salts.   A value of $\gamma \cong 2$ frequently is used as a typical value.

Since the thermal properties of major interest here are those of elemental solids, the following discussions are limited to the vibrational properties of a single species of particles in the acoustical mode.

5.1.3.2 The vibrating string.   Now consider the vibrations in a *homogeneous*, one-dimensional line of identical particles of linear density, ρ.   Only longitudinal waves will be involved; that is, the motion of each particle in the string will be parallel to the line itself.   This behavior is shown in Fig. 5-4. Let x be the coordinate of a given line element, and its displacement from its equilibrium position be μ.   Then the strain, ε, or fractional change in length is $\partial \mu/\partial x$.   If F is the force producing the strain, the elastic modulus, or stiffness, is given by

$$c_x = \frac{F}{\varepsilon} \tag{5-17}$$

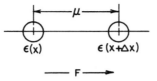

Figure 5-4 Displacement of a particle in a longitudinal wave along a string.

Now consider the forces on an element of the line, Δx. At one end of the element the strain is $\varepsilon(x)$ and is $\varepsilon(x + \Delta x)$ at the other end. The strain at $x + \Delta x$ is

$$\varepsilon(x + \Delta x) = \varepsilon(x) + \frac{\partial \varepsilon}{\partial x} \Delta x \tag{5-18}$$

The strain was given by

$$\varepsilon = \frac{\partial \mu}{\partial x} \tag{5-19}$$

From this, by differentiation,

$$\frac{\partial \varepsilon}{\partial x} = \frac{\partial^2 \mu}{\partial x^2} \tag{5-20}$$

This is substituted into Eq. (5-18) to give

$$\varepsilon(x + \Delta x) = \varepsilon(x) + \frac{\partial^2 \mu}{\partial x^2} \Delta x \tag{5-21}$$

The net strain on the element is

$$\varepsilon = \varepsilon(x + \Delta x) - \varepsilon(x) = \frac{\partial^2 \mu}{\partial x^2} \Delta x \tag{5-22}$$

From Eqs. (5-17) and (5-22), the force acting on the line element is

$$F = \varepsilon c_x = c_x \frac{\partial^2 \mu}{\partial x^2} \Delta x \tag{5-23}$$

The force on the line element also is given by

$$F = ma = \rho \, \Delta x \frac{\partial^2 \mu}{\partial t^2} \tag{5-24}$$

since its mass is a linear function of x and is given by $\rho \, \Delta x$. Equations (5-23) and (5-24) are equated, and, since the length increment vanishes, may be written as

$$\frac{c_x}{\rho} \frac{\partial^2 \mu}{\partial x^2} = \frac{\partial^2 \mu}{\partial t^2} \tag{5-25}$$

Let

$$\frac{c_x}{\rho} = A^2 \tag{5-26}$$

Then

$$A^2 \frac{\partial^2 \mu}{\partial x^2} = \frac{\partial^2 \mu}{\partial t^2} \tag{5-27}$$

Equation (5-27) is d'Alembert's equation (1747) for vibrating strings. It should be noted that

$$A = \left(\frac{c_x}{\rho}\right)^{1/2} = c_s \tag{5-28}$$

This is the Newtonian expression for the velocity of sound, c_s (longitudinal waves), in a homogeneous medium.

A solution to Eq. (5-27) is given by

$$\mu = \exp[i(At + x)] \tag{5-29}$$

Equation (5-29) can be shown to be a solution of Eq. (5-27) by differentiation and substitution.

The exponents of Eq. (5-29) will be modified for future use. By Eq. (5-28) and the substitutions shown sequentially in Eqs. (5-32),

$$A = c_s = \frac{\bar{\mathbf{k}}}{\bar{\mathbf{k}}} c_s = \frac{\bar{\mathbf{k}}}{\bar{\mathbf{k}}} \lambda v = \frac{\bar{\mathbf{k}}}{\bar{\mathbf{k}}} \frac{2\pi}{\bar{\mathbf{k}}} v = \frac{\bar{\mathbf{k}}\omega}{\bar{\mathbf{k}}^2} = \frac{\omega}{\bar{\mathbf{k}}} \tag{5-30}$$

and

$$x = \frac{\bar{\mathbf{k}}x}{\bar{\mathbf{k}}} \tag{5-31}$$

Thus, Eq. (5-29) can be expressed, using Eqs. (5-30) and (5-31), as

$$\mu = D \, \exp[i\bar{\mathbf{k}}(At + x)] = D \, \exp(i\bar{\mathbf{k}}At + i\bar{\mathbf{k}}x)$$

$$\mu = D \, \exp\left[i\bar{\mathbf{k}} \frac{\omega}{\bar{\mathbf{k}}} t + i\bar{\mathbf{k}} \frac{\bar{\mathbf{k}}}{\bar{\mathbf{k}}} x\right] = D \, \exp[i\omega t + i\bar{\mathbf{k}}x] \tag{5-32}$$

$$\mu = D \, \exp[i(\omega t + \bar{\mathbf{k}}x)]$$

where D is a constant. This will be the general form employed for the displacement induced by traveling longitudinal waves in the string. This expression will be applied later, as Eq. (5-39), to derive the properties of waves in a one-dimensional array of ions.

5.1.3.3 Nonhomogeneous strings.   Some of the properties of the wave motion on a line of identical particles will now be derived.   In this case, the string is no longer homogeneous.   Here the wave motion is similar to the elastic waves on a homogeneous string when the wavelength is much larger than the distance between the particles.   An examination also will be made of the properties of waves of shorter wavelength.

The kind of displacement of interest is indicated in Fig. 5-5.   Only nearest neighbor interactions will be considered.   The force acting on the nth particle is

$$F_n = \beta(\mu_{n+1} - \mu_n) - \beta(\mu_n - \mu_{n-1}) \tag{5-33}$$

in which β is the spring, or force, constant, μ_i are the displacements, and the quantities in the parentheses give the changes in the bond length between the designated particles.   On a macroscopic scale, the line of identical particles is treated as an independent string that has a linear mass whose density is given by

$$\rho = \frac{M}{d} \tag{5-34}$$

where the mass is given by M and d is the equilibrium distance between the particles. The force required to stretch a single bond between the particles is

$$F = \beta(\mu_n - \mu_{n-1}) = \beta\varepsilon d \tag{5-35}$$

where ε is the unit strain.   The elastic modulus for this nonhomogeneous string is

$$\frac{F}{\varepsilon} = \beta d \tag{5-36}$$

Equation (5-17) gave this modulus for the homogeneous string as

$$\frac{F}{\varepsilon} = c_x$$

Equating these two relationships gives

$$\beta d = c_x \tag{5-37}$$

This equation will be useful in simplifying Eq. (5-53) later.

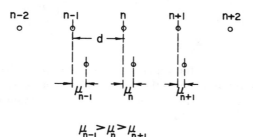

Figure 5-5 Equilibrium positions and displacements of particles in a longitudinal wave in a string.

Now, using Newton's second law, $ma = F$, equating Eqs. (5-24) and (5-33), and collecting the terms in Eq. (5-33),

$$M\frac{d^2\mu}{dt^2} = \beta(\mu_{n+1} + \mu_{n-1} - 2\mu_n) \tag{5-38}$$

It is helpful to find a solution to Eq. (5-38) similar to that obtained for traveling waves in the homogeneous string. However, since discrete particles are now involved, the quantity nd must be used instead of x in the exponent of Eq. (5-32), and Eq. (5-32) is written as

$$\mu_n = \exp[i(\omega t + \bar{\mathbf{k}}nd)] \tag{5-39}$$

It will be shown that a solution to Eq. (5-38) is given by

$$-\omega^2 M = \beta(e^{i\bar{\mathbf{k}}d} + e^{-i\bar{\mathbf{k}}d} - 2) \tag{5-40}$$

This relationship will be used to define the properties of the waves under consideration here.

Upon differentiation of Eq. (5-39),

$$\frac{d\mu}{dt} = i\omega \, \exp[i(\omega t + \bar{\mathbf{k}}nd)]$$

and

$$\frac{d^2\mu}{dt^2} = -\omega^2 \, \exp[i(\omega t + \bar{\mathbf{k}}nd)] \tag{5-41}$$

This expression is multiplied through by M to give an equation equivalent to Eq. (5-38). Equating these two relationships gives

$$M\frac{d^2\mu}{dt^2} = -M\omega^2 \, \exp[i(\omega t + \bar{\mathbf{k}}nd)] = \beta(\mu_{n+1} + \mu_{n-1} - 2\mu_n) \tag{5-42}$$

Now, (5-39) may be substituted for the quantities within the parentheses on the right side of Eq. (5-42) to obtain

$$(\mu_{n+1} + \mu_{n-1} - 2\mu_n) = e^{i\omega t}[e^{i\bar{\mathbf{k}}(n+1)d} + e^{i\bar{\mathbf{k}}(n-1)d} - 2e^{i\bar{\mathbf{k}}nd}] \tag{5-43}$$

Substituting Eq. (5-43) in Eq. (5-42) gives

$$-M\omega^2 e^{i\omega t} \cdot e^{i\bar{\mathbf{k}}nd} = e^{i\omega t}[e^{i\bar{\mathbf{k}}(n+1)d} + e^{i\bar{\mathbf{k}}(n-1)d} - 2e^{i\bar{\mathbf{k}}nd}]\beta \tag{5-44}$$

The factor $\exp(i\omega t)$ vanishes. And, when this expression is divided through by $\exp(i\bar{\mathbf{k}}nd)$,

$$-M\omega^2 = (e^{i\bar{\mathbf{k}}d} + e^{-i\bar{\mathbf{k}}d} - 2)\beta \tag{5-45}$$

This equation will be used to obtain relationships between ω and $\bar{\mathbf{k}}$ and, consequently, to describe the ways in which these factors affect the waves in the nonhomogeneous string.

The quantity within the parentheses of Eq. (5-45) can be simplified in the following way:

$$(e^{i\bar{k}d} + e^{-i\bar{k}d} - 2) = (e^{i\bar{k}d/2} - e^{-i\bar{k}d/2})^2 \tag{5-46}$$

Letting $x = \bar{k}d/2$ and using the exponential expression for the sine of x gives

$$\sin x = \frac{1}{2i}(e^{ix} - e^{-ix}) \tag{5-47}$$

and, when squared, this becomes

$$\sin^2 x = -\tfrac{1}{4}(e^{ix} - e^{-ix})^2$$

Or, when substituting for x,

$$4\sin^2 x = -(e^{i\bar{k}d/2} - e^{-i\bar{k}d/2})^2 = -(e^{i\bar{k}d} + e^{-i\bar{k}d} - 2) \tag{5-48}$$

This is now substituted back into Eq. (5-45) to obtain

$$M\omega^2 = 4\beta \sin^2 \frac{\bar{k}d}{2}$$

Or

$$\omega = \pm\left(\frac{4\beta}{M}\right)^{1/2} \sin \frac{\bar{k}d}{2} \tag{5-49}$$

This is the general expression for the angular frequency as a function of its wave vector. Equation (5-49) is now used to obtain maximum values for $\bar{k}$ and ω. Upon differentiation, this gives

$$\frac{d\omega}{d\bar{k}} = \left(\frac{4\beta}{M}\right)^{1/2} \frac{d}{2} \cos \frac{\bar{k}d}{2} = 0 \tag{5-50}$$

if a maximum exists. Thus, $\cos(\bar{k}d/2)$ must equal zero, since the other factors do not equal zero. This can only be the case for $\bar{k} = \pi/d$, or

$$\cos\left(\frac{\pi}{d} \cdot \frac{d}{2}\right) = \cos \frac{\pi}{2} = 0$$

So, for this condition, it follows that

$$\bar{k}_{max} = \pm\frac{\pi}{d} \tag{5-51}$$

By means of $\bar{k}_{max} = 2\pi/\lambda_{min}$, Eq. (5-51) gives $\lambda_{min} = 2d$, in agreement with Eq. (5-14). This, in turn, can be used to determine ω_{max} when $\bar{k}_{max}$ is used in Eq. (5-49):

$$\omega_{max} = \pm\left(\frac{4\beta}{M}\right)^{1/2} \sin\left(\frac{\pi}{d} \cdot \frac{d}{2}\right) = \pm\left(\frac{4\beta}{M}\right)^{1/2} \tag{5-52a}$$

For ω_{min}, Eq. (5-49) is equated to zero:

$$\omega_{min} = \pm\left(\frac{4\beta}{M}\right)^{1/2} \sin\left(\frac{\bar{k}d}{2}\right) = 0 \tag{5-52b}$$

Since the coefficient of Eq. (5-52b) does not equal zero, it follows that

$$\sin\frac{\bar{k}d}{2} = 0, \qquad \frac{\bar{k}d}{2} = 0, \pi, \ldots, \qquad \bar{k}_{min} = 0, \frac{2\pi}{d}, \ldots \tag{5-52c}$$

Equations (5-51) and (5-52) are the bases for the descriptions of the acoustical mode given in Section 5.1.3.1.

Equation (5-49) and the results given by Eqs. (5-52a) and (5-52b) permit a description of the behavior of ω as a function of $\bar{k}$, as shown in Fig. 5-6. Values of $\bar{k}$ larger in magnitude than $\pm(\pi/d)$ only repeat wave motion already described within that range because of the periodic behavior of Eq. (5-49). If a value of $\bar{k} > \pi/d$ is obtained, it is treated by subtracting an integral multiple of $\bar{k} = 2\pi/d$ from it to bring it back within the range $\bar{k}_{max} = \pm\pi/d$. This is analogous to the reduced zone given by Fig. 4-17. Thus, all values of $\bar{k}$ can be represented within this range.

These waves are waves in which all the particles are moving in synchronization. This behavior, as previously noted, is known as the acoustical mode of vibration, and the wave vector range between $\pm\pi/d$ is known as the first Brillouin zone. This is analogous to the linear Brillouin zone shown in Fig. 4-16(c). Representations of this kind are commonly used to picture other physical phenomena, as indicated here, and are based on the concept of the reciprocal lattice. The case developed here, and shown in Fig. 5-6, actually represents a linear reciprocal lattice, since the behavior of ω is given in terms of the wave vector that is a function of the reciprocal of the spacing of the ions on the vibrating string.

The foregoing discussion is based on strings of identical particles. What happens when strings composed of two or more species of particles constitute a string? The kinds of vibrations that result are most readily described by a diatomic string or linear lattice. Each species may be treated as an independent string of particles. The two types of particles are considered as being uniformly spaced on a single string at a distance a apart, and the spacing between like particles is $2a = d$. In a way similar to

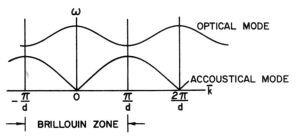

Figure 5-6 One-dimensional variation of the angular velocity of particles with wave vector. The acoustical mode is for a string of like particles. The optical mode is for strings containing two species of particles alternately positioned on a string, where d is the distance between two particles of the same species.

that given for identical particles, it can be shown that

$$\omega^2 = \beta\left(\frac{1}{m} + \frac{1}{M}\right) \pm \left[\left(\frac{1}{m} + \frac{1}{M}\right)^2 - \frac{4\,sin^2\,\bar{k}a}{mM}\right]^{1/2} \quad (5\text{-}53)$$

in which m and M are the masses of the particles. Thus, for a given combination of particles, the frequency is a double-valued function of the wave vector. In other words, the frequencies split into two branches; these are known as the *acoustical* and *optical branches*.

The acoustical mode of oscillations takes place in the range of sound frequencies in the string. In this case, both species of particles oscillate in synchronization just like a string of identical particles. However, when the particles are oppositely charged, the electric field of electromagnetic radiation can cause each species to oscillate independently like separate strings that are out of phase with each other. This is the optical mode of oscillation.

The following results are obtained from Eq. (5-53) for two species of particles. At $\bar{k} = \pm\pi/d$, for the case where $M > m$,

$$\omega_{\text{optical}} = \left(\frac{2\beta}{m}\right)^{1/2} \quad (5\text{-}53\text{a})$$

and

$$\omega_{\text{acoustical}} = \left(\frac{2\beta}{M}\right)^{1/2} \quad (5\text{-}53\text{b})$$

Also, at $\bar{k} = 0$,

$$\omega_{\text{optical}} = \left[2\beta\left(\frac{1}{m} + \frac{1}{M}\right)\right]^{1/2} \quad (5\text{-}53\text{c})$$

and

$$\omega_{\text{acoustical}} = 0 \quad (5\text{-}53\text{d})$$

These results have been treated as being superimposed on Fig. 5-6 to show the similarity between the acoustical mode of oscillation and that for identical particles. It will be understood that strings of identical particles will not manifest optical modes of vibrations. Note also will be made that a gap occurs in the frequency range at $\bar{k} = \pm\pi/d$. This separates the allowed frequencies in the sound range from those within the electromagnetic range. Frequencies within this forbidden range correspond to imaginary, or complex, values for $\bar{k}$. This means that waves with such frequencies are absorbed, or damped out, and cannot propagate.

Consideration will now be given to waves in strings of identical particles with small values of $\bar{k}$ (those with long wavelengths). For this case, where the sine is small, Eq. (5-49) becomes

$$\omega \cong \pm\left(\frac{4\beta}{M}\right)^{1/2} \frac{\bar{k}d}{2} \quad (5\text{-}54)$$

When the factor $d/2$ is brought under the square root sign,

$$\omega \cong \pm \left(\frac{4}{4} \beta d \frac{d}{M} \right)^{1/2} \bar{k} \tag{5-55}$$

It will be apparent that Eqs. (5-34) and (5-37) may now be used here to give

$$\omega \cong \left(c_x \cdot \frac{1}{\rho} \right)^{1/2} \bar{k} \tag{5-56}$$

Now, using Eq. (5-28), this becomes simply

$$\omega \cong c_s \bar{k} \tag{5-57}$$

Thus, where $\bar{k}$ is small, the wavelength is very much greater than the distance between the particles. Here the string of particles can be treated as a homogeneous line. This results from the fact that large numbers of particles are involved in the long wavelengths, and such waves are not affected by the discrete structure of the string. It will be shown later that the group and phase velocities are equal in this case.

As the frequency is increased, the wavelength shortens and each wave contains fewer numbers of particles. Now the shortest wavelength will be determined by equating Eq. (5-51) to the definition of $\bar{k}_{max}$:

$$\bar{k}_{max} = \frac{\pi}{d} = \frac{2\pi}{\lambda_{min}}$$

or

$$\lambda_{min} = 2d \tag{5-58}$$

in agreement with Eq. (5-14).

The picture of the composition of the short waves under these conditions is quite different from that of the long waves. In both cases the spacing between the oscillating particles remains the same, but now the wavelength is of the same order of magnitude as the particle spacing. Hence, each high-frequency wave is made up of relatively few oscillating particles. In this case the group and phase velocities are no longer equal. See Section 1.6.1.

First consider long waves, made up of many oscillating particles, by starting with Eq. (5-30) in the form

$$\omega = v_g \bar{k} \tag{5-59}$$

where v_g is the group velocity. Upon differentiation, this becomes

$$\frac{d\omega}{d\bar{k}} = v_g \tag{5-60}$$

Another expression for $d\omega/d\bar{k}$ can be obtained from Eqs. (5-54), (5-37), and (5-28):

$$\omega = \left(\frac{4\beta}{M} \right)^{1/2} \sin \frac{\bar{k}d}{2} = \left(\frac{4c_s/d}{\rho d} \right)^{1/2} \sin \frac{\bar{k}d}{2} = \frac{2}{d} v \sin \frac{\bar{k}d}{2} \tag{5-61}$$

This, upon differentiation, gives

$$\frac{d\omega}{d\bar{\mathbf{k}}} = \frac{2}{d} \cdot \frac{d}{2} \mathbf{v} \cos \frac{\bar{\mathbf{k}}d}{2} \tag{5-62}$$

Then, equating Eqs. (5-62) and (5-60),

$$\mathbf{v}_g = \mathbf{v} \cos \frac{\bar{\mathbf{k}}d}{2} \tag{5-63}$$

If this relationship is correct, the phase and group velocities should be equal at long wavelengths (small $\bar{\mathbf{k}}$). Under these conditions, Eq. (5-62) becomes, for $\bar{\mathbf{k}} \to 0$,

$$\mathbf{v}_g = \mathbf{v} \cos \frac{\bar{\mathbf{k}}d}{2} \cong \mathbf{v} \cos 0 = \mathbf{v} \tag{5-63a}$$

So Eq. (5-63a) verifies the statement made earlier: the long waves composed of many particles have identical phase and group velocities [Eq. (1-15)]. This permits their treatment as quantized wave packets.

For the case of the shortest possible wavelength, $\bar{\mathbf{k}}_{max} = \pi/d$ [Eq. (5-51)]. In this limiting case, Eq. (5-62) becomes

$$\mathbf{v}_g = \mathbf{v} \cos \frac{\pi}{d} \cdot \frac{d}{2} = 0$$

Here the group velocity does not exist and $\mathbf{v}_g \neq \mathbf{v}$, so "wave packets" cannot be present.

If c_s is the velocity of sound along the string of particles, then, by Eqs. (5-57) and (5-61), for $\bar{\mathbf{k}}_{max}$,

$$\mathbf{v} = \frac{\omega}{\bar{\mathbf{k}}} = \frac{1}{\bar{\mathbf{k}}} \cdot \frac{2}{d} c_s \sin \frac{\bar{\mathbf{k}}d}{2} = \frac{d}{\pi} \cdot \frac{2}{d} c_s \sin \frac{\pi}{d} \cdot \frac{d}{2}$$

Then, since $\sin \pi/2 = 1$,

$$\mathbf{v} = \frac{2c_s}{\pi} \sin \frac{\pi}{2} = \frac{2c_s}{\pi}$$

Reintroducing this along with Eq. (5-51) into Eq. (5-57),

$$\omega_{max} = \mathbf{v}\bar{\mathbf{k}}_{max} = \frac{2c_s}{\pi} \cdot \frac{\pi}{d} = \frac{2c_s}{d} \tag{5-64}$$

This gives ω_{max} in terms of the velocity of sound along the string of particles; as such, it represents the upper limit of frequencies.

5.1.4 The Born–von Kármán Theory

The Born–von Kármán (1912) model represents an important advance in the attempt to overcome the major failing of the Einstein approach; it provides a means for accounting for particle–particle interactions. Like the other models, it assumed that the internal energy of a solid resided in the ions and no account was made of the contributions of the electrons. In this case, however, the solid is treated as an assembly of *coupled*, simple-harmonic oscillators. This concept represented an important advance. And, as in Einstein's model, Eq. (2-67) was used for the average energy of the ion. But the big difference here is that each ion was considered to oscillate at its own frequency, v_i. Considering that each oscillating ion has three degrees of freedom and that the solid consists of N ions, the internal energy of the solid is determined by the sum of the energies of all the oscillating ions:

$$U = \sum_{i=1}^{3N} \frac{hv_i}{\exp(hv_i/k_BT) - 1} + \text{zero point energy} \tag{5-65}$$

The zero point energy (Sections 1.6.6 and 5.1.8) may be neglected because it is a constant for a given elemental solid, and it will vanish in the calculation of the heat capacity [Eq. (5-1)].

If ω, the angular frequency, is used instead of v in Eq. (2-67), where $\omega = 2\pi v$, then the average energy of an oscillating ion is given by

$$U = \frac{\hbar\omega}{\exp(\hbar\omega/k_BT) - 1} \tag{5-66}$$

where $\hbar = h/2\pi$. If the number of vibrational modes of the ions were known, the product of this quantity and the average energy of each of the ions would give the average internal energy of the solid. Assume, then, that there are $N(\omega)\,d\omega$ vibrational modes in the frequency range between ω and $\omega + d\omega$. This is known as the *density of vibrational modes*. Now, if the number of oscillating ions is sufficiently large, it is possible to integrate over the range of angular frequencies instead of summing as in Eq. (5-65). This gives the internal energy of the solid as

$$U = \int_{\omega_{\min}}^{\omega_{\max}} \frac{\hbar\omega}{\exp(\hbar\omega/k_BT) - 1} \, N(\omega)\,d\omega \tag{5-67}$$

This has simplified the problem, but it is now necessary to define the distribution function $N(\omega)$. To do this, it will be necessary to examine the oscillations of the coupled particles comprising the Born–von Kármán model of a solid.

The behavior of the strings of ions previously derived may now be applied. Specifically, it is necessary to obtain an expression for $N(\omega)\,d\omega$ of Eq. (5-67). This can be obtained as a function of $\bar{k}$ by starting with its reexpression, by means of the chain rule, as

$$N(\omega)\,d\omega = \frac{dn}{d\bar{k}} \cdot \frac{d\bar{k}}{d\omega} \cdot d\omega \tag{5-68}$$

This is the product of three factors: the number of modes per range of $\bar{k}$, the variation of $\bar{k}$ with ω, and the range of ω.

The first of these factors may be obtained with the help of Fig. 5-7. The relationship between the wavelength and the fixed length L of the string is, in which n

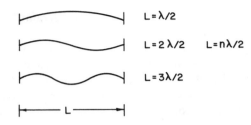

Figure 5-7 Waves in a homogeneous string of length L.

is the number of half-wavelengths,

$$\lambda = \frac{2L}{n} \tag{5-69}$$

Equation (5-69) may be rewritten by expressing the wavelength as a function of the wave vector to get

$$\frac{2L}{n} = \frac{2\pi}{\bar{\mathbf{k}}} \tag{5-69a}$$

This, in turn, results in

$$\bar{\mathbf{k}} = \frac{n\pi}{L} \tag{5-70}$$

Then, upon differentiation, the first factor required for Eq. (5-68) is obtained as

$$\frac{d\bar{\mathbf{k}}}{dn} = \frac{\pi}{L} \quad \text{or} \quad \frac{dn}{d\bar{\mathbf{k}}} = \frac{L}{\pi} \tag{5-70a}$$

The second factor is obtained by starting with Eqs. (5-49) and substituting (5-23) in it:

$$\omega = \left(\frac{4\beta}{M}\right)^{1/2} \sin \frac{\bar{\mathbf{k}}d}{2} = \omega_{\max} \sin \frac{\bar{\mathbf{k}}d}{2}$$

This is reexpressed in the form of

$$\frac{\omega}{\omega_{\max}} = \sin \frac{\bar{\mathbf{k}}d}{2}, \qquad \frac{\bar{\mathbf{k}}d}{2} = \sin^{-1} \frac{\omega}{\omega_{\max}}, \qquad \bar{\mathbf{k}} = \frac{2}{d} \sin^{-1} \frac{\omega}{\omega_{\max}}$$

Then, upon differentiation, the second term to be used in Eq. (5-68) is found as

$$\frac{d\bar{\mathbf{k}}}{d\omega} = \frac{2}{d(\omega_{\max}^2 - \omega^2)^{1/2}} \tag{5-71}$$

Equation (5-68) may now be rewritten by substituting Eqs. (5-70) and (5-71) into it to give the density of vibrational states as

$$N(\omega)\,d\omega = \frac{L}{\pi} \cdot \frac{2}{d(\omega_{\max}^2 - \omega^2)^{1/2}}\,d\omega \tag{5-72}$$

This may now be used in Eq. (5-67) to obtain the Born–von Kármán expression for the internal energy of a solid as

$$U = \frac{2L\hbar}{\pi d} \int_0^{\omega_{max}} \frac{\omega \, d\omega}{[\exp(\hbar\omega/k_B T) - 1][\omega_{max}^2 - \omega^2]^{1/2}} \tag{5-73}$$

where the upper limit is given by Eq. (5-64). The internal energy of a solid obtained in this way is the sum of the energies of parallel linear arrays of coupled ions acting as standing waves. The differention of this equation with respect to temperature will give an expression for the heat capacity of a solid so constituted.

It is very difficult to determine the heat capacity of real solids using Eq. (5-73) because of uncertainties involved with the function for ω. It will be recalled [Eq. (5-49)] that the force constants (β_i) for the various directions in a solid must be known to determine ω_{max}. This information rarely is available.

A less intricate expression for approximating the internal energy of a solid than that given by Eq. (5-73) can be obtained by means of a less sophisticated approximation for $N(\omega)$. This can be treated more readily than that given by Eq. (5-72). Starting with Eq. (5-69), for one set of normal modes, and recalling that $\lambda v = c_s$,

$$\lambda = \frac{2L}{n} = \frac{c_s}{v}$$

Treating c_s as a constant, this is rearranged to give

$$n = \frac{2L}{c_s} v \quad \text{and} \quad dn = \frac{2L}{c_s} dv$$

The density of states is obtained as a function of ω by the use of $dv = d\omega/2\pi$. This results in

$$dn = dN(\omega) = \frac{2L}{c_s} \cdot \frac{d\omega}{2\pi} = \frac{L}{\pi c_s} d\omega \tag{5-74}$$

The substitution of this function into Eq. (5-67) gives, for one set of normal modes,

$$U_1 \cong \frac{L\hbar}{\pi c_s} \int_0^{\omega'_{max}} \frac{\omega \, d\omega}{\exp(\hbar\omega/k_B T) - 1} \tag{5-75}$$

When all the normal modes are taken into account, the internal energy of a solid based on the Born–von Kármán model may be approximated by

$$U \cong \frac{3L\hbar}{\pi c_s} \int_0^{\omega'_{max}} \frac{\omega \, d\omega}{\exp(\hbar\omega/k_B T) - 1} \tag{5-76}$$

The upper limit of this integral is obtained by integrating Eq. (5-74) and rearranging it to obtain

$$\omega'_{max} = \frac{N\pi c_s}{L} \tag{5-77}$$

It will be noted that this upper limit is different from that given by Eq. (5-64), the corresponding upper limit for Eq. (5-73).

Equation (5-76) is readily treated for high temperatures. Here the first two terms of the series are used for the exponential term in the denominator. The simplified expression is integrated and gives $U \cong 3Nk_BT$; this gives $C_v \cong 3R$, in agreement with the findings of Einstein and Dulong and Petit.

Equation (5-76) also is used for the low-temperature approximation of internal energy. It may be expressed as

$$U \cong \frac{3L}{\pi c_s} \int_0^{\omega'_{max}} \frac{\hbar\omega \, d\omega}{\exp(\hbar\omega/k_BT) - 1} = \frac{3Lk_BT}{\pi c_s} \int_0^{\omega'_{max}} \frac{(\hbar\omega/k_BT) \, d\omega}{\exp(\hbar\omega/k_BT) - 1} \qquad (5\text{-}78)$$

This integral is found, from the tables, to express the internal energy as

$$U \cong \frac{3Lk_BT}{\pi c_s} \ln \frac{\exp(\hbar\omega/k_BT)}{\exp(\hbar\omega/k_BT) - 1} \Big|_0^{\omega'_{max}} \qquad (5\text{-}79)$$

The derivative of this function with respect to temperature gives the heat capacity.

The expression for C_V approximated in this way gives C_V as being a linear function of temperature; C_V approaches zero as T approaches zero. The curve for C_V derived from Eq. (5-73) shows the same behavior, but has a slightly smaller slope than that derived from Eq. (5-78).

Thus, the heat capacity approximated from Eqs. (5-76) and (5-78) approaches the proper limits for both 0 K and for relatively high temperatures. It fails in that it does not correspond to the experimentally observed behavior that C_V varies as T^3 in the very low range of temperatures. The same comments also apply to C_V derived from Eq. (5-73), the Born–von Kármán equation. These failures result from the fact that the interactions of the vibrating particles comprising solids are much more complex than those capable of description by the oscillations of strings of particles.

5.1.5 The Debye Theory

The Debye model (1912) overcomes most of the difficulties inherent in the prior attempts to describe the heat capacity of solids. Here the solid is treated as an isotropic, homogeneous medium instead of an assembly of oscillating particles or strings of particles. Again, the electron energy is neglected. This solid, thus, is considered as being a homogeneous, monatomic substance. The oscillating ions constitute standing waves that conform to Eq. (1-15) and may be quantized, or treated as phonons (Sections 5.1.2 and 5.1.7). Discrete energy states, which are the equivalent of quantized standing waves, can be propagated within this solid (Section 5.1.7). This model can be considered to treat the solid as though it was pulsating with such standing waves. In this respect, such an approach gives a simpler, yet more realistic, representation than the Born–von Kármán model. Here an integration is made over all the quantized waves in the solid treated as normal vibrational modes, rather than as a summation over strings of ions, to obtain the internal energy of the solid. The limitation of $3N$ normal modes takes the discrete crystalline structure into account because it considers each of the ions as participating in the quantized waves.

It will be recalled from prior discussions that it is possible to have any number of normal modes of vibration, depending on the number of vibrating particles. When complex vibrations exist, they can be represented by the addition of normal modes. In addition, as previously noted, when the phase and group velocities are equal [Eqs. (1-15) and (5-64)], the waves may be quantized and treated as discrete energy states.

The internal energy, U, of the solid is dependent on the average energy of each quantized standing wave [Eq. (2-67)] and the total number of such waves between v and $v + dv$, known as the *density of frequency states*, or $N(v)\,dv$. Thus, the total internal energy of a solid, based on this model, is

$$U = \int_0^{v_{\max}} \frac{hv}{\exp(hv/k_B T) - 1} \, N(v)\,dv \tag{5-80}$$

Here the upper limit is taken as $v_{\max}$, but the redundancy that occurs beyond $3N$ normal vibrational modes must be taken into account. The lower limit is taken as being zero. In actuality, it is greater than zero; but for low energies λ is large, v is small, and only negligible errors are introduced. This will be discussed later.

The density of states must be derived. The total number of standing waves must be equal to $3N$. So it follows that

$$\int_0^{v_{\max}} N(v)\,dv = 3N \tag{5-81}$$

The task now is to count the number of waves in the solid to determine $N(v)\,dv$. This is accomplished by starting as follows. Consider standing waves in a very small cube of the solid, the length of whose side is a, as in Fig. 5-8. From this figure it is seen that

$$\frac{\lambda}{2} = x \cos \theta_x$$

Since all three dimensions of the solid must be considered, similar expressions must be obtained for the y and z directions. In each of these expressions, $\cos \theta$ is the direction cosine. Since standing waves are being considered, all the waves must be such that they consist of an integral number of half-wavelengths, regardless of the direction of

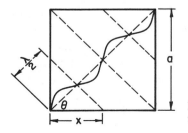

Figure 5-8 Schematic diagram of a wave in a Debye solid.

their propagation within the solid.   If, for convenience, l_i designates the direction cosines, and the n_i are integers, then, summarized in tabular form,

$$a = \frac{\lambda}{2l_x} n_x, \qquad l_x = \frac{n_x \lambda}{2a}, \qquad l_x^2 = \frac{n_x^2 \lambda^2}{4a^2}$$

$$a = \frac{\lambda}{2l_y} n_y \qquad l_y = \frac{n_y l}{2a} \qquad l_y^2 = \frac{n_y^2 \lambda^2}{4a^2}$$

$$a = \frac{\lambda}{2l_z} n_z \qquad l_z = \frac{n_z \lambda}{2a} \qquad l_z^2 = \frac{n_z^2 \lambda^2}{4a^2}$$

Recalling that the l_i are direction cosines and that the sum of their squares equals unity, the addition of the squared terms gives

$$\frac{\lambda^2}{4a^2} (n_x^2 + n_y^2 + n_z^2) = 1$$

or

$$\frac{4a^2}{\lambda^2} = n_x^2 + n_y^2 + n_z^2 \tag{5-82}$$

Thus, the volume of the solid and the waves have been quantized by Eq. (5-82).   Each phonon is described by its coordinates in the volume of the solid.

The distance from the origin to any point in the volume of the solid is obtained from Eq. (5-82) as

$$r = (n_x^2 + n_y^2 + n_z^2)^{1/2} = \frac{2a}{\lambda} \tag{5-83}$$

All points corresponding to the nodes of the standing waves in the solid must be on the surfaces of spheres whose radii are given by Eq. (5-83).   The velocity of sound is given by

$$c_s = \lambda v \tag{5-84}$$

When this is substituted into Eq. (5-83) for λ,

$$r = \frac{2av}{c_s} \tag{5-85}$$

Now consider all the vibrational modes with frequencies between v and $v + dv$; $N(v)\,dv$ of such modes includes this range of frequencies by definition.   These modes must be contained within the volume formed between the surfaces of two spheres whose radii are $2av/c_s$ and $2a(v + dv)/c_s$, respectively.   Thus, in the first octant (the other octants represent redundant modes),

$$dV = \frac{4\pi r^2 \, dr}{8} \tag{5-86}$$

Equation (5-85) and its derivative with respect to frequency are substituted into Eq. (5-86) to obtain

$$dV = \frac{\pi}{2} r^2 \, dr = \frac{\pi}{2} \left(\frac{2av}{c_s}\right)^2 \cdot \left(\frac{2a}{c_s} \, dv\right) \tag{5-87}$$

The number of modes in dV is just $N(v) \, dv$, so Eq. (5-87) becomes

$$N(v) \, dv = \frac{4\pi a^3}{c_s^3} v^2 \, dv \tag{5-88}$$

or

$$N(v) \, dv = \frac{4\pi V}{c_s^3} v^2 \, dv \tag{5-88a}$$

since $V = a^3$. Equation (5-88) is the expression for the density of vibrational states, or the number of quantized waves per unit infinitesimal volume, for the Debye model.

Now consider the types of waves that are present within the solid, with the aid of Fig. 5-9. The oscillation directions of the ions are given by the double-headed arrows. The oscillations parallel to both the direction of propagation and to the x axis constitute longitudinal waves. The oscillations of ions parallel to the y and z axes are perpendicular to the direction of propagation and constitute transverse waves. Thus, for the longitudinal waves, using Eq. (5-88a),

$$N(v)_{L_x} = \frac{4\pi V v^2}{c_{L_x}^3} \tag{5-88b}$$

and for the transverse waves

$$N(v)_{T_y} = \frac{4\pi V v^2}{c_{T_y}^3} \tag{5-88c}$$

and

$$N(v)_{T_z} = \frac{4\pi V v^2}{c_{T_z}^3} \tag{5-88d}$$

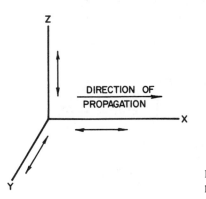

Figure 5-9 Directions of oscillations of particles in a Debye solid.

in which c_{L_x}, c_{T_y} and c_{T_z} are the velocities of sound in the longitudinal and two transverse directions, respectively.

Since the solid originally was assumed to be a homogeneous, isotropic medium, it may also be assumed that

$$c_{T_y} \cong c_{T_z} \cong c_T \tag{5-89}$$

In other words, the velocities of the transverse waves are assumed to be independent of their direction within the lattice. This gives, when included in Eq. (5-88a),

$$N(v) = 4\pi V v^2 \left(\frac{1}{c_L^3} + \frac{2}{c_T^3} \right) \tag{5-90}$$

Now, if the further approximation is made, on the same basis as Eq. (5-89), that the longitudinal and transverse waves have the same velocities, then the velocity of sound in any of these directions is given by

$$c_L \cong c_T \cong c_s \tag{5-91}$$

This approximation simplifies Eq. (5-90):

$$N(v) = 4\pi V v^2 \left(\frac{3}{c_s^3} \right) \tag{5-92}$$

It will be shown later that the first of these assumptions as given by Eq. (5-89) is reasonable, but the second, Eq. (5-91), is poor and limits the accuracy of the model. However, when Eq. (5-92) is used, one obtains

$$\int_0^{v_{max}} N(v)\,dv = \frac{12\pi V}{c_s^3} \int_0^{v_{max}} v^2\,dv = \frac{4\pi V v_{max}^3}{c_s^3} \tag{5-93}$$

Equation (5-93) could present difficulties. In a continuum, v could vary between zero and infinity. However, v_{max} is determined by λ_{min}; it cannot correspond to a wavelength less than twice the interionic distance [Eqs. (5-14) and (5-58)]. This determines the maximum value for v in the absence of any constraints. But the upper limit can only reach some cutoff frequency because no more than $3N$ normal modes of vibration are possible. Thus, the upper limit must correspond to the $3N$ modes and no more. Hence, $v_{max} = v_c$, some cutoff frequency. Therefore,

$$\int_0^{v_c} N(v)\,dv = 3N \tag{5-94}$$

Thus, using Eq. (5-93) and integrating,

$$\frac{12\pi V}{c_s^3} \int_0^{v_c} v^2\,dv = \frac{4\pi V v_c^3}{c_s^3} = 3N \tag{5-95}$$

From this the cutoff frequency is found to be

$$v_c = \left(\frac{3N}{4\pi V} \right)^{1/3} \cdot c_s \tag{5-96}$$

and the minimum wavelength

$$\lambda_{\min} = \frac{c_s}{v_c} = \left(\frac{4\pi V}{3N}\right)^{1/3} \tag{5-97}$$

It will be noted that the fraction V/N in Eq. (5-97) is the volume occupied by one of the oscillating ions, so $\lambda_{\min}$ cannot equal zero.   Thus, both $\lambda_{\min}$ and v_c are determined by the number of normal modes.

The velocity of sound and the volume can be eliminated as factors in these expressions.   Starting with Eq. (5-97), after rearranging, cubing, and inverting, it becomes

$$\frac{1}{c_s^3} = \frac{1}{v_c^3} \cdot \frac{3N}{4\pi V} \tag{5-98}$$

This may be substituted into Eq. (5-92) to get

$$N(v) = 12\pi V v^2 \cdot \frac{1}{v_c^3} \cdot \frac{3N}{4\pi V}$$

or, upon simplification,

$$N(v) = \frac{9Nv^2}{v_c^3} \tag{5-99}$$

This is an approximation of the Debye function for the density of states.   When this is applied to Eq. (5-80), the Debye expression for the internal energy of a solid is obtained as follows:

$$U = \int_0^{v_c} \frac{hv}{\exp(hv/k_B T) - 1} \, N(v)\, dv \tag{5-80}$$

This becomes the internal energy of a mol of a solid when N_A is Avogadro's number:

$$U = \frac{9N_A}{v_c^3} \int_0^{v_c} \frac{hv^3 \, dv}{\exp(hv/k_B T) - 1} \tag{5-100}$$

Before using this equation to determine the heat capacity, Eq. (5-100) will be changed to a more tractable form.   A temperature, the *Debye temperature*, θ_D, is defined by equating the classical and quantum expressions for energy:

$$k_B \theta_D = hv_c \tag{5-101a}$$

or

$$\theta_D = \frac{hv_c}{k_B} \tag{5-101b}$$

where k_B is Boltzmann's constant.   This effectively defines θ_D as the maximum temperature at which the quantum mechanics need be used. The following relationship is derived from Eq. (5-101) for later use:

$$\frac{1}{\theta_D} = \frac{k_B}{h v_c}, \qquad \frac{1}{\theta_D^3} = \frac{k_B^3}{h^3 v_c^3} \tag{5-101c}$$

and

$$\frac{1}{v_c^3} = \frac{h^3}{\theta_D^3 k_B^3} \tag{5-101d}$$

Now let the exponent in Eq. (5-100) be given by

$$x = \frac{h v}{k_B T} \tag{5-102}$$

Then

$$v = \frac{x k_B T}{h}, \qquad v^3 = \frac{x^3 k_B^3 T}{h^3} \tag{5-102a}$$

Differentiating Eq. (5-102) gives

$$dx = \frac{h}{k_B T} \, dv \quad \text{or} \quad dv = \frac{k_B T}{h} \, dx \tag{5-103}$$

By means of these relations, Eq. (5-100) is reexpressed as

$$U = 9 N_A \left(\frac{h^3}{\theta_D^3 k_B^3} \right) \int \frac{h[(x^3 k_B^3 T^3)/h^3][(k_B T)/h]}{e^x - 1} \, dx \tag{5-104}$$

or, upon simplification,

$$U = 9 N_A k_B \frac{T^4}{\theta_D^3} \int \frac{x^3 \, dx}{e^x - 1} \tag{5-105}$$

The limits of this integral must be determined. Using Eqs. (5-102) and (5-101), the upper limit is obtained as

$$x = \frac{h v}{k_B T}, \qquad \theta_D = \frac{h v_c}{k_B}, \qquad x_c = \frac{\theta_D}{T} \tag{5-106}$$

As noted previously, the lower limit is approximated by setting it equal to zero. But it must be remembered that it cannot be less than that frequency corresponding to a wavelength equal to the maximum wavelength. Such frequencies correspond to very low energies, so Eq. (5-105) may be written simply as

$$U = 9R \frac{T^4}{\theta_D^3} \int_0^{x_c = \theta_D/T} \frac{x^3}{e^x - 1} \, dx \tag{5-107}$$

recalling that $N_A k_B = R$. The heat capacity of the solid is obtained from the differentiation of Eq. (5-107):

$$C_V = \frac{\partial U}{\partial T} = 9R\left\{\frac{4T^3}{\theta_D^3}\int_0^{\theta_D/T}\frac{x^3\,dx}{e^x-1} + \frac{T^4}{\theta_D^3}\frac{d}{dT}\left[\int_0^{\theta_D/T}\frac{x^3\,dx}{e^x-1}\right]\right\} \qquad (5\text{-}108)$$

Equation (5-108) is used and the indicated differentiation is performed.

$$\frac{T^4}{\theta_D^3}\frac{d}{dT}\left[\int_0^{\theta_D/T}\frac{x^3\,dx}{e^x-1}\right] + \frac{T^4}{\theta_D^3}\frac{d}{dT}\left[\int_0^{\theta_D/T}\frac{(\theta_D/T)^3\cdot(-\theta_D/T^2)\,dT}{e^{\theta_D/T}-1}\right]$$

$$\frac{T^4}{\theta_D^3}\frac{d}{dT}\int_0^{\theta_D/T}\frac{x^3\,dx}{e^x-1} + \frac{T^4}{\theta_D^3}\cdot\frac{(\theta_D^4/T^5)(-1)}{e^{\theta_D/T}-1} = \frac{\theta_D}{T}\cdot\frac{-1}{e^{\theta_D/T}-1} \qquad (5\text{-}109)$$

Thus, Eq. (5-108) is reduced to

$$C_V = 9R\left[\frac{4T^3}{\theta_D^3}\int_0^{\theta_D/T}\frac{x^3\,dx}{e^x-1} - \frac{\theta_D}{T}\cdot\frac{1}{e^{\theta_D/T}-1}\right] \qquad (5\text{-}110)$$

This is a general expression for the heat capacity of a solid given by the Debye model.

If this expression is correct, it should give the classical value of $3R$ at high temperatures. So, for large T, the fraction under the integral sign in Eq. (5-110) becomes, when the first two terms of a series approximation are used for e^x,

$$\frac{x^3}{e^x-1} \cong \frac{x^3}{1+x-1} = x^2$$

And the second term within the brackets becomes

$$\frac{\theta_D}{T}\cdot\frac{1}{e^{\theta_D/T}-1} \cong \frac{\theta_D}{T}\cdot\frac{1}{1+\theta_D/T-1} = \frac{\theta_D}{T}\cdot\frac{T}{\theta_D} = 1$$

The approximate expression for high temperatures is reduced to

$$C_V \cong 9R\left(\frac{4T^3}{\theta_D^3}\int_0^x x^2\,dx - 1\right)$$

When integrated and θ_D/T is substituted for x,

$$C_V \cong 9R\left(\frac{4T^3}{\theta_D^3}\cdot\frac{x^3}{3} - 1\right) = 9R\left(\frac{4T^3}{\theta_D^3}\cdot\frac{\theta_D^3}{3T^3} - 1\right)$$

$$C_V \cong 9R\left(\frac{4}{3} - 1\right) = 3R$$

Thus, the Debye model gives correct results for high temperatures.

Equation (5-110) must be evaluated to determine its behavior at low temperatures. Consider the second term in the brackets of Eq. (5-110) as T approaches zero:

$$\lim_{T\to 0}\left[\frac{-\theta_D}{T}\cdot\frac{1}{e^{\theta_D/T}-1}\right] = 0$$

This is so because $\theta_D/T\to\infty$ as $T\to 0$, but $\exp(-\theta_D/T)\to 0$ as $T\to 0$, so the second fraction approaches 0 as $T \to 0$ and their product approaches 0. The value of the integral in Eq. (5-110) is $\pi^4/15$. So, at low temperatures,

$$C_V = 9R\left(\frac{4T^3}{\theta_D^3}\frac{\pi^4}{15}\right) = \frac{12}{5}\pi^4\left(\frac{T}{\theta_D}\right)^3 R \qquad (5\text{-}111)$$

This is known as the T^3 law and predicts that the heat capacity of a solid should approach zero as T^3 approaches zero. This agrees with the observed experimental behavior in the temperature range of about $\theta_D/20$ for many solid materials. It also removes the difficulty experienced by the Einstein model, where C_V approached zero much too quickly. This treatment is applicable only to nonconducting materials since electron effects have been excluded.

The Debye function, Eq. (5-110), may be approximated for various temperature ranges by a function of the form

$$C_V \cong \beta\left(\frac{T}{\theta_D}\right)^\alpha \qquad (5\text{-}110\text{a})$$

Here β is a function approximating Eq. (5-110) and α usually is an integer. Equation (5-110a) is helpful in the approximation of the influence of the temperature variation of C_V on other physical properties in which it plays a role.

The Debye temperature is an important bench mark because it defines that temperature above which the vibrational energy is large enough to give average amplitudes or ionic displacements ($\bar{x}$) such that, where K is the "spring constant",

$$\tfrac{1}{2}K\bar{x}^2 = \text{mean potential energy} = \tfrac{1}{2}k_B T \qquad (5\text{-}112)$$

so classical mechanics can be employed. Below this temperature, quantum mechanics *must* be used. This results from the way in which θ_D is defined [Eq. (5-101)]. It also explains the partial success of the classical approach [Eq. (5-8)]. The data of Dulong and Petit were taken at temperatures at or near θ_D, thus accounting for their results.

Further insight into θ_D may be obtained from Eq. (5-101). Noting that $v_c = c_s/\lambda_c$ and substituting this into the expression for θ_D gives

$$\theta_D = \frac{hc_s}{k_B\lambda_c} \qquad (5\text{-}113)$$

It is apparent that θ_D depends on the velocity of sound in the solid. This is not a constant, but is dependent on such factors as the temperature, the lattice spacing, or its density, and the way in which the phonons, or quanta of lattice vibrational energy, are transmitted and/or scattered by the lattice. In addition, the simplifying assumptions and approximations must also be considered. Thus, the Debye temperature is not exactly a constant; it also varies slightly depending on the approach used in its experimental determination, as well as the range of temperatures in which it is determined. Equation (5-111) provides one way to determine θ_D; it also may be calculated from measurements of electrical resistivity. Typical variations in the Debye temperature are shown in Fig. 5-10. The treatment given here is applicable only to nonconductors and to the ion cores of a metallic lattice, since the electron effects are

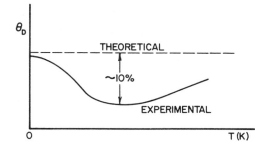

Figure 5.10 Schematic representation of the variation of the Debye temperature of a solid as a function of temperature; $T \ll \theta_D$.

not included. θ_D actually represents C_V at 96% of its limit when all factors are considered.

The theory treats θ_D as being constant with temperatures, as shown in the figure as a broken line.   The experimentally determined values for this parameter for copper show a variation of about 10%.   Other substances show different ranges of variations, but most appear to have the same general type of variations of θ_D with T as are shown in the figure.

The magnitude of θ_D appears to influence the way in which some metals behave mechanically.   For example, both gold and lead have low values for θ_D and have low recrystallization temperatures.   Both of these are extremely malleable and re-crystallize during working at room temperature.   This behavior can be explained by

TABLE 5-1   DEBYE TEMPERATURES AND HEAT CAPACITIES OF SOME ELEMENTS

Element	$\theta_D(K)$	C_p	Element	$\theta_D(K)$	C_p	Element	$\theta_D(K)$	C_p
Li	400	4.95	Mn	400	6.71	Sn (gray)	260	—
Be	1000	5.21	Fe	420	5.30	Sn (white)	170	6.30
B	1250	4.62	Co	385	5.93	Sb	200	6.03
C (diamond)	1860	3.18	Ni	375	6.16	La	132	6.65
Ne	63	4.97	Cu	315	5.86	Pr	74	6.45
Na	150	6.71	Zn	234	6.07	Gd	152	7.03
Mg	318	5.88	Ga	240	6.24	Ta	225	6.43
Al	394	5.82	Ge	360	6.22	W	310	5.97
Si	625	5.91	As	285	5.89	Pt	230	6.19
Ar	85	4.97	Zr	250	6.92	Au	170	6.03
K	100	7.12	Mo	380	5.67	Hg	100	6.50
Ca	230	6.28	Pd	275	6.21	Tl	96	6.29
Ti	400	6.00	Ag	215	5.70	Pb	88	6.12
V	390	5.67	Cd	120	6.19	Bi	120	6.10
Cr	460	5.24	In	129	6.50	Th	100	6.29

θ_D: calculated to agree with experimental data in the temperature range at which C_V is approximately half of the value of Dulong and Petit.

C_p: cal/g-atom/K at 298 K.

1 cal = 4.187 J.

θ_D after J. de Launay, *Solid State Physics*, vol. 2, p. 233, F. Seitz and D. Turnbull, eds., Academic Press, New York, 1956. C_p after K. K. Kelley, Bureau of Mines Bulletin 584, U.S. Government Printing Office, Washington, D.C. 1960.

Eq. (5-112) on the basis that their ionic displacements in the neighborhood of room temperature are comparable to those of other metals at elevated temperatures; hence their relative ease of working and low recrystallization temperatures. In other words, working gold and lead in the neighborhood of room temperature constitutes hot work.

Einstein (1911) noted a relationship between the characteristic temperature and mechanical behavior, in this case the compressibility. This relationship is

$$\theta_E = \frac{13.25 \times 10^{-4}}{A^{1/3} \rho^{1/6} \chi^{1/2}}$$

in which A is the atomic weight, ρ is the density, and χ is the compressibility. This permits θ_E to be related to other mechanical properties. Another empirical formula by Lindemann may also be used to approximate the Debye temperature:

$$\theta_D \cong C \left[\frac{T_M}{A \cdot V^{2/3}} \right]^{1/2}$$

where C is a constant (ranging from about 115 to 140), T_M is the melting point in kelvins, and V is the atomic volume. Some experimental values for θ_D are given in Table 5-1.

5.1.6 Comparison of Models

The same fundamental relationship was employed in each theory of heat capacity to obtain a basic function for the internal energy of a solid. This may be expressed as

$$U = \int \text{average energy of an oscillator} \times \text{density of states}$$

Each theory employed essentially the same expression, Eq. (2-67), for the average energy of an oscillator. Thus, the theories may be compared by examining the difference in the ways the densities of states were counted. The Einstein model used just one frequency, v_0, for all the independently oscillating ions of the solid. The solid is considered as being composed of coupled oscillators in the Born–von Kármán treatment. The original Debye approach assumed that the oscillating ions could be treated as constituting quantized, standing waves in the solid. The effects of these hypotheses on the resultant densities of states of these models are shown graphically in Fig. 5-11. The density of states for the Einstein model is just the most probable distribution of frequencies about v_0. Those for the Born–von Kármán and Debye theories are shown, respectively, as schematic plots of Eqs. (5-72) and (5-99). The singularity in Eq. (5-72) at $\omega_{max} = 2\pi v_{max} = 2\pi v_c$ should be noted.

The Debye theory is of greatest interest here because it provides the best description of the heat capacity of elemental solids. The curve of the density of states for the original Debye model is derived from the assumption that v_c, or the cutoff frequency, is the same for both longitudinal and transverse waves. The densities of

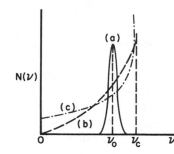

Figure 5-11   Densities of states for vibrational states for (a) the Einstein model, (b) the original Debye model, and (c) the Born–von Kármán model. Note the singularity in the Born–von Kármán model at $\omega = 2\pi v_c$.

states for the longitudinal and transverse waves, however, are different [Eqs. (5-88b, c, d)], because for most cases $c_T < c_L$. If these differences are taken into account as in Fig. 5-12a, then the use of Eqs. (5-88) can attempt to explain the observed behavior somewhat more realistically; but the assumption of a common velocity of sound leading to the same cutoff frequency for all vibrational modes is incorrect.

The effect of the Born modification to overcome this difficulty may be shown by starting with the general form of Eqs. (5-88), where c now is the velocity of sound in the longitudinal or in either of the two transverse modes, expressed as

$$N(v) = \text{constant } \frac{v^2}{c^3} \tag{5-114}$$

This essentially is the same as treating Eqs. (5-88c) and (5-88d) as being identical. A minimum wavelength is assumed to be common to each of these vibrational modes in a way analogous to Eq. (5-58). This results in a different cutoff frequency for each mode, $v_{max} = c(3N/4\pi V)^{1/3}$, using Eq. (5-97), since $c_T < c_L$. The net effect of the Born approach is to provide different expressions for $N(v)_T$ and $N(v)_L$ [Eq. (5-114) and Fig. 5-12b]. It is apparent that the original Debye approximation for $N(v)$, based on $c_T = c_L$, would be reasonably good only when the resultant density of states lies between the two distributions shown in the figure and when the cutoff frequency falls between those of each type of wave. The original Debye expression for $N(v)$ provides a good approximation only at very low temperatures because it closely approximates the densities of states proposed by Born.

The bases for the Born approach may be summarized as follows. It will be recalled that $N(v)$ was determined from the way in which the standing waves fit into

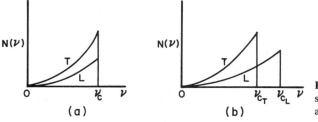

Figure 5-12   Densities of vibrational states (a) for the original Debye model and (b) as modified by Born.

the solid cube. In addition, it previously was noted in this section and in Eq. (5-58) that λ cannot be less than two interionic spacings. Thus, one such minimum wavelength sets the maximum frequency limit for each mode. The *actual* minimum wavelength and its corresponding cutoff frequency for each type of vibration are determined by $3N$ normal modes, respectively. These factors, along with the respective velocities of sound, constitute the bases for the Born modification.

If the Born modification [Fig. 5-12(b)] is used, where the cutoff frequencies for transverse and longitudinal waves are taken into account, the model, while still oversimplified, provides a closer representation of observed behavior than does the unmodified Debye model. Better approximations may be obtained when the two transverse modes, Eqs. (5-88c) and (5-88d), are treated independently and with different values of c_T.

It must also be recalled that the Debye model assumed an ideal solid. Impurities and imperfections exist in even the best single-crystal specimens. In polycrystalline materials, the deviation from the ideal is compounded by grain boundaries that add further complications. Furthermore, if the solid is a metal, the role of the valence, or conduction, electrons has been neglected here. See Section 5.1.9. This factor must be considered, especially at low temperatures. Nonetheless, the Debye approximation provides the best model for the behavior of the heat capacity of lattices at low temperatures.

5.1.7 Ions Treated as Harmonic Oscillators

Each ion within a solid is relatively restricted in its motion. These ions may be considered to oscillate in a simple harmonic mode about a mean, or equilibrium, position. This approximation is best at very low temperatures. The problem of determining the probability of finding such a particle will be approached both from the classical and quantum mechanical methods. This will provide an additional application of Schrödinger's equation and will provide another basis for understanding lattice vibrations that will be of use in explaining the thermal properties of solids.

First, consider the classical case as sketched in Fig. 5-13. The particle of mass m is acted on by a potential $V(x) = \frac{1}{2}Kx^2$ and a restoring force $F = -Kx$.

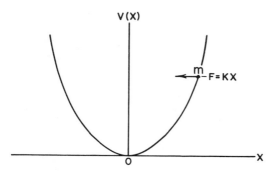

Figure 5-13 Factors acting on a simple harmonic oscillator.

The classical solution starts with Newton's second law,

$$F = ma = m \frac{d^2x}{dt^2} \qquad (5\text{-}115)$$

and equating the two expressions for F gives

$$-Kx = m \frac{d^2x}{dt^2} \qquad (5\text{-}116a)$$

or

$$\frac{d^2x}{dt^2} + \frac{Kx}{m} = 0 \qquad (5\text{-}116b)$$

The equation for the position, x, of the oscillating particle is given by

$$x = x_m \cos(2\pi vt) \qquad (5\text{-}117)$$

where x_m is its maximum displacement and v is its frequency.   The first and second derivatives of Eq. (5-117) are

$$dx = x_m(-2\pi v) \sin(2\pi vt)\, dt \qquad (5\text{-}118a)$$

and

$$d^2x = x_m(-4\pi^2 v^2) \cos(2\pi vt)\, dt^2 \qquad (5\text{-}118b)$$

At $t = 0$, $x = x_m$ and, since $\cos 0 = 1$,

$$\frac{d^2x}{dt^2} = -4\pi^2 v^2 x_m \qquad (5\text{-}119)$$

Equation (5-119) can be used with Eq. (5-116a) to determine the spring constant, K,

$$\frac{d^2x}{dt^2} = -\frac{Kx}{m} = -4\pi^2 v^2 x \qquad (5\text{-}120)$$

and it is found that

$$K = 4\pi^2 v^2 m \qquad (5\text{-}121)$$

The potential energy may be reexpressed by the substitution of Eq. (5-121) into $V(x)$. Thus,

$$V(x) = \tfrac{1}{2}Kx_m^2 = \tfrac{1}{2}(4\pi^2 v^2 m)x_m^2 \qquad (5\text{-}122)$$

The solution of Eq. (5-122) for x_m gives the general classical expression

$$x_m = \pm \frac{1}{\pi v}\left(\frac{V(x)}{2m}\right)^{1/2} \qquad (5\text{-}123)$$

This is the classical limit for the position of the oscillating particle.   Based on the classical theory, there is zero probability of finding the particle beyond $\pm x_m$.   This now will be explained.

The classical probability, P, of finding the particle in the region dx (Fig. 5-14) is proportional to the time spent by the particle in that region, or

$$P \, dx = \beta \, dt \qquad (5\text{-}124a)$$

where β is a constant of proportionality. Or

$$P = \beta \frac{dt}{dx} = \beta \frac{1}{\mathbf{v}} \qquad (5\text{-}124b)$$

where $\mathbf{v}$ is the velocity of the particle. However, as x approaches x_m, $\mathbf{v}$ slowly approaches zero. Since $\mathbf{v}$ approaches zero slowly, P approaches infinity slowly; this means that the area under the curve is finite. Since P becomes very large as x approaches $\pm x_m$, there is zero probability of finding the particle beyond $\pm x_m$. This verifies the prior statement that, when treated classically, the particle always will be found within the potential well.

Now consider the quantum mechanic approach to this problem. Starting with Eq. (3-34) and the expression for $V(x)$ [Eq. (5-122)], the Schrödinger equation for the oscillating particle is

$$\frac{\partial^2 \Psi(x)}{\partial x^2} + \frac{8\pi^2 m}{h^2} [E - 2\pi^2 v^2 m x^2] \Psi(x) = 0 \qquad (5\text{-}125)$$

Now consider the wave function

$$\Psi_0(x) = \exp(-\alpha x^2) \qquad (5\text{-}126)$$

as a trial solution to Eq. (5-125), where $\alpha = 2\pi^2 m v / h$. Its derivatives are

$$\frac{\partial \Psi_0(x)}{\partial x} = -2\alpha x \exp(-\alpha x^2)$$

and

$$\frac{\partial^2 \Psi_0(x)}{\partial x^2} = -2\alpha \exp(-\alpha x^2) + 4\alpha^2 x^2 \exp(-\alpha x^2) \qquad (5\text{-}127)$$

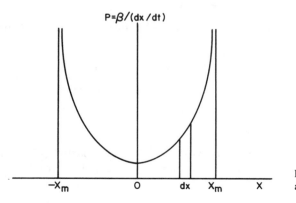

Figure 5-14 Quantized energy levels of a simple harmonic oscillator.

or, when Eq. (5-126) is substituted into Eq. (5-127),

$$\frac{\partial^2 \Psi_0(x)}{\partial x^2} = -2\alpha\Psi_0(x) + 4\alpha^2 x^2 \Psi_0(x) = -(2\alpha - 4\alpha^2 x^2)\Psi_0(x) \qquad (5\text{-}128)$$

Now, substituting for α,

$$\frac{\partial^2 \Psi_0(x)}{\partial x^2} = -\left(2 \frac{2\pi^2 mv}{h} - 4 \frac{4\pi^4 m^2 v^2}{h^2} x^2\right) \Psi_0(x) \qquad (5\text{-}129)$$

$$\frac{\partial^2 \Psi_0(x)}{\partial x^2} = -\frac{4\pi^2 m}{h^2} [hv - 4\pi^2 mv^2 x^2]\Psi_0(x) \qquad (5\text{-}130)$$

$$\frac{\partial^2 \Psi_0(x)}{\partial x^2} = -\frac{8\pi^2 m}{h^2} \left(\frac{hv}{2} - 2\pi^2 mv^2 x^2\right) \Psi_0(x) \qquad (5\text{-}131)$$

Or

$$\frac{\partial^2 \Psi_0(x)}{\partial x^2} + \frac{8\pi^2 m}{h^2} \left(\frac{hv}{2} - 2\pi^2 mv^2 x^2\right) \Psi_0(x) = 0 \qquad (5\text{-}132)$$

Equation (5-132) is identical to Eq. (5-125), and $\Psi_0(x)$ is a solution to Eq. (5-125) if

$$E = E_0 = \frac{hv}{2} \qquad (5\text{-}133)$$

Equations (5-125) and (5-132) are now the same and are identical to the Schrödinger time-independent equation. Thus, E_0 is an eigenvalue and $\Psi_0(x)$ is an eigenfunction of Schrödinger's equation.

In a similar way, using the wave function

$$\Psi_1(x) = x \exp(-\alpha x^2) \qquad (5\text{-}134)$$

it can be shown that

$$E_1 = \tfrac{3}{2}hv \qquad (5\text{-}135)$$

A third eigenfunction,

$$\Psi_2(x) = (1 - 4\alpha x^2) \exp(-\alpha x^2) \qquad (5\text{-}136)$$

will give an eigenvalue

$$E_2 = \tfrac{5}{2}hv \qquad (5\text{-}137)$$

The selection of the eigenfunctions, Eqs. (5-126), (5-134), and (5-136), is not arbitrary but may be accomplished by direct methods. These methods are beyond the scope of this text. The eigenvalues found thus far may be summarized as follows:

$$E_0 = \tfrac{1}{2}hv = (0 + \tfrac{1}{2})hv$$

$$E_1 = \tfrac{3}{2}hv = (1 + \tfrac{1}{2})hv$$

$$E_2 = \tfrac{5}{2}hv = (2 + \tfrac{1}{2})hv$$

or, generalizing,

$$E_n = (n + \tfrac{1}{2})hv \qquad (5\text{-}138)$$

Here n may take on the value of zero to give the lowest energy of the oscillating particle. This condition, $n = 0$, will be discussed as the zero-point energy in Section 5.1.8. It will be noted that the oscillating particle can take on only those energy values that differ from each other by exactly hv. This is shown in Fig. 5-15. Thus, changes in the energy of oscillation can occur only by discrete amounts; $|\Delta E| = hv$. Each such quantum of energy is called a phonon (Sections 5.1.2 and 5.1.5).

As discussed in Section 5.1.3.1, each oscillating particle is responsible for a normal vibrational mode in the lattice of a crystalline solid. Thus, the findings summarized by Eq. (5-138) can be regarded as a way of considering that the energy of a vibrating particle of a given mass and frequency is the equivalent of that of a normal vibrational mode of that oscillator, provided that its mass and frequency are unchanged. This leads to the treatment of the quantized vibrational modes, such as that given in Section 5.1.5 for the Debye theory, also as being considered for convenience, as a "gas" of normal modes, or as a "phonon gas," or of quanta in the solid. The properties of such "gases," then, may be used to describe the thermal properties of the solid.

The eigenfunctions of Eq. (5-125) are of interest. These are shown in Fig. 5-16. First, consider the extent of these functions in contrast to the behavior indicated in Fig. 5-14. Note that the tails extend considerably beyond $\pm x_m$. This condition could not exist in the classical treatment [Eq. (5-124b)]. It means that a small but real probability exists of finding the particle anywhere outside the well between $\pm x_m$ and $\pm$ infinity. As n becomes considerably larger, the tails shrink; the probability of finding the particle outside the well diminishes correspondingly. In other words, it becomes increasingly probable that the particle will be in the well.

Ψ_0 represents the wave function corresponding to the lowest energy state, the zero-point energy or ground state, E_0. One maximum is present in $\Psi_0{}^*\Psi_0$. The eigenvalue E_1 is the first excited state; the probability density $\Psi_1{}^*\Psi_1$ has two maxima. The second excited state is given by the eigenvalue E_2 obtained from Ψ_2, and $\Psi_2{}^*\Psi_2$

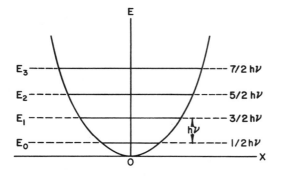

Figure 5-15 Probable position of a harmonic oscillator obtained by classical means. The probability scale is compressed.

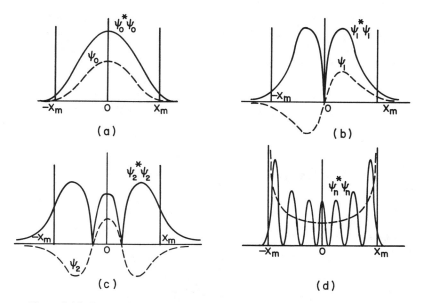

Figure 5-16 Normalized wave functions and probability densities for a harmonic oscillator as a function of the quantum number, n. The probability-density scale is compressed.

has three maxima. The probability density $\Psi_n{}^*\Psi_n$ has $n + 1$ maxima. Thus, as n becomes sufficiently large, the number of maxima becomes large and $\Psi_n{}^*\Psi_n$ approaches the classical case as the limit, in agreement with the correspondence principle. This is shown schematically in Fig. 5-16(d), where the broken curve indicates the average probability.

It is important to emphasize that discrete energy states for oscillating particles or normal modes no longer need to be assumed, such as was done in order to explain black-body radiation. The discrete eigenvalues derive naturally from the solutions of Eq. (5-125).

5.1.8 Zero-Point Energy

Classical physicists considered that the ions in a solid were motionless, or "frozen in," at 0 K. It now is known that this is not so. Solutions to Schrödinger's equation [Eq. (5-133)] can demonstrate that the ionic vibrations in a solid at the absolute zero are considerable. This is in agreement with the conclusions derived from Eqs. (1-22) and (1-24). Since the ions are not at rest at 0 K, their energies must be taken into account. This may be approximated in the following way.

It was shown that the energy of an oscillator is given by

$$E_n = (n + \tfrac{1}{2})h\nu \qquad\qquad (5\text{-}138)$$

This expression must be used because the classical equation for the energy of an oscillating ion is invalid at temperatures below θ_D [Section 5.1.5 and Eq. (5-101a)]. The ground state, or lowest energy state, of an oscillating ion, with a frequency v, is given for the condition $n = 0$ as

$$E_0 = \tfrac{1}{2}hv \tag{5-139}$$

The zero-point energy of a solid is then obtained, by summing over all the ions, as

$$U_0 = \int_0^{v_c} \tfrac{1}{2}hvN(v)\,dv \tag{5-140}$$

Using the original Debye approximation [Eq. (5-99)] for $N(v)$, Eq. (5-140) becomes

$$U_0 \cong \int_0^{v_c} \frac{1}{2}hv \cdot \frac{9N_A}{v_c^3}\, v^2\, dv$$

or, upon integration,

$$U_0 \cong \frac{9}{2}\frac{N_A h}{v_c^3}\int_0^{v_c} v^3\, dv = \frac{9}{8}N_A h v_c \tag{5-141}$$

Thus, the zero-point energy of a solid is a function of the nonzero cutoff frequency of ions of which it is composed.

Equation (5-141) may be reexpressed by use of Eq. (5-101a), $hv_c = k_B\theta_D$. Thus, the least possible internal energy of the solid is approximately given as

$$U_0 \cong \tfrac{9}{8}N_A k_B \theta_D = \tfrac{9}{8}R\theta_D \tag{5-142}$$

This energy is appreciable and is very close to the energy of an ideal gas ($\tfrac{3}{2}RT$) in the neighborhood of room temperature, since the θ_D of many elements is about 400 K.

On the basis of classical mechanics, it would have been expected that the lattice of the solid would be perfect and that all the ions would have been at rest on their lattice sites at the absolute zero. That this is not the case is apparent from Eqs. (5-141) and (5-142), the zero-point energy. Thus, while the internal energy of a solid is least at 0 K, it is quite high by classical standards. *There is no state in which the internal energy of a solid can be zero.*

Since energy is a relative quantity, the lowest possible energy of a solid is taken to represent that state in which both all the ions and all the electrons of that solid attain a unique lowest energy state at zero kelvins. For certain thermodynamic purposes, such a condition may be *assigned* as being the zero of energy. This is not at variance with classical thermodynamics. When the zero-point energy is used as a reference for energy measurements, energy differences, or changes, in the energy actually are being measured; thus, the zero-point energy cancels out.

In a corresponding way, the entropy of a substance may be arbitrarily designated as being equal to zero for this set of conditions at zero kelvins. In other words, the zero of entropy may be considered to be constituted by that state in which all the ions and electrons are in an optimum, but still imperfect, configurational order that corresponds to the lowest energy state.

5.1.9 Electron Heat Capacity

The nearly free electrons in metals contribute to the internal energy of solids as described in Sections 2.7, 3.12.4, 4.5 and 4.5.1. This energy component must be taken into account along with that of the ion cores as described by Eq. (5-107).

As discussed in Section 4.4, only a small fraction of the available electrons can absorb energy. This is shown schematically in Fig. 5-17 for a normal metal. That fraction, $f(N)$, of the total number of electrons, N, that can absorb energy is thermally promoted from states in area 1 to states in area 2 in the figure. The excited electrons will have energies up to about $2k_BT$ above E_F. These, as shown in Section 2.7, constitute approximately 1%, or less, of all the available electrons. This causes the "hypotenuse" of area 2 to be very steep. (Areas 1 and 2 in Fig. 5-17 are greatly enlarged for illustrative purposes.) Approximations now are made, in terms of the energy ranges involved, that the average fraction of excited electrons is

$$\langle f(N) \rangle \cong \frac{3}{4} \cdot \frac{2k_BT}{E_F} = \frac{3}{2} \cdot \frac{k_BT}{E_F} \tag{5-143}$$

Then the average number of thermally promoted electrons is given by

$$\langle N(T) \rangle = N\langle f(N) \rangle \cong \frac{3Nk_BT}{2E_F} = n(E_F) \tag{5-144}$$

The notation $n(E_F)$ frequently is used to designate the number of electrons per unit volume with energies at or near E_F. Such electrons are considered to behave as "nearly free" electrons (Section 2.7). Each such electron, therefore, has an average energy of close to $\frac{3}{2}k_BT$ above E_F. Thus, the total average increase in the thermal energy of the electrons is given by the product of $\langle N(T) \rangle$ and the energy of one excited electron:

$$U(T)_e \cong \langle N(T) \rangle \cdot \frac{3}{2}k_BT \cong \frac{3Nk_BT}{2E_F} \cdot \frac{3}{2}k_BT = \frac{9}{4}\frac{Nk_B^2T^2}{E_F} \tag{5-145}$$

Since Eq. (5-145) provides an approximation for the internal energy contribution made by the electrons, it can be used to determine their increment of the heat capacity.

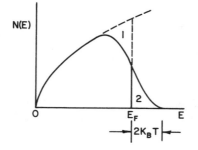

Figure 5-17 Density of states used to approximate the electron heat capacity of normal metals, $T \gg 0$ K. Energy range E to E_F greatly expanded.

Thus, by differentiating Eq. (5-145) with respect to T,

$$C_{V,e} = \frac{\partial U(T)_e}{\partial T} \cong \frac{9}{2} \frac{N k_B^2 T}{E_F} = A\gamma T \qquad (5\text{-}146)$$

A more accurate calculation by Seitz gives the electron component of the heat capacity as

$$C_{V,e} \cong \frac{\pi^2}{2} \frac{N k_B^2 T}{E_F} = A\gamma T, \qquad \gamma = \frac{\pi^2}{2} \cdot \frac{N k_B^2}{E_F} \qquad (5\text{-}147)$$

The coefficient $\frac{9}{2}$ of Eq. (5-146) is very close to $\pi^2/2$ of Eq. (5-147). Both of these equations contain approximations. The effects of some of these are given in Section 5.1.10.2. It is for this reason that γ is considered to be the same constant of proportionality in both Eqs. (5-146) and (5-147) for a given metal. The parameter A, a constant for each normal, elemental metal, compensates for these approximations. Also included in A are the variations in the electron heat capacity that occur from metal to metal. This parameter has been called the *conduction coefficient*.

The preceding discussion for normal metals is based on the availability of electrons in a partially filled valence band. The outer electron configurations of transition elements are considerably different from those of normal metals; they are characterized by incomplete, inner d levels that overlap partially filled, outer s levels in the solid state, as noted in Sections 3.11 and 4.5 and shown in Figs. 4-14 and 4-15. The unfilled states in the overlapping bands, known as holes, can act as positively charged particles and engage in conduction processes.

This behavior is taken into account by considering the extent of the unfilled states in the overlapping bands in contrast to the approach used for normal metals. This may be rationalized by considering that, if the d bands of transition elements were filled, they would be normal metals, not transition elements. Thus, their properties may be considered in terms of the extent to which their d bands remain unfilled. In effect, this provides an equally useful model, in terms of the density of states of the positive holes, rather than in terms of the electrons. This places the zero of the density of states at the top of the d band, at E_0 in Fig. 5-18, rather than at zero. This can be considered to be a mirror image of the density of states for the d electrons.

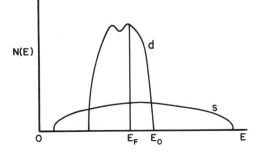

Figure 5-18 Density of states for a transition element showing the hybridized states resulting from s–d overlap.

Both methods describe the same condition: The density of states of electrons describes the extent to which the container is filled; that of holes describes the extent to which the same container remains unfilled.

Now, based on the foregoing model and Fig. 5-18, the electron contribution to the internal energy of a transition element is given, in terms of the corresponding number of holes, as

$$U(T)_h \cong \langle N(T) \rangle \cdot \frac{3}{2} k_B T \cong \frac{N k_B T}{E_0 - E_F} \cdot \frac{3}{2} k_B T = \frac{3}{2} \frac{N k_B^2 T^2}{E_0 - E_F} \tag{5-148}$$

where $E_0 - E_F$ is the energy range of the d level holes. Thus, the heat capacity contribution of the electrons is given approximately as

$$C_{V,h} \cong \frac{3}{2} \frac{N k_B^2 T}{E_0 - E_F} = A' \gamma' T \tag{5-149}$$

or more closely by

$$C_{V,h} \cong \frac{\pi^2}{6} \frac{N k_B^2 T}{E_0 - E_F} = A' \gamma' T, \qquad \gamma' = \frac{\pi^2}{6} \frac{N k_B^2}{E_0 - E_F} \tag{5-150}$$

As was noted for Eqs. (5-146) and (5-147), the coefficients of Eqs. (5-149) and (5-150) are very similar, and A' and γ' serve the same purposes for transition elements as A and γ do for normal metals.

5.1.10 Heat Capacity of Metals

The total internal energy of a metal is the sum of the energies residing both in its ion cores and in its valence electrons. This may be expressed as

$$U = U_e + U_i \tag{5-151}$$

The energy contained in the electrons is, where $U(0)_e$ is the zero-point energy,

$$U_e = U(0)_e + U(T)_e \tag{5-152}$$

The energy associated with the ion cores is given correspondingly by

$$U_i = U(0)_i + U(T)_i \tag{5-153}$$

Now, Eq. (5-151) may be written, using Eqs. (5-152) and (5-153), as

$$U = U(0)_e + U(T)_e + U(0)_i + U(T)_i \tag{5-154}$$

The zero-point energies are constants and vanish upon differentiation with respect to T, so the heat capacity of a metal is given by

$$C_V = \frac{\partial U}{\partial T} = \frac{\partial U(T)_e}{\partial T} + \frac{\partial U(T)_i}{\partial T} = C_{V,e} + C_{V,i} \tag{5-154a}$$

Equation (5-154a) is the general expression for the heat capacity of a metal.

5.1.10.1 Normal metals. The heat capacity of a normal metal is obtained from Eq. (5-154a) by means of Eq. (5-147) for $N = N_A$, Avogadro's number, as

$$C_V = C_{V,e} + C_{V,i} \cong \frac{\pi^2}{2} \frac{N_A k_B^2 T}{E_F} + C_{V,i} \qquad (5\text{-}155)$$

This may be reexpressed as

$$C_V = C_{V,e} + C_{V,i} \cong \frac{\pi^2}{2} N_A k_B \cdot \frac{k_B T}{E_F} + C_{V,i} \qquad (5\text{-}156)$$

For temperatures equal to or higher than θ_D, $C_{V,i} = 3R = 3N_A k_B$ (Section 1.1), so Eq. (5-156) may be written as

$$C_V = C_{V,e} + C_{V,i} \cong \frac{\pi^2}{2} N_A k_B \cdot \frac{k_B T}{E_F} + 3N_A k_B \qquad (5\text{-}157)$$

Upon dividing Eq. (5-157) through by the ionic component,

$$\frac{C_V}{C_{V,i}} \cong \frac{C_{V,e}}{C_{V,i}} + 1 = \frac{\pi^2}{6} \frac{k_B T}{E_F} + 1 \qquad (5\text{-}158)$$

Equation (5-158) provides a means for the approximation of the relative contribution of electrons to the heat capacity of a normal metal. Thus, if $k_B T$ is approximated as being about 0.03 eV near room temperature and E_F as being about 6 eV, then the component of the heat capacity arising from the electrons is about 0.8% of the total. The net result is that electrons impart a small, positive slope to C_V and that $C_V \cong 3R$ at elevated temperatures, in agreement with the observed properties.

The results provided by Eq. (5-158) also indicate the reason for the failure of the Drude–Lorentz model; it included *all* the valence electrons. In the analysis presented here, $f(N)$ represents only about 1%, or less, of all the valence electrons. This removes the Drude–Lorentz dilemma.

At *very low temperatures*, of the order of $T/\theta_D < 0.02$, $C_{V,e} \gg C_{V,i}$. Thus, while the electron component is small, it can constitute the major portion of C_V. Thus, $C_{V,e}$ can be of importance in some cryogenic applications. This behavior may be expressed by means of Eq. (5-154), using Eqs. (5-147) and (5-111), where the latter is written as $C_{V,i} = BT^3$, to obtain

$$C_V = A\gamma T + BT^3 \qquad (5\text{-}159)$$

This equation enables the use of heat capacity measurements, at these very low temperatures, to provide considerable insight into the specific electron behavior of an element by means of the conduction coefficients A or A' of Eqs. (5-147) and (5-150), respectively. This is accomplished by plotting C_V/T versus T^2 as in Fig. 5-19. The intercept of such a plot at 0 K is $A\gamma$ or $A'\gamma'$, and the slope is given by B. It is significant that the conduction coefficients may differ significantly from unity. The smaller portions of such deviations arise from the simplifying assumptions in the derivations of Eqs. (5-147) and (5-150). The major component results from those differences in

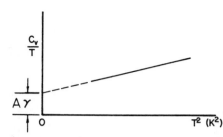

Figure 5-19 Electron heat capacity for normal metals at very low temperature ($T \ll \theta_D$).

electron behavior that are to be expected when comparing the properties of different metals. The transition metals may show much larger values for their conduction coefficients than do the normal metals. Both of these are discussed in the next section.

5.1.10.2 Transition elements.

In a way similar to that used to obtain Eq. (5-158) it is found that, for transition metals,

$$\frac{C_V}{C_{V,i}} \cong \frac{\pi^2}{18} \cdot \frac{k_B T}{E_0 - E_F} + 1 \qquad (5\text{-}160)$$

Using $E_0 - E_F \cong 1\,\mathrm{eV}$ and $k_B T \cong 0.03\,\mathrm{eV}$, it is approximated that the carriers, electrons and/or holes, contribute about 1.6% of the internal energy and heat capacity, or about twice that in normal metals.

It was noted in discussing Eqs. (5-147) and (5-159) that the conduction coefficients compensate for the simplifying assumptions used to arrive at Eqs. (5-147) and (5-150) and also to account for differences in the electron behaviors of metals. Perhaps the major simplifying assumption in these equations is that they are based on the concept of "nearly free" carriers, electrons or holes. Implicit in this simplification is the assumption that the mass of a carrier, electron or hole, is $|m_e|$, the mass of a free electron. This is an oversimplification because the carrier mass is affected by carrier–carrier effects (Section 4.3), by carrier reactions with the "phonon gas" (Section 5.1.7), and by the influence of the periodic potential of the lattice on the carrier (Sections 4.5–4.7). Corrections for the combined effects of these factors are included in the conduction coefficients. These may be expressed as, where m_{th} is the thermal effective mass,

$$A \text{ or } A' = \frac{m_{th}}{m} = \frac{\gamma \text{ (observed)}}{\gamma \text{ (free)}} \qquad (5\text{-}161)$$

If factors such as those noted here had no effect on the properties of a metal, its conduction coefficient would be unity. Thus, where the values of A or A' are found to be reasonably close to unity, such values are considered to represent primarily the effects of the major simplifications. See Table 5-2. However, where the values of A or A' are considerably greater than unity, they are considered as being essentially representative of more complex electron behaviors than those included in the models represented by Eqs. (5-147) and (5-150).

TABLE 5-2 EXPERIMENTAL AND FREE ELECTRON VALUES OF ELECTRONIC HEAT CONSTANT γ OF METALS

Li	Be											B	C	N
1.63	0.17													
0.749	0.500													
2.18	0.34													
Na	Mg											Al	Si	P
1.38	1.3					Observed γ in mJ mol^{-1} K^{-2}						1.35		
1.004	0.992					Calculated free electron γ in mJ mol^{-1} K^{-2}						0.912		
1.26	1.3					m_{th}/m = (observed γ)/(free electron γ)						1.48		
K	Ca	Sc	Ti	V	Cr	Mn(γ)	Fe	Co	Ni	Cu	Zn	Ga	Ge	As
2.08	2.9	10.7	3.35	9.26	1.40	9.20	4.98	4.73	7.02	0.695	0.64	0.596		0.19
1.668	1.511									0.505	0.753	1.025		
1.25	1.9									1.33	0.85	0.58		
Rb	Sr	Y	Zr	Nb	Mo	Tc	Ru	Rh	Pd	Ag	Cd	In	Sn(w)	Sb
2.41	3.6	10.2	2.80	7.79	2.0	-------	3.3	4.9	9.42	0.646	0.688	1.69	1.78	0.11
1.911	1.790									0.645	0.948	1.233	1.410	
1.26	2.0									1.00	0.73	1.37	1.26	
Cs	Ba	La	Hf	Ta	W	Re	Os	Ir	Pt	Au	Hg(α)	Tl	Pb	Bi
3.20	2.7	10.0	2.16	5.9	1.3	2.3	2.4	3.1	6.8	0.729	1.79	1.47	2.98	0.008
2.238	1.937									0.642	0.952	1.29	1.509	
1.43	1.4									1.14	1.88	1.14	1.97	

From compilations kindly furnished by N. Phillips and N. Pearlman to C. Kittel. Reprinted, by permission, from Charles Kittel, *Introduction to Solid State Physics*, © 1976 by John Wiley & Sons, Inc., New York.

It should be noted that alloying elements in solid solution in a given metal may affect its conduction coefficient. The properties of alloys, including electrical, thermoelectric, and magnetic, may be sensitive to these changes.

5.2 THERMAL EXPANSION

All solids show dimensional changes as the temperature varies. Almost all solids expand as a function of increasing temperature. This behavior can be explained by considering the oscillations of the ions of which the solid is composed. As the temperature increases, the amplitudes of the vibrations of the ions become greater and the dimensions of the solid increase correspondingly.

If two ions are treated as classical, simple-harmonic oscillators, but one is fixed at the origin while the other oscillates, then the condition sketched in Fig. 5-20 should prevail. According to classical ideas, the particles would be at rest at a distance a_0 apart at 0 K. As the temperature increased, the particle would oscillate about a_0, its mean, or equilibrium, position; the average amplitude of this oscillation would be the same for all directions. Since the curve is symmetrical, it is probable that, for a given set of conditions, as many ions would be found with positions greater than a_0 as those with positions less than a_0. The net result is that no thermal expansion can be predicted by the classical approach because of the postulated symmetry of ionic oscillation. Thus, the curve of Fig. 5-20 cannot represent the observed conditions.

In order that the observed behavior be accounted for, the amplitudes of the ionic oscillations must be such that their net average is greater than a_0 at a given temperature. Therefore, it follows that thermal expansion can be described only by amplitudes of ionic oscillators that must be asymmetric about a_0, as shown in Fig. 5-21. The average lattice position of an oscillating ion, as a function of temperature, is given by the line connecting the midpoints of the ranges of vibration such as are indicated for the four temperatures shown on the deduced curve.

The classical expression, $V(x) = \frac{1}{2}Kx^2$, cannot describe the preceding behavior, since it is symmetric about a_0. A first approximation of the actual behavior may be obtained by the inclusion of an anharmonic, or repulsion, term to obtain another classical expression that describes the asymmetric vibrational amplitudes more realistically. This is given, for the same set of conditions as for the symmetric

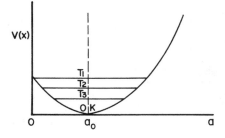

Figure 5-20 Average amplitudes of an oscillating particle as a function of temperature treated classically $(T_1 > T_2 > T_3 \gg 0 \text{ K})$.

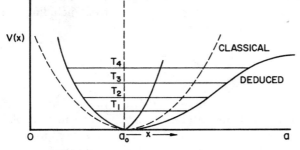

Figure 5-21 Asymmetric vibrational amplitudes, x, required to explain observed thermal expansion.

oscillations, as

$$V(x) = Lx^2 - Mx^3 \tag{5-162}$$

where x is the amplitude, or the displacement from a_0, and the coefficients L and M are related to the bonding energies of the ions. The effect of this anharmonic term, Mx^3, is shown schematically in Fig. 5-22.

The average value of the displacement, $\langle x \rangle$, from a_0 is obtained by means of Eq. (2-23) as

$$\langle x \rangle = \frac{\displaystyle\int_{-\infty}^{\infty} x \exp[-V(x)/k_B T]\, dx}{\displaystyle\int_{-\infty}^{\infty} \exp[-V(x)/k_B T]\, dx} = \frac{\alpha}{\beta} \tag{5-163}$$

This will be simplified as follows. First, consider the numerator of Eq. (5-163). This is more specifically expressed by the substitution of Eq. (5-162) as

$$\alpha = \int_{-\infty}^{\infty} x e^{-V(x)/k_B T} dx = \int_{-\infty}^{\infty} x e^{-(Lx^2 - Mx^3)/k_B T} dx \tag{5-164}$$

or as

$$\alpha = \int_{-\infty}^{\infty} \left(x e^{-Lx^2/k_B T} \cdot e^{Mx^3/k_B T} \right) dx \tag{5-165}$$

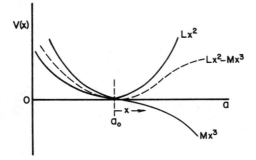

Figure 5-22 Model used to approximate the average amplitude, x, for the calculation of thermal expansion.

The second exponential factor in Eq. (5-165) may be expressed as the first two terms of a series, so that

$$\alpha = \int_{-\infty}^{\infty} xe^{-Lx^2/k_BT}\left(1 + \frac{Mx^3}{k_BT}\right) dx \qquad (5\text{-}166)$$

or expanding Eq. (5-166) results in

$$\alpha = \int_{-\infty}^{\infty} xe^{-Lx^2/k_BT}dx + \frac{M}{k_BT}\int_{-\infty}^{\infty} x^4 e^{-Lx^2/k_BT}dx \qquad (5\text{-}167)$$

The first term in Eq. (5-167) may be written as

$$\int_{-\infty}^{\infty} xe^{-Lx^2/k_BT}dx = 2\int_{0}^{\infty} xe^{-Lx^2/k_BT} dx \qquad (5\text{-}168)$$

because it is a symmetric function. This means that when Eq. (5-168) is used in Eq. (5-163) the positive and negative portions of the integral will cancel each other. It, therefore, requires no further consideration. The second integral in Eq. (5-167) is given by the general expression

$$\int_{0}^{\infty} x^{2n}e^{-ax^2}dx = \frac{1\cdot3\cdot5\cdots\cdots(2n-1)}{2^{n+1}a^n}\cdot\left(\frac{\pi}{a}\right)^{1/2}$$

So, for $n = 2$ and $a = L/k_BT$,

$$\frac{2M}{k_BT}\int_{-\infty}^{\infty} x^4 e^{-Lx^2/k_BT}dx = \frac{M}{k_BT}\left[\frac{3}{4}\left(\frac{k_BT}{L}\right)^{5/2}\pi^{1/2}\right] \qquad (5\text{-}169)$$

Consider now the denominator of Eq. (5-163). Using only the first term in $V(x)$, it is found from tables that

$$\beta = \int_{-\infty}^{\infty} e^{-V(x)/k_BT}dx \cong \int_{-\infty}^{\infty} e^{-Lx^2/k_BT} dx = \left(\frac{\pi k_BT}{L}\right)^{1/2} \qquad (5\text{-}170)$$

when the anharmonic term is neglected. Equations (5-169) and (5-170) are substituted into Eq. (5-163) to give, *for temperatures higher than* θ_D,

$$\langle x \rangle = \frac{\alpha}{\beta} \cong \frac{\frac{3}{4}M[(k_BT)^{3/2}/L^{5/2}]\cdot\pi^{1/2}}{[(k_BT)^{1/2}/L^{1/2}]\cdot\pi^{1/2}} = \frac{3}{4}M\frac{k_BT}{L^2} \qquad (5\text{-}171)$$

Equation (5-171) also may be written, for $T \geqslant \theta_D$, as

$$\langle x \rangle \cong \frac{3}{4}\frac{M}{L^2}k_BT = \frac{3}{4}\frac{M}{L^2}\langle U \rangle \qquad (5\text{-}172)$$

where $\langle U \rangle$ is the average energy. Using Eq. (5-66), this becomes, *for temperatures less than* θ_D,

$$\langle x \rangle \cong \frac{3}{4}\frac{M}{L^2}\frac{\hbar\omega}{\exp(\hbar\omega/k_BT) - 1} \qquad (5\text{-}173)$$

At temperatures above θ_D, Eq. (5-173) can be approximated as being the same as Eq. (5-171) in the same way as for Eq. (2-68). Upon differentiation with respect to temperature, Eq. (5-171) becomes

$$\frac{\partial \langle x \rangle}{\partial T} \cong \frac{3}{4} \frac{M}{L^2} k_B \tag{5-174}$$

Now, using the original equilibrium interionic distance, a_0, as a reference dimension and dividing Eq. (5-174) through by it give the temperature coefficient of linear expansion as

$$\alpha_L \cong \frac{\partial \langle x \rangle}{a_0 \partial T} = \frac{3}{4} \frac{M}{L^2} \frac{k_B}{a_0} \tag{5-175}$$

for $T \geqslant \theta_D$.

The approximation given by Eq. (5-175) does not expressly show α_L to be a function of temperature. However, it will be recalled that the coefficients L and M are related to the bonding energies; these are inverse functions of temperature. Thus, to a first approximation, Eq. (5-175) is linear with temperature. The coefficient of linear expansion is, in fact, a small, approximately linear function of temperature at temperatures above θ_D. Equation (5-175) gives a reasonable approximation of the observed behavior. See Tables 5-3 and 5-4.

For temperatures below θ_D, Eq. (5-171) cannot be used [Eqs. (5-101) and (5-112)]. Equation (5-173) can be approximated for low temperatures by

$$\langle x \rangle \cong \frac{3}{4} \frac{M}{L^2} \hbar \omega \exp \left(-\frac{\hbar \omega}{k_B T} \right) \tag{5-176}$$

Following the same procedure as for Eq. (5-175) results in

$$\alpha_L \cong \frac{\partial \langle x \rangle}{a_0 \partial T} = \frac{3}{4} \frac{M k_B}{L^2 a_0} \left(\frac{\hbar \omega}{k_B T} \right)^2 \exp \left(-\frac{\hbar \omega}{k_B T} \right) \tag{5-177}$$

Thus, according to Eq. (5-177), α_L will approach zero exponentially as T approaches zero in much the same way as the Einstein expression for C_V [Eq. (5-13)]. This approximation is theoretically incorrect for low T, where α_L should vary in the same way as C_V, that is, as T^3.

The relationship of α_L to C_V may be shown most readily by reference to Eqs. (5-171) and (5-173). In both cases, $\langle x \rangle$ is directly proportional to the internal energy of the solid. Thus, for the entire range of temperatures in which the material remains in the solid state,

$$\langle x \rangle \cong \text{constant} \; \langle U \rangle \tag{5-178}$$

The relative average amplitude of oscillation is obtained by dividing Eq. (5-178) through by a_0:

$$\frac{\langle x \rangle}{a_0} \cong \frac{\text{constant}}{a_0} \langle U \rangle \tag{5-179}$$

TABLE 5-3 SOME SELECTED TEMPERATURE COEFFICIENTS OF THERMAL EXPANSION FOR ELEMENTS NEAR 20°C (cm/cm/°C × 10⁶)

Element	α_L	Element	α_L
Aluminum	23.6[a]	Platinum	8.9
Antimony	8.5 to 10.8[b]	Rhenium	6.7[h]
Beryllium	11.6[c]	Rhodium	8.3
Bismuth	13.3	Ruthenium	9.1
Cadmium	29.8	Selenium	37
Carbon (graphite)	0.6 to 4.3[a]	Silicon	2.8 to 7.3
Chromium	6.2	Silver	19.68[i]
Cobalt	13.8	Tantalum	6.5
Copper	16.5	Tellurium	16.75
Gold	14.2	Thorium	12.5[j]
Hafnium	519[d]	Tin	23[i]
Indium	33	Titanium	8.4
Iron	11.76[e]	Tungsten	4.6
Lead	29.3[f]	Uranium	6.8 to 14.1
Magnesium	27.1[g]	Vanadium	8.3[k]
Molybdenum	4.9[a]	Zinc	39.7[l]
Nickel	13.3[c]	Zirconium (α)	5.88
Palladium	11.76		

[a]20° to 100°C	[f]17° to 100°C	[i]0° to 100°C
[b]20° to 60°C	[g]In basal plane,	[j]25° to 1000°C
[c]25° to 100°C	24.3° in c direction	[k]23° to 100°C
[d]20° to 200°C	[h]20° to 500°C	[l]20° to 250°C
[e]25°C		

Source: Abstracted from *Metals Handbook*, Vol. 1, American Society for Metals, Metals Park, Ohio, 1961.

The differentiation of Eq. (5-179) with respect to temperature gives

$$\alpha_L \cong \frac{1}{a_0} \frac{\partial \langle x \rangle}{\partial T} \cong \text{constant} \frac{\partial \langle U \rangle}{a_0 \, \partial T} \tag{5-180}$$

And, since $\partial \langle U \rangle / \partial T \equiv C_V$, Eq. (5-180) gives the relationship being sought:

$$\alpha_L \cong \text{constant } C_V \tag{5-181}$$

Equation (5-181) is not limited by any temperature range as was Eq. (5-171), but is valid for all temperatures at which the material is a solid.

This result is verified by the thermodynamic expression, based on the Helmholtz free energy, that

$$\alpha_L = \frac{\gamma C_V}{3BV} \cong \text{constant } C_V \tag{5-182}$$

Here γ is the Grüneisen constant (Section 5.1.3.1), B is the bulk modulus (the

TABLE 5-4 TYPICAL THERMAL PROPERTIES OF SOME SELECTED COMMERCIAL ALLOYS NEAR ROOM TEMPERATURE:

Material	Coefficient of Linear Expansion ($\%/°C \times 10^4$)	Thermal Conductivity (K cal/sec m^2 °C)
SAE 1020	12.2	0.23
Gray cast iron	12.1	0.20
304 Type stainless steel	17.3	0.08
3003 Aluminum alloy	23.2	0.68
380 Aluminum alloy	20.1	0.42
Copper	16.7	1.70
Yellow brass	18.9	0.52
Aluminum bronze	16.6	0.31
Copper–beryllium	16.7	0.05
Solder (50 Pb–50 Sn)	23.6	0.20
Constantan	14.6	0.10
Titanium (commercial pure)	8.8	0.08

Abstracted from *Handbook of Chemistry and Physics*, pp. D171–172, 56th ed., CRC Press, Boca Raton, Fla. 1975.

reciprocal of the compressibility), and V is the volume. The latter are only relatively small functions of temperature compared to C_V, and γ is independent of temperature. This confirms the proportionality between α_L and C_V given by Eq. (5-181).

C_V is the most important factor in Eqs. (5-180) and (5-182). Thus, the curve of α_L as a function of temperature must be expected to be very similar to that of C_V. This is the case since both properties arise from the way in which the oscillations of the ions are affected by temperature. Many materials approximate this behavior (Fig. 5-23).

Metals show similar dependencies on C_V. Since the models used here are based on homogeneous, isotropic, monatomic solids in which electron effects were omitted, deviations from these ideal conditions would be expected to influence the temperature coefficients of expansion of metals and alloys. Electron effects, including magnetic changes, impurities, imperfections, grain boundaries, internal stresses, ordering, clustering, and the presence of other phases, as well as phase transformations, all affect α_L just as they do C_V. The behavior of pure copper is shown in Fig. 5-24.

Very pure, nearly perfect, single crystals of metals with cubic crystal lattices show anisotropic variations in α_L that are very apparent, the values being identical along the cube edges. Noncubic crystals show pronounced differences in α_L for each principal lattice direction. This generally results in larger anisotropic effects than usually are shown by cubic lattices. The data in Table 5.3, with one exception, are for polycrystalline elements. These data represent "smeared," average values of α_L, in most cases, because of the random orientations of the grains as described in Section 4.6.

The anisotropy of α_L may be shown by considering the least complicated case, that of single, noncubic crystals with three mutually perpendicular axes. This is done

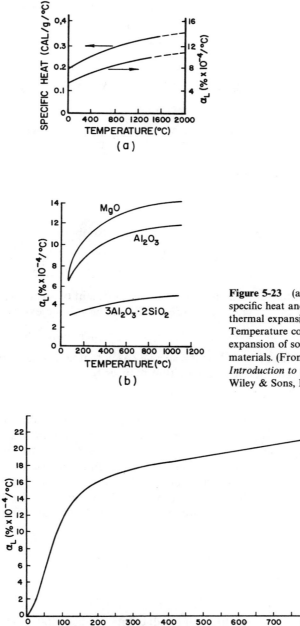

Figure 5-23   (a) Similarity of behavior of specific heat and temperature coefficient thermal expansion of Al_2O_3. (b) Temperature coefficient of thermal expansion of some common insulating materials. (From W. O. Kingery, *Introduction to Ceramics*, p. 470, John Wiley & Sons, Inc., New York, 1960.)

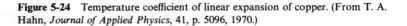

Figure 5-24   Temperature coefficient of linear expansion of copper. (From T. A. Hahn, *Journal of Applied Physics*, 41, p. 5096, 1970.)

by considering the volume changes of the crystal in terms of its dimensions. Thus, the volume of the crystal is

$$V = L_x L_y L_z$$

And a change in its volume is

$$\Delta V = L_x L_y \Delta L_z + L_x L_z \Delta L_y + L_y L_z \Delta L_x$$

Then the relative change in volume is

$$\frac{\Delta V}{V} = \frac{L_x L_y \Delta L_z}{L_x L_y L_z} + \frac{L_x L_z \Delta L_y}{L_x L_y L_z} + \frac{L_y L_z \Delta L_x}{L_x L_y L_z}$$

Or, upon simplification,

$$\frac{\Delta V}{V} = \frac{\Delta L_z}{L_z} + \frac{\Delta L_y}{L_y} + \frac{\Delta L_x}{L_x}$$

This equation now is divided through by ΔT to obtain

$$\frac{\Delta V}{V \Delta T} = \frac{\Delta L_z}{L_z \Delta T} + \frac{\Delta L_y}{L_y \Delta T} + \frac{\Delta L_x}{L_x \Delta T} \tag{5-183}$$

Each of the three terms is the temperature coefficient of linear expansion of an edge of the crystal. Thus, the temperature coefficient of volume expansion of the crystal is

$$\alpha_V = \alpha_{L_x} + \alpha_{L_y} + \alpha_{L_z} = \alpha_{100} + \alpha_{010} + \alpha_{001} \tag{5-184}$$

using crystal directions. For the special case of cubic crystals, in which the edges of the unit cells are identical, Eq. (5-184) reduces to

$$\alpha_V = 3\alpha_{100} \tag{5-185}$$

The properties of gallium, with an orthorhombic unit cell, serve as a good example of anisotropy of α_L. Its values may be expressed as $\alpha_{100} \cong 0.7\alpha_{001}$ and $\alpha_{010} \cong 1.9\alpha_{001}$. Data such as these may not be used for polycrystalline materials. The expansion characteristics of such polycrystalline materials are affected by their thermal and mechanical histories and the resultant grain sizes and orientations. And, for the same reasons, the expansion properties of any polycrystalline material should not be considered as being representative of single-crystal materials.

5.3 THERMAL CONDUCTIVITY

5.3.1 Phonon Contribution

In this discussion of the thermal conductivity of elemental insulators, the transfer of thermal energy is accomplished by the ionic oscillations in the solid. The amplitudes of these vibrations increase as a function of temperature. An ion at the surface, oscillating about its equilibrium position, with an amplitude corresponding to that of the given temperature, will increase its amplitude when the ambient temperature is increased. This induces periodic forces to act on adjacent, internal neighbors; their amplitudes increase and the energy is transferred by the phonons. This mechanism proceeds throughout the entire solid, conducting the thermal energy through it. A continuous flow of heat will occur in this way until thermal equilibrium is reached. When a temperature difference exists across a solid, the vibrational amplitudes of the oscillating ions gradually diminish across the solid, from the hotter to the cooler surface; the flow of thermal energy will continue as long as the temperature difference exists.

Consider the transfer of thermal energy by the flow of a "phonon gas" (Section 5.1.7) crossing the xy plane (Fig. 5-25) that is at energy level E_0. Each phonon arriving at the plane will have an average mean free path λ, make a solid angle θ with the z axis, and have a mean energy

$$\langle E \rangle = E_0 + \lambda \cos \theta \, \frac{\partial E}{\partial z} \tag{5-186}$$

The total energy added by all phonons crossing the plane is

$$\langle E \rangle - E_0 = \Delta E = \lambda \cos \theta \, \frac{\partial E}{\partial z} \cdot \text{flux} \tag{5-187}$$

The flux, or the number of phonons that cross unit area per unit time, within the solid angle between θ and $\theta + d\theta$, is determined with the help of Fig. 5-26. Assume that N phonons, each with an average velocity $\mathbf{v}$, pass across the surface of the hemisphere in unit time. The area of the hemisphere of unit radius is 2π. The circumference of the element of area is $2\pi r = 2\pi \sin \theta$. The area of this element is $2\pi \sin \theta \, d\theta$. The number of phonons that intercept this element of surface then is dN. The ratio of dN/N will equal the ratio of the areas involved. Thus

$$\frac{dN}{N} = \frac{2\pi \sin \theta \, d\theta}{2\pi}$$

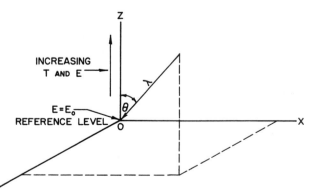

Figure 5-25 Model for the calculation of the energy of a phonon of wavelength λ at a reference plane.

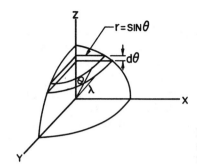

Figure 5-26 Model for the calculation of phonon flux showing one-quarter of the reference hemisphere.

or

$$dN = N \sin \theta \, d\theta \qquad (5\text{-}188)$$

The component of the velocity of the phonons parallel to the z axis is $\mathbf{v}_z = \mathbf{v} \cos \theta$. The flux is given by

$$df_z = dN\mathbf{v}_z \qquad (5\text{-}189)$$

The flux parallel to the z axis (perpendicular to the xy plane) is

$$df_z = dN\mathbf{v} \cos \theta \qquad (5\text{-}190)$$

Or, using Eq. (5-188), the expression for dN, in Eq. (5-190) gives

$$df_z = N\mathbf{v} \sin \theta \cos \theta \, d\theta \qquad (5\text{-}191)$$

Now, assuming that as many phonons in the "gas" travel in the positive direction as in the negative, then

$$df_z = \frac{N}{2} \mathbf{v} \sin \theta \cos \theta \, d\theta \qquad (5\text{-}192)$$

The substitution of Eq. (5-192) into Eq. (5-187) results in

$$\Delta E = \frac{1}{2} N\mathbf{v}\lambda \frac{\partial E}{\partial z} \int_0^\pi \cos^2 \theta \sin \theta \, d\theta \qquad (5\text{-}193)$$

The integral is integrated by parts, as follows:

$$\int_0^\pi \cos^2 \theta \sin \theta \, d\theta = -\tfrac{1}{2} \cos^2 \theta - \tfrac{1}{2} \int_0^\pi \cos^2 \theta \sin \theta \, d\theta$$

Then, transposing,

$$\tfrac{3}{2} \int_0^\pi \cos^2 \theta \sin \theta \, d\theta = -\tfrac{1}{2}[\cos^3 \theta]_0^\pi$$

Clearing and changing the limits gives the desired result:

$$2 \int_0^{\pi/2} \cos^2 \theta \sin \theta \, d\theta = -\tfrac{2}{3}[\cos^3 \theta]_0^{\pi/2}$$

$$= -\tfrac{2}{3}[0 - 1] = \tfrac{2}{3}$$

This is substituted into Eq. (5-193) to obtain the increase in energy at the reference plane as

$$\Delta E = \frac{1}{3} N v \lambda \frac{\partial E}{\partial z} \tag{5-194}$$

This provides the basis for the determination of the expression for the thermal conductivity of nonconductors.

The thermal energy transferred between two planes perpendicular to the z axis at temperatures T_2 and T_1, respectively, where $T_2 > T_1$, is given by

$$\Delta E = \kappa \frac{T_2 - T_1}{z_2 - z_1}$$

where the constant of proportionality, κ, is the *thermal conductivity*.   Or, in differential form,

$$\Delta E = \kappa \frac{\partial T}{\partial z}, \qquad \kappa = \Delta E \frac{\partial z}{\partial T} \tag{5-195}$$

Now, Eqs. (5-194) and (5-195) may be equated.

$$\Delta E = \frac{1}{3} N v \lambda \frac{\partial E}{\partial z} = \kappa \frac{\partial T}{\partial z} \tag{5-196}$$

This may be simplified by reexpressing two of the factors in Eq. (5-194) as

$$N \frac{\partial E}{\partial z} = N \frac{\partial E}{\partial T} \cdot \frac{\partial T}{\partial z}$$

and noting that, for $N = N_A$,

$$N_A \frac{\partial E}{\partial T} = C_V$$

Thus, Eq. (5-196) may be written as

$$\Delta E = \frac{1}{3} C_V v \lambda \frac{\partial T}{\partial z} = \kappa \frac{\partial T}{\partial z} \tag{5-197}$$

By inspection of Eq. (5-197), it is seen that the thermal conductivity of the ions in the lattice is given by

$$\kappa = \tfrac{1}{3} C_V v \lambda = \kappa_L \tag{5-198}$$

The thermal conductivity of solids varies widely as a function of temperature. It is apparent from prior discussions that each of the three factors in Eq. (5-198) is related to the others. However, it appears that each factor assumes a different degree of influence in various ranges of temperature, as discussed next.

In the range of temperatures up to about $\theta_D/20$, the normal vibrational modes in solids are independent and behave as described in Section 5.1.3.1, and the phonons may be treated as ideal gas particles (Section 5.1.7). Here the wave vectors are additive vectorially, energy is conserved, and the flux is transferred efficiently with minimum decay. Such wave vectors represent relatively long wavelengths that are not affected by crystal imperfections or grain boundaries. Phonon interactions of this kind are called *normal processes*.

Starting at temperatures of about $\theta_D/20$, the anharmonic component of the oscillations becomes increasingly operative as the temperature increases. This causes increasing interactions between the normal modes, and a temperature is reached at which they no longer may be treated as being independent. The interacting phonons may be scattered and form new phonons with wave vectors generally opposite in direction and smaller in magnitude than the sum of the original vectors. The flux is transferred very inefficiently under these conditions. This kind of phonon–phonon scattering is called an *umklapp process*.

The transition from normal to umklapp processes occurs at the condition at which the probability of the scattering of a phonon becomes appreciable. This probability is described in terms of the phonon scattering cross section, σ_p. The probability that a phonon will "collide" with, or represent a "target" for, another phonon will vary as its cross-sectional area. Thus, if a phonon is approximated as being spherical,

$$\sigma_p \propto r_p^2 \qquad (5\text{-}199)$$

where r_p is the phonon "radius." And, because the anharmonic effects are responsible for the scattering [Eq. (5-162)],

$$\sigma_p \propto r^2 \propto M^2 \qquad (5\text{-}200)$$

Thus, the scattering cross section is a function of the bond strength as measured by M.

At very low temperatures, the wavelengths are very long and the number of phonons per unit volume, n_p, is correspondingly small. As the temperature increases, λ decreases and n_p increases accordingly. In addition, as n_p increases the greater is the probability of phonon–phonon scattering. This also has the effect of diminishing λ. These effects are summarized by the inverse relationship

$$\lambda \propto \frac{1}{n_p \sigma_p} \qquad (5\text{-}201)$$

Or, using Eq. (5-200),

$$\lambda \propto \frac{1}{n_p M^2} \qquad (5\text{-}202)$$

Thus, Eq. (5-198) can be reexpressed as

$$\kappa_L \propto \frac{C_V \mathbf{v}}{n_p M^2} \tag{5-203}$$

The anharmonic factor, M, in Eqs. (5-200) and (5-203) has a strong influence on thermal conductivity. In addition to its effect in the denominator of Eq. (5-203), it strongly influences the phonon velocity, $\mathbf{v}$. Where the bond strength is high, M is large, oscillation amplitudes are smaller, and $\mathbf{v}$ is relatively smaller. Since it is approximated that $M \propto 1/T$, increasing temperature decreases M and permits larger amplitudes of vibration, which in turn result in larger values of $\mathbf{v}$, greater degrees of anharmonicity, and greater numbers of phonons, n_p. The net result of increasing temperature on M is to diminish thermal conductivity. At a selected temperature a material with a given value of M will have a higher thermal conductivity than a similarly bonded material that has a lower value of M.

In the temperature range $0 < T < \theta_D/20$, λ is large, essentially constant, and limited by the dimensions of the solid. Decreases in temperature below about $T = \theta_D/20$ do not increase λ because the surfaces of the solid do not reflect phonons effectively. Thus, n_p is essentially constant. The ratio $\mathbf{v}/M^2$ also is essentially constant in this range. Thus, using Eq. (5-111) in Eq. (5-203), results in the variation

$$\kappa_L \propto C_V \propto T^3, \qquad 0 < T < \frac{\theta_D}{20} \tag{5-204}$$

The approximation given by Eq. (5-204) is in general agreement with the observed results (Fig. 5-1).

At higher temperatures, in the range of about $\theta_D/20 < T < \theta_D/10$, it is approximated that $C_V \propto T/\theta_D$ [Fig. 5-1(b)], that $\mathbf{v}/M^2 \propto T^{1/2}$, and that $\lambda \propto 1/T$ so that $n_p \propto T$. This gives

$$\kappa_L \propto T \cdot T^{1/2} \cdot T^{-1} = T^{1/2}, \qquad \frac{\theta_D}{20} < T < \frac{\theta_D}{10} \tag{5-205}$$

This also is in qualitative agreement with the dependence shown in Figs. 5-1 and 5-27.

Two competing processes appear to occur as the temperature varies in the neighborhood of $T = \theta_D/10$. Here, C_V increases with temperature while λ decreases and n_p increases. These opposing trends appear to equal each other close to this temperature and a maximum occurs in κ_L. At higher temperatures, the rate of decrease in λ is greater than the corresponding increase in C_V, causing κ_L to decrease with increasing temperatures.

At the higher temperature range of about $\theta_D/10 < T < \theta_D/2$, the ratio $\mathbf{v}/M^2$ is approximated as being constant because the higher temperatures decrease M^2 and $\mathbf{v}$ decreases simultaneously as a result of increased umklapp reactions. C_V is considered as varying approximately as $T^{1/2}$ (Fig. 5-1). Then, since $C_V \propto T^{1/2}$ and $U \propto T^{3/2}$

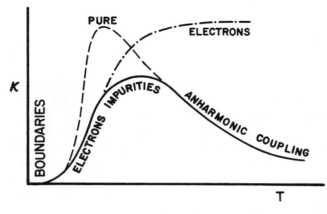

Figure 5-27 General behavior of the thermal conductivity of solids. Pure, crystalline insulators are indicated by the broken curve. The dot–dash curve is based on electron scattering by phonons. (From R. E. B. Makinson, *Proceedings Comb. Philos. Soc.*, 34, p. 474, 1938.)

[Eq. (5-1)], it follows that $n_p \propto T^{3/2}$. This combination of temperature variations gives

$$\kappa_L \propto T^{1/2} \cdot T^{-3/2} = T^{-1}, \qquad \frac{\theta_D}{10} < T < \frac{\theta_D}{2} \qquad (5\text{-}206)$$

At still higher temperatures, $T > \theta_D/2$, C_V approaches its limiting value of $3R$, so it may be approximated as being constant. The ratio of v/M^2 again is approximated as being constant for the reasons given previously. In this high-temperature range, n_p is considered to approach a relatively large limiting value. This is because of the high degree of phonon–phonon interactions. These conditions result in the approximation that

$$\kappa_L \cong \text{constant}, \qquad T > \frac{\theta_D}{2} \qquad (5\text{-}207)$$

This result may be combined with Eq. (5-206) to give

$$\kappa_L T \cong \text{constant}, \qquad T > \frac{\theta_D}{10} \qquad (5\text{-}208)$$

The hyperbolic relationship may be approximated by Eq. (5-208) because Eq. (5-207) can be considered to represent the small asymptotic value of Eq. (5-206).

5.3.2 Electron Contribution

In addition to the thermal energy transfer provided by the oscillating ion cores, the "nearly free" valence electrons of metals make the major contribution to the thermal conductivity, κ_e. Thus, for metals, the thermal conductivity is the sum of these factors:

$$\kappa = \kappa_L + \kappa_e \qquad (5\text{-}209)$$

The electron component is derived in the same way as is Eq. (5-198), the result being

$$\kappa_e = \tfrac{1}{3}C_{V,e}\mathbf{v}(E_F)\lambda(E_F) \tag{5-210}$$

where $C_{V,e}$ is the heat capacity of the "nearly free" valence electrons and $\mathbf{v}(E_F)$ and $\lambda(E_F)$, respectively, are the velocity and wavelength of an electron with energy at or near E_F. See Sections 4.2, 4.4, and 5.1.9. Using Eq. (5-147), Eq. (5-210) is reexpressed for normal metals as

$$\kappa_e \cong \frac{\pi^2}{6}\cdot\frac{Nk_B^2 T}{E_F}\;\mathbf{v}(E_F)\lambda(E_F) \tag{5-211}$$

and for transition metals, by means of Eq. (5-150), as

$$\kappa_e \cong \frac{\pi^2}{18}\cdot\frac{Nk_B^2 T}{E_0 - E_F}\;\mathbf{v}(E_F)\lambda(E_F) \tag{5-212}$$

At very low temperatures, very near 0 K, n_p of Eq. (5-201) is extremely small. It was noted that the phonon wavelength is very long in this range, so scattering of electrons by phonons is small. Electron scattering in this temperature range is caused primarily by impurity ions in solid solution in the metal. It must be emphasized that impurities are always present. Consider, for example, a mol of a very pure metal, which usually has impurity contents greater than 1 ppm. The number of impurities, then, is greater than $10^{23} \times 10^{-6} = 10^{17}$ impurities per mol. Thus, impurities, not the phonons, scatter the electrons, or

$$\lambda(E_F) \propto \frac{1}{n_i} \tag{5-213}$$

where n_i is the impurity concentration. As a consequence, κ_e [Eqs. (5-211) and (5-212)] varies as T. [See Section 6.0 and Eq. (6-29).]

In the range of temperatures above about $\theta_D/10$, the only variable in Eq. (5-211) is T; and, in virtually all cases, the same is true for Eq. (5-212). Here, the umklapp processes effectively decrease the phonon wavelengths, and electron–phonon scattering plays the predominant role. So, for $T \geqslant \theta_D/10$,

$$\lambda \propto \frac{1}{n_p} \tag{5-214a}$$

and

$$n_p \propto T \tag{5-214b}$$

so

$$\lambda \propto \frac{1}{T} \tag{5-214c}$$

Then Eqs. (5-211) and (5-212) are virtually constant with increased temperatures, as shown schematically in Fig. 5-27.

It is interesting to compare the relative roles of κ_e and κ_L. This is most readily seen by comparing Eqs. (5-198) and (5-210) and by noting that the coefficients vanish and that $C_V/C_{V,e} \cong 10^2$ [Eq. (5-158)]:

$$\frac{\kappa_L}{\kappa_e} = \frac{C_V \mathbf{v}\lambda}{C_{V,e}\mathbf{v}(E_F)\lambda(E_F)} \cong 10^2 \frac{\mathbf{v}\lambda}{\mathbf{v}(E_F)\lambda(E_F)} \tag{5-215}$$

Here the phonon velocity (the velocity of sound) $\cong 10^5$ cm/s, $\lambda \cong 10^{-7}$ cm, $\mathbf{v}(E_F) \cong 10^8$ cm, and $\lambda(E_F) \cong 10^{-6}$ cm. The substitution of these values in Eq. (5-215) results in

$$\frac{\kappa_L}{\kappa_e} \cong 10^2 \frac{10^5 \times 10^{-7}}{10^8 \times 10^{-6}} = 10^{-2}$$

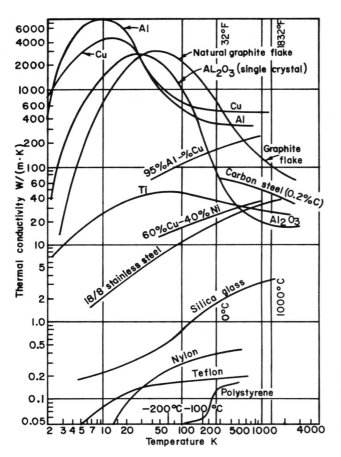

Figure 5-28 Thermal conductivities of some typical engineering materials. (From Z. D. Jastrzebski, *The Nature and Science of Engineering Materials*, p. 504, John Wiley & Sons, Inc., New York, 1976.)

or

$$\kappa_e \cong 100\kappa_L \qquad (5\text{-}215a)$$

The large disparity between κ_e and κ_L explains why the normal metals conduct heat so much better than elements bonded in any other way; the conduction by phonons alone is so very inefficient. The electron component in transition elements is approximately one-half that of the normal metals because the valence electrons are scattered by the unfilled states in the incompleted d band. And where significant alloying additions are made to either normal or transition metals, $\lambda(E_F) \cong \lambda$, so the electron component of κ approaches that of the lattice component. This explains the comparatively low thermal conductivity of many metallic alloys.

5.3.3 Commercial Materials

The very large number of thermal insulating materials used in engineering applications makes it impossible to provide detailed information here; these data are available in handbooks. The thermal conductivities of some typical materials are

TABLE 5-5 THERMAL CONDUCTIVITIES OF SOME SELECTED CERAMIC AND ORGANIC MATERIALS

Single Crystals	Thermal Conductivity (cal/s) (cm²) (°C/cm)		
	37.8°C (100°F)	93.3°C (200°F)	148.9°C (300°F)
Silicon carbide	0.21	0.21	0.20
Periclase	0.11	0.09	0.08
Spinel MgO·Al₂O₃	0.03	0.03	0.02
Quartz (c axis)	0.03	0.02	0.02
Quartz (basal plane)	0.01	0.01	0.01
Fluorite	0.02	0.02	0.01
Polycrystalline Materials			
BeO (pure, hot pressed)	0.52	0.43	0.38
MgO (spec. pure)	0.09	0.08	0.07
ThO₂ (hot pressed)	0.04	0.03	0.03
PbO	0.007	0.005	0.004
Organic Materials[a]	K (near room temperature)		
Polyethylene	0.002		
Rubber	0.007		
Urethane foam	0.0001		

[a]Abstracted from A. G. Guy, *Essentials of Materials Science*, p. 241, McGraw-Hill Book Co., New York, 1976.

Abstracted from *Handbook of Chemistry and Physics*, 56th ed., p. E5, CRC Press, 1975.

shown in Fig. 5-28. The high conductivities of aluminum and copper are to be expected because of the transport properties of their nearly free valence electrons [Eq. (5-215a)]. The behavior of graphite is explained by the trigonal $s-p$ hybridized bonding. The weak bond between the basal planes is readily available for conduction (see Section 3.12.3). The comparably high properties of the single Al_2O_3 crystal result from phonon conduction due to its unusual purity and crystalline perfection.

The effects of "impurities" in the materials shown in the figure are the result of intentionally added alloying elements. These are shown by the properties of 95% Al–5% Cu and 60% Cu–40% Ni when compared to those of the respective pure elements, Al and Cu. The comparison of the properties of the 18-8 stainless steel with those of the 0.2C steel also shows a significant contrast that results from large differences in alloying contents; both are alloys of transition elements.

The organic polymeric materials show the lowest thermal conductivities because of the strong covalent bonding involved in these materials; the energy transfer occurs almost entirely by phonons in these virtually noncrystalline solids.

Data for some typical ceramic and organic materials are given in Table 5-5. Also see Table 5-4.

5.4 BIBLIOGRAPHY

BUTTS, A., *Metallurgical Problems*, McGraw-Hill, New York, 1943.

DARKEN, L. S., and GURRY, R. W., *Physical Chemistry of Metals*, McGraw-Hill, New York, 1953.

DEKKER, A. J., *Solid State Physics*, Prentice-Hall, Englewood Cliffs, N.J., 1959.

KITTEL, C., *Introduction to Solid State Physics*, Wiley, New York, 1966.

POLLOCK, D. D., *Physical Properties of Materials for Engineers*, Vol. I, CRC Press, Boca Raton, Fla., 1982.

RICHTMEYER, F. K., KENNARD, E. H., and LAURITSEN, T., *Introduction to Modern Physics*, McGraw-Hill, New York, 1955.

SOKOLNIKOFF, I. S., and REDHEFFER, R. M., *Mathematics of Physics and Modern Engineering*, McGraw-Hill, New York, 1958.

SPROULL, R. L., *Modern Physics*, Wiley, New York, 1956.

5.5 PROBLEMS

5.1. Calculate and obtain handbook data in the literature to verify Eq. (5-7) for $T > 100°C$.

5.2. Verify Kopp's rule for several inorganic compounds by using heat capacity data for their components at both 100° and 500°C. Check the results with data for these compounds from the literature.

5.3. Why does the rule of Dulong and Petit agree with the experimental data for $T \gtrsim 300$ K?

5.4. Explain why Eq. (5-58) must be the case.

5.5. Show that the assumption given by Eq. (5-89) is unrealistic for real crystals.

5.6. Show how the Heisenberg uncertainty principle validates Eq. (5-139).

5.7. Use Eq. (5-158) to estimate the percentage of the electron contribution to the heat capacity of lithium and copper at room temperature. Account for any significant difference between these values.

5.8. Use an equation of the general form of Eq. (3-75) to approximate the influence of factors affecting the coefficient of thermal expansion of solids.

5.9. Why should α_L be expected to be highly anisotropic in crystals of low symmetry?

5.10. Why should the electron component of thermal conductivity be expected to be much greater than the phonon component?

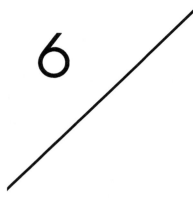

6 / ELECTRICAL CONDUCTIVITY

6.0 FUNDAMENTAL RELATIONSHIPS

Electrical conductivity is approached by considering the effects of an applied electric field, $\bar{E}$, on the electron momentum states in a solid, as shown in Fig. 6-1. This two-dimensional picture is based on the model shown in Fig. 4-4(a) for electrons near $E_F(\mathbf{p})$, with $\mathbf{p} \cong \mathbf{p}(E_F)$.

The field causes a slight net shift in the momenta of electrons to slightly higher levels. Those excited electrons at the high levels, lune A, are scattered, give off a phonon (lattice vibration) and drop back to the lower levels at or near lune B. This process results in a small net flow of electrons. The asymmetry of the momentum space can be approximated by the shift of the center of the momentum sphere to some position $d\mathbf{p}$. The drift velocity induced by the electric field is very small compared to the average velocity (Section 4.2), so $d\mathbf{p}$ is very small. The electrons are considered to behave elastically, since interactions or scattering mechanisms are not included.

The force acting on an electron is given, from Eqs. (3-4) and from electrostatics, as

$$F = ma = \frac{\partial \mathbf{p}}{\partial t} = e\bar{E} \tag{6-1}$$

By means of the de Broglie relationship and the expression for the wave vector, it is

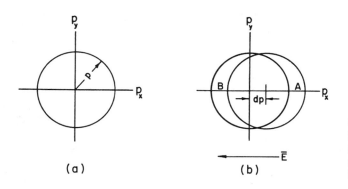

Figure 6-1 Effect of an external electric field on the momentum states in a metal. (a) In the absence of a field. See Figs. 4-1 and 4-4(a). (b) After the application of the electric field.

found that, by multiplying the numerator and denominator by 2π,

$$\mathbf{p} = \frac{h}{\lambda} = \frac{h}{2\pi} \cdot \frac{2\pi}{\lambda} = \frac{h}{2\pi}\,\bar{\mathbf{k}} \tag{6-2}$$

The differentiation of Eq. (6-2) with respect to time gives

$$\frac{d\mathbf{p}}{dt} = \frac{h}{2\pi}\frac{\partial\bar{\mathbf{k}}}{dt} \tag{6-3}$$

The substitution of Eq. (6-3) into Eq. (6-1) results in the force acting on the electron as

$$F = \frac{\partial\mathbf{p}}{\partial t} = \frac{h}{2\pi}\frac{\partial\bar{\mathbf{k}}}{\partial t} = \bar{E}e$$

and the corresponding change in its wave vector as

$$\partial\bar{\mathbf{k}} = \frac{2\pi e\bar{E}}{h}\,\partial t \tag{6-4}$$

This represents the displacement of the center of the Fermi sphere in $\bar{\mathbf{k}}$ space as a result of the application of the electric field. The shift of the center of the sphere in $\mathbf{p}$ space (Fig. 6-1) also may be determined from the differentiation of Eq. (6-2) to give the relationship between the changes in momenta and wave vectors:

$$\partial\mathbf{p} = \frac{h}{2\pi}\,\partial\bar{\mathbf{k}} \tag{6-5}$$

Then the substitution of Eq. (6-4) gives

$$\partial\mathbf{p} = \frac{h}{2\pi}\frac{2\pi e\bar{E}}{h}\,\partial t = e\bar{E}\,\partial t \tag{6-6}$$

Actually, the electron interacts with other electrons, phonons, impurities, imperfections, and so on, in the lattice and is scattered. If the time between such interactions is $\tau(E_F)$, the relaxation time of electrons with $E \cong E_F$, and the change in the velocity (drift velocity) of the electron is $\Delta \mathbf{v}$, then from Eqs. (3-3a) and (6-6),

$$\Delta \mathbf{v} = \frac{\Delta p}{m} \cong \frac{\bar{E}\tau(E_F)}{m} \tag{6-7}$$

$$\mu \equiv \frac{\Delta \mathbf{v}}{\bar{E}} \cong \frac{e\tau(E_F)}{m} \tag{6-7a}$$

where m is the effective mass of the electron and the approximation is made that $\tau(E_F) \cong \partial t$. Since $\tau(E_F)$ is of the order of 10^{-14} s, this approximation is fair. Equation (6-7) leads directly to the mobility, μ. This is defined as the drift per unit field, or $\mu \equiv \Delta \mathbf{v}/\bar{E} = e\tau(E_F)/m$. It will be observed that $\Delta \mathbf{v}$ is a constant for a given $\bar{E}$. Only electrons near $E_F(\mathbf{p})$ are involved; and $\Delta \mathbf{p} \ll \mathbf{p}(E_F)$. Thus, $\tau(E_F)$ and $\Delta \mathbf{v}$ are very small. Equations (6-18) and (6-19) verify this. It will be recognized from Lenz's law that $\Delta \mathbf{v}$ is opposite in direction to $\bar{E}$ because the charge on the electron is negative.

Basic to the preceding analysis is the idea that the electric field acting on an electron, as given by Eq. (6-1), does *not* cause it to accelerate indefinitely. The previously noted interactions and lattice defects have the effect of setting up an opposing force, F_L, that prevents this. This force is expressed, using the previous approximation that $\tau(E_F) \cong \Delta t$, as

$$F_L \cong -\frac{1}{\tau(E_F)} m \Delta \mathbf{v} \tag{6-8}$$

So, using Eqs. (3-4), (6-1), and (6-8), the forces acting on an electron may be used to obtain the equation for the motion of the electron as

$$m \frac{d(\Delta \mathbf{v})}{dt} \cong -e\bar{E} - \frac{1}{\tau(E_F)} m \Delta \mathbf{v} \tag{6-9}$$

Now the influence of F_L may be more fully understood. Assume that the electron flow has reached its steady state. If the external field is shut off, $\bar{E} = 0$ and Eq. (6-9) is reduced to

$$m \frac{d(\Delta \mathbf{v})}{dt} \cong -\frac{1}{\tau(E_F)} m \Delta \mathbf{v} \tag{6-10}$$

Noting that m vanishes, Eq. (6-10) may be rewritten as

$$\frac{d(\Delta \mathbf{v})}{\Delta \mathbf{v}} \cong -\frac{dt}{\tau(E_F)} \tag{6-11}$$

Upon integration, Eq. (6-11) becomes

$$\ln[\mathbf{v}(t) - \Delta \mathbf{v}] \cong \ln \left[\frac{\mathbf{v}(t)}{\Delta \mathbf{v}} \right] = -\frac{t}{\tau(E_F)}$$

where $\mathbf{v}(t)$ and $\mathbf{v}$ are the velocities at times t and $t = 0$, respectively. Thus

$$\mathbf{v}(t) \cong \Delta \mathbf{v} e^{-t/\tau(E_F)} \tag{6-12}$$

Here t is the elapsed time after the field has been shut off. So, for $t = \tau(E_F)$, $\mathbf{v}(t) = \Delta\mathbf{v}/e \cong \Delta\mathbf{v}/2.7$ of its value at the instant prior to removing the field. The previously noted interactions and imperfections in the lattice cause $\Delta\mathbf{v}$ to decay exponentially. This is the basis for the approximation used for Eq. (6-7). It may also be seen, from Eq. (6-12), that the larger the value of $\tau(E_F)$, the longer will be the time required for the electron to return to an unexcited state (the *decay time*). The decay time of normal metals is of the order of 10^{-12} s.

The *relaxation time* is useful in defining the *mean free path of an electron*, that is, the distance that it travels between "collisions", interactions, or scatterings. Thus, most simply, the mean free path is given by

$$L(E_F) = \tau(E_F)\mathbf{v}(E_F) \tag{6-13}$$

This implies that the electron has given up its energy of excitation in one scattering event. In cases in which more than one such event is required for $\Delta\mathbf{v} \cong 0$, that is, where more than one scattering interaction is involved in changing the direction of $\Delta\mathbf{v}$ so that it becomes random,

$$L(E_F) \cong \frac{\tau(E_F)\mathbf{v}(E_F)}{n_s} \tag{6-14}$$

Here n_s is the number of such events required to randomize the direction of $\Delta\mathbf{v}$.

It may now be appreciated why an applied electric field, $\bar{E}$ in Eq. (6-10), is responsible for the displacement of the momentum states, shown in Fig. 6-1(b), that gives rise to a given $\Delta\mathbf{v}$. In the absence of an external field, the momentum states are symmetric [Fig. 6-1(a)], so their average velocity in any direction is zero and $\Delta\mathbf{v} = 0$. An asymmetry corresponding to that shown in Fig. 6-1(b) also occurs in $\bar{\mathbf{k}}$ space. This shift, given by Eqs. (6-4) and (6-5), displaces the Fermi surface, $E_F(\bar{\mathbf{k}})$, as shown in Fig. 4-4(c). These asymmetrical electron states (excited states) that result from the external field give a nonzero $\Delta\mathbf{v}$ that constitutes a current flow.

The current density is given, in terms of the number of excited electrons per unit volume with $E \cong E_F$, $n(E_F)$, as

$$j = n(E_F)e\,\Delta\mathbf{v} \tag{6-15}$$

Here $n(E_F)$ may be approximated as shown by Eq. (5-144) and Section 5.1.9. Then, using Eq. (6-7), for $\Delta\mathbf{v}$,

$$j = \frac{n(E_F)e^2\bar{E}\tau(E_F)}{m} \tag{6-16}$$

The electrical conductivity is determined from Ohm's Law by equating it to Eq. (6-16):

$$j = \sigma\bar{E} = \frac{n(E_F)e^2\bar{E}\tau(E_F)}{m}$$

Then, since $\bar{E}$ vanishes, and using the definition of mobility given by Eq. (6-7a), the conductivity is given by

$$\sigma = \frac{n(E_F)e^2\tau(E_F)}{m} = \frac{1}{\rho} = n(E_F)e\mu \tag{6-17}$$

where ρ is the electrical resistivity.

The relaxation time, $\tau(E_F)$, may be taken, most simply, using Eq. (6-13). as

$$\tau(E_F) = \frac{L(E_F)}{\mathbf{v}(E_F)} \tag{6-18}$$

in which $L(E_F)$ is the average distance traveled by an electron between interactions (mean free path), and $\mathbf{v}(E_F)$ is the average electron velocity with $E \cong E_F$. This may be used in Eq. (6-9) to obtain another expression for the electrical conductivity as

$$\sigma(E_F) = \frac{n(E_F)e^2 L(E_F)}{m\mathbf{v}(E_F)} = n(E_F)e\mu(E_F) \tag{6-19}$$

The magnitude of $\mathbf{v}(E_F)$ is about 10^8 cm/s (Section 4.2). Here the mean free path in normal metals is found to be of the order of about 10^2 interionic spacings, whereas in the Drude–Lorentz theory it is of the order of the interionic distance.

Further consideration must be given to Eq. (6-17) because of the simplification employed in the use of Eq. (6-13) to represent $\tau(E_F)$ as in Eq. (6-18). Only those electrons in the degenerate "gas" (Sections 2.4, 2.6, 2.7, and 3.11) with energies in the narrow range of $E \cong E_F$ can accept the energy increment induced by the applied electric field. It, therefore, is considered highly probable that more than one scattering event is required for the randomization of $\Delta\mathbf{v}$ [Eq. (6-14)]. Thus, a more realistic expression for the relaxation time is obtained from Eq. (6-14) as

$$\tau(E_F) \cong \frac{n_s L(E_F)}{\mathbf{v}(E_F)} \tag{6-20}$$

Then the electrical conductivity is given by reexpressing Eq. (6-19) as

$$\sigma(E_F) \cong \frac{n(E_F)e^2}{m} \cdot \frac{n_s L(E_F)}{\mathbf{v}(E_F)} \tag{6-21}$$

and the mobility as given by Eq. (6-7) is now expressed by

$$\mu(E_F) \cong \frac{e\tau(E_F)}{m} = \frac{n_s e L(E_F)}{m\mathbf{v}(E_F)} \tag{6-22}$$

However, when the density of nondistinguishable particles is so low that the probability of their interaction is negligibly small, they may be considered to singly occupy states in state space as described in Section 2.2. The Pauli exclusion principle (Section 3.9) does not become operative, and the particles may be treated as a nondegenerate "gas". Also see Sections 2.4, 2.6, 2.7, and 3.11. Under these conditions the independent, noninteracting particles may be treated by means of classical statistics (Sections 2.1 and 2.2).

Since electrons do not interact under these conditions, they may be considered as being independent and behaving as "ideally free" electrons in contrast to the "nearly free" degenerate electrons with $E \cong E_F$ described previously. These "ideally free" electrons do not influence one another, so *all* of them can enter into such physical processes as electrical conduction. Thus, the relaxation time must be given as

$$\tau \cong \frac{\langle n_s \rangle \langle \lambda \rangle}{\langle \mathbf{v} \rangle} \tag{6-23}$$

the conductivity as

$$\sigma \cong \frac{\langle n \rangle e^2}{m} \cdot \frac{\langle n_s \rangle \langle \lambda \rangle}{\langle \mathbf{v} \rangle} \tag{6-24}$$

and the mobility as

$$\mu \cong \frac{\langle n_s \rangle e \langle \lambda \rangle}{m \langle \mathbf{v} \rangle} \tag{6-25}$$

where $\langle n_s \rangle$, $\langle \lambda \rangle$, $\langle \mathbf{v} \rangle$, and $\langle n \rangle$ represent the averages of the number of collisions, the mean free path, the velocity, and the number of electrons per unit volume, respectively, all obtained by considering the entire conductor by means of classical statistics. This frequently is done by means of Eq. (2-87). Other reductions to the classical statistics are given in Sections 2.4 and 2.6.

The relationships given by Eqs. (6-23), (6-24), and (6-25) hold for intrinsic or lightly doped semiconductors. Here the carrier concentrations are sufficiently low to enable their treatment by classical means. In these cases, in contrast to the behavior of metals as approximated by Eq. (5-215a), the nondegenerate "electron gas" does not contribute as greatly to the thermal conductivity as does the degenerate "electron gas" characteristic of normal metals. This leads directly to the observation that the factor N [Eq. (5-211)] for semiconductors must be very much less than that for normal metals. Thus, the relative contribution to the thermal conductivity of semiconductors that is made by the ions is considerably greater than that shown by Eq. (5-215a). See Section 6.3.1.2.

The effectiveness of electron transport processes in semiconductors can be approximated by starting with Eq. (5-198). Here the specific heat, $c_V = \langle n \rangle k_B$, where $\langle n \rangle$ is the average number of electrons per cubic centimeter, rather than $C_V = N_A k_B$. Thus, for the present case,

$$\kappa = \langle n \rangle k_B \langle \mathbf{v} \rangle \langle \lambda \rangle \tag{6-26}$$

Now, obtaining the ratio of Eq. (6-26) to Eq. (6-24), where $\langle n_s \rangle = 1$, gives

$$\frac{\kappa}{\sigma} = \langle n \rangle k_B \langle \mathbf{v} \rangle \langle \lambda \rangle \cdot \frac{m \langle \mathbf{v} \rangle}{\langle n \rangle e^2 \langle \lambda \rangle} = \frac{k_B m \langle \mathbf{v} \rangle^2}{e^2} \tag{6-27}$$

Classical statistics give $\langle \mathbf{v} \rangle = (8 k_B T / \pi m)^{1/2}$, so this gives the Wiedemann–Franz ratio as

$$\frac{\kappa}{\sigma} = \frac{k_B m}{e^2} \cdot \frac{8 k_B T}{\pi m} = \frac{8}{\pi} \left(\frac{k_B}{e}\right)^2 T \tag{6-28}$$

or

$$\frac{\kappa}{\sigma T} = \frac{8}{\pi} \left(\frac{k_B}{e}\right)^2 = L_0 \cong 2.5 \left(\frac{k_B}{e}\right)^2 \tag{6-29}$$

Here L_0 is the Lorenz number for semiconductors. It has been found that the coefficient $8/\pi$ may in practice vary from about 1.5 to about 2.5 and that a value of 2 represents a reasonable approximation.

In a similar way, for the degenerate electron case, the Wiedemann–Franz ratio for normal metals is found by means of Eqs. (5-210) and (6-21) as

$$\frac{\kappa}{\sigma T} = \frac{\pi^2}{3} \left(\frac{k_B}{e}\right)^2 = L_0 \cong 3.3 \left(\frac{k_B}{e}\right)^2 \tag{6-30}$$

The theoretical values of L_0 as calculated from Eq. (6-30) are 2.72×10^{-13} esu/deg^2, 2.45×10^{-8} watt-Ω/deg^2 and 5.85×10^{-9} (cal-Ω)/(s-deg^2). A good empirical value for some normal metals and alloys near 300 K is $L_0 \cong 5.7 \times 10^{-9}$ (cal-Ω)/(s-deg^2).

Equation (6-30) gives L_0 as a constant. In reality, L_0 is a strong function of composition and temperature. In the case of "pure" normal metals L_0 approaches zero as T^2 approaches zero. Small quantities of alloying additions can have strong effects on this temperature dependence. Wide ranges in $L_0 = f(T)$ behavior occur for many alloy systems. No unified theory as yet appears to have been proposed to explain this. At best, it now appears that L_0 as given by Eq. (6-30) lies within a *limiting range* of values for a given alloy system at temperatures above about 800 K.

Equations (6-29) and (6-30) basically arise from the assumption of identical wavelengths and, therefore, of identical relaxation times for electrons involved in electrical and thermal conduction. This condition results from the external factors responsible for the small dynamic departures of the Fermi sphere from its ground-state symmetry (Figs. 5-17 and 6-1).

In electrical conduction, at $T > \theta_D$, $\bar{E}$ changes the state of an electron at or near the Fermi surface to another equivalent state; its energy remains virtually unchanged, but its vectorial direction is changed significantly by the scattering. Thus, the Fermi surface may be considered to retain its spherical symmetry. The electron energy changes (phonon absorption or emission) during the transport of thermal energy along a gradient, for the same condition of $T > \theta_D$, are small with respect to the energy range above E_F (Fig. 5-17), and the scattering is elastic [Eq. (5-207)]. Here, again, the Fermi surface retains its spherical symmetry. Thus, under these conditions, the wavelengths and relaxation times for electrical and thermal conductivity are approximately the same [Eqs. (5-210) and (6-18)]. This explains the previously noted observed conformity to Eq. (6-30) at elevated temperatures.

At comparatively lower temperatures, $T < \theta_D$, the electron–lattice scattering is much less effective than for $T > \theta_D$. The electron wavelengths and relaxation times are comparatively long. The spherical asymmetry of the Fermi surface caused by $\bar{E}$ is not removed as effectively as is the case for electron–phonon scattering in thermal conduction Eqs. (5-214). As a consequence, the wavelengths and relaxation times may no longer be considered as being the same; indeed, they are considerably different for each conduction process. This difference increases with decreasing temperatures. Thus, the theoretical values given for Eq. (6-30) cannot be expected to be valid at $T < \theta_D$.

The behavior of semiconductors [Eq. (6-29)] is expected to show greater deviations than those described for normal metals. This results from the small number of nondegenerate electrons involved and the major role relegated to the lattice in thermal conduction in these materials. Behavior approximately inverse to that given by Eq. (5-213) is typical of semiconductors.

6.1 NORMAL METALS AND THEIR ALLOYS

Normal metals and some of their alloys are good conductors because their nearly "free electrons" only partly fill their valence bands and the corresponding Brillouin zones [Sections 3.12.4, 4.2, 4.5.1, and 4.6]. These electrons have comparatively high drift velocities and, for elemental metals, have relatively long mean free paths in the lattice. The ions in the lattice are in constant motion about their equilibrium positions. And, as described in Section 5.1.7, their quantized oscillations may be considered to constitute a "phonon gas". The phonons obstruct the motion of the electrons and determine their mean free paths and relaxation times. These electron–phonon "collisions" are known as *scattering*. As the temperature increases, the "density" of the "phonon gas" increases and more efficient scattering of the electrons occurs; the phonons become better "targets" for electrons. See Section 5.3.1. This offers more resistance to the flow of electrons. In terms of Eqs. (6-8), (6-18), and (6-19), $L(E_F)$ decreases and $v(E_F)$ remains virtually unchanged, so the relaxation time, $\tau(E_F)$, decreases with increasing temperature. This diminishes the electrical conductivity and increases its reciprocal, the electrical resistivity, as functions of temperature.

Both the electrons and the ions in the lattice, therefore, affect the electrical resistivity and its behavior as a function of temperature. As would be expected, these properties change with alloying and reflect fundamental changes in the solid state. In fact, early investigators of phase equilibria, Tamman and Kurnakov to cite two, made extensive use of such physical properties in their investigations. These properties have continued to be employed to monitor and/or detect changes in the solid state.

The engineering applications of alloys for electrical components are extensive. These include such applications as "thick" metallic films for use in temperature measurement or in integrated circuits, for precision, wire-wound resistors and potentiometers, resistance thermometers, and for heating elements for ovens and furnaces, to cite a few.

The ideas presented here will provide a basis for understanding the solid-state phenomena reflected by the physical properties of materials used for such applications and for explanations of the behaviors of commercially available alloys.

Not present

6.1.1 Dilute Alloys

In the case of normal elemental metals, the Brillouin zone is partly filled and, using the spherical approximation for the Fermi level [Section 4.7 and Fig. 4-4(c)], may be considered as not affecting the electron transport process. The zone boundaries, thus, are considered not to affect the electrical conductivity of such metallic solids [Section 4.7 and Figs. 4-18 and 4-22]. The electrical resistivity must, therefore, result from other interactions [Eqs. (6-8) and (6-9)].

Almost all the pure elemental metals show extended ranges of virtually linear resistivity versus temperature behavior at temperatures above about 0.2 of the Debye temperature. The normal elements show temperature coefficients of resistivity of about 0.4%/°C. Values of this property for some of the transition elements are exceptional and may be higher than this and can range up to nearly 0.7%/°C for very pure nickel (see Table 6.1). This relative uniformity of behavior permits the generalization of the electrical resistivity of most normal metals by means of a "universal" curve such as is shown in Fig. 6-2.

In Fig. 6-2, the data, ρ_T, are normalized with respect to the electrical resistivity at the Debye temperature, θ_D (Section 5.1.5). A brief review is given here. These data

TABLE 6-1 ELECTRICAL RESISTIVITIES AND TEMPERATURE COEFFICIENTS OF SOME ELEMENTS NEAR ROOM TEMPERATURE (20°C)

Element	ρ^a	α^b	Element	ρ^a	α^b
Aluminum	2.6548	4.29	Molybdenum	5.2 (0°C)	5.3
Antimony	39.0 (0°C)	3.6	Nickel	6.84	6.92
Beryllium	4	25	Palladium	10.8	3.77
Bismuth	106.5 (0°C)	5.6	Platinum	9.85	3.927
Cadmium	6.83 (0°C)	4.2	Plutonium	141.4 (107°C)	−2.08 (107°C)
Calcium	3.91 (0°C)	4.02 (0°C)	Rhenium	19.3	3.95
Carbon (graphite)	13.75 (0°C)	—	Rhodium	4.51	4.3
Chromium	12.9 (0°C)	3	Silicon (impure)	10 (0°C)	—
Cobalt	6.24	5.3	Silver	1.59	4.1
Copper	1.6730	4.3	Sodium	4.69	—
Germanium (impure)	46	—	Sulfur (yellow)	2×10^{23}	—
Gold	2.35	3.5	Tantalum	13.5	3.83
Indium	8.0 (0°C)	5	Tellurium	4.35×10^5 (23°C)	—
Iodine	1.3×10^{15}	—	Thallium	15 (0°C)	—
Iridium	5.3	3.93	Thorium	15.7 (25°C)	3.8
Iron	9.71	6.51	Tin	11 (0°C)	3.64
Lead	20.648	3.68	Titanium	42	3.5
Lithium	9.35	5	Tungsten	5.3 (27°C)	4.5
Magnesium	4.45	3.7	Uranium	11 (0°C)	2.1 (27°C)
Mercury	98.4 (50°C)	0.97	Zinc	5.916	4.19

$^a\Omega$-cm $\times 10^6$.
$^b\Omega/\Omega$-deg $\times 10^3$.
Abstracted largely from *Metals Handbook*, Vol. 2, American Society for Metals, Metals Park, Ohio, 1979.

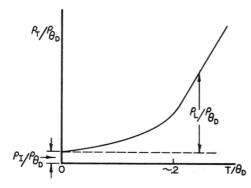

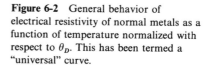

Figure 6-2 General behavior of electrical resistivity of normal metals as a function of temperature normalized with respect to θ_D. This has been termed a "universal" curve.

are plotted versus T/θ_D. The temperature θ_D is that temperature at which the heat capacity of a metal reaches 96% of its asymptotic value. As such, it may be used as a reference temperature. The heat capacity is defined as the rate of change of the internal energy of a mol of a solid with respect to temperature [Eqs. (5-1) and (5-2)]. This may be determined either for the case of constant volume, C_V, or for constant pressure, C_P. For the case of metals, it may be approximated that $C_V \cong C_P$, with an error that usually is less than about 8% [Eq. (5-7)]. The rule of Dulong and Petit gives $C_P \cong 6$ cal/g atom/deg for elemental metals at temperatures at and above room temperature. This value divided by Avogadro's number gives, as noted in Section 5.1.1,

$$\frac{6 \text{ cal/g atom/deg}}{6.02 \times 10^{23} \text{ atoms/g atom}} \cong 1 \times 10^{-23} \text{ cal/atom/deg}$$

Thus, the Debye temperature is that temperature at which the atoms have an approximately constant rate of change of energy of oscillation with temperature. The normalization of the data with respect to θ_D permits the properties of many metals to be shown on the same plot; it provides a common basis for the comparison of the same relative effects of oscillation energies on their physical properties. It is for this reason that curves of this type have been called "universal" curves.

At cryogenic temperatures, very much lower than θ_D, ρ varies as T^5. Electron–electron scattering is important in this range (Section 4.3). The T^5 behavior may be altered by impurities. Some metals may show minima in this range of very low temperatures.

The intercept at the ordinate of Fig. 6-2 gives important clues respecting some of the factors affecting electrical resistivity. For example, it is found that both the impurity content and, to a much lesser extent, the degree of crystalline imperfection of the materials significantly affect this parameter, the residual resistivity, in addition to the electron–electron scattering (Section 4.3). The magnitude of the intercept diminishes as the purity and crystalline perfection improve. The International Practical Temperature Scale makes use of this behavior in the definition of the purity of the platinum standard reference material in terms of its residual resistance ratio. This is given as an average value of $R_{273}/R_0 \cong 3500$.

The electrical properties of the alloys discussed subsequently will be considered at temperatures at which the resistivity has a linear temperature dependence, well above those where T^5 behavior or minima are present. This will simplify the analyses, since the lattice interactions involved may be treated more simply.

6.1.2 Matthiessen's Rule

Matthiessen (1862) showed that the slopes of the curves of resistivity versus temperature ($\Delta\rho/\Delta T$) of well-annealed, very dilute, solid-solution alloys containing about 3 At%, or less, of alloying elements, were virtually the same as that of the annealed, pure (unalloyed) base element, but were offset from that of the base. This behavior is shown in Fig. 6-3. In other words, the resistivity intercepts of the curves for the alloys increase with increasing alloy or impurity content, but their slopes remain the same:

$$\frac{\Delta\rho(C_A)}{\Delta T} = \frac{\Delta\rho(C_B)}{\Delta T} = \frac{\Delta\rho(C_0)}{\Delta T} \tag{6-31}$$

Here C_0 denotes the "pure" metal and C_A and C_B are the compositions of any two suitable binary alloys. This behavior is the basis for *Matthiessen's rule*. It now is generally expressed as

$$\rho = \rho_I + \rho_L \tag{6-32}$$

for a given temperature. Here ρ is the resistivity of the metal or alloy, ρ_I is that part of the resistivity induced primarily by the presence of alloying elements or impurities, and ρ_L is the component of the resistivity caused by the scatter of the electrons by the phonons in the lattice (Section 5.3.1).

It was noted previously that the intercept of Fig. 6-2 was primarily affected by

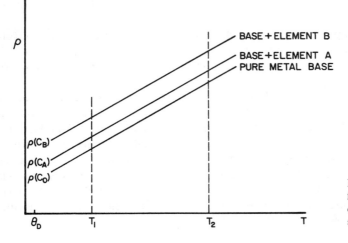

Figure 6-3 Electrical resistivity as a function of temperature for annealed, dilute ($<3At\%$), random, binary solid solutions of metals.

the impurity or alloy content. This is the physical manifestation of ρ_I, which is also known as the *residual resistivity*. The linear portion of the curve, or the lattice component, ρ_L, is also known as the *ideal resistivity*. The imperfection contribution to the residual resistivity also may affect ρ_L slightly, provided that annealing does not occur at higher temperatures.

Another way of stating Eq. (6-32) is to say that the electrical resistivity of a metal or an alloy results from the sum of the electron-scattering effects of these two factors. This frequently is expressed on the basis that the relaxation time (the time between successive scattering events) is a function of the probability that an electron will be scattered [Eqs. (5-199) and (5-201)]. The total scattering of an electron, then, can be approximated simply in terms of Eqs. (6-32) and (6-17) as

$$\rho = \frac{m^*}{n(E_F)e^2\tau(E_F)} = \frac{m^*}{n(E_F)e^2}\left(\frac{1}{\tau(E_F)_I} + \frac{1}{\tau(E_F)_L}\right) \tag{6-33}$$

where the subscripts are the same as those for Eq. (6-32). From this, the total relaxation time is found to be

$$\frac{1}{\tau(E_F)} = \frac{1}{\tau(E_F)_I} + \frac{1}{\tau(E_F)_L} \tag{6-34}$$

Here all the impurity and lattice effects specifically included in Eq. (6-32) are now lumped in the two relaxation times. These parameters are considered as being measures of scattering. This also is discussed in the first section of this chapter and in 6.1.3. The approach used here is somewhat oversimplified, but is sufficient for present purposes.

It would appear to be inconsistent with Eq. (6-34) that several dilute alloys, each containing different alloying elements, should have the same slope as the base, or "pure", metal (Fig. 6-3). The differences in their residual resistivities can be understood simply by considering that $\tau(E_F)_I$ varies with the kind and amount of impurity ions and is the predominating factor in Eq. (6-33) at very low temperatures; the small scattering, small values of n_p and σ_p in Eq. (5-201) result in long wavelengths and correspondingly large values for $\tau(E_F)_L$ [Eq. (6-13)] that are negligible in Eq. (6-34). This factor, $\tau(E_F)_I$, is a function of composition and not of temperature. It is responsible for the offsets between the curves of the alloys and the pure, base metal as shown in Fig. 6-3. But why should $\tau(E_F)_L$ be virtually unaffected by the various dilute alloys in a solid solution in a given base metal at elevated temperatures as evidenced by their unchanged slopes?

To answer this question, it is necessary to examine the factors in Eq. (6-19) in greater detail in the temperature range close to or above θ_D. Under this condition, both the Einstein and Debye models for lattice vibrations give similar and reasonable results. The Einstein model is used here because of its greater simplicity. Each vibrating ion in the lattice may be considered to be a simple harmonic oscillator with a constant frequency. This theory gives good agreement with experiment in the temperature range being considered here (Section 5.1.2). The potential energy of such a particle is given by

$$\text{PE} = \tfrac{1}{2}M\omega^2\langle x\rangle^2 = \tfrac{1}{2}M\cdot 4\pi^2 v^2\langle x\rangle^2 \tag{6-35}$$

in which ω and v are the angular frequency and linear frequency, respectively, M is the mass of the ion, and $\langle x\rangle$ is its average displacement, or amplitude. Each ion will oscillate within a sphere of radius $\langle x\rangle$. The average potential energy is one-half of the thermal energy for $T \geqslant \theta_D$. This may be equated with Eq. (6-35) to give

$$2\pi^2 M v^2\langle x\rangle^2 = \tfrac{1}{2}k_B T \tag{6-36}$$

An electron with a given mean free path, $L(E_F)$, must have a 100% probability of being scattered by an oscillating ion. Thus, the scattering cross section of the ion, or the cross section of the sphere in which it oscillates, $A(E_F)$, may be defined by

$$n(E_F)L(E_F)A(E_F) = 1 \quad \text{or} \quad A(E_F) = \frac{1}{n(E_F)L(E_F)} \tag{6-37}$$

where $n(E_F)$ is the number of electrons per unit volume that enter into the conduction process [Eq. (5-144)]. $A(E_F)$ is the average effective area swept out by an oscillating ion and will vary as $\langle x\rangle^2$; both have the same dimension of area. This is the same approach as used for Eq. (5-199). Equation (6-37) is substituted in Eq. (6-36) to include such scattering. This is done by starting with the substitution of $A(E_F)$ for $\langle x\rangle^2$ in Eq. (6-36) to get

$$2\pi^2 M v^2 A(E_F) \cong \tfrac{1}{2}k_B T \tag{6-38}$$

When Eq. (6-38) is rearranged for $A(E_F)$, and using Eq. (6-37), it is found that

$$A(E_F) = \frac{1}{n(E_F)L(E_F)} = \frac{k_B T}{4\pi^2 M v^2} \tag{6-38a}$$

This is solved to give the mean free path as

$$L(E_F) = \frac{4\pi^2 M v^2}{n(E_F)k_B T} \tag{6-39}$$

The numerator and denominator of the fraction are multiplied by $k_B h^2$, and

$$L(E_F) = \frac{4\pi^2 k_B}{n(E_F)h^2}\cdot\frac{M h^2 v^2}{k_B^2 T} \tag{6-40}$$

Since Eq. (6-35) is applicable at or near θ_D, the relationship $\theta_D = hv/k_B$, [Eq. (5-101b)] may be used in Eq. (6-40). This simplifies Eq. (6-40), giving

$$L(E_F) = \frac{4\pi^2 k_B}{n(E_F)h^2}\cdot\frac{M\theta_D^2}{T} \tag{6-41}$$

Equation (6-41) now may be substituted into Eq. (6-19):

$$\sigma = \frac{n(E_F)e^2 L(E_F)}{m^*\mathbf{v}(E_F)} = \frac{n(E_F)e^2}{m^*\mathbf{v}(E_F)}\cdot\frac{4\pi^2 k_B}{n(E_F)h^2}\cdot\frac{M\theta_D^2}{T}$$

Simplification results in

$$\sigma = \frac{4\pi^2 k_B e^2}{h^2 m^* \mathbf{v}(E_F)} \cdot \frac{M\theta_D^2}{T} \qquad (6\text{-}42)$$

It will be recognized that the denominator of Eq. (6-42) contains the product $m^*\mathbf{v}(E_F)$; this is the momentum of an electron at or near the Fermi surface, denoted by $\mathbf{p}(E_F)$. In normal metals, the Brillouin zones either are partly filled or zone overlap occurs. In each of these cases, the electrons may be approximated as being "nearly free" (Sections 4.6 and 4.7). The momentum of such an electron may be given by $\mathbf{p}(E_F) = (2m^* E_F)^{1/2}$. Now, using Eq. (3-25) for the energy of a free electron, slightly modified to fit this case, and assuming a spherical Fermi surface that is unaffected by the zone walls results in

$$\mathbf{p}(E_F) = (2m^* E_F)^{1/2} = \left[2m^* \frac{h^2 \bar{\mathbf{k}}(E_F)^2}{8\pi^2 m^*} \right]^{1/2} = \left[\frac{h^2 \bar{\mathbf{k}}(E_F)^2}{4\pi^2} \right]^{1/2} = \frac{h\bar{\mathbf{k}}(E_F)}{2\pi} \qquad (6\text{-}43)$$

This expression is substituted for the momentum term in Eq. (6-42) to give

$$\sigma = \frac{4\pi^2 k_B e^2}{h^2} \cdot \frac{2\pi}{h\bar{\mathbf{k}}(E_F)} \cdot \frac{M\theta_D^2}{T} \qquad (6\text{-}44)$$

or, in terms of the electrical resistivity,

$$\rho = \frac{h^3 \bar{\mathbf{k}}(E_F) T}{8\pi^3 e^2 M\theta_D^2 k_B} = \frac{\hbar^3 \bar{\mathbf{k}}(E_F) T}{e^2 M\theta_D^2 k_B}, \qquad \hbar = \frac{h}{2\pi} \qquad (6\text{-}45)$$

The rate of change of the resistivity with temperature is obtained from the derivative of Eq. (6-45); this is

$$\frac{\partial \rho}{\partial T} = \frac{\hbar^3 \bar{\mathbf{k}}(E_F)}{e^2 M\theta_D^2 k_B} \qquad (6\text{-}46)$$

It will be recalled that E_F is a function of the electron:ion ratio [Eq. (4-9)]. The small alloying additions (<3 At%) will result in negligibly small changes in E_F when compared with that of the pure metal, since the change in $(N/V)^{2/3}$ [Eq. (4-9)] is very small. Consequently, $\bar{\mathbf{k}}(E_F)$ remains essentially unchanged [Eq. (3-25)]. All the other factors in Eq. (6-46) may be treated as constants; M and θ_D may also be considered to be virtually unchanged by the small quantities of alloying ions in solid solution in the metal. Therefore, since all the factors in Eq. (6-46) are essentially constant, the slopes of the resistance–temperature curves of all these very dilute alloys are correspondingly constant and must be the same as that of the "pure" base metal as given in Eq. (6-31) and shown in Fig. 6-3; this accounts for Matthiessen's observations.

The residual resistivities, being primarily a result of the impurity or alloy contents, shift the curves from that of the unalloyed metals and account for the differences in the resistivities of the various dilute alloys at a given temperature. These explain the differences between $\rho(C_0)$, $\rho(C_A)$, and $\rho(C_B)$ at a given temperature in Fig. 6-3. And, since each dilute alloy has essentially the same temperature dependency (slope) as the base metal [Eq. (6-46)], their curves are parallel.

It will also be observed that both Eqs. (6-45) and (6-46) can account for anisotropic variations of these properties in crystals. This behavior is taken into consideration by the inclusion of the wave vector in both equations. See Section 4.7.

When very accurate measurements of the electrical resistivity are made over a temperature range of several hundred degrees, it is found that very small departures from linearity exist. These small deviations occur both above and below the nominal linear relationship given by Eq. (6-45). The resultant curves of resistivity, or resistance, versus temperature are slightly S-shaped in character. This behavior is taken into consideration in the calibration of highly accurate resistance thermometers. The linear behavior described here is very good for use in the majority of engineering applications.

6.1.3 Electrical Resistivity, Higher-Concentration Binary Alloys

Two-component, nominally random solid solutions of normal metals, with alloy concentrations greater than $3 \, At\%$, represent the least complicated case to illustrate the mechanisms responsible for the resistivities of the alloys containing more concentrated amounts of alloying elements. The assumption again is made that the Brillouin zone is incompletely filled and that the spherical Fermi surface is sufficiently far from the zone boundaries as to be unaffected by them. This approximation is permissible since the contents of the alloys being considered here are well within the limits of solid solubility [Sections 4.2 and 4.7]. However, the other approximations employed to derive Eq. (6-45), to explain the properties of very dilute alloys, are not valid in the present case. The mechanisms responsible for the resistivities of the more concentrated solid solutions are considerably more complex than those noted for the dilute solid solutions.

One way to explain the behavior of this class of alloys is to begin by examining Eq. (6-17). The most significant factor in this equation, with respect to the alloys being considered here, is the relaxation time. Beginning with Eq. (6-18) and using Eq. (6-37), this factor can be reexpressed in the following way:

$$\tau(E_F) = \frac{L(E_F)}{\mathbf{v}(E_F)}, \qquad L(E_F) = \frac{1}{n(E_F)A(E_F)_T}$$

so that

$$\tau(E_F) = \frac{1}{n(E_F)\mathbf{v}(E_F)A(E_F)_T}, \qquad A(E_F)_T = \frac{1}{n(E_F)\mathbf{v}(E_F)\tau(E_F)} \qquad (6\text{-}47)$$

Here $A(E_F)_T$ is defined as the total scattering cross section; this is discussed more completely later. When Eq. (6-47) is substituted into Eq. (6-17), it is found that

$$\sigma(E_F) = \frac{n(E_F)e^2\tau(E_F)}{m^*} = \frac{n(E_F)e^2}{m^*} \cdot \frac{1}{n(E_F)\mathbf{v}(E_F)A(E_F)_T}$$

This reduces to

$$\sigma(E_F) = \frac{e^2}{m^* v(E_F) A(E_F)_T} \quad \text{or} \quad \rho = \frac{m^* v(E_F) A(E_F)_T}{e^2} \tag{6-48}$$

As noted earlier, solid solutions with alloy concentrations well within the limits of solid solubility may be approximated as having spherical Fermi surfaces that are unaffected by proximity to the Brillouin zone boundaries. Thus, $m^*/m \cong 1$ and the mass may be considered to be that of a free electron. See Section 4.6 [Eq. (4-34)].

Alloy additions to the base metal will usually increase E_F, since, according to Eq. (4-9), it is a function of the electron : ion ratio. However, the percentage increase in E_F will be relatively small, since E_F for normal metals is quite large prior to any alloying. See Table 4-1. So it may be assumed that such alloy additions will induce only negligible changes in $v(E_F)$; such changes may be neglected in this approximation.

The remaining factor in Eq. (6-48) most significantly affected by the ions in random solid solution is $A(E_F)_T$. The changes in the other factors previously discussed are relatively small compared to those induced in the total scattering cross section. This factor includes the scattering effects of both the host and the impurity ions, because each of these is contained in $\tau(E_F)$ [Eq. (6-34)]. This is implicit in Eqs. (6-17) and (6-19), upon which this analysis is based. $A(E_F)_T$ contains another source of scattering that was not included in Eq. (6-34); its effects were small and did not require consideration in the case of the very dilute alloys. In the case of the more concentrated alloys, however, changes in the periodic potential within the lattice become very significant; they can cause a high degree of electron scattering and must be taken into account. So, to provide a more accurate and complete relationship, Eq. (6-34) must be rewritten, as follows, to include this factor:

$$\frac{1}{\tau(E_F)} = \frac{1}{\tau(E_F)_I} + \frac{1}{\tau(E_F)_L} + \frac{1}{\tau(E_F)_e} \tag{6-49}$$

The last term gives that component of the total relaxation time that is due to changes in the periodic potential of the lattice; it results from relatively large numbers of alloying ions, whose charges are different from that on the host ions, that occupy substitutional sites in the host lattice.

In a way corresponding to the derivation of Eqs. (6-34), (6-47), and (6-49), the total scattering cross section may be approximated to be

$$A(E_F)_T = A(E_F)_I + A(E_F)_L + A(E_F)_e \tag{6-50}$$

or as

$$A(E_F)_T = A'(E_F)_L + A(E_F)_e, \quad A'(E_F)_L = A(E_F)_I + A(E_F)_L \tag{6-50a}$$

If the lattice were perfect and contained none of the usual lattice imperfections, including the lattice distortions resulting from impurity ions and the disregistry at grain boundaries, the only resistance to electron flow would be that caused by the thermal oscillations of the host ions (the "phonon gas"); the resistivity would be quite low at relatively low temperatures. However, all these sources of lattice distortion are present in real polycrystalline alloys; they disrupt the spatial periodicity of the lattice and, consequently, scatter electrons. All such scattering effects are included in $A'(E_F)_L$.

Alloying ions in solid solution in a host lattice usually have different charges from those of the host ions. These differences disrupt the periodic electrical potential of the lattice and contribute to electron scattering. This effect is contained in $A(E_F)_e$.

Rather than attempt to determine the total effects of these two scattering parameters, it is more convenient to consider the *changes* in each of these factors relative to the annealed, unalloyed metal at a given temperature. Such changes are obtained more readily by experiment and explained more simply by analyses. So, for these reasons, Eq. (6-50a) usually is expressed as isothermal scattering changes given by

$$\Delta A(E_F)_T = \Delta A'(E_F)_L + \Delta A(E_F)_e \qquad (6\text{-}50\text{b})$$

The electron scattering effects of local changes in the periodic potential of the lattice that result from electrical charge differences between the ions will be approximated first. The host ions each have a charge of $Z_\alpha e$, where Z_α is the valence of the host atom and e is the charge on an electron. Each alloy, or foreign, ion has a charge of $Z_\beta e$, where Z_β is its atomic valence. The difference in the charge between an impurity ion and a host ion is $(Z_\beta - Z_\alpha)e$; this disrupts the periodic electrical potential in the lattice and constitutes a major source of electron scattering. The electron scattering will be proportional to the square of the difference of these charges, or

$$\Delta A(E_F)_e \propto (Z_\beta - Z_\alpha)^2 \qquad (6\text{-}51\text{a})$$

per impurity atom present in the host lattice. On an atomic percent basis, this scattering is given by the Rutherford model [Eq. (6-135a)] as

$$\Delta A(E_F)_e = k_a(Z_\beta - Z_\alpha)^2 \Delta C \qquad (6\text{-}51\text{b})$$

in which k_a is a constant of proportionality and ΔC is atom percent of the foreign or alloy ions present in solid solution in the host lattice.

The scattering effect of ion-size differences may be expressed in terms of Vegard's rule by which the lattice parameter of solid solutions is given as a linear function of the concentration of the solute. This is accurate for many solid solutions; it was observed first in the behavior of ionic crystals. This is shown schematically in Fig. 6-4. Here element D represents the host lattice in which ions of elements E or F are present on random lattice sites. If element E has a larger ionic "diameter" than D, the lattice parameter of the alloy would be expected to increase with increasing amounts of E in random solid solution in D. If the ionic "diameter" of element F is smaller than that of D, the lattice parameter of the solution would be expected to decrease with increasing amounts of F. In either event, an alloy ion on a host lattice site will distort or warp a volume of the host lattice that surrounds it. This results from the fact that no two species of ions have exactly the same "diameters". Larger alloy ions cause a bulging, while smaller ions bring about a contraction of the surrounding host lattice.

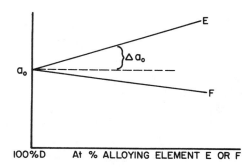

Figure 6-4 Vegard's rule for the variation of the lattice parameter, a_0, as a function of temperature for dilute, random, binary solid solutions of metals.

Thus, the kind and amount of distortion that occurs will depend on the relative sizes of the host and alloy ions; this governs the changes in the spatial periodicity of the host lattice and explains Vegard's rule. These effects are included in $A(E_F)_I$ in Eq. (6-50). Such effects are lumped with those of any other imperfections (such as dislocations, grain boundaries, vacancies, and interstitials) normally present in the host lattice. All the factors present within the host lattice that impede electron flow are considered to contribute to the scattering of electrons and all are included in $\Delta A'(E_F)_L$ of Eq. (6-50a) for simplicity.

The change in the lattice parameter (lattice periodicity) at a given temperature, using Vegard's rule, can be expressed by

$$\Delta a_0 = \alpha \, \Delta C \tag{6-52}$$

where α is the rate of change of the lattice parameter with respect to composition and a_0 is the lattice parameter. The effect of this behavior on the scattering cross section is given by

$$\Delta A'(E_F)_L = k_b \, \Delta C \tag{6-53}$$

where the constant slope, k_b, includes the effects of the lattice imperfections, as well as those of the distortions induced by alloy ions in the host lattice. $\Delta A'(E_F)_L$ usually is considerably smaller than $\Delta A(E_F)_e$. Equations (6-51) and (6-53) can be added, as indicated by Eq. (6-50a), to give

$$\Delta A(E_F)_T = k_b \, \Delta C + k_a (Z_\beta - Z_\alpha)^2 \, \Delta C$$

or

$$\Delta A(E_F)_T = [k_b + k_a (Z_\beta - Z_\alpha)^2] \, \Delta C \tag{6-54}$$

Then, since Eqs. (6-50) and (6-50a) are directly related to Eqs. (6-33) and (6-34) by means of Eq. (6-41), Eq. (6-54) is substituted into the differential of Eq. (6-48), where $A(E_F)_T$ is the variable under consideration, to give

$$\Delta \rho = [k_2 + k_1 (Z_\beta - Z_\alpha)^2] \, \Delta C \tag{6-55}$$

This is the *Linde equation* (1931) for the change in the electrical resistivity of an annealed, binary alloy as a function of the alloying element present in an annealed, random solid solution at a constant temperature.

If the quantity $\Delta\rho/\Delta C$ is plotted against $(Z_\beta - Z_\alpha)^2$, the resultant curve, for solutes of a given period in the Periodic Table, should be linear (Fig. 6-5). The slope of the line is k_1 and the intercept is k_2. For a given solvent, the values of k_1 and k_2 are valid for solutes from a given period in the Periodic Table, including the transition elements of that period, provided that they are soluble in the host lattice. Where the solute and solvent ions are of the same period, k_2 becomes quite small, usually approaching zero. Different sets of values of k_1 and k_2 hold for solute elements of different periods in a given base. The general behavior is shown in Fig. 6-5 and Table 6-2.

When more than one alloying element is present in solid solution in the base, Eq. (6-55) holds for each of the individual constituents, provided that the limit of solid solubility is not exceeded by the combined alloy additions present in the host lattice in the temperature range of interest. The effects of each of the several alloying elements are additive, and the change in the resistivity of the annealed base metal, at a constant temperature, is given by the sum of the resistivity changes of all of the alloying elements present:

$$\Delta\rho = \Sigma\,\Delta\rho_i \qquad\qquad (6\text{-}56\text{a})$$

and the resistivity of the alloy is

$$\rho = \rho_0 + \Delta\rho = \rho_0 + \Sigma\,\Delta\rho_i \qquad\qquad (6\text{-}56\text{b})$$

where $\Delta\rho_i$ is the resistivity change caused by one of the several alloying elements. Again, care must be taken to ensure that all the alloy components are in solid solution in the base lattice and that the alloy is thoroughly annealed.

The Linde equation [Eq. (6-55)] used in conjunction with Eq. (6-56b), constitutes one of the most useful means for the design of commercial engineering alloys

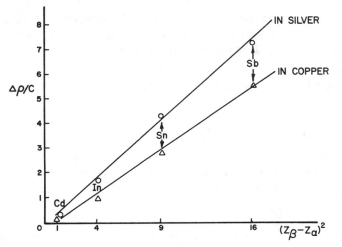

Figure 6-5 Effects of valence on the change in electrical resistivity per atom percent of alloying element in annealed, dilute, random, binary solid solutions of metals at 18°C. See Table 6-2.

TABLE 6-2 CHANGE IN RESISTIVITY PER ATOM PERCENT OF ALLOYING ELEMENT IN Au, Ag, AND Cu

Alloying Element	Solvent Material[a]		
	Au	Ag	Cu
Cu	0.485	0.068	—
Ni	1.00	—	1.25
Co	6.1	—	6.4
Fe	7.66	—	9.3
Mn	2.41	—	2.83
Cr	4.25	—	—
Ti	14.4	—	—
Ag	0.38	—	0.14
Pd	0.407	0.436	0.89
Rh	4.2	—	4.40
Au	—	0.38	0.55
Pt	1.02	1.59	2.51
Ir	—	—	6.1
Zn	0.96	0.62	0.335
Ga	2.2	2.28	1.40
Ge	5.2	5.52	3.75
As	—	8.46	6.8
Cd	0.64	0.382	0.21
In	1.41	1.78	1.10
Sn	3.63	4.32	2.85
Sb	—	7.26	5.45
Hg	0.41	0.79	1.00
Tl	—	2.27	—
Pb	—	4.64	—
Bi	—	7.3	—

[a]In microohm-centimeter per atom percent at 18°C.
After J. O. Linde, *Annalen der Physik*, 5 Folge, B and 15, p. 239, 1932.

for special electrical applications. These equations also give insight into the properties of commercially available alloys. Also see Section 6.1.4.

The understandings provided by Eqs. (6-55) and (6-56b) are somewhat limited because they are restricted to isothermal properties. Additional knowledge may be obtained by applying Eq. (6-45) to multicomponent alloys in a way analogous to that used to obtain Eq. (6-33). This results in

$$\rho \cong \frac{\hbar^3 T}{e^2 \theta_D^2 k_B} \left[f_0 \frac{\bar{\mathbf{k}}(E_F)_0}{M_0} + f_1 \frac{\bar{\mathbf{k}}(E_F)_1}{M_1} + f_2 \frac{\bar{\mathbf{k}}(E_F)_2}{M_2} + \cdots \right] \qquad (6\text{-}57)$$

where the f_i are in mol fractions, the subscript zero refers to the host ion, and the integers each represent one of the several alloying ions. Here θ_D is the Debye temperature of the given, multicomponent alloy. Equation (6-57) may be simplified to read

$$\rho = \frac{\hbar^3 T}{e^2 \theta_D^2 k_B} \, \bar{\mathbf{k}}(E_F)_T \sum_{i=0}^{n} f_i \frac{1}{M_i} \tag{6-58}$$

In this case, $\bar{\mathbf{k}}(E_F)_T$ is the total electron wave vector; it results from *all* the scattering effects of the alloying and host ions and the imperfections in the host lattice. Where the alloy concentration is relatively high, the scattering effects of lattice defects, such as imperfections, dislocations, and grain boundaries, are quite small compared to those caused by alloying. Large numbers of alloy ions in solution strongly decrease the mean free path, or wavelength, of the electrons and increase $\bar{\mathbf{k}}(E_F)_T$ primarily as a result of the disruption of the electrical periodicity of the lattice. This correspondingly increases the resistivity at a given temperature. The larger the amounts of alloy ions in solid solution, the greater will be the resistivity at the given temperature, provided that solid solubility is maintained. The effects of increases in alloy concentration appear to be very much larger than the phonon effects of increasing temperature in most, but not all, solid-solution alloys.

The combined effects of alloy concentration and temperature are most readily understood by examining the relationships between the scattering parameter, $A(E_F)_T$, the wave vector, $\bar{\mathbf{k}}(E_F)_T$, and the relaxation time, $\tau(E_F)$ and their influences on electrical resistivity. Starting with Eq. (6-37) and recalling that $\bar{\mathbf{k}}(E_F)_T = 2\pi/L(E_F)_T$,

$$A(E_F)_T = \frac{1}{n(E_F)L(E_F)_T} = \frac{\bar{\mathbf{k}}(E_F)_T}{2\pi n(E_F)} \tag{6-59}$$

The factor $n(E_F)$ is only slightly affected by composition and temperature because both Nk_B and k_B are very much smaller than E_F (Section 5.1.9 and Fig. 5-13). Thus, $A(E_F)_T$ may be approximated as being proportional to $\bar{\mathbf{k}}(E_F)_T$ or

$$A(E_F)_T = \text{constant}_1 \, \bar{\mathbf{k}}(E_F)_T \tag{6-60}$$

The relaxation time may be more closely associated with $A(E_F)_T$ and $\bar{\mathbf{k}}(E_F)_T$ by means of Eq. (6-47). This is substituted into Eq. (6-17) as follows:

$$\rho = \frac{m}{n(E_F)_T e^2 \tau(E_F)} = \frac{m}{n(E_F)e^2} \cdot n(E_F)\mathbf{v}(E_F)A(E_F)_T \tag{6-61}$$

The factor $n(E_F)$ vanishes and the equation becomes

$$\rho = \frac{m}{e^2} \, \mathbf{v}(E_F)A(E_F)_T \tag{6-62}$$

It will be recalled that $\mathbf{v}(E_F)$ is of the order of 10^8 cm/s (Section 4.2). Any changes in this factor as a result of changes in composition or temperature would constitute only very small percentage changes. This permits the approximation that $\mathbf{v}(E_F)$ is essentially constant and may be treated as such along with the factors m and e. The result is that Eq. (6-62) may be written as

$$\rho \cong \text{constant}_2 \, A(E_F)_T \tag{6-63a}$$

and, using Eq. (6-60), the electrical resistivity may be given by

$$\rho \cong \text{constant}_3\, \bar{k}(E_F)_T \tag{6-63b}$$

The inverse relationship between $\tau(E_F)$ and $A(E_F)_T$ may be shown by Eq. (6-47). When the same approximations are made for $n(E_F)$ and $v(E_F)$ as were made previously, this may be expressed as

$$\tau(E_F) = \frac{1}{n(E_F)v(E_F)A(E_F)_T} \cong \frac{\text{constant}_4}{A(E_F)_T} \tag{6-64}$$

The inverse relationship shown in Eq. (6-64) also holds for $\bar{k}(E_F)_T$ because of Eq. (6-60). Again, by means of Eq. (6-17), using Eq. (6-64),

$$\rho = \frac{m}{n(E_F)e^2} \cdot \frac{1}{\tau(E_F)} \cong \frac{m}{n(E_F)e^2} \cdot \frac{A(E_F)_T}{\text{constant}_4} \cong \text{constant}_5\, A(E_F)_T \tag{6-64a}$$

so Eqs. (6-63a) and (6-64a) are essentially the same.

In the more concentrated alloys, the effects of increases in alloy content on resistivity will be greater than those caused by increases in temperature. This is based on the relative influence of the factors in Eq. (6-50). The alloy additions have large effects on $A(E_F)_I$ and $A(E_F)_e$. The very presence of alloying ions, whose sizes, masses, and charges inevitably are different from those of the host atom, will cause greater changes than are possible by increases in temperature on the "phonon gas", since n_p approaches a large, limiting value, [Section 5.3 and Eq. (5-207)]. The lattice distortions and resultant obstructions to electron flow caused by these factors are much more effective scatterers than those resulting primarily from virtually any vibrational effects caused by temperature alone as represented by $A(E_F)_L$ in nondilute alloys. Thus, the increase in the scattering of electrons by phonons will be small compared to the scattering caused by the presence of alloy ions. Consequently, the slopes of the curves of resistivity or resistance versus temperature of such alloys would be expected to be small and to decrease with alloy content. In fact, a general rule is that the greater the alloy content in a random solid solution, the smaller will be the slope of such curves. This effect is described in more detail in Section 6.1.4.

In certain cases in which the appropriate alloy content is sufficiently large and such that increases in temperature cause the Fermi surface to begin to overlap into the next Brillouin zone, the wavelengths, or mean free paths, of the electrons begin to increase slightly. This has the effect of slightly decreasing $\bar{k}(E_F)_T$ as the overlap increases with increasing temperature, thus correspondingly decreasing the resistivity. Some alloys of this kind show a flat, nearly parabolic type resistivity versus temperature relationship. These have maxima at the temperatures at which the overlapping of the zones just becomes appreciable. Some such alloys also show minima at about 300° to 400°C, where $\bar{k}(E_F)_T$ begins to increase again. The temperatures at which the maxima and minima occur and the curve shapes can be controlled by varying the composition of the alloy.

One other solid-state reaction requires mention here. All the alloys considered thus far have been random solid solutions. Where short-range ordering takes place

TABLE 6-3 TEMPERATURE COEFFICIENTS OF ELECTRICAL RESISTANCE OF SOME SELECTED METALS AND ALLOYS

Typical Composition[a]	$\dfrac{\Delta R}{R\,\Delta T}$ ($\Omega/\Omega\,°C \times 10^6$)	Typical Composition[a]	$\dfrac{\Delta R}{R\,\Delta T}$ ($\Omega/\Omega\,°C \times 10^6$)
99.99 Cu	$+4270$ ($0°-50°C$)	99.8 Ni	$+6000$ ($20°-35°C$)
98 Cu–2 Ni	$+1500$ ($0°-100°C$)	71 Ni–29 Fe	$+4500$ ($20°-100°C$)
94 Cu–6 Ni	$+800$ ($0°-100°C$)	80 Ni–20 Cr	$+85$ ($-55°-100°C$)
89 Cu–11 Ni	$+400$ ($0°-100°C$)	75 Ni–20 Cr–3 Al +	
78 Cu–22 Ni	$+300$ ($0°-100°C$)	Cu or Fe	±20 ($-55°-100°C$)
55 Cu–45 Ni	±40 ($20°-100°C$)		
87 Cu–13 Mn	±15 ($15°-35°C$)	60 Ni–16 Cr–24 Fe	$+150$ ($20°-1000°C$)
83 Cu–13 Mn–4 Ni	±10 ($15°-35°C$)	35 Ni–20 Cr–45 Fe	$+350$ ($20°-100°C$)

[a]Weight percent.
From "Magnetic Electrical and Other Special-Purpose Materials," *Metals Handbook, Properties and Selection*, Vol. 1, 8th ed., T. Lyman, Ed., American Society for Metals, Metals Park, Ohio, 1961, pp. 779–864.

(Section 6.1.7), the calculated value for $\Delta\rho$ [Eqs. (6-55) and (6-56)] will be larger than the experimental value by approximately 5%. Where long-range ordering occurs, this difference will be very much greater. The previously noted equations should not be used for this case since they are derived on the implicit assumption of random occupation of lattice sites. This decrease in the resistivity comes about because relatively large fractions of the alloy atoms now are situated on preferred lattice sites; they no longer occupy random lattice sites. This increases the degree of lattice regularity and effectively decreases $\bar{k}(E_F)_T$ in Eq. (6-58). This explains the decrease in resistivity observed when any ordering occurs. Alloys showing this behavior normally are not used in electrical instrumentation because of potential electrical instability.

Important exceptions to this behavior are some of the commercial quaternary Ni–Cr–Al–X alloys (Table 6-3). The mechanisms responsible for their properties have not, as yet, been resolved. Both short-range ordering and clustering have been proposed as explanations of their anomalous increases in resistivity and decreases in their temperature coefficients of resistance. It has also been suggested that ferromagnetic, ferrimagnetic and antiferromagnetic transitions, shown to be responsible for other electrical anomalies in other, similar alloys, might influence the observed behavior of these alloys, depending on their composition. See Section 6.1.7.

6.1.4 Temperature Coefficients of Resistivity

The temperature coefficients of electrical resistivities of alloys must be known if resistance materials are to be used properly in engineering applications. This is the case because either Joule or ambient heating will change the temperature of the component and, consequently, its resistance. Therefore, it is very helpful if this property can be approximated closely when more accurate data are unavailable. It turns out that this can be done quite simply by the means derived here.

As previously noted, Matthiessen's original findings are valid only for relatively dilute alloys. However, these may be used to obtain an approximation of the temperature dependence of resistance alloys ($>3\,\text{At}\%$ alloying elements). Starting with

$$\frac{\Delta\rho(C_A)}{\Delta T} = \frac{\Delta\rho(C_0)}{\Delta T} \tag{6-31}$$

and multiplying the left side of Eq. (6-31) by $\rho(C_A)/\rho(C_A)$ and the right side by $\rho(C_0)/\rho(C_0)$, one obtains

$$\rho(C_A)\frac{\Delta\rho(C_A)}{\rho(C_A)\Delta T} = \rho(C_0)\frac{\Delta\rho(C_0)}{\rho(C_0)\Delta T} \tag{6-65}$$

Since the *temperature coefficient of electrical resistivity* is defined by

$$\alpha \equiv \frac{\Delta\rho}{\rho\,\Delta T} \qquad (\Omega\text{-cm}/\Omega\text{-cm}\cdot{}^{\circ}\text{C}) \tag{6-66}$$

Eq. (6-65) becomes

$$\rho(C_A)\alpha_A = \rho(C_0)\alpha_0 \tag{6-67}$$

The properties of the "pure" base materials, $\rho(C_0)$ and α_0, are known so that

$$\rho(C_A)\alpha_A = \rho(C_0)\alpha_0 = K_M \tag{6-68}$$

Equation (6-68) is a hyperbolic equation where K_M is the *Matthiessen constant*. Data for some metals and alloys are given in Table 6-3.

The relationships between α_A and $\alpha_R = \Delta R/R\,\Delta T$, the *temperature coefficient of electrical resistance*, are important. It is useful to note that $\alpha_A \cong \alpha_R$, with a maximum error of 0.2%, for $\Delta T \leqslant 100^{\circ}\text{C}$ and $\alpha_L \leqslant 20\ \mu\text{m/m}^{\circ}\text{C}$. This includes virtually all commercial metals and alloys with the exception of aluminum. This relationship *is not valid* for applications intended for use over wider temperature ranges.

Equations (6-55), (6-56), and (6-68) can be employed to predict the electrical resistivity and temperature coefficient of nearly all annealed solid-solution alloys. At any given temperature, the resistivity of an alloy is given, based on Eq. (6-56), by

$$\rho(C_A) = \rho(C_0) + \Delta\rho(C_A) \tag{6-69}$$

Since $\Delta\rho(C_A)$ is given by Eq. (6-55), it can be employed in Eq. (6-69) to give the resistivity of the annealed solid-solution alloy at the temperature of interest. This, in turn, can be substituted into Eq. (6-68) to compute its temperature coefficient of resistivity. Calculations for α made in this way give good approximations for those alloys that have linear, or nearly linear, resistivity–temperature characteristics.

The temperature coefficient of electrical resistivity is very sensitive to small quantities of impurity elements present in solution in a "pure" metal. This may be demonstrated in an uncomplicated way by considering the effect of a small, binary addition to a base metal. The temperature coefficients of such alloys may be determined from Eq. (6-68) rearranged as

$$\alpha_A = \frac{K_M}{\rho_A} \tag{6-68a}$$

Equation (6-68a) is differentiated to obtain

$$d\alpha_A = -\frac{K_M}{\rho_A^2} \, d\rho_A \tag{6-70}$$

Expressions for ρ_A and $d\rho_A$ are obtained from Eqs. (6-55) and (6-56) as follows:

$$\rho_A = \rho_0 + \Delta\rho_A = \rho_0 + [K_2 + K_1(Z_\beta - Z_\alpha)^2]\Delta C_A \tag{6-71}$$

Equation (6-71) may be reexpressed more simply for a given alloying element as

$$\rho_A = \rho_0 + K_3 C_A, \qquad K_3 = K_2 + K_1(Z_\beta - Z_\alpha)^2 \tag{6-72a}$$

Then, upon differentiation of Eq. (6-72a),

$$d\rho_A = K_3 \, dC_A \tag{6-72b}$$

since ρ_0 is a constant and vanishes. In cases in which the influence of the impurity content is such that its presence in small amounts has the effect that $\Delta\rho_A \geqslant \rho_0$, Eq. (6-72a) may be approximated as

$$\rho_A \cong K_3 C_A \tag{6-72c}$$

Equations (6-72b) and (c) are substituted into Eq. (6-70) to give

$$d\alpha_A \cong -\frac{K_M}{K_3^2 C_A^2} \cdot K_3 \, dC_A = -\frac{K_M}{K_3 C_A^2} \, dC_A \tag{6-73a}$$

or

$$\frac{d\alpha_A}{dC_A} \cong -\frac{\text{constant}}{C_A^2}, \qquad C_A > 0 \quad \text{and} \quad \text{constant} = \frac{K_M}{K_3} \tag{6-73b}$$

On the basis of Eq. (6-73b) it is expected that the rate of change of α will be large and will decrease rapidly with composition. This is especially true for small alloying or impurity contents of less than one atom percent. This high sensitivity is consistent with Eq. (6-70) and with the experimental data.

The integration of Eq. (6-73b) results in an effective reexpression of Eq. (6-68):

$$\alpha_A \cong \frac{\text{constant}}{C_A} \tag{6-74a}$$

or

$$\alpha_A C_A \cong \text{constant} \tag{6-74b}$$

The sensitivity noted for Eq. (6-73b) results from the hyperbolic nature of Eq. (6-74b). This is shown schematically in Fig. 6-6, where the effect of small impurity or alloying contents is emphasized by the steep initial slope of the curve.

Equations (6-73b) and (6-74b) were derived to illustrate the powerful influence of alloying or impurity elements on α. These equations are in good agreement with the observed behavior, but are most accurate for solute elements with strong influences upon $\Delta\rho$. This is because of the simplifying approximation made to obtain Eq. (6-72c).

A more general relationship may be obtained for the effect of any small alloy or impurity contents on the temperature coefficient of electrical resistivity.

This may be shown by beginning with Eq. (6-67), using a simplified notation:

$$\alpha\rho = \alpha_0\rho_0 \tag{6-75}$$

in which the subscript zero denotes the properties of the "pure" metal. Good practice requires that the factors used for the calculation of α be determined at fixed reproducible temperatures. The properties at $0°$ and $100°C$ are commonly used for this purpose. Thus, $\Delta T = 100°C$, and ρ is the resistivity of the material being tested at $0°C$. The selection of such a reference temperature permits the consideration of ρ as a function only of composition [Eqs. (6-55) and (6-56a)].

The previously noted hyperbolic nature of Eqs. (6-73b) and (6-74b) may be expressed as

$$\alpha = \alpha_0 f(C) \tag{6-76}$$

where $f(C)$ is a hyperbolic function of composition in atom percent.

For $C \leqslant 1 \, \text{At}\%$ it is found that

$$\alpha \cong \alpha_0 - \frac{C}{\gamma(1 + C)} \tag{6-77a}$$

Or

$$\alpha \cong \alpha_0 - \frac{C}{\gamma + \gamma C} \tag{6-77b}$$

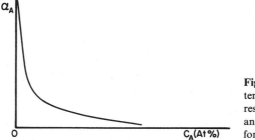

Figure 6-6 Schematic diagram of the temperature coefficient of electrical resistance as a function of composition of annealed, dilute, random, solid solutions for a specific temperature range.

in which $\gamma = 1/\alpha_0$. Equations (6-77a) and (b) are valid for annealed metals in which the small alloy or impurity contents are in solid solution. These equations have added utility because they do not require the determination of empirical coefficients. The value of α_0 may be obtained from Table 6-1 or from handbooks. They also clearly demonstrate the strong effect of small amounts of alloying or impurity elements on the temperature coefficient of electrical resistance.

Wider application of Eqs. (6-77a) and (b) may be obtained when *equivalent composition*, rather than atom percent, is used. This parameter is calculated by multiplying the atom percent of an impurity or alloy element by its valence. For example, 0.20 atom percent of antimony, valence 5, is the equivalent of 0.25 atom percent of tin, valence 4. In each case the products give the same equivalent composition of 1.00 equivalent atom percent.

The importance of valence is indicated in Eqs. (6-51b) and (6-55). These equations can only show the effects per atom percent of various individual elements and thus provide an indirect basis for comparison. The use of equivalent compositions places alloy or impurity elements on a common basis and permits the effects of many such elements to be expressed by a single equation or by a single curve. The use of this parameter in effect provides a generic means for the comparison of the effects of such elements.

This approach makes it possible to obtain a good estimate of the total impurity content. When Eq. (6-77a) is rewritten as

$$C \cong \frac{\alpha_0}{\alpha} - 1 \tag{6-78}$$

it is only necessary to obtain a good measurement for α in order that C may be calculated; α_0 is known. The value for C obtained in this way is the approximate, *total* impurity content in equivalent atom percent. This is not a method of chemical analysis. It can detect *total* impurity contents of the order of 0.02 equivalent atom percent.

Another practical application of Eqs. (6-77b) or (6-78) may be in their use to define the purity of metal by specifying a minimum acceptable value for α. This constitutes a quality control because it defines a maximum allowable total impurity content without the necessity of requiring chemical analyses. Characterizations of this kind can be useful when the costs of the chemical analyses of several impurity elements, present in small quantities, may not be warranted.

Impurity contents of the order of 0.02 equivalent atom percent may be unacceptable for some standards-grade materials. One example of this is that of the reference platinum, Pt-67. The electrical properties of this material constitute one of the bases for the International Practical Temperature Scale. This material may be defined by an average temperature coefficient of electrical resistance of $0.0039269 \, \Omega/\Omega/K$ for the $0°$ to $100°C$ temperature range that was described previously. The purity of nickel intended for use in resistance thermometry also may be defined in a similar way. For example, this may be controlled by specifying $\alpha_R = 0.0065 \, \Omega/\Omega/°C$, minimum, for the conditions noted. It will be understood that highly accurate temperature measurement and control as well as electrical measurement techniques must be used to characterize materials for these precision applications. The major problem in the establishment of a suitable curve or equation for α versus C resides primarily in the difficulties inherent in the required chemical analyses and not in the physical testing.

The value of α is lowered by cold working a given alloy. Cold working increases ρ, and, as may be seen from Eq. (6-68a), α is diminished accordingly. However, such increases in ρ are limited (see Section 6.1.9); this limits the change in α.

Data in the literature for the electrical properties of cold-worked metals and alloys are sometimes subject to question because stress relief may have occurred at room temperature in the interval between cold working and electrical measurement. In addition, in some cases the heat generated in the metal during cold working may be sufficient to induce stress relief, recovery, or even recrystallization, if the rate of working is sufficiently severe. Autogeneous processes of these kinds may occur at or below room temperature.

6.1.5 Application to Phase Equilibria

Electrical properties were among the earliest means employed to determine the ranges of constitutional phases as functions of composition and/or temperature. A series of isothermal plots of these electrical properties, or their changes, versus composition continue to be used for such work. The changes in the slopes of such plots can demark phase boundaries as functions of composition and temperature. The use of two of such properties for this purpose is described here.

Equations (6-55) and (6-56b) can be employed to verify phase relationships both for binary and pseudobinary systems. Here, isothermal electrical measurements are made of the properties of the equilibrated alloys. The only variable is that of composition. Both of these equations hold only for alloying compositions within the limits of solid solubility. Beyond the limits of solid solubility, a change in slope occurs because the resistivity then follows the law of mixtures. The composition at which the slope changes is that of the phase boundary. This is shown in Fig. 6-7(a). The hyperbolic behavior of the temperature coefficient of resistivity, noted in Eq. (6-68) and Fig. 6-6, can be used to verify the resistivity data. However, the demarcation between phase fields is much more difficult to establish by means of this property. Both methods, as described here, are applicable only in the solid state.

Where complete solid insolubility exists between two components, or two phases, the relationships are as shown in Fig. 6-7(b). Here a linear relationship exists across the entire diagram because this case represents a mixture of two components.

In the case of isomorphous binary systems (complete solid solubility), the resistivity isotherm is a linear function of composition for about 25% to 30% of the range beyond each "pure" component. It reaches a maximum in the neighborhood of 50% of each component [Fig. 6-7(c)].

Nordheim's rule may be used to approximate the location of such maxima. This

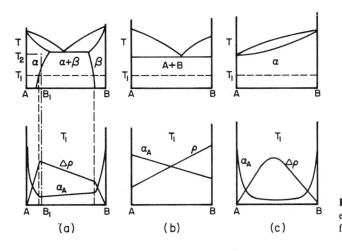

Figure 6-7 Correlation between some electrical properties and phase equilibria for some common metallic systems.

rule gives the composition dependence of the isothermal resistivity as

$$\rho \cong AC(1 - C) = AC - AC^2 \tag{6-79}$$

in which A is an empirical constant. The differentiation of Eq. (6-79) gives

$$\frac{d\rho}{dC} \cong A - 2AC \tag{6-80}$$

And, where a maximum exists, Eq. (6-80) may be equated to zero:

$$A - 2AC \cong 0 \tag{6-80a}$$

The constant A vanishes and

$$C = 0.5 \cong 50 At\% \tag{6-80b}$$

Relatively few of the maxima occur exactly at $50\,At\%$ as indicated by Eq. (6-80b). Most maxima lie well within the range of about $50 \pm 10\,At\%$. The composition of a maximum may be located more closely if data for some of the dilute binary alloys of each terminal component are available. Statistical methods are used to fit these data to an equation of the form

$$\rho \cong AC(1 - BC) \tag{6-81}$$

The introduction of the second empirical constant, B, in Eq. (6-81) reduces the symmetry of Eq. (6-79). This small asymmetry is responsible for the closer approximation provided by Eq. (6-81) as compared to that provided by Eq. (6-79). Here the composition of the maximum is given by $C \cong 0.5B^{-1}$, where B usually lies between about 0.8 to 1.3, depending on the alloy system.

On the basis of the foregoing, it can be seen that data from well-annealed alloys, suitably equilibrated and measured at various isotherms, can reveal phase fields and boundaries as functions of composition and temperature. This technique is limited to temperatures at which the alloys are in the solid state.

These techniques also have been employed to investigate more complex alloy systems. Consider an alloy of n components. The alloy base will consist of $n-1$ components that are maintained at constant ratios. The composition and properties of this base then constitute one terminal of a pseudobinary system. Alloy additions of the single component being investigated are made to the fixed, complex, terminal-alloy base. Any suitable combination of the $n-1$ components may be used as the terminal alloy. In employing this approach, it is imperative to maintain the $n-1$ components of the base as closely to their nominal ratios as possible.

Isothermal plots of the electrical properties of these multicomponent alloys are made in the same ways as described for binary alloys. The assembly of several such plots, taken at different isotherms, can be made to constitute mappings of constitutional phases as functions of both composition and temperature.

It should be emphasized that these electrical measurements can provide unambiguous data only when the metallurgical specimens approach equilibrium at the temperatures at which the tests are made.

6.1.6 Precipitation Effects

Many engineering alloys that derive desirable sets of other physical or mechanical properties from precipitation phenomena must be used with caution, if at all, in electrical instrumentation applications. The reason for this is that small changes in the conditions of the precipitates, which may occur as a function of time, may cause relatively large changes in the electrical properties, even at room temperature.

As previously noted (Section 6.1.3), the presence of alloying elements in solid solution increases the electrical resistivity of a metal. The precipitation of such ions, whether they are in the form of the pure component, a terminal solid solution, or an intermetallic phase, would be expected to cause a decrease in the electrical resistivity of the alloy. This frequently does occur, but the resistivity of an equilibrated two-phase alloy may be greater or less than that of the single-phase alloys bounding a two-phase field. The slope as a function of composition depends on the properties of the two alloys of which the two-phase mixture is composed.

Reference to Fig. 6-7(a) shows that precipitation of a second phase from a supersaturated solid solution can either increase or decrease the resistivity of the alloy compared to the respective terminal solid solutions. Here the A-rich, two-phase alloys show a decrease in resistivity. The B-rich alloys show the opposite. Both cases assume equilibrium microstructures.

An alloy of composition B_1, in Fig. 6-7(a), that had been rapidly quenched (see later) from temperature T_2 would represent a metastable, supersaturated, non-equilibrium solid solution. Its higher resistivity will diminish with time, the rate depending on the kinetics of the precipitation mechanism and the temperature at which the alloy is maintained. The resultant properties of the alloy will vary depending on the elapsed time and temperature, the type of precipitate, its precipitation mode, and the degree of departure of the alloy from equilibrium. For example, small particles precipitating within a grain will impart different effects when compared to similar precipitation at the grain boundaries. In addition, if these particles are smaller than the equilibrium size, the electrical properties of the alloy will change if and when they are allowed to grow so that they approach equilibrium conditions. The composition of these particles also influences the electrical properties of the alloy.

If, however, the precipitates are coherent (have nonequilibrium compositions and crystal lattices that have close registries with that of the host lattice) in nature, added effects are noted. The mechanisms involved in the precipitation of such phases are considered to follow the Guinier–Preston (G–P) theory (see Fig. 6-8). As the incipient, nonequilibrium particles nucleate from the supersaturated solid solution and grow, they add a large strain component (coherency strain) to the lattice, which increases $\bar{k}(E_F)_T$. The alloy content of the host lattice simultaneously becomes slightly depleted. The lattice coherency strain energy overwhelmingly predominates over the depletion effects, and it becomes larger as the G–P zones continue to grow and to increase the lattice distortion. This strain component reaches a maximum that is indicated by the maximum resistivity. At this point, the particles commence to approach both their equilibrium composition and lattice structure and begin to lose their coherency with the host lattice. When this occurs, the strain energy starts to diminish, the host lattice becomes less distorted, and the decreased resistivity reflects this. As the particles continue to grow, they behave increasingly more like noncoherent particles until they show noncoherent behavior and approach equilibrium size and composition. They finally become dispersed particles of the precipitating phase in the host matrix. The electrical resistivity decreases relatively rapidly as most of the lattice strain is relieved; it then asymptotically approaches its equilibrium value.

The detection of maximum coherency is difficult to observe by most methods of detection, including electron microscopy. Electrical resistance, or resistivity, data are ideal for the purpose of determining aging times and temperatures, since they readily and economically reflect the precipitation process. Maximum hardness, yield, and tensile strengths occur in precipitation-hardening alloys when the number of G–P zones is greatest and the resistivity is maximum. This resistometric method is used

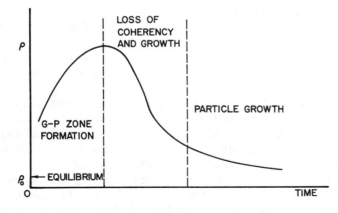

Figure 6-8 Schematic diagram of isothermal changes in electrical resistivity of a supersaturated, age-hardening alloy as a function of aging time.

commercially for the accurate determination of the thermal treatments of such alloys. It is widely used for the treatment of aluminum alloys in aircraft construction. These alloys frequently are aged for times just beyond that of the maximum shown in Fig. 6-8. This is done to improve their resistance to corrosion at the expense of relatively small decreases in yield or tensile strength. Aluminum alloys that have been given this treatment are designated as being in the T7 condition.

The combination of electrical conductivity and hardness measurements has been shown to provide insight into the thermal treatments to which age-hardenable aluminum alloys have been subjected and into their resulting metallurgical structures. *Nucleation diagrams*, which are analogous to the time–temperature–transformation diagrams for steels, may be constructed from such data. These diagrams show the isothermal nucleation and growth of the G–P zones and their transformation to the equilibrium precipitate as functions of aging temperatures and times after quenching from the solution temperature. Such diagrams could make it possible to optimize the thermal treatments of other age-hardenable alloys, as well as those of aluminum. They could also serve as bases for the quality control of the heat treatments of such alloys that have been made under commercial production conditions.

It should be noted that quenching a metallic specimen will also induce lattice strains. The degree of such strains will depend on the modulus of elasticity of the material as a function of temperature, the magnitude of the temperature differential, and the rate of the quench. Material strained in this way will also show an increase in resistivity; this increment will be metastable and will decrease with time, even at room temperature.

Electrical resistivity measurements are also capable of detecting very small amounts of microsegregation at cryogenic temperatures. One such case involved the identification of highly dispersed, very small precipitates of tin in bismuth. These were not detected by an electron microscope equipped for x-ray fluorescence analyses. Electrical resistivity measurement showed the precipitate to be tin, but was not able to determine their geometry or size.

6.1.7 Order–Disorder, Clustering

In previous sections the influences of alloying ions were approximated on the assumption that they occupied random sites on the host lattice. This is not always applicable. In some cases the alloy and host ions are sited so that they form regular, periodic arrays. The extent to which this ordering occurs within a lattice has strong influences on the physical, chemical, and mechanical properties of the alloy.

The effects of alloying ions in random solid solution are dramatically illustrated by their near absence when a long-range ordering transformation occurs within a given alloy. In this case, the alloy ions occupy preferred sites over long distances in the host lattice, rather than random ones. Such a lattice, also known as a *superlattice*, has a much greater degree of spatial periodicity, or regularity, than the same alloy in the random state. The same is true for the periodicity of the electrical potential within the lattice.

As mentioned at the end of Section 6.1.3, ordering of the alloy ions in the host lattice increases the spatial and electrical regularity of the lattice and can result in a large decrease in resistivity because large decreases can occur in $\bar{k}(E_F)_T$. This is shown clearly in the classic Cu–Au system in Fig. 6-9. The quenched alloys, many of which are metastable, show the behavior previously noted in Fig. 6-7(c) as being typical of disordered or random, substitutional solid solutions. When these alloys are aged and become ordered, large decreases in the electrical resistivity are observed. The maximum changes occur at 25% and 50% Au in Cu. Superlattices are formed at these compositions. At the 25% alloy composition, nearly all the Au atoms occupy corner sites in the FCC lattice of the copper. At the 50% alloy composition, the Au atoms may be considered to occupy body-centered sites, with Cu atoms at the corners of the unit cell, or vice versa. This actually is a CsCl-type lattice. Both ordered compositions represent much higher degrees of lattice spatial regularity and uniformity of electrical periodicity than those of random, or disordered, substitutional solid solutions. Under these inherently uniform conditions, $\bar{k}(E_F)_T$ becomes considerably smaller because the scattering by the more regular, ordered lattice is much less than that of a random solid solution. From Eq. (6-58), it can be seen that the electrical resistivity would decrease. The total wave vector decreases because the decreased scattering results in longer effective electron wavelengths. Alloys of the Cu–Au system other than the two discussed here have structures with varying degrees of order and disorder.

The short-range ordering mentioned in Section 6.1.3 concerns the case in which the ions take preferred sites over comparatively short distances in the lattice. This situation is most readily pictured in terms of the nearest neighbors of a given, reference ion. Thus, short-range ordering may be considered as that situation in which the reference ion has more unlike ionic neighbors. The net result is that only very small volumes of the lattice have high degrees of spatial and electrical periodicities. These have the same qualitative effects that were described for long-range order, but the magnitudes usually are, as cited in Section 6.1.3, very much smaller.

The case in which the reference ion has more ionic neighbors of the same kind as

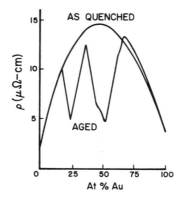

Figure 6-9 Effects of disorder and order on the electrical resistivity of Cu–Au alloys. (From C. S. Barrett, *Structure of Metals*, p. 288, McGraw Hill Book Co., New York, 1952.)

itself is known as *clustering*. This condition also results in small volumes of the lattice with high degrees of spatial and electrical periodicities. The effects of clustering on electrical resistivity may be considered to be approximately the same as those described for short-range ordering.

It also will be noted that any significant decreases in the resistivity caused by ordering or clustering will be accompanied by corresponding increases in the temperature coefficient of resistivity [Eq. (6-68)].

Alloys showing such metastable behaviors normally are not used for electrical instrumentation or for electronic components. As previously noted (Section 6.1.3), the quarternary Ni−Cr−Al−X alloys (Table 6.3) constitute an important exception to this behavior. Their electrical properties are ideal for use in some precision electrical resistors.

6.1.8 Allotropic Changes

Many elemental metals and their alloys possess different crystal lattices known as *allotropes*. The lattice type present depends on the ambient temperatures and/or pressures. Such solid-state transformations are first-order phase changes. Each allotrope has different properties because each lattice is different. This results in a different Brillouin zone for each allotrope.

When a metal or an alloy undergoes an allotropic change, $\bar{k}(E_F)_T$ changes because the lattice type changes. This is reflected in a change in electrical resistivity. The resulting discontinuity of the curve of electrical resistivity as a function of temperature can thus be used to mark such a phase boundary.

Such measurements must be made using very slow heating and cooling rates. If this is not the case, nonequilibrium effects can mask the transition; the discontinuity may be suppressed. Hysteresis effects can also affect such determinations. For example, if an alloy contains elements with slow diffusion rates in the temperature range of interest, the transition could occur in heating over a spread of temperatures. Upon cooling such alloys, hysteresis effects are also observed. If the same slow absolute values of heating and cooling rates are employed, the allotropic change may

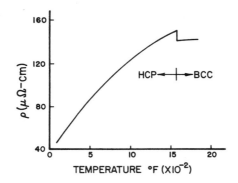

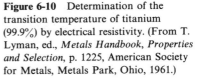

Figure 6-10 Determination of the transition temperature of titanium (99.9%) by electrical resistivity. (From T. Lyman, ed., *Metals Handbook, Properties and Selection*, p. 1225, American Society for Metals, Metals Park, Ohio, 1961.)

be approximated as starting halfway between the beginnings of the transformations noted upon heating and cooling. The slower the heating and cooling rates, the more accurate will be the determination of the transformation temperature. A typical allotropic change as determined by electrical resistivity is shown in Fig. 6-10.

6.1.9 Effects of Deformation

The crystal lattices of cold-worked metals are deformed by such treatments and may be highly stressed and imperfect, depending on the kind, rate, and amount of deformation. They will remain in this condition, to a large extent, unless thermal treatments are applied. The resulting, irregular spatial and electrical periodicities of such lattices are very effective in scattering electrons. This condition also results in distorted Brillouin zones, accounting for the increased anisotropy of properties; these factors can also cause large increases and variations in $\bar{k}(E_F)_T$.

The working, or plastic deformation, of a metal or alloy below its recrystallization temperature or below a phase reaction isotherm usually results in a highly stressed microstructure that consists of fine, distorted grains with orientations that are influenced by the kind, rate, amount, and direction of the working. In addition, the structures of the small grains are highly imperfect and contain many types of lattice imperfections. Metals treated in this way would be expected to show more highly anisotropic electrical and other physical properties than they do in the annealed condition because of the lattice distortions, which result in warped or distorted Brillouin zones.

The electrical resistivity of metallic materials subjected to such treatments would be expected to show greater resistivities than those in the annealed condition. This is anticipated because $\bar{k}(E_F)_T$ [Eqs. (6-45) or (6-58)] would be expected to be increased by the condition of the highly imperfect and distorted lattices of the grains, as well as by the increased quantity of grain boundary material.

The increase in resistivity induced by cold working is not as great as one might think. It appears that a maximum increase exists for a given metallic material. For example, very pure metals show increases of the order of up to about 5%. The maximum increase in the resistivity of commercially pure metals lies between about 4% and 8%. Alloys may show resistivity increases of from about 10% to 25%. The data indicate that purity greatly affects this behavior. It is well known that the presence of alloying, or impurity, elements increases the recrystallization temperature. The purest material has the lowest recrystallization temperature; its physical properties are more susceptible to change by thermal means than are those of less pure materials. The mechanical energy that is converted to heat during the working process could be sufficient to recrystallize or to induce some degree of recovery or stress relief in the purer materials during the working process. The less pure, or alloy, materials could have been subject to some lesser degree of recovery or stress relief; these would show correspondingly higher resistivity increments upon annealing, in agreement with the observed data. Such autogeneous thermal effects can obscure the true effects of cold working on the electrical properties. Working that is suitably performed at cryogenic temperatures can reveal virtually all the effects of cold working.

The changes in the electrical properties that are induced by cold working may be eliminated by suitable thermal treatments. If such treatments are not performed, the internal stresses will relieve themselves, even at room temperature. Studies have shown that the stresses induced in annealed wires during the fabrication of resistors or of potentiometers for electrical instrumentation will continue to show decreasing resistances for indefinitely long periods of time at room temperature. This electrical instability is a measure of the stress relief that is occurring. The best solution to this problem is to fully anneal the cold-worked material. However, as is the case for many components, such as resistors, a full anneal would destroy any coatings or insulation materials present on the surface of the part. In this case the maximum temperature of any thermal treatment is determined by the nature of the surface material. Thermal treatments at temperatures as low as $100°$ to $150°C$ for 24 to 48 hours can have the effect of greatly reducing the electrical instabilities that result from the autogeneous relief of internal stress.

It is considered that some of the scatter in the published data for the effects of cold working on electrical properties may be explained as resulting from some degree of stress relief. Unless the electrical measurements are made immediately after working, the measurement will contain an error that is a function of the elapsed time at room temperature.

Dimensional instability also arises from the relief of internal stresses in the same way as electrical instability. This may be eliminated by suitable thermal treatments.

6.1.10 Commercially Available Alloys: General

Large numbers of alloys intended for electrical applications are commercially available. Alloys for applications in electrical and electronic instrumentation are melted and processed very carefully to maintain their compositions within narrow limits and to avoid impurities and contaminations. Many alloy producers provide very similar alloys, each under its own trade name. Some of the more commonly used alloy types are discussed here.

Some of the copper–nickel (approximately 2% to 20% Ni) alloys have low resistivities. These are offered to provide ranges of resistivities and temperature coefficients for specific applications and sometimes to carry relatively high currents. Here it becomes necessary to design the resistors to help dissipate the Joule heat. This heating effect must also be taken into account in the determination of the resistances of the components of the circuits at the temperatures of operation (Section 6.1.4). Relatively large changes in resistance can occur when the temperature coefficient of resistance is large. These changes must be taken into account for the components to be able to perform the desired functions in the circuit under operating conditions.

The commercial *constantans* (about 55–65 Cu, 45–35 Ni) are used in precision resistors. Their use is based not only on their medium resistivities, but primarily on their relatively flat, nearly parabolic resistance–temperature characteristics over temperatures from $-55°$ to $105°C$.

Many constantans have very flat, nearly parabolic resistance–temperature curves at low temperatures that have maxima in the neighborhood of room temperature. These properties are a result of Brillouin zone overlap (see Section 6.1.3). The curve shapes and locations of the maxima may be controlled by manipulating the compositions of this class of alloys. Each alloy within the nominal composition range differs somewhat in properties. Most constantans contain small but significant quantities of Fe, Mn, Co, Si, and other alloying elements, depending on the source. These alloys almost invariably are purchased in the annealed condition. It is customary to stress relieve electrical components made from them after fabrication (Section 6.1.9).

The small temperature coefficients of resistance, resulting from the flat resistance–temperature curves, make constantans desirable for use as electrical components that will not change appreciably in resistance calibration as the ambient temperature changes in the vicinity of room temperature. However, the large thermoelectric power of constantans versus copper must be considered in applying these alloys to components for very accurate electrical and electronic instrumentation. The constantans can form small thermocouples with the copper connecting wires. These can create small, extraneous parasitic voltages and currents when temperature differentials exist throughout the electrical circuitry. This can result in the introduction of errors in high-accuracy devices intended to measure very small voltages or currents. One method for the elimination of this problem is to enclose and to thermostatically regulate such temperature-sensitive circuits.

Manganin, like constantan, is a generic term for a class of alloys for electrical resistors. Manganin-type alloys usually are solid solutions of copper, manganese, and nickel with small amounts of other alloy and impurity elements. Many manganin-type alloys, like constantans, are characterized by very flat, nearly parabolic resistance–temperature curves whose maxima are in the neighborhood of room temperature. The temperature coefficients of many manganins *over relatively narrow temperature ranges* on either side of the maxima can be smaller than those of constantans. Manganins have a decided advantage over constantans in that their thermoelectric powers versus copper are quite small. This minimizes the potential for those extraneous sources of error cited for constantans. Manganins also can demonstrate a very high degree of electrical stability when properly stress relieved and protected from strains (see Section 6.1.9) and injurious atmospheres.

The maximum points (peak temperatures) and curve shapes of the parabolic resistance–temperature relationships of manganins also are a result of zone overlap and may be controlled by variations in alloy composition. This is useful in order that the peak temperatures of resistors lie close to their temperatures of operation. Thus, the effects of any change in the ambient temperature of the instrumentation will be minimized, and its calibration will remain virtually unaffected. The temperature coefficients of commercial manganins are generally less than $\pm 0.00001 \ \Omega/\Omega/°C$ for a $10°C$ interval on either side of the peak. The peak temperatures of these manganins lie in the range from $20°$ to $30°C$. Peak temperatures close to $25°C$ frequently are specified for manganins intended for use in dc measuring devices. When instrumentation is designed to operate at higher temperatures, the chemical content of the alloy is changed to shift the peak temperature to the desired value. Shunt manganin, which carries high currents and consequently heats considerably in use, usually has its peak temperature in the range from 45° to 65°C, depending on the amount of heating to be generated.

The preceding combinations of properties have lead to the almost universal use of manganin in the fabrication of very high accuracy resistors, slide-wires (potentiometers), and other components for electrical measuring and control devices. Manganin alloys always should be purchased in the fully annealed condition. The components made from them should be stress relieved after fabrication.

The Ni–Cr–Al based alloys have resistivities of about three times that of manganins. They also have nearly parabolic resistance–temperature curves. This means that accurate resistors of higher resistance values can be made to occupy smaller volumes. Their temperature coefficients of resistance are less than ± 0.00002 $\Omega/\Omega/°C$ over the range from $-50°$ to $+100°C$. The average thermoelectric powers of this class of alloys versus copper are small. This combination of properties is excellent for accurate resistors. However, the electrical stability of this class of alloys can be very sensitive to mechanical straining and thermal treatments. They also have the drawback that highly corrosive fluxes are required to produce acceptable soldered or brazed joints. If such fluxes are not completely removed, they continue to corrode and to cause electrical instability. Components made from these alloys should also be stress relieved after fabrication.

The alloys based on the compositional types of Ni–Fe–Cr, Ni–Cr, Fe–Ni–Cr, and Fe–Cr are most widely used for high-temperature service, such as for electric resistance heating elements. (They have relatively high temperature coefficients of electrical resistance and are not ordinarily used for precision electrical resistance applications.) Many of these alloys, especially those containing large amounts of nickel and chromium, should only be used in oxidizing atmospheres. Typical applications of these alloys are their use as furnace components and heating elements.

Such applications should be made with caution. Ni–Cr alloys of this class will undergo selective attack of their chromium components when the partial pressure of the oxygen in the furnace atmosphere drops significantly below its normal atmospheric value. This attack is known as *green rot*. Most of these alloys deteriorate rapidly in reducing atmospheres and in those containing even every small amounts of sulfur. These susceptibilities become most apparent when these alloys are used as heating elements.

6.2 SUPERCONDUCTORS

The behaviors of the electrical resistances of normal metals and alloys in general are described in the preceding sections. These materials decrease in resistivity with decreasing temperature and approach their residual resistivity as T^5 approaches 0 K. Certain other materials do not show this behavior; their resistances decrease with

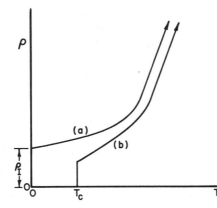

Figure 6-11 Typical electrical resistivity at cryogenic temperatures: (a) normal metals and (b) type I superconductors.

temperature and then more or less abruptly drop to zero at various critical or transition temperatures below about 24 K, depending on the material. They show resistanceless, or superconducting, properties at these low temperatures. Since $\rho = \bar{E}/j$, this means that their electric fields are zero at any point in the solid in the superconducting state. This phenomenon was first discovered by Heike Kamerlingh-Onnes (1911) while working with mercury at cryogenic temperatures. Such behavior is shown schematically in Fig. 6-11.

The absence of electrical resistance in the material enables currents to flow for very long times; hence the designation *superconductor*. The maximum theoretical electrical resistance has been approximated to be less than 10^{-23} Ω-cm. Under certain conditions, the decay time of the supercurrent can be expected to be of the order of 10^5 years. Shorter decay times are observed in other superconducting materials because of the effects of irreversible magnetic fluxes within the superconductor; these are described subsequently. For comparison purposes, the decay time of a current in a normal metal is of the order of 10^{-12}s [Eq. (6-12)].

6.2.1 General Properties

The temperature at which the electrical resistance abruptly goes to zero, in the absence of a magnetic field, is the *transition temperature*, T_c. Some superconducting metallic elements, alloys, intermetallic compounds, and semiconductors discontinuously revert to normal behavior as the magnetic field reaches a critical value, H_c, at temperatures below T_c; these materials are known as type I superconductors. Electrical conducting materials that do not show superconducting properties are designated as being *normal* in the sense that they show residual resistivities.

The transition temperatures of some elements are given in Table 6-4. The electrical resistivities of these elements tend to be high at normal temperatures, indicating the presence of strong electron–ion–phonon interactions (Sections 4.3 and 6.2.2). Superconducting elements are infrequent or not present in groups 1A, 1B, 2A, 3A, 5B, 6B, 7B, and 8B of the Periodic Table. Many elements in the remaining groups (2B, 3B, 4A, 4B, 5A, 6A, and 7A) are superconductors. The elements belonging to

TABLE 6-4 SUPERCONDUCTING PROPERTIES OF SOME SELECTED SUPERCONDUCTING ELEMENTS

Element	ρ^*	$T_c(K)$	H_0 (Oe)	Element	ρ^*	T_c (K)	H_0 (Oe)
W	5.3	0.012	1070	Th (α)	15.7	1.37	162
Be	4	0.026	—	Pa	10.8	1.4	—
Ir	5.3	0.14	19	Re	19.3	1.7	193
Hf (α)	35.1	0.165	—	Tl	15	2.4	171
Ti (α)	42	0.39	56	In	8	3.4	293
Ru	7.6	0.49	66	Sn (β)	11	3.7	309
Cd	6.8	0.52	30	Hg	98	4.15	412
Os	9.5	0.65	65	Ta	13.5	4.48	830
U (α)	30	0.68 (?)	—	V	24.8	5.3	1020
Zr (α)	45	0.55	47	La (β)	61.5	5.9	1600
Zn	5.9	0.85	52	Pb	20.6	7.2	803
Mo	5.2	0.92	98	Tc	18.5	8.2	1410
Ga	15	1.09	59	Nb	15	9.2–9.4	1950
Al	2.6	1.19	99				

[a]In units of microohms-centimeter near room temperature.

After E. M. Savitskii et al., *Superconducting Materials*, p. 82, © Plenum Press, New York, 1973, by permission.

these groups have relatively high electrical resistivities arising primarily from electron–phonon interactions. See Tables 6-1 and 6-4 and Sections 6.1 and 6.1.1. This is important in the generation of Cooper pairs, as described later.

While superconductivity does not appear to be an inherent atomic property, it has been demonstrated empirically that the average number of valence electrons per atom must lie between two and eight for superconductivity to occur; a complex relationship has been used to approximate the effect of the electron:ion ratio on the superconductivity transition temperature. The transition temperatures of alloys composed of elements that are superconductors with other superconductor elements, or composed of superconducting elements alloyed with normal metals, may be higher or lower than those of the superconductors involved. Some of these alloys can show superconducting transitions that occur gradually over a range of temperatures, rather than at a specific temperature; such transitions depend on the strength of any internal or external magnetic fields. Materials showing this behavior are known as type II superconductors. These transition ranges may be affected by such factors as prior metallurgical history, stress, and purity.

Several intermetallic compounds show superconductivity. Some of those with what were considered to be high values of T_c are given in Table 6-5. Most of these contain at least one constituent that is a superconductor. In the small number of compounds that do not contain an elemental superconductor, one of the constituents occupies a place in the Periodic Table next to a superconductor. The lattice type is important because of the role of lattice vibrations in the formation of Cooper pairs. The sodium chloride, nickel arsenide, β-tungsten, and σ-phase crystal structures are frequently found. Ceramics with high T_c (~ 120 K) are currently being investigated.

TABLE 6-5 SUPERCONDUCTIVITY OF SELECTED COMPOUNDS

Compound	T_c (K)	Compound	T_c (K)
Nb_3Sn	18.05	V_3Ga	16.5
Nb_3Ge	23.2	V_3Si	17.1
Nb_3Al	17.5	$Pb_1Mo_{5.1}S_6$	14.4
NbN	16.0	Ti_2Co	3.44
$(SN)_x$ polymer	0.26	La_3In	10.4

Reprinted, by permission, from Charles Kittel, *Introduction to Solid State Physics*, © 1976 by John Wiley & Sons, Inc., New York.

When a metal or alloy becomes superconducting, no change occurs in its crystal structure, nor does recrystallization take place. In the absence of a magnetic field, no latent heat is evolved during the transition. However, very small changes in lattice parameter and volume do occur. The lattice becomes slightly more perfect. These changes are of the order of about one part in 10^7. The molar entropy change resulting from this is about $10^{-3}R$. Observable changes in elastic properties and thermal expansion have not as yet been noted, despite theoretical predictions that these properties should be very slightly affected.

It is of significance that no Thomson effect is generated below T_c. When electrons flow in a normal conductor in a thermal gradient, an electrical potential is set up along that conductor. This electrical potential is the Thomson effect; it results from the entropy of the conduction electrons and is the basis for explaining thermoelectricity. The absence of this effect at $T \leqslant T_c$ is important evidence for considering that the electrons must be highly ordered and that no electron entropy changes occur in type I superconductors below T_c or in type II materials below H_{C1}. See Figs. 6-12(a), 6-15, and Section 6.2.7.

It is of interest to consider how the analyses given earlier in this chapter might be used in an attempt to explain superconductivity. Consider Eq. (6-16) written, using $\tau(E_F)$ of Eq. (6-7) in the integrated form, as

$$j = \frac{n(E_F)e^2\bar{E}}{m}\tau(E_F) = \frac{n(E_F)e^2\bar{E}}{m}\Delta t \tag{6-16a}$$

In the absence of any constraints, such as those noted in previous sections, j would increase without limit as a function of Δt *as long as $\bar{E}$ was operative*. This condition may also be expressed by Eq. (6-17), again using Eq. (6-7) as was done previously, in the form

$$\lim_{\Delta t \to \infty} \sigma = \lim_{\Delta t \to \infty} \left(\frac{n(E_F)e^2}{m}\Delta t \right) = \infty \tag{6-17a}$$

Thus, unlimited electrical conductivity (zero electrical resistivity) would be expected in

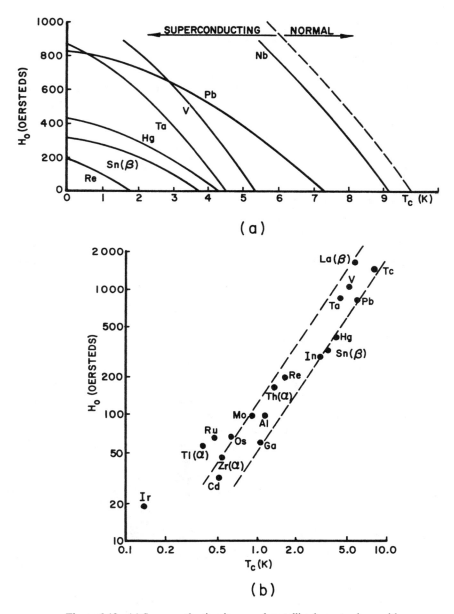

Figure 6-12   (a) Superconduction in several metallic elements along with a schematic diagram. (b) Plot of log H_0 versus T_c. (Data from E. M. Savitskii, and others, *Superconduction Materials*, p. 82, Plenum Press, New York, 1973.)

this case. However, this condition cannot explain superconductivity because its basis is contrary to the experimentally observed conditions of $0 < j \ll \infty$ and $0 < \sigma \ll \infty$ for $\bar{E} = 0$ and *not* for $\bar{E} > 0$.

If one is tempted to increase $\bar{E}$, the concomitant magnetic field within the conductor will increase and reach a critical value at which the superconductor will transform back to the normal state (Section 6.2.4). Where $\bar{E}$ is increased to a degree less than that causing reversion, another undesirable effect will take place. This may be shown by using Eq. (6-7a) in Eq. (6-19):

$$\sigma = n(E_F)e\mu = n(E_F)e \cdot \frac{\Delta \mathbf{v}}{\bar{E}} \qquad (6\text{-}19a)$$

So, by increasing $\bar{E}$ (assuming that $\Delta \mathbf{v}$ remains constant), the conductivity is decreased; the mobility is decreased by increasing $\bar{E}$ as a result of increased electron scattering, a factor that has not been considered up to this point. Electron scattering may not be neglected even at $0\,\mathrm{K}$. In fact, as is shown later, electron–phonon interactions play a very important part in explaining the mechanisms responsible for superconductivity at very low temperatures.

While the present theory provides good explanations for the mechanisms involved in superconductivity, it as yet is unable to predict those elements, alloys, or compounds that will show this property. It is for this reason that some of the descriptions that follow are empirical.

6.2.2 Theoretical Background

Some of the early ideas are presented to lead up to and to help to clarify the modern theory. It is of interest to note that over 50 years elapsed between the discovery of superconductivity and the development of the present theory.

One early model, known as the two-fluid theory, assumed that some of the valence electrons were in the ground state and were considered to behave as a superfluid. The remaining valence electrons were considered to behave normally. These two groups of electrons were treated as behaving like two interpenetrating fluids. This approach was able to provide explanations for the electromagnetic properties, the diamagnetism, and the zero resistivity, but it was unable to explain the superfluid behavior of some of the electrons.

F. London (1935) first suggested that electron mechanisms involving both electron coupling and an energy gap were fundamental to the understanding of superconductivity. This energy gap* subsequently was considered to separate the coupled, superfluid valence electrons in the lowest states from those in higher, excited, but normal levels. London's model was based on the concept that a high degree of uniformity of the momentum of electrons was required to explain superconductivity. Thus, it was thought that the superfluid, coupled electrons all possessed the same

---

*It should be noted that one type of superconductor does not appear to have an energy gap. Such "gapless" superconductors are not described here.

average momentum throughout the superconductor, regardless of whether or not a subcritical field was present.   Essentially, all these electrons were regarded as being in the same momentum state.   And, in conformance with experiment, these electrons required discrete energies for promotion to the excited states.   This constituted the basis for considering that an energy gap must exist between the superfluid electrons and those in the higher, normal states.   In effect, this is a macroscopic model of two electron fluids where many superfluid electrons in a single, low-energy level (bosons), in apparent disagreement with the exclusion requirement, are separated by an energy gap from normal electrons that conform to exclusion (fermions).   These concepts are the bases for that part of the present theory in which paired electrons, all with the same net momenta, account for the distribution of momentum; all the paired particles now must be considered to have the same momentum, as described later.

Other investigators had shown that superconductivity does not derive from the electron configurations of atoms, but is closely related to electron reactions with phonons.   This was demonstrated by early work that clearly indicated T_c to be an inverse function of the square root of the isotopic mass of Hg.   This led J. Bardeen and others to the idea that superconductivity might result from electron interactions with phonons, because phonon frequencies show a similar mass dependency.

This behavior of phonon frequencies, at very low temperatures, may be shown in an elementary way by starting with an expression for the velocity of sound in the superconductor as given by

$$c_s = \lambda v$$

where λ and v are the phonon wavelength and frequency, respectively.   The Newtonian expression for c_s is given by

$$c_s = \left(\frac{E_y}{\rho}\right)^{1/2} \tag{6-82}$$

Here E_y is the elastic modulus and ρ is the density of the material.   These two expressions for c_s may be equated to find

$$v = \frac{1}{\lambda}\left(\frac{E_y}{\rho}\right)^{1/2} \tag{6-83}$$

The density is reexpressed as mass/volume $= M/V$, and this is substituted into Eq. (6-83) to obtain

$$v = \frac{1}{\lambda}\left(\frac{E_y V}{M}\right)^{1/2} \tag{6-84}$$

This may be rewritten using the wave number, $k = 1/\lambda$, as

$$v = k(E_y V)^{1/2}(M)^{-1/2} \tag{6-85}$$

For any isotope of a given element, the factors k, E_y, and V will be constant, and the phonon frequency will be proportional to $M^{-1/2}$.   And, since T_c also varies as $M^{-1/2}$,

T_c is expected to vary with the phonon frequency. This provides further insight into the prior statement (Section 6.2.1) regarding the role of phonons, as well as that of isotopic mass. It turns out that electron–ion–phonon reactions are crucial in superconduction.

The idea that electron–ion–phonon interactions could be responsible for superconductivity might seem to contradict the mechanism described in Section 6.1.2 for the ideal resistivity of normal metals. It is now known that the electrons undergoing such interactions with ions and phonons at very low temperatures can have their characteristics changed. What is the process and how are the properties of the electrons changed? The answers reside in the mechanisms of electron–pair formation and in the properties of these pairs. This poses another question. Why should identically charged particles be attracted to each other in such pair formation when they normally would be expected to show electrostatic repulsion?

These questions may be resolved by first considering an electron with $E \cong E_F$ moving through a metallic lattice at very low temperatures. When the electron passes very close, within about 10^{-8} cm, to one of the ions, its negative charge induces a repulsion in those electrons in the filled shells of the ion that surround, or screen, its nucleus (Section 4.3). This effectively deforms and slightly unbalances the normally uniformly distributed charges around its nucleus for an instant. It leaves the ion slightly polarized with a momentarily partially unscreened positive charge. The ion then is weakly attracted to the electron and it undergoes a very small, transient lattice displacement that is the equivalent of the absorption of a phonon. This mechanism can continue as the electron proceeds through the lattice and interacts with many ions along its path. The absorbed phonons (or the displacements of the ions that are caused by the passage of the electron) are emitted by the ions and, in turn, are also transmitted through the lattice. The electrons involved in this process must be considered to be quasiparticles (Section 4.3) because of their repeated, mutual reactions with the filled electron shells of the ion cores.

These reactions now are used to describe the mechanisms involved in forming an electron pair. Consider an electron at $T \cong 0$ K, with $E \cong E_F$ and wave vector $\bar{k}_A$ that is closely approaching an ion. The ion becomes slightly polarized, is attracted to the electron, undergoes a small displacement, and then emits the energy of the displacement as a phonon with a wave vector $+\bar{k}(v)$. Then another electron with wave vector $\bar{k}_B$, whose spin and equal momentum are opposite to those of the first electron, also is attracted to the previously polarized ion. The second electron simultaneously repels other electrons and emits a phonon with a wave vector of $-\bar{k}(v)$ equal in magnitude and with a direction opposite to that of the first phonon. This is the equivalent of the absorption of $+\bar{k}(v)$ by the second electron. The emissions of the phonons, which resulted from the interactions of the two electrons with the ion, reduce the nominal energies of each electron to a value of $E \cong E_F - E_b/2$ and result in a small attraction (or less repulsion) between the two electrons; this constitutes the bonding energy, E_b, of an electron pair. Their mutual electrostatic repulsion energy is diminished by an amount E_b, a degree that permits them to form pairs.

The minimum energy required to return the paired electrons to normal states is E_b. Thus, using E_F as a reference, an energy gap of $E_g \cong 2E_b$ must exist between the paired states and the lowest normal states. The formation of many pairs is reflected by E_g since all their bonding energies are included in E_b. This energy gap is formed within the valence band and includes an energy range $E_g \cong E_F \pm E_b$ that formerly, in the normal state, constituted a portion of the quasicontinuum. The energy gap divides the valence band of the superconductor at $T \cong 0\,\mathrm{K}$ into two parts. All states with energies up to and including $E_F - E_g/2$ are filled; all states with energies equal to or greater than $E_F + E_g/2$ are empty. The resultant energy spectrum is similar to that shown in Fig. 4-16(b), but with important differences. First, the gap appears *within* the valence band and not between bands. Second, the departure from parabolic behavior in the energy range just below $E_F - E_g/2$ is much more pronounced than that indicated for the normal state. This large departure from normal behavior is necessitated in order to accommodate the electrons that formerly occupied states within the gap. Third, the energy spectrum of the unoccupied states at and above $E_F + E_g/2$ shows a similar deviation. [It will be noted that this deviation is opposite to that shown at the bottom of the second band for normal materials in Fig. 4-16(b).] This deviation is required to account for the range of vacated states that now constitutes the energy gap. The postulated energy spectrum and gap has been confirmed by absorption studies using far-infrared radiation. The evidence for this is the very strong absorption that begins at a frequency corresponding to the lower edge of the gap and continues with increasing frequency over a range of frequencies that corresponds to E_g. This absorption is greatly decreased at the top of the gap, where it approaches values expected of the normal states.

Now consider the energy spectrum of the filled states just below the energy gap. Its slope is given by the derivative of Eq. (3-26). This, along with Eq. (6-2), provides an expression for the velocity of such states, using the simplifying assumption $E_F \cong E_F - E_b/2$, since E_b is of the order of $10^{-3}\,\mathrm{eV}$:

$$\frac{dE}{d\bar{\mathbf{k}}} = \frac{\hbar^2 \tilde{\mathbf{k}}(E_F)}{4\pi^2 m^*} = \frac{\hbar^2 \mathbf{k}(E_F)}{m^*} = \frac{\hbar^2}{m^*} \cdot \frac{\mathbf{p}(E_F)}{\hbar} = \frac{\hbar}{m^*}\,\mathbf{p}(E_F) = \frac{\hbar}{m^*}\,m^*\mathbf{v}(E_F) = \hbar\mathbf{v}(E_F)$$

But $dE/d\bar{\mathbf{k}} \cong 0$ at or near the lower edge of the gap, so translational velocity, $\mathbf{v}(E_F)$, of these electrons is zero. However, again referring to Eq. (6-2), both $\mathbf{p}(E_F)$ and $\bar{\mathbf{k}}(E_F)$ may be large [Figs. 4-4(a) and (c)]. These disparate properties have significant implications for superconduction.

This pair formation may be expressed vectorially using the notation used previously. The vector path of the first electron that undergoes this reaction is

$$\bar{\mathbf{k}}_A + \bar{\mathbf{k}}(v) = \bar{\mathbf{k}}(A, v) \qquad (6\text{-}86)$$

and that of the second electron is

$$\bar{\mathbf{k}}_B - \bar{\mathbf{k}}(v) = \bar{\mathbf{k}}(B, v) \qquad (6\text{-}87)$$

in which $\bar{\mathbf{k}}(A, v)$ and $\bar{\mathbf{k}}(B, v)$ are the respective electron wave vectors subsequent to the interactions with the ion. The emission of the positive phonon caused by the first electron and the negative phonon by the second electron may be regarded as the absorption of the first phonon by the second electron. This can be expressed by adding the vectorial paths given by Eqs. (6-86) and (6-87) to obtain

$$\bar{\mathbf{k}}_A + \bar{\mathbf{k}}_B = \bar{\mathbf{k}}(A, v) + \bar{\mathbf{k}}(B, v) \tag{6-88}$$

The paired electrons, known as *Cooper pairs*, are said to be correlated, and the phonon absorbed in this process, $\bar{\mathbf{k}}(v)$, is called a *virtual* phonon. The wave vectors $\bar{\mathbf{k}}_A$ and $\bar{\mathbf{k}}_B$ must be equal and opposite in sign for this reaction to take place.

The electron-pairing process has been described here as occurring in the neighborhood of a single ion because it represents the least complicated case. But the initial interaction of the first electron with the ion may be such that conditions are unfavorable for the absorption of the phonon by a second electron at the time that it was generated. This would be the case if the second electron had the wrong direction or spin, if it was too different in momentum, or if it was not sufficiently close to the reacting ion when the phonon was emitted. Under these conditions, the phonon generated by the first electron would not be absorbed by the second electron, but would be transmitted through the lattice. Its passage would continue until a third electron, with suitable properties, reacted with some other ion in the lattice just as it was interacting with and being deformed by the phonon that had been generated by the first electron–ion reaction. An electron pair then would have been created by electron–phonon interactions with two very widely separated ions. These reactions may occur as far as approximately 10^3 to 10^4 interionic distances apart. Thus, the coherence length, or effective diameter or distance between members of an electron pair, may be of the same magnitude. The production of many electron pairs, therefore, is a process involving all electrons with $E \cong E_F$ and all the ions in the lattice.

Field theory analyses by H. Frölich (1950) showed that mutual reactions between electrons and phonons, similar to those just described, can result in attractive interactions among pairs of electrons with energies near E_F. The optimum conditions for this to occur were found to be that the two electrons should have different spins, have equal but opposite wave vectors and momenta, and not be separated by more than about 10^{-8} cm. This led to models based on the treatment of electron pairs acting as individual pseudomolecules.

J. Bardeen suggested that if the application of the concept of bound electron pairs was applied to quasiparticles it could be an important factor in superconductivity theory. It was shown later by L. N. Copper (1956) that if there is an attractive interaction such as that described by Frölich a pair of quasiparticles should form a bound state, regardless of the weakness of the attraction. At $0\,K$, the ground state would contain paired quasielectrons. Such pairs now are known as Cooper pairs.

The paired quasielectrons, each of opposite spin, have zero net spin as a Cooper pair. As a particle, each Cooper pair, thus, may be treated as a boson. Unlike the normal, or unpaired, quasielectrons, which as fermions must obey the Pauli exclusion principle and follow the Fermi–Dirac statistics (Sections 2.5 and 2.6), the Cooper pairs obey the Bose–Einstein statistics (Section 2.4). This difference (Section 2.8) is very important, since large numbers of identical Cooper pairs as bosons may occupy the same state, an important condition for superconduction, and fermions are forbidden more than single-state occupancy.

6.2.3 Present Theory

Normal metals are described by the way in which the quasielectrons as fermions occupy single states near E_F [Sections 4.5 and 4.6 and Figs. 4-14(a) and 4-16]. In contrast, the Cooper pairs formed in superconductors *all* condense as bosons into a *single* energy state at 0 K. This state lies slightly below the highest state occupied by the remaining electrons in the valence band. These remaining, normal electrons retain their original behavior as fermions and occupy states as required by the Pauli exclusion principle and by the Fermi–Dirac statistics. The Cooper pairs are formed by those electrons that formerly occupied states within about $\pm k_B T_c$ of E_F. Approximately 0.01% of the nearly free electrons are involved. Their presence in a single, condensed state results in a range of disallowed energies of about 10^{-3} to 10^{-2} eV and thus forms an energy gap, E_g, within the valence band between the highest state filled by the remaining normal valence electrons and the empty outer states. E_F lies at the center of this gap at 0 K. The single state occupied by all the Cooper pairs lies at about $k_B T_c$ below E_F. This is shown in Figs. 6-13(a) and (b).

The original work by J. Bardeen, L. N. Cooper, and J. R. Schrieffer (1957), now known as the BCS theory, required that each of the paired particles have opposite spin and that all Cooper pairs have both zero net spin and the same total momentum. The sum of the wave vectors of any Cooper pair [Eq. (6-88)] is the same as that of all other similarly bound pairs. The bound pairs of all the quasielectrons, all occupying the same state, overlap each other because the "distance between pairs" is much less than

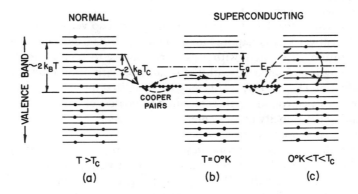

Figure 6-13   Model for the transition from the normal state (a) to the formation of Cooper pairs and the creation of an energy gap in the superconducting state (b). (c) The dynamic equilibrium between the Cooper pairs and normal states for $0 K < T < T_c$. (From C. M. Hurd, *Electrons in Metals*, p. 207, John Wiley & Sons, New York, 1975.)

the previously noted large "size" of a pair. It is estimated that the density of centers of masses of Cooper pairs per unit volume, as defined by the probable coherence length, is approximately 10^6. This extremely high degree of overlap between pairs is such that the densely associated pairs cannot show independent motion; they are capable only of collective action. This is in agreement with London's prior theory of paired particles all having the same momentum. In addition, it is in contrast to the models based on individual pseudomolecules; these could not possess the collective properties required of Cooper pairs.

This behavior also explains the absence of the Thomson effect in superconductors noted in Section 6.2.1. The high degree of order of the Cooper pairs, all with the same net wave vectors and momenta and all residing in a single energy state, may be regarded as being equivalent to zero electron entropy in the temperature range $T \leqslant T_c$ (Section 5.1.8). Thus, the Thomson effect must be absent for $T \leqslant T_c$ because it is a function of electron entropy change.

When a current flows in a superconductor, each Cooper pair must move in the same direction as the current with the same velocity as all other pairs; all must have the same energy and wave vector. Some of the pairs may split up as a result of thermal excitation. When this occurs, the individual electrons may jump across the gap to fill low-lying energy states just above the gap in the valence band. The remaining Cooper pairs stay together and resist separation. All pairs must have the same momentum when an electric field is applied because all have the same drift velocity. The resulting flow of the pairs essentially constitutes the superconducting current. And, since all the pairs are required to have the same collective properties, the scattering of some of the Cooper pairs cannot affect the momentum of any of the other pairs. The momentum of all other pairs remains unchanged as the current continues to flow. Thus, the scattering is ineffective as a source of adversely affecting the properties of the Cooper pairs, as shown later, and the current will flow indefinitely as long as $T < T_c$.

Because the density of the Cooper pairs is so very high, only a correspondingly small drift velocity is required for them to generate a considerable current. The small momentum associated with this corresponds to a very much larger de Broglie wavelength as compared with the mean free path of an electron involved in normal conduction [Eq. (6-19)]. This may be visualized more readily when it is recalled that each quasielectron in a pair is required to possess equal and opposite velocities. The resulting net velocity at the center of mass of each of the densely associated pairs may then be very small when current flows.

Other important properties may be explained by de Broglie's equation, [Eq. (1-17)] $\lambda = h/p = h/m\mathbf{v}$, where λ now has macroscopic dimensions because of the very small velocity of a pair. This single de Broglie wavelength describes all of the Cooper pairs because all of them have the same set of dynamic properties; it extends macroscopically over all the pairs. This very long de Broglie wave is basically responsible for the previously noted ineffectiveness of scattering. In general, a wave will be scattered when its wavelength is about the same or smaller than the dimension of the scattering source. The very long de Broglie wave is much larger than such scattering sites as impurities and most lattice imperfections that affect normal electrons (Sections 6.1.1–6.1.3). Thus, the pairs are unaffected by such nonthermal scattering sources. This explains the absence of such scattering and the persistence of an unattenuated, resistanceless current. It is now apparent why some of the early work on superconductivity considered that this might result from the presence of a superfluid. All the paired carriers are bosons occupying a single energy state with the same momentum and with an extremely long, macroscopic de Broglie wavelength. Unpaired quasielectrons are also present, in addition to the Cooper pairs; these have microscopic wavelengths and may be scattered. Thus, the simultaneous presence of both fermions and bosons may be considered to be the quantum mechanic equivalent of the early, two-fluid theory.

Continuous scattering of the Cooper pairs can occur by thermal means [Fig. 6-13(c)]. This mechanism and its effects may be explained most clearly by beginning with a brief review. As the temperature is decreased, the quasielectrons in normal states in the range $\pm k_B T_c$ of E_F form Cooper pairs that constitute bosons at $T = T_c$. This causes an abrupt transition from the normal state to the superconducting state. The net result is the formation of an energy gap of $E_g \cong 2k_B T_c$ in the valence band as $T \rightarrow 0\,\mathrm{K}$, an energy range equivalent to about 10^3 to 10^4 times that of a single Cooper pair. At $0\,\mathrm{K}$, all the Cooper pairs are in the single, superconducting ground state that is at approximately $k_B T_c$ below the uppermost filled, normal fermion state. The normal particles (fermions), in contrast, must occupy individual states singly, or two electrons of opposite spin may occupy a state, as required by the principle of exclusion (Sections 2.3.2, 2.5, and 3.9).

The energy gap, E_g, unlike that of a semiconductor, is mobile and is not a function of E_F; E_F remains unaffected. E_g is temperature dependent because the number of pairs as represented by E_b is temperature dependent ($E_g \cong 2E_b$) and has no definite location in $\bar{\mathbf{k}}$ space. As the temperature decreases, E_g increases monotonically from zero at T_c to its maximum at $0\,\mathrm{K}$. E_F is at $E_g/2$ at $T = 0\,\mathrm{K}$ and the gap edges approach E_F as $T \rightarrow T_c$.

Both of these parameters, E_g and E_F, are shifted in $\bar{\mathbf{k}}$ space (and momentum space) as a function of temperature when current flows for $0 < T < T_c$. This is analogous to that described by Eq. (6-4) and arises from a dynamic equilibrium between the number of paired and unpaired particles at a given temperature. At $T \ll T_c$, the thermal scattering, or splitting, of pairs is very small. However, when an equilibrium is established at a given temperature $T_1 < T_c$, which may be considered to closely approximate that at $0\,\mathrm{K}$, the thermal scattering increases and a smaller number of pairs and net flow of the paired quasiparticles exist for that temperature; this is equal to $\rho_s \mathbf{v}_p$, where ρ_s corresponds to the density of the superfluid in the two-fluid model (the density of the paired quasiparticles) and $\mathbf{v}_p$ is their rate of flow. As $T \rightarrow T_c$, the number of pairs is decreased by thermal scattering, $E_g \rightarrow 0$, and $\rho_s \rightarrow 0$. This results only from the thermally induced splitting up of the paired particles and not from scattering, nor from an "evaporation" of pairs. Below T_c, the current continues to flow, despite any scattering, for the reasons given previously. At a higher temperature, T_2, $T_1 < T_2 < T_c$, thermal excitations of the Cooper pairs to quasiparticle states reduce the current until another equilibrium is reached between the Cooper pairs and the normal particles and new values of ρ_s and $\mathbf{v}_p$ are established. Increasing T causes increasing pair-to-normal quasiparticle excitations until, at T_c, the parameters E_g and ρ_s vanish and normal behavior returns.

The thermal excitation, or splitting, of the Cooper pairs at $0 < T < T_c$ results in an equilibrium number, or ρ_s, of pairs at a given temperature. Thus, even if significant amounts of nonthermal scattering of the Cooper pairs could occur, the equilibrium number of these pairs would be reestablished simultaneously, and current changes would not be observed.

It should be noted that the application of a voltage (electric field) to a superconductor induces both the Cooper pairs and the normal particles to transport the current within it. The pairs, all moving with the same velocity in the same direction as the electric field, behave as though they short-circuit that part of the current carried by the normal particles that are moving in the direction opposite to the field. The superconducting current is the net result of these two flows. This explains the early models based on two interpenetrating fluids.

An upper limit to the way in which the pairs may be scattered is by forcing them to exceed the velocity of sound in the solid, c_s. The phonons that generate the Cooper pairs have a maximum velocity of c_s. If the pairs are forced to travel at a velocity $\mathbf{v} \geqslant c_s$, additional pair production cannot proceed. The dynamic equilibrium with quasiparticles determines the number of pairs at a given temperature. Increasing $\mathbf{v} > c_s$ for a fixed number of carriers is the equivalent of increasing the current density, since $j = ne\mathbf{v}$ [Eq. (6-15)]. But the pair-production mechanism can no longer operate under this condition, so the number of pairs decreases to zero and the material reverts to the normal state. See Section 5.1.3.3 and Eq. (6-96).

Reversion to the normal state also may be caused by magnetic fields as indicated by Fig. 6-12(a). This results because field strengths greater than H_c split the Cooper pairs. In this case, each member of a Cooper pair is subject to a force that is perpendicular to the motion of the particle and is proportional to the flux density of the field. This force, known as the *Lorentz force*, induces curvatures in the paths of the particles without changing their energies. When the Lorentz force becomes sufficiently large, the Cooper pairs are disrupted and revert to unpaired particles and superconduction vanishes. See Section 6.2.6.

6.2.4 Effects of Internal Electric Fields

Consider a suitable long, cylindrically shaped superconductor in an electric circuit such that a uniform electric field, $\bar{E}$, will be set up in it when the circuit is completed. The creation of $\bar{E}$ causes all the Cooper pairs to begin to flow in a direction opposite to that of $\bar{E}$; all the pairs move simultaneously with the same acceleration. This is given by, where the charge and mass of a pair are $2e$ and $2m$, respectively,

$$a = \frac{F}{m} = \frac{2e\bar{E}}{2m} = \frac{e\bar{E}}{m} \qquad (6\text{-}89)$$

where m is the mass of a quasielectron. The establishment of the electric field actually is a function of time, so this will affect the current density in the superconductor. Thus, starting with the general expression for the current density,

$$j = ne\mathbf{v} = \frac{n_q}{2} \cdot 2e \cdot \mathbf{v} = n_q e\mathbf{v} \tag{6-90}$$

where n_q is the number of quasielectrons per unit volume and $\mathbf{v}$ is the drift velocity of a Cooper pair; upon differentiation with respect to time, Eq. (6-90) becomes, using Eq. (6-89),

$$\frac{dj}{dt} = ne\frac{d\mathbf{v}}{dt} = n_q ea = n_q e \cdot \frac{e}{m}\frac{d\bar{E}}{dt} = \frac{n_q e^2}{m}\frac{d\bar{E}}{dt} \tag{6-91}$$

Thus, the rate of increase in the current density varies directly as the rate of establishment of $\bar{E}$; this, as previously noted, is a function of time.

The current creates a solenoidal magnetic field, H, within the superconductor. H increases with time because j increases with time [Eq. (6-91)]. In its turn, H simultaneously induces an electric field, $\bar{E}_I$, and a current, j_I. These oppose $\bar{E}$ and j, respectively. The induced current, j_I, creates a magnetic field, H_I, that opposes H. The net result is that, since $\bar{E}_I$ and H_I are equal and opposite to $\bar{E}$ and H, respectively, both the electric and magnetic fields within the superconductor are equal to zero. And, since $\bar{E} = 0$ and $\rho = 0$, resistanceless current flow is present.

This state will continue as long as the Cooper pairs continue to flow after the establishment of $\bar{E}$ and the resistivity of the specimen is zero; that is, the specimen remains in the superconducting state. In addition, the current, j_I, must be concentrated within a *very thin* volume adjacent to the surface of the cylindrical specimen. Such a current-containing volume is necessary because it induces a magnetic field external to the specimen and not inside it. Thus, when the steady state has been reached within the specimen, j_I is constant and both $\bar{E}$ and H_I are equal to zero. The thin volume in which the current is concentrated is characterized by its thickness, λ, a parameter known as the *penetration depth* (see Section 6.2.5).

The expulsion of the electric field, $H_I = 0$, constitutes ideal diamagnetic behavior. Since $H_I = 0$, the magnetization is $-M = H$ and the magnetic susceptibility is $M/H = -1$. Thus, except for the very small, outer volume containing both the concentrated current and the magnetic field, the superconductor is a perfect diamagnetic material.

In the steady state, the Cooper pairs move with a constant drift velocity (zero acceleration). Here the momentum of the Cooper pairs is acquired during the interval in which $\bar{E}$ was established originally. This may be expressed, for the time interval over which the momentum and the electric field change from zero to their respective steady-state values of $\mathbf{p}$ and $\bar{E}$, using the same approach as that used for Eq. (6-91), as

$$\frac{m\mathbf{v}}{\mathbf{p}} \propto \bar{E} \tag{6-92a}$$

so that

$$\mathbf{v} \propto \frac{\mathbf{p}\bar{E}}{m} \qquad\qquad (6\text{-}92\mathrm{b})$$

The substitution of Eq. (6-92b) into Eq. (6-90) results in a current density of

$$j \cong n_q e\,\frac{\mathbf{p}\bar{E}}{m} \qquad\qquad (6\text{-}93)$$

within the thin, outer layer of the superconductor known as the penetration depth and designated by λ.

6.2.5 Effects of External Magnetic Fields

Now consider the effect of the application of an external field, H_{Ex}, to the specimen described in the previous section. This causes a solenoidal electric field to form within the specimen that generates an electric current of density j_{I1}. This current has a solenoidal magnetic field H_I associated with it that is equal and opposite to H_{Ex}. The current, j_{I2}, created by H_I offsets j_{I1}. Thus, H_{Ex} is virtually entirely expelled from the volume of the specimen by H_I (Fig. 6-14). H_{Ex} is confined to the very thin volume of material adjacent to the surface of the specimen, where it is balanced by the induced current that flows within it. This behavior is the basis for the term penetration depth.

In summary, not only does its electrical resistance go to zero, but a type I superconductor also is virtually a perfect diamagnet. When a magnetic field of less than the critical field H_c is applied to a superconductor, a current is induced in it in the direction of the applied field. The field created by the induced current exactly balances the applied field in almost all the volume of the superconductor. The net effective field in this volume is zero. This is known as the *Meissner effect* (Fig. 6-14). A very small surface layer, previously described by the penetration depth, remains in which the applied field is not completely canceled. Thus, all superconductors reject virtually all the flux from their interiors and react like nearly perfect diamagnets. Under certain conditions, some superconductors, type II, may show mixed behavior and permit appreciable flux penetration (Section 6.2.7).

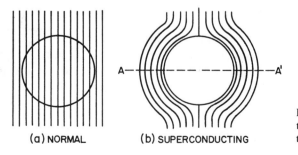

(a) NORMAL   (b) SUPERCONDUCTING

Figure 6-14 (a) Magnetic flux in a transverse field for a normal metal; (b) the Meissner effect in a superconductor.

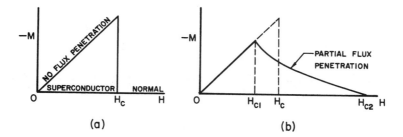

Figure 6-15 Magnetization curves for superconductors: (a) type I, and (b) type II.

The macroscopic effect of an external magnetic field on a superconductor is indicated in Fig. 6-14. Here the field has been forced out of the interior of the material. As a result, it will be noted that the field intensity is greatest at $A-A'$ in the figure. An increase in the field intensity at the surface of the specimen can cause H_{c1} to be reached in the volumes of the material adjacent to the concentrated field. These volumes, therefore, can revert to the normal state while the remainder of the bulk material remains superconductive. A mixture of superconductive and normal regions may exist under these circumstances (Section 6.2.7). The extent of the normal material increases as the field strength increases beyond H_{c1} [Fig. 6-15(b)]. Transition to the completely normal state occurs with increasing H when the upper critical field, H_{c2}, is reached. Superconducting materials showing this kind of behavior are called type II, or hard, or high-field superconductors. The magnitude of the upper critical field, H_{c2}, may be of the order of 100 times the value of H_c. Solenoids made of hard superconductors have provided fields greater than 100 kG.

6.2.6 Critical Magnetic Field

The centers of masses of Cooper pairs carrying the current in the superconducting state all have the same momentum and, consequently, the same drift velocity. The energy of such pairs is obtained from this as

$$E = \tfrac{1}{2} \cdot 2m \cdot \mathbf{v}^2 = m\mathbf{v}^2 \tag{6-94}$$

since the mass of a pair is that of two quasielectrons. Where no current flows in the superconductor, that is $\bar{E} = 0$ as in Eq. (6-93), $\mathbf{v}$ and E both equal zero. Thus, the energy increment of each Cooper pair in motion above the ground state is given by Eq. (6-94). This permits an insight into the maximum permissible drift velocity of a pair. It results from the fact that the energy increment may not exceed the bonding energy. Thus, the onset of reversion to the normal state may be expressed as

$$m\mathbf{v}^2 \geqslant E_b = \frac{E_g}{2} \tag{6-95}$$

and the limiting drift velocity as

$$\mathbf{v}_c = \left(\frac{E_g}{2m}\right)^{1/2} \tag{6-96}$$

And, when Eq. (6-96) is substituted into Eq. (6-90), the critical current density is obtained as

$$j_c = n_q e \left(\frac{E_g}{2m}\right)^{1/2} \tag{6-97}$$

The magnitudes of Eqs. (6-96) and (6-97) may be obtained by using $E_g \cong 10^{-3}\,\text{eV}$, $m \cong 9.1 \times 10^{-28}\,\text{g}$, $n_q \cong 10^{18}\,\text{cm}^{-3}$, and $e \cong 1.6 \times 10^{-19}\,\text{C}$. These result in $\mathbf{v}_c \cong 1 \times 10^6\,\text{cm s}^{-1}$ and $j_c \cong 2 \times 10^5\,\text{A}\cdot(\text{cm}^2)^{-1}$.

Reversion to the normal state will also occur at the critical field strength as approximated by

$$H_c \cong \lambda j_c \tag{6-98}$$

The penetration depth is given by

$$\lambda = \left(\frac{m}{n_q e^2 \mu_0}\right)^{1/2} \tag{6-99}$$

where μ_0 is the permeability of vacuum (free space). Equations (6-99) and (6-97) are substituted into Eq. (6-98) to obtain

$$H_c \equiv \lambda j_c = \left(\frac{m}{n_q e^2 \mu_0}\right)^{1/2} \cdot n_q e \left(\frac{E_g}{2m}\right)^{1/2} \tag{6-100a}$$

$$H_c \cong \left(\frac{m}{n_q e^2 \mu_0} \cdot \frac{n_q^2 e^2 E_g}{2m}\right)^{1/2} = \left(\frac{n_q E_g}{2\mu_0}\right)^{1/2} \tag{6-100b}$$

Again using the data given for illustrative purposes, $H_c \cong 10^4\,\text{A}\cdot\text{m}^{-1}$.

The reaction now represented by Eq. (6-98) was observed first by Kamerlingh-Onnes (1913). He showed that an electric current in a wire would eliminate its superconductivity if the current exceeded a certain amount. Silsbee (1916) explained this by demonstrating that the field induced by the current was responsible for this, rather than the current itself.

The properties $\mathbf{v}_c$ and j_c [Eqs. (6-96) and (6-97), respectively] are directly related to each other and have the same temperature dependence, since each is a function of $E_g^{1/2}$ and $E_g \cong 2k_B T_c$ (Section 6.2.3). Thus, H_c also is expected to be a function of temperature. This is given by

$$H_c = H_0 \left[1 - \left(\frac{T}{T_c}\right)^2\right] \tag{6-101}$$

where H_0 is the critical field at $0\,\text{K}$ and T_c is the superconducting transformation temperature in the absence of a magnetic field. Equation (6-101) delineates the permissible $H_c - T$ regimes for superconduction as shown in Fig. 6-12(a).

Another interesting relationship between H_0 and T_c is shown in Fig. 6-12(b). In general, the higher that T_c is found to be, the higher H_0 will be. Most of the data given in Table 6-4 are in agreement with this relationship. It is natural to consider superconductors for use in electric power transmission. However, this would be self-defeating because the high electric and magnetic fields set up by the currents in the cables would ensure that they would revert to the normal state rather than be superconducting [Eq. (6-98)]. It appears that this application might be feasible if superconductor materials with values of T_c of approximately 10 to 20 times those of presently known materials (Table 6-5) were available. See Sections 6.2.8 and 6.2.9.

6.2.7 Transition Temperature

The superconducting transition temperature, as given by the BCS theory, is

$$T_c = 1.14\theta_D \exp[-1/AN(E_F)] \tag{6-102}$$

Here θ_D is the Debye temperature, $N(E_F)$ is the density of states within $\pm k_B T$ of E_F, and A is a positive, particle–lattice-pair-interaction coefficient that is related to E_b. The factors in Eq. (6-102) may be considered on an intuitive basis.

As discussed in Section 6.2.2, $T_c \propto M^{-1/2}$. The Debye temperature can be shown to have a dependency similar to that of Eq. (6-85) by starting with the relationship

$$\theta_D = \frac{hc_s}{k_B\lambda_c} \tag{5-113}$$

where λ_c is the phonon wavelength corresponding to the phonon cutoff frequency. Now, using the same approach as was used to obtain Eq. (6-85),

$$\theta_D = \frac{hc_s}{k_B\lambda_c} = \frac{h}{k_B\lambda_c}\cdot\left(\frac{E_y V}{M}\right)^{1/2} \tag{6-103}$$

And, for the reasons given previously,

$$\theta_D = \text{constant}_1 \, M^{-1/2} \tag{6-104}$$

The previously noted variation of T_c may be expressed as

$$T_c \cong \text{constant}_2 \, M^{-1/2} \tag{6-105}$$

Thus, the close relationship between T_c and θ_D is obtained by dividing Eq. (6-105) by Eq. (6-104) to eliminate the factor $M^{-1/2}$:

$$\frac{T_c}{\theta_D} \cong \text{constant}_3 \tag{6-106a}$$

This may be shown in a way similar to that used in Eq. (6-102) as

$$T_c \cong \text{constant}_3 \, \theta_D \tag{6-106b}$$

It has been shown that θ_D depends on c_s [Eq. (5-113)]. As such, it is affected by such factors that include temperature, lattice spacing, and the ways in which phonons are transmitted or scattered, or both, by the lattice (Section 5.1.7). All these variables also play important parts in the pair-production process (Section 6.2.2). The higher the temperature of θ_D, the more uniform will be the lattice vibrations at very low temperatures. This may be restated as the greater the degree of anharmonicity (umklapp) of the phonons (Section 5.3.1), the less likely they are to participate in the pair-production process. A more uniform "phonon gas" has a greater probability for Cooper pair production than does one involving anharmonics. The factor A represents the attractive interaction required for pair formation. It must be positive for superconduction, and, being related to E_b, it has a strong effect on T_c. The number of states available to enter into the pairing process is taken into account by $N(E_F)$. In this situation, $N(E_F)$ does not drop discontinuously to zero at $E = E_F$ as is the case for fermions at $T = 0\,\text{K}$ [Fig. 4-3(c)]. Instead, as discussed in Section 2.4, the particles that can form or act as bosons obey the Bose–Einstein statistics; these show a considerable distribution of energies about the range $\mu = E_F$ for $T \to 0\,\text{K}$ [Eq. (2-65)]. This distribution is similar to that shown for the density of states of fermions for $T > 0\,\text{K}$ in Fig. 4-3(c). This is the reason why $N(E_F)$ is significant in the range $\pm k_B T_c$ at these very low temperatures. Thus, the major factors involved in the generation of Cooper pairs are included in Eq. (6-103).

6.2.8 Penetration Depth

The generation and the roles of the penetration depth, λ, are discussed in the previous sections. The Cooper pairs in the superconducting state are confined to the very small, outer volume of the material whose average thickness is defined by λ. This influences the properties of bulk materials. The number of Cooper pairs may be approximated by

$$2n_q \propto \left[1 - \left(\frac{T}{T_c} \right)^4 \right] \tag{6-107}$$

And, since λ is a function of n_q [Eq. (6-99)], it is expected to be temperature dependent. The relationship between λ and T is

$$\left[\frac{\lambda(0)}{\lambda(T)} \right]^2 = \left[1 - \left(\frac{T}{T_c} \right)^4 \right]^{-1} \tag{6-108}$$

in which $\lambda(0)$ and $\lambda(T)$ are the penetration depths at 0 and $T < T_c$, respectively. Values of λ for various superconductors are of the order of about 10^{-6} to 10^{-5} cm. Those of a normal dielectric metal are about 5×10^{-6} cm.

The relationships given by Eqs. (6-107) and (6-108) are used for penetration analyses. Measurements of λ, thus, can evaluate n_q and, consequently, v_c, j_c, and H_c [Eqs. (6-96)–(6-98)]. This also permits the calculation of E_g (Sections 6.2.2 and 6.2.3).

As noted in Section 6.2.4, a perfect superconductor would be entirely diamagnetic throughout its volume. That is, the induced magnetic field, B, would be zero at any point in the superconductor. However, field penetration does occur and magnetic induction does take place in the outer layer. Measurements of B obtained in this way permit the calculation of λ for penetration analyses.

The penetration depth can be used to explain the different behaviors of types I and II superconductors. The penetration depth, in addition to the previously noted factors, depends on the material and is a function of the coherence length (Sections 4.3 and 6.2.2). Materials with type I properties have Cooper pairs with coherence lengths larger than λ. These paired particles are unaffected by the field penetration. Consequently, they show the relatively uncomplicated behavior indicated in Fig. 6-15(a). Where the coherence length of the Cooper pairs of a material is smaller than λ, type II behavior is shown. When this situation prevails, it permits the disruption of the members of a pair by the magnetic field and that part of the outer volume of the material where this take place reverts to normal properties. This accounts for the mixed behavior between H_{c2} and H_{c1} in Fig. 6-15(b), shown as partial flux penetration.

6.2.9 Applications and Present Status

Important questions still remain to be answered by the theory. The behaviors of the elements continue to be treated empirically. The mechanisms responsible for their behaviors in alloys and in intermetallic compounds have yet to be described; the prediction of the superconducting properties of such materials is not yet possible. Thus, the continuing search for materials with higher values of T_c remains essentially empirical and is based on general properties such as those outlined in the foregoing sections.

Superconducting materials are finding increasingly wider applications as windings on ferromagnetic cores to form very high field magnets. These are used for scientific and some industrial applications. Other uses include heat valves, radiation detection, microwave cavities, and computer memories. One of the more spectacular applications is their use to levitate very high speed bullet trains to eliminate rail friction. The use of these materials for electric power transmission was noted earlier (Section 6.2.6).

The costs of refrigeration of the power transmission cables will play an important part should this application be made. Prior to 1987, this factor also appears to have adversely affected the use of superconducting components in electronic circuits. This could be minimized if T_c could be increased considerably. The difficulty may be appreciated by considering an approximate form of Eq. (6-102): $T_c \cong \theta_D e^{-\alpha}$. It has been shown that $|\alpha| \geqslant 2$ is characteristic, a condition apparently limited by E_b. But, to obtain $T_c \cong 100\,\text{K}$, $|\alpha| \cong 1.6$ is required; this is not compatible with the presently accepted mechanism for the formation of Cooper pairs in materials such as are shown in Tables 6-4 and 6-5.

A new class of ceramic materials has recently (1987) been discovered that overcomes the prior limitation of T_c to very low temperature. A number of ternary oxides containing rare-earth elements have been reported to have values of

$T_c > 90$ K.* Up until recently, materials with compositions of the types Nb_3M or V_3M, with β-tungsten lattice types in which M is a normal metal, have maximum values of $T_c \cong 18$ K (Table 6-5).

The basis of the new class of materials was the finding of oxides, such as $LiTi_2O_4$ and $BaPb_{1-x}Bi_xO_3$, with values of T_c between 13 and 14 K. These are unusual because their low electron densities cannot account for their values of T_c in terms of the present theory [Eq. (6-102)]. Theoretical considerations showed that materials of these kinds with intense electron–phonon interactions that also could manifest a transition from the superconducting state to a bipolaron insulating state should have a high value of T_c. (A *bipolaron* state is one in which two quasielectrons become highly associated with lattice sites by being effectively trapped in the potential wells that result from their reactions with the lattice.) This ultimately led to materials of the Ba–La–Cu–O system that had values of T_c from 23 to 30 K. The application of a pressure of 12 kbars to one of these increased T_c to 52 K. The value of T_c from a material of a similar system, Sr–La–Cu–O, was shown to be 40 K. It then was found that values of T_c ranging from 80 to 100 K were present in members of the Y–Ba–Cu–O system in which the superconducting phase is $YBa_2Cu_3O_{9-y}$. It turned out that this material is one of a general group of superconducting materials with compositions of the type $MBa_2Cu_3O_{9-y}$, in which M is a rare-earth or a transition element. These have $T_c > 90$ K. Percent indications are that materials with $T_c \sim 240$ K may be possible.

These new materials undergo the transition to the superconducting state over relatively large temperature ranges and thus are of type II. This behavior has led to measurements of their electrical resistivities at temperatures at which ρ is 90%, 50%, and 10% of the values extrapolated from their ranges of normal behavior. The 50% point, known as the midpoint T_c, is used as an estimate of their actual values of T_c. This value is conservative and is considerably lower than that at which superconduction is first observed.

The T_c of these materials appears to be a strong function of the sintering parameters and their compositions. The compacting pressure, sintering time, temperature, cooling rate, and the atmosphere under which this is performed determine the composition factor y in the composition types noted previously.

Other materials, such as $La_{2-x}Sr_xCuO_{4-y}$, have the additional variable of chemical content. Here $0.05 < x < 0.4$. It has been shown that the composition $x = 0.15$ gives the most abrupt transition (over a range from 1.5 to 2 K) and the highest value of T_c. Some of the compositions of these materials outside of the superconducting composition ranges are semiconductors.

These new materials give rise to questions regarding the ways in which the BCS theory (Section 6.2.3) must be modified to explain their properties. Among these are: What is the mechanism of the production of electron pairs? Do these have different configurations? Does the pairing involve more than two electrons? Is the phonon frequency limited to that of the Debye frequency as indicated by Eq. (5-113)? Can

---

*See A. Khurana, *Physics Today*, p. 17, April 1987, for a review of the status of this work.

pair production take place at higher frequencies than the Debye frequency? If this is not the case, how can the necessary high pair production, required for high values of T_c, be explained? How do some of these materials show semiconduction?

Several theoretical models for the properties of these oxides have been proposed, none of which has gained wide acceptance. Despite this, the very large increases in the values of T_c that were observed in so short a time (early in 1987) ensure that a technological revolution, at least equal to that caused by semiconductors, is in the near future.

6.3 SEMICONDUCTORS

Semiconductors are good examples of the theoretical and practical applications of modern electron theory. Their electrical resistivities vary in the range from about 10^{-2} to 10^9 Ω-cm. This range is intermediate between those of good conductors (10^{-6} to 10^{-5} Ω-cm) and insulators (10^{14} to 10^{22} Ω-cm). The pure elements and compounds discussed here are insulators at 0 K; semiconduction processes occur in many materials at relatively elevated temperatures. Most of these materials are covalently bonded. Their numbers of nearest neighbors are given by the $(8 - N)$ rule. The compound semiconductors mainly consist of combinations of ions from groups III and V and II and VI of the Periodic Table, but many other covalently bonded compounds also have desirable properties. These materials are intrinsic semiconductors in the "pure" state. They have negative temperature coefficients of electrical resistivity.

Modifications of the electrical properties of these materials by means of dopants (extrinsic semiconductors), junctions, barriers, surface effects, and so on, have led to the development of many types of devices that have had wide application and have revolutionized electronic circuitry. Only those properties relating to their electrical conductivity are considered here. See Chapters 8 and 9.

6.3.1 Intrinsic Semiconductors

The elements C (diamond), Ge, Si, and Sn (α) are covalently bonded and crystallize according to the $(8 - N)$ rule. Thus, all their valence electrons are tightly bound in the bond-sharing process between nearest neighbors. This sharing provides each of the ions with completed, outer electron levels (Section 3.12.2). In other words, the valence bands of these elements are completely filled at 0 K (Section 4.5 and Fig. 4-14). The next outermost band, the *conduction band*, is completely empty of electrons at 0 K. These two outermost bands are separated by a relatively small energy gap, E_g, of the order of $k_B T$; the materials behave as insulators as long as this condition prevails. As the temperature is raised above absolute zero, increasing numbers of electrons are promoted across the gap, $k_B T > E_g$, into the formerly empty conduction band. Holes (absences of electrons) are created simultaneously in the states the electrons vacate in the valence band. Semiconduction becomes apparent when appreciable numbers of such carriers are generated.

It should be noted that suitable electromagnetic radiation ($E = hv \geqslant \frac{3}{2}E_g$, Section 6.3.2) may also excite electrons from the valence to the conduction band. Materials having appropriate energy gaps, which enable them to respond to exposure to portions of the spectrum, are known as *photoconductors*. Thus, intrinsic semiconductors may be considered to be a class of insulators whose band structures render them susceptible to property changes that may be induced by external energy.

A vacant state, or hole, remains when external influences excite an electron from the top of a nearly filled valence band. Another excited electron, from a slightly different, next-lower state, fills this hole. In so doing, it leaves behind an unoccupied state at its original energy level. After a series of such reactions, the promotion of a series of electrons to successively higher energy states results in the corresponding displacement of the hole to lower-energy states. The particles move in opposite directions. The hole has a positive charge [Eq. (6-157)] and, consequently, is expected to behave in a manner opposite to that of an electron.

It will be recalled (Section 3.12.3) that the covalent bond of interest here consists of pairs of electrons of opposite spin that are shared by nearest neighbors in the lattice. The formation of a hole consists of the splitting of one such bond pair. The remaining member of the pair then constitutes a partial or dangling bond between the adjacent lattice neighbors. The dangling bond (remaining electron) is readily transmitted from host ion to host because of the indistinguishability of each corresponding state in the lattice. The transfer of the dangling bond in the lattice is the equivalent of the displacement of the hole.

As increasing numbers of electrons occupy states in the conduction band, the material becomes increasingly conductive, because the many available vacant states in the conduction band permit these promoted electrons to behave in a nearly free way. The holes also serve as carriers. Since this carrier-generation process increases with increasing temperatures, the electrical resistivity decreases correspondingly; therefore, these materials have negative temperature coefficients of electrical resistivity. Simply restated, increasing the temperature promotes more electrons from the valence to the conduction band and simultaneously creates more holes in the valence band. The greater the number of these mobile carriers of electricity, the greater the conductivity and the smaller the resistivity of the semiconductor. Thus, the temperature coefficient of electrical resistance is negative.

The numbers of carriers involved in this conduction mechanism is very much less than those active in metallic conduction. The number of carriers responsible for the electrical conductivity of semiconductors is about 10^{-7} of the number of electrons engaged in conduction in normal metals.

6.3.1.1 Intrinsic conduction mechanism. The promotion of electrons to the conduction band simultaneously creates holes in the valence band. The vacated states in the valence band may be treated as particles that have a positive charge equal in magnitude to that of an electron and engage in the conduction mechanism.   In the case of intrinsic semiconductors, the number of holes per unit volume, n_h, is equal to the number of electrons per unit volume, n_e, that are promoted to the conduction band because of their simultaneous generation.   This process is called *pair production*.

The electrical conductivity, σ, for the case of electrons in a metallic conductor was given by Eq. (6-17).   This also can be expressed generally as

$$\sigma = ne\mu = \frac{1}{\rho} \qquad (6\text{-}109)$$

Here ρ is the electrical resistivity and μ, the mobility, is given by Eq. (6-7a) as

$$\mu = \frac{e\tau}{m^*} \qquad (6\text{-}7a)$$

in which m^* is the effective mass (Sections 4.6 and 6.3.2).

Since both electrons and holes are carriers and contribute to the conductivity of a semiconductor, the effects of both must be considered:

$$\sigma = \sigma_e + \sigma_h \qquad (6\text{-}110)$$

Or, using Eq. (6-109), the contributions of both types of carriers are taken into account by

$$\sigma = n_e e\mu_e + n_h e\mu_h \qquad (6\text{-}111)$$

It is necessary to obtain expressions for n_e and n_h in order to calculate σ and its temperature dependence.   These relationships may be obtained by determining the number of states for each carrier in each band (the area under the curve) in a way analogous to that shown in Fig. 4-3(c).   Thus, an expression must be obtained for the density of electron states in the conduction band, $N(E)_c$, and for the density of holes in the valence band, $N(E)_h$.   This is done with the help of Fig. 6-16.

The density of states in the conduction band is given by

$$n(E)_c \, dE = \int_{E_c}^{\infty} N(E) f(E) \, dE \qquad (6\text{-}112)$$

This is the integrated product of Eqs. (4-5) and (2-86); see Fig. 4-3(c).   Since a large number of states is available for the relatively few electrons [Eq. (2-55)], it may be considered that the width of the energy span is large with respect to $k_B T$; so it may be

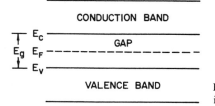

Figure 6-16   Simple band structure of an intrinsic semiconductor.

approximated, actually using the Boltzmann tail [Eq. (2-87)], that

$$f(E) \cong e^{-(E-E_F)/k_B T} \tag{6-113}$$

When spin is included, the factor $N(E)$ for the electrons in the conduction band in Eq. (6-112) becomes, using Eq. (4-5),

$$N(E)\, dE = \frac{4}{h^3}\, (2m_e^*)^{3/2}(E - E_c)^{1/2}\, dE \tag{6-114}$$

Here, m_e^* is the effective mass of an electron at the bottom of the conduction band. The concept of effective mass was introduced in Section 4.6 and is discussed in more detail in the next section. The factor V is equal to unity when Eq. (4-5) is used here; it does not appear explicitly in Eq. (6-114) because n_e and n_h are the numbers of carriers per unit volume. Equations (6-113) and (6-114) are substituted into Eq. (6-112) to give the density of states in the conduction band as

$$N(E)_c\, dE \cong \frac{4\pi}{h^3}\, (2m_e^*)^{3/2} \int_{E_c}^{\infty} e^{-(E-E_F)/k_B T}(E - E_c)^{1/2}\, dE \tag{6-115}$$

The upper limit of infinity may be used because only a negligible number of states are filled at energies much above E_c as a result of the very rapid decrease in $f(E)$ for $E > E_c$. In other words, only a very small fraction of the states available for occupancy in the conduction band are filled, and these are close to the lower conduction band edge, E_c.

When $E_c - E_F$ is small, or, in other words, when it may be approximated that $E_c \cong E_F$, the integral can be expressed in the general form given next, since for this case $E - E_F \cong E - E_c$:

$$\int_0^{\infty} e^{-u} u^{1/2}\, du = \frac{\pi^{1/2}}{2}$$

Equation (6-115) is modified, as shown next, to conform to this general form. Upon integration, it is found, by letting $u = E - E_F \cong E - E_c$, that

$$N(E)_c\, dE \cong \frac{4\pi}{h^3}\, (2m_e^*)^{3/2} \int_{E_c}^{\infty} e^{-(E-E_F)/k_B T}(k_B T)^{1/2}\, \frac{(E - E_c)^{1/2}}{(k_B T)^{1/2}}\, k_B T\, \frac{dE}{k_B T}$$

$$= \frac{4\pi}{h^3}\, (2m_e^* k_B T)^{3/2}\, \frac{\pi^{1/2}}{2}$$

Then, when the expression for $N(E)_c\, dE$ is simplified, the total number of electrons per unit volume in the conduction band is given by

$$n_e \cong N(E)_c f(E) = 2 \left(\frac{2\pi m_e^* k_B T}{h^2}\right)^{3/2} \cdot e^{(E_F - E_c)/k_B T} \tag{6-116}$$

Since intrinsic behavior is being considered, the probability of finding a hole, or an unoccupied state in the valence band, is determined by the number of electrons that have left it and entered the conduction band. This probability is given by $[1 - f(E)]$.

The derivation of the density of states of holes in the valence band parallels that given previously for the electrons in the conduction band. Thus,

$$N(E)_h \, dE = \int_0^{E_v} N(E)[1 - f(E)] \, dE \tag{6-112a}$$

In this case the relatively few unoccupied states are very close to the valence band edge, E_v, and $[1 - f(E)]$ very rapidly becomes negligible as the interior of the valence band is approached. Thus, the density of states for the holes in the valence band is the mirror image of that for the electrons in the conduction band, and it may be approximated that

$$1 - f(E) \cong e^{-(E_F - E)/k_B T} \tag{6-113a}$$

Following the same reasoning as given previously, that $E_F - E_v$ is small and that $(E_F - E) \cong (E_v - E)$, it is found that

$$N(E)_h \, dE \cong \frac{4\pi}{h^3} (2m_h^*)^{3/2} \int_0^{E_v} e^{-(E_F - E)/k_B T} (E_v - E)^{1/2} \, dE \tag{6-114a}$$

in which m_h^* is the effective mass of a hole. Following the previous treatment given for n_e and performing the integration, the number of holes per unit volume in the valence band is obtained as

$$n_h \cong N(E)_h[1 - f(E)] \cong 2 \left(\frac{2\pi m_h^* k_B T}{h^2} \right)^{3/2} \cdot e^{(E_v - E_F)/k_B T} \tag{6-117}$$

The quantities n_e and n_h are in the range of about 10^{18} to $10^{19}/\text{m}^3$ in the neighborhood of room temperature. These densities are very small compared to the density of electrons, about $10^{25}/\text{m}^3$, in a normal metal.

Equations (6-116) and (6-117) can be equated, since $n_e = n_h$ in "pure" intrinsic semiconductors. Noting that the coefficients vanish, this equivalence gives

$$\exp[(E_F - E_c)/k_B T]m_e^{*3/2} \cong \exp[(E_v - E_F)/k_B T]m_h^{*3/2} \tag{6-118}$$

It should be observed at this point that m_e^* and m_h^* have been assumed to be single-valued scalar quantities. These actually are tensor values, known as the density-of-states effective masses (Section 6.3.2). However, this simple assumption is reasonable for the essentially one dimensional analysis given here.

When the logarithms of both sides of Eq. (6-118) are taken and the fractions are cleared, it is found that

$$E_F - E_c + \tfrac{3}{2} \cdot k_B T \ln m_e^* \cong E_v - E_F + \tfrac{3}{2} \cdot k_B T \ln m_h^*$$

This reduces to

$$2E_F \cong E_v + E_c + \frac{3}{2} \cdot k_B T \ln \frac{m_h^*}{m_e^*} \tag{6-119}$$

Another way of arriving at the relationships between the various energies involved is to consider the *effective widths* of the valence and conduction bands to be small with respect to the energy gap between them. This approach includes only those states in the respective bands in which the numbers of carriers are significant. This is a reasonable approximation because of the relatively small numbers of carriers that are involved near each band edge. The number of electrons per unit volume in the conduction band may be approximated closely by using the Boltzmann tail given by Eq. (2-87). Thus,

$$\frac{n_e}{N(E)_c} = f, \qquad n_e \cong \frac{N(E)_c}{\exp[(E_c - E_F)/k_B T]} \tag{6-120}$$

This is permissible since $(E_c - E_F)/k_B T \gg 1$. And, in the same way, the density of holes in the valence band is approximated by

$$n_h \cong \frac{N(E)_v}{\exp[(E_v - E_F)/k_B T]} \tag{6-121}$$

In each of these equations the total numbers of states per unit volume are identical and equal to N. From Eqs. (6-120) and (6-121), the total number of carriers per unit volume is

$$n_e + n_h \cong \frac{N(E)_c}{\exp[(E_c - E_F)/k_B T]} + \frac{N(E)_h}{\exp[(E_v - E_F)/k_B T]} \cong 2N \tag{6-122}$$

Simplifying this equation and using the first two terms of the series expansion for e^x in each denominator, gives, since $N(E)_c = N(E)_h = N$,

$$\frac{1}{(E_c - E_F)/k_B T + 1} + \frac{1}{(E_v - E_F)/k_B T + 1} \cong 2$$

This may be simplified further to read

$$\left(\frac{E_c - E_F + k_B T}{k_B T} + \frac{E_v - E_F + k_B T}{k_B T} \right)^{-1} \cong 2 \tag{6-123}$$

since $k_B T$ is much less than either $E_c - E_F$ or $E_v - E_F$. And finally, where E_g is the width of the energy gap, and the approximation is made that $k_B T/2 \to 0$, the Fermi energy is given by

$$E_F \cong \frac{E_v + E_c}{2} = \frac{E_g}{2} \tag{6-124}$$

Therefore, the Fermi level lies at the center of the energy gap between the valence and conduction bands (Fig. 6-16).

The substitution of Eq. (6-124) into Eq. (6-119) gives

$$E_F \cong \frac{E_g}{2} + \frac{3}{4} k_B T \ln \frac{m_h^*}{m_e^*} \tag{6-125}$$

It will be observed that Eq. (6-125) is the same as Eq. (6-119). When the bands are exactly symmetrical, $m_h^* = m_e^*$ and Eq. (6-125) gives the result, since $\ln 1 = 0$, that the

Fermi energy lies at the center of the energy gap:

$$E_F \cong \frac{E_g}{2} \tag{6-126}$$

Equation (6-126) is the same as Eq. (6-124). The approximation of the equivalence of the effective masses could have been used directly in Eq. (6-119) to obtain the same result given by Eq. (6-126).

It should be noted that m_h^* generally is larger than m_e^* (see Section 6.3.2). However, even when m_h^*/m_e^* is significantly different from unity, the temperature dependence of E_F is small because Eqs. (6-119) and (6-125) contain the logarithm of this ratio. Thus, in general, E_F has a small, positive slope as a function of temperature. Silicon has $m_h^*/m_e^* < 1$ and therefore has a small, negative slope and is an exception to the more general behavior.

The identical Eqs. (6-124) and (6-126) facilitate the simplification of Eq. (6-116) as

$$n_e = 2\left(\frac{2\pi m_e^* k_B T}{h^2}\right)^{3/2} \cdot \exp\left(-\frac{E_g}{2k_B T}\right) \tag{6-127a}$$

and Eq. (6-117) as

$$n_h = 2\left(\frac{2\pi m_h^* k_B T}{h^2}\right)^{3/2} \cdot \exp\left(-\frac{E_g}{2k_B T}\right) \tag{6-127b}$$

The information is now available for the more complete understanding and use of Eq. (6-111). Equations (6-127a) and (6-127b) may be substituted into Eq. (6-111) to obtain

$$\sigma = 2\left(\frac{2\pi m_e^* k_B T}{h^2}\right)^{3/2} \cdot \exp\left(-\frac{E_g}{2k_B T}\right)e_e\mu_e - 2\left(\frac{2\pi m_h^* k_B T}{h^2}\right)^{3/2} \cdot \exp\left(-\frac{E_g}{2k_B T}\right)e_h\mu_h \tag{6-128}$$

The minus sign appears because the sign of the charge on a hole is opposite to that on an electron. Since these charges have the same absolute value, Eq. (6-128) can be expressed as

$$\sigma = 2\left(\frac{2\pi m^* k_B T}{h^2}\right)^{3/2} \cdot |e| \cdot \exp\left(-\frac{E_g}{2k_B T}\right) \cdot (\mu_e - \mu_h) \tag{6-129}$$

if it is assumed again that $m_e^* \cong m_h^* = m^*$.

When the first three factors of Eq. (6-129) are included in a single parameter, A, since $T^{3/2}$ is small compared to the exponential term, this equation may be expressed in logarithmic form as

$$\ln \sigma = \ln A - \frac{E_g}{2k_B T} + \ln(\mu_e - \mu_h) \tag{6-130a}$$

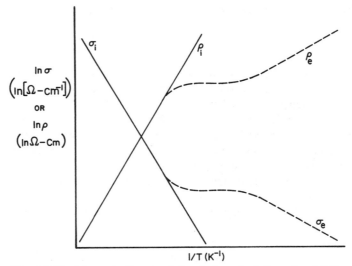

Figure 6-17   Schematic diagram of electrical conductivity and resistivity
of semiconductors. Intrinsic behavior is shown by the solid curves, and
extrinsic behavior by the broken curves.

or, using Eq. (6-126), as

$$\ln \sigma = \ln A - \frac{E_F}{k_B T} + \ln(\mu_e - \mu_h) \qquad (6\text{-}130\text{b})$$

If the temperature dependencies of the mobilities of the electrons and holes are
not considered explicitly, and these mobilities are included along with the other
factors in A in a new parameter σ_0, then the conductivity may be approximated, using
Eq. (6-124), as

$$\ln \sigma \cong \ln \sigma_0 - \frac{E_F}{k_B T} = \ln \sigma_0 - \frac{E_g}{2k_B T} \qquad (6\text{-}131)$$

And, since ρ is the reciprocal of σ, the electrical resistivity may be approximated
similarly by

$$\ln \rho \cong \ln \rho_0 + \frac{E_F}{k_B T} = \ln \rho_0 + \frac{E_g}{2k_B T} \qquad (6\text{-}131\text{a})$$

Schematic plots of these equations are given in Fig. 6-17.

The solid curves in the figure show the negative type of temperature dependence
indicated earlier for the intrinsic behavior of semiconductors. The broken curves
show the corresponding properties of extrinsic materials that are discussed later in
Section 6.3.3.

6.3.1.2 Variation of mobility with temperature. The temperature dependence of the mobilities of the carriers was neglected in the derivation of Eq. (6-131). A complete presentation of carrier mobility is beyond the scope of this work. Simple descriptions are given here.

Semiconductors. The relatively few electrons promoted to the conduction band are readily accommodated in the large number of available states. Thus, they satisfy the condition of Eq. (2-55) and may be treated as classical, noninteracting, nondegenerate particles. The mobilities of these electrons, therefore, can be approximated by means of Eq. (6-25). In addition, $\langle n_s \rangle$ can be taken as unity in the same way as for Eq. (6-13). This gives the mobility as

$$\mu = \frac{\langle n_s \rangle e \langle \lambda \rangle}{m \langle \mathbf{v} \rangle} = \frac{e}{m} \cdot \frac{\langle \lambda \rangle}{\langle \mathbf{v} \rangle} \tag{6-132a}$$

And, when e/m is considered to be approximately constant, because the electrons occupy the lowest levels in the conduction band (Section 6.3.2), the temperature variation of μ may be obtained from Eq. (6-132a) as

$$\mu \propto \frac{\langle \lambda \rangle}{\langle \mathbf{v} \rangle} \tag{6-132b}$$

Here $\langle \lambda \rangle \propto 1/T$ [Eq. (5-214c)], and, as classical particles, $\langle \mathbf{v} \rangle \propto T^{1/2}$ so that

$$\mu_e \propto \frac{T^{-1}}{T^{1/2}} = T^{-3/2} \tag{6-133}$$

The mobility of holes is significantly different from that of electrons because of their disparate nature (Sections 6.3.1). A hole, while treated as a particle, actually is a readily transferrable, dangling bond in a covalently bonded lattice. It may be considered that the scattering mechanism of holes as charged carriers in a "pure", elemental, intrinsic semiconductor results primarily from lattice imperfections; the relative effects of impurities and any consequent changes in the periodic potential of the lattice are expected to be small [Eq. (6-50)]. However, assume that such composite scattering centers may be represented by spheres of radius r_s with a scattering cross section of $\Delta A_T \propto r_s^2$. Thus $r_s \propto T$ [Eq. (5-175)], and $r_s^2 \propto T^2$, so $\langle \lambda \rangle \propto 1/r_s^2 \propto 1/T^2$. These permit the temperature variation of μ_h to be obtained from Eq. (6-132b) as

$$\mu_h \propto \frac{\langle \lambda \rangle}{\langle \mathbf{v} \rangle} = \frac{T^{-2}}{T^{1/2}} = T^{-5/2} \tag{6-134}$$

The temperature dependence of μ for extrinsic semiconductors is somewhat more complex than that of intrinsic materials. Here the electron scattering is caused primarily by the ionized impurities (dopants) in a way analogous to that described in Section 6.13. [Eq. (6-51a)]; in that case, only the electrical perturbations required consideration because the other factors were essentially constant. In the present instance, the temperature variation of $\langle n_s \rangle$ must be taken into account. This is obtained, by use of the Rutherford model, as

$$\langle n_s \rangle \propto \frac{1}{\Delta A_T} \propto \left(\frac{\varepsilon m^* \langle \mathbf{v} \rangle^2}{(Z_\beta - Z_\alpha)e^2} \right)^2 \tag{6-135a}$$

Here ε is the relative dielectric constant of the host material. Thus, for a given dopant ion, $m^*\varepsilon/(Z_\beta - Z_\alpha)e^2 \cong$ constant, so

$$\langle n_s \rangle \propto \langle \mathbf{v} \rangle^4 \tag{6-135b}$$

The holes also are treated as classical particles so that, as in Eq. (6-134), $\langle \mathbf{v} \rangle \propto T^{1/2}$. However, in this case, as indicated by Eq. (5-123), λ is inversely proportional to the impurity concentration; it is not a function of temperature. Thus, from Eq. (6-132a),

$$\mu \propto \frac{\langle n_s \rangle \langle \lambda \rangle}{\langle \mathbf{v} \rangle} \propto \frac{\langle n_s \rangle}{\langle \mathbf{v} \rangle} = \frac{\langle \mathbf{v} \rangle^4}{\langle \mathbf{v} \rangle} \tag{6-136a}$$

$$\mu \propto \langle \mathbf{v} \rangle^3 \propto T^{3/2} \tag{6-136b}$$

See Section 6.3.

Metals.   The temperature variation of the mobility of electrons as degenerate particles is considerably different from that in semiconductors, where they are nondegenerate. This may be shown by reference to Eq. (6-22):

$$\mu(E_F) \cong \frac{n_s e L(E_F)}{m v(E_F)} \propto n_s L(E_F) \tag{6-137}$$

Here $e/[mv(E_F)]$ is treated as a constant because $m \cong m^*$ (Section 6.3.2), and the variation of $\mathbf{v}(E_F)$ with T is negligible [Eq. (2-85) and Section 4.2]. So, since $n_s \propto T$ and $L(E_F) \propto 1/T$,

$$\mu(E_F) \propto T \cdot T^{-1} = \text{constant} \tag{6-138}$$

It is interesting to note that the same result as is given by Eq. (6-138) is also obtained for metals in the temperature range in which $C_V \propto T^3$ [Eq. (5-111)]. Here $U \propto T^4$ [Eq. (5-1)]. If the energy of a phonon is roughly approximated as $E = h\nu \cong k_B T$ [Eq. (5-138)], then the number of phonons $n_p \propto n_s \propto T^4/T = T^3$ for this kind of electron–phonon scattering. And, since $L(E_F) \propto 1/n_s \propto T^{-3}$, $\mu(E_F)$ is constant for this case, too.

In the range of about $\theta_D/20 < T \ll \theta_D$, the electron–phonon scattering will vary as the number of phonons. In turn, this varies inversely as the average phonon energy:

$$n_s \propto n_p \propto \frac{1}{E} \tag{6-139}$$

The phonon energy is given by $E = \mathbf{p}^2/2m$, so $E \propto \mathbf{p}^2$ and

$$n_s \propto \frac{1}{\mathbf{p}^2} \tag{6-140}$$

The phonon energy may also be expressed as $E = \frac{1}{2}mc_s^2$, or $2E/c_s = mc_s = \mathbf{p}$. And, in this temperature range where the variation in c_s is small, again assuming $E \cong k_B T$,

$$E \propto \mathbf{p} \propto T \tag{6-141}$$

Thus, using Eq. (6-141) in Eq. (6-140),

$$n_s \propto \frac{1}{T^2} \tag{6-142}$$

Now, substituting Eq. (6-142) along with the prior approximation for $L(E_F) \propto T^{-3}$ into Eq. (6-137) gives the electron–phonon scattering as

$$\mu(E_F) \propto T^{-2} \cdot T^{-3} = T^{-5} \tag{6-143}$$

In the temperature range $T \geqslant \theta_D$, $L(E_F) \propto 1/T$. In this regime, the average energy of an electron is such that a single scattering event with a phonon is sufficient to randomize its direction; thus, $n_s \cong 1$. For these conditions, Eq. (6-137) is reduced to

$$\mu(E_F) \propto L(E_F) \propto T^{-1} \tag{6-144}$$

Thus, for high temperatures, $\mu(E_F)$ approaches a limiting asymptotic value similar to that of Eq. (6-138), and $L(E_F)$ behaves similarly.

6.3.2 Electron Behavior and Effective Mass

The masses of electrons and holes were taken into consideration in the previous sections and in Section 4.6. The effective masses of these particles give a more accurate representation than that of nearly free particle behavior in certain situations. In the case of normal metals, where the Fermi levels are well below the tops of their valence bands, the effective mass is approximately equal to the nearly free electron mass. Thus, the relationships given in previous chapters for metals are essentially correct. However, in the case of semiconductors, where electrons and holes are, respectively, very close to the bottom of the conduction band or very near to the top of the valence band, their effective masses must be taken into account. Effective masses are designated by m^*, as in previous discussions.

The behavior of electrons close to the top of the band or very near the zone boundary will be examined first. Here the electron is considered as a wave packet. The group velocity is obtained from Eq. (1-15) for one dimension, now using the wave vector, $\bar{\mathbf{k}}_x$, instead of the wave number, k. Thus, as shown in Section 1.6.1,

$$\mathbf{v} = \mathbf{v}_g = \frac{dv}{dk} = \frac{\partial \omega}{\partial \bar{\mathbf{k}}_x} \tag{6-145}$$

Rewriting Eq. (1-5), $E = hv = h\omega/2\pi$, its differentiation with respect to $\bar{\mathbf{k}}_x$ results in

$$\frac{\partial E}{\partial \bar{\mathbf{k}}_x} = \frac{h}{2\pi} \frac{\partial \omega}{\partial \bar{\mathbf{k}}_x} \tag{6-146}$$

Rearranging Eq. (6-146) gives

$$\frac{\partial \omega}{\partial \bar{k}_x} = \frac{2\pi}{h} \frac{\partial E}{\partial \bar{k}_x} \tag{6-146a}$$

The substitution of Eq. (6-146a) into Eq. (6-145) provides a more convenient expression for the velocity of an electron:

$$\mathbf{v} = \frac{2\pi}{h} \frac{\partial E}{\partial \bar{k}_x}, \qquad \frac{\partial E}{\partial \bar{k}_x} = \frac{h\mathbf{v}}{2\pi} \tag{6-147}$$

Thus, the velocity of an electron depends on the slope and, consequently, on its position on the E versus $\bar{k}_x$ curve [Fig. 4-16(b)]. An expression needed for $\partial E/\partial \bar{k}_x$ is obtained from the derivative of Eq. (3-26) as

$$\frac{\partial E}{\partial \bar{k}_x} = \frac{h^2 \bar{k}_x}{4\pi^2 m} \tag{6-148}$$

Now, the velocity of an electron with E and $\bar{k}_x$ coordinates that lie on the parabolic portion of this essential curve is obtained by the substitution of Eq. (6-148) into Eq. (6-147):

$$\mathbf{v} = \frac{2\pi}{h} \frac{h^2 \bar{k}_x}{4\pi^2 m} = \frac{h\bar{k}_x}{2\pi m} \tag{6-149}$$

Therefore, in the range in which an electron is approximated as being nearly free, its velocity is a linear function of $\bar{k}_x$ [Fig. 6-18(a)], since the other factors are constants for this case.

However, an electron with coordinates at and beyond the inflection point on the E versus $\bar{k}_x$ curve, near the top of a band (close to the zone boundary) as shown in Fig. 4-16(b), may no longer be treated as being nearly free because $\partial E/\partial \bar{k}_x$ decreases and approaches zero at the critical values of $\bar{k}_x$ that occur at $\pm n\pi/a$. The velocity [Eq. (6-149)], therefore, also approaches zero between the inflection points and $\pm n\pi/a$. This is shown in Fig. 6-18(b).

The acceleration of an electron, a, is obtained from the derivative of Eq. (6-147)

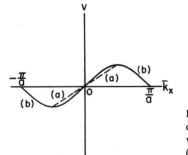

Figure 6-18 Schematic diagram of electron velocity as a function of the wave vector: (a) for a nearly free electron; (b) influence of the zone boundary.

with respect to time:

$$a = \frac{d\mathbf{v}}{dt} = \frac{2\pi}{h} \frac{\partial}{\partial t} \left(\frac{\partial E}{\partial \overline{\mathbf{k}}_x} \right) \tag{6-150}$$

The indicated operation is performed and reexpressed using the chain rule as

$$a = \frac{2\pi}{h} \frac{\partial^2 E}{\partial \overline{\mathbf{k}}_x \, \partial t} = \frac{2\pi}{h} \frac{\partial^2 E}{\partial \overline{\mathbf{k}}_x^2} \frac{\partial \overline{\mathbf{k}}_x}{\partial t} \tag{6-151}$$

An expression for $\partial \overline{\mathbf{k}}_x / \partial t$ is needed to simplify Eq. (6-151). This is obtained by considering the effect of an applied electric field, $\overline{E}$ on an electron. If the field influences the electron for a time dt, and its velocity is $\mathbf{v}$, the distance that it travels in that time is $\mathbf{v}\,dt$. Hence, using the product of the force on the electron and the distance traveled, its energy may be expressed as

$$dE = e\overline{E}\mathbf{v}\,dt \tag{6-152}$$

Equation (6-152) may be reexpressed by rewriting dE/dt, again using the chain rule, as

$$\frac{dE}{dt} = \frac{\partial E}{\partial \overline{\mathbf{k}}_x} \cdot \frac{\partial \overline{\mathbf{k}}_x}{\partial t} = e\overline{E}\mathbf{v} \tag{6-152a}$$

An expression for $\partial E / \partial \overline{\mathbf{k}}_x$ is obtained from Eq. (6-147). This is substituted into Eq. (6-152a) with the result being

$$e\overline{E}\mathbf{v} = \frac{h}{2\pi} \mathbf{v} \frac{\partial \overline{\mathbf{k}}_x}{\partial t} \tag{6-153}$$

The factor $\mathbf{v}$ vanishes and the rearrangement of Eq. (6-153) will provide the expression for $\partial \overline{\mathbf{k}}_x / \partial t$ being sought:

$$\frac{\partial \overline{\mathbf{k}}_x}{\partial t} = \frac{2\pi}{h} e\overline{E} \tag{6-153a}$$

The substitution of Eq. (6-153a) into Eq. (6-151) gives the acceleration of an electron as

$$a = \frac{2\pi}{h} \cdot \frac{\partial^2 E}{\partial \overline{\mathbf{k}}_x^2} \cdot \frac{2\pi}{h} e\overline{E} = \frac{4\pi^2}{h^2} \cdot \frac{\partial^2 E}{\partial \overline{\mathbf{k}}_x^2} \cdot e\overline{E} \tag{6-154}$$

A nearly free electron that is neither close to the top of the band nor close to the zone wall, that is, one that lies on the parabolic portion of the primary E versus $\overline{\mathbf{k}}_x$ curve (as in a normal metal), will have its acceleration determined by the use of the derivative of Eq. (6-148) in Eq. (6-154). For these conditions, Eq. (6-148) becomes

$$a = \frac{4\pi^2}{h^2} \cdot \frac{h^2}{4\pi^2 m} \cdot e\overline{E} = \frac{e\overline{E}}{m} \tag{6-155}$$

so that the acceleration of a nearly free electron is constant for a given electric field [Fig. 6-19(a)].

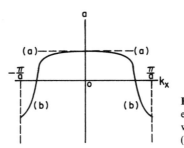

Figure 6-19 Schematic diagram of electron acceleration as a function of the wave vector: (a) for a nearly free electron; (b) influence of the zone boundary.

The acceleration of an electron near the top of the band is more complicated; it may not be treated as being "nearly free" and it is more complex, since $\partial^2 E/\partial \bar{\mathbf{k}}_x^2$ [Eq. (6-154)] no longer remains constant. This factor is constant and positive in the parabolic portion of the E versus $\bar{\mathbf{k}}_x$ curve ("nearly free"). Near the top of the band, $\partial^2 E/\partial \bar{\mathbf{k}}_x^2$ no longer is constant, but becomes zero at the inflection points, and then is negative from the inflection points to the zone boundaries. The acceleration, accordingly, decreases from a positive value, goes through zero, and then becomes negative [Fig. 6-19(b)].

The effective mass of an electron may be determined from an examination of Eq. (6-154). Recalling that $a = F/m$ and $F = e\bar{E}$, it can be seen from Eq. (6-154) that the effective mass is given by

$$m^* \propto \frac{1}{a} = \frac{h^2}{4\pi^2}\left(\frac{\partial^2 E}{\partial \bar{\mathbf{k}}_x^2}\right)^{-1} \tag{6-156}$$

And, since m^* varies as $1/a$, m^* may be positive or negative and may be larger or smaller in magnitude than that of a free electron in the $\bar{\mathbf{k}}_x$ range at and beyond the inflection points.

An inspection of Fig. 4-16(b) or 4-17 and Eq. (6-148) shows that $\partial^2 E/\partial \bar{\mathbf{k}}_x^2$ is positive and constant in the parabolic part of the curve and $m^* \cong m$, the mass of a nearly free electron. Another way to see this is by means of the second derivative of the equation for the energy of a *free* electron [Eq. (3-26)]. This operation results in

$$\frac{\partial^2 E}{\partial \bar{\mathbf{k}}_x^2} = \frac{h^2}{4\pi^2 m} \tag{6-156a}$$

The substitution of Eq. (6-156a) into Eq. (6-156) gives the previously noted result:

$$m^* = \frac{h^2}{4\pi^2}\left(\frac{\partial^2 E}{\partial \bar{\mathbf{k}}^2}\right)^{-1} = \frac{h^2}{4\pi^2}\cdot\frac{4\pi^2 m}{h^2} = m \tag{6-156b}$$

This is shown in Fig. 6-20(a). (See Table 6-6.)

At the inflection points of the curve of E versus $\mathbf{k}_x$, the second derivative equals zero. This produces singularities [Eq. (6-156)]. Beyond this, as the zone boundary is approached, the second derivative is negative and m^* is negative. The singularities at the inflection points actually are represented by maxima and minima, rather than by

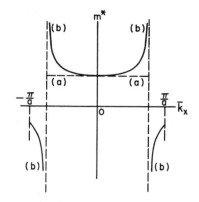

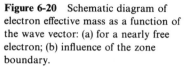

Figure 6-20 Schematic diagram of electron effective mass as a function of the wave vector: (a) for a nearly free electron; (b) influence of the zone boundary.

infinite masses. This results from the Heisenberg uncertainty principle (Section 1.6.6). Since neither the wave vector nor the energy can be known exactly, average values of m^* must be determined over suitable ranges of $\Delta \bar{\mathbf{k}}$ and ΔE. These uncertainties give rise to the maxima or minima.

The role of the hole in the valence band, actually a short-hand construct for describing the behavior of the electrons in the valence band, is most readily pictured by first considering a completely filled band. In the absence of external energy, the gap restricts the electrons to the filled states where they are distributed symmetrically as shown in Figs. 4-4(a) and (c) and in Fig. 6-1(a). This symmetry means that the effects of any electron with given values of $\mathbf{p}$ (or $\mathbf{v}$) or $\bar{\mathbf{k}}$ will be balanced exactly by those of another electron of equal but opposite values of these parameters. Thus, even when the electrons respond to external influences, their effects mutually cancel each other; the net effect is zero.

This situation changes considerably when even a very small fraction of the electrons in the valence band is promoted to the conduction band; the symmetry now is imperfect because some of the remaining electrons are no longer balanced by others. The resulting asymmetry is analogous to that shown in Fig. 6-1(b), and the net effect is no longer zero. The holes describe this net effect. This may be seen by observing the influence of just one vacant state in the valence band.

If the contribution to the current made by a single electron, whose velocity is $\mathbf{v}_i$, is

$$I_i = e\mathbf{v}_i \tag{6-157a}$$

then the current that would be constituted by *all* the states, *n*, in the completely filled valence band is

$$I = e \sum_i^n \mathbf{v}_i = 0 \tag{6-157b}$$

This must be so because of the symmetry of the velocities; the summation must equal zero for this case and no net current flows. Now, if only one unoccupied state is present in the valence band and corresponds to that of an electron of velocity $\mathbf{v}_j$, then the current provided by all the valence electrons is given by

$$I = e \sum_{i \neq j}^{n} \mathbf{v}_i = e \sum_{i}^{n} \mathbf{v}_i - e\mathbf{v}_j \tag{6-157c}$$

So, when Eq. (6-157b) is considered in Eq. (6-157c), the current is reduced to that determined by the hole:

$$I = -e\mathbf{v}_j \tag{6-157d}$$

Here, since the charge on the electron is negative, the direction of the current is positive and the charge on the hole is positive.

The utility of the construct known as a hole, and treated as a real particle, is that it permits the behaviors of the many electrons in the valence band to be described conveniently by those of relatively very few vacant states treated as fictitious particles.

The curve of E versus $\bar{\mathbf{k}}_x$ for holes in a nearly filled valence band may be considered to be the mirror image of that for electrons (Fig. 6-21), because their charges, energies, and masses are opposite in sign. Thus, $\bar{\mathbf{k}}_x$ for a hole equals zero at the top of the valence band in contrast to the wave vector of an electron that would have been maximum at $\pm \pi/a$. Furthermore, the energies of the holes increase parabolically downward, symmetric about $\bar{\mathbf{k}}_x = 0$, as the magnitudes of their wave vectors increase. This gives the sign of $\partial^2 E/\partial \bar{\mathbf{k}}_x^2$ for a hole as opposite to that for an electron. As a result of this, the effective mass of a hole is opposite in sign to that of a corresponding electron on any given portion of the curve [Eq. (6-156)].

In the case of insulators, where the filled valence bands and empty conduction bands are separated by relatively large energy gaps, electrical conduction is not usually possible [Eq. (6-157b)]. Neither the application of normal electric fields nor normal temperatures can promote electrons across the gap. Thus, under most conditions, the effective mass of electrons in insulators need not be considered. However, in certain cases photons of electromagnetic radiation with energies of about $\frac{3}{2}E_g$ can promote valence electrons to the conduction band, and some of these

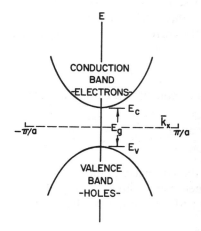

Figure 6-21 Schematic diagram for the energies of a small number of electrons at the bottom of the conduction band and of an equivalent number of holes at the top of the valence band as functions of the wave vector.

materials, normally considered as insulators, can behave as photoconductors in the same ways as semiconductors (Section 6.3.1).

The application of external energy to a semiconductor results in the utilization of a *part* of the applied energy in the promotion of some of the electrons from the valence to the conduction band. The carriers remaining in the valence band and the lattice are also affected. Consider the case in which an electric field is applied parallel to and in the $+\bar{k}_x$ direction [Fig. 6-18(b)]. Let one unfilled state with a $-\bar{k}_x$ coordinate be present very close to the top, or edge, of the valence band. An electron in the next highest state, under the influence of the field, would be expected to occupy the unfilled state; its energy and its velocity component parallel to the field should be decreased in the direction opposite to the field (Lenz's law). This electron behaves as expected, fills the hole, and causes the hole to move into the state it vacated. Thus, the hole effectively moves from a higher state to the next lower state; it moves in the same direction as the applied field, as would be expected of a particle with a positive charge. Its velocity is positive because its value of $\partial E/\partial \bar{k}_x$ is opposite in sign to that of electrons [Eq. (6-149)]. Its acceleration and effective mass are positive because $\partial^2 E/\partial \bar{k}_x^2$ is positive [Eq. (6-156)].

The electron that moved into and occupied the original hole in the band will have its negative velocity decreased because it obeyed Lenz's law [Fig. 6-18(b)]. This results since Eq. (6-148) is negative and correspondingly affects Eq. (6-149). It will have a negative effective mass because $\partial^2 E/\partial \bar{k}_x^2$ is negative [Eq. (6-156) and Fig. 6-20(b)]. A subsequent occupation of the newly created hole by a second electron, from another slightly lower state, transports the hole to that slightly lower state. It will be observed again that motions of the holes mirror and actually determine those of the electrons [Eq. (6-157d)].

Following the same reasoning, the velocity of the electron will be slightly less negative, its acceleration will be slightly more negative, and its effective mass will be slightly less negative. In a corresponding way, the velocity of the hole will increase slightly, its acceleration will be greater, and its effective mass will be slightly more positive.

This description is incomplete until the effective mass of an electron is considered in terms of the reflection of the electron at the Brillouin zone boundary. When the Fermi level is close to $\bar{k}_c$ [Eq. (4-31)], the electrons with energies near E_F will be accelerated by an applied electric field. Some of these electrons will have an effective wave vector $\bar{k}_x \geqslant \bar{k}_c$ and, therefore, will be reflected back into the zone (Section 4.6). The reversal of the motion of the electron changes the sign and magnitude of $\bar{k}_x$. The direction of motion is opposite to that which normally would have been expected to have been induced by the field. The construct of effective mass is employed because it readily accounts for the changes in the behavior of electrons near the top of the band. This simplifies matters by neglecting other effects of external influences and by only considering that the electron has changed its mass so that its motion remains in conformity with Newton's laws. The change in the mass of the electron is a function of $\bar{k}$. The reversal of the sign of $\bar{k}$ and the other corresponding changes in the velocity, acceleration, and mass of the electron are properties associated with negative effective mass.

In reality, no electrons have negative masses. This is a convenient physical construct. Newton's laws have not been violated, since only the reaction of the electron to the external influence has been considered; other factors, including the effect of lattice forces on the electron, have been neglected. The effective mass is actually a crystal-wave interaction; it is not an inertial effect. It is necessary to be concerned with negative effect masses only for those electrons in states near the top of the band; this is the region in which the electrons cannot be treated as being nearly free because the primary E versus $\bar{k}_x$ curve undergoes inflection and deviates from its initial, parabolic shape. States with lower E, $\bar{k}_x$ coordinates, where the energies lie on the parabolic portion of this curve, may be considered to be nearly free and $m^* \cong m$ [Eq. (6-156b)].

It will be appreciated that the essentially one dimensional treatment given here represents the simplest case. Real lattices are both three dimensional and anisotropic. In one such case, assuming an orthorhombic lattice with ellipsoidal energy surfaces, the effective mass may be obtained from Eq. (3-48) by letting $L_i = \lambda_i/2$ and recalling that $\bar{k}_i = 2\pi/\lambda_i$, giving, where x, y, and z lie on orthogonal axes,

$$E = \frac{h^2\bar{k}_x^2}{8\pi^2 m_x} + \frac{h^2\bar{k}_y^2}{8\pi^2 m_y} + \frac{h^2\bar{k}_z^2}{8\pi^2 m_z} \tag{6-158}$$

and the effective mass is given by

$$\frac{1}{m^*} = \frac{1}{m_x^*} + \frac{1}{m_y^*} + \frac{1}{m_z^*} \tag{6-159}$$

This is known as the *density-of-states* effective mass. In the most general case, the density-of-states effective mass must be given by a nine-component tensor. However, Eq. (6-156) can be used where the Fermi energy surfaces may be approximated reasonably well as being spherical. Therefore, m may be used in place of m^* [Eq. (6-156b)] in the equations based on "nearly free" electrons (given in Chapter 4 and earlier in this chapter) without introducing appreciable errors, where E_F is used to describe the properties of metals and metallic alloys whose valence bands do not approach maximum filling.

Normal metals have partially filled valence bands in which the occupied states lie on the parabolic portion of the primary curve of E versus $\bar{k}$, with E_F well below the inflection point. Their Fermi surfaces may be approximated as being spherical as in Fig. 4-4(c) (Section 4.4). This is the equivalent of the approximation that $m^* \cong m$ and that the electrons are "nearly free". See Table 6-6.

Transition elements have narrow, incompletely filled d bands that overlap broad, incompletely filled s bands and hybridization or mixed s–d behavior results (Section 3.11). An unfilled quasicontinuum exists above the highest filled hybridized state. This permits the assumption that the valence electrons are "nearly free" and the approximation that $m^* \cong m$. This is not as good an approximation as that for normal metals. Even though s–d hybridization is present, the d states are more strongly bound to the nucleus than are the s states; so this approximation must be used with caution.

TABLE 6-6 EFFECTIVE MASSES OF ELECTRONS
IN SOME NORMAL METALS

Metal	m^*/m	Metal	m^*/m
Li	1.19	Cu	0.99
Na	1.0	Ag	1.01
K	0.99	Au	1.01
Rb	0.97		
Cs	0.98		

After L. A. Girifalco, *Statistical Physics of Materials*, p.
108, John Wiley & Sons, Inc., New York, 1973.

It will be appreciated from the previous discussions that properties involving the mass of the carrier must reflect the nature of the band structure of the crystalline solid. This is particularly true of semiconductors, where both electrons and holes are involved in different bands (Fig. 6-22). This figure gives the approximate variations in the band structure of silicon for two different crystallographic directions.

The effective masses of the carriers in some of the more common semiconductor materials are given in Table 6-7. The contrast between the values given here and those in Table 6-6 is striking.

Bands separated by relatively small energy gaps, as is the case for many semiconductors, have large curvatures in their curves for E versus $\bar{k}$ [Fig. 4-16(b)] adjacent to the zone boundaries. This causes electrons in states at or near the top of the valence band to possess small, negative values for m_e^*. Carriers in such states are represented by the holes by means of m_h^* [Eq. (6-157d)]. States at the bottom of the conduction band have small, positive values of m_e^*, as shown in Table 6-7.

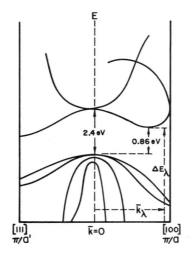

Figure 6-22 Approximate band structure of silicon as a function of wave vector.

TABLE 6-7   PROPERTIES OF SOME LIGHTLY DOPED SEMICONDUCTORS AT 300 K

Type	Semiconductor	E_g (eV)	μ_e (cm²/V s)	μ_h (cm²/V s)	m_e^*/m	m_h^*/m
IV	C	5.3	1,800	1,600	—	—
	Si	1.1	1,350	475	0.23	0.12
	Ge	0.7	3,900	1,900	0.02	0.08
	SiC	2.8	400	50	0.60	1.20
III–V	AlS	2.2	180	—	—	—
	AlP	3.0	80	—	—	—
	AlSb	1.6	200	420	0.30	0.40
	BN	4.6	—	—	—	—
	BP	6.0	—	300	—	—
	GaAs	1.4	8,500	400	0.07	0.09
	GaP	2.3	110	75	0.12	0.50
	GaSb	0.7	4,000	1,400	0.20	0.39
	InAs	0.4	33,000	460	0.03	0.02
	InP	1.3	4,600	150	0.07	0.69
	InSb	0.2	80,000	750	0.01	0.18
II–VI	CdS	2.6	340	18	0.21	0.80
	CdSe	1.7	600	—	0.13	0.45
	CdTe	1.5	300	65	0.14	0.37
	ZnS	3.6	120	5	0.40	—
	ZnSe	2.7	530	16	0.10	0.60
	ZnTe	2.3	530	900	0.10	0.60
IV–VI	PbS	0.4	600	200	0.25	0.25
	PbSe	0.3	1,400	1,400	033	0.34
	PbTe	0.3	6,000	4,000	0.22	0.29
II–IV	Mg₂Ge	0.7	530	110	—	—
	Mg₂Si	0.8	370	65	—	0.46
	Mg₂Sn	0.4	210	150	—	—
II–V	Cd₃As₂	0.1	—	15,000	0.05	—
	CdSb	0.5	300	1,000	0.16	0.10
	Zn₃As₂	0.9	—	10	—	—
	ZnSb	0.5	10	350	0.15	—

After H. F. Wolf, *Semiconductors*, p. 33, Wiley-Interscience, New York, 1971.

6.3.3 Roles of Impurities (Extrinsic Semiconductors)

The effects of impurities in solid solution in semiconductors are directly related to the ways in which they enter into the bonding mechanism of the host materials and affect their band structures. Semiconductors frequently containing very small, controlled amounts of intentionally added impurities, or alloying elements, are known as extrinsic semiconductors. Such additions are called *dopants*.

It will be recalled that the elements that crystallize in the diamond-cubic lattice have valences of four; this makes possible the tetrahedral bonding discussed in Section 3.12.3 and shown in Figs. 3-10(a) and (b). An ion from group V of the Periodic Table will have one more valence electron than is required for this type of bonding. Since the valence band of the semiconductor is completely filled, this fifth electron must occupy a state in the gap. The extra electron constitutes a local negative charge in the lattice that is loosely associated with the impurity ion; this electron can be excited to the conduction band when an external influence is such that k_BT is sufficient to promote it there from a state in the gap, leaving a positive impurity ion behind. The four remaining valence electrons completely satisfy the bonding requirements of the host lattice and behave in the same way as those electrons binding the host ions. Such ions are called *donors* because they contribute electrons to the conduction process without affecting the bonds of the host lattice, neglecting local lattice distortion [Eq. (6-52)]. Materials doped in this way are called *n-type* semiconductors.

Ions from group III lack one electron for the required, saturated, covalent bonding. An electron from one of the semiconductor ions adjacent to it in the lattice will "orbit" about both the group III ion and the ion with which it originally was associated. This induces a positive charge on the group III ion. The resulting induced positive charge acts as a potential well, and this can capture the orbiting electron from the nearby semiconductor ion. Thus, one of the neighboring covalently bonded host ions loses an electron in this process; it becomes incompletely bonded, and the trivalent ion simultaneously becomes negatively charged. The original lack of an electron on the trivalent (group III) ion has caused an absence of an electron, or a dangling bond or a hole, to transfer to the host semiconductor ion, leaving the trivalent dopant ion negatively charged. The dangling bond, or hole, now may be transferred from one host semiconductor ion to another in the valence band without requiring the application of additional energy, because each such corresponding state is indistinguishable from the first state (Section 6.3.1). The mobility of a hole governed by this mechanism, thus, is expected to be smaller than that of a comparatively "free electron" in the conduction band. This is shown by a comparison of Eqs. (6-134) and (6-133) and by Table 6-7. The motion of the hole through the lattice in the valence band contributes to the conduction process. Ions that create such positive holes are known as *acceptors*. Semiconductors containing such dopants are called *p-type* because the extrinsic conduction is a result of the positive holes [Eq. (6-157d)].

Another class of dopants can act either as donors or acceptors. These are known as *amphoteric* dopants. For example, copper may act as either of these in Ge, depending on its degree of ionization. Another way in which amphoteric behavior can occur is by the addition of a tetravalent ion to a III–V compound. If it substitutes for a group V ion, it will behave as an acceptor; it will act as a donor if it replaces a group III ion. A third mechanism for amphoteric behavior depends on the position of the dopant ion in the host lattice. It may act as an acceptor or as a donor depending on whether it is on a substitutional or in an interstitial site.

The general effect of the introduction of any of these types of dopants into the semiconductor material is to create new energy states. These levels may lie anywhere in the gap between the top edge of the valence band and the bottom edge of the conduction band. In any event, the doped semiconductor will be an insulator at 0 K.

It will remain an insulator until, at higher temperatures or energy increments, some of the electrons from donor levels, which occupy states just below the conduction band edge, can absorb the thermal or other energy and jump up to the bottom of the conduction band. Such electrons behave as though they were "nearly free" because many vacant states are available to them. The relatively small number of electrons thus excited to the conduction band produces an increase in the electrical conductivity (Fig. 6-17) because increasing numbers of mobile carriers become available. This continues up to a limiting condition, the *exhaustion range*, which is discussed next. Such materials are *n*-type semiconductors because the principal carriers are electrons.

The acceptor dopants behave in a way similar to that of the donors. The capture of an electron from a semiconductor ion will occur only when the added external energy $(k_B T)$ is sufficient to excite an electron from a state at or near the top of the valence band up to an acceptor state that lies within the gap, but that is sufficiently close to the top edge of the valence band for the exchange to take place. Increasing numbers of holes are created in the valence band as this process continues with increasing thermal or other energies. Since the number of holes increases, so does the electrical conductivity (Fig. 6-17). This mechanism also can continue up to a limiting range, the *exhaustion range*. Materials of this kind are *p*-type semiconductors because the majority carriers are holes. Simple schematic band models of *n*- and *p*-type semiconductors are shown in Fig. 6-23.

It should be noted that the positive charges on donor ions are not holes and that the negative charges on acceptor ions are not electrons. These are the charges on those dopant ions that have engaged in the giving up or the taking on of electrons. Therefore, the two conduction mechanisms may be considered as resulting from *ionization processes* that provide the majority carriers in each case. The ionized dopant ions can move only by diffusion in the lattice of the semiconductor. Since this effect is zero or negligibly small under most normal conditions, the dopant ions may be considered as not entering into the conduction mechanism. However, dopant migration must be taken into account in the design and operation of high-density chips.

Both types of impurity levels are localized in the energy gap and are finite in number. The exhaustion ranges, which were indicated previously, are reached when

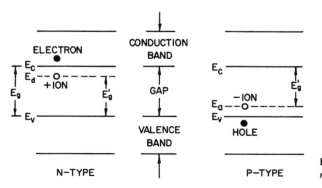

Figure 6-23 Schematic band models for *n*- and *p*-type semiconductors.

virtually all the donors, or acceptors, become ionized. Here the carrier concentration is approximately equal to the dopant concentration. No more carriers are available for excitation from the dopants, a condition sometimes known as *saturation*. This occurs at relatively low temperatures or increments of energy because the energy levels of the dopant ions usually are relatively close to the edges of the conduction or valence bands. These energy differences are small compared to the energy range of the gap of the intrinsic material. At temperatures or energies below the onset of the exhaustion or saturation ranges, the numbers of intrinsic carriers of either type are very small compared to the numbers of extrinsic carriers. After exhaustion occurs, the extrinsic electrical conductivity changes very slowly with increasing energy until a temperature or an energy is reached at which the added external energy is sufficient to promote increasing numbers of electrons from close to the top edge of the valence band into states low in the conduction band; increased intrinsic behavior, then, is responsible for the further increase in electrical conductivity (Section 6.3.1). Here the number of intrinsic carriers may equal or exceed the number of extrinsic carriers. Greater numbers of both electrons and holes now participate in the conduction process when both mechanisms are operative (Fig. 6-17).

In the case of intrinsic behavior, E_F was approximated as being at $E_g/2$ [Eq. (6-124) or (6-126)]. The reason for this is that the probability of finding a carrier, either an electron or a hole, is 0.5 at that location. However, in the case of extrinsic materials, this is no longer true because of the presence of dopant states of either kind within the energy gap. The Fermi level for either type of dopant can be determined from the number of carriers of a given kind per unit volume in the given band. In the case of donors, this number of electrons is obtained by use of the Boltzmann tail [Eq. (2-87)] to give

$$n_e \cong N(E)_c \exp\left[-\frac{E_c - E_{F(d)}}{k_B T} \right]$$

(6-160)

or

$$\ln n_e \cong \ln N(E)_c - \frac{E_c - E_{F(d)}}{k_B T}$$

(6-160a)

where n_e and $N(E)_c$ are the density of electrons per unit volume and the density of states in the conduction band, respectively, and $E_{F(d)}$ is the Fermi level in an *n*-type material. Similarly,

$$n_h \cong N(E)_v \exp\left[-\frac{E_v - E_{F(a)}}{k_B T} \right]$$

(6-161)

or

$$\ln n_h \cong \ln N(E)_v - \frac{E_v - E_{F(a)}}{k_B T}$$

(6-161a)

in which n_h, $N(E)_v$, and $E_{F(a)}$ have the meanings for p-type material that correspond to those given for Eq. (6-160). Referring back to intrinsic behavior and using Eq. (6-116) to determine the density of states in the conduction band, it is found that

$$N(E)_c \cong n_i \exp\left[-\frac{E_F - E_c}{k_B T}\right] \tag{6-162}$$

where n_i denotes the density of intrinsic carriers per unit volume. Equation (6-162) is substituted into Eq. (6-160) to give

$$n_e \cong n_i \exp\left[-\frac{E_F - E_c}{k_B T}\right] \exp\left[-\frac{E_c - E_{F(d)}}{k_B T}\right]$$

This simplifies to

$$n_e \cong n_i \exp\left[-\frac{E_F - E_{F(d)}}{k_B T}\right]$$

or

$$E_{F(d)} - E_F \cong k_B T \cdot \ln \frac{n_e}{n_i} \tag{6-163}$$

And, in a similar way, but using Eq. (6-117), it is found that

$$E_F - E_{F(a)} \cong k_B T \cdot \ln \frac{n_h}{n_i} \tag{6-164}$$

Equations (6-163) and (6-164) give the effects of dopants and temperature on $E_{F(d)}$ and $E_{F(a)}$ based on the intrinsic Fermi level. The effects of dopant concentration, at constant temperature, are shown in Fig. 6-24.

The effects of both dopant concentration and temperature are shown in Fig. 6-25. In this figure the low concentrations of carriers are of the order of about $10^{14}/cm^3$ and the high concentrations are about $10^{18}/cm^3$. At low dopant concentrations and at low temperatures, $E_{F(d)}$ and $E_{F(a)}$ lie between the impurity states and the edges of either the conduction or the valence bands, respectively. As the temperature increases, increasing amounts of extrinsic behavior take place, decreasing the supply of dopants of either type that are available for ionization, and both $E_{F(a)}$

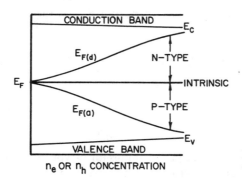

Figure 6-24 Isothermal effect of dopant concentrations on the Fermi energy.

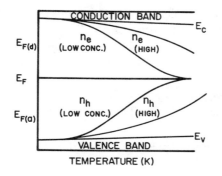

Figure 6-25   Effects of dopant concentrations and temperature on the Fermi energy.

and $E_{F(d)}$ begin to approach the intrinsic Fermi level. As exhaustion is approached, this trend increases until saturation occurs, and at that temperature at which intrinsic behavior begins to predominate, both $E_{F(d)}$ and $E_{F(a)}$ become identical to E_F. Materials with high dopant concentrations have high exhaustion temperatures and, therefore, approach this limiting condition at higher temperatures than do materials with low dopant concentrations.

Approximations of the energy levels of the donors and acceptors may be made based on their ionization energies in solid solution in the covalently bonded semiconductor crystal. Consider a group III or V ion, with either one less or one more electron than a host ion, respectively. The charge difference, either a positive hole or a negative electron, represents a carrier that is not tightly bound to the dopant ion, but exists as a relatively distant "cloud" around it; this "cloud" is spread over a relatively large number of host ions (say Si or Ge) that surround the dopant ion. The dopant ion-carrier configuration is analogous to that of a hydrogen atom. If, for example, a phosphorous ion (valence 5) is in solution in Si (four of the valence electrons form the strong, paired, covalent bonds that are required by its four Si nearest neighbors), the average "distance", or "radius", of the fifth valence electron is relatively far from the P ion. This comes about because the charge on the "orbiting" carrier (the fifth electron in this illustration) polarizes the host ions within the electrically neutral, spherical volume enclosed by its "orbit". This polarization partially offsets the charge on the dopant ion, so the "orbiting" carrier has a "radius" of the order of about 200 to 300 ion diameters. This distance is known as the *Debye length*. Under this set of conditions, the volume of the host lattice surrounding the dopant ion (P ion) may be approximated as being continuous and to possess a uniform dielectric constant because it reacts macroscopically with respect to the carrier (fifth valence electron). Thus, the extra carrier (electron) is treated as polarizing the lattice around the dopant ion (P ion) and as being unaffected by the discrete nature of the volume of the lattice between itself and the dopant ion (P ion).

The problem of the calculation of the ionization energy of the extra carrier now is similar to the calculation of the bonding energy of an electron in the case of the Bohr hydrogen atom (Section 1.5). In the preceding illustration, the potential energy of the extra electron is reduced because the surrounding volume of polarized silicon ions decreases the positive potential field of the P ion; hence its very large "radius" as compared to that of the hydrogen atom; so

$$PE = -\frac{e^2}{\varepsilon r} \tag{6-165}$$

where ε is the relative dielectric constant of the semiconductor material (Si in this discussion) and r is the Debye length. The energy of the extra carrier (electron) is

$$E_d = PE + KE = -\frac{e^2}{\varepsilon r} + \tfrac{1}{2}mv^2 \tag{6-166}$$

An expression for r can be obtained from equating the electrostatic and mechanical forces acting on the carrier:

$$F = \frac{e^2}{\varepsilon r^2} = ma = \frac{mv^2}{r}$$

to obtain the Debye length as

$$r = \frac{e^2}{\varepsilon m v^2} \tag{6-167}$$

An expression for the velocity may be obtained by first considering that the extra electron travels in a closed path in the host lattice about the P ion. Using the Bohr relationship that $\mathbf{p}(\theta) = nh/2\pi$ results in

$$\oint \mathbf{p}(\theta)\, d\theta = 2\pi\mathbf{p}(\theta) = nh = 2\pi mvr \tag{6-168}$$

in which $\mathbf{p}(\theta)$, the momentum, is constant since there is no torque. From Eqs. (6-167) and (6-168),

$$nh = 2\pi mvr = 2\pi mv\,\frac{e^2}{\varepsilon m v^2} = \frac{2\pi e^2}{\varepsilon v}$$

And, solving for $\mathbf{v}$,

$$\mathbf{v} = \frac{2\pi e^2}{\varepsilon n h} \tag{6-169}$$

Now, substituting Eq. (6-169) into Eq. (6-167) gives

$$r = \frac{e^2}{\varepsilon m} \cdot \frac{\varepsilon^2 n^2 h^2}{4\pi^2 e^4} = \frac{\varepsilon n^2 h^2}{4\pi^2 e^2 m} \tag{6-170}$$

Equations (6-169) and (6-170) are substituted back into Eq. (6-166) to obtain

$$E_d = -\frac{e^2}{\varepsilon} \cdot \frac{4\pi^2 e^2 m^*}{\varepsilon n^2 h^2} + \frac{m^*}{2} \cdot \frac{4\pi^2 e^4}{\varepsilon^2 n^2 h^2}$$

This reduces to

$$E_d = -\frac{4\pi^2 e^4 m^*}{\varepsilon^2 n^2 h^2} + \frac{2\pi^2 e^4 m^*}{\varepsilon^2 n^2 h^2} = -\frac{2\pi^2 e^4 m^*}{\varepsilon^2 n^2 h^2} \tag{6-171}$$

Here m^* has been substituted for m for the reasons given in the previous section. Equation (6-171) is essentially the same relationship that was obtained from the Bohr model of the hydrogen atom (Section 1.5), in which the atomic number $z = 1$ and in which an implicit squared relative dielectric constant of unity is used because the electron is traveling in free space. Here E_d is the binding energy of the extra carrier (electron). The denominator of Eq. (6-171) explicitly contains the square of the relative dielectric constant of the medium in which it is traveling because $\varepsilon \gg 1$ for these materials. When appropriate substitutions are made, Eq. (6-171) becomes, in terms of the Bohr model,

$$E_d = -\frac{13.6 m^*}{\varepsilon^2 m} \text{ eV} \tag{6-171a}$$

The importance of ε now may be appreciated. It is seen that the inclusion of the relative dielectric constant in Eq. (6-171a) significantly lowers the bonding energy of the extra carrier (electron) considerably below that given for hydrogen; it is the major factor in the determination of the ionization energy of the dopant. It is interesting to note that E_d is independent of the dopant ion; it is a function of the relative dielectric constant of the host material and the effective mass of the carrier. The Debye length, or the "radius" of the extra carrier, is about two orders of magnitude greater than that of the H atom, so the carrier most probably is quite far from the donor or acceptor ion. Therefore, any local lattice effects introduced by the dopant ion do not affect the extra carrier (electron); it only "sees" the surrounding volume of host material. This is the reason why the relative dielectric constant, ε, a macroscopic quantity, may be used in this analysis.

If the value of ε for Si is taken as being about 12 and that for $m^*/m \cong 0.5$, then $E_d \cong 0.047$ eV. This is in good agreement with the observed values for P donor states in Si. These lie about 0.045 eV below the bottom edge of the conduction band.

Another important and interesting observation is that, despite the fact that Eq. (6-171a) was derived for a donor ion, it also holds for acceptor ions as well. In fact, this equation could have been derived for an acceptor ion (group III) instead of for a group V ion using a positive hole as the carrier instead of an electron. This is shown by the fact that the observed value for boron (valence 3) acceptor states is $E_a \cong 0.045$ eV above the top edge of the valence band in Si. Similar calculations for Ge, using values of $\varepsilon \cong 16$ and $m^*/m \cong 0.8$, give $E_d = E_a \cong 0.043$ eV. The corresponding experimental values for P and B in solution in Ge are $E_d \cong 0.044$ eV and $E_a \cong 0.045$ eV, respectively. For the case of a III–V compound doped with a group IV ion, which, as discussed earlier, can show amphoteric behavior, using a value of $\varepsilon \cong 12.5$ and an average value of $m^*/m \cong 0.3$ gives $E_d \cong 0.026$ eV below the conduction band. This also is in excellent agreement with the experimental value for Si acceptor levels in GaAs.

The relatively small values of E_d and E_a compared to those of E_g (see Table 6-7), which result primarily from ε, explain the onset of extrinsic semiconduction at temperatures so very much lower than that of the beginning of appreciable intrinsic behavior; the close proximity of the impurity levels to the edges of the valence or conduction bands requires comparatively little thermal energy for the carrier transitions (Fig. 6-17). This low-energy response of dopants also explains the smaller slope of the extrinsic range as compared to that of the intrinsic range shown in the figure. Dopant levels such as those described here are designated as being shallow levels. Those levels farther from band edges are classed as deep levels.

The behavior just described is that for relatively dilute concentrations of dopants. As the dopant concentration is increased, a condition is reached at which the Debye radii begin to overlap. The impurity levels then start to interact, hybridize, and form an impurity band within the gap. The presence of such a band permits conduction more like that of metals than of semiconductors.

6.3.4 Materials

The elemental semiconductors belong to group IV of the Periodic Table. These materials are insulators at 0 K (Sections 4.5.1, 4.6, and 6.3.1). Under usual conditions, they become semiconductors as increasing energy, generally as thermal energy or as suitable electromagnetic radiation, promotes electrons across the gaps to their conduction bands. Some values of the energy gaps of semiconductor materials are given in Table 6-7.

Diamond has the largest energy gap of any of the semiconducting elements and it is an insulator below about 1000 K. It shows intrinsic behavior above this temperature. The most widely used elemental semiconductor materials are Si and Ge. Si shows intrinsic behavior at about 200°C. Ge becomes intrinsic at about 100°C. Each of the two "pure" elements are virtual insulators below the temperatures given. The reason for the differing behaviors is that the values of E_g of each of these elements are significantly different. Diamond has the largest value of E_g, and Ge has the smallest; that of Si lies between these, but is much closer to that of Ge than to that of diamond. The electrical properties of these elements at temperatures below that at which intrinsic behavior occurs are largely determined by any very small amounts of impurities that are always present. Si and Ge almost invariably are used with controlled amounts of dopants. See Section 6.3.3 and Table 6-7.

The difference between the ranges of the energy gaps of Si and Ge determines that intrinsic Ge will show a greater conductivity at a given temperature than will Si. More carriers, both electrons and holes, will be available for conduction in the Ge than in the Si at the given temperature because of the smaller energy gap between its valence and conduction bands. However, this difference provides a practical advantage favoring Si. The preparation of essentially intrinsic Si is less difficult than that required for Ge; more impurity ions can be tolerated in states within its E_g range, because of its wider energy gap, before it shows any significant extrinsic behavior.

A large number of compounds also demonstrate semiconductor properties. These usually (but not necessarily) have an electron:ion ratio of four (Table 6-7 and

Fig. 3-10). Such compounds are covalently bonded and frequently crystallize in the zinc-blende [similar to Figs. 3-10(a) and (b) in which an S ion has four nearest Zn neighbors] and wurtzite lattices [Fig. 3-10(c)]; they tend to have relatively large values for E_g. This makes it possible to dope them and, in effect, to tailor them by manipulating E_a or E_d for many applications in various devices, including those involving junctions. See Section 6.3.3.

Many compound semiconductors show higher carrier mobilities than the elemental materials (Section 6.3.1.2). This is useful because a given concentration of carriers will provide a higher conductivity in these compounds than in the elemental materials. It also will be noted that the materials cited in Table 6-7 vary largely in their values of E_g. They are useful because they permit more latitude in the selection and manipulation of materials for use in the specific application to various devices than do Ge or Si.

Some of the compounds may have varying degrees of covalent and ionic bonding (Section 3.12.3). Generally, the larger the *difference* in the electronegativities of the components, the greater will be the probability of ionic bonding. The degree of covalent bonding determines their applicability to semiconductor devices. In other words, the greater the amount of ionic bonding, the greater will be their tendency toward ionic behavior; such compounds tend to act as insulators rather than as semiconductors (Sections 3.12.1 and 3.12.3).

A valence of four is not always required. Some compounds belong to IV–VI, II–IV, and II–V types (Table 6-7). Other compounds, not shown in the table, are of the V–VI, V–VIII, and III–VI types. These generally, but not always, have narrow energy gaps; many of the V–VI compounds have values of E_g from about 1.0 to 1.6 eV. Those with the smaller energy gaps are useful for devices that operate in the infrared region.

Two important factors must be given consideration in compound semiconductors when intrinsic conduction is being sought. The first is that the stoichiometric ratios of the ions must be maintained as exactly as possible. The second is that their purity must be extremely high. Variations in the ratios of the ions would have the same effects as the intentional doping of the compound; the compound could be either *n*- or *p*-type depending on which ion was in excess of the required stoichiometric ratio. Impurities present in such compounds also act as donors and acceptors and cause extrinsic behavior where intrinsic properties may be desired. These factors make it more difficult to manufacture pure single crystals of intrinsic, compound semiconductors than elemental semiconductors. However, these factors also make it possible to produce extrinsic, compound semiconductors of either type simply by adjusting the relative amounts of the constituent ions, without the introduction of dopants.

Metal oxides can show semiconductor properties when an excess of the metal ion is present. Zinc oxide can show *n*-type properties when this situation exists. The electrical conductivity of these crystals may be varied by controlling the amount of excess cation.

Extrinsic semiconductor properties also are shown by defect oxides when the cation (metal ion) can have more than one state of ionization. Since the given cation is present in the lattice with more than one level of ionization (with more than one valence), either cation or anion vacancies must be present in the crystal to preserve the required electrical neutrality. The electrons may migrate between those cations with different degrees of ionization (difference valences). In some cases, the resulting vacancies may permit ionic conductivity to contribute to that of the usual carriers. However, this is an exception rather than the rule. Metal oxides in which the cation has only a single ionization state (only one possible valence) and in which there is no excess of cations act as insulators (Section 6.4).

The defect oxides are not quite stoichiometric in composition. Wüstite, nominally FeO, is a good example of this. Its actual composition is such that the ratio O : Fe is slightly greater than unity. This results because the valence of iron can be either $+2$ or $+3$; it has two ionization states. Assuming an exact stoichiometric composition, each O^{2-} ion should have an Fe^{2+} ion as its nearest neighbor. However, some of the ions are ionized to the Fe^{3+} state instead of being at the Fe^{2+} ionization level. For each pair of Fe^{3+} sites present in the lattice, one of the Fe^{2+} sites must be vacant to maintain the electrical neutrality. This is the basis for the designation of its lattice as a defect structure, as well as for its nonstoichiometry. The extra electron made possible by the two ionization states of iron causes "FeO" to behave as an extrinsic semiconductor. Since the metallic ion is present with more than one level of ionization, the extra electrons may migrate between the Fe ions with differing ionization or valence states and give rise to extrinsic behavior (Section 6.3.3).

As previously noted, metal oxides of cations that have single ionization states and in which there is no excess number of cations act as insulators. Their filled valence bands are separated by large gaps from their conduction bands (Sections 4.5 and 4.6). As an example of this, stoichiometric ZnO is an insulator. However, ZnO with an excess of Zn^{2+} ions shows n-type semiconducting properties; these act as n-type dopants, as described in Section 6.3.3. This is the basis for one method of xerography in which papers with such coatings are employed.

6.3.5 External Variables

The resistance–temperature characteristics of semiconductors are shown schematically in Fig. 6-17. The large, negative slopes of intrinsic materials demonstrate their high sensitivity to changes in temperature. The corresponding responses of extrinsic materials to temperature are similar to, but smaller than, those of intrinsic semiconductors and take place over lower ranges of temperatures. The influence of dopants on these properties may be shown by the use of Eqs. (6-131), (6-163), (6-164) and (6-171a).

The temperature coefficient of the electrical resistivity of a semiconductor is obtained by starting with the reciprocal of Eq. (6-109).

$$\rho = (ne\mu)^{-1} \tag{6-109a}$$

The derivative of Eq. (6-109a), taken with respect to temperature, is

$$\frac{d\rho}{dT} = \frac{1}{e}\left[-\frac{n(d\mu/dT) + \mu(dn/dT)}{n^2\mu^2} \right] = \frac{1}{e}\left(-\frac{1}{n\mu^2}\frac{d\mu}{dT} - \frac{1}{n^2\mu}\frac{dn}{dT} \right) \tag{6-172}$$

Equation (6-172) is substituted into Eq. (6-66) along with Eq. (6-109a) to obtain

$$\alpha = \frac{1}{\rho}\frac{d\rho}{dT} = ne\mu \cdot \frac{1}{e}\left(-\frac{1}{n\mu^2}\frac{d\mu}{dT} - \frac{1}{n^2\mu}\frac{dn}{dT} \right) \tag{6-173}$$

which, upon simplification, gives the temperature coefficient as

$$\alpha = -\frac{1}{\mu}\frac{d\mu}{dT} - \frac{1}{n}\frac{dn}{dT} \tag{6-174}$$

in the absence of strain.

In general, for intrinsic semiconductors, increasing the temperature increases the probability that a carrier will be scattered by the lattice and decreases both the relaxation time and the mobility; $d\mu/dT$, therefore, will be negative [Eqs. (6-133) and (6-134)]. The numbers of carriers will increase with increasing temperature, and dn/dT will be positive and much larger than $|d\mu/dT|$. The net result is that α will be negative because the second term in Eq. (6-174) predominates.

At temperatures up to and approaching that at which saturation begins to become manifest, extrinsic materials will have negative temperature coefficients. This results because both $d\mu/dT$ [Eq. (6-136)] and dn/dT are positive and Eq. (6-174) is negative.

In the exhaustion range, the numbers of carriers are virtually constant as a function of temperature; dn/dT will be very small, positive, and approximately constant. Increasing the temperature decreases μ, and $d\mu/dT$ becomes more negative, as described previously. In this case, α will be slightly positive because of the slightly greater influence of the first term of Eq. (6-174).

At moderate temperatures, conduction occurs by both electrons and holes in intrinsic and extrinsic materials. As the temperature is lowered, the resistivity normally increases because the rate of decrease of the number of carriers resulting from the filling of donor states, the emptying of acceptor states, and/or the return of electrons from the conduction band to the valence band is greater than the increase in mobility (Section 6.3.1.2). In other words, the number of nearly free carriers diminishes faster than their mobility increases as the impurity levels and the valence band refill [Eqs. (6-133) and (6-134)]. Any impurity scattering will limit the mobility at very low temperatures [Eq. (6-136)].

In some extrinsic materials the hole contribution to the conductivity [Eq. (6-111)] can be larger than that due to electrons, even though their mobility usually is more limited. The result is that the resistivity ceases to increase and shows only a small, positive slope with decreasing temperature. For example, Ge can show this behavior below 20 K, depending on its purity.

Another mechanism may take place at low temperatures when the numbers of donors and acceptors are small, that is, when the material is nearly intrinsic. An ionized pair consisting of an acceptor and a donor ion can be considered as being analogous to an ionized hydrogen molecule if they are close enough in the lattice. The application of a voltage (external electric field) will cause the electron to jump to the other impurity ion; a small degree of conductivity can result from repetitions of this mechanism.

The large, negative temperature coefficients of electrical resistance of many semiconductors make them ideal as sensors for the measurement and control of temperature. Devices of this kind are known generically as *thermistors*. These temperature-sensing elements may be made from either single-crystal or from polycrystalline materials. The kinds and amounts of dopants used in these detectors depend on the temperature ranges of their intended applications (Fig. 6-25). As an example silicon containing a dopant concentration of about 10^{14} impurity atoms/cm^3 becomes intrinsic at about 500 K. Thus, a thermistor intended for use at higher temperatures would be based on a semiconductor with a correspondingly higher dopant concentration. Similarly, a thermistor intended for medical or biological use could employ much lower dopant concentrations (Section 6.3.3).

The ways in which the electrical connections are made to any semiconducting material, including thermistor elements, are very important. As an example, assume that such a welded connection is made with a small-diameter, normal-metal wire, such as copper. If the metallurgical joint contains high, sharp, or abrupt concentration gradients of the materials that have been welded together, a rectifying junction will be formed. This is undesirable because it will permit the current to flow only in one direction; the joint will behave in the same way as in a metal-semiconductor rectifying diode. Such behavior may be avoided by ensuring that the concentration gradient of the connecting material in the volume of the semiconductor adjacent to the joint is relatively low, gradual, or shallow rather than steep, sharp, or abrupt. This usually is accomplished by either of two general methods. One of these is to use connecting wires whose alloy components have high diffusion coefficients in the semiconductor. Joints made with such combinations of materials normally will have gradual concentration gradients of the connecting material in the joint at the semiconductor. The sharp concentration gradients are avoided and the current will flow in both directions without rectification. Connections of this kind are known as *ohmic connections* or as ohmic contacts. Gold connections to semiconductors, in the form of fine wires or thick surface films, are used widely because of the high diffusion coefficient of gold in silicon. A variant of this approach is to make use of specially alloyed connecting wires. These materials contain alloying elements, such as In, Sn, and Au, that diffuse rapidly into the semiconductor during welding. This technique has the advantage of inherently providing the metallurgical conditions necessary for the presence of gradual concentration gradients of metal ions in the semiconductor at a lower cost than that of gold. It has the disadvantage of increasing the electrical resistivity of the connecting wires [Section 6.1.3 and Eqs. (6-55), (6-56a), and (6-56b)]. Another method is to very highly dope the small, local volume of the semiconductor to form an "impurity band" where the heat-affected zone of the welded joint is to be formed. The introduction of metal ions from the connector into such a small volume already containing a high dopant concentration, of the order of 10^{18}/cm^3, will have no appreciable rectification effects; it will be ohmic because of the formation of the "impurity band" (Section 6.3.3).

Some additional physical properties of semiconductors that must be considered in the design of thermistors include the desired temperature coefficient of electrical resistivity in the temperature range of interest, the electrical resistance, the heat capacity, the thermal dissipation, and the power required. The thermal and electrical properties of the encapsulating materials also must be taken into account.

Mixtures of sintered iron oxides that normally contain wüstite or that have been treated to contain prescribed amounts of "FeO" are commonly used as thermistor elements. Other mixtures of sintered oxides have also been used. The most important of these are manganese oxide–nickel oxide combinations. NiO doped with Li_2O is also used. Here the replacement of a Ni ion is accomplished by two Li ions. This requires the presence of a vacancy on an O ion lattice site. In turn, this determines the electrical properties (Section 6.3.4).

Difficulties sometimes are encountered in the manufacturing of large numbers of thermistors from defect-oxide semiconductors; these may show large variations in their electrical properties within a given lot or from batch to batch. In industrial applications in which these thermistors may be thermally cycled to relatively high temperatures, some defect-oxide thermistors may show unpredictable instabilities and poor reproducibilities of temperature readings.

In addition to the use of thermistors for the measurement and control of temperature, other applications include their employment as means for the compensation of temperature-induced resistance changes in electrical and electronic circuitry and in such devices as vacuum gages, fire-alarm systems, and current regulators.

Semiconductor resistors are used widely because they are reliable and inexpensive. These are available with either linear or nonlinear temperature coefficients. Those with linear negative coefficients find use in the compensation of temperature-induced resistance changes in electrical and electronic circuitry. Those with nonlinear coefficients have been used in such applications as for overload controls.

Another class of semiconductor resistors is voltage sensitive and is marketed under many trade names, one of which is the Varistor (Western Electric). These are made from sintered SiC particles and ceramic binders. Their properties may be varied by the use of proprietary additions and by adjustments in compacting pressure and in sintering times and temperatures. Metallized contacts are provided for the connecting wires. The adjacent SiC particles are considered to act as rectifiers. This enables them to behave as voltage-sensitive resistors. The voltage dependence of the current that occurs in rectification permits them to act as though their resistance varies with the voltage. The SiC resistors are sturdy, have little reaction to overloading, and are not excessively expensive.

Variable resistors of another kind may be made by combining a photoconductor (Sections 6.3.1 and 6.3.2) and a source of illumination. The number of carriers is a function of the intensity of the illumination for a given frequency range of irradiation. Thus, variations in intensity correspondingly affect the electrical properties. The illumination from an incandescent lamp passes through a fixed aperture before reaching the photoconductor. The intensity of the illumination is varied by voltage changes in the circuit controlling the filament of the lamp. The same results may be obtained by using constant illumination and employing electromechanical means to vary the opening of the aperture. Polycrystalline CdS usually is used for these devices. The dark current (current in the absence of illumination) is small, the sensitivity to the visible portion of the spectrum is good, and the range of change in resistance is broad and reasonably stable when properly encapsulated.

6.4 ELECTRICAL INSULATORS

Materials classed as electrical insulators are either ionically or covalently bonded (Sections 3.12.1 and 3.12.3). These also are known as dielectric materials. Their valence bands are completely filled and these are separated from their empty, outer conduction bands by large energy gaps ($\gg k_B T$) (Sections 4.5, 4.5.1, 4.6, and 4.7 and Fig. 4-15]. Consequently, abnormal energies generally are required to promote electrons from the topmost levels of the valence band to the lowest levels in the conduction band. And, since virtually no electrons are available to transport electrical or thermal energy under these conditions, these materials are very poor conductors of electrical or thermal energy. Any conduction that does occur in pure materials of these kinds must therefore be a result of lattice mechanisms.

A brief review of the natures of these bond types is given here as a convenient basis for explaining the properties of these classes of materials. Ionic bonding can be considered to result from the strong electrostatic attraction between oppositely charged ions. One of the most commonly used examples of this type of bond is that of rock salt, or sodium chloride. The outer electron configuration of the Na atom is that of $2p^6 3s$ (Table 3-3). This array may be considered as being that of a neon core with an additional $3s$ valence electron. Since sodium is a normal metal, the $3s$ valence electron is relatively weakly bound to its core. In contrast to this, the outer electrons of the isolated chlorine atom are arrayed as $3s^3 3p^5$. This is a high-energy, unstable configuration, since one electron is lacking to complete its outer $3p$ levels. Its energy can be lowered considerably by completing the outermost levels and forming the stable $3p^6$ (argonlike) state. (The strong tendency of the chlorine atom to lower its energy in this way is the reason for its high degree of chemical activity.) When sodium and chlorine atoms react, the weakly bound $3s$ electron is *transferred* from the sodium atom to the unoccupied $3p^6$ level of the chlorine atom. This causes the chlorine atom to have an electron configuration that is the same as that of argon. The net result of the transfer of the $3s$ electron is to induce the sodium atom to become a much smaller, positively charged, neonlike ion because of the loss of its valence electron; simultaneously, the chlorine atom becomes a considerably larger, negatively charged, argonlike ion because of its gain of the electron. Both ions have acquired the stable electron structures of the noble gases. The electron acquired by the chlorine atom can be considered to be in a very deep potential well because of the large decrease in the energy of the chlorine atom that occurs upon its conversion to the ionic state. The energies required to reverse this process (remove the extra electron) are quite high, being of the order of about 6 to 10 eV/mol. Thus, all the electrons involved in this process are very tightly bound in their respective ions; they cannot engage in transport processes under the influence of normal energies ($k_B T$). The valence bands of such ionically bonded substances may be considered to be completely filled. And, because of their high bonding energies, a large energy gap must be considered to lie between their valence and conduction bands (see Section 3.12.1).

Covalent bonding occurs when the atoms involved have completed inner electron shells and have four or more outer, valence electrons. In this situation the valence electrons from the atoms involved in the bonding are mutually *shared* in pairs by adjacent atoms. Thus, the outer shells of each of the resulting ions have electron configurations similar to those of the noble gases. Their valence bands are completely filled. This type of bonding in crystalline semiconductors is described in Sections 3.12.3, 6.3.1, and 6.3.3. The highly localized electrons shared by the ions result in a bonding that also lowers the energies of the atoms involved; the bonding electrons may be considered to lie in a comparatively deep potential well of the order of about $1\frac{1}{4}$ to $4\,\text{eV/mol}$. As such, these shared electrons completely fill the valence band, have a high degree of directionality, and are immobile. And, while not as tightly bound as in the ionic case, they are unable to engage in conduction processes when normal energies are applied. Thus, as is the case for ionic bonding, covalently bonded materials also have comparatively large energy gaps between their valence and conduction bands (see Section 3.12.3).

It has been considered that pure ionic and pure covalent bonding rarely exist and that at least some degree of mixed bonding prevails. It will be recalled (Section 3.10) that the sizes of the "orbits" of the outer electrons of an atom usually are based on the maxima of their probability densities. These may be considered to be a measure of atomic size. However, the probability-density "clouds" have significant values at distances considerably farther in space beyond the maxima. Thus, when two ions are at an equilibrium distance from each other, as in a crystal, these "clouds" will overlap to some degree. In cases where the bonding is principally ionic, the interpenetration of the clouds can result in some sharing of electrons. The net result of this is to introduce at least a small degree of covalency to the ionic bonding. Similarly, some amounts of electron transfer may result in some degree of ionic bonding in covalent materials. Thus, the conductivities of dielectric materials may have electronic or ionic components. In some cases, both factors may be operative.

Other crystals may show varying degrees of mixed ionic and covalent bonding. Approximations of the amount of ionic character may be estimated by means of several semiempirical relationships. These usually are given as functions of the electronegativity differences of the component atoms. The electronegativities are the single-bond strengths of the atoms, usually as expressed in units of $(\text{eV})^{1/2}$. They may be considered as measures of the tendencies of atoms to attract an electron. The reliability of approximations of the amount of ionic bond character decreases as the degrees of covalent or metallic bonding increase.

The interpenetrations of the electron "clouds" not only are responsible for the mixed bonding described here, but also are partly responsible for the nonintegral valences noted in Section 3.11. In addition, this "cloud overlap" also contributes to variations in the "sizes" of ions. This parameter will vary when the degree of ionization of the solute ion changes, as well as with the amount of the given solute ions present and also with the quantities and species of other solute ions in the solution. Atomic or ionic "diameter" will also change with allotropic changes and with order–disorder reactions in metallic solid solutions, as well as with the formation of compounds with various other species. Atomic size, or diameter, usually is taken as the distance of closest approach of atoms in the elemental condition; this can be quite variable, as indicated here, for all other cases. *The atomic "diameter" is not a constant.*

Where organic molecules are formed, the bonds between the component atoms usually are assumed to be purely covalent. This also is the case where long molecular chains, or polymers, are formed. However, the intermolecular bonding in molecular crystals or between the random molecular chains in polymeric materials results primarily from weak, electrostatic, intermolecular attractions. This type of bonding is known as van der Waals, London, or molecular bonding (Section 3.12.5). Briefly, this type of bond arises from the mutually induced polarizations of adjacent molecules; these result in the formation of oppositely charged dipoles. The weak molecular bond is a consequence of the attraction between such molecular dipoles. Bond strengths of this kind range from about 0.5 to 0.005 eV/mol. The low melting points of molecular crystals attest to such low bond energies.

All these dielectric materials show some degree of electrical conductivity, however small. The electrical properties of metals and semiconductors are described in subsections beginning with 6.1 and 6.3, respectively. Most materials obey Ohm's law. However, as the strength of the electric field is increased, the conductivity of dielectrics becomes a strong, nonlinear function of the field strength. Solid materials of these kinds may undergo lasting damage or be destroyed when subjected to sufficiently high fields.

6.4.1 Ionic Conduction

Electrical conduction in essentially pure ionic crystals can result from the motion, or diffusion, of the component ions in the lattice. This is called *intrinsic conductivity* and generally is important at higher temperatures. Conduction also may result from the presence of impurity ions present in the lattice. This is called *extrinsic conductivity* and may occur at relatively low temperatures. In an analogous way, the conductivity observed in most molecular crystals and in polymeric materials usually results from the extrinsic behavior caused by weakly bound impurity ions or easily ionizable components present in these materials.

The bonding energies of ionic crystals are high, approximately 6 to 10 eV/mol. This indicates that an external electric field of the order of 10^6 to 10^8 V/cm should be required for conduction by charged ionic carriers to take place in "perfect" ionic crystals. However, experience shows that electric fields of much lower strengths than these can cause small currents to flow in real ionic solids. All crystalline solids are known to contain lattice imperfections, and virtually none are perfectly pure. These factors have important influences in diffusion processes in the solid state. It is logical

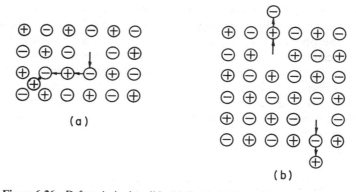

Figure 6-26   Defects in ionic solids: (a) Frenkel defects; (b) Schottky defects.

to assume that the presence of such impurities and imperfections could expedite the motion of ionic carriers and account for their properties. These ideas naturally lead to explanations of ionic conductivity in terms of the impurities and the lattice defects present in these substances.

Two of the types of thermally induced lattice defects employed to explain this behavior are Frenkel and Schottky defects (Fig. 6-26). A Frenkel defect is caused by the movement of an ion from its lattice site to an interstitial position; two point defects are created: a vacancy and an interstitial. It is most probable that such vacancies and interstitials are created as a result of a series of jumps involving several ions, rather than by a single jump of just one ion. A Schottky defect is the result of the migration of an ion from its lattice site to the surface; thus a vacancy is created. It is much easier to visualize the formation of these defects as resulting from a multiplicity of jumps than is the Frenkel mechanism. Since electrical neutrality must, on the average, be maintained both on the surface and within the volume of the crystal, Schottky defects must be created in pairs of opposite sign. It has been shown that the concentration of Schottky defects is very much greater than that of Frenkel defects in ionic crystals.

The transport of charged carriers can occur by the jumping of ions into adjacent lattice vacancies induced by either of these defect-producing mechanisms. Transport of this kind can only occur when adjacent vacancies are available. In a way analogous to that given for electrons and holes in Section 6.3.2, the movements of the ions may be described by the motions of the vacancies. Ionic motion consists of a series of ion jumps from lattice sites to adjacent, vacant lattice sites. Each ion jump to a vacancy leaves its last lattice site unoccupied. The vacancies thus "move" in directions opposite to those of the ions, and their motions are mirror images of ionic motions.

The vacancies may be considered to control the ion jumps. The probability of an ion jump will vary directly with the density of the vacancies. This enables properties based on ion jumps, and consequently ionic conduction, to be described in terms of the densities of the vacancies.

6.4.1.1 Statistical treatment. It will be recalled that interionic spacings are determined by the energy minima (Section 4.5 and Figs. 4-5 and 4-11). The depths of the potential energy wells in which the ions lie constitute barriers between the adjacent ions. An ion, and consequently a vacancy, moves through the lattice by jumping over the potential barrier between its site and a vacant site next to it. All these barriers are identical, and each lattice site for a given species is the equivalent of every other lattice site. In the absence of an external electrical field, the probability that thermal motion will cause an ion to jump across a barrier, or a vacancy to move, may be expressed as

$$p^* = A \exp(-E_b/k_B T) \tag{6-175}$$

where the height of the barrier that must be overcome (activation energy) is E_b. The coefficient A is primarily a function of the frequency of an ion nearest to the vacancy, because the probability that an ion will make a jump is a function of its frequency [See Eq. (6-181)]. When a weak, external electric field is applied, the probability that a vacancy will jump from one site to the next in a direction parallel to the field is given approximately by

$$p_t^* \cong p^* \frac{\bar{E}ea}{k_B T} \tag{6-176}$$

where a is the distance between lattice sites and the energy (force × distance) is $\bar{E}ea \ll k_B T$. The electric polarization for one jump of a singly charged ion is ea. This results in a current density, using Eq. (6-176), of

$$j = np_t^* ea \cong np^* \frac{\bar{E}e^2 a^2}{k_B T} \tag{6-177}$$

in which n is the number of vacancies per unit volume (density of vacancies). Using Ohm's law and Eq. (6-177) and noting that $\bar{E}$ vanishes, the ionic conductivity is obtained as

$$\sigma = \frac{j}{\bar{E}} \cong np^* \frac{e^2 a^2}{k_B T} \tag{6-178}$$

The ionic mobility is obtained by analogy with Eq. (6-109) and, using Eq. (6-178),

$$\sigma = ne\mu \cong np^* \frac{e^2 a^2}{k_B T} \tag{6-179}$$

so that the mobility is obtained as

$$\mu \cong p^* \frac{ea^2}{k_B T} \tag{6-180}$$

The coefficient A of Eq. (6-175) is a function of the frequency of the jumping ion, rather than a constant, because jumping depends on energy, a function of frequency. This may be appreciated by starting with the assumption that the ions of each of the species situated on an ideal ionic lattice have the same frequency v_0. This simplifying assumption is the basis for the Einstein model of the heat capacity of insulators (Section 5.1.2). It is most accurate in the temperature range at and above θ_D [Section 5.1.5 and Eq. (5-101b)]. This ideal lattice serves as a reference for real lattices that contain many types of imperfections, including vacancies and interstitials. An ion closest to a vacancy created by one of its own species will be subject to a smaller vibrational restoring force than one properly surrounded by next-nearest neighbors as in the ideal lattice because it is acted on by the attractive forces of one less nearest neighbor. An examination of Eq. (5-122) shows that, for a simple harmonic oscillator, the restoring force is

$$F(x) = -\frac{d}{dx} V(x) = -Kx = -4\pi^2 v^2 m x$$

So, if the amplitude of oscillation of an ion, x, is primarily a function of temperature, any smaller value of $F(x)$ will require a corresponding decrease in the frequency, v. In the present case v, the frequency of all the 12 like ions closest to the vacancy in a NaCl-type lattice, will be decreased. This effect is most pronounced in those crystalline directions between the 12 ions of the same species and the vacancy. The six nearest neighbors of the other ionic species will be affected similarly.

When calculations are made of the entropy of all the possible jumps in this imperfect ionic crystal, it is found that the coefficient of Eq. (6-175) is given by

$$A' = \left(\frac{v_0}{v}\right)^3 = \left(\frac{hv_0}{hv}\right)^3$$

for an ion located adjacent to a vacancy. However, not all such ion jumps, or configurational changes, can contribute to the conductivity, as explained in Section 6.4.1.2. When this limitation is taken into account, the coefficient of Eq. (6-175) becomes

$$A = \frac{(hv_0)^3}{h^3 v^2} \tag{6-181a}$$

Further simplification of Eq. (6-181a) may be obtained by recalling that, in the temperature range of interest here, $k_B T = hv_0$, so

$$A = \frac{(k_B T)^3}{h^3 v^2} \tag{6-181b}$$

Conduction also may be considered to take place by the mobility of interstitial ions, as well as by that of the vacancies. In the case of interstitial migration, Eq. (6-175) may also be used to give the probability that an interstitial ion will be thermally activated to jump to another interstitial site, provided that suitable values of A and E_b are used. Where vacancies are responsible for the conduction, Eq. (6-175) gives the probability that an ion will occupy an adjacent vacancy and, thus, effectively cause the vacancy to move. When the energy required to create a vacancy, E_V, is greater than the energy needed to form an interstitial,

$$E_V = E_I + E_L \tag{6-182}$$

where E_I is the energy of the ion on an interstitial site and E_L is the energy of the ion on a lattice site. In this case, it is probable that conduction will take place by the migration of interstitials; its energy of formation is smaller than that of a vacancy.

The use of Eq. (6-181) is bounded by two conditions. The first of these is that relatively large numbers of interstitials or vacancies must exist in the lattice. For interstitial migration, the energy factor of the exponent of Eq. (6-175) is given by $E_T = (E_V + E_b)$; this is the total activation energy for interstitial migration. The second condition is based on a perfect lattice and the simultaneous formation of an interstitial and a vacancy within that lattice; this is the Frenkel mode. Here the total activation energy is $(E_T + E_b)/2$, since two species are created.

Where the conduction is primarily a result of vacancy migration, as is the case for Schottky defects, the energy required for the creation of a Schottky pair, E_p, must be smaller than that for the creation of an interstitial ion. This leads to the probability that conduction in ionic crystals is primarily by vacancies rather than by interstitials if the lattice is closely packed. In this case, the activation energy for a single vacancy is $(E_T + E_p/2)$, since Schottky vacancies must be made in pairs to maintain electrical neutrality.

The intrinsic conductivities for the cases just described may be summarized as follows:

1. For interstitial migration:

$$\sigma = C_1 \exp\left[-\frac{(E_V + E_b)}{k_B T} \right] \tag{6-183}$$

2. For the migration of both interstitials and vacancies:

$$\sigma = C_2 \exp\left[-\frac{(E_T + E_b)}{2k_B T} \right] \tag{6-184}$$

3. For vacancy migration:

$$\sigma = C_3 \exp\left[-\left(\frac{E_T + E_p}{2k_B T}\right) \right] \tag{6-185}$$

The coefficients C_i in these equations are given by the product of the coefficient of p^* as given by Eq. (6-181b) and the other terms in Eq. (6-179): $C_i = n(k_B Tea/v)^2/h^3$. The magnitude of C_i is about 10^6 $(\Omega\text{-cm})^{-1}$.

Where conduction takes place by more than one of the means just described, the intrinsic conductivity is given by the sum of the contributions of the participating mechanisms:

$$\sigma = \Sigma C_i \exp\left(-\frac{E_i}{k_B T}\right) \tag{6-186}$$

Here the coefficients and activation energies are those appropriate for each migrating carrier.

In the case of closely packed ionic crystals, it would be expected that intrinsic conductivity should be primarily a result of vacancy migration. This should be so since the interstices are of minimum size and, as noted previously, the concentration of Frenkel defects is very much less than that of Schottky defects. In less densely packed crystals, interstitial migration of the smaller ion, usually the cation, becomes increasingly likely as the *difference* between the "radii" of the ions increases. The resulting increasing percentage of unoccupied volume of a lattice is responsible for this.

At lower temperatures, the conductivity largely results from the migration of impurity ions in the lattice. In this case, the summation given by Eq. (6-186) would include any conduction modes contributed by the host lattice and any additional terms required to account for the extrinsic behavior.

Impurities in molecular crystals frequently can account for their conductivities. Some of these impurities may also be associated with lattice defects. When the number of lattice defects is significantly greater than the number of impurity ions, a situation analogous to that described by Eq. (6-184), for large numbers of interstitials and vacancies, may be employed to describe the conductivity.

Where interstitials are the majority carriers at relatively low temperatures, their number is comparatively small. Consequently, the number of vacancies is correspondingly small. As the temperature increases, impurity ions could migrate interstitially and increase the conductivity [Fig. 6-27(a)]. If the majority carriers are vacancies at low temperatures, impurity ions could reduce the conductivity by filling the vacancies as the temperature increases [Fig. 6-27(b)].

Other factors, in addition to those just noted, may influence the discontinuities shown in Fig. 6-27. One important effect is that the exponential value decreases with increasing temperature. Another consideration is that the conduction mechanism may change. For example, if the positive ions are the majority carriers at lower temperatures and the anions become mobile at higher temperatures, the increased number of carriers could account for the behavior as sketched in Fig. 6-27(a). It has also been considered that vacancies may be predominant in one range of temperatures and interstitials in another. This could account for the change in conductivity as shown in Fig. 6-27(b).

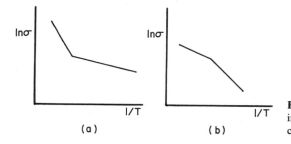

Figure 6-27   Schematic diagrams for the influences of different factors on ionic conductivity.

6.4.1.2 Free energy treatment. Another approach to the analysis of ionic conductivity is based on the change in the free energy that is required to produce a change in the crystal structure. Starting with a one-dimensional lattice, the probability of a jump by a vacancy is

$$p^* = v \exp\left(-\frac{\Delta F}{k_B T}\right) \tag{6-187}$$

where v is the frequency of the ions on either side of the vacancy and ΔF is the change in the Gibbs free energy of activation (energy to overcome the potential barrier) in the absence of an electric field [see Eq. (6-175)]. The vacancy created by the absence of a positive ion upon a lattice site has an effective charge of e. Upon application of a suitable, uniform, electric field, $\bar{E}$, the vacancy may be considered to be induced to jump from one lattice site to the next (see Section 6.4.1). The application of the field increases the barrier to vacancy motion because it increases the potential energy of the vacancy in the direction of the field by an amount $e\bar{E}a/2$, where a is the interionic distance (Fig. 6-28). This increases the depth of the potential well.

The probability that the vacancy will make a jump along the same direction as the field must take the energy barrier of $\Delta F + e\bar{E}a/2$ into consideration. This is expressed by including the additionally required energy in the exponent of Eq. (6-187):

$$\underset{\rightarrow}{p^*} = v \exp\left[-\left(\frac{\Delta F + e\bar{E}a}{2k_B T}\right)\right] \tag{6-188a}$$

since $a/2$ is the distance it must travel to reach the peak of the adjacent activation hump, so the additional energy (force × distance) is $e\bar{E}a/2$. By the same reasoning, the probability for the vacancy to jump in the direction opposite to the electric field is greater, because the barrier energy, $\Delta F - e\bar{E}a/2$, is smaller than that for a jump in the same direction as the field. The probability of a vacancy jump opposite to the field is

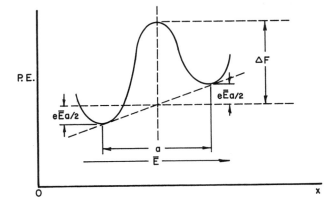

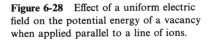

Figure 6-28 Effect of a uniform electric field on the potential energy of a vacancy when applied parallel to a line of ions.

considered by including the diminished energy of the barrier in the exponent of Eq. (6-187):

$$\underleftarrow{p^*} = v \exp\left[-\left(\frac{\Delta F - e\bar{E}a}{2k_B T}\right)\right] \tag{6-188b}$$

The average drift velocity, v_d, of the vacancies is determined by the product of the difference between the two opposing jump probabilities [Eqs. (6-188a) and (6-188b)] and the jump distance; this is given in its factored form as

$$\mathbf{v}_d = a(\underleftarrow{p^*} - \underrightarrow{p^*}) = av \exp\left(\frac{-\Delta F}{k_B T}\right)\left\{+\exp\left[+\frac{(e\bar{E}a/2)}{k_B T}\right] - \exp\left[\frac{-(e\bar{E}a/2)}{k_B T}\right]\right\}$$

This average drift velocity, v_d, is the analog of Δv in Eq. (6-7). It is more conveniently reexpressed by combining the exponentials into one hyperbolic factor:

$$\mathbf{v}_d = av \exp\left(\frac{-\Delta F}{k_B T}\right) 2 \sinh \frac{e\bar{E}a}{2k_B T} \tag{6-189}$$

At relatively low external fields, the hyperbolic function may be approximated as

$$2 \sinh \frac{e\bar{E}a}{2k_B T} \cong \frac{2e\bar{E}a}{2k_B T} = \frac{e\bar{E}a}{k_B T} \tag{6-190}$$

The substitution of Eq. (6-190) into Eq. (6-189) gives

$$\mathbf{v}_d \cong \frac{a^2 ve\bar{E}}{k_B T} \exp\left(\frac{-\Delta F}{k_B T}\right) \tag{6-191}$$

for such low applied electric fields. The mobility is obtained from this by means of the definition [Eq. (6-7)]. Noting that $\bar{E}$ vanishes, the ionic mobility is given as

$$\mu = \frac{\mathbf{v}_d}{\bar{E}} \cong \frac{a^2 ve}{k_B T} \exp\left(\frac{-\Delta F}{k_B T}\right) \tag{6-192}$$

Equations (6-191) and (6-192) are for a one-dimensional lattice. They may be applied to an NaCl-type lattice by recalling the cubic array of alternating Na^+ and Cl^- ions. A given Na^+ ion has 6 nearest Cl^- ions and 12 next-nearest Na^+ ions. Consider the central lattice position in this array to be an Na^+ vacancy. This vacancy may jump to any of the 12 surrounding Na^+ sites. The presence of the electric field does not affect the activation energy of any jump transverse to the field. In addition, such transverse jumps do not contribute to the flow of current. Only four of the eight remaining possible jumps that the Na^+ vacancy could make would contribute to the flow of current. Thus, for NaCl-type lattices, Eqs. (6-189) through (6-192) must be multiplied by a factor of 4 to convert the results for a linear lattice to those for a three-dimensional, NaCl-type lattice. Factors for the application of these equations to other types of ionic lattices may be obtained in a similar way; some of these types are given in Table 3-5. [Also see Fig. 3-8(b).]

The intrinsic conductivity of a material with an NaCl-type lattice can be described by an expression analogous to Eq. (6-111), where the vacancies of each species act as the majority carriers. The conductivity imparted by the vacancies in this case may be expressed as the sum of the contributions of each kind:

$$\sigma_v = n_+ e\mu_+ + n_- e\mu_- \tag{6-193}$$

Here the subscripts $+$ and $-$ refer to anion and cation vacancies, rspectively, and n is the number of a given type of such vacancies per unit volume.

Equation (6-192) is multiplied, as derived above, by a factor of 4 to adapt it to include the effective number of jumps in the NaCl-type lattice. The general equation for the mobility of either type of vacancy in the NaCl-type lattice is thus given by

$$\mu = \frac{4a^2 v e}{k_B T} \exp\left(\frac{-\Delta F}{k_B T}\right) \tag{6-194}$$

It will be noted again that different factors must be used to account for those vacancy jumps that can affect the mobility in other types of ionic lattices.

The substitution of Eq. (6-194) into Eq. (6-193), taking both types of vacancies into account, gives the intrinsic conductivity arising from all vacancies as

$$\sigma_v = 4n_+ e \frac{a^2 v_+ e}{k_B T} \exp\left(\frac{-\Delta F_+}{k_B T}\right) + 4n_- e \frac{a^2 v_- e}{k_B T} \exp\left(\frac{-\Delta F_-}{k_B T}\right)$$

or, upon factoring,

$$\sigma_v = \frac{4a^2 e^2}{k_B T}\left[n_+ v_+ \exp\left(\frac{-\Delta F_+}{k_B T}\right) + n_- v_- \exp\left(\frac{-\Delta F_-}{k_B T}\right)\right] \tag{6-195}$$

If N is the total number of all possible vacancy sites of either kind for a unit volume of a crystal, since $n_+ \equiv n_-$, the probable fraction of each type of vacancy, cation or anion, is given by the Boltzmann equation as

$$\frac{n}{N} = \exp\left(\frac{-F_p}{2k_B T}\right) \tag{6-196}$$

in which F_p is the Gibbs free energy required to create a Schottky pair Fig. 6-26(b). Equation (6-196) may be introduced by multiplying the numerator and denominator of Eq. (6-195) by N to obtain

$$\sigma_v = \frac{4Na^2 e^2}{k_B T}\left[\frac{n_+}{N} v_+ \exp\left(\frac{-\Delta F_+}{k_B T}\right) + \frac{n_-}{N} v_- \exp\left(\frac{-\Delta F_-}{k_B T}\right)\right] \tag{6-197}$$

Equation (6-196) now is substituted into Eq. (6-197) and, recalling that n_+ must equal n_-, the result is factored to the form

$$\sigma_v = \frac{4Na^2 e^2}{k_B T} \exp\left(\frac{-F_p}{2k_B T}\right)\left[v_+ \exp\left(\frac{-\Delta F_+}{k_B T}\right) + v_- \exp\left(\frac{-\Delta F_-}{k_B T}\right)\right] \tag{6-198}$$

Equation (6-198) includes the contributions of both types of vacancies to the intrinsic conductivity. However, the mobility of the cation vacancies would, in general, be expected to be greater than that of the anion vacancies. This results from the considerably smaller sizes of the cations as compared to the anions. The size disparity is largely a consequence of the electron transfer involved in the bonding mechanism (Sections 3.12.1 and 6.4). When $\mu_- \gg \mu_+$, then, according to Eq. (6-194), ΔF_- must be much less than ΔF_+. Thus, because of its larger negative exponent, $\exp(-\Delta F_+/k_B T)$ in Eq. (6-198) is comparatively small and may be neglected for this case. For this condition, Eq. (6-198) becomes

$$\sigma_v \cong \frac{4Na^2e^2v_-}{k_B T} \exp\left(\frac{-F_p}{2k_B T}\right) \exp\left(\frac{-\Delta F_-}{k_B T}\right) \tag{6-199}$$

Equation (6-199) is specifically intended only for the case in which the cation vacancies constitute the primary means of intrinsic conduction.

Equation (6-199) may be reexpressed in terms of enthalpy, H, and entropy, S, by using the Gibbs equations for F and ΔF:

$$F = H - TS$$

and

$$\Delta F = \Delta H - T\Delta S$$

These substitutions express the intrinsic vacancy contributions to the conductivity given by Eq. (6-199) as

$$\sigma_v = \frac{4Na^2e^2v_-}{k_B T} \exp\left(\frac{-H_p}{2k_B T} + \frac{S_p}{2k_B}\right) \exp\left(\frac{-\Delta H_-}{k_B T} + \frac{\Delta S_-}{k_B}\right)$$

Again, the subscript p refers to the thermodynamic quantities involved in Schottky pair production. Like terms are combined to give

$$\sigma_v = \frac{4Na^2e^2v_-}{k_B T} \exp\left[\frac{(S_p/2 + \Delta S_-)}{k_B}\right] \exp\left[\left(\frac{-H_p}{2} - \frac{\Delta H_-}{k_B T}\right)\right]$$

or, when both exponents are made to have $k_B T$ in their denominators,

$$\sigma_v = \frac{4Na^2e^2v_-}{k_B T} \exp\left[\frac{T(S_p/2 + \Delta S_-)}{k_B T}\right] \exp\left[\frac{-(H_p/2 + \Delta H_-)}{k_B T}\right] \tag{6-200}$$

The same restrictions hold for Eq. (6-200) as for Eq. (6-199). Equation (6-198) or its equivalent must be used when both types of vacancies have significant mobilities and make appreciable contributions to the conductivity.

6.4.2 Extrinsic Conduction

The influence of impurity ions on the electrical conductivity of ionic crystals can be more important than those of intrinsic conduction mechanisms. The same is true for molecular crystals and polymeric materials. Assume, for example, that divalent impurity cations are present on lattice sites in an NaCl-type lattice. Each such impurity ion will bond with two anions. This results in the formation of a vacant lattice site that normally would have been occupied by a monovalent cation. Electrical neutrality is maintained by the absence of that cation and the resultant lattice defect structure (see Sections 6.3.4 and 6.3.5). Thus, the number of cation vacancies per unit volume must include those induced by the impurity ions. And, since $n_- = n_+$, the density of cation vacancies is given by

$$n_- = n_+ + n_d \tag{6-201}$$

Here n_d is the number of vacancies per unit volume created by the presence of the divalent impurities. Since each divalent ion induces one vacant cation site, n_d also equals the concentration of the divalent impurities.

It is convenient to reexpress the density of vacant anion sites (number per unit volume) as

$$n_+ = n_- - n_d \tag{6-202}$$

Using Boltzmann's equation as applied to Eq. (6-202), in the same way as for Eq. (6-196), it is found that

$$\frac{n_+}{N} = \frac{n_- - n_d}{N} = \exp\left(\frac{-F_p}{2k_BT}\right) \tag{6-203a}$$

and

$$\frac{n_-}{N} = \exp\left(\frac{-F_p}{2k_BT}\right) \tag{6-203b}$$

An expression involving both types of vacancies may be obtained by the product of Eqs. (6-203a) and (6-203b). This results in

$$\frac{n_-}{N} \cdot \frac{n_- - n_d}{N} = \exp\left(\frac{-F_p}{k_BT}\right) \tag{6-204}$$

This may be rewritten in quadratic form as

$$n_-^2 - n_- n_d = N^2 \exp\left(\frac{-F_p}{k_BT}\right) \tag{6-205}$$

The density of vacancies induced by the divalent impurities is obtained from Eq. (6-205) in terms of the density of cation vacancies as

$$n_d = \frac{n_-^2 - N^2 \exp(-F_p/k_BT)}{n_-} \tag{6-206}$$

The solution of Eq. (6-205) for n_- is useful for the description of the extrinsic conductivity of this class of solids. The use of the equation for the general solution of a quadratic equation as applied to Eq. (6-205) gives

$$n_- = \frac{n_d + [n_d^2 + 4N^2 \exp(-F_p/k_BT)]^{1/2}}{2} \tag{6-207}$$

This may be reexpressed by multiplying and dividing the quantity within the brackets by n_d^2. This results in

$$n_- = \frac{n_d + \left(\dfrac{n_d^4 + n_d^2 4N^2 \exp(-F_p/k_B T)}{n_d^2}\right)^{1/2}}{2}$$

When n_d^2 is factored out of the brackets, the fraction is simplified as

$$n_- = \frac{n_d}{2} + \frac{n_d}{2}\left(1 + \frac{4N^2 \exp(-F_p/k_B T)}{n_d^2}\right)^{1/2}$$

Then, by factoring $n_d/2$, the number of cation vacancies per unit volume is found to be

$$n_- = \frac{n_d}{2}\left[1 + \left(1 + \frac{4N^2 \exp(-F_p/k_B T)}{n_d^2}\right)^{1/2}\right] \tag{6-208}$$

This relationship will be used to simplify the approximation for the extrinsic conductivity as expressed by Eq. (6-210).

The extrinsic conductivity may be expressed for this case, in the same way as for Eq. (6-193), in the form

$$\sigma_d = n_+ e\mu_+ + n_- e\mu_- \tag{6-209}$$

Equation (6-208) now is used in Eq. (6-209) to obtain

$$\sigma_d = n_+ e\mu_+ + \frac{n_d}{2}\left[1 + \left(1 + \frac{4N^2 \exp(-F_p/k_B T)}{n_d^2}\right)^{1/2}\right] e\mu_- \tag{6-210}$$

When the concentration of the divalent ions is high enough to make a significant contribution to the conductivity, the density of vacancies induced by their presence will be much greater than the density of thermally induced Schottky pairs. When this is the case, it may be approximated that

$$n_d^2 \gg 4N^2 \exp\left(\frac{-F_p}{k_B T}\right) \tag{6-211}$$

The use of the inequality given by Eq. (6-211) in Eq. (6-210) simplifies matters because the fraction within the brackets may be considered to be small enough to be neglected. Thus, Eq. (6-210) may be approximated as

$$\sigma_d \cong n_+ e\mu_+ + \frac{n_d}{2}\left[1 + (1)^{1/2}\right]e\mu_-$$

and, finally, noting that the factor 2 vanishes, the extrinsic conductivity is obtained in the form

$$\sigma_d \cong n_+ e\mu_+ + n_d e\mu_- \tag{6-212}$$

As noted previously (Section 6.4.1.1), cation mobility is much greater than that of anions in these crystals. Since $\mu_- \gg \mu_+$, it may be further approximated that

$$\sigma_d \cong n_d e\mu_- \tag{6-213}$$

Corrections must be made for n_d since only a fraction of the vacancy density, depending on the lattice type, can contribute to the conductivity. In this sense, Eq. (6-213) may be considered in a way similar to Eqs. (6-190) and (6-191). It then may be treated as described for Eq. (6-194) to account for this behavior.

It should be noted that situations analogous to that described here may occur in molecular crystals, in polymeric dielectric materials, and in some glasses. Conduction may take place when such substances for whatever reason, inadvertently or otherwise, contain impurity ions or easily ionizable molecules. This also can occur when the molecules of which these materials are composed are susceptible to ionization. Such cases permit extrinsic condution to take place, particularly in the presence of strong electric fields. Conditions of these kinds may result in appreciable leakage currents and may eventually lead to breakdown.

Equation (6-213) may be reexpressed, for divalent impurities in an NaCl-type lattice, by using the expression for mobility given by Eq. (6-194) as

$$\sigma_d \cong n_d e \frac{4a^2 v_- e}{k_B T} \exp\left(\frac{-\Delta F}{k_B T}\right)$$

or as

$$\sigma_d \cong n_d \frac{4a^2 v_- e^2}{k_B T} \exp\left(\frac{-\Delta F_-}{k_B T}\right) \qquad (6\text{-}214)$$

The total conductivity is given by the sum of the intrinsic conductivity as approximated by Eq. (6-199) and the extrinsic conductivity as obtained from Eq. (6-214). In this case, $\sigma \cong \sigma_v + \sigma_d$.

Certain types of ionic crystals show unusually high ionic conductivities, compared to those discussed previously, primarily as a result of their more complex crystal structures. Some of these lattices are layered crystal structures in which close-packed planar arrays are alternated by much less densely packed ionic planes. The alkali metal ions in the latter planes are small with respect to the other ions and can have, on a comparative basis, an extremely high degree of mobility. The Li ion has been the subject of much study because it is the smallest of this class of ions.

Two compounds of this type that have received much attention are based on $Na_2O \cdot 11Al_2O_3$ (β alumina) and $NaAl_{11}O_{17}$ (β'' alumina). Both substances actually contain more NaO than are shown by the stoichiometric formulas. The Na ions occupy sites in the less densely packed layers (sometimes called conduction planes). The O ions occupy preferred sites in the close-packed planes. This crystalline configuration permits the relatively small Na ions to have relatively high mobilities in the less densely packed planes. This results in a surprisingly high conductivity. As an example, β'' alumina can have an electrical conductivity of about 20 times that of a 0.1 molar solution of NaCl in water in the neighborhood of room temperature.

Mobilities that result in conductivities of this magnitude have caused materials of this type to be considered as excellent candidates for use as solid electrolytes in storage batteries. In addition, these compounds are very stable and the electron component of the conductivity is very small (Section 6.4). Other compounds, such as

$RbAg_4I_5$ and Li_xTiS_2, also show high conductivities. However, these materials have larger components of electron conduction than do the aluminas.

The operation of storage batteries that are based on these materials as solid electrolytes depends on the diffusion of an alkali metal ion (from the anode) across the inert, solid electrolyte where it forms a readily reversible compound with the cathode. At present, the use of Li or Li-containing anodes and such cathodes as V_2O_5, which form easily reversible compounds that permit battery recharging, are being investigated. Recharging is accomplished, in the same way as for conventional storage batteries, by the application of a reverse external current. Ions of such metals as Li, K, Rb, and Ag have been used in the aluminas to replace the Na. The Li anodes are being used with Li_xTiS_2 electrolytes and V_2O_5 cathodes.

Other ionic conductors have been based on TiS_2. Here the Li ions from anodes form layered structures by occupying preferred planes in the TiS_2 lattice. The result is a compound of composition Li_xTiS_2. The TiS_2 layers are covalently bonded, but the intercalated Li ions are weakly held by van der Waals bonds. The resulting material has a high conductivity because the relatively small, loosely bound Li ions have a high mobility on the conduction planes of the TiS_2. This material is but one of a class of compounds whose compositions are based on the general formula of TX_2, where T is a transition element and X is an atom such as S, Se, and Te.

The conduction in the layered compounds discussed here is two-dimensional because it occurs in the planes occupied by the mobile ions. Other compounds show one-dimensional conduction. For example, $LiAlSiO_4$ forms a hexagonal lattice in which the Li ions are in linear arrays perpendicular to the basal planes and, thus, the conductivity is unidirectional. An example of the other extreme is given by LiN (NaCl-type lattice). In this case the mobility of the Li ions is high in the three orthogonal directions.

Batteries using solid electrolytes, instead of aqueous solutions, are expected to have a much longer life and a storage capacity of up to about three times that of a conventional battery of the same weight. Small batteries with solid electrolytes involving lithium are widely available.

It will have been noted that extrinsic conduction in polymeric materials has been treated as being detrimental to their properties. This results from the fact that many of these materials are used primarily for their desirable properties as electrical and thermal insulators. Extrinsic conduction would degrade these properties in many applications.

Recently, however, considerable work has been devoted to greatly increasing the electrical conductivities of some of these materials. One technique for the control of the properties of these conductive polymers is based on the use of multiphase systems. In this case, conducting powders, granules, fibers, and the like are included in the polymeric matrices. The mechanical and electrical properties of composite materials of this kind may be manipulated by varying such parameters as the physical properties, sizes, geometries, degrees of dispersion, or volume fractions of the additives. Thus, comparatively low resistance electrical paths may be built into the polymeric matrices. Conduction can take place without the necessity for high electric fields or without the risk of breakdown (Section 6.4.4). Parametric variations of the additives can result in relatively large ranges of electrical conductivities.

Another approach now being investigated is one that is roughly analogous to the doping of semiconductors (Section 6.3.3). In this case, the dopants are added to such materials as polyacetylene and poly (p-phenylene) to obtain the desired conduction properties. For example, polyacetylene can be made to have semiconducting properties with low concentrations of such dopants as ClO_4^- or I_3^- or to have conduction resembling that of metals with higher concentrations of these dopants. The dopant concentrations used here are much greater than those employed in elemental and compound semiconductors. The quantities of dopants used for this purpose are of the order of several percent. Recent work appears to indicate that the physical properties of some doped polymers may be explained by treating them as having linear conduction mechanisms.

Irrespective of the methods employed to produce conductive polymers, they can provide another class of useful materials with controlled ranges of electrical conductivities that also have other desirable combinations of optical, thermal, mechanical, and chemical properties.

6.4.3 Ionic Conduction and Diffusion

When a concentration gradient exists in a solid, it, at best, represents a metastable situation. Such gradients tend to be self-eliminating, under ideal conditions of temperature and time, so that the concentrated species ultimately becomes uniformly distributed. In ionic solids, the concentrated species is transported by diffusion mechanisms. This also is called *mass* or *matter transport*. The concentration gradient is the driving force for the mass transport; the steeper the concentration gradient, the greater will be the ion flow, or the rate of mass transport. This proportionality is described by Fick's first law as

$$J = -D \frac{dC}{dx} = -D \frac{dn}{dx} \tag{6-125}$$

The factors in Eq. (6-215) may be more readily understood in the following discussions when they are considered in terms of their respective units (neglecting signs):

$$J = \frac{\text{number of ions}}{\text{cm}^2 \cdot \text{s}} = \frac{\text{cm}^2}{\text{s}} \frac{\text{number of ions/cm}^3}{\text{cm}} \tag{6-215a}$$

It also may be expressed in molar form as

$$J = \frac{\text{g-mol}}{\text{cm}^2 \cdot \text{s}} = \frac{\text{cm}^2}{\text{s}} \frac{\text{g-mol/cm}^3}{\text{cm}} \tag{6-215b}$$

Here J, the flux, is number of ions that pass across a unit area perpendicular to the direction of the concentration gradient per unit time; D, the constant of proportionality, is the *diffusion coefficient* (cm^2/s); and either dC/dx (where C is the concentration) or dn/dx is the concentration gradient in the direction of interest. The relative tendencies of the migrations of various ions may be considered on the basis of their comparative values of D at a given temperature. This important factor is expressed as a function of temperature by

$$D = D_0 \exp\left(\frac{-E}{k_B T}\right) \tag{6-216}$$

In addition, as indicated by Eq. (6-215b), D may be used as a measure of the quantity of the diffusion species (gram-moles) passing across an area of 1 cm² in 1 s across a concentration gradient unit of mol/cm³/cm. D_0 is a frequency factor similar to that in Eq. (6-175), and E is the activation energy for an ionic jump. It will be noted that Eq. (6-216) is essentially the same as Eq. (6-175).

The ions will migrate under the influence of an external electric field as well as by thermally induced diffusion. Einstein's equation is based on relating these two mechanisms of flow. This equation is derived by first considering an ion with a charge e in a constant electric field $\bar{E}$. The Boltzmann equation gives the concentration of such a diffusion ion, which migrated a distance x from its original position, as being proportional to $\exp(-e\bar{E}x/k_B T)$. The ion flows, or fluxes, induced by an applied electric field and by thermal activation are equal at equilibrium, so

$$\mu n \bar{E} + D \frac{dn}{dx} = 0 \tag{6-217}$$

Since the probable density of ions engaged in the thermally induced flow is given by

$$n = A \exp\left(\frac{-e\bar{E}x}{k_B T}\right) \tag{6-218}$$

and the differentiation of Eq. (6-218) provides the concentration gradient of the diffusing, or migrating, species as

$$\frac{dn}{dx} = -\frac{Ae\bar{E}}{k_B T} \exp\left(\frac{-e\bar{E}x}{k_B T}\right) \tag{6-219}$$

then, by substitution of Eqs. (6-218) and (6-219) in the equation for equilibrium conditions [Eq. (6-217)], it is found that

$$\mu \cdot A \exp\left(\frac{-e\bar{E}x}{k_B T}\right) \bar{E} = D \frac{Ae\bar{E}}{k_B T} \exp\left(\frac{-e\bar{E}x}{k_B T}\right) \tag{6-220}$$

The quantity $A\bar{E}\exp(-e\bar{E}x/k_B T)$ vanishes and the equation reduces to

$$\mu = \frac{De}{k_B T} \tag{6-221a}$$

or, when expressed as the ratio of the mobility to the diffusion coefficient,

$$\frac{\mu}{D} = \frac{e}{k_B T} \tag{6-221b}$$

This is the Einstein equation that relates the ion flows resulting from the influences of external electric fields and of diffusion. Both sides of Eq. (6-221b) may be multiplied by ne, n again being the number of carriers per unit volume. This gives, when use is made of Eq. (6-179),

$$\frac{\sigma}{D} = \frac{ne\mu}{D} = \frac{ne^2}{k_B T} \qquad (6\text{-}222)$$

Equation (6-222) conveniently relates the electrical conductivity to the diffusion coefficient.

Equation (6-222) does not always agree with experimental data because some jumps do not contribute to the current flow. This behavior may be taken into account in a manner similar to the way in which Eq. (6-194) was obtained from Eq. (6-192). See Sections 6.4.1.2 and 6.4.2. The agreement between the calculated and the experimental data, however, improves as the temperature increases. At relatively low temperatures, the normally low intrinsic mobility is enhanced by the presence of crystalline lattice defects. The influences of lattice imperfections are not included in Eq. (6-222) because an ideal lattice is implicit in its derivation. At higher temperatures, the mobility becomes much greater and is considerably less affected by lattice imperfections. This behavior at higher temperatures is similar to that indicated in Fig. 6-27(a), in the portion of the curve with the larger slope corresponding to Eq. (6-222). This corresponds to the range of temperatures at which ionic diffusion appreciably increases, because D increases with increasing temperatures [Eq. (6-216)].

In most dielectric materials, the direct current decays with time. That is, when these crystals are subjected to constant voltages (constant external electric fields) at intermediate temperatures, the dc current diminishes to a lower, steady-state value, with a corresponding higher residual dc resistance, R_r. This decrease in current, which occurs at a high initial rate, makes experimental measurement of the conductivity difficult. The initial flow of current also results in the formation of an internal space charge within the dielectric. This can result in a high internal electric field that also can enhance ion migration. The initial resistance is $R_i = V/I_i$, where I_i is the initial current. (The true dc resistance, R_t, which represents the inherent property of the material, may be determined from the sum of the conduction and capacitive currents.) See Eqs. (6-228) and (6-231). The initial and true resistances are rarely equal. This difference has been explained by the formation of the internal space charge that forms as a result of the additional flow of ion carriers in the material.

The space charge may be determined from the variation of the potential within the dielectric material. This potential, V_P, defines the internal field, $\bar{E}_p$, on the basis of the relationship

$$\bar{E}_p = -\frac{\partial V_p}{\partial x} \qquad (6\text{-}223)$$

Considering the ion flow in the x direction, the number of diffusing carriers per unit volume that is induced by the potential V_p is

$$n_p = \beta \, \exp\!\left(\frac{-eV_p}{k_B T}\right) \tag{6-224}$$

where β is a constant for the given material.   Equation (6-224) may be converted into a more useful form by means of the Einstein relationship [Eq. (6-221)].   Both sides of Eq. (6-221) are multiplied by V_p to give $V_p\mu/D = eV_p/k_B T$. This equivalence is used to reexpress Eq. (6-224) as

$$n_p = \beta \, \exp\!\left(\frac{-V_p\mu}{D}\right) \tag{6-225}$$

The internal potential, V_p, does not vary uniformly within the dielectric.   Some analyses make use of Poisson's equation, because it relates $\bar{V}_p$ to the charge density, rather than using Eq. (6-223).   In addition, the direction of $\bar{E}_p$ is opposite to that of the external applied field, $\bar{E}$.   This causes the migration of some of the carriers in a direction opposite to that caused by $\bar{E}$ as determined by Fick's law [Eq. (6-215)]. This is known as *back diffusion*.   The effective current density is the resultant of these two opposite flows.   This situation is described by

$$j = \sigma\bar{E} - De\,\frac{\partial n_p}{\partial x} = ne\mu\bar{E} - De\,\frac{\partial n_p}{\partial x} \tag{6-226}$$

Equation (6-225) is a solution to Eq. (6-226) when equal numbers of ion carriers flow in opposite directions.   The variation in j, at a given temperature, depends on the variation of $\partial n_p/\partial x$ in the space-charge region.   Experiments show that, as might be anticipated, smaller space charges are induced in purer materials.

Once the space charge fully develops, the dc resistivity remains constant for a short time.   Then, as diffusion progresses, the carriers are swept out of the dielectric, the resistivity approaches a much higher value, and the current decreases.   The influences of these mechanisms are shown schematically in Fig. 6-29.   A typical value of j is of the order of 10^{-12} A/cm^2.

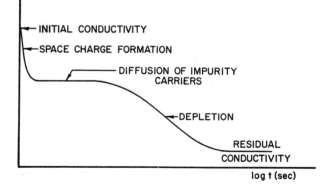

Figure **6-29**   Schematic diagram for the current density in a dielectric material as a function of time (is $\sim 10^{-12}$ A cm^{-2}).

In terms of Eq. (6-226), the initial current density is relatively high prior to the formation of a space charge. The formation of a space charge requires times of the order of a few milliseconds. Once this is formed, back diffusion takes place and the effective current density is decreased accordingly. The relatively constant current density that ensues results from the transport of impurity ions and may exist for times ranging from seconds to minutes, depending on the degree of purity of the dielectric. The transport resulting in the sweeping out of the impurity is virtually completed in times of the order of a few minutes. After this interval, the conductivity and thus the current density drop to a residual level; virtually all the readily available carriers have been removed.

A simple approach may be used to obtain an approximation of the internal potential. The field of the space charge is opposite to that of the external field. This imparts an internal potential, V_p, opposite to that of the externally applied potential, V. This effect may be taken into account by starting with Ohm's law for the initial current:

$$I_i = \frac{V}{R_i} \tag{6-227}$$

where I_i and R_i are the initial current and resistance, respectively. The residual current, considering the net effect of the applied and internal potentials, may be approximated as being given by

$$I_r \cong \frac{V - V_p}{R_t} \tag{6-228}$$

in which R_t is the true resistance. A first approximation for $V_{p(\text{max})}$ may be obtained that is based on the properties of very pure materials. The difference between Eqs. (6-227) and (6-228) is

$$I_i - I_r \cong \frac{V}{R_i} - \frac{V - V_p}{R_t} \tag{6-229}$$

For very pure materials, $R_i \cong R_t$, so Eq. (6-229) reduces to

$$I_i - I_r \cong \frac{V_{p(\text{max})}}{R_i} \tag{6-230a}$$

Then

$$V_{p(\text{max})} \cong (I_i - I_r)R_i \tag{6-230b}$$

Substituting R_i from Eq. (6-227) into Eq. (6-230a) gives

$$V_{p(\text{max})} \cong (I_i - I_r) \cdot V/I_i \tag{6-231}$$

The use of the initial and residual currents and the applied potential to determine $V_{p(\text{max})}$ makes it much easier to approximate the true resistance, R_t, using Eq. (6-228). This eliminates the necessity for attempting to make accurate measurements during the previously noted periods of rapid transitions of the electrical properties.

6.4.4 Effects of Strong Fields and Breakdown

The previous discussions show that the conductivities of dielectrics are affected by their purity. Ohm's law is obeyed up to high field strengths when V_p [Eq. (6-228)] is taken into account. The expected relationship (Ohm's law) between current density and field strength no longer holds when the energy exerted by the field approximates the thermal energy required for an ion jump: $\bar{E}ex \cong E_b$ of Eqs. (6-218) and (6-175), respectively. Field strengths of the order of 10^5 V/cm are required for this condition in "pure" crystals. Field strengths of this magnitude approach those required for breakdown.

Beyond the range of applicability of Ohm's law, the conductivity may be approximated by the empirical equation

$$\sigma \cong A \exp(B\bar{E}) \tag{6-232}$$

in which A and B are constants at a given temperature for the material being considered. The exponential coefficient B decreases with increasing temperature because the bonding energies also decrease and any space charges break down. In addition, intrinsic conductivity can also make a significant contribution to total conductivity under these conditions.

Large increases in the electrical conductivity of dielectrics would be expected when such materials are subjected to very strong fields. Almost all crystalline dielectric materials contain some impurities and some show photoelectric properties (Sections 6.3.1 and 6.3.4). This means that it is possible for a very small number of electrons to be promoted to and to occupy states in the conduction bands of such materials. Under the influence of very strong fields these few, "nearly free" electrons would be accelerated so that they could excite other electrons to the conduction band. These, in turn, excite more electrons, and so on. The process is repeated until many are promoted. This behavior is known as an *avalanche mechanism*. Such a process could result in a relatively large density of electrons in the conduction band. This also could occur in high-purity materials where electric fields of strengths of $\bar{E}ex \geqslant 3E_g/2$ would be sufficient to promote some electrons from states near the top of the valence band to low levels in the conduction band. In any event, the resulting large increase in current, caused by ionic as well as electronic conductivity, would then lead to breakdown (sometimes called *conductive breakdown* or *dielectric strength*).

A simple model for this behavior, due to Fröhlich, is outlined here for illustrative purposes. The drift velocity of an electron that has been promoted to the conduction band by an external electric field is obtained from Eq. (6-7) as

$$\mathbf{v}_d = \frac{\bar{E}e\tau(E)}{m^*} \tag{6-233}$$

in which $\tau(E)$ is the relaxation time of the electron with energy E. The energy of such an electron is, using Eq. (6-233),

$$E = \frac{m^*\mathbf{v}_d^2}{2} = \frac{m^*}{2}\left(\frac{\bar{E}e\tau(E)}{m^*}\right)^2 \tag{6-234}$$

The rate of change of the energy of the electron excited as a result of the application of the electric field is obtained by differentiating Eq. (6-234) with respect to time:

$$\frac{dE}{dt} = \frac{\bar{E}^2 e^2 \tau(E)}{m^*} \tag{6-235}$$

Here, because of the very small value of $\tau(E)$, it is approximated that $d\tau(E) \cong dt$, as in Eq. (6-8). If the portion of the energy, generated by this mechanism, that is conducted out of the dielectric as thermal losses is greater than the energy required to initiate avalanching, a very small leakage current may flow without inducing breakdown; large numbers of electrons cannot be excited to the conduction band. This condition is expressed as

$$\left| -\left(\frac{dE}{dt}\right)_{th} \right| > \frac{\bar{E}^2 e^2 \tau(E)}{m^*} = \frac{dU}{dt} \tag{6-236}$$

When $|(dE/dt)_{th}| = dU/dt$ (the rate of energy absorption by a unit volume of the dielectric) becomes high at high electric fields or at frequencies at which the absorbed energy cannot be dissipated rapidly enough, the relationship given as Eq. (6-236) no longer holds. Therefore, when the energy balance is such that the heat is removed too slowly,

$$\left| \left(\frac{dE}{dt}\right)_{th} \right| \leqslant \frac{\bar{E}^2 e^2 \tau(E)}{m^*} = \frac{dU}{dt} \tag{6-237}$$

and breakdown will occur for all energies greater than that required to promote the valence electron to the conduction band of the dielectric material.

Any condition that diminishes $\tau(E)$, for the electric field that approaches the limiting condition of Eq. (6-237), will induce greater resistance to breakdown, causing it to occur at higher field strengths. It will be recalled that increased temperature has the effect of decreasing $\tau(E)$ (Section 6.1.3). More advanced theories indicate that electron avalanching takes place at field strengths of about 80% of that given by Eqs. (6-236) and (6-237). This results in breakdown at correspondingly smaller fields than are indicated by these equations. See Table 6-8.

Another type of breakdown can take place in nonhomogeneous dielectric materials. These substances may contain volumes of material with lower electrical resistivities than the surrounding matrix. The low-resistivity volumes will conduct more leakage current than the matrix. The resistivity is decreased further as the temperature and the current increase and the localized heating increases. A temperature finally is reached at which the conduction is large enough to cause breakdown. This type of thermal breakdown may be encountered in multiphase polymeric dielectrics.

Other solid dielectric materials may break down as a result of molecular dissociation under the influence of strong electric fields. In cases where other readily ionized substances, or other impurity ions, are present in a polymeric dielectric, leakage currents may be caused in a way analogous to the extrinsic behavior described

TABLE 6-8 TYPICAL PROPERTIES OF DIELECTRIC MATERIALS

Material	Electrical Resistivity (Ω-cm)	Dielectric Strength (V-cm)	Material	Electrical Resistivity (Ω-cm)	Dielectric Strength (V/cm)
Ceramics			Fiber Reinforced Resins		
Alumina	10^{11}–10^{14}	1.6×10^4–6.4×10^5	Polyester	10^{13}–10^{14}	1.4×10^5–2×10^5
Corderite	10^{12}–10^{14}	1.6×10^4–9.9×10^4	Phenolic	10^{11}–10^{12}	0.6×10^5–1.5×10^5
Forsterite	10^{14}	9.4×10^4	Epoxy	10^{14}–10^{15}	1.4×10^5
Porcelains			Melamine	10^{10}–10^{11}	0.7×10^5–1.2×10^5
Dry process	10^{12}–10^{14}	1.6×10^4–9.5×10^4	Polyurethane	10^{11}–10^{14}	1.3×10^5–3.5×10^5
Wet process	10^{12}–10^{14}	3.5×10^4–1.6×10^5			
Zirconia	10^{13}–10^{15}	9.9×10^4–1.6×10^5	Rubbers and Elastomers		
Steatite	10^{13}–10^{15}	7.9×10^4–1.6×10^5	Polyisoprene	10^{15}	2.4×10^5
Titanates			Butadiene	10^{15}	—
(Ba, Sr, Ca, Mg, Pb)	10^8 –10^{15}	2×10^3–1.2×10^5	Styrene-butadiene	10^{14}	2.4×10^5
Titanium dioxide	10^{12}–10^{18}	3.9×10^4–8.3×10^4	Acrylonitrile butadiene	10^{10}	1.9×10^5
			Polychloroprene	10^{11}	2.0×10^5
Plastics, Resins			Isobutylene-isoprene	10^{17}	3.0×10^5
Allyl, cast	10^{13}–10^{14}	1.5×10^5	Polysulfide	10^8	1.3×10^5
Analine formaldehyde	10^{16}–10^{17}	2.5×10^5	Polymethane	10^{11}	2.0×10^5
Epoxy, cast	10^{16}–10^{17}	1.6×10^5			
Melamine formaldehyde	10^{11}–10^{14}	1.5×10^5			
Methyl melhacrylate	10^{14}–10^{15}	1.8×10^5			
Nylons	10^{13}–10^{15}	1.2×10^5–2.4×10^5			
Phenol formaldehyde	10^{12}–10^{13}	1.4×10^5			
Polyethylene	10^{12}–10^{13}	1.8×10^5			
Polystyrene	10^{17}–10^{19}	2.0×10^5			
Rubber, hard	10^{15}	1.9×10^5			
Shellac	10^9	0.8×10^5			
Vinyl chloride	10^{14}–10^{16}	2.4×10^5			

Abstracted from *Handbook of Tables for Applied Engineering Science*, 2nd ed., R. E. Boltz and B. L. Tuve, eds., CRC Press, Boca Raton, Fla., 1979.

in Section 6.4.2. If the concentration of the impurity ions is excessive, either as a result of a high, initial impurity concentration or by the continuous dissociation of the matrix or of any ionizable impurities, leakage currents will be present and breakdown can occur.

Breakdown may also be primarily mechanical in nature. The application of an external field can cause both ionic and dipolar displacements. When the field strength becomes sufficiently high, large compressive forces can be induced within dielectric materials. A point may be reached at which such forces exceed the compressive strengths of the materials. This can result in mechanical failure and in consequent breakdown.

Voids may be present in polymeric dielectric solids as a result of evaporated solvents, moisture, or improper manufacturing techniques. These voids may act as stress concentrators. Thus, mechanical failure may occur at applied fields of lower strength than is the case when such voids are absent.

6.5 BIBLIOGRAPHY

BARDEEN, J., *Physics Today*, p. 19, Jan. 1963, American Physical Society.

BEEFORTH, T. H., and GOLDSMID, H. J., *Physics of Solid State Devices*, Pion, London, England, 1970.

BIRKS, J. B., *Modern Dielectric Materials*, Academic Press, New York, 1960.

BYLANDER, E. G., *Materials for Semiconductor Functions*, Hayden, Rochelle Park, N.J., 1971.

CHIHOSKI, R. A., *Metal Progress*, 123, p. 27, 1983.

COHEN, M. M., *Introduction to the Quantum Theory of Semiconductors*, Gordon & Breach, New York, 1972.

DEAN, M., III, ed., *Semiconductor and Conventional Strain Gages*, Academic Press, New York, 1962.

DEKKER, A. J., *Solid State Physics*, Prentice-Hall, Englewood Cliffs, N.J., 1957.

EVANS, W. D. J., *Platinum Metals Rev.*, 25, p. 1, 1981.

FELDMAN, J. M., *The Physics and Circuit Properties of Transistors*, Wiley, New York, 1972.

HANSEN, M., JOHNSON, W. R., and PARKS, J. M., *J. Metals*, 3, No. 12, p. 1184, 1951.

HEREMANS, J., BOXUS, J., and ISSI, J. P., *Phys. Rev.*, B, 19, p. 3476, 1979.

HILL, N. E., and others, *Dielectric Properties and Molecular Behavior*, Van Nostrand Reinhold, New York, 1969.

HUNT, L. B., *Platinum Metals Rev.*, 24, p. 104, 1980.

HURD, C. M., *Electrons in Metals*, Wiley, New York, 1975.

KITTEL, C., *Introduction to Solid State Physics*, 5th ed., Wiley, New York, 1976.

LINDE, J. O., *Ann. Physik*, 10, p. 52, 1931; 14, p. 353, 1932; 15, 219, 1932.

MATTHIESSEN, H., and VOGT, C., *Pogg. Ann.*, 122, S.19, 1864.

MOTT, N. F., and GURNEY, R. W., *Electronic Processes in Crystals*, Dover, New York, 1964.

O'DWYER, J. J., *Theory of Electrical Conduction and Breakdown in Solid Dielectrics*, Oxford University Press, New York, 1973.

POLLOCK, D. D., *Trans. TMS-AIME*, 230, p. 753, 1964.

———, *Physical Properties of Materials for Engineers*, Vols. I, II, and III, CRC Press, Boca Raton, Fla., 1982.

———, *Electrical Conduction in Solids*, American Society for Metals, Metals Park, Ohio, 1985.

ROBINSON, A. T., and DORN, J. E., *Trans. TMS-AIME*, 191, p. 457, 1951.

ROSE-INNES, A. C., and RHODERICK, E. H., *Introduction to Superconductivity*, Pergamon Press, Elmsford, N.Y., 1969.

ROSENBERG, H. M., *The Solid State*, 2nd ed., Oxford University Press, New York, 1978.

SAUMS, H. L., and PENDLETON, W. W., *Materials for Insulating and Dielectric Functions*, Hayden, Rochelle Park, N.J., 1973.

SAVITSKII, E. M., and others, *Superconducting Materials*, Plenum Press, New York, 1973.

SOLYMAR, L., and WALSH, D., *Lectures on the Electrical Properties of Materials*, Oxford University Press, New York, 1979.

SPROULL, R. L., *Modern Physics*, 2nd ed., Wiley, New York, 1963.

STANLEY, J. K., *Electrical and Magnetic Properties of Metals*, American Society for Metals, Metals Park, Ohio, 1963.

WOLF, H. F., *Semiconductors*, Wiley-Interscience, New York, 1971.

ZHELUDEV, I. S., *Physics of Crystalline Dielectrics*, Vols. 1 and 2, Plenum Press, New York, 1971.

6.6 PROBLEMS

6.1. Estimate the mobility of a valence electron in a normal metal using mobility data from Section 4.1.

6.2. Approximate the mean free path of a valence electron of a normal metal in terms of number of lattice parameters ($a_0 \cong 3\,\text{Å}$), assuming that $n_s \cong 10$. (Select data from Sections 4.1 and 4.2.)

6.3. Approximate the relative change in the resistivity of silver for equal quantities of solutes, one of which has a valence of 2 and the other a valence of 5.

6.4. Calculate the temperature coefficient of electrical resistivity of a copper-base alloy containing $6\,\text{At}\%$ Ni. Compare the results with the data in Table 6-3 and explain any differences.

6.5. Estimate the impurity content of the platinum given in Table 6.1 with respect to a purer specimen with a temperature coefficient of $3.935 \times 10^{-3}\ \Omega/\Omega/°C$.

6.6. Explain the basis for Eq. (6-79). Why must this equation be modified as Eq. (6-81)? (*Hint*: See J. S. Dugdale, *The Electrical Properties of Metals and Alloys*, Edward Arnold, London, 1977.)

6.7. Explain the advantages of the quality control of precipitation-hardening alloys by resistometric means.

6.8. Define the optimum conditions for the use of constantans for resistance elements in precision electrical circuitry.

6.9. Approximate the ratio of the density of electrons to the density of holes for the case where $E_g = 1.4\,\text{eV}$. Explain the answer. Does this hold for any other values of E_g for the reasons just given?

6.10. Explain how the uncertainty principle may be employed to rationalize the discontinuities of the m^* versus $\bar{k}_x$ relationship at $\bar{k}_x = \bar{k}_c$.

6.11. Why is it possible to represent electrons and holes as shown in Fig. 6-21?

6.12. Explain the location of impurity states in extrinsic semiconductors.

6.13. Why may classical equations of the type given by Eqs. (6-160) and (6-161) be employed?

6.14. Explain the ways in which (a) temperature and (b) dopant concentration should be expected to influence the Fermi energy.

6.15. (a) Explain the difference between Eq. (1-11) and Eq. (6-171a).

(b) Why is this of significance in the properties of doped materials.

6.16. What is the utility of available semiconductor materials with large variations in their values of E_g?

6.17. What is the difference in the roles of dopant atoms with a valence of 4 as compared to those of valence less than 4?

6.18. (a) Describe the fundamental band structures required of all electrical insulators.

(b) Cite and explain why some classes of materials conform to these requirements.

6.19. Why do some organic polymers behave as semiconductors?

6.20. What practical use may be made of carrier diffusion currents in insulating materials?

6.21. Show how the value of E_g affects the dielectric strength of an insulating material.

7 / THERMOELECTRIC AND GALVOMAGNETIC EFFECTS

The electrical conduction discussed in Chapter 6 gave primary consideration to the flow of carriers under the influence of imposed electric fields. Other physical phenomena arise from the influences of temperature effects on magnetic effects on carriers. The former result in such thermoelectric manifestations as the Seebeck, Peltier, and Thomson effects, while the latter are evidenced by the galvomagnetic Hall, Ettingshausen, and Nernst effects.

7.1 THERMOELECTRIC PROPERTIES

7.1.1 Seebeck Effect

When two dissimilar conductors,* A and B, constitute a circuit, a current will flow as long as the junctions of the two conductors are at different temperatures. Conductor A is defined as being positive to conductor B if the electrons flow from A to B at the colder junction (Fig. 7-1). It can be shown that the Seebeck effect is related to the Peltier and Thomson effects [see Eq. (7-6)].

It must be emphasized that the Seebeck effect is solely a function of the temperature difference between the two junctions of the dissimilar, *homogeneous*

*The term conductor as used in this chapter includes metals, alloys, and semiconductors.

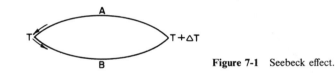

Figure 7-1   Seebeck effect.

conductors. And, where the compositions of the two thermoelements may be *assumed* to remain unchanged, the Seebeck effect is independent of all other variables, including changes in cross section, temperature distributions, and temperature gradients.

This phenomenon is the basis for thermoelectric thermometry. Combinations of any of the metallic elements and their alloys or semiconducting elements and compounds (either intrinsic or extrinsic as described in Chapter 6) will give Seebeck currents and voltages. One of the present objectives is to explain why so few pairs of conductors have been used to exploit this phenomenon, especially for the measurement and control of temperature.

The Seebeck effect is in no way related to the contact potential, or Volta effect. Contact potential is measured by the difference in work functions (Fig. 8-2), when two different metals are brought sufficiently close so that a transient electron transfer creates a common Fermi level in both. This does not require a temperature difference, and for closed circuits (Fig. 7-1), the net effect is zero (see Section 8.5).

7.1.2 Peltier Effect

When an electric current flows across a junction of two dissimilar conductors, heat is liberated or absorbed. When the electric current flows in the same direction as the Seebeck current, heat is absorbed at the hotter junction and liberated at the colder junction. The Peltier effect is defined as the reversible change in heat content when 1 coulomb crosses the junction (Fig. 7-2).

The direction in which the current flows determines whether heat is liberated or absorbed. This effect is reversible and is independent of the shape or dimensions of the materials composing the junction. It is a function of the compositions of the materials and the temperature of the junction, not of the contact (see Section 8.5).

It is interesting to note that Peltier, using a thermocouple made of antimony and bismuth, was able to freeze a droplet of water on a junction. This was the first demonstration of thermoelectric refrigeration. Peltier devices, made from suitably doped semiconductors, are widely used to cool sensitive, solid-state electronic circuitry such as that used in computers. These devices have low thermal efficiencies, but are virtually trouble free and lend themselves to simple control. In the reverse mode, junctions with large Peltier effects have been used for power generation in situations where thermal efficiency is not of primary importance. Use has been made of this to power satellites.

The Peltier cooling devices of greatest interest in thermoelectric thermometry are those that are being used to provide reference temperatures for thermocouple

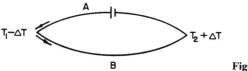

Figure 7-2   Peltier effect.

junctions.   Small, battery-powered devices are available that can automatically maintain 0°C (32°F), within very narrow limits, for long periods of time.   In most cases, their maintenance consists of battery (small dry cells) replacement.

7.1.3 Thomson Effect

The Thomson effect is defined as the change in the heat content of a single conductor of unit cross section when a unit quantity of electricity flows along it through a temperature gradient of 1 K (Fig. 7-3). This phenomenon arises because the carriers transport both the electrical and thermal energy.   Consider a single conductor [Fig. 7-3(a)], which has been heated at one point to some temperature T_2.   A thermal gradient will exist on either side of the heated point.   Two points, P_1 and P_2, of equal temperature, $T_1 < T_2$, will be found on either side of T_2.   If current flows in a circuit that includes the single conductor, the temperatures at P_1 and P_2 will change [see Fig. 7-3(b)].   These changes are a result of the motion of the electrons with respect to the direction of the temperature gradient.   The electrons flowing past P_1 will absorb energy in moving against the temperature gradient and increase their potential energy. The electrons flowing in the same direction as the thermal gradient (toward decreasing temperatures at P_2) will give up energy and thus decrease their potential energy.

Heat (thermal energy) accordingly will be absorbed at P_1, where the electron flow is opposite to the heat flow.   Heat will be liberated at P_2 where the electron flow

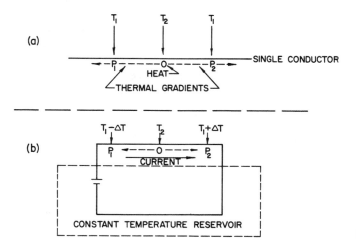

Figure 7-3   Thomson effect.

is the same as the heat flow. These changes in the heat content of the conductor are the Thomson effects and constitute the basis for the statement that the Thomson coefficient, σ, is the specific heat of electricity.

In Figure 7-3, the Thomson effects are equal and opposite and cancel each other. This illustrates why dissimilar metals must be used in thermocouples. The law of homogeneous conductors, which states that a thermoelectric current cannot be maintained solely by the application of heat to a single homogeneous conductor, regardless of any cross-sectional variations, has its origin in this cancellation. However, this condition is ideal for extension wires connecting thermocouples to measuring devices; no extraneous emf is added to the circuit when the ends of such connecting wires are at the same temperatures. When most dissimilar metals are paired to form a thermocouple, the Thomson effects are unequal, and a voltage and a flow of current results.

7.1.4 Thermodynamics of Thermoelectricity

The understanding of the thermodynamic interdependencies among the thermoelectric effects is important not only to have clear ideas of the nature of the basic phenomena, but because the quantum mechanic treatments are based on these relationships. Thermodynamics provides the interrelationships and quantum mechanics gives the mechanisms. The thermodynamic approach given here essentially is that of Thomson and represents the least complicated way of showing these relationships.

Neglecting small conduction losses and Joule heating, a typical thermoelectric circuit can be considered to very closely approximate a reversible heat engine in the following way. The irreversible (Joule) losses can be approximated as follows. The current in most thermoelectric circuits is of the order of 10^{-3} A. The electrical resistance of such circuits usually is minimized in order to achieve maximum sensitivity. This can be approximated to be considerably less than $10\,\Omega$. Thus the irreversible heat loss will be very much less than 10^{-5} W. Losses of this magnitude may be neglected and thus permit the treatment of thermoelectric properties as being thermodynamically reversible. Thomson, aware that this was not strictly rigorous, used this approximation.

Consider a circuit of two dissimilar metals, A and B of Fig. 7-1, where the colder junction is at temperature T and the hotter junction is at $T + \Delta T$. Both of these temperatures are maintained by heat reservoirs. The electromotive force (emf) generated in this circuit is the Seebeck effect, E_{AB}. The thermoelectric power is defined as the change in emf per kelvin, or dE_{AB}/dT. Then the electrical energy is given by

$$E_{AB} = \frac{dE_{AB}}{dT}\,\Delta T \qquad (7\text{-}1)$$

It was noted previously that the Peltier effect describes changes in the heat contents of the junctions and that the Thomson effect describes changes in the heat contents of the individual conductors. These thermal effects may be expressed as follows:

Peltier effects (at the junctions): $\qquad$ (7-2a)
    Heat absorbed at the hotter junction $= P_{AB}(T + \Delta T)$
    Heat liberated at the colder junction $= -P_{AB}(T)$

Thomson effects (within the conductors): $\qquad$ (7-2b)
    Heat absorbed in conductor $B = \sigma_B(\Delta T)$
    Heat liberated in conductor $A = -\sigma_A(\Delta T)$

Since the thermoelectric circuit can be considered to approximate a reversible heat engine, the thermal and electrical energies may be equated:

$$\frac{dE_{AB}}{dT}\,\Delta T = P_{AB}(T + \Delta T) - P_{AB}(T) + (\sigma_B - \sigma_A)\Delta T \qquad (7\text{-}3)$$

When both sides of this equation are divided by ΔT,

$$\frac{dE_{AB}}{dT} = \frac{P_{AB}(T + \Delta T) - P_{AB}(T)}{\Delta T} + (\sigma_B - \sigma_A) \qquad (7\text{-}4)$$

The fraction on the right is the only term containing the quantity ΔT. It will be recognized as being in the form of a difference quotient, which, for the condition where ΔT approaches zero, gives the instantaneous rate of change of the Peltier effect with respect to temperature:

$$\underset{\Delta T \to 0}{\text{Lim}} \left[\frac{P_{AB}(T + \Delta T) - P_{AB}(\Delta T)}{\Delta T} \right] = \frac{dP_{AB}}{dT} \qquad (7\text{-}5)$$

When Eq. (7-5) is substituted into Eq. (7-4), the *fundamental theorem of thermoelectricity* is obtained:

$$\frac{dE_{AB}}{dT} = \frac{dP_{AB}}{dT} + (\sigma_B - \sigma_A) \qquad (7\text{-}6)$$

This equation in effect gives the electrical Seebeck effect as the algebraic sum of the thermal Peltier and Thomson effects in terms of *thermal energies*. Equation (7-6) is the basis for the statement made earlier that the Seebeck effect is related to the Peltier and Thomson effects (Section 7.1.1).

7.1.4.1 Reversibility and entropy.

The approximation made earlier (Section 7.1.4) that thermoelectric properties may be considered as being thermodynamically reversible simplifies their treatment. This permits the net change in the entropy of the surroundings of a thermocouple to be approximated as being equal to zero. While this is not rigorous (a factor considered by Thomson, who employed this concept), it avoids the necessity to account for irreversible effects and gives results that are in excellent agreement with experimental findings. Thus, it is possible to describe the properties of thermocouples in a way that is based on the net entropy changes of their surroundings. These surroundings are represented by the thermal reservoirs in the following analysis.

Place two additional reservoirs at the midpoints of conductors A and B. Both of these central reservoirs are maintained at a temperature that is the mean of those at the hotter and colder junctions (Fig. 7-4). These will provide a measure of the average change in the entropy of the surroundings of each of the conductors.

Allow a unit quantity of electricity to flow through the circuit. The assumption of reversibility requires that the net change in the entropy, ΔS, of the reservoirs at the junctions and along the conductors will be zero. Thus, the net change in the entropy of the surroundings of the thermocouple is given by

$$\Delta S \simeq \frac{-P_{AB}(T+\Delta T)}{T+\Delta T} + \frac{P_{AB}(T)}{T} - \frac{\sigma_B(\Delta T)}{T+(\Delta T/2)} + \frac{\sigma_A(\Delta T)}{T+(\Delta T/2)} = 0 \qquad (7\text{-}7)$$

When the first two terms of Eq. (7-7) are are multiplied by $\Delta T/\Delta T$, it may be expressed as

$$\Delta S \simeq \left[\frac{-\dfrac{P_{AB}(T+\Delta T)}{T+\Delta T} + \dfrac{P_{AB}(T)}{T}}{\Delta T} \right] \Delta T - \frac{\sigma_B(\Delta T)}{T+(\Delta T/2)} + \frac{\sigma_A(\Delta T)}{T+(\Delta T/2)} = 0 \qquad (7\text{-}8)$$

In the limit as ΔT approaches zero, the quantity within the brackets is $-(d/dT)(P_{AB}/T)$. This is substituted into Eq. (7-8) to give

$$\Delta S \simeq - \frac{d}{dT}\left(\frac{P_{AB}}{T}\right)\Delta T - \frac{\sigma_B(\Delta T)}{T+(\Delta T/2)} + \frac{\sigma_A(\Delta T)}{T+(\Delta T/2)} = 0 \qquad (7\text{-}9)$$

The Thomson effect is defined as the change in heat content for a gradient of 1 K. Thus, $\Delta T = 1$ K. Since T is much greater than 1 K, $T + \Delta T/2 = T + \frac{1}{2} \simeq T$. Under these conditions, it may be approximated that Eq. (7-9) may be written simply as

$$\frac{d}{dT}\left(\frac{P_{AB}}{T}\right) \simeq \frac{\sigma_A}{T} - \frac{\sigma_B}{T} \qquad (7\text{-}10)$$

The indicated derivative is taken and Eq. (7-10) becomes

$$\frac{T(dP_{AB}/dT) - P_{AB}}{T^2} \simeq \frac{\sigma_A}{T} - \frac{\sigma_B}{T} \qquad (7\text{-}11)$$

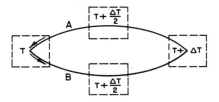

Figure 7-4 Thermodynamic model for calculating the thermoelectric power of a thermocouple. The thermoelectric circuit is that of Fig. 7-1 with the addition of thermal reservoirs at the midpoint of each conductor.

After multiplication through by T and rearrangement, Eq. (7-11) becomes

$$\frac{P_{AB}}{T} \simeq \frac{dP_{AB}}{dT} + \sigma_B - \sigma_A \qquad (7\text{-}12)$$

Equation (7-12) actually describes the entropy change of a thermoelectric-junction because P_{AB} is the change in the heat content of the junction and, when divided by the absolute temperature, gives (by definition) the change in the entropy of the junction at the given temperature.

The significance of entropy in thermoelectricity is emphasized in the following sections.

7.1.4.2 Peltier effect.   Equation (7-12) relates the Peltier effect to the Thomson effects. As such, it can be helpful in the selection of materials for use in power generation or refrigeration. This equation is rearranged, for clarity, in the form

$$\frac{dP_{AB}}{dT} = -\frac{P_{AB}}{T} + (\sigma_B - \sigma_A) \qquad (7\text{-}12a)$$

Where a maximum Peltier effect exists, $dP_{AB}/dT = 0$. Thus,

$$P_{AB} = (\sigma_B - \sigma_A)T \qquad (7\text{-}12b)$$

Equation (7-12b) is useful because the Thomson coefficients of a pair of thermoelements can be determined both more readily and more accurately than the Peltier effect of their junction. Or, where data are available for σ_A and σ_B, optimum combinations of thermoelements may be selected by means of simple calculations.

Use is now made of the fundamental theorem [Eq. (7-6)] to simplify Eq. (7-12), which upon substitution beomes

$$\frac{P_{AB}}{T} = \frac{dE_{AB}}{dT} \qquad (7\text{-}13)$$

Thus, the thermoelectric power of a thermocouple is a direct measure of the change in entropy of a thermoelectric junction and may be used as such.

Equation (7-13) may be rewritten as

$$P_{AB} = \frac{dE_{AB}}{dT}\, T \qquad (7\text{-}14)$$

Equation (7-14) is important in describing the behavior of Peltier devices. Combinations of thermoelements with large Peltier effects may be used for power generation or for refrigeration. The thermal efficiency is low in either case.

7.1.4.3 Thomson effect.   Upon differentiation with respect to temperature, Eq. (7-14) becomes

$$\frac{dP_{AB}}{dT} = \frac{dE_{AB}}{dT} + T\frac{d^2E_{AB}}{dT^2} \qquad (7\text{-}15)$$

This may be rearranged to read

$$\frac{dP_{AB}}{dT} - \frac{dE_{AB}}{dT} = T\frac{d^2E_{AB}}{dT^2} \tag{7-15a}$$

The fundamental theorem may also be rewritten as

$$\frac{dP_A}{dT} - \frac{dE_{AB}}{dT} = -(\sigma_B - \sigma_A) \tag{7-6a}$$

Equations (7-6a) and (7-15a) may be equated to give

$$T\frac{d^2E_{AB}}{dT^2} = -(\sigma_B - \sigma_A) \tag{7-16}$$

or they may be rearranged to read

$$\frac{d^2E_{AB}}{dT^2} = \frac{\sigma_A - \sigma_B}{T} \tag{7-16a}$$

Upon integration, Eq. (7-16a) becomes

$$\frac{dE_{AB}}{dT} = \int_0^T \frac{\sigma_A - \sigma_B}{T}\,dT = \int_0^T \frac{\sigma_A}{T}\,dT - \int_0^T \frac{\sigma_B}{T}\,dT \quad \mu V/{}^\circ C \tag{7-17}$$

Equation (7-17) is integrable because the quantities σ/T are entropy, and, by reason of the third law of thermodynamics, they approach zero as the temperature approaches zero. On this basis, the thermoelectric power of a thermocouple may be considered to be the difference between the entropies of the two legs from which it is formed. It is also of importance to note that the thermoelectric power of a thermocouple, at a given temperature, may be expressed in terms of the Thomson coefficients of its components.

7.1.5 The Concept of Absolute EMF

The separation of Eq. (7-17) into two integrals is the basis for the concept known as absolute thermoelectric power. The derivation of Eq. (7-17) is the proof of the statement that the thermoelectric power of a thermocouple is the algebraic sum of the absolute thermoelectric powers of its components. This may be written as

$$\frac{dE_{AB}}{dT} = S_A - S_B \tag{7-18}$$

where

$$S_A = \int_0^T \frac{\sigma_A}{T}\,dT \quad \text{and} \quad S_B = \int_0^T \frac{\sigma_B}{T}\,dT \tag{7-18a}$$

If the absolute thermoelectric power (ATP) of one thermoelement is known and the thermoelectric power of the couple has been determined experimentally, the ATP of the unknown leg may be calculated from Eq. (7-18).   This approach is very important because it permits the study of the characteristics of individual thermoelements in terms of their inherent properties, without recourse to a reference component.   Virtually all the analysis in the following sections are based on this concept.

The element lead (Pb) has been used for thermoelectric reference purposes.   By this it is meant that it has been employed as a standard thermoelement, with established thermoelectric properties, to serve as one component of a thermocouple. The absolute thermoelectric properties of any other thermoelectric materials may then be determined by the use of Eq. (7-18) as just described.   The reason for the use of lead is that its ATP is relatively small compared to most other elements or alloys. Thus, when lead is used for this purpose, the emf of the thermocouple is largely due to the Thomson effect of the other thermoelement.   The use of lead for this purpose is limited by its relatively low melting point.   At present it is used primarily for reference purposes at low temperatures (Table 7-1).

Very pure platinum is the most widely used reference thermoelement.   Its high melting point and thermoelectric stability in oxidizing atmospheres make it more useful than lead.   The thermoelectric properties of platinum have been well established (Table 7-2).

The reference temperature for the absolute thermoelectric scale is 0 K [Eq. (7-17)].   Any known, convenient, easily reproducible temperature may be used as a practical reference temperature.   The melting point of ice at 1 atmosphere pressure, 0°C, is the most commonly used practical reference temperature.   The ice point was selected because of its universal availability and ease of accurate reproducibility.

TABLE 7-1 ABSOLUTE THERMO-ELECTRIC POWER OF LEAD (μV/°C)

Temperature, K	S_{Pb}
10	-0.43_4
50	-0.77_4
100	-0.86_5
153.2	-1.02
193.2	-1.10_5
253.2	-1.21
273.2	-1.15
293.2	-1.27_5

Abstracted from W. B. Pearson, "The Effects of Chemical Impurities and Physical Imperfections on the Thermoelectricity of Metals," *Ultra-High-Purity Metals*, p. 237, American Society for Metals, Metals Park, Ohio, 1962.

Data for intermediate temperatures are given in the reference.

TABLE 7-2   ABSOLUTE THERMOELECTRIC POWER (ATP) OF SOME ELEMENTS (μV/°C)

Temperature, K	Cu	Ag	Au	Pt	Pd	W	Mo
100	1.19	0.73	0.82	4.29	2.00	—	—
200	1.29	0.85	1.34	−1.27	−4.85	—	—
273	1.70	1.38	1.79	−4.45	−9.00	0.13	4.71
300	1.84	1.51	1.94	−5.28	−9.99	1.07	5.57
400	2.34	2.08	2.46	−7.83	−13.00	4.44	8.52
500	2.83	2.82	2.86	−9.89	−16.03	7.53	11.12
600	3.33	3.72	3.18	−11.66	−19.06	10.29	13.27
700	3.83	4.72	3.43	−13.31	−22.09	12.66	14.94
800	4.34	5.77	3.63	−14.88	−25.12	14.65	16.13
900	4.85	6.85	3.77	−16.39	−28.15	16.28	16.86
1000	5.36	7.95	3.85	−17.86	−31.18	17.57	17.16
1100	5.88	9.06	3.88	−19.29	−34.21	18.53	17.08
1200	6.40	10.15	3.86	−20.69	−37.24	19.18	16.65
1300	6.91		3.78	−22.06	−40.27	19.53	15.92
1400				−23.41	−43.30	19.60	14.94
1600				−26.06	−49.36	18.97	12.42
1800				−28.66	−55.42	17.41	9.52
2000				−31.23	−61.48	15.05	6.67
2200						12.01	4.30
2400						8.39	2.87

From N. Cusak and P. Kendall, *Proc. Phys. Soc.*, 72, 289, (1958).

The platinum reference standard for thermoelectric thermometry in the United States, Pt-67, is defined in the National Bureau of Standards Monograph 125.

The absolute thermoelectric powers, S in units of μV/°C, as used here should not be confused with the Seebeck coefficients, $S(T)$, also in units of μV/°C, such as are given in the National Bureau of Standards Monograph 125.   As derived in Eq. (7-17), S denotes the rate of change with temperature of the Thomson voltage in a *single conductor* when one of its ends is at 0 K (-273°C) and its other end is at any other temperature, T K (T°C).   In contrast to this, $S(T)$, also in units of μV/°C, refers to the thermoelectric power of a thermocouple formed by the conductor and the reference platinum, Pt-67, when the reference junction of this *thermocouple* is at 0°C ($+273$ K) and its measuring junction is at any other positive or negative temperature, T°C.   As noted, temperature–voltage data, such as that given as $E(T)$ by National Bureau of Standards, may be converted to the absolute scale by means of Eq. (7-18) using the well-established absolute values for very pure platinum, such as those given by J. Nystrom, *Arkiv Mathemetic Fysik*, Vol. 34A, No. 27, 1948, p. 1, or by those in Table 7-2.

7.1.6 Laws of Thermoelectric Circuits

If two wires of the same homogeneous conductor are used to form a thermoelectric circuit, the resultant emf will be zero.   This results from the fact that S_A and S_B in Eq. (7-18) are identical.   This also is discussed in Section 7.1.3.

Another consequence of Eq. (7-18) is that when no temperature difference exists between the ends of a homogeneous conductor, even though temperature gradients may exist between its ends, the net emf across the conductor will be zero.* Any number of such conductors may be connected in series with the same results. Indeed, such a series can be made to constitute the measuring junction of a thermocouple without affecting its calibration. This behavior is summarized in the law of intermediate conductors, which states that the sum of the ATPs of dissimilar conductors is zero when their ends are at the same uniform temperature. See Section 8.5.

A second law of the same name states that the emf of two thermocouples AC and CB, whose junctions are at the same temperatures, may be expressed [Eq. (7-18)] as $dE_{AB}/dT = S_A - S_C + S_C - S_B = S_A - S_B$. Separate integration gives the algebraic sum $E_{AB} = E_{AC} + E_{CB}$, in which the thermoelectric effects of component C are eliminated. Platinum common legs are used in this way to pair thermoelements.

A fourth result has been termed the law of successive temperatures. Based on Eq. (7-18), this may be expressed as

$$E_{AB} = \int_{T_0}^{T_1} (S_A - S_B)\,dT + \int_{T_1}^{T_2} (S_A - S_B)\,dT + \int_{T_2}^{T_3} (S_A - S_B)\,dT = \int_{T_0}^{T_3} (S_A - S_B)\,dT$$

$$(7\text{-}18b)$$

In other words, the emf of a thermocouple composed of homogeneous conductors can be measured, or expressed, as the sum of its emfs over successive temperature intervals. This is important in the calibration of thermocouples and in establishing their emf–temperature characteristics very accurately over wide ranges of temperature. Equation (7-18b) is also important in understanding the role of extension wires.

7.1.7 Application to Real Thermoelements

Consideration will now be given to the way in which the concept of ATP is applied to real thermoelements. The thermoelectric power of a thermocouple was shown to be given by

$$\frac{dE_{AB}}{dT} = S_A - S_B \qquad (7\text{-}18)$$

Therefore, the ATP of each thermoelement of a thermocouple must be considered in the description of the thermoelectric properties of the thermocouple. Two limiting cases of pairs of thermoelements will be considered that will include the behavior of most thermocouples. The general case is that of two thermoelements with different slopes as functions of temperature (Fig. 7-5).

The thermoelectric power of a thermocouple composed of elements A and B may be read directly from Fig. 7-5 for any given temperature. According to Eq.

*See Sections 8.5 and 8.5.1.

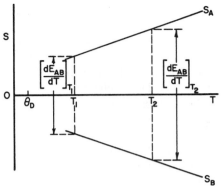
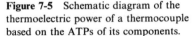

Figure 7-5 Schematic diagram of the thermoelectric power of a thermocouple based on the ATPs of its components.

(7-18), the thermoelectric power at any temperature is just the algebraic sum between S_A and S_B at that temperature, as shown in the figure.

The respective ATPs of the thermoelements A and B are given by

$$S_A = C_1 + m_A T \tag{7-19a}$$

and

$$S_B = C_2 + m_B T \tag{7-19b}$$

where C_1 and C_2 are constants and the slopes are given by m_A and m_B. The thermoelectric power of the thermocouple thus formed by B versus A is, from Eq. (7-18),

$$\frac{dE_{AB}}{dT} = C_3 + (m_B - m_A)T \quad \text{or} \quad dE_{AB} = [C_3 + (m_B - m_A)T]\,dT \tag{7-20}$$

where $C_3 = C_2 - C_1$. Thus, Eq. (7-20) may be considered to represent an element of area between the curves. From this it can be seen that the emf generated by the thermocouple is the area between the two curves integrated over the temperature range between the reference and measuring junctions. This also illustrates the arbitrary nature of the practical reference junction. If the reference junction is maintained at a known temperature, T_0, the emf of this couple can be found by integrating Eq. (7-20) to obtain

$$E_{AB} = \int_{T_0}^{T} [C_3 + (m_B - m_A)T]\,dT \tag{7-21}$$

Upon integration, this may be expressed as

$$E_{AB} = C_3(T - T_0) + \tfrac{1}{2}(m_B - m_A)(T^2 - T_0^2) \tag{7-21a}$$

The emf of this couple thus contains a term involving the difference of the squares of the temperatures; that is, it will be a parabolic function. This is not the most desirable case because of its explicit nonlinearity.

It should be noted that institutions such as the National Bureau of Standards make use of power series in $(T - T_0)$ for the descriptions of the emf–temperature characteristics of thermocouples. This usually is done in degrees Celsius with $T_0 = 0°C$. The power series may contain as many as two to about six terms. The number of terms depends on the shape of the emf–temperature curve in the temperature interval over which the emf is being expressed and the degree of precision desired. Those terms with cubic or higher exponents are, at present, considered to be useful curve-fitting parameters for those thermocouples whose components obey Eqs. (7-19a) and (7-19b).

The most desirable situation is one in which the ATPs of two thermoelements A and D are parallel functions of temperature. In this case, the slopes will be equal $(m_A = m_D = m)$, and the ATPs are given by

$$S_A = C_1 + mT \tag{7-22a}$$

and

$$S_D = C_4 + mT \tag{7-22b}$$

Then $S_D - S_A = C_4 - C_1 = C_5$, another constant. Or

$$\frac{dE_{AD}}{dT} = C_4 - C_1 = C_5 \tag{7-23}$$

since the temperature-dependent terms containing the equal slopes, m, cancel each other. And the emf of the ideal couple formed by thermoelements A and D is expressed by

$$E_{AD} = \int_{T_0}^{T} C_5 \, dT = C_5(T - T_0) \tag{7-24}$$

where T_0 again is the reference temperature.

The emf of couple A–D is a linear function of the temperature difference. This response is considered by manufacturers of electrical measuring devices to be ideal behavior when contrasted to that shown by couple A–B [Eq. (7-21a)]. Components normally used in the measurement of emf, such as slidewires, potentiometers, and electrical circuits that are based on linear behavior are more readily made and calibrated than those based on nonlinear responses of sensing devices. This permits the production of more accurate and more economical temperature-measuring and control devices.

This requirement is one of the reasons for the relatively few combinations of metals and alloys in common use as thermoelements. It does not, as yet, appear to be possible to obtain alloys with identical values of m_A and m_B as required by the ideal case. While no two thermocouple alloys have exactly the same slopes, some pairs of thermoelements are available that have relatively small differences between their slopes. This causes the second term of Eq. (7-21a) to be small compared to the linear term. None of these thermocouples have ideally linear emf–temperature properties. To a good first approximation, such thermocouples can be considered to approach a linear emf–temperature characteristic.

It should be noted that a linear emf–temperature characteristic may be sacrificed when other thermoelectric properties are more desirable. Considerable work at the National Research Council of Canada has been devoted to the study of a Pt–Au thermocouple for use between 0° and 1000°C. The ATP of the Pt component is in reasonable agreement with Eq. (7-19a), but that of Au follows Eq. (7-19b) only up to about 400 K. It shows relatively flat, parabolic ATP–temperature properties for $T > 400$ K. Thus, the equation for this Pt–Au thermocouple, which corresponds to Eq. (7-21a), contains significantly large quadratic and cubic terms; as such, its emf output is larger and its emf–temperature characteristic is highly nonlinear as compared to the standard Pt–Pt$_{90}$Rh$_{10}$ thermocouple. However, since the proposed thermocouple is intended primarily for use in standards laboratories, its nonlinearity is acceptable because its larger thermoelectric power permits more precise temperature measurements than does the standard thermocouple. It is also of prime significance that the Pt–Au couple has shown greater stability than the Pt–Pt$_{90}$Rh$_{10}$ thermocouple. The large increase in stability arises from the fact that the Au component appears to be much more stable than does the Pt$_{90}$Rh$_{10}$ leg.

On the basis of the foregoing, it is seen that the temperature-dependent terms of Eqs. (7-19a) and (b) are of fundamental importance in the understanding of the thermoelectric properties of thermocouple elements. Subsequent sections are devoted to different ways of explaining the electron mechanisms that attempt to describe their behavior.

7.1.8 Relation to Heat Capacity

The thermodynamic analyses given in the preceding sections provide means for describing the observed thermoelectric properties of individual thermoelements and of the thermocouples that may be formed by them. They do not, however, provide a model that can explain the mechanisms responsible for their behaviors. This can be accomplished only by understanding the roles of the carriers in thermoelectric behavior.

In the discussion of the Thomson effect (Sections 7.1.3 and 7.1.4.3), note was made of the absorption and release of energy by the electrons as they moved along a thermoelement either against or with the direction of the flow of thermal energy. It was also shown why the Thomson coefficient has been termed the specific heat of electricity. This concept is the basis for the analysis developed in this section; these combine thermodynamics and quantum mechanics.

This approach affords a means to introduce electron physics into thermodynamics in an uncomplicated way. The heat capacity (heat required to raise the temperature of 1 mol of material 1 K) will be used rather than the specific heat (heat required to raise 1 g of material 1 K) so that the derivations will be easier to follow. The result will be expressions corresponding to the temperature-dependent terms of Eqs. (7-19a) and (b). Then, in addition to the ability to describe thermoelectric properties thermodynamically, it will be possible to explain them in terms of the physical concepts presented in Chapter 4.

7.1.8.1 Normal metals.   Normal metals have been defined as those with complete inner electron shells and three or less outermost valence electrons (see Section 3.11). This is the simplest case because the valence electrons of this class of metals cannot interact with any unfilled states in inner levels since none are present.

Consider a single, monovalent, metallic conductor in a known, constant, temperature gradient and in which no current flows. The difference in the Gibbs free energy, F, across the conductor is given by

$$\Delta F = V \Delta P - S \Delta T \qquad (7\text{-}25)$$

where V is the volume, P the pressure, S the entropy, and T the absolute temperature. At constant pressure (1 atmosphere), $\Delta P = 0$. Then

$$\Delta F = -S \Delta T \quad \text{or} \quad \frac{\Delta F}{\Delta T} = -S \qquad (7\text{-}26)$$

And the gradient in free energy in the conductor being held in the temperature gradient from T_1 to T_2 is, from the second law of thermodynamics,

$$\frac{\Delta F_1}{\Delta T} = -S_1 = -\int_{T_1}^{T_2} \frac{C_{P1}}{T} \, dT \qquad (7\text{-}27)$$

where C_{P1} is the heat capacity of the conductor at constant pressure. The relationships indicated by Eq. (7-27) demonstrate the significance of the heat capacity and entropy.

Now allow an electric current to flow in the conductor while maintaining it in the original temperature gradient. The heat content of the conductor will change because of the Thomson effect, and the gradient in free energy now becomes

$$\frac{\Delta F_2}{\Delta T} = -S_2 = -\int_{T_1}^{T_2} \frac{C_{P2}}{T} \, dT \qquad (7\text{-}28)$$

The difference between the two conditions given by Eqs. (7-28) and (7-27) is

$$\Delta \left(\frac{\Delta F}{\Delta T} \right) = \frac{\Delta F_2}{\Delta T} - \frac{\Delta F_1}{\Delta F} \simeq \frac{dF}{dT} = \int_{T_1}^{T_2} \frac{\Delta C_P}{T} \, dT \qquad (7\text{-}29)$$

Equation (7-29) can be simplified if the approximation, given in Section 5.1, is made that

$$C_P \simeq C_V \qquad (5\text{-}6)$$

where C_V is the heat capacity of the conductor at constant volume. This approximation introduces an error that usually is much less than 10%, because the atomic volume and the compressibility of a normal metallic conductor do not change with the flow of current. The differentiation of Eq. (5-6) gives a relationship that will be seen to simplify this analysis:

$$\Delta C_P \simeq \Delta C_V \qquad (7\text{-}30)$$

Use now is made of Eq. (5-155) in which the coefficient of the electron component of the heat capacity is given more generally by Eq. (5-147). The substitution of these equations into Eq. (5-6) gives

$$C_P \simeq C_V \simeq 3N_A k_B + A \frac{\pi^2}{2} \frac{n N_A k_B^2 T}{E_F} \tag{7-31}$$

Here N_A is Avogadro's number, k_B is Boltzmann's constant, A is a conduction coefficient that accounts for the thermal effective mass of the electrons (Sections 5.1.10.1 and 5.1.10.2), n is the valence of the metal, T is the absolute temperature, and E_F is the Fermi energy (Fig. 5-17).

If the valence electrons are assumed to be "nearly free," *neglecting any magnetic effects*, the heat capacity of a normal metal in the absence of any current flow can be approximated by Eq. (5-8) for $T \geqslant \theta_D$ as

$$C_{V1} \simeq 3N_A k_B + A \frac{\pi^2}{2} \frac{n N_A k_B^2 T}{E_F} \tag{7-32}$$

In the case where the current is flowing in the conductor, the *electron flow no longer is random*, and, in the limit, the heat capacity can be approximated to be (Section 5.1.10.1), assuming *random electron spin* orientations,

$$C_{V2} \simeq 3N_A k_B \tag{7-33}$$

The substitution of Eqs. (7-32) and (7-33) into Eq. (7-30) gives

$$\Delta C_P \simeq \Delta C_V \simeq C_{V2} - C_{V1} \simeq 3N_A k_B - 3N_A k_B - A \frac{\pi^2}{2} \frac{n N_A k_B^2 T}{E_F} \simeq C_{V,e}$$

$$\Delta C_P \simeq - A \frac{\pi^2}{2} \frac{n N_A k_B^2 T}{E_F} \simeq C_{V \text{ (carrier)}} \simeq C_{V,e} \tag{7-34}$$

Equation (7-34) is substituted into Eq. (7-29). It will be noted that the factor T vanishes and that

$$\frac{dF}{dT} \simeq - \int_{T_1}^{T_2} A \frac{\pi^2}{2} \frac{n N_A k_B^2}{E_F} dT \tag{7-35}$$

The integration of Eq. (7-35) between the limits $T_1 = 0$ and $T_2 = T$ results in

$$\frac{dF}{dT} \simeq - A \frac{\pi^2}{2} \frac{n N_A k_B^2 T}{E_F} \tag{7-36}$$

If this change in the free energy results only from the difference in the conditions caused by current flow, that is, the Thomson effect, it may be assumed that this very closely approximates a reversible thermodynamic process (Section 7.1.4.1). By analogy to the thermodynamics of reversible cells, it is found that

$$\Delta F = - n \mathscr{F} \Delta E \tag{7-37}$$

Here $\mathscr{F}$ is Faraday's number and ΔE is the emf across the cell. Since $\mathscr{F} = N_A e$, Eq.

(7-37) may be expressed as

$$\Delta F = -nN_A e\, \Delta E \tag{7-38}$$

When both sides of this equation are divided by ΔT,

$$\frac{\Delta F}{\Delta T} = -nN_A e \frac{\Delta E}{\Delta T}$$

or, for very small intervals,

$$\frac{dF}{dT} = -nN_A e \frac{dE}{dT} \tag{7-39}$$

Equations (7-36) and (7-39) are equated to give

$$-nN_A e \frac{dE}{dT} \simeq -A \frac{\pi^2}{2} \frac{nN_A k_B^2 T}{E_F}$$

Noting that the factors n and N_A vanish, it is found that

$$\frac{dE}{dT} = S \simeq A \frac{\pi^2}{2} \frac{k_B^2 T}{eE_F} \quad \mu V/^\circ C \tag{7-40}$$

It will be observed that S actually is given by the heat capacity of a single carrier divided by its charge. See Eqs. (5-155) and (7-34). The approximation made for Eq. (7-33) limits the application of Eq. (7-40) to those metals and alloys in which any magnetic effects resulting from cooperative electron spin are absent. In other words, Eq. (7-40) is applicable to paramagnetic and diamagnetic materials in the absence of magnetic fields.

For practical purposes, Eq. (7-40) as given here is the equivalent of the temperature-dependent terms of Eqs. (7-19a) and (b). Equation (7-40) provides a means for understanding thermoelectric behavior that the other equations do not give. The factors A and E_F permit the construction of a model to explain thermoelectric properties in terms of electron physics for $T \geqslant \theta_D$. The corresponding thermodynamic equations given earlier in this chapter do not lend themselves to this.

7.1.8.2 Transition Metals.

Because of the assumptions implicit in Eq. (7-32), Eq. (7-40) can be expected to be valid only for normal metals. In the case of transition elements such as Fe, Co, and Ni, there is a positive hole contribution to the heat capacity. See Section 5.1.9 and Fig. 4-15(d). This effect arises from the incompletely filled d states of such atoms. In this case, where $(E_0 - E_F)$ represents the energy range of unoccupied states, or holes, of the incompletely filled portion of the d band, and E_0 is the energy of the top of the d band, Eq. (5-150) is used to express the heat capacity:

$$C_{V,h} \simeq A' \frac{\pi^2}{6} \frac{N_A k_B^2 T}{(E_0 - E_F)} \tag{5-150}$$

This equation is the basis for an approach that parallels that given for normal metals in the preceding section and leads to

$$S \simeq A' \frac{\pi^2}{6} \frac{k_B^2 T}{e(E_0 - E_F)} \quad \mu V/°C \tag{7-41}$$

for transition elements. Equation (7-41) is negative because the carriers are opposite in sign to those of Eq. (7-40). As was noted for Eq. (7-40), Eq. (7-41) is just the heat capacity of a d-level hole, the carrier in this case, divided by its charge. It should be noted that $(E_0 - E_F)$ as used in the above equations is not a simple difference; accurate determinations of this quantity include $E_F = f(T)$.

For those transition elements or their alloys in which the number of d-level holes per atom is equal to or greater than that of nickel, the variation of the ATP with temperature is large and negative in slope. In this case, $(E_0 - E_F)$ is virtually unaffected by temperature and may be approximated as being constant with a high degree of confidence.

This approximation does not apply to high-concentration alloys of normal metals in random solid solution in transition elements in which the d bands approach maximum filling. Those nickel-based alloys that approach an electron : atom ratio that is the equivalent of 60 At% of a monovalent, normal metal to 40 At% of Ni, such as constantans, show that $(E_0 - E_F)$ is affected by temperature, as discussed in detail in later sections dealing with commercial constantans.

It must be noted that Eq. (7-41) was derived in essentially the same way as Eq. (7-40). Since random spin orientations are implicit and all magnetic effects were neglected in arriving at Eq. (7-40), these effects consequently are not included in Eq. (7-41). *This means that Eq. (7-41) is not valid for those transition elements and those of their alloys in the temperature ranges over which they show any of the types of magnetic behaviors that arise from aligned electron spins.* However, Eq. (7-41) is valid for this class of metals and alloys at temperatures above their magnetic transition temperatures or their Debye temperatures, whichever is the higher; their spins are essentially random in this range and they are paramagnetic. The influence of this factor on commercially available thermoelements is discussed in detail in sections that include the nickel-based commercial alloys.

Those transition elements and their alloys that show only paramagnetic properties, such as platinum and its alloys, are reasonably well described by Eq. (7-41).

7.1.8.3 Semiconductors.

The carriers in nondegenerate semiconductors may be treated classically (Sections 2.3.1 and 2.7). Thus, at any given temperature, $E = nk_B T$. However, n_x, the number of extrinsic carriers per unit volume, varies with temperature, so using $E(T)$ as a reference,

$$E = E(T) + E(n_{ext}, T)$$

where $n_{ext} \to 1$ is the *fraction* of carriers operative at a given temperature, and

$$\frac{1}{n_x} \frac{\partial E}{\partial T} \simeq k_B \left[1 + \frac{\partial n_{ext}}{\partial T} T \right] \simeq C_{V \text{ (carrier)}} \tag{7-42}$$

Now, applying the results of Eqs. (7-40) and (7-41) to Eq. (7-42) for the case of semiconductors,

$$S = \frac{C_{V \text{ (carrier)}}}{n_x e_{\text{carrier}}} \simeq \frac{k_B}{e}\left[1 + \frac{\partial n_{\text{ext}}}{\partial T} T\right] \tag{7-43}$$

The relationship given in Eq. (7-43) anticipates Eq. (7-104). Use is now made of Eqs. (6-162), now written in general terms to conform to Eq. (7-43). These equations are combined as

$$n_{\text{ext}} \simeq \frac{n_x}{N(E)_x} \simeq \exp\left[\frac{-(E_F - E_{BE})}{k_B T}\right] \tag{7-44}$$

Here $N(E)_x$ is the density of states of a given extrinsic carrier, in its respective band, E_{BE} is the energy of the band edge (either E_c or E_v), and $E_{F \text{ (ext)}}$ is the Fermi energy of the extrinsic carriers. For simplification, let

$$\exp\left[\frac{-(E_F - E_{BE})}{k_B T}\right] = \exp\left(\frac{-\Delta E}{k_B T}\right) \tag{7-45}$$

so that the derivative of Eq. (7-44) with respect to temperature is

$$\frac{\partial n_{\text{ext}}}{\partial T} \simeq \frac{1}{N(E)_x} \cdot \frac{\partial n_x}{\partial T} = \frac{(\Delta E/k_B T^2)\exp(-\Delta E/k_B T)}{\exp(-\Delta E/k_B T)} = \frac{\Delta E}{k_B T^2} \tag{7-46}$$

The specific notation used in Eqs. (6-160) and (6-161) now is substituted successively into Eqs. (7-45) and (7-46). Thus, for n-type material,

$$\frac{\partial n_x}{\partial T} \simeq \frac{1}{N(E)_c} \cdot \frac{\partial n_x}{\partial T} = \frac{E_c - E_F}{k_B T^2} \tag{7-47a}$$

and similarly, for p-type material,

$$\frac{\partial n_{\text{ext}}}{\partial T} \simeq \frac{1}{N(E)_v} \cdot \frac{\partial n_x}{\partial T} = \frac{E_F - E_v}{k_B T^2} \tag{7-47b}$$

Equations (7-47a) and (b) are now substituted into Eq. (7-43) with the results for n-type material being

$$S_{e^-} \simeq \frac{k_B}{e}\left[1 + \frac{E_c - E_F}{k_B T}\right] \quad \text{mV/°C} \tag{7-48a}$$

and that for p-type material

$$S_h \simeq \frac{k_B}{e}\left[1 + \frac{E_F - E_v}{k_B T}\right] \quad \text{mV/°C} \tag{7-48b}$$

When the charges on the carriers are taken into consideration, Eq. (7-48a) will be negative and Eq. (7-48b) will be positive. In both cases the ATP of the materials will be large because the energy differences in the numerators are large with respect to $k_B T$.

It will be noted that the temperature variation of the carrier densities given by Eqs. (7-47a) and (b) are the only variables in Eqs. (7-48a) and (b). As such, experimental measurements of S provide indexes of carrier densities and carrier sign. In contrast, electrical conductivity measurements only provide measures of carrier density and not of carrier sign [see Eq. (6-129)].

In addition, the numerical factors in Eqs. (7-48a) and (b) should be written as $1 + x$, for reasons similar to those given in Section 5.1.10.2. Thus,

$$S_e \simeq \frac{k_B}{e}\left[(1 + x) + \frac{E_c - E_F}{k_B T}\right] \tag{7-48c}$$

and

$$S_h \simeq \frac{k_B}{e}\left[(1 + x) + \frac{E_F - E_v}{k_B T}\right] \tag{7-48d}$$

In the case in which the carrier scattering is by phonons, $x \simeq 1$. And where the ionized impurities scatter the carriers, $x \simeq 3$. This large difference in the values of x is the reason why extrinsic semiconductors are preferred for use in Peltier devices [Eq. (7-14)] and for thermoelectric power generation or refrigeration applications [Eq. (7-18)].

7.1.9 Quantum Mechanic Treatments

The derivations given in the preceding sections employed thermodynamics to introduce quantum mechanic relationships. This was based on electron contributions to heat capacity. The end results were uncomplicated derivations of the thermoelectric properties of normal and transition elements and extrinsic semiconductors. Having provided this background, theories based solely on quantum mechanics are derived in these sections. These result in general equations that can be applied to both types of elements. The findings obtained from all models are compared and shown to be in close agreement. The Mott and Jones theory, based on electrical conductivity, is used to show an interesting relationship between ATP and heat capacity. All the models can be used to explain why most thermocouples in common use in thermometry are made of transition elments or of alloys of transition elements.

7.1.9.1 Relation to internal potentials.

The derivations of Eqs. (7-40) and (7-41) are based on a combination of thermodynamic and quantum mechanic analyses. The same results may be obtained from a more general, uncomplicated quantum mechanic approach that is based on the internal potentials in a conductor.

Consider an *isolated* conductor, such as just one of the conductors in Fig. 7-1, along which a temperature difference ΔT is created and that thus contains a temperature gradient dT/dx. The kinetic energy of the carriers in the hotter portion of the conductor is greater than that of the carriers in its cooler portion. The more energetic carriers will group at the cold end and create a potential difference E_{pd} between its ends. This effect may be expressed as

$$S_{pd} = \frac{\partial E_{pd}}{\partial T} \tag{7-49}$$

This has been termed the *drift component,* or the volumetric component, of the ATP. Thus, the temperature difference can be considered to exert a "pressure" on the "gas" of carriers. This pressure may be approximated by use of the gas laws, and assigning $V = 1$, as

$$PV \simeq nk_BT \tag{7-50}$$

$$P \simeq \frac{n}{V} k_BT = \frac{2}{3} n\langle E \rangle \tag{7-51}$$

so that n is the number of carriers per unit volume and $\langle E \rangle$ is their average energy.

The potential difference between the ends of the conductor is a result of the electric field arising from the difference in the carrier densities. At equilibrium, this electric field, $\bar{E}_{pd}$, is equal to the carrier "pressure". This may be expressed as

$$ne\bar{E}_{pd}\,\partial x = \partial P \tag{7-52a}$$

and be rearranged as

$$ne\bar{E}_{pd} = \frac{\partial P}{\partial x} \tag{7-52b}$$

Using the chain rule, Eq. (7-52b) may be rewritten as

$$ne\bar{E}_{pd} = \frac{\partial P}{\partial T} \cdot \frac{\partial T}{\partial x} \tag{7-53}$$

so that

$$ne\bar{E}_{pd} \cdot \frac{\partial x}{\partial T} = \frac{\partial P}{\partial T} \tag{7-54}$$

or

$$ne\,\frac{\partial E_{pd}}{\partial T} = \frac{\partial P}{\partial T} \tag{7-54a}$$

$$\frac{\partial E_{pd}}{\partial T} = \frac{1}{ne}\,\frac{\partial P}{\partial T} \tag{7-55}$$

since $\bar{E}_{pd} \cdot \partial x/\partial T = \partial E_{pd}/\partial T$. Now, using Eq. (7-55) in Eq. (7-49),

$$S_{pd} = \frac{\partial E_{pd}}{\partial T} \simeq \frac{1}{ne}\,\frac{\partial P}{\partial T} \tag{7-56}$$

Another source of potential difference exists in conductors as a result of the temperature differences between their ends. In normal metals, $E_F(T)$ decreases very slightly with increasing temperature [Eq. (2-84)]. This effect is more pronounced in *n*-type extrinsic semiconductors [Eq. (6-163)], and a corresponding change also occurs in *p*-type materials [Eq. (6-164)]. In any of these cases, the values of E_F will be different at the hot and cold ends of the conductors. This difference constitutes another potential within the conductor. In this situation, even very small differences in E_F are significant, because E_F is *not constant* along the conductor, but varies along the temperature gradient. The simplest case for the description of this behavior, that of normal metals, may be seen by the use of Eq. (2-84). The potential difference, E_f, created by the variation of $E_F(T)$ is given by

$$edE_f \, \partial T = -\partial E_F(T) \, dT \tag{7-57a}$$

so that

$$S_f = \frac{dE_f}{dT} = -\frac{1}{e} \frac{\partial E_F(T)}{\partial T} \tag{7-57b}$$

The absolute thermoelectric power of a conductor is the result of the effects of both of these internal potential differences, so it may be given by the sum of Eqs. (7-56) and (7-57b). Thus, the general expression for ATP of an isolated conductor in the absence of current flow is

$$S = S_{pd} + S_f = \frac{1}{ne} \frac{\partial P}{\partial T} - \frac{1}{e} \frac{\partial E_F(T)}{\partial T} \tag{7-58a}$$

or

$$S = \frac{1}{e} \left[\frac{1}{n} \frac{\partial P}{\partial T} - \frac{\partial E_F(T)}{\partial T} \right] \tag{7-58b}$$

7.1.9.2 Normal metals. Equation (7-58b) may now be applied to obtain an expression for normal metals. The first term needed for Eq. (7-58b) is obtained from Eq. (7-51) and using the approximation that

$$\langle E \rangle \simeq \frac{3}{5} E_F(0) \left[1 + \frac{5\pi^2}{12} \left(\frac{k_B T}{E_F(0)} \right)^2 \right] \tag{7-59}$$

so that

$$P \simeq \frac{2}{3} n \langle E \rangle \simeq \frac{2}{3} n \cdot \frac{3}{5} E_F(0) \left[1 + \frac{5\pi^2}{12} \left(\frac{k_B T}{E_F(0)} \right)^2 \right] \tag{7-60}$$

Equation (7-60) is simplified to obtain

$$P \simeq n \left[\frac{2}{5} E_F(0) + \frac{2}{5} E_F(0) \cdot \frac{5\pi^2}{12} \left(\frac{k_B T}{E_F(0)} \right)^2 \right]$$

$$P \simeq n \left[\frac{2}{5} E_F(0) + \frac{\pi^2}{6 E_F(0)} (k_B T)^2 \right] \tag{7-61}$$

Since $E_F(0)$ is a constant and vanishes upon differentiation, the first term of Eq. (7-50) is obtained by differentiating Eq. (7-61):

$$\frac{1}{n} \frac{\partial P}{\partial T} \simeq \frac{1}{n} \cdot n \left[0 + \frac{2\pi^2}{6 E_F(0)} (k_B^2 T) \right] = \frac{\pi^2}{3 E_F(0)} k_B^2 T \tag{7-62}$$

Differentiating Eq. (2-85), the second term for Eq. (7-58b) is obtained as

$$\frac{\partial E_F(T)}{\partial T} \simeq -\frac{\pi^2}{6}\frac{k_B^2 T}{E_F(0)} \tag{7-63}$$

The use of Eqs. (7-62) and (7-63) in Eq. (7-58b) gives the ATP of normal metals as

$$S \simeq \frac{\pi^2 k_B^2 T}{3eE_F(0)}\left(1 + \frac{1}{2}\right) \tag{7-64a}$$

or

$$S \simeq \frac{\pi^2}{2}\frac{k_B^2 T}{eE_F(0)} \tag{7-64b}$$

It will be noted that Eq. (7-64b) is the same as Eq. (7-40), except that the numerical coefficient now is shown to arise from three sources. It will be recalled that conduction coefficients (Sections 5.1.9, 5.1.10, 5.1.10.1, and 5.1.10.2) take into account some of the variations between the electron behaviors and approximations used to describe their properties. This situation also exists in the derivation of Eq. (7-64b). One such approximation resides in the application of classical results [Eq. (7-51)] to degenerate particles (Chapter 2). This approximation, introduced in Eq. (7-60), affects Eq. (7-62) and finally is reflected in the factor $(1 + \frac{1}{2})$ in Eq. (7-64a). In addition, it also may be approximated that $E_F(0) = E_F(T) = E_F$ and that any small errors so introduced are also included in the conduction coefficient. Thus, for the same reasons given for Eqs. (5-147), (5-161), and (7-32), Eq. (7-64b) should be given as

$$S \simeq A\,\frac{\pi^2}{2}\frac{k_B^2 T}{eE_F} \quad \mu V/^{\circ}C \tag{7-65}$$

7.1.9.3 Transition metals. The thermoelectric behavior of transition metals is presented in terms of the number of vacant states as described in Section 7.1.8.2. Here E_0 is used as a reference energy, rather than zero, and the holes are treated as the mirror images of the filled electron states in a way similar to that discussed in Section 6.3.2 for the holes in the nearly filled valence bands of semiconductors. The average energy of the holes is approximated as

$$\langle E_h\rangle \simeq \frac{3}{5}\left[E_0 - E_F(0)\right]\left[1 + \frac{5\pi^2}{12}\left(\frac{k_B T}{E_0 - E_F(0)}\right)^2\right] \tag{7-66}$$

The substitution of Eq. (7-66) into Eq. (7-51) results in

$$P \simeq n\left[\frac{2}{5}\left[E_0 - E_F(0)\right] + \frac{\pi^2}{6[E_0 - E_F(0)]}(k_B T)^2\right] \tag{7-67}$$

and from this, Eq. (7-56), as applied to transition elements, is

$$\frac{1}{ne}\frac{\partial P}{\partial T} \simeq \frac{\pi^2}{3e[E_0 - E_F(0)]}\cdot k_B^2 T \tag{7-68}$$

In a corresponding way, Eq. (7-57b) is approximated for transition elements by starting with the temperature dependence of the vacant states as

$$E_0 - E_F(T) \simeq [E_0 - E_F(0)] \left[1 + \frac{\pi^2}{12} \left(\frac{k_B T}{E_0 - E_F(0)} \right)^2 \right] \qquad (7\text{-}69)$$

so that Eq. (7-57b) is, for transition metals,

$$-\frac{1}{e} \frac{\partial [E_0 - E_F(T)]}{\partial T} \simeq -\frac{\pi^2}{6e[E_0 - E_F(0)]} \cdot k_B^2 T \qquad (7\text{-}70)$$

The substitution of Eqs. (7-70) and (7-68) into Eq. (7-58a) gives the desired result as

$$S \simeq \frac{\pi^2}{3} \frac{k_B^2 T}{e[E_0 - E_F(0)]} \cdot \left(1 - \frac{1}{2} \right) \qquad (7\text{-}71\text{a})$$

or

$$S \simeq \frac{\pi^2}{6} \frac{k_B^2 T}{e[E_0 - E_F(0)]} \qquad (7\text{-}71\text{b})$$

And, for the same reasons given for Eq. (7-65), Eq. (7-71b) may be expressed simply as

$$S \simeq A' \frac{\pi^2}{6} \frac{k_B^2 T}{e[E_0 - E_F]} \quad \mu V/{}^\circ C \qquad (7\text{-}72)$$

Because of the way in which the holes are treated, the sign of the charge on a hole is opposite to that on an electron, so the equations for S given here are negative in sign. In addition, they are subject to the same limitations imposed by random spin orientations as those cited for Eqs. (7-40) and (7-41).

7.1.9.4 Alternate derivation of numerical coefficients.

Greater insight into the numerical coefficients of Eqs. (7-64a) and (7-71a) may be obtained by another approach. The carriers were considered to have average energies, $\langle E \rangle$, in the neighborhood of $E_F(T)$. Here, attention is directed to the carriers in the range $\Delta E \simeq k_B T$ above E_F. This energy range is given by the second term of Eq. (2-84) rewritten as

$$\Delta E \simeq \frac{\pi^2}{12 E_F(0)} (k_B T)^2 \propto E^2 \qquad (7\text{-}73)$$

Using Eq. (4-8) to approximate the number of carriers in the small range of ΔE,

$$\Delta N \propto E^{1/2} \qquad (7\text{-}74)$$

Then

$$\frac{\Delta E}{\Delta N} \propto E^2 \cdot E^{-1/2} \simeq E^{3/2} \qquad (7\text{-}75)$$

for a spherical Fermi surface. Equation (7-75) may be expressed more generally as

$$\frac{\Delta E}{\Delta N} \propto E^{1+x} \tag{7-76}$$

Then the derivative of Eq. (7-76) is taken with respect to ΔT and divided by the charge on a carrier to give

$$\frac{1}{e}\frac{d}{d\,\Delta T}\left[\frac{\Delta E}{\Delta N}\right] \propto \frac{1}{e}(1 + x)E^x_{E \simeq E_F} \tag{7-77}$$

since $E = f(T)$. This gives the variation of the energy with temperature or the heat capacity of one carrier divided by its charge. As is shown more concretely in Section 7.1.11, Eq. (7-77) defines the ATP. Thus, Eq. (7-77), another expression for S, provides the coefficients being sought.

For the case of normal metals, $x \simeq \frac{1}{2}$ and the numerical coefficient of Eq. (7-77) is in agreement with that of Eq. (7-64a). The factor E^x corresponds to the energy factor $k_B^2 T/E_F(0)$ in Eq. (7-64a). When the transition metals are treated as previously noted, $x = -1/2$, in agreement with Eq. (7-71a), and the factor E^x corresponds to $k_B^2 T/[E_0 - E_F(0)]$.

7.1.9.5 Semiconductors.

The derivation of the ATP of nondegenerate semiconductors also is based on the carrier "pressure" arising from the temperature gradient. Here, since the condition that $n_i/N \ll 1$, which was used to obtain Eq. (2-40) [and restated as Eq. (2-55)], is obeyed so that the classical approach may be used. This leads to the direct use of Eq. (7-51). Thus, differentiating Eq. (7-51) with respect to T gives

$$\frac{\partial P}{\partial T} = nk_B + T\frac{k_B\,\partial n}{\partial T} = k_B\left[n + \frac{T\,\partial n}{\partial T}\right] \tag{7-78}$$

Then the equivalent of Eq. (7-56) is found by dividing Eq. (7-78) by ne:

$$\frac{1}{ne}\frac{\partial P}{\partial T} = \frac{k_B}{e}\left[1 + T\frac{\partial n}{n\,\partial T}\right] = \frac{k_B}{e}\left[1 + T\frac{\partial \ln n}{\partial T}\right] \tag{7-79}$$

Now, Eq. (2-54) is used to obtain the equivalent of Eq. (7-57b). This is accomplished in the same way that was used to obtain Eq. (6-116). This is the case since the condition $n_i/N \ll 1$ permits classical treatment. In the present instance, the chemical potential, μ, of Eq. (2-64) actually is the Fermi energy, E_F, as explained in Section 2.7. [This was noted in the use of Eq. (6-113) to obtain Eq. (6-116).] So, for present purposes,

$$f_{MB} = \frac{1}{e^{(E-\mu)/k_B T}} \simeq f \simeq \frac{1}{e^{(E-E_F)/k_B T}} \tag{7-80}$$

So that, in the same way as for Eq. (6-116), the number of carriers per unit volume is given by

$$n = \frac{n}{V} \simeq N(E)f \simeq 2 \left[\frac{2\pi m k_B T}{h^2} \right]^{3/2} \cdot e^{(E - E_F)/k_B T} \tag{7-81}$$

since $V = 1$. Equation (7-81) is now expressed in logarithmic form as

$$\ln n = \ln 2 + \ln \left[\frac{2\pi m k_B T}{h^2} \right]^{3/2} + \frac{E_F}{k_B T} \tag{7-82}$$

so that it may be rewritten, letting $E = k_B T$, as

$$\frac{E_F}{k_B T} = - \ln 2 - \ln \left[\frac{2\pi m k_B T}{h^2} \right]^{3/2} + \ln n \tag{7-83}$$

This simplifies to give the energy of the carriers, in terms of E_F, as

$$E_F = k_B T \ln \frac{n h^3}{2(2\pi m k_B T)^{3/2}} \tag{7-84}$$

The equivalent of Eq. (7-57b) is obtained as

$$\frac{1}{e} \frac{\partial E_F}{\partial T} = \frac{k_B}{e} \cdot \frac{\partial}{\partial T} \left[T \ln \frac{n h^3}{2(2\pi m k_B T)^{3/2}} \right] \tag{7-85}$$

The derivative required by Eq. (7-85) is simplified if the logarithmic function is treated as follows:

$$\frac{1}{k_B} \frac{\partial E_F}{\partial T} = \ln \frac{n h^3}{2(2\pi m k_B T)^{3/2}} + T \frac{\partial}{\partial T} \{\ln(n h^3) - \ln[2(2\pi m k_B T)^{3/2}]\} \tag{7-85a}$$

Then

$$\frac{1}{k_B} \frac{\partial E_F}{\partial T} = \ln \frac{n h^3}{2(2\pi m k_B T)^{3/2}} + T \left[\frac{h^3 \partial n}{n h^3} - \frac{2(2\pi m k_B)^{3/2} \cdot \frac{3}{2} T^{1/2}}{2(2\pi m k_B)^{3/2} \cdot T^{3/2}} \right] \tag{7-85b}$$

Use is now made of Eq. (7-84) to simplify the logarithmic term, the result being

$$\frac{1}{k_B} \frac{\partial E_F}{\partial T} = \frac{E_F}{k_B T} + T \left(\frac{\partial n}{n} - \frac{3}{2T} \right) \tag{7-86a}$$

Or

$$\frac{1}{k_B} \frac{\partial E_F}{\partial T} = \frac{E_F}{k_B T} + T \frac{\partial \ln n}{\partial T} - \frac{3}{2} \tag{7-86b}$$

Equation (7-86b) is substituted into Eq. (7-85) to obtain

$$\frac{1}{e} \frac{\partial E_F}{\partial T} = \frac{k_B}{e} \left[\frac{E_F}{k_B T} + T \frac{\partial \ln n}{\partial T} - \frac{3}{2} \right] \tag{7-87}$$

Equations (7-79) and (7-87) are substituted into Eq. (7-58a) to obtain

$$S = \frac{k_B}{e}\left[1 + T\frac{\partial \ln n}{\partial T} - \frac{E_F}{k_B T} - T\frac{\partial \ln n}{\partial T} + \frac{3}{2}\right] \qquad (7\text{-}88)$$

The derivatives of the logarithmic terms vanish and Eq. (7-88) simplifies to

$$S = \frac{k_B}{e}\left[\frac{5}{2} - \frac{E_F}{k_B T}\right] \quad \text{mV/}^{\circ}\text{C} \qquad (7\text{-}89)$$

Or, again using Eq. (7-84), Eq. (7-89) becomes

$$S = \frac{k_B}{e}\left[\frac{5}{2} + \ln\frac{2(2\pi m k_B T)^{3/2}}{nh^3}\right] \quad \text{mV/}^{\circ}\text{C} \qquad (7\text{-}90)$$

Equations (7-89) and (7-90) are general expressions for the ATP of semiconductors. In the case of n-type materials, the charge of e is negative and S is negative. Similarly, Eqs. (7-89) and (7-90) are positive for p-type semiconductors. On the basis of a rough approximation using Eq. (7-89), it is seen that the ATP of semiconductors is expected to be very much greater than that of metals or their alloys simply because $E_F/k_B T \gg 1$. In fact, the ATP of these materials is of the order of millivolts/$^{\circ}$C as compared to that of microvolts/$^{\circ}$C for metals.

Because of the assumption of ideal gas behavior implicit in the use of Eq. (7-51) to obtain Eq. (7-78), it is considered that Eq. (7-90) should not be used for $T < \theta_D$. In addition, for the reasons cited for the previous expressions for ATP [Eqs. (7-40), (7-41), (7-65), and (7-72)], Eq. (7-90) should more properly be given as

$$S \simeq \pm\frac{k_B}{e}\left[(2 + x) + \ln\frac{2(2\pi m k_B T)^{3/2}}{nh^3}\right] \qquad (7\text{-}90\text{a})$$

where x may be obtained in a way similar to that indicated by Eq. (7-77). And, for convenience, Eq. (7-90a) may be reexpressed, using Eq. (7-89), as

$$S \simeq \pm\frac{k_B}{e}\left[(2 + x) - \frac{E_F}{k_B T}\right] \qquad (7\text{-}90\text{b})$$

7.1.10 Relation to Electrical Conductivity

The Mott and Jones treatment of thermoelectricity outlined here is a quantum mechanic approach that is based on electrical conductivity. Consider a rod of unit cross section in an electric field, $\bar{E}$, containing an electric current j, a heat current Q, and a temperature gradient $\partial T/\partial x$ (see Sections 7.1.3 and 7.1.4). The energy in the rod per unit volume per unit time is

$$U = \bar{E}j - \frac{\partial Q}{\partial x} \qquad (7\text{-}91)$$

The electric and heat currents are given, respectively, by

$$j = ne\mathbf{v}_x \tag{7-92}$$

and

$$Q = nE\mathbf{v}_x \tag{7-93}$$

where n is the number of electrons with wave vector $\bar{\mathbf{k}}$ in the element of wave vector space $d\bar{\mathbf{k}}_x d\bar{\mathbf{k}}_y d\bar{\mathbf{k}}_z$. The factor n is given by

$$n = 2f \frac{d\bar{\mathbf{k}}}{(2\pi)^3} \tag{7-94}$$

where f is the Fermi function [Eq. (2-86)]. The factor e in Eq. (7-92) is the charge on an electron, E in Eq. (7-93) is the energy of an electron in state $\bar{\mathbf{k}}$, and $\mathbf{v}_x$ is the velocity of an electron in the temperature gradient.

When the effects of the heat and the current flows on the Fermi function are taken into account, the results are substituted into Eq. (7-92) and (7-93) and then evaluated in the very narrow range of energies close to E_F. This results in the general equation for ATP of a thermoelement as

$$S \simeq \frac{\pi^2}{3} \frac{k_B^2 T}{e} \frac{\partial}{\partial E} \left[\ln \sigma(E) \right]_{E=E_F} \tag{7-95}$$

where $\sigma(E)$ is the electrical conductivity.* In its simplest form, this is given by

$$\sigma(E) \simeq n(E_F) e^2 \tau(E_F)/m \tag{7-96}$$

Here $n(E_F)$ is the number of carriers per unit volume with energies near E_F, $\tau(E_F)$ is their relaxation time, and m is the effective electron mass. Since $\tau(E_F)$ is the predominant variable in Eq. (7-96), Eq. (7-95) provides an additional means for examining carrier mobility as shown by Eqs. (6-19) and (6-7a).

Equation (7-95) is valid for pure elements and alloys *above* the Debye characteristic temperature. The relationship is correct to the first order in $k_B T/E_F$.

An approximation can be made in the range $E \simeq E_F$, where x includes the effects of $\bar{\mathbf{k}}_x$ in terms of $\tau(\bar{\mathbf{k}}_x)$, the relaxation time as a function of the wave vector, such that

$$\sigma(E) \simeq \text{constant } E^x \tag{7-97}$$

The close kinship between Eqs. (7-97) and (7-76) becomes apparent when the interrelationships between σ, $\bar{\mathbf{k}}$, τ, n, and E are taken into account. When Eq. (7-97) is substituted into Eq. (7-87), the constant vanishes as a result of the differentiation and

$$S \simeq \frac{\pi^2}{3} \frac{k_B^2 T}{eE_F} \cdot x = \frac{2.45 \times 10^{-2} T}{E_F} \cdot x \quad \mu\text{V}/°\text{C} \tag{7-98}$$

---

*See D. D. Pollock, *Thermoelectricity, Theory, Thermometry, Tool,* American Society for Testing and Materials, Philadelphia, PA, ASTM STP 852, 1985, for a complete derivation.

Equation (7-98) is the reduced Mott and Jones equation for the ATP of elements other than semiconductors and transition elements.

7.1.10.1 Normal metals.

Since more experimental work has been done on the noble metals than on most other normal metals, they frequently are used to illustrate the properties of this class of elements. The monovalent noble metals have experimentally determined values for $x \simeq -\frac{3}{2}$ [see Eq. (7-75)]. The inclusion of this value in Eq. (7-98) gives the ATP for these elements: Cu, Ag (and Au up to 400 K) as

$$S \simeq -\frac{\pi^2}{2} \frac{k_B^2 T}{e E_F} \quad \mu V/°C \tag{7-99}$$

It should be noted that $x \simeq -3/2$ was obtained from experiment, in contrast to the coefficient derived for Eq. (7-64a). When the negative charge on the electron is taken into account, Eq. (7-99) becomes positive.

7.1.10.2 Transition elements.

The preceding relationships were derived for metals in which the conduction is by electrons. They are based primarily on the extent to which the valence band is filled. In the case of the transition metals, conduction may be considered to be largely by the d-shell holes (unoccupied states) that lie above the Fermi level. (See Sections 3.11, 4.5, 5.1.9, 5.1.10.2, and prior sections in this chapter.) This must be taken into account in assessing the behavior of the factor [$\ln \sigma(E)$] in Eq. (7-95) if it is to be applied to transition elements.

Since the d bands of the transition elements are incompletely filled with electrons, their behavior can be described by considering the extent to which their d levels remain unoccupied. Thus, the behavior of the unoccupied states, or holes, can be used to portray the d-band behavior. On this basis it is approximated that the density of d-level holes is given by

$$N_d(E) \propto (E_0 - E_F)^{1/2} \tag{7-100}$$

where E_0 is the highest energy level in the d band. The use of this parabolic relationship permits the treatment of the holes in a way similar to that used for the electrons in normal metals.

When it is assumed that the factor $N_d(E)$ is the major variable in $\sigma(E)$, and the probability of s to d electron transitions is taken into account in terms of $N_d(E)$ in Eq. (7-95), then the ATP of transition elements and their alloys is given by approximating that

$$\ln \sigma(E) \propto \ln N_d(E) \propto \ln(E_0 - E_F)^{1/2} \tag{7-101a}$$

so

$$S \simeq -\frac{\pi^2}{6} \frac{k_B^2 T}{e(E_0 - E_F)} \quad \mu V/°C \tag{7-101b}$$

Since $N_d(E)$ decreases sharply as E_F approaches E_0, it would be expected that the ATP should be large and negative compared to Eq. (7-99). The sign of e in Eq. (7-101b) is taken to be positive since $(E_0 - E_F)$ represents a range of empty states, or holes, rather than electrons.

It must be noted again that none of the cooperative electron spin behaviors responsible for any of the magnetic effects were considered in the derivation of Eqs. (7-95) and (7-101b). Therefore, for the same reasons cited previously for the applicability of Eqs. (7-40), (7-41), and (7-72), Eq. (7-101b) is *not valid for transition elements that show any magnetic behaviors arising from aligned spins or for any alloys of such transition elements below any of their magnetic transition temperatures.* Equation (7-101b) is valid for such materials at temperatures above their transition temperatures at which they show paramagnetic properties. These equations are applicable to diamagnetic and paramagnetic materials in the absence of external magnetic fields.

7.1.10.3 Semiconductors.
In this case, the energy surfaces in $\bar{\mathbf{k}}$ space are considered to be nonspherical, and, at temperatures at which the valence band is almost but not quite filled, their energies are given by

$$E \simeq E_0 - \frac{h^2}{8\pi^2 m} \beta \bar{\mathbf{k}}^2 \qquad (7\text{-}102)$$

where E_0 is the energy level at the top of the nearly filled valence band. And, when Eq. (7-102) is taken into account in the factor $\ln \sigma(E)$ of Eq. (7-95), it is found, for hole conduction, where $(E_0 - E_F) \ll k_B T$, that

$$S_h \simeq -\frac{\pi^2}{3} \frac{k_B^2 T}{e(E_0 - E_F)} \cdot \frac{3 + x}{2} \qquad (7\text{-}103)$$

Here the minus sign accounts for the sign of the charge on a hole being opposite to that of an electron. In addition, x is positive, so S is positive.

The strong dependence of $\sigma(E)$ on $\bar{\mathbf{k}}$ [Eqs. (6-53) and (6-65)], which is affected strongly by a nonspherical Fermi surface, is reflected in the numerical coefficient of Eq. (7-103).

7.1.11 Thermoelectric Power and Electron Heat Capacity

It is instructive to compare the relationships for the electron and hole contributions to the heat capacity of solids with the corresponding equations for the ATP of the solids. For the noble metals, using Eq. (5-147),

$$C_{V,e} = \frac{\pi^2}{2} \frac{n N_A k_B^2 T}{E_F} \qquad (5\text{-}147)$$

and, neglecting any conduction coefficients for purposes of comparison,

$$S \simeq -\frac{\pi^2}{2} \frac{k_B^2 T}{e E_F} \qquad (7\text{-}40), (7\text{-}64\text{b}), (7\text{-}99)$$

For transition metals, using Eq. (5-150),

$$C_{V,h} = \frac{\pi^2}{6} \frac{nN_A k_B^2 T}{E_0 - E_F} \tag{5-150}$$

and, on the same basis as before,

$$S \simeq -\frac{\pi^2}{6} \frac{nN_A k_B^2 T}{e(E_0 - E_F)} \tag{7-41), (7-71b), (7-101b}$$

It will be recalled that Eqs. (7-40) and (7-41) were derived from the changes in the heat capacities of solids that occur as a result of the Thomson effect [Eq. (7-34)]. As can be seen from the preceding relationships, the ATP of a metal is given by the heat capacity of a carrier, electron or hole, divided by its charge:

$$S = \frac{C_{V\,\text{(carrier)}}}{e_\text{carrier}} \tag{7-104}$$

The relationship given by Eq. (7-104) provides important insight into the thermoelectric behavior of many alloys. It also provides an accurate, inexpensive means for the determination of the carrier signs and contributions to the heat capacities of metals and their alloys.

The relationship given by Eq. (7-104) also holds for nondegenerate semiconductors. Here the carriers may be treated classically as in Section 7.1.8.3. Thus, the heat capacity per carrier is obtained from Eq. (7-42) as

$$C_{V\,\text{(carrier)}} \simeq \frac{1}{n_x} \frac{\partial E}{\partial T} \simeq k_B \left[1 + \frac{\partial n_\text{ext}}{\partial T} T \right] \tag{7-105}$$

Thus, in the same way as for Eq. (7-43),

$$S = \frac{C_{V\,\text{(carrier)}}}{n_x e_\text{carrier}} \simeq \frac{k_B}{e} \left[1 + T \frac{\partial n_\text{ext}}{\partial T} \right] \tag{7-106}$$

Use now is made of Eq. (7-83) to expand Eq. (7-106). Equation (7-84) is reexpressed as

$$\ln n_\text{ext} \simeq \ln 2 + \ln \left(\frac{2\pi m k_B T}{h^2} \right)^{3/2} + \frac{E_F(0)}{k_B T} \tag{7-107}$$

Then, upon differentiation with respect to T, Eq. (7-107) becomes

$$\frac{\partial n_\text{ext}}{\partial T} \simeq \frac{3}{2} \frac{1}{T} - \frac{E_F(0)}{k_B T^2} \tag{7-108}$$

And, when Eq. (7-108) is multiplied through by T, the second term of Eq. (7-106) is obtained as

$$T \frac{\partial n_\text{ext}}{\partial T} \simeq \frac{3}{2} - \frac{E_F(0)}{k_B T} \tag{7-109}$$

Then, using Eq. (7-109) in Eq. (7-105), the contribution to the heat capacity made by a single carrier is

$$C_{V,\,\text{carrier}} \simeq k_B \left[1 + \frac{3}{2} - \frac{E_F(0)}{k_B T} \right] \qquad (7\text{-}110)$$

so the ATP is given by the substitution of Eq. (7-110) into Eq. (7-104) as

$$S = \frac{C_{V\,(\text{carrier})}}{e_{\text{carrier}}} \simeq \pm \frac{k_B}{e} \left[\frac{5}{2} - \frac{E_F(0)}{k_B T} \right] \qquad (7\text{-}111)$$

The results given by Eq. (7-111) are the same as those given by Eq. (7-89). Thus, Eq. (7-111) provides another confirmation of Eq. (7-104). For the same reasons given for Eq. (7-90a), the numerical terms in Eqs. (7-110) and (7-111) should be written as $(2 + x)$.

7.1.12 Comparison of Models

It is interesting and instructive to compare the results obtained in the preceding sections with the help of Table 7-3. Here the general equations are provided for each method for the derivation of thermoelectric properties.

All three approaches give the same expressions for the ATP of normal metals. The parameters A and x, respectively, vary as the way in which electrons contribute to the heat capacity (Sections 5.10.1 and 5.10.2) and the way that they contribute to electrical conductivity [Sections 6.0, 6.3.1.2, 7.1.10.1, and 7.1.10.2 and Eqs. (7-95) and (7-96)]. In other words, these parameters depend on the electron models used to describe the classes of solids (Section 3.12.4). All these methods involve approximations. A comparison of the findings for normal metals shows that

$$A \simeq \tfrac{2}{3} x \qquad (7\text{-}112)$$

And, on the same bases, it is shown for transition metals that

$$A' \simeq -1 \qquad (7\text{-}113)$$

if the charge on the hole [Eq. (7-41)] is not taken into consideration. The series of equations for semiconductors [Eqs. (7-48), (7-90), and (7-103)] also provides similar results when the parameter x is taken into account specifically for each model (Sections 6.0, 6.3.1.1 and 6.3.1.2).

7.1.13 Comparison of Thermoelectric Powers

An examination of the data in Table 7-3 shows why almost none of the noble metals are used as thermoelements at elevated temperatures. It also provides the reason why most of the thermoelements in common use for thermometric purposes are transition elements or alloys of transition elements.

A comparison of the ATP of these two classes of metals can be based on their Fermi energies. The noble metals have values of E_F of approximately 6 eV. The

TABLE 7-3 SUMMARY OF THERMOELECTRIC RELATIONSHIPS

Thermoelement	Heat Capacity	Internal Potentials	Electrical Conductivity (Mott and Jones)
General equations	(7-104): $S \simeq \dfrac{C_{V \text{ (carriers)}}}{N_A n e}$	(7-56): $S_{pd} \simeq \dfrac{\partial \bar{E}_{pd}}{\partial T} = \dfrac{1}{ne}\dfrac{\partial P}{\partial T}$ (7-57b): $S_f \simeq \dfrac{\partial E_f}{\partial T} = -\dfrac{1}{e}\dfrac{\partial E_F(T)}{\partial T}$ (7-58b): $S \simeq S_{pd} + S_f$	(7-96): $\sigma(E) \simeq n(E_F)e^2\tau(E_F)/m$ (7-95): $S \simeq \dfrac{\pi^2}{3}\dfrac{k_B^2 T}{e}[\ln \sigma(E)]_{E=E_F}$
Normal metals	(7-34): $C_{V,e} \simeq A\,\dfrac{\pi^2}{2}\dfrac{n N_A k_B^2 T}{E_F}$ (7-40): $S \simeq A\,\dfrac{\pi^2}{2}\dfrac{k_B^2 T}{eE_F}$	(7-64b): $S \simeq A\,\dfrac{\pi^2}{2}\dfrac{k_B^2 T}{eE_F}$	(7-98): $S \simeq \dfrac{\pi^2}{3}\dfrac{k_B^2 T}{eE_F}x, \qquad x \simeq \dfrac{3}{2}$
Transition metals	(5-150): $C_{V,h} \simeq A'\,\dfrac{\pi^2}{6}\dfrac{N_A k_B^2 T}{e(E_0 - E_F)}$ (7-41): $S \simeq A'\,\dfrac{\pi^2}{6}\dfrac{k_B^2 T}{e(E_0 - E_F)}$	(7-71b): $S \simeq A'\,\dfrac{\pi^2}{6}\dfrac{k_B^2 T}{e(E_0 - E_F)}$	(7-101b): $S \simeq -\dfrac{\pi^2}{6}\dfrac{k_B^2 T}{e(E_0 - E_F)}$
Semiconductors	(7-42): $C_{V\text{ (carrier)}} \simeq k_B\left[1 + \dfrac{\partial n_{\text{ext}}}{\partial T}\,T\right]$ (7-48a)	(7-90b): $S \simeq \pm\dfrac{k_B}{e}\left[(2+x) - \dfrac{E_F}{k_B T}\right]$ to (7-48c): $S \simeq \pm\dfrac{k_B}{e}\left[(1+x) \pm \dfrac{E_F - E_{BE}}{k_B T}\right]$	(7-103): $S_h \simeq \dfrac{\pi^2}{3}\dfrac{k_B^2 T}{e(E_0 - E_F)}\cdot\dfrac{3+x}{2}$

transition elements and their dilute alloys have values of $(E_0 - E_F)$ of about 1 eV. On this basis alone the ATP of a transition element is expected to be at least twice that of a noble thermoelement. A thermocouple made of transition metals or their alloys would give an emf considerably greater than that of a noble metal thermocouple at a given temperature. In addition, most of the transition thermoelements have reasonably linear emf–temperature characteristics (see Section 7.1.7).

The large ATPs of semiconductors result from the fact that the numerators of the energy terms in Eqs. (7-48a), (7-48b), and (7-90b) are very much larger than $k_B T$. A similar situation exists for Eq. (7-103), where the nearly filled valence band is such that $(E_0 - E_F) \ll k_B T$. The net result of this condition is that the ATPs of intrinsic semiconductors are larger than those of metals by a factor of about 10^2; those of extrinsic semiconductors are larger by a factor of about 10^3.

7.1.14 Influence of the Fermi Level

It was indicated previously that an understanding of the behavior of the Fermi level is necessary to complete the models for ATP because it is the most important factor in the relationships summarized in Table 7-3. The discussions in the previous sections assumed that the thermoelements were in the ideal state; it was considered that the metals had perfect, stress-free, isotropic crystal lattices and that they contained no impurities. Questions thus arise as to the effects of temperature, composition, stress, and so on, on the Fermi level and the resulting influences on the ATP of the material being considered. These factors have important influences not only on thermoelectric properties, but also on the stability of thermoelements.

7.1.14.1 Effect of temperature. It has been stated [Eq. (2-84)] that the Fermi level as a function of temperature is given by

$$E_F(T) = E_F(0)\left[1 - \frac{\pi^2}{12}\left(\frac{k_B T}{E_F(0)}\right)^2 \right] \qquad (2\text{-}85)$$

where $E_F(T)$ is the Fermi energy at temperature T and $E_F(0)$ is that at 0 K. The value of $k_B T$ is very much smaller than $E_F(0)$ at all temperatures normally encountered by metallic thermoelements. Thus, when this very small fraction is squared, the second term within the brackets is extremely small compared to unity. So, for most purposes, since

$$E_F(T) \simeq E_F(0) \qquad (7\text{-}114)$$

the Fermi level may be treated as being independent of temperature. In other words, the negligibly small changes in the Fermi level with temperature may be neglected, and the Fermi level of a given metallic alloy may be regarded as being essentially constant for most purposes. Some important exceptions to this are given by Eqs. (7-89) and (7-90) and Section 7.1.14.3.

7.1.14.2 Effects of alloying elements in solid solution in noble metals.   Now consider the effects of dilute solid solutions on the Fermi energy of noble metals. To do this, the density-of-states curve for the pure, noble metal is examined first [Fig. 7-6(a)]. This figure shows that the band is half-filled (see Sections 3.9, 4.1, and 4.7). This must be the case because the noble metal atoms have one outer, s, electron per atom, and the band can accommodate two electrons per atom. The Fermi energy of the "pure", noble metal is shown by E_{F_0} in Fig. 7-6(a). The electron:atom ratio equals unity.

Now consider a dilute, substitutional, solid solution in which the alloying, or added, normal element has a greater number of outer, or valence, electrons than the noble metal in whose lattice it is dissolved. If the added element has a valence of Z_β, each such atom in solution will add $Z_\beta - 1$ additional electrons to the lattice for each noble atom it replaces. The electron:atom ratio will be greater than unity. When this is the case, the Fermi level would be expected to rise [Eq. (4-9)]. This is shown schematically as $E_{F_1} > E_{F_0}$ in Fig. 7-6(b). The assumption is made that the shape of the band, or zone, is virtually unaffected by the presence of dilute amounts of alloying atoms occupying sites in the lattice of the noble metal [Figs. 4-19 and 4-22]. The valence electrons from the alloying atoms are considered to occupy states in the valence band of the host lattice. This is known as the *rigid-band model*.

The situation in which the added element contains fewer valence electrons than the host atom is shown schematically in Fig. 7-6(c). Here the solute atoms will add fewer electrons than would have been the case if the lattice sites had been occupied by the original noble metal atoms. Each such substitutional solute atom present will create a deficit of $1 - Z_\beta$ electrons. The electron:atom ratio will be less than unity. Thus, the Fermi level would be expected to decrease. This is shown schematically as $E_{F_2} < E_{F_0}$ in Fig. 7-6(c), again assuming that the band, or zone, remains unchanged.

In either of these cases, the increase or decrease in the Fermi level is a function of the amount of solute element in substitutional solid solution in the noble metal. These changes in E_F may be calculated by the use of Eq. (4-9).

7.1.14.3 Effects of alloying elements in solid solution in transition metals.   The differences in behavior make it convenient to consider this class of alloys as being composed of two subgroups: dilute and concentrated solid solutions.

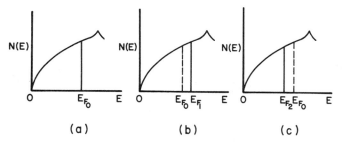

Figure 7-6 Densities of states for dilute alloys of monovalent normal metals.

Dilute Solutions.   The assumptions are again made that the rigid-band model may be used and that dilute amounts of alloying elements in substitutional solid solution in the lattices of transition metals do not significantly change the shape of the bands of the "pure" elemental lattice.   In the case of transition metal solvents, account must be taken of the hybridized s and d states.   The overlap of these bands is largely responsible for their properties (Sections 3.11 and 7.1.9.3).   A typical schematic diagram for the density of states of a transition element is given in Fig. 7-7(a).   Here the s–d band overlap is indicated clearly.   The energy range from E_{F_0} to E_0 is the range of d-level holes, or absence of electrons.   As approximate examples, Fe has about 2.2 such holes per atom, Co has 1.7, and Ni has 0.6 in the crystalline state (see Section 3.11).

When normal atoms with completed inner electron levels, such as noble metals, are in substitutional solution in the lattices of transition metals, their valence electrons will tend to fill the d-level holes of the transition atoms.   This decreases the energy span at the top of the d band: $(E_0 - E_{F_1}) < (E_0 - E_{F_0})$ as shown by comparing Figs. 7-7(a) and (b).

Transition elements in substitutional solid solution in the lattices of other transition elements can have the effect of adding holes to those already present.   This occurs when the number of unoccupied d states of the alloy atom is larger than that of the host atom.   The energy range at the top of the band increases and $(E_0 - E_{F_2}) > (E_0 - E_{F_0})$.   This is shown by a comparison of Figs. 7-7(a) and (c).   Figures 7-7(b) and (c) represent the application of the rigid-band model to alloys of transition elements.

In a given *dilute* alloy, $(E_0 - E_F)$ is very nearly constant.   This comes about because E_F is virtually constant with temperature [Eq. (7-114)], and $(E_0 - E_F)$ is reasonably large, that is, very much greater than $k_B T$.   It will be seen that small changes in E_F will cause only small changes in $N_d(E)$ (Fig. 7-7).   Thus, any minor variations in $N_d(E)$ in Eq. (7-100) will be negligible when included in Eq. (7-95).   This is true for concentrations in the range of up to about 20 At%.   The verification of this behavior is demonstrated in Sections 7.1.19.1, 7.1.19.2, 7.1.20, and 7.1.22.1 through

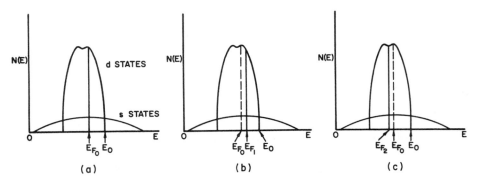

Figure 7-7   Densities of states for dilute alloys of transition metals.

7.1.22.4, where it is used to explain the thermoelectric properties of many of the thermoelements in common use.

Concentrated Solutions of Noble Metals in Transition Elements.   In concentrated, substitutional, solid-solution alloys of noble metals in the lattices of transition metals such as Cu–Ni, Ag–Pd, and Au–Pd, in which the ratio of the noble to the transition metal approaches 60 : 40, the d levels approach maximum filling using the rigid-band model.   This is shown schematically in Fig. 7-8.   This maximum filling of the hybridized d band implies that $(E_0 - E_F)$ is quite small.   In this case, the small variations in E_F as a function of temperature cannot be neglected, as was done for the dilute alloys, because they may now constitute an appreciable percentage of $(E_0 - E_F)$.   More importantly, it will be recalled (Section 7.1.10.2) that the approximation for the ATP of the transition elements was based on the density of states, $N_d(E)$ of the d band [Eq. (7-100)].   As indicated in Fig. 7-8, a small change in E_F, in the range in which $E \to E_F$, causes a relatively large change in $N_d(E)$ as a result of the very large slope in the range $E_F \to E_0$.   Thus, the small variations in E_F must be considered, and $(E_0 - E_F)$ *cannot be regarded as being a constant for any given concentrated alloy.*

Another departure from the prior discussions resides in the fact that it is no longer possible to assume that the shapes of the s or d bands are virtually unaffected by the presence of the solute atoms, as was the case for dilute solutions.   This model can only attempt to show the extent to which the d levels appear to be unfilled; it cannot provide information regarding the rest of the band (see Sections 7.1.8.2, 7.1.9.3, and 7.1.10.2).

The band structure is more involved than it appears to be on first consideration. The complexity arises from the overlapping s and d bands and the resulting hybridization of the s and d states.   The distributions of the electrons between the s and d states may be approximated statistically.   One simplifying approach has been to consider that the electrons of solute and solvent atoms share common s and d bands in alloys such as are under consideration here.   This leads to the rigid-band model shown in Fig. 7-8 and discussed earlier in this section.

The rigid-band model permits a readily visualized approximation of the electron

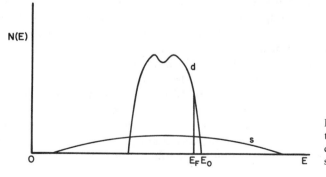

Figure 7-8   Schematic representation of the densities of states for annealed, concentrated, substitutional solid solutions in transition elements.

behavior of alloys as the d band fills. When E_F approaches E_0 of the shared d band, the value of the Fermi function [Eq. (2-86)] decreases rapidly and becomes very small. This means that the probability of finding any additional electrons (from the presence of increased amounts of normal metal alloying ions in solid solution in the lattice of the transition element) in the shared d band rapidly decreases and becomes very small. The probability of finding such electrons in the shared s states increases accordingly and becomes relatively large because of the s–d band overlap. Thus, a critical electron concentration is reached at which the shared d levels become filled to a maximum degree. It appears that such d levels do not become completely filled. At electron concentrations greater than that of the critical, or maximum, d-level filling, the electron behavior becomes increasingly more s-like as more electrons are added. As increasing s-like behavior occurs, the magnitude of $(E_0 - E_F)$ increases and the ATP decreases. These effects are also discussed in Sections 7.1.18.2 and 7.1.23, and the following.

In cases involving transition solutes where

$$(E_0 - E_F)_{\text{solute element}} > (E_0 - E_F)_{\text{solvent element}}$$

the presence of such solute elements has the effect of causing

$$(E_0 - E_F)_{\text{alloy}} > (E_0 - E_F)_{\text{pure solvent}}$$

This behavior increases with increasing concentrations of solutes up to a limiting concentration. Its effect is to cause the ATP of alloys to become increasingly less negative as a function of the solute concentration. The ATPs of such alloys may show positive values of ATPs at concentrations at which conduction by d-level holes predominates over that of the valence electrons. This behavior is limited by a maximum addition of such d-level holes to the hybridized d band. Thus, in a way analogous to that described for the maximum acquisition of electrons, only a critical, or maximum, number of holes may be added to the d band. This is another way of saying that a limited number of electrons may be removed from the d band without modifying its structure. This is discussed in Section 7.1.18.2.

The strong influence of chemical composition on E_F and therefore on $(E_0 - E_F)$ makes it imperative that extreme care be taken in the commercial melting and fabrication of thermocouple alloys (Section 7.1.16). On the bases of the mechanisms described here and in the preceding sections, it should be apparent that what appear to be innocuous amounts of impurity, or "tramp", elements, which may be picked up during normal commercial melting and fabrication processes, can act cooperatively to affect E_F and $(E_0 - E_F)$ and result in the production of off-calibration materials. Tramp elements of the order of 0.01 wt%, or less, have become an increasingly important problem in thermocouple irons (Type JP thermoelements). This problem can be more serious in platinum and its thermoelectric alloys. It has long been common practice to guard against the surface contamination of these materials by very small amounts of iron (usually of the order of a few parts per million). In this case, the iron contamination takes place during the fabrication of the wires.

7.1.15 Effects of Stress or Cold Working

The effects of stress or mechanical working are considered in terms of the lattice distortions that they cause. The resulting lattice disarray warps, or distorts, the bands and the Brillouin zones and increases the scatter of electrons (see Sections 3.12.4 and 4.4–4.6). This results in changes in the Fermi level and is observed easily by changes in the ATP.

The types of stress, their magnitude, and their distribution have long been known to affect the thermoelectric properties in a general way. The simplest case, that of nearly homogeneous, uniaxial stress effects, has been described in the literature. Unfortunately, the stresses normally encountered in practice are much more complex than this.

The residual stress distributions that are usually present in metals and alloys after cold working or straining are mixed tensile and compressive, with at least some degree of triaxiality. Under such conditions, the changes in the Fermi level are very difficult, if not impossible, to assess, in terms of the stress distributions, on a fundamental basis. The intricacy of the variables and their distributions in anisotropic, polycrystalline metals and alloys are such that their interactions make the separation of their effects extremely complex. Their lumped effects have been ascribed as representing different "states". The differences between such states are considered to depend on the specific methods of working and the annealing regimes to which the metal had been subjected. It usually is not possible to improve on such a description.

This becomes increasingly apparent when the deformation textures of metals and alloys are examined using pole figures obtained by x-ray analyses. These vary widely, depending on the method and amount of deformation. Under such conditions, even a moderately stressed or cold-worked wire will give an emf relative to annealed wire of identical composition. This indicates that the Fermi level has changed as a result of the stored energy in the deformed lattice. In other words, the law of homogeneous conductors does not hold for this case (Sections 7.1.3 and 7.1.6).

The improper handling and/or installation of thermocouples in practical applications can result in calibration errors that originate from the introduction of internal stresses. Thermocouple wires as furnished by their manufacturers normally are in their optimum conditions for reliable service. The inadvertent introduction of stresses commonly results from the straightening, bending, or twisting of the larger-gage wires or from the pulling finer-gage wires through tight conduits. It is relatively easy to exceed their yield strengths. The presence and subsequent relief of such stresses both at ambient and at elevated temperatures will cause thermoelectric instabilities (Section 7.1.16).

The presence of residual internal stresses and the accompanying stored energy can be eliminated by suitable annealing. Thermocouple materials must be annealed at temperatures for times that will result in virtually stress free, recrystallized, equiaxed grain structures. Such treatments usually are performed in protective atmospheres to avoid any chemical reactions or composition changes in the thermoelements. The virtually random orientations of the vast number of grains that constitute a properly annealed thermoelement make it possible to consider it as being isotropic and to use the equations for normal metals and transition elements and their alloys as summarized in Table 7-3.

It must also be remembered that, even under optimum annealing conditions, polycrystalline metals and alloys are far from being composed of perfect grains; they will contain dislocation densities of about $10^8/cm^2$. So even under the best of conditions the thermoelectric properties will be influenced by the presence of local lattice imperfections and grain boundaries.

7.1.16 Major Causes of Thermoelectric Instability

The descriptions of the effects of composition in previous sections have been based on ideally pure, homogeneous, stress-free, random, isotropic solid solutions. These are considered to be at thermodynamic equilibrium over the entire temperature ranges of their applications. It is not possible to achieve this state of purity or perfection even under the best of laboratory conditions. The problems are magnified under the conditions of commerical production. Here, melts of the order of 1000 kg are not uncommon, as compared to research-sized melts of about 1 kg. However, the ideas presented in the prior sections make it possible to understand and to deal with most of the practical problems.

In the first place, none of the metallic components of the alloys are perfectly pure when they are added to the metals; all contain some small amounts of impurity elements. Very high purity alloy components are not available in the quantities needed for commercial production. Even if they were available, their costs would increase the prices of thermocouples to a level that would make their commercial use prohibitive. Thermocouple manufacturers, therefore, make use of available alloying ingredients, with known impurity contents, and take these into account in the formulation of the various alloys (Section 7.1.14.3). This problem is regularly kept under reasonable control.

A much more difficult problem is that of making homogeneous wires. It is relatively easy to achieve virtually homogeneous melts of these alloys, especially when they are small and made in high-frequency, induction furnaces in vacua or in protective atmospheres. The problems of chemical nonhomogeneity have their origins not so much in the melting as in the solidification process. In normal, commercial practice, the as-cast structures of ingots consist mainly of dendrites (relatively large crystals with treelike shapes). The chemical compositions of the dendrites vary from their centers to their surfaces; the centers contain more higher-freezing alloy, while their outer portions are enriched in lower-freezing alloy. In addition, on a much larger scale, similar chemical variations occur within an ingot from its surface to its center and from its bottom to its top. This chemical variation is known as *segregation*. Such segregation is decreased by the thermal homogenization treatments of the ingot prior to its reduction to billets. Additional homogenization occurs during the hot working of the billets down to rods and during the intermediate annealing in the wire-reduction process. These thermal and mechanical treatments very largely reduce the chemical nonhomogeneity, but it is thermodynamically impossible to eliminate it completely in reasonable times under commercial conditions.

Any small chemical segregations present in thermoelements will tend to minimize themselves when the thermoelement is heated. Significant changes can occur by diffusion processes of the kinds noted in Section 6.4.3, in the absence of electric fields, because of the low bonding energies of metals. The extent to which these take place depends on the temperatures at the measuring junctions, the thermal gradients in the wires, the local composition differences, and the times of exposure. The magnitudes and signs of the resulting changes in thermoelectric properties are not predictable because the factors just noted are highly variable with respect to a given lot of material as well as between applications. This behavior is one of the most important factors affecting the accuracy of thermoelectric measurements.

Instabilities in thermoelectric properties resulting from chemical nonhomogeneities may become apparent after a few thermal excursions. These usually are readily discernible and the thermocouples can be replaced. Of much greater importance are those changes that take place at very slow rates over long periods of time. These often are difficult to detect over short periods, but their cumulative effects during continuous, long-time use at elevated temperatures, if not monitored, can have disastrous consequences.

Good practice requires the periodic checking of the accuracy of thermocouples and their replacement. The changes noted make it desirable to test thermocouples in position, without changing the extent of their insertions. When they must be removed for testing, they should be replaced at their original insertion distances. Shorter insertions can result in changes in readings and instabilities. This problem may be avoided by the periodic replacement of thermocouples after times that *specific experiences* have shown them to be approaching the limits of acceptable accuracies. The useful life of a thermocouple will vary with each application.

The effects of compositional changes in thermoelements resulting from some environmental factors are given in Sections 7.1.19.3, 7.1.22.4, and 7.1.22.5. It is considered that failure to account for the influences of such factors probably constitutes the largest single source of instability and error in thermocouples.

Neutron irradiation will affect the thermoelectric properties of all the standard thermocouples (Table 7-4). These undergo changes in their thermoelectric properties as a result of two simultaneous processes. One of these changes is caused by transmutation effects. The resultant changes in composition, even when small, can have significant effects (Section 7.1.14.3). The other change is caused by neutron-induced lattice imperfections and distortions. Its effect on a thermoelement is analogous to that caused by cold work (Section 7.1.15). Type K thermocouples appear to be the least sensitive to neutron irradiation. This is explained in Section 7.1.20.

Thermocouples should be replaced immediately whenever any doubt arises regarding their accuracy. The price of thermocouples is small compared to the costs of rectification of some of the problems that can arise from erroneous temperature readings. Spare thermocouples should always be readily available for the immediate replacement of those in service.

The presence of internal stresses in thermoelements, either as a result of incomplete annealing or of their inadvertent introduction, will affect their properties (Section 7.1.15). This factor also is important in incompletely annealed, sheathed thermocouples that have been swaged or cold drawn down to small diameters. In some cases these are continuously annealed during manufacturing, and the actual times and temperatures experienced by the thermoelements can be insufficient to produce recrystallized, equiaxed, nearly stress free grain structures. When such annealing times are too short, thermoelectric instabilities will result (Sections 7.1.14.3 and 7.1.15). Their thermoelectric properties will change as functions of the amount of residual stress, temperature, temperature gradient, and time.

This effect frequently is neglected in dealing with extension wires (Section 7.1.22.6). Any internal stresses in the wires will slowly relieve themselves at room temperature over long periods of time. The result is a continuous drift in the thermoelectric properties of the extension wires and corresponding changes in the accuracy of the temperature readings.

Many extension wires are exposed to temperatures of about 200°C at the point at which they are connected to the thermoelements. This relatively high temperature and the consequent thermal gradients in the wires will not only tend to relieve some of the internal stresses, but will also affect the chemical nonhomogeneities. Simultaneous changes of these kinds are also readily apparent in the inaccuracy of the temperature readings and should be minimized. The thermal gradients cannot be avoided, but the internal stresses in the extension wires should be kept as low as possible by careful manipulation during their installation, as well as during their use.

7.1.17 Thermoelectric Behavior of Alloys

The ATP of alloys will be discussed in a general way. The alloy groupings previously employed will be used here.

7.1.17.1 Alloys of noble metals.

In preceding discussions, it was assumed that the shape of the zone was virtually unaffected by the presence of alloying elements in dilute solid solution. From this it was shown that the Fermi level is a function of the composition of the solution [Section 7.1.14.2 and Eq. (4-9)]. As discussed in Section 4.2, the composition determines the way in which the Fermi level is affected because it determines the electron:atom ratio [Eq. (4-9)]. Under these conditions, the Fermi level will rise or fall depending on whether the electron:atom ratio increases or decreases. As indicated in Table 7-3, all the thermoelectric relationships show that S is expected to be sensitive to variations in the Fermi level. Thus, electrical conductivity as a function of the Fermi level, $\sigma_e(E_F)$, is expected to be proportional to the electrical conductivity as a function of composition, $\sigma_e(C)$.

$$S \simeq \text{Constant}_1 \frac{\partial}{\partial E} [\ln \sigma_e(E)]_{E=E_F} \qquad (7\text{-}115)$$

When the proportionality between $\sigma_e(E_F)$ and $\sigma_e(C)$ is taken into account, it may be approximated [Eq. (7-95)] that

$$S \simeq \text{Constant}_2 \frac{\partial}{\partial C} [\ln \sigma_e(C)] \tag{7-115a}$$

When the indicated operation is performed and note is made that the electrical resistivity, $\rho_e(C)$, is the reciprocal of the electrical conductivity,

$$S \simeq \text{Constant}_3 \, \rho_e(C) \frac{d\sigma_e(C)}{dC} \tag{7-116}$$

where both factors are functions of C, the amount of alloying element in the solid solution in units of atom percent.

For silver-based alloys, the constant in Eq. (7-116) has been found to be approximately

$$\text{Constant}_3 \simeq 0.3(7 - Z_\beta) \tag{7-117}$$

where Z_β is the valence of the solute element. Equations (7-116) and (7-117) take into account the valence and amount of the alloying element in the solution. They also permit the approximation of the ATP of alloys from other electrical data.

It should be noted that Eq. (7-116) gives negative values for S. This results from the fact that $d\sigma_e(C)/dC$ is negative (Section 6.1.4). This effect is shown schematically in Fig. 7-9.

In Fig. 7-9, the ATP of the "pure" noble metal is shown to be small and positive. Increasing amounts of alloying additions cause the alloys of the noble metal to show increasingly more negative values of S. This follows because of the ways in which alloying elements in solid solution affect the variable factors in Eq. (7-116). The electrical resistivity, $\rho_e(C)$, increases with composition, and $d\sigma_e(C)/dC$ consequently becomes more negative.

Data show that noble atoms in solution in other noble metals cause only a slight decrease in ATP. This behavior is understandable on the basis that the electron:atom ratio remains unchanged and that small changes in the shape of the band or zone give rise to the observed, small ATP. Equation (7-116) may also be used

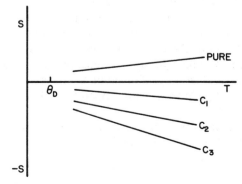

Figure 7-9 Schematic diagram for the effects of annealed, dilute, binary alloying additions in substitutional solid solution in monovalent metals; $C_1 < C_2 < C_3$.

to explain this. The change in the resistivity is expected to be small. Consequently, the change in conductivity as a function of composition would also be small. Under these conditions, S would be expected to be small and negative.

Multivalent solute elements in substitutional solution in lattices of noble metals should increase the Fermi level as shown in Fig. 7-6(b). This increase in E_F should cause a decrease in the ATP [Eqs. (7-40), (7-64b), and (7-98)].

Transition elements in substitutional solution in the lattices of noble metals should give the greatest changes in the ATP. These elements, with partially filled d levels, are strong scattering centers [Section 6.1.3]. They therefore exert large effects on $\rho_e(C)$ and $d\sigma_e(C)/dC$ [Eq. (7-116)].

In terms of equations of the type of Eq. (7-40), the presence of alloying atoms of the transition elements in solid solution in the lattices of noble metals should cause electrons that would normally occupy s states to occupy some of the vacant d states belonging to the transition atoms. This effectively decreases the number of s electrons per atom and lowers the Fermi level, as shown in Fig. 7-6(c). A larger (more negative) ATP results. It would be expected that the more d "orbitals" that are available, the greater would be the change in ATP.

7.1.17.2 Alloys of multivalent normal metals.

Substitutional normal metal alloy solutes in the lattices of other multivalent normal metals, whose inner electron shells are completed, behave similarly to noble metal solid solutions. Solute atoms with valences less than that of the solvent metal lower E_F and increase the ATP. Solute atoms with valences greater than that of the solvent metal increase E_F and thus decrease the ATP [Eqs. 7-6(b) and 7-6(c)].

The presence of transition ions in substitutional solid solution in the lattices of normal multivalent solvents also results in behavior similar to that noted for noble metal solvents. These solutes provide d "orbitals" for the s and p electrons of the solvent atom and effectively lower the value of E_F. This is evidenced by increases in the magnitude of the thermoelectric power.

The effectiveness with which a normal metal atom is able to contribute electrons to affect E_F cannot always be based on its valence in the ground state (see Section 3.10). For example, magnesium in the ground state has a valence of two. In contrast to this, when Mg is in solid solution in copper-based alloys, it appears to have an effective valence of about 0.6. This may vary somewhat, depending on the kinds and quantities of other ions also present in the solid solution. Such a small effective valence has very important practical applications, because Mg is frequently used to deoxidize melts of copper-based thermocouple alloys. A small excess of any deoxidant is always allowed to remain in order to ensure the maximum removal of oxygen from the melt. The small amount of residual Mg in the solid solution has an almost negligible influence on E_F, and consequently on the ATP, because of its small effective valence.

Aluminum is another commonly used deoxidant for copper-based alloys. It has an effective valence of about 2.85 in these alloys. This is nearly five times as influential in the contribution of valence electrons as is Mg. Thus, a normal, given amount of residual Mg would have an inconsequential effect on the ATP of an alloy, while the same amount of Al would have an appreciable effect on the ATP of the alloy. Influences such as these must be taken into account in the manufacture of thermocouple alloys. Failure to do so will result in thermocouple elements that do not lie within the nationally accepted specifications. This effect is especially apparent in the higher temperature ranges. Refer to the discussion on residual elements in Section 7.1.14.3.

7.1.18 Alloys of Transition Elements

Alloys of transition elements are conveniently described by considering two general classifications. One of these concerns normal metal solute elements, those with completed inner electron shells, including the noble metals. The second case deals with transition elements in solid solution in other transition elements.

7.1.18.1 Normal metal solute atoms.   In this case, the valence electrons from the normal metal solute atoms tend to occupy hybridized d-level "orbitals" of the alloy and to effectively decrease $(E_0 - E_F)$. The ATP, consequently, becomes larger, or more negative. The more effective a given solute atom is in contributing its valence electrons to the d-level orbitals of the alloy, the greater will be this effect (Section 7.1.14.3).

7.1.18.2 Transition metal solutes.   The class of solute atoms may again be divided into two categories for convenience. The first of these consists of those transition atoms with relatively large numbers of available d "orbitals". This group includes such elements as V, Cr, Mo, and W. In solution in other transition elements, they effectively increase the value of $(E_0 - E_F)$ by sharing the common s–d levels (see Section 7.1.14.3). In so doing, they cause the thermoelectric power to become smaller, or less negative.

The other category of solutes may be represented by such elements as Ni, Pt, and Pd whose d levels are more nearly filled. Again, the influence of such elements must be considered in relation to the degree to which they affect the hybridized s and d levels of the alloy. If the solute atom has fewer d orbitals available than the solvent atom, it will decrease the value of $(E_0 - E_F)$ of the alloy and cause the ATP to become larger, or more negative [Eqs. (7-41), (7-71b), and (7-101b)]. If the solute atom has more d "orbitals" than the solvent atom, the value of $(E_0 - E_F)$ of the alloy will increase; the ATP of such an alloy will become smaller, or more positive. This is verified in Sections 7.1.19.2 and 7.1.22.1.

As discussed in Section 7.1.14.3, when the electron concentration approaches that of the maximum filling of the d band, the factor $(E_0 - E_F)$ cannot be treated as a constant. When the electron concentration exceeds this critical maximum, increasing s-like behavior occurs, $(E_0 - E_F)$ increases, and the magnitude of the ATP decreases.

The work of Wang and others* on binary alloys of nickel illustrates the effects of alloying elements on the d band. Their conclusions are quoted here:

> 1. The absolute thermoelectric power of binary alloys of nickel is dependent on the $(s + d)$ electron concentration of the alloy. If the addition of a transition solute atom causes an increase in the electron concentration when compared to the matrix nickel, the absolute thermoelectric power of the alloy becomes more negative than that of nickel. On the other hand, if the addition of the solute atom causes a decrease in electron concentration, the absolute thermoelectric power of the alloy becomes more positive than that of nickel.

> 2. The positive maxima of S of the binary alloys of nickel and a transition element occur at a common electron concentration of about 9.6 $(s + d)$ electrons per atom.

These conclusions are in agreement with the preceding discussion. The case in which the alloy additions increase the electron concentration and cause a more negative ATP is the equivalent of considering that the electrons from the solute atoms cause an effective decrease in $(E_0 - E_F)$ of the alloy as compared to that of nickel. The converse is true for solute atoms that decrease the electron concentration of the alloy. In this case, the solute atoms effectively increase the value of $(E_0 - E_F)$ and result in alloys with ATPs that are more positive (less negative) than that of nickel.

The sign of the positive maxima, noted in the second conclusion, is anomalous. It is suggested by these authors that it may be a result of the predominance of hole conduction. It is of interest to note that the solutes that show the greatest anomalous behavior are Cr, Mo, and W, elements that have relatively large numbers of unoccupied d states. Fe and Co also have this effect, but to a lesser degree than do Cr, Mo, and W.

The appearance of the maxima is thought to result from the mechanism noted in Section 7.1.14.3. The maxima appear to be a result of maximum depletion of d-level electrons. According to these investigators, the maximum number of unfilled d states occurs at an electron concentration of about 9.6 $(s + d)$ electrons per atom. Thus, the value of $(E_0 - E_F)$ is largest at this concentration and S is most positive. Conduction by d-level holes apparently makes an important contribution in this range.

It might be considered that if nickel has about $0.6\,s$ electrons per atom, and that if the $3d$ level can accommodate 10 electrons, the preceding $(s + d)$ concentration would indicate that the maximum number of unfilled $3d$ orbitals that can be induced in nickel by another transition element is one unoccupied d state per nickel atom. This would result in a value for $(E_0 - E_F)$ that is just the equivalent of one $3d$ hole. This is a highly oversimplified picture. It is doubtful that the two-band model can be employed and that the hybridized s electrons of the solute atoms can be neglected. It can only be said with certainty that this behavior occurs when the solute atoms possess significantly greater numbers of unoccupied d states than does the solvent transition element.

*Wang, T.-P., and others, *Acta Metallurgica*, Vol. 14, p. 649, 1966.

7.1.19 Standard, Low-Alloy Thermocouples

The American National Standards Institute and the American Society for Testing and Materials have standardized the most commonly used types of thermocouples as given in Table 7-4. Additional information on the performances of these thermocouples is available in ASTM STP 470B, *Manual on the Use of Thermocouples in Temperature Measurement.*

The American National Standards Institute–American Society for Testing and Materials designations given in Table 7-4 are generic. All thermocouple materials that reproduce the emf–temperature characteristics of a given type as defined by the National Bureau of Standards* are equally acceptable.

All the thermoelements in Table 7-4, with the exception of copper, are transition elements or alloys of transition elements. See Sections 7.1.7, 7.1.8.2, 7.1.9.3, and 7.1.10.2. Copper will not be treated here since the properties of the noble elements have been presented in Sections 7.1.8.1, 7.1.9.3, and 7.1.10.1. Constantans (Cu–Ni) are treated separately because they are high-concentration alloys (Sections 7.1.8.2, 7.1.18.2, and 7.1.23–7.1.23.3).

The approximate compositions of this class of transition thermoelements are given in Table 7-5. The ATPs of these thermoelements are given in Table 7-6; their calculated properties are given in Table 7-7. These alloys are discussed in groups according to their major constituents.

TABLE 7-4 STANDARD THERMOCOUPLES IN MOST COMMON USE

ANSI–ASTM Thermocouple Designation	Positive Component	Negative Component	Temperature Range, °C[a]
Type S	$Pt_{90}Rh_{10}$ (SP)[b]	Pt (SN)	$-50-1767$
Type R	$Pt_{87}Rh_{13}$ (RP)	Pt (RN)	$-50-1767$
Type B	$Pt_{70}Rh_{30}$ (BP)	$Pt_{94}Rh_6$ (BN)	$0-1820$
Type J	"Iron" (JP)	Constantan (JN)[c]	$-210-1200$
Type K	Nickel –Cr alloy (KP)	Nickel –Al alloy (KN)	$-270-1372$
Type E	Nickel –Cr alloy (EP)	Constantan (EN)	$-270-1000$
Type T	Copper (TP)[d]	Constantan (TN)	$-270- 400$

[a]R. L. Powell, and others, *Thermocouple Reference Tables Based on the IPTS-68*, (U.S.), Monograph 125, 1974. These are not the ranges recommended by the ANSI–ASTM specifications.

[b]ANSI–ASTM thermoelement (thermocouple component) designation.

[c]Constantan is a generic name for Cu–Ni alloys.

[d]Usually electrolytic, tough-pitch copper.

*Name changed to National Institute of Standards and Technology (NIST), August 23, 1988.

TABLE 7-5 NOMINAL COMPOSITIONS OF TRANSITION THERMOELEMENTS (ATOM PERCENT)

ANSI Designation: Element:	RN, SN Pt	BN $Pt_{94}Rh_6$	SP $Pt_{90}Rh_{10}$	RP $Pt_{87}Rh_{13}$	BP $Pt_{70}Rh_{30}$	Ni	EP, KP Chromel[a]	KN Alumel[a]	Nicrosil	Nisil	JP Iron[b]
Pt	99.99+	89.2	82.6	78.0	55.2						
Rh		10.8	17.4	22.0	44.8						
Fe						0.005			0.01	0.01	99.5
Co						0.013					
Cu						0.004					
Ni						99.9	89.1	90.4	81.9 min	90.8 min	0.08
Cr							10.3		15.2	0.03 max	0.01
Mn								3.0			0.2
Si						0.014	0.5	2.4	2.8	8.8	0.2
Al								4.2			
C						0.06			0.01 max	0.1 max	
Mg									0.2	0.2	

[a]Chromel and Alumel are registered trademarks of the Hoskins Manufacturing Co. These have been selected as illustrations because of the author's greater familiarity with them. The trademarks and manufacturers of equally acceptable alloys are: EP, KP: Tophel, Wilbur B. Driver Co; T-1, Driver-Harris Co.; ThermoKanthal KP, Kanthal Corp.; and KN: Nial, Wilbur B. Driver Co., T-2, Driver-Harris Co.; ThermoKanthal KN, Kanthal Corp.

[b]Also ThermoKanthal JP, Kanthal Corp.

TABLE 7-6 CALCULATED ABSOLUTE THERMOELECTRIC POWERS OF TRANSITION THERMOELEMENTS ($\mu V/°C$)[a]

ANSI* Designation:	RN, SN Pt	BN $Pt_{94}Rh_6$	SP $Pt_{90}Rh_{10}$	RP $Pt_{87}Rh_{13}$	BP $Pt_{70}Rh_{30}$	Ni	EP, KP Chromel[b]	KN Alumel[b]	Nicrosil	Nisil	JP Iron[c]
Temp., °C											
100	-7.3	-1.4_6	-0.8_6	-0.8_4	-1.1_2	-22.1_0	$+20.8_2$	-20.1_3	$+10.5_5$	-17.2_0	$+10.4_6$
200	-8.7	-2.3_4	-1.5_1	-1.3_6	-1.4_5	-24.2_0	$+21.1_4$	-19.5_5	$+11.0_2$	-18.5_6	$+8.1_6$
300	-10.9	-4.2_3	-3.1_7	-2.9_1	-2.7_8	-26.2_0	$+20.1_6$	-20.5_3	$+10.2_5$	-20.8_8	$+4.6_9$
400	-12.3	-5.3_9	-4.1_6	-3.9_9	-3.4_3	-25.9_3	$+19.6_0$	-21.3_9	$+9.9_9$	-22.4_4	$+2.0_3$
500	-13.7	-6.6_4	-5.2_4	-4.7_7	-4.1_5	-26.0_2	$+18.7_1$	-22.5_7	$+9.4_9$	-24.0_0	$+0.3_2$
600	-15.3	-8.1_2	-6.5_8	6.0_5	-5.1_3	-27.0_3	$+17.3_8$	-24.1_2	$+8.6_4$	-25.7_1	-2.3_9
700	-16.8	-9.5_1	-7.8_4	7.1_8	-6.0_4	-28.3_7	$+15.9_7$	-25.6_4	$+7.7_6$	-27.2_7	-3.8_0
800	-18.4	-11.0_2	-9.2_3	-8.4_8	-7.0_8	-30.0_9	$+14.3_5$	-27.2_4	$+6.7_1$	-28.8_6	—
900	-19.9	-12.4_3	-10.5_2	-9.7_0	-8.0_3	-31.7_8	$+12.7_4$	-28.7_3	$+5.6_9$	-30.2_7	—

[a]Calculations based on data from R. L. Powell, and others, *Thermocouple Reference Tables Based on the IPTS-68*, National Bureau of Standards, (U.S.), Monograph 125, 1974, and N. A. Burley, and others, *The Nicrosil versus Nisil Thermocouple: Properties and Thermoelectric Reference Data*, National Bureau of Standards, (U.S.), Monograph 161, 1978, and values given for platinum by J. Nystron, *Arkiv Mathematic Fysik, 34A*, No. 27, 1, 1948.

[b]See footnote a, Table 7-5.

[c]See footnote b, Table 7-5.

*Name changed to National Institute of Standards and Technology (NIST), August 23, 1988.

TABLE 7-7 CALCULATED PROPERTIES OF TRANSITION THERMOELEMENTS USING EQ. (7-101b)

ANSI Designation:	RN, SN Pt	BN $Pt_{94}Rh_6$	SP $Pt_{90}Rh_{10}$	RP $Pt_{87}Rh_{13}$	BP $Pt_{70}Rh_{30}$	Ni	EP, KP Chromel[c]	KN Alumel[c]	Nicrosil	Nisil	Iron[d]
Property											
$\Delta S/\Delta T$[a]	-1.5_2	-1.3_5	-1.2_1	-1.1_1	-0.8_6	-1.5_3	-1.5_1	-1.4_7	-0.9_5	-1.5_7	-2.2_1
$(E_0 - E_F)$[b]	0.8_0	0.9_0	1.0_1	1.1_0	1.4_2	-0.8_0	0.8_1	0.8_3	1.2_8	0.7_8	0.5_8

[a] $\mu V/°C \times 10^2$ (actual units are $V/°C^2 \times 10^{-8}$).
[b] eV.
[c] See footnote a, Table 7-5.
[d] See footnote b, Table 7-5.

The thermocouples composed of the alloy combinations in common use evolved largely as a result of selection based on practical experience and requirements. Most of them possess three important thermoelectric properties. They have nearly linear emf–temperature characteristics (Section 7.1.7) and they show relatively large thermoelectric powers (Section 7.1.13). In addition, these materials provide reproducible and stable temperature readings *when they are used properly.* One important reason for their electrical stability is that all these alloy thermoelements are relatively stable substitutional solid solutions.

Metallurgical factors responsible for their wide use are concerned with the relative ease of production of these alloys. The compositions of these alloys can be controlled to be within very narrow limits during the melting and casting phases of their manufacture (Section 7.1.14.3). In addition, methods have been developed for the reduction of the ingots to wires so that compositional changes are minimized during fabrication and the residual stresses in the finished wires are at the lowest possible levels (Section 7.1.15).

These thermoelectric and metallurgical qualities combine to make thermocouples a highly cost effective means for temperature measurement and control.

7.1.19.1 Platinum and platinum alloys.

It is expected that this group of alloys should show linear absolute thermoelectric properties as functions of temperature, since the differentiations of Eqs. (7-41), (7-71b), or (7-101b) give

$$\frac{dS}{dT} \simeq -\frac{\pi^2 k_B^2}{6e(E_0 - E_F)} \tag{7-118}$$

It will be recalled from Section 7.1.14.3 that $(E_0 - E_F)$ is virtually constant for dilute alloys. Therefore, Eq. (7-118) is a constant for a given alloy of this type, and S is expected to be a linear function of T with a negative slope.

The ATPs of this group of alloys are given as functions of temperature in Table 7-6 and in Fig. 7-10, in which the data in Table 7-6 and in Fig. 7-10 agree with the behavior predicted by Eq. (7-118), since the curves are linear and negative and have negative slopes. This linearity confirms the prior statements (Sections 7.1.8.2, 7.1.9.3, 7.10.2 and 7.1.14.3) that $(E_0 - E_F)$ is essentially constant for the pure element and for dilute, substitutional, solid solutions of other atoms, of the order of 30%, in the lattice of the transition element. It was also shown that the small variations in E_F have such relatively minor effects on $N_d(E)$ [Eq. (7-100)] in Eq. (7-96) that these may be neglected when considering the ATPs of these dilute alloys. Thus, $\ln \sigma_e(E)$ in Eq. (7-95) is not a significant function of temperature, and $(E_0 - E_F)$ is virtually constant.

7.1.19.2 Effect of rhodium additions.

The addition of rhodium to platinum has the effect of decreasing the magnitude of the ATP. The alloys become increasingly positive with respect to platinum; they are less negative. Rhodium atoms have more *d*-level holes than do platinum atoms. They effectively increase the values of $(E_0 - E_F)$ of their alloys with platinum as compared to that of pure platinum. The

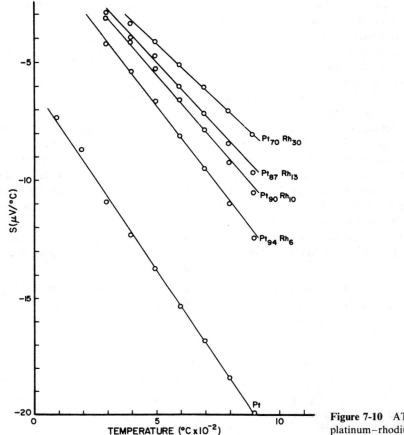

Figure 7-10 ATP of platinum and platinum–rhodium alloys.

data in Table 7-7 confirm this and the behavior described in Sections 7.1.14.3 and 7.1.18.2.

The effective range of $(E_0 - E_F)$ of platinum is increased approximately 0.01_3 eV per atom percent of rhodium present in solid solution in the alloys. This effect may be described by the approximation

$$(E_0 - E_F)_{\text{Pt-Rh alloy}} \simeq 0.8_1 + 0.01_3 C \quad \text{eV} \tag{7-119}$$

where C is the atom percent of rhodium in solution in the alloy.

If a d-band width of about 2.8 eV is assumed for platinum, then values for E_F are obtained that are reasonably consistent with the approximation $E_F \simeq \frac{3}{5} E_0$. This gives $(E_0 - E_F) \simeq (E_0 - \frac{3}{5} E_0) \simeq \frac{2}{5} E_0 \simeq 1$ eV and is another indication that the data for the dilute Pt-based and Ni-based alloys in Table 7-7 are in reasonable agreement with the theory.

Type BP alloys contain nearly 45 At% of Rh and, thus, being concentrated solid solutions, do not follow this approximation for E_F. However, they do obey the approximation given by Eq. (7-119). The high Rh content adds holes to the d band and widens the energy range between E_0 and E_F, and these alloys behave similarly to dilute alloys as shown in Fig. 7-7. The result is that these alloys may also be considered to have essentially constant values of $(E_0 - E_F)$. This is confirmed by the constant slope shown in Fig. 7-10 for the $Pt_{70}Rh_{30}$ alloy.

7.1.19.3 General comments on platinum-based thermocouples.

It will be noted that the curves for these alloy thermoelements (Fig. 7-10) are not quite parallel to each other or to that of platinum. This means that the emfs of the thermocouples that they form with platinum, or with each other, are not exactly linear functions of temperature, as discussed in Section 7.1.7.

Platinum and its alloy thermoelements should not be exposed to vacuum or to reducing atmospheres. In such atmospheres as those containing hydrogen, carbon monoxide, methane, and organic vapors, including those originating from other organic sources or materials, these thermoelements tend to react with and to accelerate the decomposition of the normally stable, refractory ceramics that are used to insulate the thermoelements. The metallic elements of the decomposing ceramics react with platinum and its alloys and usually form stable intermetallic compounds; some solid solutions also may be formed. Readily reducible metal oxides in the furnace atmosphere must be avoided for this reason. In any event, the compositions of the thermoelements are changed permanently. Their Fermi levels and, consequently, the ATPs are irreversibly altered. Similar reactions can also take place between the platinum-based alloys and metallic protection tubes in atmospheres other than reducing atmospheres. Thermocouples of this class should always be sheathed in stable, ceramic protection tubes. These assemblies may be inserted into metallic tubes to guard against possible breakage, but the platinum-based alloys must not be permitted to come into contact with the metallic protection tubes.

Once such chemical reactions start, their continuation results in large progressive errors in the temperature readings. Reactions with ceramics may start at temperatures as low as 400°C. The inaccuracy in the temperature reading will begin to be evident at this point and will increase both with temperature and time at temperature.

Furnace atmospheres containing nonmetallic elements such as phosphorus, sulfur, or arsenic or those in which zinc, cadmium, mercury, or lead are present, either as elemental vapors or as easily reducible compounds, will also react with the platinum-based alloys. The effects of these reactions are similar to those just described.

Platinum-based thermocouples are very stable, provide highly reproducible emfs, and are more reliable than most other thermocouples when properly used in uncontaminated oxidizing atmospheres. They are highly resistant to oxidation. Degrading effects of the kinds noted do not take place. These characteristics have led to the adaption of the $Pt-Pt_{90}Rh_{10}$ (Type S) thermocouple as the means for defining the International Practical Temperature Scale between 630.74°C and the freezing point of gold (1064.43°C).

None of these thermocouples should be exposed to neutron irradiation. The thermoelements containing rhodium change their thermoelectric properties because of the relatively high susceptibility of the transmutation of this element to palladium.

The effective lives of these thermocouples are limited by two factors. Long-time exposures at elevated temperatures can result in the gradual pickup of contaminants. These impurities will diffuse into the wires. This can lead to gradual drifts (instabilities) in their thermoelectric properties that eventually cause them to exceed acceptable limits of error. These long-time, high-temperature regimes can also result in secondary grain growth (also known as secondary recrystallization). Under certain conditions, some of the grains grow by absorbing neighboring grains. The resultant microstructure is one of abnormally large grains. Grain sizes that approach the sizes of the diameters of the wires are not uncommon. (This is the basis of one method of growing single crystals.) Such very large grain sizes render the wires very susceptible to mechanical failure.

Thermocouple Types S and R may be used for short times up to 1769°C, the melting point of platinum. They may be used continuously at temperatures up to 1400° to 1450°C when they are used properly, as noted previously.

The thermoelectric powers of Type B thermocouples are too small for use at temperatures below about 450°C. Their maximum optimum temperature for prolonged operation is about 1700°C, but they may be used intermittently at temperatures as high as 1820°C under suitable conditions.

7.1.20 Nickel and Nickel-Based Alloys

Nickel and its dilute alloys, at temperatures above about 600°C, show thermoelectric properties similar to those of platinum-based alloys (Figs. 7-11 and 7-12). However, the data for these materials show considerable deviations from linearity at temperatures in the range from about 250° to 500°C. For example, considerable attention has been devoted to the idea that the thermoelectric deviations of Type KP (Ni –Cr) alloys could be a result of short-range ordering.[1] However, the x-ray diffraction patterns of alloys of this type fail to show any superlattice lines that would confirm the presence of such ordering.[2] Moreover, such a mechanism cannot explain the thermoelectric properties of elemental nickel that are more anomalous than those of Type KP alloys in this temperature range (Sections 7.1.21, 7.1.21.3, and 7.1.21.4).

It is possible to explain the behavior of both elemental nickel and its thermoelectric alloys by means of a common approach. This is based on the changes in electron spin that occur as a result of any of the magnetic transformations that take place when these materials are heated through a range of temperatures that includes their Curie or Néel temperatures, T_c, T_N, or T_{FN}, respectively, depending on the alloy.

---

[1] R. L. Powell and others, *Thermocouple Reference Tables Based on IPTS-68*, National Bureau of Standards (U.S.), Monograph 125, U.S.G.P.O., Washington, D.C., 1974.

[2] R. Nordheim and N. J. Grant, *J. Institute of Metals, 82*, p. 440, 1953–54.

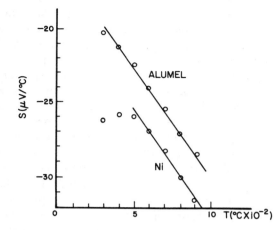

Figure 7-11 ATP of nickel and Alumel (type KN). (From D. D. Pollock and C. P. Conard II, *Trans. Met. Soc. AIME,* *227*, 478, 1963.)

Another advantage of this approach is that it is possible to correlate the thermoelectric deviations with anomalies in other physical properties, which occur within the same temperature regimes and which have been explained by this means. The thermoelectric properties of elemental nickel, which cannot be explained by ordering, are derived in Section 7.1.21.

Rosenberg cites data on the specific heat and thermal expansion of pure nickel that indicate that some of the electronic effects of relatively strong, short-range, magnetic interactions are present at temperatures just below T_c and persist up to above 500°C.[3] This is considerably above the T_c of nickel, which usually is given as 358°C.[4] Such behavior may be explained by a mechanism, treated by Weiss[5] and others, which is presented in a complete but more elementary way by Morrish.[6] It is

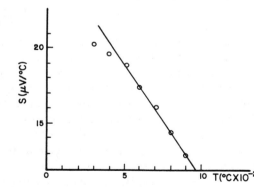

Figure 7-12 ATP of chromel (type KP). (From D. D. Pollock and C. P. Conard II, *Trans. Met. Soc. AIME, 227,* 478. 1963.)

[3]S. J. Rosenberg, *Nickel and Its Alloys,* National Bureau of Standards (U.S.), Monograph 106, U.S.G.P.O., Washington, D.C., 1968.

[4]E. M. Wise, in *Metals Handbook,* 8th ed., Vol. 1, American Society for Metals, Metals Park, Ohio, 1961, p. 217.

[5]P. R. Weiss, *Phys. Rev., 74,* p. 1493, 1948.

[6]A. H. Morrish, *Physical Principles of Magnetism,* Wiley, New York, 1966, pp. 271, 287.

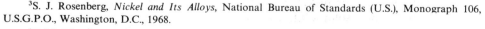

considered that the molecular field, which acts on and aligns the spin magnetic moments in ferromagnetic materials, is rapidly diminished by the increasingly larger amplitudes of the atomic oscillations as T_c is approached and exceeded. The large decrease of the internal field in passing through this range of temperatures results in increasingly random spin orientations. Since ions with parallel spin orientations represent a lower energy and entropy configuration than those with random spin, a fraction of ions with parallel spins will continue to exist as small clusters in the neighborhood of T_c. The number of such clusters and their size appear to approach zero at temperatures significantly greater than T_c. In the temperature range in which such a fraction exists, small numbers, or clusters, of neighboring ions with parallel spins produce small, local internal fields. The magnetic moments of these small groups of ions are parallel to their spins and are independent of the average surrounding magnetization. The persistence of some of these groups, known as *spin clusters*, at temperatures ranging from below T_c to above 500°C has been used as a means for explaining the anomalies in the specific heat and the thermal expansion of nickel.[7] Calculations based on this mechanism are in good agreement with the thermoelectric deviation of Ni (see Section 7.1.21). This model may also be applied to antiferromagnetic and ferrimagnetic alloys.

Chromium ions in substitutional solid solution in the nickel lattice, as in Type KP alloys, have the effect of adding holes to the Ni d band (Sections 7.1.14.3 and 7.1.18.2). They also act to decrease the ratio of half the interatomic distance of closest approach, r_a, to the "radius" of the d "orbits", r_d. The decrease in the r_a/r_d ratio decreases the average magnetic moment per ion in the alloy.[7] This suppresses the T_c of Type KP alloys to below −100°C. Above this temperature the spins appear to be oppositely oriented, so the net magnetic moment is nearly zero and the alloys behave as antiferromagnetic. This condition diminishes with increasing temperature as T_N is approached. Clusters with a small spin imbalance form below T_N. These react in the same way as those described for ferromagnetic materials. These are described in Sections 7.1.21.1 and 7.1.21.4.

A first approximation of these effects indicates that the T_N of Type KP alloys is about 300°C. The decreased effect of spin clusters and the difference between this T_N and the T_c of nickel are reflected in the thermoelectric properties of the Type KP alloy as shown in Fig. 7-12. Thus, its deviations from linearity in the range from about 300° to 500°C are smaller than those of Ni. This approach can also explain the anomalous changes in electrical resistivity reported by Nordheim and Grant to occur between about 280 and 430°C[2] (see Section 7.1.21.3). The value of T_N, obtained from other data, is within this range, and recourse to ordering is not required to explain this behavior. This is derived in detail in Section 7.1.21.4.

The properties of Type KN alloys may also be explained by spin clusters in the same way as for nickel (Section 7.1.21.1). These alloys contain Mn, Si, and Al in substitutional solid solution in the Ni lattice. The Mn ions behave similarly to the Cr ions, as described previously. They add holes to the $3d$ band and appear to decrease

---

[7]D. D. Pollock, *Physical Properties of Materials for Engineers*, Chapters 9 and 12, CRC Press, Boca Raton, Fla., 1982. Also see Section 10.2.1.

the effective number of spin clusters. The Si and Al atoms, with valences of 4 and 3, respectively, contribute their valence electrons directly to the $3d$ band and are very effective in decreasing the average number of unbalanced spins per nickel atom; this is especially true of Si. This results in a decrease in the effective fraction of spin clusters in the vicinity of T_c and a drop in T_c. It is also reflected in the low values of T_c of about 150° to 200°C, depending on composition, as well as in the high degree of thermoelectric linearity of Type KN alloys at and above about 400°C.

This mechanism may be tested by examining the effects of small amounts of alloying additions on the magnetic properties of other Ni-based alloys. One of these is the cast nickel alloy CZ-100.[8] Its nominal composition is quite close to those of some of the commercially available Type KN alloys.[9] The magnetic properties of alloy CZ-100 vary inversely with its silicon content (2 Wt% max.), and it is only slightly magnetic near room temperature.[8] The thermoelectric properties of Nisil, containing nearly 9 At% Si, also indicate that its magnetic transformation has been suppressed (Section 7.1.22.3).

The magnetic effects under consideration here are not trivial. Morrish has shown that they constitute a large percentage of the heat capacity of nickel below, at, and beyond T_c.[10] The thermoelectric significance of this is shown by Eq. (7-104). Thus, any analytic expression for the description of the ATPs of nickel and its alloys, valid over the entire temperature ranges of their applications, must include factors that account for the differences in electron behavior both below and above T_c, T_N, or T_{FN}, depending on the alloy. Equation (7-104) provides a means for the fundamental evaluation of such an analysis.

It was noted in Section 7.1.8.2 that the electron effects responsible for magnetic properties and their contribution to the heat capacity were not taken into account in the derivation of Eq. (7-41). The derivation of Eq. (7-72) (Section 7.1.9.3) also assumed random spin orientations. The same is true for the Mott and Jones analysis (Section 7.1.10) in the use of Eq. (7-95) to derive Eq. (7-101b). None of these equations is valid when ferromagnetic, antiferromagnetic, or ferrimagnetic effects are present.

The decreasing deviations from linearity and the linear thermoelectric behavior of nickel as the temperature increases to above 500°C indicate that the spin clusters act as if they are progressively dissipated by the increased amplitudes of the lattice vibrations as the temperature increases. It appears that all the spins have become random at and above this temperature. The transition from residual parallel spin to random spin provides the basic explanation for the thermoelectric properties below and above about 500°C. Random spin is implicit in Eqs. (7-41), (7-71b), and (7-101b); this is the reason why these equations provide reasonable descriptions of the behavior of nickel at temperatures above 500°C.

As in the case of platinum and its alloys, nickel and its alloys also show linear

[8]D. E. Wenschhof, R. B. Herchenroeder, and C. R. Bird, in *Metals Handbook*, 9th ed., Vol. 3, American Society for Metals, Metals Park, Ohio, 1980, p. 164.

[9]N. A. Burley and others, *Nicrosil versus Nisil Thermocouple: Properties and Thermoelectric Reference Data*, National Bureau of Standards (U.S.), Monograph 161, U.S.G.P.O., Washington, D.C., 1978.

[10]A. H. Morrish, *Physical Principles of Magnetism*, Wiley, New York, 1966, p. 177.

thermoelectric properties (above 500°C) with negative slopes. This provides confirmation of the statements made in Sections 7.1.8.2, 7.1.14.3, and 7.1.18.2. It will be noted (Table 7-5) that most of the nickel-based alloys contain about half, or less, of the alloy content present in the platinum-based alloys. Both of these alloy groups verify the approximation for the essential constancy of the value of $(E_0 - E_F)$ for dilute solid solutions made in the previously noted sections.

Both Type KP and Type KN alloys are relatively insensitive to *small* variations in composition or in impurity content (Sections 7.1.14.3 and 7.1.16). This is not so much a consequence of the quantities of their alloy contents as it is a result of the intersections of E_F with $N_d(E)$ at relatively flat portions of the density-of-states curve (Fig. 7-7). Equation (7-100) remains virtually unaffected by such small variations of $N_d(E)$ and, consequently, Eq. (7-101b) is unchanged (Sections 7.1.10.2, 7.1.14.3, and 7.1.18.2).

The similarities between the values for $(E_0 - E_F)$ and the thermoelectric slopes of nickel and platinum are consistent with the evidence that both of these elements lack approximately the same number of electrons in their d bands and that their d bands extend over nearly equal energy ranges.[11,12] For illustrative purposes, such similarity of behavior should make this pair of thermoelements desirable for use at temperatures above 500°C. The data (Table 7-7) suggest that a Pt–Ni thermocouple could approach the ideal as represented by Eq. (7-24). One convenient way of using such a thermocouple is to design the temperature measuring/control instrument with a suppressed zero. This automatically accounts for the emf that the thermocouple shows when its reference junction is at 0°C and its measuring junction is at 500°C. This instrument would begin to indicate and/or control at temperatures of 500°C or higher. The thermoelectric power of this thermocouple would be much larger than those of the other platinum-based thermocouples, and its emf would be very close to being a linear function of temperature between 500° and 1400°C (Section 7.1.7). It would have all the desirable thermoelectric properties noted for platinum-based alloys noted in Section 7.1.19.3. However, the performance of a Pt–Ni thermocouple would be limited by the properties of the nickel thermoelement. When used in air at temperatures above 1000°C, it would be unstable and short lived as a result of oxidation.

The nonlinear behavior of the ATP of nickel at low temperatures could also be minimized if another alloy thermoelement could be found that demonstrated nearly parallel thermoelectric properties (Section 7.1.7). Such attempts have not won wide acceptance. Pure nickel is not normally employed as a thermoelement for this reason.

[11]J. Callaway, in *Solid State Physics*, F. Seitz and D. Turnbull, eds., Vol. 7, Academic Press, New York, 1958, p. 182.

[12]N. F. Mott and H. Jones, *Theory and Properties of Metals and Alloys*, Dover, New York, 1958, p. 199.

7.1.21 Effects of Magnetic Transitions on Nickel and Its Alloys

A general description of the thermoelectric behavior of nickel and its important thermoelectric alloys is given in Section 7.1.20. The discussion in that section is based on the idea that electron spin clusters can provide a single means for the explanation of the thermoelectric properties of nickel and its thermoelectric alloys. Such an explanation conforms with those presently accepted for the other physical properties of these materials.[1]

This approach is adopted in order to provide a unified thermoelectric model for these materials, without recourse to short-range lattice order–disorder, for which no x-ray evidence has been presented.[2,3]   See Section 7.1.21.3.

The transition elements and their alloys are expected to show the linear thermoelectric properties expressed by

$$S \simeq -A' \frac{\pi^2 k_B^2 T}{6e(E_0 - E_F)} \quad \mu\text{V/deg)} \quad (7\text{-}41), (7\text{-}71b), (7\text{-}101b)$$

Platinum and its alloys show good agreement with this equation (Fig. 7-10). But nickel and some of its thermoelectric alloys show negative deviations, ΔS, from the linear behavior of ATP predicted by the indicated equations. These occur in the range of about 250° to above 500°C (Figs. 7-11–7-14). These materials do, however, conform to the predicted linear behavior at temperatures above this range. In the case of nickel, the Curie temperature, T_c, lies well within the temperature range in which the thermoelectric deviations are observed.

The heat capacities of *nickel and other magnetic materials also show deviant behaviors in the ranges of temperatures that include their transformation temperatures*, T_{tr}. The anomalous heat capacity takes the form of relatively large, continuous increases in the heat capacity that begin at temperatures well below T_{tr} and continue to increase up to T_{tr}. A discontinuity occurs at T_{tr}. The aberrant behavior then smoothly diminishes and disappears at temperatures considerably above T_{tr}. This increment in heat capacity resulting from magnetic effects is denoted by C_M.[4]

Spin clusters have been used to explain this heat-capacity anomaly.[5-8]   These

[1] D. D. Pollock, *Proposed Mechanism for the Thermoelectric Properties of Nickel and Some of Its Alloys near the Curie Temperature*, in *Temperature, Its Measurement and Control in Science and Industry*, Vol. 5, Part 2, J. F. Schooley, ed., American Institute of Physics, New York, 1982, p. 1115.

[2] R. Nordheim, and N. J. Grant, *J. Inst. Met.* (London), *82*, p. 440, 1953–54.

[3] N. A. Burley and others, *The Nicrosil versus Nisil Thermocouple: Properties and Thermoelectric Reference Data*, National Bureau of Standards (U.S.), Monograph 161, U.S.G.P.O., Washington, D.C., 1978, pp. 2, 5.

[4] D. D. Pollock, *Physical Properties of Materials for Engineers*, Vol. II, CRC Press, Boca Raton, Fla., 1982, p. 127 and Section 10.2.3.

[5] H. Bethe, *Proc. Roy. Soc.* (London), *A-150*, pp. 552, 1935.

[6] R. E. Peierls, *Proc. Roy. Soc.* (London), *A-154*, pp. 207, 1936.

[7] P. R. Weiss, *Phys. Rev.*, *74*, p. 1493, 1964.

[8] A. H. Morrish, *Physical Principles of Magnetism*, Wiley, New York, 1966, pp. 271, 287.

analyses may be described in a simple way by considering that at temperatures below, but close to, T_{tr}, the volumes of aligned spins (domains in ferromagnetic materials) break down into very small volumes; these may be considered to be the remnants of the original volumes or domains. Each of these consists of clusters of atoms that still retain their aligned spins. The atoms on the lattice sites surrounding the clusters have random spins. The numbers and sizes of the spin clusters diminish and appear to approach zero at temperatures considerably above T_{tr} (see Sections 10.2 and 10.2.3 and Fig. 10.26).

The present theory is developed starting with the basic relationship for ATP given previously by

$$S = \frac{C_{carrier}}{e_{carrier}} \qquad (7\text{-}104)$$

The differentiation of Eq. (7-104) results in

$$\Delta S = \frac{\Delta C_{carrier}}{e_{carrier}} = \frac{C_{M\,(carrier)}}{e_{carrier}} \qquad (7\text{-}120)$$

Here the change in the heat capacity, $\Delta C_{carrier}$, is considered to result from, and to be equivalent to, one electron spin contribution to the anomalous heat capacity, or $C_{M\,(carrier)}$ (Section 7.1.11).

C_M may be approximated on the basis of the molecular field theory as[9]

$$C_M = -\frac{3JN_A k_B}{2(J+1)} \cdot g(T) \qquad (7\text{-}121)$$

Here J is the total quantum number, N_A is Avogadro's number, k_B is Boltzmann's constant, and $g(T)$ is the temperature variation. According to this theory, C_M is considered to result from the spontaneous spin orientations that are induced by the molecular field. The function $g(T)$ causes C_M to increase continuously up to T_{tr}, in reasonable agreement with the observed behavior. However, $g(T)$ discontinuously drops to zero at T_{tr}. This implies that all aligned spins should become completely random at and above T_{tr}.

This does not agree with the observed behavior. The experimental results show the discontinuity at T_{tr}, but the decrease in C_M is much less abrupt and extends over a relatively wide range of temperatures above T_{tr}. This leads to the conclusions that some spin alignment must persist above T_{tr} and that $g(T)$ is incorrect in this range.

Thus, another temperature dependence is needed that more closely describes the observed behavior in the range of temperatures in which the molecular field theory breaks down. In the range of temperatures at and above T_{tr}, in effect using T_{tr} as a reference temperature, for $J = \frac{1}{2}$ at $T \simeq T_{tr}$, Eq. (7-120) becomes[9]

$$\Delta S = \frac{C_{M\,(carrier)}}{e_{carrier}} = -\frac{C_M}{Z_\alpha N_A e} = -\frac{3R}{2Z_\alpha N_A e} \cdot f(T) \quad \mu V/deg \qquad (7\text{-}122)$$

[9]A. H. Morrish, *Physical Principles of Magnetism*, Wiley, New York, 1966, pp. 271.

in which Z_α is the number of carriers per atom, R the gas constant, and $f(T)$ the temperature dependence for $T \geqslant T_{tr}$. A first-order estimation of this function can be obtained from

$$f(T) \simeq N(T)z(T) \tag{7-123}$$

Since Eq. (7-123) is the product of the approximate temperature variations of the number of clusters, $N(T)$, and the effective number of electron states per cluster, $z(T)$, it provides a reasonable approach to the temperature dependence of the number of electrons involved in clustering.

The function $f(T)$ is approximated by starting with the estimation of the fraction of electrons engaged in clustering. Using T_{tr} as a reference temperature, this may be approximated by

$$f_e \simeq \frac{n(T_{tr})}{n(T)} \tag{7-124}$$

in which $n(T_{tr})$ and $n(T)$ are the numbers of electrons involved in clustering at T_{tr} and T, respectively.[10] The temperature variation of the number of clusters, $N(T)$, may be approximated from Eq. (7-124) and the energy range of the available electrons, $N(E)$:

$$N(T) \simeq N(E)f_e \tag{7-125}$$

Equation (7-125) is based on the approximation that the average number of clusters at a given temperature is expected to be proportional to the number of electrons available for clustering at that temperature.

The factors in Eq. (7-124) may be obtained from Eq. (4-8)[11]

$$N(T) = \text{constant}_1\, VE^{3/2} = \text{constant}_2\, VT^{3/2} \tag{7-126}$$

since all of the other factors in Eq. (4-8) are constant. For a given magnetic metal or alloy, Eq. (7-126) may be written as

$$n(T) \simeq \text{constant}_3\, T^{3/2} \tag{7-127a}$$

or as

$$n(T_{tr}) \simeq \text{constant}_3\, T_{tr}^{3/2} \tag{7-127b}$$

Equations (7-127a) and (b) are valid and may be expressed as given because $n(T)$ varies as $N(T)/V$ which, in turn, varies as $T^{3/2}$. The constants vanish when Eqs. (7-127a) and (b) are substituted into Eq. (7-124) for final use in Eq. (7-125) as f_e.

For a given half-band, using the Sommerfeld equation, $N(E)$ in Eq. (7-125) varies as $E^{1/2}$ [Eq. (4-5)]. Since E is directly proportional to temperature, for nearly free electrons, $N(E)$ may be considered to vary as $T^{1/2}$ in Eq. (7-125).

The substitution of the expressions for $N(E)$ and f_e, obtained as indicated

[10]C. Kittel, *Introduction Solid State Physics*, 3rd ed., Wiley, New York, 1966, p. 460.

[11]A. H. Morrish, *Physical Principles of Magnetism*, Wiley, New York, 1966, p. 299.

previously, into Eq. (7-125) results in

$$N(T) \simeq T^{1/2} \frac{T_{tr}^{3/2}}{T^{3/2}} = \frac{T_{tr}^{3/2}}{T} \tag{7-128}$$

Equation (7-128) serves as a first approximation of the temperature dependence of the number of clusters for use in Eq. (7-123).

The second factor in Eq. (7-123), the temperature variation of the effective number of electrons in a cluster, remains to be developed. This factor may be measured by the effective number of near neighbors in a cluster. This is the case since only the unbalanced spins in a half-band need be considered. And, within a given cluster, the number of these spins is proportional to the number of atoms in that cluster.[12] The effective number of atoms in a cluster will vary as the number of nearest neighbors in a cluster. Thus, the effective number of spins in a cluster may be approximated from the effective number of nearest neighbors, z. This may be obtained by starting with

$$\frac{J_e}{k_B T} \simeq \ln \frac{z}{z-2} \tag{7-129}$$

in the neighborhood of T_{tr}.[13] Equation (7-129) may be reexpressed in exponential form as

$$\exp\left(\frac{J_e}{k_B T}\right) \simeq \frac{z}{z-2}$$

The exponential term is closely approximated by the first two terms of a series:

$$1 + \frac{J_e}{k_B T} \simeq \frac{z}{z-2}$$

This expression simplifies to

$$z J_e - 2 J_e \simeq 2 k_B T$$

Then, upon differentiation, treating J_e as a constant because of the spin alignment, it is found that

$$dz \simeq \frac{2k_B}{J_e} dT \tag{7-130}$$

The temperature variation of the number of states per cluster, then, may be approximated as being proportional to

$$z(T) \simeq \frac{2k_B}{J_e} \int_T^{T\infty} dT \tag{7-131}$$

[12]C. M. Hurd, *Electrons in Metals*, Wiley-Interscience, New York, 1975, p. 27.

[13]A. H. Morrish, *Physical Principles of Magnetism*, Wiley, New York, 1966, p. 290.

The upper limit, T_∞, is defined as that temperature at which all the clusters appear to vanish. According to the molecular field theory, T_∞ should equal T_{tr}. However, experiments show that T_∞ is always greater than T_{tr}.

The number of nearest neighbors in a cluster depends on the lattice type. This factor is taken into account, in a way analogous to that in Eq. (7-126), by including the number of atoms in a unit cell, N_0, in Eq. (7-131):

$$z(T) \simeq \frac{2k_B}{J_e N_0} \int_T^{T_\infty} dT \tag{7-132}$$

It is shown later that the coefficient $2k_B/(J_e N_0)$ is given more conveniently by Eq. (7-138). Also see Eq. (10-199a) and Table 10-4.

The proposed temperature dependence may now be obtained by the substitution of Eqs. (7-128) and (7-132) into Eq. (7-123):

$$f(T) \simeq N(T)z(T) \simeq \frac{T_{tr}^{3/2}}{T} \cdot \frac{2k_B}{J_e N_0} \cdot \int_T^{T_\infty} dT \tag{7-133}$$

The approximate deviation of the ATP of nickel from linearity is finally obtained by the substitution of Eq. (7-133) into Eq. (7-122) as

$$\Delta S \simeq -\frac{3R}{2Z_\alpha N_A e} \cdot \frac{T_{tr}^{3/2}}{T} \cdot \frac{2k_B}{J_e N_0} \int_T^{T_\infty} dT \quad \mu\text{V/deg} \tag{7-134}$$

The following data were used in Eq. (7-134) to describe the thermoelectric deviations of "pure" nickel: $Z_\alpha \simeq 0.54$ carriers/atom, $T_{tr} = T_c = 631$ K $\simeq 633$ K, and $k_B/J_e \simeq 1/200$.[14-16] The data show (Fig. 7-11) that linear behavior is fully established at 600°C, so T_∞ is taken as 873 K. These data reduce Eq. (7-134) to

$$\Delta S \simeq -3.8(873 - T)\frac{(633)^{3/2}}{T} \times 10^{-4} \quad \mu\text{V/deg} \tag{7-135}$$

The values of ΔS, calculated by means of Eq. (7-135), are given in Table 7-8 along with the experimental data.

The calculated values for ΔS for the "pure" nickel are in good agreement with the experimental data for temperatures at and above T_c. This is considered to confirm the simplified approximations of the temperature dependencies given by $N(T)$ and $z(T)$; these were made expressly for this temperature range.

The good agreement for ΔS at 573 K, 58 K below T_c, was unexpected. It appears that Eq. (7-134) permits *short* extrapolations, without the introduction of excessive errors, in the range $T \leqslant T_c$. In addition, it is also of interest to note that, at T_c, Eq. (7-132) gives the number of nearest neighbors in a cluster as being proportional to $z(T)N_0 \simeq 6$, or to one-half those in a ferromagnetic domain within a FCC lattice.

[14]C. Kittel, *Introduction to Solid State Physics*, Wiley, New York, 1966, p. 581.

[15]M. Fallot, *Ann. Phys. (Paris)*, 6, p. 305, 1936.

[16]A. H. Morrish, *Physical Principles of Magnetism*, Wiley, New York, 1966, p. 291.

TABLE 7-8 CALCULATED AND EXPERIMENTAL
VALUES OF ΔS FOR NICKEL $-\Delta S$ (μV/deg)

T (K)	Calculated	Experimental
573	$3._2$	3.5
633	$2._3$	2.4
673	$1._8$	1.8
773	$0._8$	0.7
873	0	0

This is in reasonable agreement with the maximum value of C_M and its curve shape at the discontinuity at T_c. The number of nearest neighbors must decrease for $T_c \leqslant T \leqslant T_\infty$. In this range, $z(T)N_0$ diminishes from less than six nearest neighbors and approaches zero as $T \to T_\infty$; it conforms with the observed behavior of C_M. The agreements of Eq. (7-134) with the experimental data in the ranges $T \leqslant T_c$ and $T_c \leqslant T \leqslant T_\infty$ are a consequence of its continuous nature; it shows no discontinuity at $T = T_c$.

Considering the complexity of the mechanism and the elementary approximations used for its description, the results provided by this model indicate that it may be considered to provide a reasonable explanation of the effects of spin clusters on the ATP of nickel.

The electron effects responsible for magnetic phenomena and their contribution to heat capacity are not taken into account in the derivation of Eqs. (7-41) and (7-71b). The same is true for the Mott and Jones analysis [Eq. (7-101b)]. None of these analyses is valid when any magnetic effects are present.

The decreasing thermoelectric deviations and the linear behavior of nickel as the temperature increases to above 500°C indicate that the spin clusters dissipate progressively as the temperature increases. The net result is that all the spins appear to have become random above this temperature. The transition from parallel spin, as spin clusters, to random spin provides the basis for the explanation of the thermoelectric properties both below and above about 500°C. Random spin is implicit in Eqs. (7-41), (7-71b), and (7-101b); this is the reason why they provide good descriptions of the ATP of nickel at temperatures above 500°C.

This model is based on the anomalous behavior of the heat capacity in the neighborhood of the Curie temperature. It thus provides an explanation for the thermoelectric deviations of nickel that is consistent with and conforms to that generally accepted for its heat capacity and thermal expansion.

7.1.21.1 Alumel.   The spin-cluster model has been shown to provide a good description of the thermoelectric deviations of nickel (see Section 7.1.21). Its application to Alumel extends it to the more complex case involving the effects of alloy additions to the nickel base.

The general relationship for the thermoelectric deviations from Eqs. (7-41),

(7-71b), or (7-101b) are given by Eq. (7-134) as

$$\Delta S = -\frac{3R}{2Z_\alpha N_A e} \cdot \frac{T_{tr}^{3/2}}{T} \cdot \frac{2k_B}{J_e N_0} \int_T^{T\infty} dT \qquad (7\text{-}134)$$

Data for values of J_e for many thermoelectric alloys do not appear to be available, and this limits the utility of Eq. (7-134). This difficulty may be eliminated by the use of a relationship between J_e and T_{tr} that is given by [see Eq. (10-199a)

$$J_e = \frac{3k_B T_{tr}}{2N_0 S(S+1)} \qquad (7\text{-}136)$$

Equation (7-136) may be simplified by recalling that spin alone is responsible for magnetic properties and that $S = \frac{1}{2}$ for this case. This substitution in Eq. (7-136) results in

$$J_e = \frac{2k_B T_{tr}}{N_0} \qquad (7\text{-}137)$$

The rearrangement of Eq. (7-137) reduces the coefficient of the integral in Eq. (7-134) to

$$\frac{2k_B}{J_e N_0} = \frac{1}{T_{tr}} \qquad (7\text{-}138)$$

Equation (7-138) makes it unnecessary to obtain values for J_e, and Eq. (7-134) is simplified by the use of $1/T_{tr}$ as the coefficient of the integral. Upon substitution of Eq. (7-138) and integration of Eq. (7-134), the *thermoelectric deviations of ferromagnetic origin* are given by

$$\Delta S \simeq -\frac{3R}{2Z_\alpha N_A e} \cdot \frac{T_c^{1/2}}{T} \cdot (T_\infty - T) \quad \mu V/deg \qquad (7\text{-}139)$$

Then, when substitutions are made for the constants in Eq. (7-139), it is found that

$$\Delta S \simeq -\frac{8.129 \times 10^{-2}}{Z_\alpha} \cdot \frac{T_c^{1/2}}{T} \cdot (T_\infty - T) \quad \mu V/deg \qquad (7\text{-}140)$$

Equations (7-134), (7-139), and (7-140) are general expressions for the effects of magnetic spin clusters on thermoelectric deviations.

The parameters required for Eq. (7-140) were obtained as follows. The Curie temperature of Alumel ($T_c = T_{tr}$) is given as 170°C, or 443 K.[2] The value for Z_α was obtained from a nominal analysis of Alumel (in atom percent) of Ni: 90.4, Mn: 3.0,

[1] D. D. Pollock, *Physical Properties of Materials for Engineers*, Vol. II, Boca Raton, Fla., 1982, p. 130.

[2] R. L. Powell and others, cited by Burley, N. A., and others, *The Nicrosil versus Nisil Thermocouple: Properties and Thermoelectric Data*, National Bureau of Standards (U.S.), Monograph 161, U.S.G.P.O., Washington, D.C., 1978, p. 7.

Si:2.4 and Al:4.2, with corresponding valences of 0.54, 2, 4, and 3, respectively.[3] These data give $Z_\alpha = 0.77$ carriers per atom. T_∞ was taken as 773 K, since linear thermoelectric behavior appears to be well established at and above 500°C. The calculated values for ΔS [Eq. (7-140)] are shown in Table 7-9.

The ATP of Alumel is shown graphically in Fig. 7-11. Its deviations from linearity become more clearly evident when its ATP is expressed algebraically. The empirical equation of best fit for the linear portion of this curve, from $t = 500°$ to 900°C, is given by

$$S' \simeq -15.14 - 0.015t \; (°C) \quad (vV/°C) \tag{7-141}$$

Equation (7-141) has an average error of less than 0.5%. This equation was used to extrapolate the thermoelectric properties of Alumel below 500°C. These data, S', also are included in Table 7-9.

The calculated values of the ATP of Alumel were obtained from the algebraic sums of the extrapolated values of S' [Eq. (7-141)] and the deviations induced by ferromagnetic effects, ΔS [Eq. (7-140)]. These data are presented in Table 7-9 along with the experimental values of the ATP of Alumel. The differences between the observed and calculated values of S appear to be of the same order of accuracy as that given for Eq. (7-141).

This degree of conformity for Alumel is better than the good agreement previously obtained for "pure" nickel; this had been determined by means of Eq. (7-134). See Table 7-8. The calculations for the properties of elemental nickel included a value for k_B/J_e that had been approximated from several sets of experimental data. The improved agreement obtained here is considered to result from the use of Eq. (7-138). However, irrespective of the different methods used for the calculations, the spin-cluster model provides results that are in good accord with the experimental data in both cases.

The electron spin influences responsible for magnetic effects and their influence on the heat capacity were not taken into account in the derivation of Eq. (7-41) or (7-71b). The effects of these factors were also neglected in the Mott and Jones analysis that is based on electrical conductivity [Eq. (7-101b)]. The calculated values of ΔS for

TABLE 7-9   CALCULATED AND EXPERIMENTAL DATA FOR ALUMEL

t (°C)	T (K)	S' Eq. (7-141) (μV/°C)	ΔS Eq. (7-140) (μV/°C)	$S_{calc.}$ $S' + \Delta S$ (μV/°C)	$S_{exp't.}$ (μV/°C)
170	443	−17.7	−1.7	−19.4	−19.4
200	473	−18.1	−1.4	−19.5	−19.6
300	573	−19.6	−0.8	−20.4	−20.5
400	673	−21.1	−0.3	−21.4	−21.4
500	773	−22.6	0	−22.6	−22.6

[3]C. Kittel, *Introduction to Solid State Physics*, 3rd ed., Wiley, New York, 1966, pp. 462, 581.

both nickel (Section 7.1.21) and Alumel substantiate the proposition that such electron influences are responsible for the nonlinear thermoelectric properties of the materials at temperatures below T_∞. The apparent absence of such electron behavior at temperatures higher than T_∞ explains the conformance of these materials to the thermoelectric models cited previously.

The decreasing degree of nonlinearity of the ATP of Alumel as the temperature approaches 500°C is considered to indicate that the effects of the spin clusters are being progressively dissipated by the increasing amplitudes of the lattice vibrations. The net result is that the effects of all the clusters appear to have vanished at $T_\infty = 500°C$. The linear ATP of Alumel above this temperature then conforms to that described by the previously noted theoretical models because any remaining spin clusters may be regarded as being uncorrelated and their thermoelectric influence effectively eliminated.[4]

The use of the anomalous heat capacity as a basis for this model has significant consequences. It constitutes a unified model for the explanation of ΔS for nickel, as well as Alumel. And, of prime importance, it provides results that are in conformance with other physical properties, with deviations in the vicinity of T_c, that have been explained by spin clusters.[5,6]

7.1.21.2 Nicrosil.

The chromium content of Nicrosil suppresses the ferromagnetic behavior of nickel. This alloy is considered to show antiferromagnetic behavior at room temperature and, as the temperature increases, to go through a Néel transition beyond which temperature it is paramagnetic. Very strong evidence for this behavior is given by the observation of magnetic susceptibility peaks in Ni–Cr alloys whose Cr contents are greater than that required to suppress ferromagnetism in Ni.[1]

Antiferromagnetic behavior provides the most probable explanation of the thermoelectric properties of these high-chromium alloys. Averbach's model may be extended to explain this behavior.[2] These alloys are considered to consist of *oppositely oriented* clusters of ions with parallel spins in a lattice matrix of ions with random spins. Such an antiparallel array of spin clusters results in a negligibly small magnetic susceptibility at or near room temperature. Here, the "correlation", or molecular field interaction, between clusters is sufficiently strong so that any resultant

[4]B. L. Averbach, "Spin Clusters in Iron near the Curie Temperature" in *Magnetic Properties of Metals and Alloys*, American Society for Metals, Metals Park, Ohio, 1959, p. 280.

[5]D. D. Pollock, *Physical Properties of Materials for Engineers*, Vol. II, CRC Press, Boca Raton, Fla., 1982, p. 127.

[6]A. H. Morrish, *Physical Principles of Magnetism*, Wiley, New York, 1966, p. 287.

[1]R. J. Weiss and K. J. Tauer "A Metallurgical Slide Rule for Determining Magnetic Properties of 3d Transition Alloys," in *Theory of Alloy Phases*, American Society for Metals, Metals Park, Ohio, 1956, pp. 290–300.

[2]B. L. Averbach, "Spin Clusters in Iron near the Curie Temperature," in *Magnetic Properties of Metals and Alloys*, American Society for Metals, Metals Park, Ohio, 1959, pp. 280–287.

magnetization is quite small and difficult to detect. Increasing temperatures above the neighborhood of room temperature decrease the correlation between the antiparallel arrayed clusters and cause small, increasing degrees of spin imbalance between clusters. Consequently, as the temperature is increased, the alloys gradually show increasing magnetizations, reach relatively small, maximum magnetizations, and therefore show susceptibility peaks. They then show a typical Curie–Weiss decay of magnetization. This antiferromagnetic behavior is virtually independent of external fields.[3,4] The maxima occur as cusps or peaks at the Néel temperature, T_N. At temperatures above T_N, the magnetizations decrease continuously as the correlations between clusters diminish. At these higher temperatures, the alloys are increasingly paramagnetic as a result of the increasingly diminishing correlations between clusters.

Since magnetic susceptibility peaks have been reported in other Ni-based alloys with Cr compositions greater than that required to suppress ferromagnetism in Ni, it is considered that antiferromagnetism most probably is present in similar thermoelectric alloys. It has been shown that the spin-cluster model, as applied to an antiferromagnetic transition, can explain the thermoelectric properties of Nicrosil.

The theoretical approach given in Section 7.1.21 is used as a basis for this model. Here T_N is used as a reference temperature, instead of T_{tr}, in Eqs. (7-122), (7-124), (7-127b), and (7-128). Equation (7-128) becomes, for the present case,

$$N(T) \simeq \frac{T_N^{3/2}}{T} \tag{7-142}$$

This inverse relationship is in agreement with the observed behavior of antiferromagnetic materials for $T \geqslant T_N$.[5]

The temperature variation of the number of electrons per cluster is expected to vary in a way paralleling that occurring in ferromagnetic alloys for $T \geqslant T_c$. This follows since both types show Curie–Weiss behavior at temperatures at and above their transformation temperatures.[6] This function is proportional to the number of near neighbors in a cluster and is approximated by the substitution of Eq. (7-138) in Eq. (7-132) as

$$z(T) \simeq \frac{1}{T_N} \int_T^{T_\infty} dT \tag{7-143}$$

Here T_∞ is again used as that temperature at which the effects of the spin-cluster correlations appear to vanish.

[3]A. J. Dekker, *Solid State Physics*, Prentice-Hall, Englewood Cliffs, N.J., 1957, p. 483.

[4]B. D. Cullity, *Introduction to Magnetic Materials*, Addison-Wesley, Reading, Mass., 1972, p. 159.

[5]B. D. Cullity, *Introduction to Magnetic Materials*, Addison-Wesley, Reading, Mass., 1972, pp. 159–161.

[6]B. D. Cullity, *Introduction to Magnetic Materials*, Addison-Wesley, Reading, Mass., 1972, pp. 97 and 158.

The substitutions of Eqs. (7-142) and (7-143) into Eqs. (7-123) and (7-122) give

$$\Delta S \simeq -\frac{3R}{2Z_\alpha N_A e} \cdot \frac{T_N^{1/2}}{T} \cdot (T_\infty - T) \quad \mu\text{V/deg} \qquad (7\text{-}144\text{a})$$

When the constants in Eq. (7-144a) are taken into account,

$$\Delta S \simeq -\frac{8.129 \times 10^{-2}}{Z_\alpha} \cdot \frac{T_N^{1/2}}{T} \cdot (T_\infty - T) \quad \mu\text{V/deg} \qquad (7\text{-}144\text{b})$$

Equations (7-144a) and (7-144b) are general expressions *for the effects of antiferromagnetic spin clusters* on the thermoelectric deviations from linearity.

The unavailability of data for the magnetic properties of Nicrosil makes it necessary to estimate T_N from other physical properties. This may be done by means of the temperature coefficient of linear expansion, α_L, and its close relationship to the heat capacity, C_V.[6,7] This is given by

$$\alpha_L = \frac{\gamma C_V}{3BV} \qquad (5\text{-}182)$$

where γ is Gruneisen's constant, B the bulk modulus, and V the volume.[8] Gruneisen's constant is virtually temperature independent.[8,9] The temperature variations of B and V are quite small compared to that of C_V. Thus, α_L may be considered to vary with temperature in the same way as does C_V (see Section 5.2). And, as previously indicated [Eq. (7-120)], C_V is sensitive to electronic changes in a solid. *Thus, data on α_L may be regarded as reflecting electronic changes in a solid in the same way as C_V.* As applied to the antiferromagnetic behavior of Nicrosil, it is expected that the curve of a second-order transformation, such as that of α_L as a function of temperature, would show an inflection point at T_N with discernable differences in slope below and above T_N.[6] Data by Moore and others for Nicrosil show this behavior with the inflection point occurring at approximately 600 K.[10] The Néel temperature of Nicrosil was selected on this basis as being about 600 K.

The remaining data required for Eq. (7-144b) were obtained as follows. The value for Z_α was calculated from a nominal Nicrosil analysis, in atom percent, of Ni: 81.9, Cr: 15.2, and Si: 2.8, with corresponding valences of 0.54, 1, and 4, respectively.[11,12] These data give $Z_\alpha \simeq 0.71$ carriers per atom. Since linear ther-

[7]A. H. Morrish, *Physical Principles of Magnetism*, Wiley, New York, 1965, p. 462.

[8]C. Kittel, *Introduction to Solid State Physics*, 3rd ed., Wiley, New York, 1966, pp. 183–184.

[9]A. J. Dekker, *Solid State Physics*, Prentice-Hall, Englewood Cliffs, N.J., 1957, p. 33.

[10]J. P. Moore and others, "Nicrosil II and Nisil Thermocouple Alloys: Physical Properties and Behavior During Thermal Cycling to 1200 K," Technical Report ORNL-TM-4954, Oak Ridge National Laboratory, Oak Ridge, Tenn., 1975.

[11]C. Kittel, *Introduction to Solid State Physics*, 5th ed., Wiley, New York, 1976, p. 467.

[12]D. D. Pollock, *Physical Properties of Materials for Engineers*, Vol. I, CRC Press, Boca Raton, Fla., 1982, p. 71.

TABLE 7-10 CALCULATED AND EXPERIMENTAL DATA
FOR NICROSIL

(°C)	(K)	S' Eq. (7-145) (μV/°C)	ΔS Eq. (7-144b) (μV/°C)	$S_{\text{calc.}}$ $S' + \Delta S$ (μV/°C)	$S_{\text{exp'.}}$ (μV/°C)
300	573	11.4	−1.0	10.4	10.3
327	600	11.1	−0.8	10.3	10.3
400	673	10.4	−0.4	10.0	10.0
500	773	9.5	0	9.5	9.5

moelectric properties appear to be well established at 500°C, T_∞ was taken as 773 K. The calculated values of ΔS [Eq. (7-140a)] are given in Table 7-10.

The ATP of Nicrosil is given graphically in Fig. 7-13. The empirical equation of best fit for the linear portion of this curve, from $t = 500$ to 900°C, is given by

$$S' \simeq 14.20 - 0.0094t \quad \mu\text{V/deg} \qquad (7\text{-}145)$$

Equation (7-145) has an average error of $\pm 0.5\%$. Extrapolated data from this equation were used to calculate the thermoelectric properties of Nicrosil at temperatures at and below 500°C. These data, S', also are included in Table 7-10.

The calculated data are in good agreement with the experimental values. The comments on the omission of spin-cluster effects in the derivation of Eqs. (7-41), (7-71b), and (7-101b) that were made in Sections 7.1.21 and 7.1.21.1 are also applicable here. The inclusion of these spin influences, as manifested by ΔS [Eqs. (7-144a) and (b)] substantiates the hypothesis that antiferromagnetic influences are responsible for the nonlinear thermoelectric properties of Nicrosil at temperatures below $T_\infty \simeq 773$ K.

The decreasing departures from thermoelectric linearity, as Nicrosil is progressively heated from T_N to T_∞, are considered to result from a decrease in the effective correlation between the antiparallel clusters, or in decreases in their effective sizes and numbers, or in all of these.[2] In any case, the influence of spin clusters appears to vanish at T_∞. The effective absence of this factor at temperatures higher than T_∞ explains the agreement of the experimental values of the ATP of Nicrosil with those obtained by means of the previously noted equations for transition metals and alloys. As a consequence, the spin-cluster model provides a unified explanation for the thermoelectric deviations of Nicrosil, as well as for nickel and Alumel (Sections 7-1-21 and 7-1-21-1), which is closely related to the anomalies in the other physical properties that occur at T_N or at T_c.[7]

7.1.21.3 The question of ordering in nickel–chromium alloys.* It has been considered that the thermoelectric deviations of some of the nickel–chromium alloys may be explained by means of a short-range, lattice order–disorder

*The major portion of the material in this section was published originally as D. D. Pollock, "Short-Range Ordering in Nickel–Chromium Thermocouple Alloys," *Journal of Testing and Evaluation*, Vol. 11, No. 3, May 1983, pp. 225–226, 266.

mechanism.[1-3] If such a reaction did take place in these alloys, in the range from about $200°$ to $500°C$, it could provide a means for accounting for the observed behavior in the absence of other physical changes.

Almost all the thermoelectric alloys in common use are random solid solutions. This implies that it is highly probable that the alloy atoms occupy host lattice sites in a way dictated by pure chance. A binary (two-component), solid-solution alloy composed of elements A and B is presented here to illustrate this. The compositions of alloys of such a system are represented by the mol fractions (atom fractions) of the two components as f_A and f_B, respectively; $f_A + f_B = 1$. In the case of a random solid solution, the probability that an A atom will have a B nearest neighbor is

$$P_B = \frac{f_B}{f_A + f_B} = 1 \quad \text{or} \quad \frac{P_B}{f_B} = 1$$

In other words, for the case of a random solid solution, the probability that an A atom will have a B atom as a nearest neighbor is equal to the mol fraction of the B atoms present.

When *short-range order* takes place, some of the B atoms will occupy preferred rather than random sites. Then an A atom will have more B atoms as nearest neighbors than it would have in the random case. When this occurs, $P_B/f_B > 1$. The greater the difference between P_B/f_B and unity, the greater will be the degree of short-range ordering. This difference is small for short-range ordering. It may be expressed as

$$s = 1 - \frac{P_B}{f_B}$$

Here s is the short-range order coefficient. This may also be considered as the degree to which the occupation of a given, nearest-neighbor lattice site differs from that predicted by chance. For an ideally random solid solution, $s = 0$. Where $P_B/f_B > 1$, as just described, s is negative and indicates more unlike B nearest neighbors. Where the A atom has fewer nearest-neighbor B atoms than is predicted by chance (that is, when the A atom has more A nearest neighbors than are normally probable), $P_B/f_B < 1$ and s is positive. This condition is known as *clustering*. The spin-cluster mechanism, described in Section 7.1.21, gets its name from this because the early models were based on analogies with atomic clustering.[4,5] Either of these conditions can have pronounced effects on the electrical properties of alloys.[6]

[1]N. A. Burley and others, *The Nicrosil versus Nisil Thermocouple: Properties and Thermoelectric Reference Data*, National Bureau of Standards (U.S.), Monograph 161, U.S.G.P.O., Washington, D.C., 1978, p. 5.

[2]J. P. Moore and others, *Nicrosil II and Nisil Thermocouple Alloys: Physical Properties and Behavior During Thermal Cycling to 1200 K*, ORNL-TM 4954 (August 1975), Oak Ridge, Tenn.

[3]T. G. Kollie and others, *Temperature Measurement Errors with Type K (Chromel vs. Alumel) Thermocouples Due to Short-Ranged Ordering in Chromel*, ORNL-TM 4862 (March 1975), Oak Ridge, Tenn.

[4]H. Bethe, *Proc. Roy. Soc. (London)*, *A-150*, p. 552, 1935.

[5]R. E. Peierls, *Proc. Roy. Soc. (London)*, *A-154*, p. 207, 1936.

[6]D. D. Pollock, *Physical Properties of Materials for Engineers*, Vol. II, CRC Press, Boca Raton, FL, 1982, pp. 13, 17, 55.

The x-ray diffraction patterns of random solid solutions show lines that represent diffractions by an "average" atom of the given composition. This is a consequence of the random arrays of A and B atoms on the lattice planes. However, when short-range order is present, diffraction by the atoms on the nonrandom lattice sites causes the appearance of diffuse background patterns. These are in addition to the lines given by the x-ray diffractions from the planes of the random solid solution. Precise examinations and analyses of the diffuse patterns are employed to determine the average predisposition for an atom to have unlike nearest neighbors. Such diffraction techniques are considered to be a more reliable means for the determination of short-range ordering in a crystal lattice than single sets of other physical data that are not supported by additional evidence.

The references to the mechanism of short-range order as being responsible for the behaviors of some of the Ni–Cr thermoelectric alloys appear to have had their origin in the work of Nordheim and Grant.[7] These investigators considered that a superlattice or long-range order (atoms occupying preferred lattice sites over relatively long distances in the crystal) was most unlikely. They *postulated* that short-range ordering takes place in the composition range near that of Ni_3Cr (25 At% Cr) at temperatures below 800° to 900°C. This was done to explain experimentally observed changes in the relative electrical resistance.[6] The only alloy reported by these investigators, containing less than 22.2 At% Cr, and whose analysis approximated that of the nominal Chromel composition contained 10.9 At% Cr. The changes in its relative electrical resistance, which occurred during stepwise isothermal aging at temperatures between 288° and 424°C after water quenching from 980°C, appear to constitute the basis for ascribing the thermoelectric deviations of Chromel to short-range order. No x-ray evidence was presented to verify this. It should be noted that models based either on atomic clustering or electron spin clustering may be used to account for these electrical changes, as is indicated later.

The controversy regarding the lattice structure and the resultant thermoelectric behavior of Ni–Cr alloys arises because of the similar atomic numbers of Ni and Cr. The net result of this is "that the most intense reflection of a superlattice line would be only 0.7% that of the fundamental line."* The difficulties in resolving this question have been described by Fenton[8] and Taylor and Hinton.[9] The latter investigators attempted to solve this problem by replacing 20% of the Cr atoms in an alloy by aluminum atoms. However, since the definite observation of ordering only occurs when Al is present in these alloys, Nordheim and Grant considered that short-range order is probable for Ni-based alloys containing 25 At% Cr; *they did not discuss the mechanism responsible for the properties of the alloy with 10.9 At% Cr. Their primary concern was with alloys containing from 22.2 to 35.7 At% Cr.*

Baer showed that ordering occurs close to the Ni_2Cr(33.3 At% Cr) com-

---

[7]R. Nordheim and N. J. Grant, *J. Inst. Met.* (*London*), *82*, p. 440, 1953–54.

*T. G. Kollie, R. L. Anderson, and D. L. McElroy, personal communication, June 24, 1982.

[8]A. W. Fenton, *Proc. I.E.E.* (*London*), *116*, p. 1277, 1969.

[9]A. Taylor and K. G. Hinton, *J. Inst. Met.* (*London*), *81*, p. 169, 1953.

position.[10,11] This was demonstrated to take place between 25 and 36 At% Cr, with a maximum occurring at 33.7 At% Cr and 580°C. This work has been verified by other investigators.[12-15]

The findings of Roberts and Swalin also confirm the work cited.[16] However, their results are of greater significance with respect to the controversy regarding the thermoelectric properties of Ni–Cr alloys. Both neutron diffraction and dilatometric studies were made on Ni–Cr alloys containing 24.9 and 28 At% Cr. No evidence of ordering was detected. It will be noted that the compositions of these alloys are very close to the Ni_3Cr composition that was considered by Nordheim and Grant. Hawkins has incorporated the above data in a revised Cr–Ni phase diagram.[17] Since ordering does not take place at the Ni_3Cr composition, it would appear to be unlikely that short-range order similar to that of Ni_2Cr would occur in Chromel or Nicrosil; these alloys contain less than one-half of the Cr present in Ni_2Cr.

Burley and co-workers have comprehensively surveyed the literature on the thermoelectric instability of Ni–Cr alloys.[1] They have described Burley's own work on short-term thermoelectric instability[18] and that of Fenton[8] as "much circumstantial evidence of a hypothesis that the short-term emf changes are due to short-range ordering in the Ni–Cr atomic structure of the Type KP thermoelement." More recently, Kollie and co-workers also have published important thermoelectric work on this.[3] These investigators also state that "it is reasonable to assume on theoretical grounds that short-ranged order does occur in Chromel, even though it has not been detected by crystallographic techniques."[19] It should be noted that it is also equally theoretically reasonable, in the absence of crystallographic evidence to the contrary, to ascribe the thermoelectric deviations of these alloys as being a result of either atomic clustering or spin clusters; the explanations given by these authors for the influence short-range order upon emf also apply as well to either atomic or spin clustering.

Kollie and others* have proposed that synchrotron radiation be employed to examine and resolve this question. It is also of interest to note that Burley and others

[10]G. Baer, *Naturewissenshaften*, *43*, p. 298, 1956.

[11]G. Baer, *Z. Mettalk.*, *49*, p. 1025, 1958.

[12]Yu. A. Bagaryatskii and Yu. D. Tiapkin, *Dokl. Akad. Nauk SSSR*, *122*, p. 806, 1958.

[13]N. V. Semenova, *Fiz. Metal. i. Mettaloved*, *6*, p. 1017, 1958.

[14]N. V. Semenova, *Phys. Metals Metallog. (USSR)*, *6(6)*, p. 57, 1958.

[15]H. Schuller and P. Schwab, *Z. Mettalk.*, *51*, p. 81, 1960.

[16]B. W. Roberts and R. A. Swalin, *Trans. AIME*, *209*, p. 845, 1957.

[17]D. T. Hawkins, *Metals Handbook*, 8th ed., Vol. 8, American Society for Metals, Metals Park, Ohio, 1973, p. 291.

[18]N. A. Burley, *Cyclic Thermo-E.M.F. Drift in Nickel–Chromium Thermocouple Alloys Attributable to Short-Range Order*, Australian Defence Scientific Service, Defence Standards Laboratories Report 353, 1970.

[19]T. G. Kollie and others, *Rev. Sci. Instrum.*, *46*, p. 1447, 1975.

*T. G. Kollie, R. L. Anderson, and D. L. McElroy, personal communication, June 24, 1982.

have considered the possibility that the thermoelectric anomalies of the alloys in question may be of magnetic origin. See Section 7.1.21.

Since short-range atomic ordering has not been established as being responsible for the thermoelectric properties of Ni–Cr alloys, this mechanism may not arbitrarily be considered to be the only basis for their explanation. Chromel and Nicrosil, two very important nickel-based thermoelectric alloys, have approximate analyses of 11 and 15 At% Cr, respectively. These alloys contain only very small, trace amounts of aluminum. It is, therefore, considered that short-range, lattice order–disorder transformations do not necessarily take place in these thermoelectric alloys, nor do they constitute the sole rationale for their behavior.

It will be noted that the thermoelectric properties of nickel (Section 7.21.1), which cannot undergo any ordering transition, are explained by the spin-cluster model. The same spin-cluster mechanism also describes the deviant properties of Alumel (Section 7.1.21.1) and Nicrosil (Section 7.1.21.2) in which ordering does not occur. The same is true for Chromel, as is shown here and in Section 7.1.21.4 and for Nisil (Section 7.1.21.5). The common factor in all these materials is the presence of a magnetic transition. The spin-cluster model provides a unified basis for the explanations of the thermoelectric behaviors of all these materials when the appropriate magnetic influences are taken into account.

7.1.21.4 Chromel. Some of the attempts to explain the thermoelectric deviations of Chromel by means of short-range atomic ordering are summarized in Section 7.1.21.3. The spin-cluster model, used for the explanation of the deviations of Nicrosil (Section 7.1.21.2) is applied here to the properties of Chromel; it also is most probable that it shows an *antiferromagnetic transition.*

The expression used to describe the *effects of antiferromagnetic spin clusters* on the thermoelectric deviations from linearity is

$$\Delta S \simeq -\frac{8.129 \times 10^{-2}}{Z_\alpha} \cdot \frac{T_N^{1/2}}{T} \cdot (T_\infty - T) \qquad (7\text{-}144\text{b})$$

The value of $T_N \simeq 573$ K was obtained from the calibration changes noted by Burley and others.[1] This is in good agreement with the data of Kollie and others.[2] Both of these investigations indicate that reversible thermoelectric changes become apparent on heating to about 300°C. The other data required for Eq. (7-144b) were obtained as follows. The value for Z_α was calculated from a nominal analysis of Chromel, in atom percent, of Ni: 89.1, Cr: 10.3, and Si: 0.5, with corresponding valences of 0.54, 1, and 4,

[1]N. A. Burley and others, "The Nicrosil versus Nisil Thermocouple: A Critical Comparison with the ANSI Standard Letter-Designated Base-Metal Thermocouples," in *Temperature, Its Measurement in Science and Industry,* Vol. 5, J. F. Schooley, ed., American Institute of Physics, New York, 1982, pp. 1159–1166.

[2]T. G. Kollie and others, "Temperature Measurement Errors with Type K (Chromel vs. Alumel) Thermocouples Due to Short-Ranged Ordering in Chromel," *Rev. Sci. Instrum.,* 46, 11, pp. 1447–1461, 1975.

respectively.[3,4] These data give $Z_\alpha \simeq 0.61$ carriers per atom. Since linear thermoelectric properties appear to be well established at 600°C, T_∞ was taken as 873 K. The calculated values of ΔS [Eq. (7-144b)] are given in Table 7-11.

The ATP of Chromel is given graphically in Fig. 7-12. The empirical equation of best fit for the linear portion of this curve, from $t = 600$ to 900°C, is given by

$$S' \simeq 26.89 - 0.0157t \quad (\mu V/\text{deg}) \tag{7-146}$$

The average error of Eq. (7-146) is $\pm 0.7\%$. Extrapolated data from this equation were used to calculate the thermoelectric properties of Chromel at temperatures at and below 600°C. These data, S', also are included in Table 7-11. Calculated values of S have an average error that is slightly less than that of Eq. (7-146).

The good agreement of the data provided by the spin-cluster model for the explanation of the thermoelectric deviations from linearity of Chromel confirms the hypothesis that an antiferromagnetic transformation is responsible for its behavior. This agreement, in addition to that which this model provided for nickel, Alumel, and Nicrosil, in the preceding sections, provides further confirmation that the influence of spin clusters in alloys undergoing such transformations explains their thermoelectric behaviors. Equations (7-144b) and (7-140) also explain the transient reversible aberrations of Type K thermocouples below about 600°C.

7.1.21.5 Nisil.   Nisil is reported to have a "Curie temperature" of 290 K.[1] It will be noted that the thermoelectric deviations of Nisil (Fig. 7-13) are positive. These deviations are opposite in sign to those of elemental nickel and of Alumel (Fig. 7-11), which are known to be ferromagnetic. It thus would appear that the electron

TABLE 7-11 CALCULATED AND EXPERIMENTAL DATA FOR CHROMEL

t (°C)	T (K)	S' Eq. (7-146) ($\mu V/°C$)	ΔS Eq. (7-144b) ($\mu V/°C$)	$S_{calc.}$ $S' + \Delta S$ ($\mu V/°C$)	$S_{exp't.}$ ($\mu V/°C$)
200	473	23.8	−2.8	21.0	21.1
300	573	22.2	−1.7	20.5	20.2
400	673	20.6	−1.0	19.6	19.6
500	773	19.0	−0.4	18.6	18.7
600	873	17.5	0	17.5	17.4

[3]C. Kittel, *Introduction to Solid State Physics*, 5th ed., Wiley, New York, 1976, p. 467.

[4]D. D. Pollock, *Physical Properties of Materials for Engineers*, Vol. I, CRC Press, Boca Raton, Fla., 1982, pp. 71–72.

[1]N. A. Burley and others, *The Nicrosil versus Nisil Thermocouple: Properties and Thermoelectric Reference Data*, National Bureau of Standards (U.S.) Monograph 161, U.S. G.P.O., Washington, D.C., 1978, p. 39.

behavior responsible for the properties of Nisil must be significantly different from those of nickel and Alumel.

The simplest way to explain this behavioral difference is to consider that the spontaneous magnetization of Nisil results from *ferrimagnetic behavior*. This is accounted for by the presence of strongly correlated, or coupled, atomic dipoles within the spin clusters. The effective atoms within a cluster may be considered to occupy sites on two sublattices, each of which has unequal, antiparallel spins. This accounts for the observed magnetic properties and for the effects of silicon noted in Sections 7.1.20 and 7.1.22.3. Morrish cites seven possible spin configurations that may explain this behavior for the case where the fractions of atoms on each sublattice are equal, as would be expected in a normal, random solid solution.[2] This can occur when the host element is present in different ionic states, where alloying elements that are present can be in different ionic states, when different internal, or crystal, fields act on the sublattices, or when the sublattices contain unequal numbers of ions with unequal magnetic moments, but where the average of both numbers is representative of the composition of the alloy.

The exchange integral, J_e, is another important factor in this model. Its significance results from the fact that it determines the sign of ΔS [Eqs. (7-134), (7-138), and (7-139)]. In the cases of any ferromagnetic materials such as nickel and Alumel, J_e is positive and ΔS is negative, in ageement both with the experimental thermoelectric and magnetic data. However, the observed values of ΔS for Nisil are positive. This can only be the case if J_e is negative. This condition also is a fundamental requisite for ferrimagnetic behavior.[3] Thus, unbalanced, antiparallel spins within the spin clusters must be used to explain the thermoelectric behavior of Nisil.

The theoretical approaches given in Sections 7.1.21 and 7.1.21.2 are employed here. In the present case, the ferrimagnetic Néel temperature, T_{FN}, must be used as the reference, rather than T_c or T_N in Eqs. (7-122), (7-124), (7-127b), (7-128), and (7142). Equation (7-142) is written, for this case, as

$$N(T) \simeq \frac{T_{FN}^{3/2}}{T} \qquad (7\text{-}147)$$

This equation is valid because ferrimagnetic behavior above T_{FN} shows essentially the same Curie–Weiss decay as antiferromagnetic behavior above T_N.[4] This fact and the negative value for J_e cause Eqs. (7-137) and (7-138) to become negative. Thus, Eq. (7-137) becomes

$$-J_e = \frac{2k_B T_{FN}}{N_0} \qquad (7\text{-}137\text{a})$$

[2] A. H. Morrish, *Physical Principles of Magnetism*, Wiley, New York, 1965, p. 487.

[3] C. Kittel, *Introduction to Solid State Physics*, 3rd ed., Wiley, New York, 1966, p. 475.

[4] A. H. Morrish, *Physical Principles of Magnetism*, Wiley, New York, 1965, p. 493.

or

$$\frac{2k_B}{J_e N_0} = -\frac{1}{T_{FN}} \tag{7-148}$$

Then, upon substitution of Eqs. (7-147) and (7-148) in Eq. (7-132), using Eq. (7-138),

$$z(T) \simeq \frac{2k_B}{J_e N_0} \int_T^{T_\infty} dT \simeq -\frac{1}{T_{FN}} \int_T^{T_\infty} dT \tag{7-149}$$

and $f(T)$ [Eq. (7-123)] becomes

$$f(T) \simeq N(T)z(T) \simeq -\frac{T_{FN}^{1/2}}{T} \cdot (T_\infty - T) \tag{7-150}$$

Finally, the substitution of Eq. (7-150) into Eq. (7-122) results in

$$\Delta S \simeq -\frac{3R}{2Z_\alpha N_A e} f(T) = -\frac{3R}{2Z_\alpha N_A e} \cdot \left[-\frac{T_{FN}^{1/2}}{T}(T_\infty - T) \right] \quad \mu V/deg \tag{7-151}$$

Or, substituting for the constants, Eq. (7-151) becomes

$$\Delta S = +\frac{8.129 \times 10^{-2}}{Z_\alpha} \cdot \frac{T_{FN}^{1/2}}{T} \cdot (T_\infty - T) \tag{7-152}$$

Equations (7-151) and (7-152) are general expressions for the *effects of ferrimagnetic spin clusters* on thermoelectric deviations. The positive sign of ΔS [Eq. (7-152)], which is in agreement with the experimental data, could only have been obtained by considering the spin clusters to be ferrimagnetic in nature; atomic ordering cannot explain this behavior.

The value given by Burley and others as 290 K is used for T_{FN}.[1] The other data required for Eq. (7-152) were obtained as follows. The value of Z_α was calculated from a nominal analysis of Nisil, in atom percent, of Ni: 90.8, Si: 8.8, and Mg: 0.2, with corresponding valences of 0.54, 4, and 0.6, respectively.[5,6] These data give $Z_\alpha \simeq 0.84$ carriers per atom. Since linear thermoelectric properties are well established at 300°C, T_∞ was taken as 573 K. The calculated values for ΔS [Eq. (7-152)] are given in Table 7-12.

The ATP of Nisil is shown graphically in Fig. 7-13. The empirical equation of best fit for the linear portion of this curve from $t = 300$ to 900°C, is

$$S' \simeq -16.16 - 0.0158t \quad \mu V/°C \tag{7-153}$$

The average error of Eq. (7-153) is $\pm 0.3\%$. Extrapolated data from this equation were used to calculate the thermoelectric properties of Nisil at temperatures at and below 300°C. These data, S', also are included in Table 7-12.

The good agreement of the calculated data with the experimental results for the

[5] C. Kittel, *Introduction to Solid State Physics*, 5th ed., Wiley, New York, 1976, p. 467.

[6] C. Kittel, *Introduction to Solid State Physics*, 5th ed., Wiley, New York, 1976, p. 176.

TABLE 7-12   CALCULATED AND EXPERIMENTAL DATA FOR NISIL

t (°C)	T (K)	S' Eq. (7-153) (μV/°C)	ΔS Eq. (7-152) (μV/°C)	$S_{calc.}$ $S' + \Delta S$ (μV/°C)	$S_{exp't.}$ (μV/°C)
17	300	−16.4	+1.6	−14.8	−14.8
100	373	−17.7	+0.9	−16.8	−16.8
200	473	−19.3	+0.4	−18.9	−18.9
300	573	−20.9	0	−20.9	−20.9

deviations of the thermoelectric properties of Nisil from linearity verify the hypothesis that ferrimagnetic spin clusters are responsible for this behavior. It has thus been shown, here and in the preceding sections, that the spin-cluster model accurately explains the thermoelectric anomalies of nickel and those of all the nickel-based thermoelectric alloys in common use. Such a common basis for the explanation of the properties of these materials cannot be obtained from a model based on atomic ordering, short-range or otherwise.

7.1.22 Base Metal Thermocouple Alloys

Some of the properties of nickel-based alloys are discussed in Section 7.1.20. These are given in more detail here along with data on Type JP thermoelements. The properties of type JN alloys (constantans) are given in Sections 7.1.23 to 7.1.27.

7.1.22.1 Type KN alloys. Alumel (trademark of the Hoskins Manufacturing Co.) is selected for illustrative purposes here because of the author's greater familiarity with it and because of the large amount of published information on its properties. Equally acceptable Type KN alloys are listed in Table 7-5.

The effects of alloy composition on the thermoelectric linearity of Type KN alloys were discussed in Section 7.1.21.1. The net band-structure effect of the three alloying elements (Mn, Si, and Al) in substitutional solid solution in the nickel lattice is to induce a slight increase in the value of $(E_0 - E_F)$ as compared to that of nickel (see Sections 7.1.14.3 and 7.1.18.2). The elements Si and Al have completed inner electron shells, but do not appear to contribute four and three electrons, respectively, as would be expected.[1] The number of states that a manganese atom can accommodate in its $3d$ band when it is in substitutional solid solution in the nickel lattice is yet to be determined.[1,2] Without attempting to describe the effects of the electron interactions of this combination of elements on the band structure, except to note that the electrons from the Al and Si atoms appear to nearly balance the $3d$ holes

[1]W. H. Taylor, *Acta Metallurgica*, Vol. 2, 1954, p. 684.

[2]G. V. Raynor, in *Progress in Metal Physics*, Vol. 1, Butterworth Scientific Publications, London, 1949, p. 1.

from the Mn additions, the net result is that the energy range as described by $(E_0 - E_F)$ is about 0.0_3 eV greater than that of nickel (Section 7.1.14.3). This accounts for the facts that the ATP of Alumel is positive with respect to nickel and that it has a smaller value of $\Delta S/\Delta T$ than does nickel (Table 7-7).

The roles of the valence electrons of these alloying elements are better understood in terms of their effects on the magnetic increment of the heat capacity and its influence on the ATP. In this instance, calculations based on their nominal valences give results that are in very good agreement with the experimental data. This is shown in Section 7.1.21.1.

7.1.22.2 Type KP alloys.

Chromel (trademark of the Hoskins Manufacturing Co.) has been selected for illustrative purposes here for the same reasons given in Section 7.1.22.1 for Alumel. In addition, the compositional variations between the other equally acceptable, commercially available Type KP alloys are relatively small.

The ATP of Chromel is anomalous in that it is positive in sign. This is believed to be a result of the predominance of d-band hole conduction (Section 7.1.18.2).[3] Its departure from thermoelectric linearity at temperatures below about 500°C is discussed in Section 7.1.20. The slope of its linear thermoelectric properties, at temperatures higher than 500°C, is negative and is in agreement with Eq. (7-118). The value of $(E_0 - E_F)$ for Chromel is very close to that of Alumel (Table 7-7). The result is that both of these thermoelements show nearly parallel thermoelectric properties at temperatures above about 500°C. Even though the thermoelectric deviations of Chromel are larger than those shown by Alumel at temperatures below 500°C these are roughly parallel. This degree of parallelism between the ATPs of the two thermoelements reduces some of the effects of their nonlinearity and permits their application as a practical thermocouple [Eq. (7-20), Section 7.1.7].

Wang and others suggest that the anomalous behavior of Type KP alloys may be a result of the predominance of holes.[3] If it is assumed that the conduction mechanism is that resulting from an electron–hole interaction, the scattering mechanism may be described by the squared difference of the effective charges on the solute and solvent atoms.[4] The shift in the ATP relative to that of nickel may be given in a manner analogous to that of Linde[5] and Robinson and Dorn[6] by an equation of the form

$$\frac{\Delta S}{C} \simeq K_2 + K_1 (Z_\beta - Z_\alpha)^2 \qquad (7\text{-}154)$$

See Eqs. (6-51b), (6-135a), and (7-158). Here ΔS is the average displacement of the absolute thermoelectric power relative to that of nickel, C is the atom percent of the alloying element, K_1 and K_2 are constants, Z_β is the effective charge on the solute ion,

[3] T. P. Wang, C. D. Starr, and N. Brown, *Acta Metallurgica*, *14*, p. 649, 1966.

[4] D. D. Pollock, *Transactions of the Metallurgical Society*, A.I.M.E., *224*, p. 892, 1962.

[5] J. O. Linde, *Annalen der Physik*, *14*, p. 319, 1933.

[6] A. T. Robinson and J. E. Dorn, *Transactions of the Metallurgical Society*, A.I.M.E., *191*, p. 457, 1951.

and Z_α is the effective charge on the solvent ion. In the present case, since chromium and nickel belong to the same period in the periodic table, $K_2 \simeq 0$. Using Raynor's values[2] for the maximum negative valences of nickel and chromium (Ni: 0.6; Cr: 4.66), K_1 is found to be 0.24_4 $\mu V/°C/At\%$ Cr in substitutional solid solution in the nickel lattice. This value is similar to that found by Mott[7] and confirms the assumption of the predominance of holes in the conduction mechanism of these Ni–Cr alloys.

It is interesting to note that Eq. (7-154) is useful in approximating the interaction between chromium and silicon ions in Nicrosil alloys (Section 7.1.22.3).

While the d-level holes (negative valences) can explain the sign of the ATP of Type KP alloys, calculations based on the normal valences of Ni and Cr give results that do explain the thermoelectric deviations of these alloys from linearity. The theoretical model used to describe these is given in Section 7.1.22.2 along with these calculations.

7.1.22.3 Nicrosil–Nisil thermocouples.

The nicrosil–Nisil thermocouple represents a relatively recent development. It was devised in response to an increasing demand for a base metal thermocouple that would provide reliable, continuous readings in the range from 1000°C to 1300°C in oxidizing atmospheres. Prior to this development, thermoelectric measurements in this temperature range normally had been considered to be given best by the platinum-based thermocouples.

The Nicrosil–Nisil thermocouple meets the requirements noted, is more oxidation resistant, and has a higher degree of thermoelectric stability than other nickel-based thermocouples.[8] The nominal analyses and thermoelectric properties of these thermoelements are given in Tables 7-5–7-6, 7-7, and Fig. 7-13.

The thermoelectric behavior of Nicrosil is similar to that of type KP alloys (Fig. 7-12). The fundamental explanations given in Sections 7.1.20 and 7.1.22.2 also apply to Nicrosil alloys. It will be noted that the deviations of the thermoelectric properties of Nicrosil from linearity are smaller than those of type KP alloys at temperatures below 500°C. The increased chromium content (approximately 15.2 At% as compared to about 10.3 At%), as well as the very large increase in the silicon present, have resulted in decreases in the numbers of residual spin clusters in the temperature range from about 200° to about 600°C. A model for this is given in Section 7.1.21.2.

When Eq. (7-154) is used to calculate the average displacement of the absolute thermoelectric power of Nicrosil relative to that of nickel, it is found that one silicon atom effectively neutralizes the positive displacement effect of two chromium atoms. This agrees with the strong effects noted for silicon in Section 7.1.20.

The high chromium and silicon contents of Nicrosil also are responsible for its increased oxidation resistance. The formation of a thin intermediate layer of SiO_2 between the alloy and the outer oxides acts as an impermeable, protective shield

---

[7] N. F. Mott, *Proceedings of the Royal Society*, Series A, *156*, p. 368, 1936.

[8] N. A. Burley and others, *The Nicrosil versus Nisil Thermocouple: Properties and Thermoelectric Reference Data*, National Bureau of Standards (U.S.), Monograph 161, U.S.G.P.O., Washington, D.C., 1978.

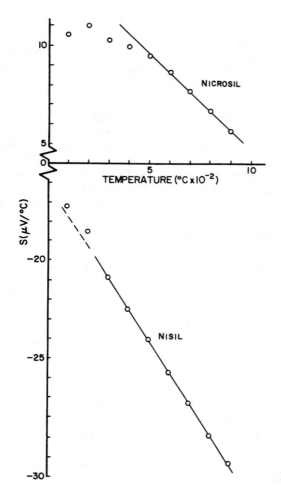

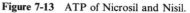

Figure 7-13   ATP of Nicrosil and Nisil.

against further oxidation. This barrier is made more impermeable to the diffusion of oxygen by the presence of a nominal 0.1 Wt% Mg. The Si–Mg combination is responsible for the long useful life and for the increased thermoelectric stability of Nicrosil at elevated temperatures.[8]

The thermoelectric properties of Nisil are considerably more linear than those of type KN alloys. As noted in Section 7.1.20, the high silicon content has caused T_{FN} to be manifested at about 290 K (17°C).[8] Small, positive deviations from thermoelectric linearity below 300°C (Fig. 7-13) are explained as shown in Section 7.1.21.5.

Nisil, virtually a binary Ni–Si alloy, also acquires its oxidation resistance and its thermoelectric stability from a very stable, impermeable, surface film of SiO_2 that also is enhanced by the presence of a nominal 0.1 Wt% Mg. The oxidation reactions and kinetics of both Nisil and Nicrosil are reviewed in detail elsewhere.[8]

It will be noted from the data in Table 7-7 that the slopes of the absolute thermoelectric powers of Nicrosil and Nisil are considerably different. Thus, the emf of this thermocouple has a greater degree of thermoelectric nonlinearity as a function of temperature than does that of the Type K thermocouple (Section 7.1.7). Its emf output is also somewhat smaller than that of the type K thermocouple. These factors are compensated for by its greater stability, higher temperature range of operation, and longer useful life because of its greater oxidation resistance. The Nicrosil–Nisil thermocouple has been recently accepted as a standard thermocouple by the National Bureau of Standards and the American Society for Testing and Materials.

7.1.22.4 General comments on nickel-based thermocouples.
None of the nickel-based thermocouples in common use should be used in reducing atmospheres or in those containing even very small amounts of sulfur. These materials are stable in oxidizing atmospheres because their surfaces are protected by thin, continuous, adherent, protective oxide films. These protective, complex-oxide films are destroyed in reducing atmospheres. Once this occurs, the alloying elements become subject to selective attack by the atmospheres, and the compositions of the alloys are depleted. The result is that $(E_0 - E_F)$ changes and, consequently, the ATPs change accordingly. Large deviations from calibration are to be expected under these conditions.

Similar undesirable behavior will take place in oxidizing atmospheres when the partial pressure of the oxygen in the atmosphere falls below a critical level. This changes the chromium compositions of the Ni–Cr alloys. Such behavior is known as "green rot".

Extended exposures to high temperatures in vacua also have undesirable effects on Type K thermocouples. Chromium depletion occurs in the type KP thermoelements and changes their calibration.

Nickel and virtually all its alloys are extremely susceptible to attack by very small amounts of sulfur, even at moderate temperatures. The resultant intergranular attack not only changes the thermoelectric properties of these alloys, but causes them to become very brittle. Early thermoelectric and mechanical failures are common under these conditions.

A survey of the oxidation reactions and kinetics of Type K thermocouple alloys is given in NBS Monograph 161.[8] These alloys are more oxidation resistant at temperatures above 500°C than are the thermocouple Types E, J, or T. This is one important reason for the extensive application of Type K thermocouples at high temperatures and for their frequent preference over other base metal thermocouples.

7.1.22.5 Type JP thermoelements.
The ATP of thermocouple iron* as a function of temperature is given in Fig. 7-14. The data are given only up to 700°C. Beyond this temperature, the effects of both the magnetic and allotropic trans-

---

*It is frequently, and incorrectly, assumed that any commercial iron, such as ARMCO Iron[R], may be used for thermoelectric purposes. Thermoelectric iron actually is an alloy containing up to nine alloying elements, depending on its source. These are present in very small amounts, and can have very strong effects on its thermoelectric properties.[10]

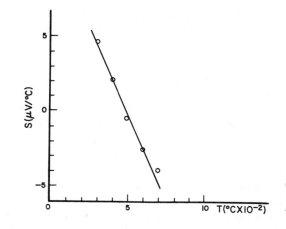

Figure 7-14 ATP of thermocouple iron (type JP). (From D. D. Pollock and C. P. Conrad II, *Trans. Met. Soc. AIME, 227,* 478, 1963.)

formations affect this property (above about 750°C).[9]   The thermoelectric behavior of iron is intermediate in character between those of Chromel and nickel or Chromel and platinum. It is linear and has a negative slope. Its ATP is positive at low temperatures and becomes negative at temperatures above about 500°C. This mixed behavior probably indicates a predominance of hole conduction at the lower temperatures, with increasing amounts of electron character as the temperature increases. Above 500°C, electron conduction appears to predominate. It will be noted that the calculated value for $(E_0 - E_F)$ is considerably smaller than those for platinum and nickel. The mixed electron–hole behavior, the small value for $(E_0 - E_F)$, and the fact that the data are given below the Curie temperature, T_c, indicate that Eqs. (7-41), (7-71b), or (7-101b) should not be used in this temperature range (see Sections 7.1.8.2, 7.1.9.3, and 7.1.10.2). They may be used at temperatures above the allotropic phase change.

Finch has shown that relatively small amounts of impurity or alloying elements may have large effects on the emf of iron.[10]   It is, therefore, expected that the impurities present in the typical sample, noted in Table 7-5, would have a significant effect on the value of $(E_0 - E_F)$, (Section 7.1.14.3). Sufficient data are as yet not available in the literature for the evaluation of the individual effects of these impurities.

Type JP thermoelements may be used in oxidizing, reducing, or inert atmospheres and in vacua. They are much less sensitive to strain than are Type KP alloys. The maximum recommended temperature of their application is 760°C. Repeated heating through the lower transformation temperature, A_1, changes the thermoelectric properties of some Type JP thermoelements. Exposure to atmospheres containing sulfur can embrittle some Type JP thermoelements. Subzero applications usually are not recommended because some Type JP thermoelements

[9]G. A. Moore and T. A. Shives, in *Metals Handbook*, 8th ed., Vol. 1, American Society for Metals, Metals Park, Ohio, 1961, p. 1206.

[10]D. I. Finch, U.S. Patent 2,323,759, 3 Aug. 1943.

appear to undergo a ductile–brittle transition and all require protection against aqueous corrosion. The limitations of Type J thermocouples are largely those imposed by the JP component.

7.1.22.6 Extension wires.

Extension wires are used to connect the actual thermoelements to the temperature-measuring/controlling devices. They also enter into and can affect the thermoelectric circuit. This may be shown simply by reference to Eq. (7-18b), rewritten to illustrate a typical pair of extension wires A and B, as

$$E_{AB} = \int_{T_0}^{T} (S_A - S_B) \, dT = \int_{T_0}^{T_1} (S_A - S_B) \, dT + \int_{T_1}^{T} (S_A - S_B) \, dT \qquad (7\text{-}155)$$

The integral from T_0 to T_1 represents the thermoelectric contribution of the extension wires, and the integral from T_1 to T represents that of the thermocouple.

It is important that the output of the extension wires (the integral from T_0 to T_1) match the specified values for that of the thermocouple over the range T_0 to T_1 within reasonable limits. If the match is exact, no measurement error will be introduced by the extension wires. The poorer this match is, the greater will be the error in the temperature reading that is introduced by the extension wires.

Extension wires are available for the standard thermocouples that, in most cases, closely match their thermoelectric properties between 0° and 200°C. In the majority of applications, it is unnecessary for these extension materials to have matching thermoelectric properties beyond that temperature.

In the case of the standard, base metal thermocouples, alloys of compositions similar to those of the thermoelements are used for the extension wires. However, where any of the noble metal thermocouples are considered, the cost of such an approach is frequently prohibitive. Consequently, base metal extension wires are commercially available for the noble metal thermocouples that satisfy Eq. (7-155) over the small range of temperatures to which these wires are normally exposed. Extension wires for many of the less commonly used types of thermocouples are available from their manufacturers or from their distributors.

Because extension wires need only match the thermoelectric properties of thermocouples over limited ranges of temperature, the P or N components can have absolute thermoelectric properties that are considerably different from those of the measuring thermoelements. However, when such wires are suitably paired, their thermoelectric properties are made to match those of the thermocouple within the small temperature range to which the extension wires are normally exposed. This makes it extremely important that the extension wires be purchased as pairs and not singly.

The factors responsible for instability noted in Section 7.1.6 also apply to extension wires. These too frequently overlooked sources of errors in extension wires may be eliminated by reasonable care in their installation and use.

Note should also be made of the fact that the thermoelectric measuring/control instruments are usually near room temperature. This has two important consequences. First, the device must be at a reasonably uniform internal temperature.

Thus, the various metals and alloys in its circuitry do not add spurious emfs to the thermocouple output (Section 7.1.6). And, second, the difference between the temperature of the instrument and the 0°C reference temperature must be taken into account. This can be done by electrical circuitry within the measuring device that accounts for the temperature (automatic reference junction compensation) within the instrument. Another way to do this is to attach Peltier devices (Sections 7.1.2 and 7.1.4.2), which operate at 0°C within reasonably narrow limits, to the ends of the thermocouples. These maintain the reference junctions of the thermocouples at the reference temperature. Two virtually identical wires (such as commercially pure copper) then connect the reference junctions of the thermocouple to the instrument to complete the circuit. The connecting wires add no emf to the circuit since their Thomson effects cancel (Section 7.1.6).

7.1.23 High-Alloy Thermoelement Theory

The standard alloys discussed in Sections 7.1.19 and 7.1.20 are relatively dilute, solid-solution alloys of transition elements that, with one exception, contain a maximum of about 20 atom percent of alloying additions. The alloys considered here contain up to about 60 atom percent of noble metal, substitutional additions to the lattice of the transition metal base. Three alloy systems of this kind have been studied: Cu–Ni, Ag–Pd, and Au–Pd.[1-5] Each of these systems shows complete solid solubility (with no intermediate phases) below the solidus.

The Cu–Ni based alloys, known generically as *constantans*, are typical of this alloy class. These are presented in detail as being representative of the three systems since they have won wide acceptance and the Ag–Pd and Au–Pd alloys have not. The negative components of the Platinel® thermocouples, $Au_{65}Pd_{35}$, are an exception.

7.1.23.1 Electron theory for binary alloys.
In the case of the dilute substitutional solid solutions of transition elements, it was pointed out that the parameter $(E_0 - E_F)$ remained virtually constant as a function of temperature (see Sections 7.1.14.3, 7.1.18.1, 7.1.19.1, and 7.1.20). This behavior no longer holds when the Fermi level approaches the top of the d band. Here the variation of $(E_0 - E_F)$ with temperature must be considered (see Sections 7.1.14.3 and 7.1.18.2).

A better understanding of the thermoelectric properties of copper–nickel alloys may be obtained by examining the rigid-band model long used to explain their ferromagnetic and paramagnetic properties. This will clarify the effects of copper variations on the hybridized d bands of the alloys.

Copper and nickel both crystallize in FCC lattices. Their Goldschmidt radii

[1] F. E. Bash, *Trans. TMS-AIME*, p. 2409, 1919.

[2] W. Geibel, *Z. an. Chem. 69*, p. 38, 1910.

[3] W. Geibel, *Z. an. Chem., 70*, p. 240, 1911.

[4] D. D. Pollock, *Trans. TMS-AIME, 224*, p. 892, 1962.

[5] D. D. Pollock, *Acta. Met., 16*, p. 1453, 1968.

and electronegativities are very nearly equal.[6]   The atomic number of nickel is 28; that of copper is 29.   Thus, to a first approximation, copper and nickel atoms are very much alike, except for the extra electron, and form a continuous series of FCC solid solutions with no intermediate phases (an isomorphous system).[7]   These elements do differ in one significant aspect, particularly in the crystalline state.   The configurations of their s and d electrons show important differences.   Nickel lacks about 0.6 electrons per atom in its d shell and has about 0.6 electrons per atom in its s levels.[8,9,10] Copper has all its d levels filled and has one, outer s electron per atom.   The densities of states of these elements are shown schematically in Figs. 7-15 and 7-16.

The absence of 0.6 electrons in the $3d$ levels of nickel and the consequent spin imbalance account for its ferromagnetic behavior.[10]   As increasing amounts of copper are alloyed with nickel, the added s electrons from the copper atoms progressively fill states in the hybridized $3d$ bands of the alloys, and the ferromagnetic properties of the Cu–Ni alloys decrease in a regular way.   At about 60 At% copper, the d bands are maximumly filled, the $3d$ spins are virtually balanced, and the ferromagnetic behavior vanishes.   This is shown in Fig. 7-17.   Alloys with copper concentrations equal to or greater than this are paramagnetic.

Verification of the effect of copper and other solutes on the ferromagnetic properties of nickel is given in detail elsewhere.[10]   In summary, the data show that the decrease of ferromagnetism in nickel alloys and its disappearance are linear functions of the valence of the alloying element in substitutional solid solution in the nickel lattice.   As examples, the compositions of binary nickel-based alloys at which ferromagnetic properties cease to be present are 30 At% for zinc (valence 2), 20 At% for aluminum (valence 3), 15 At% for each of silicon and tin (each with a valence of 4),

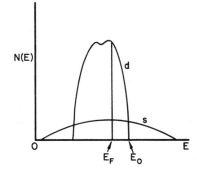

Figure 7-15   Schematic diagram of the density of states of nickel.

[6]F. Laves, in *Theory of Alloy Phases*, American Society for Metals, Metals Park, Ohio, 1956, p. 124.

[7]F. H. Rhines, *Phase Diagrams in Metallurgy*, McGraw-Hill, New York, 1956, p. 2.

[8]N. F. Mott and H. Jones, *Theory of the Properties of Metals and Alloys*, Dover, New York, 1958, p. 197.

[9]C. Kittel, *Introduction to Solid State Physics*, Wiley, New York, 1956, p. 259, 330–333.

[10]D. D. Pollock, *Physical Properties of Materials for Engineers*, Sections 9.9 and 9.10, CRC Press, Boca Raton, Fla., 1982.

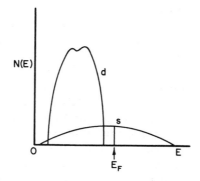

Figure 7-16 Schematic diagram of the density of states of copper.

and 12 At% for antimony (valence 5). The limiting composition for the presence of ferromagnetism in each of these alloy systems is 60 equivalent At% (Section 6.1.4.) As the number of valence electrons of the alloying element increases, the unoccupied, hybridized d states of the alloys are filled more rapidly. This is the equivalent of stating that $(E_0 - E_F)$ approaches zero more rapidly and with smaller amounts of alloying elements in substitutional solid solution in the nickel lattice as the valences of the alloying elements increase. (See Sections 10.2.7 and 10.2.8.)

Thus, as copper is added in substitutional solid solution in the lattice of nickel, the holes in the alloy d band become increasingly filled and $(E_0 - E_F)$ decreases. Theoretically, the maximum filling of the alloy d band should occur from about 55 to

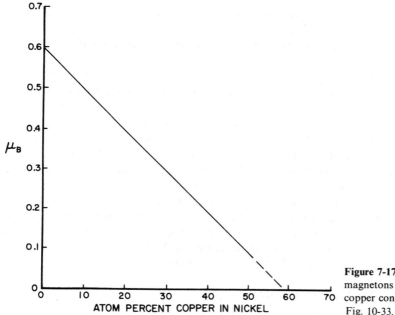

Figure 7-17 Number of Bohr magnetons per atom as a function of copper content in Ni–Cu alloys. Also see Fig. 10-33.

60 At% copper.[11]   The parameter $(E_0 - E_F)$ should be minimum at an alloy of that critical composition.   Beyond that specific composition, increasing amounts of s-like electron behavior occur, and $(E_0 - E_F)$ increases.[5,12]   This behavior is also discussed in Sections 7.1.14.3 and 7.1.18.2.

The hybridized alloy d band does not become completely filled.   If this did occur, $(E_0 - E_F)$ would approach zero as the copper content was increased.   The result would be abnormally large values for S [Eqs. (7-41), (7-71b) and (7-101b)].   The experimental results from alloys in this composition range show no evidence that $(E_0 - E_F)$ approaches zero.[4]   In addition, such very large increases in S do not correspond to observed heat capacity data and would contradict Eq. (7-104).

The variation of $(E_0 - E_F)$ as a function of composition, at a given temperature, therefore should show a minimum in the critical composition range of the maximum filling of the hybridized alloy d levels, that is, from about 55 to 60 At% copper.   The influence of such a variation in $(E_0 - E_F)$ should be manifested as a maximum in the ATP as a function of composition at the constant given temperature.

7.1.23.2 Effects of ternary alloying elements.   The addition of normal metallic elements, such as aluminum, in substitutional solid solution in the lattice of the copper–nickel base should induce effects similar to those described in the previous section.   In this instance, each aluminum atom would be expected to contribute three valence electrons to any available hybrid d levels.   On the other hand, any transition elements in substitutional solid solutions in the lattice of the copper–nickel base would be expected to contribute unoccupied states to the d band of the copper–nickel alloy base.   Cobalt may be used to illustrate this.   Each cobalt atom would be expected to supply nearly three times as many d-level holes as does nickel; it has about 1.7 holes per atom as compared to about 0.6 holes per nickel atom.   Thus, ternary cobalt additions would be expected to increase the range of $(E_0 - E_F)$ of the hybridized d band. (See Sections 7.1.14.3 and 7.1.18.2.)

The critical, binary, copper–nickel alloy, containing from about 55 to 60 At% copper, would show the maximum ATP because its d levels would be maximumly filled.   This is the same as saying that the value of $(E_0 - E_F)$ would be at a minimum.   The presence of small amounts of *any* additional substitutional ternary, alloying element, either a normal metal or a transition metal, would be expected to cause a decrease in the ATP of the Cu–Ni alloy of the critical composition.   This would result either from the presence of excess valence electrons or from an increase in the number of vacant, hybrid, d-level states, depending on the ternary alloy atom being considered.   The presence of extra electrons contributed by a normal metal atom would result in more s-like behavior, increase $(E_0 - E_F)$, and therefore decrease the ATP.   An increase in the number of unoccupied d states contributed by a transition atom also should increase $(E_0 - E_F)$ and decrease the ATP.

Such influences of ternary addition atoms on the ATP of the critical, binary,

[11]N. F. Mott and H. Jones, op cit., p. 314.

[12]N. F. Mott and H. Jones, op. cit., pp. 248 and 267.

copper–nickel alloy, in which the d band is maximumly filled and which shows the minimum $(E_0 - E_F)$, can be described by a scattering parameter, $q(E)$. This parameter can be used to account for the effects of alloying elements in this critical solid solution by modifying Eqs. (7-41), (7-71b), and (7-101b) as follows:

$$S = -\frac{1.22 \times 10^{-2}T}{(E_0 - E_F) + q(E)} \quad \mu V/^{\circ}C \qquad (7\text{-}156)$$

The scattering parameter $q(E)$ defined in this way may be considered to represent the displacement between the value of $(E_0 - E_F)$ for any ternary alloy and that of the critical, unalloyed, binary copper–nickel base with maximumly filled d levels. Since any alloy addition element in substitutional solid solution in the critical alloy should increase $(E_0 - E_F)$ of such a ternary alloy, $q(E)$ must always be positive.

7.1.23.3 Effect of temperature upon $(E_0 - E_F)$. The effects of the overlap of the hybridized s and d bands of transition elements were discussed in Section 7.1.14.3. If the electrons with energies close to E_F are considered to be "nearly free", an approximation can be made regarding the behavior of $(E_0 - E_F)$ as a function of temperature.

If $(E_0 - E_F)$ is of the order of $k_B T$, as is the case when the d band approaches maximum filling at the critical composition, electrons with energies near E_F, which would normally occupy d states, could be promoted readily to s states. If such electrons are considered to be "nearly free", they may be approximated as obeying the classical gas laws; each would have an average energy of $\frac{3}{2}k_B T$.[13] Thus, as the temperature was increased, more electrons would vacate the d band and would occupy s levels. One way to visualize this is to consider that the s band would be displaced to slightly lower energies, to the left in Fig. 7-15, to permit the d to s resonance of the hybrid electrons. As this occurred, the energy range between the Fermi level and the top of the d band should vary as $\frac{3}{2}k_B T$. Or

$$(E_0 - E_F) \propto \tfrac{3}{2}k_B T \qquad (7\text{-}157)$$

This is also discussed in Sections 7.1.24 and 7.1.26.

7.1.24 ATP of Concentrated Binary Solid Solutions

It was predicted, in the discussion in Section 7.1.23.2, that the thermoelectric power as a function of composition should pass through a maximum between about 55 and 60 At% copper. This is verified by the data[4] shown in Fig. 7-18. Geibel reported similar behavior.[2,3] The maximum thermoelectric power is shown to occur at approximately 57 At% copper. This falls nicely within the range predicted by Mott and Jones.[11] The maxima observed by Geibel occurred between 40 and 50 At% of the noble element. It is thought that this difference could have been caused by the presence of impurity atoms in the alloys that were not taken into consideration.

[13]L. S. Darken and R. W. Gurry, *Physical Chemistry of Metals*, McGraw-Hill, New York, 1953, p. 8.

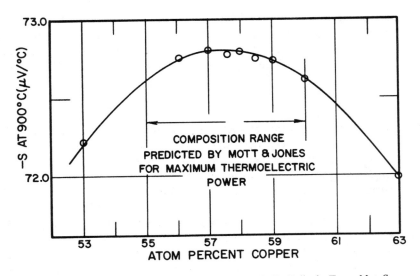

Figure 7-18   ATP of Cu–Ni alloys at 900°C. (From D. D. Pollock, *Trans. Met. Soc. AIME, 224,* 892, 1962.)

Equation (7-101b) was used to calculate the values of $(E_0 - E_F)$ for these binary alloys. The predicted minimum is shown in Fig. 7-19.

It will be noted that the minimum in Fig. 7-19 and the maximum in Fig. 7-18 are rather flat. The differences in the thermoelectric properties of alloys whose compositions range from 56 to 59 At% copper are quite small. Compositions on either side of this range (below 56 and above 59 At% copper) show smaller values of ATP and larger values for $(E_0 - E_F)$. The small thermoelectric changes over this relatively large range of copper composition between 56 and 59 At% copper can have significant commercial implications. The variations of the ATPs of commercial alloys based on the center of this 56 to 59 At% range of copper will be small compared with those of

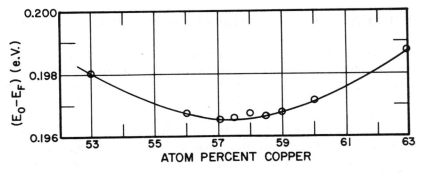

Figure 7-19 $E_0 - E_F$ of Cu–Ni alloys at 900°C. (From D. D. Pollock, *Trans. Met. Soc. AIME, 224,* 892, 1962.)

constantans based on either side of the range. Production melts based on this fact would have greater percentages of heats of constantans that meet the published tables than those that are formulated in any other way; it reduces the number of heats that must be scrapped (see Section 7.1.14.3).

The flat ranges in Figs. 7-18 and 7-19, at a constant temperature, are also reflected in the behaviors of S and $(E_0 - E_F)$ as functions of temperature (Figs. 7-20 and 7-21). Here it will be seen that, while S is nonlinear, $(E_0 - E_F)$ is a linear function of temperature. The values of S and $(E_0 - E_F)$ for alloys in the range 56 to 59 At% copper in solid solution in nickel are so close that they may be considered to be essentially the same for all practical purposes. Thus, the paraboliclike behavior of $(E_0 - E_F)$ as a function of composition, at constant temperature, and the linear behavior of $(E_0 - E_F)$ as a function of temperature, for a given composition, both agree with the predicted theoretical model (Sections 7.1.23.1 and 7.1.23.2). The temperature dependence of $(E_0 - E_F)$ is discussed in further detail in Section 7.1.26.

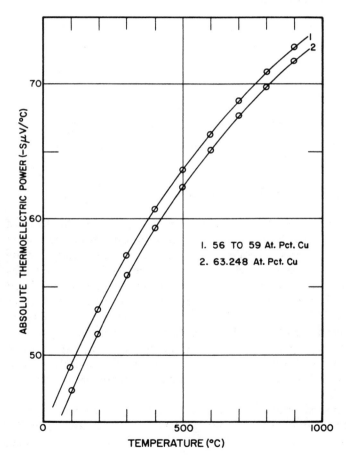

Figure 7-20 Envelope of ATP of Cu–Ni alloys as a function of temperature. (From D. D. Pollock, *Trans. Met. Soc. AIME*, *224*, 892, 1962.)

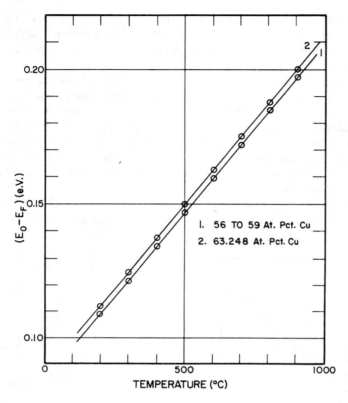

Figure 7-21 Envelope of $(E_0 - E_F)$ of Cu–Ni alloys as a function of temperature. (From D. D. Pollock, *Trans. Met. Soc. AIME, 224,* 892, 1962.)

7.1.25 ATP of Ternary Cu–Ni Solid Solutions

The effects of ternary alloying elements were determined by comparing the ATPs of alloys containing suitable alloying elements present in substitutional solid solutions in a nearly constant composition, critical copper–nickel base with the properties of the same unalloyed base.[4]   The base to which the ternary elements were added was maintained, within very narrow limits, close to the critical composition that provided the maximum ATP.   This is important because of the way in which the parameter $q(E)$ is defined (see Section 7.1.23.2).   Figure 7-22 shows the effect of a typical ternary addition element on the behavior of $(E_0 - E_F)$ as a function of temperature.

It will be noted that the displacements of the values for $(E_0 - E_F)$ are positive and vary approximately as the amount of ternary element present in substitutional solid solution in the alloy.   This agrees with the postulated theoretical model [Eq. (7-156) and Section 7.1.23.2].

It was shown that the temperature variations of $(E_0 - E_F)$ are linear, and for alloys whose ternary components are of the order of 0.5 At% or less, all are essentially parallel to that of the critical copper–nickel base.[4]   Thus, in this dilute alloy range,

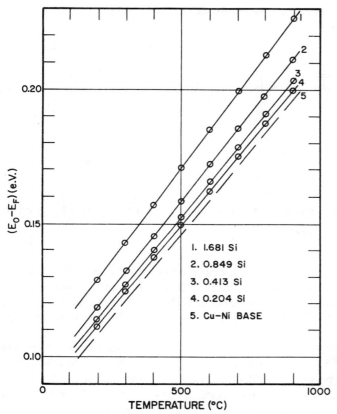

1. 1.681 Si
2. 0.849 Si
3. 0.413 Si
4. 0.204 Si
5. Cu–Ni BASE

Figure 7-22 Typical effect of a ternary alloying element on $(E_0 - E_F)$ as a function of temperature. (From D. D. Pollock, *Trans. Met. Soc. AIME*, *224*, 892, 1962.)

$q(E)$ is not a function of temperature, but is a function only of composition. Ternary additions greater than 0.5 At% do show a small effect of temperature on $q(E)$ [Section 7.1.26, Eq. (7-160)]. Ternary additions of about 0.5 At% are not trivial, since many elements present in the solid solution in copper–nickel alloys in this amount show very pronounced effects on the ATP [Sections 7.1.14.3 and 7.1.18.2].

Within the range of dilute, ternary compositions in which $q(E)$ is unaffected by temperature, the amount of the displacement of $(E_0 - E_F)$ per atom percent of the ternary element, given by $q(E)/C$, is essentially constant for the given element. Values of $q(E)/C$ are given in Table 7-13.

7.1.25.1 Scattering parameter. The copper ions in the lattice have a charge equal to $+e$, the charge on an electron. Normal metal, ternary, alloying ions have a charge of $+Z_\beta e$, where Z_β is their valence. The extra charge around such a ternary ion is $(Z_\beta - 1)e$ relative to copper. The field of this extra charge is responsible for the scattering of the electrons. The intensity of the scattering is proportional to

TABLE 7-13　VALUES FOR $q(E)/C$ ARRANGED IN PERIODIC FORM

Group Period	II	III	IV	V	VI	VII	VIII
3	Mg 0.000_5	Al 0.005_7	Si 0.013_8				
4a			Ti 0.024_6	V 0.046_5	Mn 0.0009_0	Fe 0.019_7	Co 0.010_9
5a			Zr 0.0006_6	Nb 0.045_9			
5b		In 0.003_8	Sn 0.009_8				

the square of the scattering charge.[14]　On this basis, the scattering may be given by

$$\frac{q(E)}{C} \simeq K_1 + K_2(Z_\beta - 1)^2 \tag{7-158}$$

where C is the amount of ternary alloying element in atom percent and K_1 and K_2 are constants [see Eqs. (6-51b), (6-135a), and (7-154)].　These constants are analogous to the parameters used by Linde[15] and Robinson and Dorn.[16]　This assumes that the very small quantities of ternary alloying additions have a negligible effect on the shape of the top of the d band of the critical copper–nickel base alloy.

　　When Eq. (7-158) is applied to describe the effects of the nontransition ternary additions given in Table 7-13, it is found that the constants K_1 and K_2 are valid for all solute elements of a given period in the Periodic Table that are soluable in the critical base.　The same values of these constants also accurately describe the effects of soluable ternary transition metal additions belonging to the given period.

　　In the case of ternary transition element additions to the base, the scattering mechanism is different.　Here the number of unoccupied d-shell states is used because these elements add holes.　These interact with the electrons at the highest energy levels in the hybridized d bands.　Since the binary copper–nickel alloys show that their d levels are maximumly filled at about 57 At% copper, the number of occupied, hybrid d states is taken as 0.43 per nickel atom.　This is taken into account by modifying Eq. (7-158) in the following way:

$$\frac{q(E)}{C} = K_1 + K_2[Z'_\beta - (-0.43)]^2 \tag{7-158a}$$

where Z'_β is the number of vacant d levels in an atom of the added, ternary, transition element.　A summary of the data obtained from Eqs. (7-158) and (7-158a) is given in Table 7-14.

[14]N. F. Mott and H. Jones, op. cit., p. 293.

[15]J. O. Linde, *Ann. der Physik, 14*, p. 319, 1933.

[16]A. T. Robinson and J. E. Dorn, *Trans. TMS-AIME, 191*, p. 457, 1951.

TABLE 7-14 COMPARISON OF CALCULATED AND EXPERIMENTAL VALUES FOR $q(E)/C$[a]

Period	Element	Z_β[a]	K_1[b]	K_2[c]	Calculated $q(E)/C$[d]	Experimental $q(E)/C$[d]
3	Mg	2	-0.001_2	$+0.001_7$	0.000_5	0.000_5
3	Al	3	-0.001_2	$+0.001_7$	0.005_6	0.005_7
3	Si	4	-0.001_2	$+0.001_7$	0.014_1	0.013_8
4a	Ti	4	-0.003_2	$+0.003_1$	0.024_7	0.024_6
4a	V	5	-0.003_2	$+0.003_1$	0.046_4	0.046_5
4a	Mn	3	-0.003_2	$+0.003_1$	0.009_2	0.009_0
4a	Fe	2.22[e]	-0.003_2	$+0.003_1$	0.018_6	0.019_7
4a	Co	1.71[e]	-0.003_2	$+0.003_1$	0.011_0	0.010_9
5a	Zr	4	-0.043_8	$+0.005_6$	0.006_6	0.006_6
5a	Nb	5	-0.043_8	$+0.005_6$	0.045_8	0.045_9
5b	In	3	-0.001_0	$+0.001_2$	0.003_8	0.003_8
5b	Sn	4	-0.001_0	$+0.001_2$	0.009_8	0.009_8

[a]Valence.
[b]eV per atom percent.
[c]eV per square valence difference per atom percent.
[d]eV per atom percent.
[e]Number of $3d$ holes per atom.

The good agreement between the experimental and calculated values of $q(E)/C$ has two important consequences. The first is that the constants K_1 and K_2 are shown to be valid for an entire period in the periodic table. The second is that they make it possible to predict the ATPs of alloys containing dilute ternary and higher-order alloy addition elements in substitutional solid solution in the lattice of the critical copper–nickel base (see Section 7.1.27).

7.1.26 Verification of the Temperature Dependence of $(E_0 - E_F)$

It was anticipated (Section 7.1.23.3) that $(E_0 - E_F)$ should be a linear function of temperature [Eq. (7-157)]. This was shown to be the case in Figs. 7-21 and 7-22. This linear behavior may be expressed as

$$(E_0 - E_F) \simeq B + m(E)k_B T \qquad (7\text{-}159)$$

in which B is the energy intercept, $m(E)k_B$ is the slope, and k_B is Boltzmann's constant. Since Eq. (7-101b) is valid only above the Debye temperature, no physical significance may be attributed to the constant B.

It was noted that approximations in its derivation limited the accuracy of Eq. (7-95) to the first order in $k_B T/E_F$. The values of E_F for transition elements are small compared to those of normal metals. Therefore, departures from Eq. (7-95) would be anticipated for transition elements at elevated temperatures.[17] The data show that

[17]N. F. Mott and H. Jones, op. cit., p. 311.

the values of $(E_0 - E_F)$ of all alloys of this class are linear functions of temperature over the range from 473 to 1173 K.[5] The relative error, therefore, must be a constant, because the predicted deviations at temperatures of the order of 1000 K are not observed. It would thus appear that Eq. (7-101b) provides a better description of binary copper–nickel alloys than had been anticipated originally by its authors.

The values of $m(E)$ in Eq. (7-159) for the binary copper–nickel alloys lie between 1.44 and 1.46,[5] or very close to the value of $\frac{3}{2}$ predicted by Eq. (7-157). Thus, the energy range between the Fermi level and the top of the 3d band varies approximately as $\frac{3}{2}k_B T$. This confirms the "nearly free" behavior assumed in Section 7.1.23.3.

An electron with energy within $\frac{3}{2}k_B T$ or less of the top of the hybridized 3d band would have a high probability of being found in one of the many hybrid 4s states.[18] This verifies the observation that such states probably are contained in overlapping s and d bands and that the two-band model is an oversimplification[19] (see Section 7.1.14.3).

Alloys containing less than 0.5 At% of ternary elements show very small differences between their values for $m(E)k_B$ [Eq. (7-159)] and the slope of the critical copper–nickel base. This agrees with the approximation used to define the scattering parameter $q(E)$ in Sections 7.1.23.2 and 7.1.25.

The previous discussions did not consider alloys with concentrations of ternary alloying atoms greater than 0.5 At%. These alloys show small but significant differences in their slopes of $(E_0 - E_F)$ versus temperature relative to that of the critical copper–nickel base (Fig. 7-22). This effect may be expressed empirically as

$$(E_0 - E_F) \simeq B + [m(E) + \Delta m(E)]k_B T \qquad (7\text{-}160)$$

where $m(E)k_B$ is the slope of the critical copper–nickel base and $\Delta m(E)k_B$ is the change in that slope induced by the presence of a ternary alloying element in excess of 0.5 At%. For the dilute alloys containing less than 0.5 At% of a ternary element, $\Delta m(E)$ is approximately equal to zero and Eq. (7-160) reduces to Eq. (7-159). For concentrations at which $\Delta m(E)$ becomes significant,

$$\Delta m(E) \simeq K_3(Z_\beta C) \qquad (7\text{-}161)$$

where K_3 is a constant, Z_β is the valence of the ternary alloy atom, and C is the quantity of that ternary atom present in solid, substitutional solution in the critical copper–nickel base, in atom percent. Data for this relationship are given in Table 7-15.

As noted in Section 7.1.23.2, all the ternary alloying atoms increase the value of $(E_0 - E_F)$ as compared to that of the critical copper–nickel base and can affect $\Delta m(E)$. Thus, the parameter $\Delta m(E)$ is regarded as an indicator of the changes that take place in the outer electron levels when normal metal ternary elements, those with completed inner electron bands, are added to the copper–nickel base in excess of 0.5 At%.

---

[18]S. Raimes, *Wave Mechanics of Electrons in Metals*, Wiley-Interscience, New York, 1961, p. 160.

[19]W. Hume-Rothery, in *Electronic Structure and Alloy Chemistry of the Transition Elements*, P. A. Beck, ed., Wiley, New York, 1963, p. 94.

TABLE 7-15 FACTORS FOR EQUATION (7-161)

Element	Period	$Z_\beta{}^a$	$K_3(Z_\beta)$	$\Delta m(E)/C^b$ (experimental)	$\Delta m(E)/C^b$ (calculated)
Mg	3	2	$0.005Z_\beta^2$	0.02	0.02
Al	3	3	$0.005Z_\beta^2$	0.04	0.05
Si	3	4	$0.005Z_\beta^2$	0.11	0.09
Ti	4a	4	$0.002Z_\beta^2$	0.03	0.03
V	4a	5	$0.002Z_\beta^2$	0.05	0.05
Mn	4a	3	$0.8/Z_\beta^2$	0.07	0.09
Fe	4a	2.22	$0.8/Z_\beta^2$	0.02	0.02
Zr	5a	4	$0.002Z_\beta^2$	0.02	0.03
Nb	5a	5	$0.002Z_\beta^2$	0.05	0.05
In	5b	3	$0.005Z_\beta^2$	0.03	0.04
Sn	5b	4	$0.005Z_\beta^2$	0.12	0.06

[a]Valence.

[b]Per atom percent of alloying element (average).

Increases in the value of $\Delta m(E)$ indicate that increasing amounts of s-like behavior are induced by the ternary atoms. This behavior was pictured in Section 7.1.23.3 as a small displacement of the hybrid s band to slightly lower energies to permit more d to s resonance.

Ternary, alloying, transition atoms in substitutional, solid solution in the lattice of the critical copper–nickel base have interesting effects. The presence of Co up to about 0.3 At% has a negligible effect on $\Delta m(E)$. The elements Mn and Fe affect $\Delta m(E)$ appreciably. Since these atoms add holes to the hybridized electron structures of the resultant alloys, it is expected that their effects on $\Delta m(E)$ should be different from those normal atoms that add electrons. The effects of Mn and Fe on the hybridized electron structure appear to be inversely proportional to the square of the number of d-level holes of these atoms. The negligible effect of Co probably results from the similarity of its electronic configuration to that of nickel. It also should be noted that Co additions have negligible effects on the lattice parameter of nickel.[20] It would thus appear that small, substitutional additions of Co to the lattice of the copper–nickel base leave its band and zone structures virtually unaltered.

7.1.27 Application to the Design of Copper–Nickel Thermoelements (Types EN, JN, or TN)

Equations (7-156), (7-158), (7-158a), and (7-160) make it possible to predict the ATPs of copper–nickel based solid solutions, provided that the parameter B is known. An examination of the behavior of this parameter shows that it may be expressed as an

[20]W. B. Pearson, *A Handbook of Lattice Spacings and Structure of Metals and Alloys*, Pergamon, Elmsford, N.Y., 1958, p. 777.

empirical function of the copper content, C_{Cu} (in atom percent), as

$$B = 0.0334 + 0.000314C_{Cu} \quad eV \tag{7-162}$$

It is frequently convenient to employ one temperature as being representative in expressing thermoelectric effects. In the case of copper–nickel alloys, 773 K is used often since it is midway in the 100° to 900°C range. This is permissible because the small (<0.5 At%) alloying additions normally present in substitutional solid solution in the critical base alloy induce parallel displacements of the linear $(E_0 - E_F)$ functions of the alloys from that of the critical base (Section 7.1.25). When 773 K is used and the equations noted previously are suitably combined, it is found that

$$S \simeq -\frac{9.43}{0.0129_3 + 0.000314C_{Cu} + q(E)} \quad \mu V/°C \tag{7-163}$$

Equation (7-163) is the empirical form of Eq. (7-101) for describing the ATPs of constantans at 500°C. It is valid for ternary alloys containing 56 to 59 At% of copper in solid solution in nickel and up to 0.5 At% of a third element in substitutional solid solution in the lattice of that alloy. The accuracy of Eq. (7-163) is shown in Fig. 7-23.

Practical thermoelements have been designed in this way. However, these are more commonly designed empirically by describing the thermoelectric effects of an arbitrary amount of the third element at a given temperature.[21,22] Where multiple

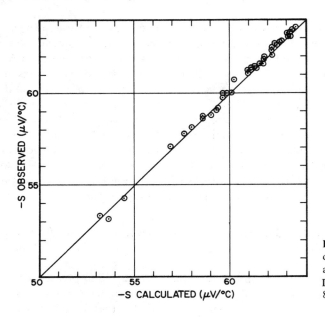

Figure 7-23 Correlation between observed and calculated values of ternary alloys of Cu and Ni at 500°C. (From D. D. Pollock, *Trans. Met. Soc. AIME, 224,* 892, 1962.)

[21]D. I. Finch, E. Korostoff, and D. D. Pollock, *Copper–Nickel Thermoelements with Controlled Voltage–Temperature Characteristics*, U.S. Patent No. 3,017,269, Jan. 16, 1962.

[22]D. I. Finch, E. Korostoff, and D. D. Pollock, *Copper–Nickel Thermocouple Alloys*, U.S. Patent No. 3,266,891, Aug. 16, 1966.

alloying additions are made to the critical copper–nickel base, the effects of small amounts of the alloying elements are additive, within their combined limit of solid solubility in the lattice of the critical copper–nickel based alloy.

It is, thus, possible to design commercial alloys with other specific sets of physical properties in addition to the required thermoelectric behavior.

7.1.28 Other Thermoelectric Devices

The Thomson thermodynamic relationships, derived in Sections 7.1 to 7.1.4 and Eq. (7-95) for metallic conductors, also hold for semiconductors; Eqs. (7-48(a) to (c), (7-90b), and (7-103) are specifically for semiconductors. It was shown that the absolute thermoelectric powers of semiconductors are much larger than those of metals. This has made their application in power generation and refrigeration desirable [Eqs. (7-48), (7-90), and (7-103) and Sections 7.1.8.3, 7.1.9.5, 7.1.10.3, and 7.1.11].

The Peltier effect (Section 7.1.2) is defined as

$$P_{AB} = \frac{\Delta Q}{I}$$

Use now is made of Eq. (7-14):

$$P_{AB} = \frac{dE_{AB}}{dT} T \qquad (7\text{-}14)$$

Equating these relations gives an expression for the thermal energy in terms of the electrical energy as

$$\Delta Q = I \frac{dE_{AB}}{dT} T \qquad (7\text{-}164)$$

This equation directly relates the thermal and thermoelectric properties. Thus, where the $A - B$ combination of thermoelements is such that it provides a large Peltier heat change, it also generates a large thermoelectric power. This means that thermoelements that are desirable for refrigeration also (in most cases) are good for power generation.

The emf generated by such a pair of thermoelements is derived from integrating Eq. (7-18):

$$E_{AB} = \int_{T_1}^{T_2} (S_A - S_B)\, dT \equiv S_{AB}\,\Delta T \qquad (7\text{-}18\text{b})$$

It will be recalled (Section 7.1.3) that the Thomson coefficient is the specific heat of electricity. Thus, the Thomson heat generated is equal to $\sigma j\, dT/dx$ and this heat will be absorbed or liberated depending on the direction of the flow of the electrons with respect to the temperature gradients in the thermoelements (Sections 7.1.3 and 7.1.4). As the temperature of the heated junction of two semiconductors is increased, the majority carriers in each leg will diffuse to the colder portions of the respective component in order to lower their potential energies (Sections 7.1.9.1 and 7.1.9.3). The

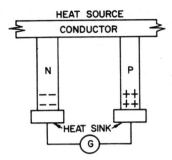

Figure 7-24 Schematic diagram of a
component of a thermoelectric generator.
The conductor and the heat sinks are
metallic. In the case of semiconducting
thermoelements, the joints are ohmic.
(Potential barriers are not present
between components.)

most efficient thermoelectric circuit is thus provided by a thermocouple consisting of
n-type and *p*-type semiconductor elements (Fig. 7-24).

The configuration shown in Fig. 7-24 is that intended for the generation of
power. Groups of thermocouples of this kind are connected in series and arranged so
that their hotter junctions are in good thermal contact with a heat source such as that
provided by nuclear fission. This technique has long been used to provide power for
satellites.

If, instead of creating a temperature difference between the junctions, current is
made to flow in the same direction as the Seebeck current, heat will be absorbed at the
former heated junction and will be liberated at the prior cooler junction. The
absorption of heat by a series arrangement of thermocouples of this kind about a
container constitutes a means of refrigeration (see Section 7.1.4.2).

In either of these applications, the electrical conductivity, σ, of the thermoelements plays an important role (Sections 6.3.1, 6.3.3, and 6.3.4). The Joule heating in
the thermoelements, I^2R, represents extraneous heating in refrigeration or a loss of
power in power generation. As such, it represents a major loss in both applications
and decreases their thermodynamic efficiencies. The thermal conductivity, κ, allows
heat to flow into the refrigerated zone or it permits heat losses from the heated
junction, decreasing the ΔT in power generation. The most desirable materials for
these purposes, therefore, should have high σ and low κ. This combination of
properties is incompatible with the Wiedemann–Franz ratio [Eq. (6-29)]; this rule
requires that κ and σ be directly proportional to each other at a given temperature:

$$\kappa = L_0 T\sigma \qquad\qquad (6\text{-}29)$$

Here, the "constant" of proportionality, L_0, is the Lorenz number. *L_0 is not a
constant for all metallic or semiconductor materials*, Section 6.0. It can vary very
widely as a function of temperature and alloy content. In some metal alloy systems,
L_0 may show a large variation of values at low temperatures and converge toward a
much smaller range of variations about the theoretical value of 5.85×10^{-9} (cal-Ω)/(s-
K^2), or 2.45×10^{-9} (W-Ω)/K^2, at temperatures above about 800 K. It is necessary to
obtain the best possible combinations of these two properties, despite their incom-
patibility and their variations, because they are of fundamental importance in
determining the efficiency of thermoelectric devices.

One indication why present theories have not been successful in the prediction of the thermodynamic efficiency of semiconductor materials has been given. Instead, a figure of merit, Z, is used to describe their properties for these applications. One such approach is given by

$$Z = \frac{S^2\sigma}{\kappa} = \frac{S^2}{\rho\kappa} \tag{7-165}$$

in which S is the absolute thermoelectric power and σ is given by Eq. (6-111) in which n_e or n_h may be approximated from Eqs. (6-127a) or (6-127b) or from (6-163) and/or (6-164). Other expressions for the figure of merit are available, but these are not as commonly used as the one given here.

The value of S of semiconductor materials depends primarily on the amount of dopant present (if any), the energy levels of the dopant states, the energy gap, and the temperature [Eqs. (7-48a) to (c), (7-90b) and (7-103)]. These factors determine the number of carriers per unit volume and, thus, are major factors influencing the heat capacity, the internal potentials, or the electrical conductivity. The conductivity, as employed in Eq. (7-95), governs the ATP. Also see Eqs. (7-48a) and (7-90b).

In the case of doped semiconductors that are used in their temperature ranges of extrinsic conduction, either Eq. (6-160) or Eq. (6-161) provides the density of carriers. The appropriate equation, depending on the type of dopant, is substituted into Eq. (6-109). This is expressed as $\ln \sigma$ and is used in Eq. (6-130) to obtain S.

Where intrinsic semiconductors are being employed, Eq. (6-129) in the logarithmic form, as in Eqs. (6-130) or (6-131), is used in Eq. (7-95) to calculate the ATP. A similar approach may be used for the case of lightly doped semiconductors when these are employed well within their intrinsic temperature ranges. In this case, the approximation sometimes can be made that the number of extrinsic carriers is very much less than the number of pairs of intrinsic carriers. This simplifies the calculations by treating the material as being intrinsic as in the manner just described. The relatively small number of extrinsic carriers is neglected.

In some cases, carriers of both origins must be taken into consideration in the calculation of S. Some of the reasons for this include such factors as the extent (width) of the energy gap, the dopant concentration, the dopant levels, and the temperature. Here the expression for the electrical conductivity that must be employed in Eq. (7-95) is the natural logarithm of the sum of the conductivities resulting from the carriers generated by both mechanisms.

The figure of merit is a function of the number of carriers. S decreases as the density of carriers, n, increases when Eq. (7-96) is substituted into Eq. (7-95). The quantity $[\partial \ln \phi(E)/\partial E]$ in Eq. (7-95) decreases as n becomes large because $\tau(\bar{k}_x)$ decreases rapidly with n. Both σ and κ generally increase with n. This variation of the parameters results in a maximum for Z as a function of n that lies approximately between 10^{18} to $10^{20}/cm^3$. Since n is a function of temperature, Z is also a function of temperature. Z usually lies in the range of about $1 \times 10^{-3}/deg$ to $3 \times 10^{-3}/deg$. Values in this range give thermal efficiencies of about 5%, although values as high as 13% have been reported. Thermal efficiencies greater than about 8% or 9% appear to be questionable since they approach the theoretical limit of thermal efficiency. The problem is one of finding the best balance of S, σ, and κ for a given application. Such a balance usually results in absolute thermoelectric powers in the neighborhood of 250 mV/°C at about 500°C.

Some materials with high values of Z, near 500°C, include $Pb_{1.05}Te_{0.95}$, Bi_2Te_3, and $Bi_2Te_{2.4}Se_{0.6}$; both Bi-based compounds are doped with CuBr. $CuGaTe_2$ has a value of $Z \simeq 3.0 \times 10^{-3}$/°C near 500°C. These are among the better materials for power generation. Bi_2Te_3 has been used for refrigeration devices. Refractory materials are used at temperatures higher than 1000°C; these have Z about 10^{-4}/°C in this temperature range.

The low thermal efficiencies of these devices make them acceptable only in remote locations or in unusual situations. These include power generation where conventional sources are unavailable, such as in arctic sites, space vehicles, or satellites. They constitute relatively long lived power sources in space applications, in which the thermal energy is provided by spontaneous nuclear fission. In addition, since no moving parts are involved, they do not affect the orbits of satellites.

Thermoelectric refrigeration is at least an order of magnitude more costly to operate and has been shown to cost more than a comparable, conventional appliance by a factor of about 20.

Despite these adverse factors, Peltier devices are being used widely to cool sensitive electronic circuitry, such as in computers, and as thermoelectric reference temperature sources. These are virtually trouble free and can be controlled relatively simply (see Sections 7.1.2 and 7.1.4.2).

7.2 GALVOMAGNETIC EFFECTS

Up to this point, current flow in solids has been examined in the absence of external magnetic fields. Here the force acting on a carrier during a steady flow of current along the conductor is

$$\bar{F}_V = e\bar{E}_x \tag{7-166}$$

The application of an externally induced magnetic field of strength $\bar{B}$, perpendicular to the conductor, induces an additional force on the carrier, known as the *Lorentz force*, that is given by

$$\bar{F}_L = e\mathbf{v}_x \times \bar{B} \tag{7-167}$$

where $\mathbf{v}_x$ is the average velocity of a nearly free electron with energy within the range of $E_F \pm 2k_BT$.

Thus, the total force acting on the carrier is obtained from the sum of Eqs. (7-166) and (7-167) as

$$\bar{F} = \bar{F}_V + \bar{F}_L = e\bar{E}_x + e\mathbf{v}_x \times \bar{B} \tag{7-168}$$

Equation (7-168) is known as the Lorentz relationship. This is valid for degenerate carriers such as those in metals and in highly doped semiconductors in which the range of velocities represented by v_x is small ($\simeq \pm 2k_BT$). Equation (7-168) does not hold for cases in which the carriers are nondegenerate, such as intrinsic or lightly doped semiconductors where the velocity range is large. Here the velocity distribution is given by the Boltzmann equation (see Sections 2.2, 2.6, and 2.8). These carriers may *not* be considered to have the same velocity, v_x, as is done for degenerate carriers.

The following sections on galvomagnetic effects are based on the influences of the Lorentz force on the carriers.

7.2.1 Hall Effect

Consider the flow of carriers in a conductor, such as that shown in Fig. 7-25, in the absence of an external magnetic field. The equipotential lines will be parallel to each other and perpendicular to the direction of current flow. Thus, when the probes are exactly opposite to each other, the voltage between the probes will be zero.

When an externally induced magnetic field is applied transverse to the conductor, such as an *n*-type semiconductor, the Lorentz force acts on each of the carriers. Where $\bar{B}$ is perpendicular to v_x, Eq. (7-167) becomes

$$\bar{F}_L = ev_x\bar{B} \qquad (7\text{-}167a)$$

Here v_x and the average energy of carriers whose energies all lie in the small range of $\frac{1}{2}mv_x^2 \simeq 2k_BT$ as in Fig. 5-17. This force causes the carrier to travel toward the bottom of the conductor (in the $-y$ direction in the figure). The density of the electrons increases at the bottom edge of the conductor. This creates an absence of electrons, an excess of positive charges, on its opposite edge. The creation of the potential difference across the conductor, V_H, the Hall voltage, induces an electric field given by

$$V_H = \bar{E}_H y, \qquad \bar{E}_H = \frac{V_H}{y} \qquad (7\text{-}169)$$

The transverse field then exerts a transverse force on each carrier equal to

$$\bar{F}_T = e\bar{E}_H \qquad (7\text{-}170)$$

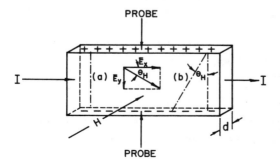

Figure 7-25 Model for the Hall effect. *H* is normal to the face of the specimen. (a) and (b) represent equipotential surfaces before and after the application of the field, respectively.

This transverse force opposes the Lorentz force. The deflection of the carriers by the Lorentz force continues until a dynamic equilibrium is reached at which $\bar{F}_T = \bar{F}_L$. This is the equivalent of setting Eq. (7-168) equal to zero. Here, using Eqs. (7-170) and (7-168),

$$e\bar{E}_H = e\mathbf{v}_x\bar{B} \qquad (7\text{-}171a)$$

or, since the charge on the carrier vanishes,

$$\bar{E}_H = \mathbf{v}_x\bar{B} \qquad (7\text{-}171b)$$

The Hall voltage is obtained by the substitution of Eq. (7-171b) into Eq. (7-169) as

$$V_H = \mathbf{v}_x\bar{B}y \qquad (7\text{-}172)$$

The Hall voltage arises because the equipotential lines no longer are perpendicular to the current flow. They now make an angle, θ_H, known as the *Hall angle*, with this direction. The probes now measure the potential difference between different equipotential lines. The Hall voltage is this potential difference.

Now, using Eq. (6-15) in the form

$$j_x = ne\mathbf{v}_x \qquad (7\text{-}173)$$

the factor $\mathbf{v}_x$ is found to be

$$\mathbf{v}_x = \frac{j_x}{ne} \qquad (7\text{-}174)$$

Equation (7-174) is substituted into Eq. (7-172) to obtain

$$V_H = \frac{1}{ne}j_x\bar{B}y = R_H j_x\bar{B}y \qquad (7\text{-}175)$$

from which the Hall coefficient, R_H, is defined as

$$R_H = \frac{1}{ne} \qquad (7\text{-}176)$$

The sign of R_H will be determined by that of the majority carrier; it will be positive for holes and negative for electrons. As previously noted for the thermoelectric relationships, Eq. (7-176) provides the sign and density of the carriers, while electrical conductivity measurements can provide the carrier density, but not their sign [Eq. (6-19)].

The Hall mobility, μ_H, may be obtained from Eq. (7-176) by multiplying its numerator and denominator by μ_H:

$$R_H = \frac{\mu_H}{ne\mu_H} = \frac{\mu_H}{\sigma_H} = \mu_H\rho_H \qquad (7\text{-}177)$$

in which σ_H and ρ_H are the Hall conductivity and resistivity, respectively. It should be noted that μ_H may not necessarily agree with the mobilities discussed in Section 6.3.1.2 (see Table 7-16).

TABLE 7-16 HALL MOBILITIES IN SELECTED
SEMICONDUCTORS

Semiconductors	μ_e^* (cm^2/V-s)	μ_h^* (cm^2/V-s)
Si	1,300 $(300/T)^{2.0}$	500 $(300/T)^{2.7}$
Ge	4,500 $(300/T)^{1.6}$	3,500 $(300/T)^{2.3}$
GaAs	8,500 $(300/T)^{1.0}$	420 $(300/T)^{2.1}$
GaP	110 $(300/T)^{1.5}$	75 $(300/T)^{1.5}$
GaSb	4,000 $(300/T)^{2.0}$	1,400 $(300/T)^{0.9}$
InP	4,600 $(300/T)^{2.0}$	150 $(300/T)^{2.4}$
InSb	78,000 $(300/T)^{1.6}$	750 $(300/T)^{2.1}$

Adapted from H. F. Wolf, *Semiconductors*, Wiley-Interscience,
New York, 1971, p. 303, with permission. See Section 6.3.1.2.
[a]cm^2/V-s.

The Hall angle may be approximated by starting with the definition of mobility given by Eq. (6-7) and rewritten as

$$\mu_H = \frac{\mathbf{v}_x}{\bar{E}_x} \tag{7-178}$$

so that the velocity component in the x direction, the direction of current flow, is

$$\mathbf{v}_x = \mu_H \bar{E}_x \tag{7-179}$$

This is substituted into Eq. (7-171b) to obtain

$$\bar{E}_H = \mu_H \bar{E}_x \bar{B} \tag{7-180a}$$

or

$$\frac{\bar{E}_H}{\bar{E}_x} = \mu_H \bar{B} \tag{7-180b}$$

And, since $\bar{E}_H$ is the field opposing the Hall field, $\bar{E}_H$ is identical to $\bar{E}_y$ in Fig. 7-25. So, for small angles, $\bar{E}_H/\bar{E}_x \equiv \bar{E}_y/\bar{E}_x \simeq \theta_H$. Thus, the Hall angle is obtained from Eq. (7-180) as

$$\theta_H \simeq \mu_H \bar{B} \tag{7-181}$$

Or, using Eq. (6-109),

$$\theta_H \simeq R_H \sigma_H \tag{7-181a}$$

The Hall coefficient may be determined directly from the experimental data. The relationship used for this is obtained from Eq. (7-175) in the form

$$V_H = \frac{R_H I \bar{B}}{d} \tag{7-182}$$

in which I is the current and d is the thickness of the specimen as in Fig. 7-25.

In the use of Eq. (7-182) to experimentally determine the R_H of metals and their alloys, considerable care must be taken to ensure accurate results. The probes must be aligned as accurately as possible so that meaningful values of V_H are obtained. In addition, V_H should be determined using a null balance; this eliminates any Joule heating and the possible creation of any transient, parasitic, thermoelectric effects. The edges of the specimen must be parallel and the thickness of the specimen must be uniform. Variations in these dimensions can affect the equipotential lines and introduce errors.

In the case of metals, n is typically large so that V_H and R_H are small. This can be partially alleviated by making d as thin as is practical and using relatively large currents [Eq. (7-182)]. These, too, can result in undesirable Joule heating and extraneous emfs, especially if d is not uniform. Errors of these kinds may be avoided by conducting such tests under conditions in which the apparatus is immersed in a constant-temperature environment, such as in a suitable oil bath. This may be very important when some semiconductors are being tested.

Unlike the case of metals, the factor n for semiconductors is relatively small and is a function of temperature [Eqs. (6-116), (6-117), (6-160), and (6-161)]. Uncontrolled temperature variations, as described previously, may introduce large uncertainties. However, V_H and R_H are large compared to those of metals; *small* errors in n can be tolerated in some semiconductor applications because of this, provided that these are known.

Hall specimens of semiconductors may be prepared by photochemical etching of thin wafers where the current leads and the potential probes are an integral part of the test specimen. Another technique involves the use of a fixture in which one probe is fixed and the other is adjustable. The latter is adjusted until a null balance is obtained. The value of θ_H may be read directly and any desired Hall data may be obtained by starting with Eqs. (7-181a) and (7-182).

Considerable data in the literature on the Hall effect are given in esu units. In this case, Eq. (7-176) is written as

$$R_H = \frac{1}{nec} \qquad (7\text{-}176a)$$

Here $c \simeq 3 \times 10^{10}$ cm/s. These esu data may be reexpressed in units of V-cm/A-G by multiplying them by a factor of 9×10^{11}, another unit frequently found in the literature.

In the case of extrinsic semiconductors (Section 6.3.3), the Hall coefficient is given by

$$R_H = \frac{\beta}{e} \cdot \frac{n_h \mu_h^2 - n_e \mu_e^2}{(n_h \mu_h + n_e \mu_e)^2} \qquad (7\text{-}183)$$

where β is a parameter that takes into account the scattering of the carriers. Where $n_h = n_e = n_i$, as is the case for intrinsic semiconductors (Section 6.3.1), Eq.

(7-183) is reduced to

$$R_H = \frac{\beta}{n_i e} \cdot \frac{\mu_h - \mu_e}{\mu_h + \mu_e} \tag{7-184}$$

And for those intrinsic semiconductors in which $\mu_h \ll \mu_e$, Eq. (7-183) may be reduced further to read

$$R_H \simeq \frac{\beta}{n_i e} \tag{7-185}$$

As noted previously, the velocity distribution used to arrive at Eq. (7-176) is valid only for metals and highly doped semiconductors in which the carriers are degenerate. Thus, for greater accuracy and generality, Eq. (7-176) should be written as

$$R_H \simeq \frac{\beta}{ne} \tag{7-186}$$

So, for degenerate carriers, at $T \leqslant \theta_D$, $\beta \simeq 1$; for $T > \theta_D$, $\beta \simeq 1.1$. For the case involving nondegenerate carriers, such as intrinsic or lightly doped semiconductors, $\beta \simeq 3\pi/8$ in Eq. (7-184) or Eq. (7-185). Here the carrier scattering is primarily caused by phonons. Where the scattering due to impurity concentration is greater than that of the lattice, in a way analogous to that described in Section 6.1.4, $\beta \simeq 5\pi/8$ in Eq. (7-183).

As previously noted, Hall-effect data are of use for theoretical and practical purposes. In contrast to electrical conductivity or resistivity measurements, these provide the sign of the majority carrier as well as their concentration and mobility. Such measurements are of great utility in the determination of the effective carrier densities and band structures of metallic alloys and of semiconductors.

Equation (7-176) provides a good representation of normal metals and their alloys; here $\beta \simeq 1$ [Eq. (7-185)]. In this case, the degenerate electrons occupy states such that E_F is unaffected by the top of the band or by the zone wall (Sections 5.1.10.1, 5.1.10.2, 6.1.3, 7.1.8.1, and 7.1.8.2). The noble and alkali metals show evidence of electron:atom ratios that are close to unity. Metals with higher valences show departures from integral valences because of the nature of the metallic bond (Sections 3.11 and 3.12.4). As is expected, those transition elements with comparatively large ranges of $(E_0 - E_F)$, such as Fe and Co, show hole conduction, while Ni, with a small value of $(E_0 - E_F)$, indicates that the majority carriers are electrons (Table 7-17).

The values of Sb and Bi are large and in agreement with Eq. (7-176a). This is in agreement with the diamagnetic mass susceptibilities of both of these elements. These are $\simeq -1 \times 10^{-6}$, an indication of nearly filled valence bands. Antimony has a positive value of R_H, while that of Bi is negative.

Considerable use is made of Hall effect data in the quality control of processing and use of semiconductor materials and devices. These provide insight into the

TABLE 7-17 HALL COEFFICIENTS OF SOME METALS

Element	$R_H(\times 10^{12})^a$	Element	$R_H(\times 10^{12})$	Element	$R_H(\times 10^{12})$
Cu	−5.5	Be	24.4	Fe	100
Ag	−8.4	Zn	3.3	Co	24
Au	−7.2	Cd	6.0	Ni	−60
Li	−17.0	Al	−3.0		
Na	−25.0				

From F. Seitz, *Modern Theory of Solids*, McGraw-Hill, New York, 1940, p. 183.
[a]V/cm-abamp-gauss.

degree of lattice perfection, the purity of intrinsic materials, the actual amounts of dopants in extrinsic materials, and the mobility of the carriers in both cases (Table 7-16). As is to be expected, the properties measured by the Hall effect may vary widely depending on the degree of crystalline perfection and the purity of the samples.

Use is made of the Hall effect in such instrumentation as magnetometers, phase meters, ac–dc and dc–ac converters, microphones, and signal generators.

7.2.2 Ettingshausen Effect

When a magnetic field is directed perpendicularly across a current-carrying conductor, a temperature gradient is formed at right angles both to the magnetic field and to the current. This transverse temperature gradient is the Ettingshausen effect.

It was noted previously that Eq. (7-168), and consequently Eq. (7-171a), hold only where the average carrier velocity v_x is representative of carriers with a relatively small range of velocities (Sections 7.2 and 7.2.1). Where nondegenerate carriers are involved, the velocity range of the carriers, as determined by the Boltzmann statistics, is much too large for v_x to be a valid representation of their velocities. Thus, in terms of Eqs. (7-168) and (7-171a), a situation will exist where the forces acting on the carriers will range from $\bar{F}_L > \bar{F}_V$ through $\bar{F}_L = \bar{F}_V$ to $\bar{F}_L < \bar{F}_V$. Where $\bar{F}_L > \bar{F}_V$, the carriers will have their paths bent down toward the bottom of the specimen in Fig. 7-25; those with $\bar{F}_L < \bar{F}_V$ will have their paths bent up toward the top of the specimen. The extra energy increment of the higher-energy carriers is emitted and absorbed as thermal energy by the lattice at the bottom of the specimen. The energy deficit of the electrons at the top of the specimen is made up by the absorption of thermal energy from the lattice. Thus, the bottom of the specimen is hotter than the top; this is the Ettingshausen effect. This effect is given by

$$\frac{\partial T}{\partial y} = P_E \bar{B} I_x \qquad (7\text{-}187)$$

where P_E is the Ettingshausen coefficient and I_x is the current.

Efforts to employ the Ettingshausen effect for cooling purposes have shown that temperatures of about 100°C below that of the ambient temperature can be achieved.

7.2.3 Nernst Effect

When a conductor that contains a temperature gradient along its length is subjected to a homogeneous magnetic field that is at right angles to the temperature gradient, the electrons that usually flow from the hotter to the colder ends of the specimen begin to be diverted by the magnetic field. They now tend to follow a circular path as a result of the Lorentz force. The radius of this path is given by

$$r = \frac{m_e \mathbf{v}}{e\bar{B}} \tag{7-188}$$

The electrons with higher velocities follow paths of larger radii than do the slower ones [Eq. (7-188)]. Thus, the front surface of the specimen (Fig. 7-25) will have a high concentration of the higher-energy electrons, while the rear surface will have a high concentration of the lower-energy electrons. This induces a transverse electric field, $\bar{E}_y$, that opposes any additional deflection of the electrons. This field is given by

$$\bar{E}_y = Q\bar{B}\frac{\partial T}{\partial x} \tag{7-189}$$

where Q is the Nernst coefficient and $\partial T/\partial x$ is the temperature gradient along the length of the specimen. This has also been called the *transverse* Nernst–Ettingshausen effect. The change in the Seebeck effect induced by a magnetic field is known as the *longitudinal* Nernst–Ettingshausen effect.

The relationship between P_E and Q, originally found by P. W. Bridgeman, is given by

$$P_E = QT\kappa \tag{7-190}$$

where κ is the thermal conductivity of the conductor in the transverse magnetic field.

7.2.4 Magnetoresistance

The application of a magnetic field transverse to the flow of current in a conductor causes a change in its electrical resistance, ΔR, as compared to the resistance, R, in the absence of such a field. This comes about because of the previously noted range of velocities of the carriers. If all the carriers had velocities such that their energies were within the narrow range of about $E_F \pm 2k_BT$, then Eq. (7-171a) would be satisfied and no change would occur in the electrical resistance. Thus, all electrons with velocities significantly different from the average $\mathbf{v}_x$ of Eq. (7-167) will have curvatures induced into their paths, while simultaneously inducing a transverse electric field that is in equilibrium with the magnetic field. The curvature of the paths of the affected electrons has the effect of decreasing the mean free paths in the absence of the magnetic field, λ, by an amount $\Delta\lambda$.

This effect may be expressed using the electrical resistivity equivalent of Eq.

(6-19) as

$$\frac{\Delta\rho}{\rho} = \frac{\Delta\lambda}{\lambda} = \frac{\Delta\mu}{\mu} \tag{7-191}$$

in which μ is the mobility. The fraction $\Delta\rho/\rho$ is called the *magnetoresistivity*. Where the applied magnetic field is small, Eq. (7-191) varies as $\bar{B}^2$, or

$$\frac{\Delta\rho}{\rho} \propto \bar{B}^2 \tag{7-192a}$$

And, where the applied magnetic field is large,

$$\frac{\Delta\rho}{\rho} \propto \bar{B} \tag{7-192b}$$

Here a large magnetic field is one that causes λ to be large with respect to the electron's "orbit". Equation (7-192a) may also be expressed as

$$\frac{\Delta\rho}{\rho} = C\mu^2\bar{B}^2 \tag{7-193}$$

where C is a constant.

A magnetoresistance effect also occurs when the magnetic field is parallel to the direction of the flow of current. This change is smaller than that induced by transverse magnetic fields.

The magnetoresistance of most metals is small and positive. The decrease in the wavelength, and in the mobility, increases the electrical resistivity (Section 6.0). Some alloys of noble and transition elements show decreased electrical resistivity under these conditions and, thus, have negative magnetoresistances (see Section 7.1.8.2). Here it appears that additional carriers become available by promotion out of the hybridized d bands. Nickel shows a positive change of about 2% (Table 7-17 and Section 7.2.1). Fully magnetically saturated ferromagnetic materials show negative magnetoresistances. Some semiconductors have large, but highly anisotropic properties.

On the basis of the foregoing sections, it is seen that combined data from Hall effect and magnetoresistance measurements can provide much insight into the band structure of crystals.

7.3 BIBLIOGRAPHY

BENEDICT, R. P., ed., *Manual on the Use of Thermocouples*, STP470B, 12, American Society for Testing and Materials, Philadelphia, 1971.

———, *Fundamentals of Temperature, Pressure and Flow Measurements*, 3rd ed., Wiley, New York, 1984.

BRIDGMAN, P. W., *Thermodynamics of Electrical Phenomena in Metals*, Dover, New York, 1961.

BRINDLEY, G. W., in *Report of a Conference on Strength of Solids*, p. 95, Physical Society, London, 1948.

BURLEY, N. A., and others, *Nicrosil versus Nisil Thermocouple: Properties and Thermoelectric Reference Data*, National Bureau of Standards, (U.S.), Monograph 161, Washington, D.C., 1978.

COHEN, M. M., *Introduction to the Quantum Theory of Semiconductors*, Gordon & Breach, New York, 1972.

CRANGLE, J., in *Electronic Structure and Alloy Chemistry of Transition Elements*, P. A. Beck, ed., p. 51, Wiley, New York, 1963.

CRUSSARD, C., in *Report of a Conference on Strength of Solids*, p. 119, Physical Society, London, 1948.

CULLITY, B. D., *Elements of X-Ray Diffraction*, Addison-Wesley, Reading, Mass., 1956.

CUSAK, N., and KENDALL, P., *Proc. Phys. Soc.* (London), *72*, 289, 1958.

DARKEN, L. S., and GURRY, R. W., *Physical Chemistry of Metals*, McGraw-Hill, New York, 1953.

DEKKER, A. J., *Solid State Physics*, Prentice-Hall, Englewood Cliffs, N.J., 1959.

DIKE, P. H., *Thermoelectric Thermometry*, Leeds and Northrup Co., North Wales, Pa., 1954.

GEIBEL, W., *Zeitschrift fur Annorganische Chemie*, Vol. 69, p. 38, 1910.

———, *Zeitschrift fur Annorganische Chemie*, Vol. 70, p. 240, 1911.

HUME-ROTHERY W., in *Electronic Structure and Alloy Chemistry of Transition Elements*, P. A. Beck, ed., Wiley, New York, 1963.

KITTEL, C., *Introduction to Solid State Physics*, 5th ed., p. 154, Wiley, New York, 1976.

MACDONALD, D. K. C., *Thermoelectricity: An Introduction to the Principles*, Wiley, New York, 1962.

MCLAREN, E. H., and MURDOCK, E. G., *Properties of Pt/PtRh Thermocouples for Thermometry in the Range 0–1100°C, I, II and III*, National Research Council, Ottawa, Canada, 1979–1983.

MOTT, N. F., and JONES, H., *Theory of the Properties of Metals and Alloys*, Dover, New York, 1958.

NYSTROM, J., *Arkiv Mathematic Fysik, 34A*, No. 27, 1, 1948.

POLLOCK, D. D., *Transactions of the Metallurgical Society*, AIME, Vol. 224, p. 892, 1962.

———, *Transactions of the Metallurgical Society*, AIME, Vol. 230, p. 753, 1964.

———, *Acta Metallurgica*, Vol. 16, p. 1453, 1968.

———, *Theory and Properties of Thermocouple Elements*, ASTM STP492, American Society for Testing and Materials, Philadelphia, 1971.

———, *Physical Properties of Materials for Engineers*, Vol. I, Section 4.1, CRC Press, Boca Raton, Fla., 1982.

———, *Physical Properties of Materials for Engineers*, Vol. I, Chapter 5, CRC Press, Boca Raton, Fla., 1982.

———, *Thermoelectricity: Theory, Thermometry, Tool*, ASTM STP 852, American Society for Testing and Materials, Philadelphia, 1985.

———, and CONARD, G. P., II, *Transactions of the Metallurgical Society*, AIME, Vol. 227, p. 478, 1963.

————, and FINCH, D. I., in *Temperature, Its Measurement in Science and Industry*, Vol. 3, Part 2, p. 237, Van Nostrand Reinhold, New York, 1962.

POWELL, R. L., and others, *Thermocouple Reference Tables Based on the IPTS-68*, National Bureau of Standards, (U.S.), Monograph 125, Washington, D.C., 1974.

REED, R. P., "Thermoelectric Thermometry—A Functional Model," in *Temperature, Its Measurement in Science and Industry*, Vol. 5, American Institute of Physics, New York, 1982.

REED-HILL, R. E., *Physical Metallurgy Principles*, Van Nostrand Reinhold, New York, 1964.

ROESER, W. F., "Thermoelectric Thermometry," *J. Appl. Physics*, 11, 388, 1940.

————, and LONBERGER, S. T., *Methods of Testing Thermocouples and Thermocouple Materials*, NBS circular 590, U.S. Government Printing Office, Washington, D.C., 1958.

SEITZ, F., *Modern Theory of Solids*, McGraw-Hill, New York, 1940.

SPROULL, R. L., *Modern Physics*, 2nd ed., Wiley, New York, 1963.

STRINGER, J., *An Introduction to the Electron Theory of Solids*, Pergamon Press, Elmsford, N.Y., 1967.

WANG, T. P., STARR, C. D., and BROWN, N., *Acta Metallurgica*, Vol. 14, p. 649, 1966.

WEISS, H., *Structure and Application of Galvomagnetic Devices*, Pergamon Press, Elmsford, N.Y., 1969.

WOLF, H. F., *Semiconductors*, Wiley-Interscience, New York, 1971.

7.4 PROBLEMS

7.1. Why is it possible to equate Eq. (7-7) to zero?

7.2. Explain the bases for the fact that the ATP of superconductors is zero.

7.3. Give a practical application of each of the laws of intermediate conductors.

7.4. Why has it not as yet been possible to achieve the ideal thermoelectric behavior indicated by Eq. (7-24)?

7.5. Why is it possible to equate Eqs. (7-36) and (7-39)?

7.6. Account for the differences between Eqs. (7-40) and (7-41).

7.7. Why are the ATPs of semiconductors inherently so much greater than those of metals?

7.8. Explain the reasons for the similarity of Eq. (5-147) to Eq. (7-40) and that of Eq. (5-150) to Eq. (7-41).

7.9. Can the major causes of thermoelectric instability be expressed by a unifying condition? If so, give this expression.

7.10. Explain the deviations from linearity of the ATPs of nickel and its alloys below about 600°C.

7.11. Rationalize the presence of the scattering factor in Eq. (7-156).

7.12. Explain Eq. (7-157).

7.13. Compare the utility of measurements of the Hall effect and of electrical resistivity in the characterization of the mobile carriers. How are these used in the quality control of semiconducting materials?

8 FACTORS AFFECTING THE PROPERTIES OF SEMICONDUCTORS

The electrical conductivity of semiconductors is discussed in Chapter 6, and the Hall effect is given in Chapter 7. These properties are but two of many that influence the application and utility of semiconductors. Other properties that are of importance in semiconductor devices are presented here. This will provide a basis for the understanding of the operation of bulk and junction devices.

8.1 THE WORK FUNCTION

The ions in a crystalline metallic solid set up a periodic potential in the solid as discussed in Section 4.5 and shown in Fig. 4.5(a). At very short distances away from a line or from a plane of the ions, the variation of this periodic potential diminishes greatly [Fig. 4-5(b)]. Thus, about half way between planes of ions, this small variation may be neglected and may be considered to be constant. A free electron requires an energy increment of $eV_0 = W$[Fig. 1-2] to escape from this potential well. (It has a maximum potential energy of $-eV_0$.) The metallic solid, thus, constitutes a potential well for the electron. The work required to remove the electron from the well is the *work function*. The more commonly used symbol ϕ will be used instead of W, as shown in the figure.

If an electron had no kinetic energy (KE) the work function would have to be equal to the energy required to remove the electron from the very bottom of the well in

Fig. 1-2. However, an electron must have energy of translation [Section 1.6.6 and Eq. (1-24)]. And, in addition, where many electrons are involved, electrons must occupy discrete states in a quasicontinuum [Sections 2.7 and 4.2]. Thus, the electrons occupy all states in the well constituted by the metal up to and including E_F [Fig. 3-4(b)]. The energy required to promote an electron out of this well, that is, the minimum energy increment from E_F to the top of the well (or the barrier of the well) is the work function, ϕ. This also has been termed the surface work function (see Section 8.2). The top of the well is taken as the reference energy and is known as the *vacuum level*.

Semiconductors may be considered to provide several wells of varying depths for electrons. Figure 6-23 is used to illustrate this. Electrons may be excited from any level starting with those at or below E_v and ranging to those at or above E_c, including any impurity levels at E_a and/or at E_d. The work function is smallest for excitations from the conduction band. If the excitations are limited to electrons in the conduction band, the equilibria required by Eqs. (6-116), (6-117), (6-160), and (6-161) would be destroyed. Thus, such excitations are replenished by other electrons from the impurity states and/or from the valence band. This replenishment requires the expenditure of energy, and, under adiabatic conditions, this is accomplished by the absorption of energy from the lattice so that the crystal becomes cooler. In the case of electron excitations solely from the valence band, equilibrium is restored by corresponding compensating transitions of electrons returning to the valence band from the conduction band and from transitions from E_a and/or E_d; the energies given up in these transitions cause the crystal to become hotter. Equilibrium can be preserved only if suitable numbers of electrons are excited from states both above and below E_F. When this is the case, E_F is taken as the average energy of all the participating electrons, prior to their ejection. Thus, the work function, or the average barrier height of the wells, again is taken as the energy increment between E_F and the top of the well. From this it is seen that E_F is the reference energy even though electrons may not necessarily occupy states in its vicinity.

The work functions of some of the elements are given in Table 8-1.

8.1.1 Effects of Surface Layers on Metals

The foregoing discussions are based on the assumption that the metallic surfaces are chemically clean. Chemically adsorbed layers on semiconductors or on metallic surfaces can strongly affect their work functions.

Molecules or atoms may adhere to metallic surfaces by means of two mechanisms. The first of these is by physical adsorption (*physisorption*). Here the adhesion results from the weak van der Waals interactions between the surface and the physisorbed species (Section 3.12.5). The small energies involved in this type of bonding [Eq. (3-88)] are readily absorbed by the crystal as phonons. The adhering molecule retains its identity because only its potential energy is slightly reduced, and the resulting bonding energies are much too small to break the intermolecular bonds of the physisorbed species. A molecule so weakly bonded oscillates within a very shallow potential well on the metal surface. The shallowness of the well and the

TABLE 8-1   WORK FUNCTIONS OF SOME METALS (eV)

Na	2.28–2.75	C	5.0	Se	5.9
K	2.22–2.30	Si	4.60–4.85	Te	4.85
Cs	1.93–2.14	Ge	4.80–5.0	Mn	4.1
Cu	4.45–4.94	Sn	4.42	Re	4.96–5.75
Ag	4.26–4.74	Pb	4.25		
Au	5.1–5.47			Fe	4.44–4.81
		Ti	4.33	Ru	4.71
Mg	3.66–3.67	Zr	4.05	Os	4.83
Ca	2.87	Hf	3.9		
Sr	2.59	Th	3.4	Co	5.0
Ba	2.7			Rh	4.98
		V	4.3	Ir	5.00–5.76
B	4.45	Nb	4.18–4.87		
Al	4.20–4.74	Ta	4.13–4.80	Ni	4.96–5.35
Ga	4.2			Pd	4.98–5.6
In	4.12	As	3.75	Pt	5.36–5.7
Tl	3.84	Sb	4.55–4.7		
		Bi	4.22	*Coatings*	
Sc	3.5			Cs	on W 1.36
Y	3.1	Cr	4.5–4.60	Ba	on W 1.56
La	3.5	Mo	4.24–4.95		
		W	4.18–4.63		
		U	3.63–3.90		

Abstracted primarily from M. B. Michaelson, in *Handbook of Chemistry and Physics*, pp. E82–83, CRC Press, Boca Raton, Fla., 1980.

The work function varies with crystalline orientation of the emitting surface and its chemical cleanliness.

relatively large amplitudes of the molecular oscillations make it highly probable that physisorbed molecules are transients on metallic surfaces. Calculated times of adhesion are about 10^{-8} and 1 s at room temperature and at 100 K, respectively.

When chemical adsorption (*chemisorption*) takes place at a metallic surface, chemical bonding takes place. Such bonding usually is covalent in character, and the strength of such bonds is many times greater than that of a van der Waals bond (Section 3.12.3). In most cases the components of the chemisorbed molecules are dissociated in the bonding with the substrate material so that the chemisorbed molecule loses its identity. And, unlike physisorbed species, the translational freedom (amplitude of oscillation) of the chemisorbed molecular fragments are severely reduced. This is to be expected when bond energies such as those given in Table 3-7 are involved.

Oxygen is commonly chemisorbed upon metallic surfaces. The oxygen molecule is dissociated into two atoms in this process. The resulting atoms bond covalently with the metallic substrate. This involves the strong localization of two valence electrons from the metal to change the electron configuration of each oxygen atom from $2p^4$ to $2p^6$ (Section 3.12.3). The bonding mechanism thus results in an outer layer consisting of doubly charged, negative oxygen ions. This layer is adjacent to a second layer just below the metallic surface that has an equal positive charge. Charges of this kind in semiconductors are called *surface states*. The electric field set up by this double layer of charge differences acts to oppose electron emission from such a surface. This is the equivalent of saying that a chemisorbed layer has the effect of increasing the work function of the metal.

Now consider the effect of a monolayer of a metal chemisorbed on a metallic substrate. Here the valence electrons of the atoms of the monolayer are more weakly bound to their atoms than are those of the substrate atoms. The metallic atoms of the monolayer give up their valence electrons to the surface atoms of the substrate, thus creating a positively charged outer monolayer of ions superimposed on a negatively charged inner layer of ions at the surface of the substrate. This double layer has the effect of facilitating electron emission from the surface layer because the polarity of the electric field, in contrast to that of the monolayer of oxygen, now has the effect of increasing the velocity and therefore the energy of those electrons that enter into it. The effective result of this double layer is to lower the work function.

A frequently utilized combination of this kind makes use of cesium on tungsten. A monolayer of cesium, $\phi \sim 2\,\text{eV}$, on tungsten, $\phi \sim 4.5\,\text{eV}$, causes a double layer that diminishes the work function of the composite to 1.36 eV. Monolayers of Ba, Ce, Th, and others have similar effects (Table 8-1).

Monolayers suitable for reducing work functions are relatively few in number. Most of the metals in Table 8-1 have $4 < \phi < 5$. Many of those with lower values of ϕ (used here as a measure of the tendency of a valence electron to remain associated with its atom) are highly reactive chemically. The number of remaining elements with suitable values of ϕ, thus, is small.

Despite the fact that vacuum tubes no longer are used as widely as they were before the advent of solid-state devices, material combinations with small work functions still find applications as cathodes in power tubes. Other applications make use of the alkali metals (Li, Na, K, Ru). These are reacted inside phototubes to form hydrides on the emitting surfaces. The hydrogen layer acts metallically and decreases their values of ϕ. Antimony–cesium cathodes are used as cathodes in photocells because this combination has a value of ϕ that readily responds to a wide range of wavelengths, including those in the infrared range.

8.2 THERMIONIC EMISSION

Here attention is directed at the effect of temperature on electron emission from chemically clean (uncontaminated) surfaces. A minimum kinetic energy of $E_F + \phi$ is required for an electron to escape from its well (Figs. 3-4 and 8-1). As previously noted, the more commonly used symbol ϕ is used here and in subsequent sections, instead of W as shown in Fig. 1-2; $\phi \equiv W$.

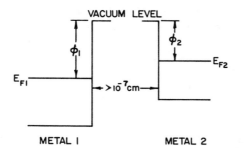

Figure 8-1 Potential wells of two metals before contact.

Since the major concern here is the kinetic energy that is required to enable the electrons to escape from the well, this property is most simply described by rewriting Eq. (4-1) as

$$E(\mathbf{p}) = \frac{1}{2m} (\mathbf{p}_x^2 + \mathbf{p}_y^2 + \mathbf{p}_z^2) \tag{8-1}$$

Considering the x direction to be perpendicular to the emitting surface, the energy of an escaping electron must be such that

$$E_{ex} = \frac{m\mathbf{v}_{ex}^2}{2} > \phi \tag{8-2}$$

where the subscript ex denotes the properties of an escaping electron.

The density of states of electrons that are traveling in the three Cartesian directions is obtained by starting with Eq. (4-2). Here, for L_i equal to unity,

$$\mathbf{p}_i = n_i \frac{h}{2} \tag{8-3}$$

Upon differentiation and rearrangement, this becomes

$$dn_i = \frac{2}{h} d\mathbf{p}_i \tag{8-4}$$

So, an element of volume, of dimensions $dn_x \cdot dn_y \cdot dn_z$, contains

$$N(\mathbf{p}) = \left(\frac{2}{h}\right)^3 d\mathbf{p}_x \cdot d\mathbf{p}_y \cdot d\mathbf{p}_z \tag{8-5}$$

states. However, all states in momentum space other than those in the first quadrant are redundant because of the squared values of $\mathbf{p}_i$ in Eq. (8-1). Thus, only one-eighth of the states represented by Eq. (8-5) enter into consideration. And, in addition, spin must be included in Eq. (8-5) as a factor of 2. These factors are included in Eq. (8-5) to obtain the desired density of momentum states:

$$N(\mathbf{p})_x = \frac{1}{8} \cdot 2 \cdot \frac{8}{h^3} d\mathbf{p}_x d\mathbf{p}_y d\mathbf{p}_z = \frac{2}{h^3} d\mathbf{p}_x d\mathbf{p}_y d\mathbf{p}_z \tag{8-6}$$

The probable number of electrons per unit area, per unit time, that contact the surface in the x direction, N_x, is based on the integration of $\mathbf{v}_x N(\mathbf{p})_x f(E)$ over all momenta, where $f(E)$ is Eq. (2-86). Since $\mathbf{p}$ is the independent variable, and $\mathbf{v}_x = \mathbf{p}_x/m$, the integration is based on $(\mathbf{p}/m)N(\mathbf{p})_x f(\mathbf{p})$, since E in $f(E)$ actually is $E(\mathbf{p})$. Thus, using Eq. (8-6),

$$N_x = \frac{2}{mh^3} \int_{-\infty}^{\infty} \int_{-\infty}^{\infty} \int_{\mathbf{p}_{ex}}^{\infty} \frac{\mathbf{p}_x}{1 + e^{(E - E_F)/k_B T}} \, d\mathbf{p}_x \, d\mathbf{p}_y \, d\mathbf{p}_z \tag{8-7}$$

The energies involved here are such that $\exp[(E - E_F)/k_B T] \gg 1$, so Eq. (8-7) may be well approximated as

$$N_x \cong \frac{2}{mh^3} \int_{-\infty}^{\infty} \int_{-\infty}^{\infty} \int_{\mathbf{p}_{ex}}^{\infty} \frac{\mathbf{p}_x}{e^{(E - E_F)/k_B T}} \, d\mathbf{p}_x \, d\mathbf{p}_y \, d\mathbf{p}_z \tag{8-8a}$$

And further simplification may be made by factoring $\exp(-E_F/k_B T)$, since it is a constant for these conditions. This results in

$$N_x \cong \frac{2e^{E_F/k_B T}}{mh^3} \int_{-\infty}^{\infty} \int_{-\infty}^{\infty} \int_{\mathbf{p}_{ex}}^{\infty} \frac{\mathbf{p}_x}{e^{E/k_B T}} \, d\mathbf{p}_x \, d\mathbf{p}_y \, d\mathbf{p}_z \tag{8-8b}$$

As noted previously, the energy term in the exponential actually is $E(\mathbf{p})$; this is given by Eq. (8-1). Equation (8-1) is substituted into Eq. (8-8b) to obtain

$$N_x \cong \frac{2e^{E_F/k_B T}}{mh^3} \int_{-\infty}^{\infty} \int_{-\infty}^{\infty} \int_{\mathbf{p}_{ex}}^{\infty} \mathbf{p}_x e^{-[(\mathbf{p}_x^2 + \mathbf{p}_y^2 + \mathbf{p}_z^2)/2mk_B T]} d\mathbf{p}_x \, d\mathbf{p}_y \, d\mathbf{p}_z \tag{8-8c}$$

The limits of the integrals for the momenta in the y and z directions include all probable values. The limits for the momenta in the x direction provide a lower limit for emitted electrons, in conformity with Eq. (8-2), to ensure that only those electrons with sufficient momenta for emission from the surface are taken into account.

The three momenta components may be considered as being independent so that they may be treated individually. The integrals for the momenta in the y and z directions are of the general form

$$\int_{-\infty}^{\infty} e^{-au^2} \, du = \left(\frac{\pi}{a}\right)^{1/2} \tag{8.9}$$

where $a = (2mk_B T)^{-1}$. And, when compensation is made for the parameter a of Eq. (8-9) in its actual form, these two integrals change Eq. (8-8c) into

$$N_x \cong \frac{2e^{E_F/k_B T}}{mh^3} (2\pi mk_B T) \int_{\mathbf{p}_{ex}}^{\infty} \mathbf{p}_x e^{-(\mathbf{p}_x^2/2mk_B T)} \, d\mathbf{p}_x \tag{8-10}$$

The integral is treated as follows:

$$\int_{\mathbf{p}_{ex}}^{\infty} \mathbf{p}_x e^{-(\mathbf{p}_x^2/2mk_B T)} d\mathbf{p}_x = -\frac{2mk_B T}{2} \int_{\mathbf{p}_{ex}}^{\infty} e^{-(\mathbf{p}_x^2/2mk_B T)} \left(-\frac{2\mathbf{p}_x}{2mk_B T} \, d\mathbf{p}_x\right)$$

so that

$$\int_{\mathbf{p}_{ex}}^{\infty} \mathbf{p}_x e^{-(\mathbf{p}_x^2/2mk_BT)} d\mathbf{p}_x = (-mk_BT) e^{-\mathbf{p}_x^2/2mk_BT)} \tag{8-11}$$

The substitution of Eq. (8-11) completes the equation for N_x [Eq. (8-10)], and noting that the factor m in Eq. (8-10) vanishes, and prior to inserting the limits,

$$N_x \cong -\frac{4\pi mk_B^2}{h^3} \cdot T^2 \cdot e^{E_F/k_BT} \cdot e^{-(\mathbf{p}_{ex}^2/2mk_BT)} \tag{8-12}$$

The exponential factor $\mathbf{p}_{ex}^2/2m = E_{ex}$ [Eqs. (8-1) and (8-2)], so Eq. (8-12) may be expressed as

$$N_x \cong -\frac{4\pi mk_B^2}{h^3} \cdot T^2 \cdot e^{-(E_F - E_{ex})/k_BT} \tag{8-13}$$

The minimum value for $E_F - E_x$ for an electron to escape is the surface work function (Section 8.1), $\phi = -E_F + E_{ex}$; so upon substitution of this value and applying the limits, Eq. (8-13) becomes

$$N_x \cong \frac{4\pi mk_B^2}{h^3} \cdot T^2 \cdot e^{-\phi/k_BT} \tag{8-14}$$

The constants composing the fractional coefficient of Eq. (8-14) frequently are denoted by A or A_0. Its theoretical value is

$$A = A_0 \cong 120\,\text{A cm}^{-2}\,\text{K}^{-2} \tag{8-14a}$$

assuming that all N_x electrons exit from the surface. Not all such electrons do so; some are considered to be reflected back into the material. This is taken into account by the use of an effective factor A^* that is defined as

$$A^* = A(1 - r) \tag{8-14b}$$

in which r is the *reflection coefficient*; $\frac{1}{2} < r < \frac{3}{4}$. In actuality, the reflection coefficient also includes the history, chemistry, and condition of the emitting surface.

8.3 THE SCHOTTKY EFFECT

Thermionic emission is enhanced by electrostatic image forces and by externally applied electric fields. This enhancement results from the decrease in the work function that is known as the *Schottky effect*.

An electron leaving the surface of a metal creates a positive charge in the volume of the metal as a result of its absence. An attractive force thus exists between the metal and the emitted electron. When the metal is considered to be a plane conductor, the force acting on the ejected electron is considered to be the same as that acting on the positive hole remaining within the metal. This approach is known as the *method of images*. It permits the treatment of the hole and the electron as being equidistant from the emitting surface. Thus, from electrostatics, the image force acting on the electron is

$$F(x) = \frac{e^2}{4\pi\varepsilon_0(2x)^2} \qquad (8\text{-}15)$$

where x is the perpendicular distance of the electron from the surface and $2x$ is the electron–hole distance, provided that the electron is several interionic distances from the surface. The dielectric constant (permittivity) of free space (vacuum) is given by ε_0.

The energy imparted to the electron by the image force is obtained by the application of Eq. (3-1) to Eq. (8-15). Thus,

$$-dV(x) = F(x)\,dx = \frac{e^2}{4\pi\varepsilon_0 \cdot 4x^2}\,dx \qquad (8\text{-}16)$$

So $V(x)$, the potential energy from the image force that is associated with the electron at a distance x from the emitting surface, is obtained by integrating Eq. (8-16) between x and infinity. Thus, the potential energy arising from the image force is the image potential

$$-V_I = -V(x) = \frac{e^2}{16\pi\varepsilon_0 x} \qquad (8\text{-}17)$$

The image potential, V_I, also has the effect of appreciably decreasing the reflection of emitted electrons by the steep change in the internal potential present at the boundary in a way analogous to Eq. (8-14b) and shown in Figs. 1-2 and 3-4. It will be noted that Eq. (8-17) does not include the energy of the electron prior to its emission from the surface.

When an external electric field, $\bar{E}$, is applied to the ejected electron, it is acted on by an additional force of $e\bar{E}$ and consequently acquires another energy component, $V(\bar{E}) = e\bar{E}x$. (Here $\bar{E}$ is applied in the x direction.) Therefore, the total potential energy acting on the electron is

$$-V = -[V_I + V(\bar{E})] = \frac{e^2}{16\pi\varepsilon_0 x} + e\bar{E}x \qquad (8\text{-}18)$$

At equilibrium, the energy components of Eq. (8-18) are equal so that

$$\frac{e^2}{16\pi\varepsilon_0 x} = e\bar{E}x \qquad (8\text{-}19a)$$

This is rearranged to give

$$\frac{e}{16\pi\varepsilon_0 \bar{E}} = x^2 \qquad (8\text{-}19b)$$

Thus, the maximum distance of an electron from the emitting surface is

$$x_{max} = \left(\frac{e}{16\pi\varepsilon_0 \bar{E}}\right)^{1/2} \qquad (8\text{-}20)$$

The image forces and the application of the external electric field decrease the work function. This is known as the Schottky effect. This decrease is obtained by substituting Eq. (8-20) into Eq. (8-18) to obtain

$$-\Delta\phi = \frac{e^2}{16\pi\varepsilon_0}\left(\frac{16\pi\varepsilon_0\bar{E}}{e}\right)^{1/2} + e\bar{E}\left(\frac{e}{16\pi\varepsilon_0\bar{E}}\right)^{1/2} \qquad (8\text{-}21a)$$

Equation (8-21a) simplifies to

$$-\Delta\phi = \frac{e^{3/2}\bar{E}^{1/2}}{(16\pi\varepsilon_0)^{1/2}} + \frac{e^{3/2}\bar{E}^{1/2}}{(16\pi\varepsilon_0)^{1/2}} \qquad (8\text{-}21b)$$

And, upon further collection,

$$-\Delta\phi = \frac{2e^{3/2}\bar{E}^{1/2}}{(16\pi\varepsilon_0)^{1/2}} = \left(\frac{e^3\bar{E}}{4\pi\varepsilon_0}\right)^{1/2} \qquad (8\text{-}21c)$$

Thus, the probability of the emission of an electron in the x direction is given as varying as

$$P_x \propto \exp-\left[\frac{(\phi-\Delta\phi)}{k_BT}\right] = \exp\left\{-\frac{\phi-(e^3\bar{E}/4\pi\varepsilon_0)^{1/2}}{k_BT}\right\} \qquad (8\text{-}22)$$

The emission current density for a single electron thermally excited in the x direction is therefore expected to behave according to

$$j_x \propto e \exp\left\{-\frac{\phi-(e^3\bar{E}/4\pi\varepsilon_0)^{1/2}}{k_BT}\right\} \qquad (8\text{-}23)$$

For N electrons excited per unit surface (area) per unit time (seconds), the emission "current density" is obtained by the use of Eq. (8-14):

$$j_x = N_x e \exp-\left[\frac{(\phi-\Delta\phi)}{k_BT}\right] = \frac{4\pi m k_B^2}{h^3}\cdot T^2\cdot\exp\left\{-\frac{\phi-(e^3\bar{E}/4\pi\varepsilon_0)^{1/2}}{k_BT}\right\} \qquad (8\text{-}24)$$

Or, using Eq. (8-14a),

$$j_x = AT^2 \exp\left\{-\frac{\phi-(e^3\bar{E}/4\pi\varepsilon_0)^{1/2}}{k_BT}\right\} \qquad (8\text{-}24a)$$

The same comments given for Eq. (8-14) also hold here. In addition, because T is involved both as a square and as an exponential factor, small variations in T can have large effects on j_x. It will be noted that an isothermal plot of the logarithm of j_x versus $\bar{E}^{1/2}$ is linear and has a positive slope; this is known as a Schottky line, or as a *Schottky plot*.

It was noted with respect to Eq. (8-15) that a lower or critical limit exists for the variable distance x. If x is at less than the critical or limiting value, the image force would approach infinity and the image potential would approach minus infinity. These difficulties are avoided by setting the critical value of x at a minimum distance equivalent to a few interionic diameters.

In general, the predicted linearity of a Schottky plot holds only above a particular value of $\bar{E}^{1/2}$ for each metal. Increasingly negative deviations from linearity occur as $\bar{E}^{1/2}$ decreases from the given initial linear value. Tungsten, for example, shows the linearity predicted by the theory at values above $\bar{E}^{1/2} \sim 10$ V/cm. But, as $\bar{E}^{1/2} \to 0$, $\ln j_x \to \phi$ in an approximately hyperbolic fashion. These departures from linearity have been considered to be a result of asperities or other variations in the emitting surface that result in nonuniformities of the work function. At the other extreme, when $\bar{E}$ becomes high ($> 10^6$ V/cm), the electrons no longer are emitted by surmounting the reduced barrier. When such high fields are applied, the electrons penetrate the barrier. This is known as *tunneling*.

8.4 FIELD EMISSION

The previous sections consider the escape of electrons *over* the potential barrier at the surface of emitter. The Schottky effect describes the increased electron emission as a result of the decrease of the barrier that occurs on application of an external electric field. The Schottky effect takes place when electric fields up to $< 10^6$ V/cm are applied. At field strengths of $\bar{E} > 10^8$ V/m, the electrons no longer need to surmount the potential barrier; they are enabled to *penetrate* through the surface barrier. Here the electric field plays the same part as does temperature in the previous sections. This effect is known as *field emission* and the mechanism is called *tunneling*. At normal temperatures, the electrons that participate in this mechanism are those in the neighborhood of E_F. It should be noted that high internal fields in the barrier zone enhance this effect (Section 8.9.1).

First consider the electrons in a potential well in the absence of an electric field as described in Section 3.7 and in Fig. 3-1. The wave function for finding an electron in region I, that originated in region II, (in the $-x$ direction) is, at the boundary of these regions, as derived from Eq. (3-36),

$$\Psi_I = A \exp\left\{\frac{2\pi}{h}[2m(V - E)]^{1/2}x\right\} \tag{8-25}$$

The electrons under consideration here are in the neighborhood of E_F, $V = V_0$ is the vacuum level, and $x = x(E_F)$. The latter is the width (thickness) of the potential barrier at $x(E_F) \cong 0$ and $E \cong E_F$. Thus, Eq. (8-25) becomes

$$\Psi_I = A \exp\left\{\frac{2\pi}{h}[2m(V_0 - E_F)]^{1/2} \cdot x(E_F)\right\} \tag{8-26}$$

From Fig. 8-1, where the barrier height is ϕ and V_0 is the vacuum level, the height of the potential barrier is

$$\phi = V_0 - E_F \tag{8-27}$$

so that the wave function for finding an electron in region I, that passed through the barrier from region II, must be

$$\Psi_I = A \exp - \left[\frac{2\pi}{h}(2m\phi)^{1/2} \cdot x(E_F) \right] \qquad (8\text{-}28)$$

The factor $x(E_F)$ permits the inclusion of the effect of the applied electric field. The electron acquires another energy component from this that permits the penetration of the boundary. This is obtained from

$$\phi(E_F) = e\bar{E}x(E_F) \qquad (8\text{-}29\text{a})$$

or

$$x(E_F) = \frac{\phi(E_F)}{e\bar{E}} \qquad (8\text{-}29\text{b})$$

The inclusion of Eq. (8-29b) into Eq. (8-28) results in

$$\Psi_I = A \exp - \left[\frac{2\pi}{h}(2m\phi)^{1/2} \cdot \frac{\phi(E_F)}{e\bar{E}} \right] \qquad (8\text{-}30)$$

The *transmission coefficient*, $\theta(E_F)$, is essentially that given by Eq. (8-30), since it considers only those electrons entering region I from region II in the $-x$ direction. This parameter may be expressed as

$$\theta(E_F)_x \propto \exp - \left[\frac{2\pi}{h}(2m\phi)^{1/2} \cdot \frac{\phi(E_F)}{e\bar{E}} \right] \qquad (8\text{-}31)$$

The number of available electrons in the energy range of interest here is given by those between E_F and $E_F + dE$, or $N(E_F)_x$, the density of such states. Thus, the emission current density, or the number of electrons in the neighborhood of E_F that traverse the barrier in the x direction per unit area per unit time, is

$$j_x = e \int_{-E_F}^{\infty} N(E_F)_x \theta(E_F)_x \, dE_x \qquad (8\text{-}32)$$

The factor $N(E_F)_x$ in Eq. (8-32) is virtually unaffected by normal temperatures. The same is true for $\phi(E_F)$ in Eq. (8-31). Thus, Eq. (8-32) is for all practical purposes independent of temperature. It can be shown that upon integration, Eq. (8-32) may be expressed as

$$j_x \propto \bar{E}^2 \theta(E_F)_x \qquad (8\text{-}33)$$

Thus, a plot of $\ln(j_x/\bar{E}^2)$ as a linear function of $(\bar{E})^{-1}$ is expected. This has been confirmed by experiment.

In the field-emission microscope, the sample consists of a sharp-tipped filament. This is placed in an evacuated chamber that contains a fluorescent screen. The high voltage between the filament and the positively charged screen accelerates the emitted electrons. These give off flashes of light upon impinging on the screen. The light patterns are characteristic of the crystalline structure of the emitter filament.

The field-ion microscope represents another application of field emission. Here, a cryogenically cooled fine "needle", with a hemispheric shaped tip, is maintained at a positive potential in a chamber containing very low pressure helium. A conducting layer on a fluorescent screen is held at a negative potential. Effective field strengths of the order of $\sim 10^8$ V/cm, or greater, may be achieved at the tip. The helium atoms react with and are ionized by the active metal ions at the needle tip. And, as positive helium ions, they are attracted to the screen and form an image of the active atoms and, therefore, of the ionic array (crystal structure) at the tip.

8.5 METAL–METAL JUNCTIONS

Consider two metals, as shown in Fig. 8-1, with a separation of many interionic distances (~ 30 to 50) between them prior to contact. The individual wells are compared by placing their vacuum levels at a common level. The situation indicated in the figure is stable as long as the metals are sufficiently far apart so that the probability density functions of their valence electrons do not overlap or significantly penetrate the opposite well (Section 3.12.4). Such penetration permits an electron to pass through the barriers and to enter the other well, a process known as *tunneling*.

As soon as the metals attain a separation of the order of a few interionic distances (<10), the situation shown in Fig. 8-1 becomes unstable because the appreciable penetration of the probability density functions permits tunneling. Electrons will be emitted by both metals by tunneling *through* the barriers and by thermal excitation *over* the barriers (thermionic emission). However, the predominating transfer of the electrons is by tunneling from metal 2 to metal 1 because $E_{F2} > E_{F1}$. The electron states in metal 2 will be depleted, while those in metal 1 will be increased. Thus, metal 1 will have an excess negative charge while metal 2 will have an excess positive charge.

The transfer of an electron from a given level in metal 2 to one in a negatively charged situation as in metal 1 at some other potential V, requires work or an energy expenditure of eV. This becomes the kinetic energy of the transferred electron. Thus, the kinetic energy of the electrons in metal 1 increases (the well is filled to a higher level), while that of the remaining electrons in metal 2 decreases (the well contains fewer electrons). The Fermi levels in the metals also must change correspondingly. A point is reached at which the Fermi energies attain a common level as shown in Fig. 8-2. The buildup of negative charges in metal 1 and the corresponding creation of positive charges in metal 2 may be considered to form an *electric double layer* that sets up an internal electric field, $\bar{E}_i$, whose polarity inhibits further electron flow from metal 2 to metal 1, but which tends to enhance electron flow from metal 1 to metal 2. The electron transport involved here occurs at rates of about 10^8 cm s^{-1} (Section 4.2) so that the process of the buildup of the double layer requires times of the order of 10^{-15} to 10^{-16} s. Thus, dynamic equilibrium in the electron flows (j_0 in the figure) is attained virtually instantaneously.

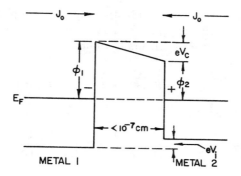
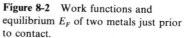

Figure 8-2 Work functions and equilibrium E_F of two metals just prior to contact.

The Fermi energies of both metals are at a common level at equilibrium so that a constant potential difference exists between the two metals. This is the *external contact potential* and is given by

$$V_c = \frac{\phi_1 - \phi_2}{e} \tag{8-34}$$

This is reasonable because the electron flow that creates the potential results from the difference in the work functions; the predominant flow came from the metal with the lower work function.

It will be noted in Fig. 8-2 that even though the Fermi energies of both metals are at the same level, *the kinetic energies of electrons near the common E_F are not the same.* As measured from the bottom of each well in Fig. 8-2, the kinetic energies of the electrons in metal 1 are greater than those in metal 2; *their respective kinetic energies are different* even though their Fermi energies become placed at the same level. Thus, at equilibrium, an *internal contact potential*, V_i, exists at the junction that is based on the kinetic energies of the electrons involved. So

$$V_i = \frac{E_{F1} - E_{F2}}{e} = -V_c \tag{8-35}$$

The situation discussed in descriptive terms may now be obtained by using the Fermi function, f [Eq. (2-86)]. The function f gives the probability of the occupancy of a state, and $1 - f$ gives the probability that a state near E_F will be unoccupied. Considering electron states with energies in the neighborhood of E_F, $(1 - f)$ may be taken as the probability that a state near E_F in one metal has been transferred to the other metal. So, using the subscripts 1 and 2 to denote metals 1 and 2, respectively, the situation on both sides of the metal 1–metal 2 interface at equilibrium is given by

$$f_1(1 - f_2) = f_2(1 - f_1) \tag{8-36}$$

Now, using Eq. (2-86), for metals 1 and 2, the distribution functions are

$$f_1 = \frac{1}{e^{(E - E_{F1})/k_B T} + 1} \quad \text{and} \quad f_2 = \frac{1}{e^{(E - E_{F2})/k_B T} + 1}$$

The use of these equations in Eq. (8-34) may be simplified by letting $x_1 = (E - E_{F1})/k_B T$ and $x_2 = (E - E_{F2})/k_B T$. The Fermi functions are substituted into Eq. (8-36) using the simplified exponents to obtain

$$\frac{1}{e^{x_1} + 1} \left(1 - \frac{1}{e^{x_2} + 1}\right) = \frac{1}{e^{x_2} + 1} \left(1 - \frac{1}{e^{x_1} + 1}\right) \tag{8-37}$$

The indicated multiplication gives

$$\frac{1}{e^{x_1} + 1} \cdot - \frac{1}{(e^{x_1} + 1)(e^{x_2} + 1)} = \frac{1}{e^{x_2} + 1} - \frac{1}{(e^{x_2} + 1)(e^{x_1} + 1)}$$

The terms containing the products in their denominators are identical and vanish, leaving

$$\frac{1}{e^{x_1} + 1} = \frac{1}{e^{x_2} + 1} \tag{8-38}$$

Equation (8-38) is true if and only if

$$e^{x_1} = e^{x_2}$$

so it must follow that

$$x_1 = x_2 \tag{8-39}$$

And, using the values represented by x_1 and x_2,

$$\frac{E - E_{F1}}{k_B T} = \frac{E - E_{F2}}{k_B T} \tag{8-40}$$

It is thus seen that Eq. (8-40) is valid only when the identity condition

$$E_{F1} \equiv E_{F2} \tag{8-41}$$

exists, that is, at equilibrium. Here the equilibrium density of electrons flowing across the junction in each direction is n_{e0}. This equilibrium flow is a result of thermionic emission and of tunneling.

Equation (8-41) represents a general situation, despite the fact that it was derived for metals. It is applicable to all solids and combinations of solids (Section 8.1).

Using Fig. 8-2 and the vacuum level as a reference, the top of the well of metal 2 lies at eV_c below ϕ_1, or

$$\phi_1 = \phi_2 + eV_c \tag{8-42}$$

This follows directly from Eq. (8-41) and confirms Eq. (8-34).

Now consider the Fermi energies of the electrons in the two wells as representing the maximum kinetic energies of their respective electrons. These have been aligned at the common vacuum level, as permitted by Eqs. (8-41) and (8-42), so that, in terms of the kinetic energies of the electrons in metal 1,

$$E_{F1} = E_{F2} + eV_i \tag{8-43}$$

This confirms Eq. (8-35) for the existence of an internal contact potential across the junction. It is useful for this to be treated as an internal electric field, $\bar{E}_i$; and, for a junction of width d,

$$\bar{E}_i = \frac{eV_i}{d} \qquad (8\text{-}44)$$

It is not possible to measure V_i by connecting a metallic junction into a circuit. It will be recalled that the creation of V_i resulted from the virtually instantaneous electron flow that acted to bring the two Fermi energies into alignment at a common energy level. Once this equilibrium is attained, the electrons at or near E_F in each metal well have the same Fermi energy as is given by Eq. (8-41). *Measurements of V_i, therefore, can be made only when the metals are separated very slightly, but are not in contact*, such as two plates of a capacitor (see Section 8.5.1). It is for this reason that note was made in Sections 7.1.1 and 7.1.2, respectively, that *the Seebeck and Peltier effects are in no way related to the contact potential*. In fact, any number of such junctions may be inserted between the active components of a thermocouple without changing its calibration, provided that all the intermediate junctions have no temperature differences between their ends (Section 7.1.6).

8.5.1 Metal–Metal Double Layers

The electron flow from metal 2 to metal 1 (Fig. 8-2) results because $\phi_2 < \phi_1$. After equilibrium is reached, metal 1 has an excess of electrons and metal 2 has a deficit of electrons. The deficit of electrons in metal 2 constitutes a positive charge at or near its surface. The two oppositely charged surfaces create an electric double layer that may be described by a potential V_i [Eq. (8-43)] or by an electric field $\bar{E}_i$ [Eq. (8-44)].

The capacitance method noted in the previous section will be used here to approximate the extent or the thickness of the double layer and its effects. Each metallic surface is considered to constitute a plate of a capacitor with a charge of Q separated from the other by a vacuum. The capacitance of such a plate of unit area is

$$C = \frac{Q}{V_i} \qquad (8\text{-}45)$$

The capacitance is also given by

$$C = \frac{\varepsilon\varepsilon_0}{d} \qquad (8\text{-}46a)$$

where ε is the relative permittivity (relative dielectric constant), ε_0 is that of the vacuum, and d is the distance between the plates. Equations (8-45) and (8-46) are equated, after setting $\varepsilon = 1$, to obtain

$$\frac{\varepsilon_0}{d} = \frac{Q}{V_i}$$

This gives the distance between the plates as

$$d = \frac{\varepsilon_0 V_i}{Q} \tag{8-46b}$$

Equations (8-44), (8-45), and (8-46b) permit an approximate description of the properties of the double layer. If it is assumed that the minimum distance between the surfaces is about one interionic distance; then, using Eq. (8-44) and approximating $d \sim 3 \times 10^{-10}$ m, the internal electric field is $\bar{E}_i \sim 3 \times 10^9$ V·m^{-1}. Now, using Eq. (8-46) with $V_i \cong 1$ V and $\varepsilon_0 \cong 8.9 \times 10^{-12}$ F·m^{-1},

$$Q \cong \frac{\varepsilon_0 V_i}{d} \cong \frac{8.9 \times 10^{-12} \times 1}{3 \times 10^{-10}} \cong 3 \times 10^{-2} \, C \cdot m^{-2} \tag{8-46c}$$

The number of electrons involved at these surfaces is obtained as the surface electron density, $\rho(j)$, by using (8-46c) and $e \cong 1.6 \times 10^{-19}$ C:

$$\rho(j) = \frac{Q}{e} \cong \frac{3 \times 10^{-2}}{1.6 \times 10^{-19}} \sim 2 \times 10^{17} \text{ electrons} \cdot m^{-2} \tag{8-46d}$$

Considering an average interionic distance of $\sim 3 \times 10^{-10}$ m, an area of 1 m^2 will contain $\sim 3 \times 10^{19}$ ions. If the ions are of monovalent metals, the number of valence electrons at the surface also will be $\sim 3 \times 10^{19}$ m^{-2}. A comparison of this result with that of Eq. (8-46d) gives an interesting insight into metal–metal double layers. This is obtained as follows: $2 \times 10^{17} \div 3 \times 10^{19} \sim 1 \times 10^{-2}$, or a maximum of about 1% of the surface electrons are involved in creating the double layer. It will be noted that this is in good agreement with the electron contribution to the heat capacity of normal metals (Section 5.1.10.1).

The thickness of the double layer is a small percentage of the mean free path of an electron in a normal metal (Section 6.0). This means that there is a negligible probability of electron scattering by the double layer. And, since the number of electrons that are involved is small, the electrical conductivity of the double layer is expected to be virtually the same as that which would be expected if the double layer did not exist. This is another reason why it is not possible to measure V_i by connecting it into an electric circuit, as noted in the previous section.

8.6 METAL–SEMICONDUCTOR JUNCTIONS

Now replace metal 2 in Fig. 8-2 with an n-type semiconductor with a work function $\phi_S < \phi_1$. In the same way as for the metal 1–metal 2 junction, electrons will flow from the semiconductor to metal 1 until an equilibrium is reached where the Fermi levels of each component reach a common level [Eq. (8-41)] and a contact potential is formed as shown in Fig. 8-3.

Using the same approximations as those in Section 8.5.1, $\rho(j) \cong 2 \times 10^{17}$ electrons·m^{-2} will transfer from the semiconductor to the metal. Using silicon, with a lattice parameter of 5.4×10^{-10}, or $\sim 5 \times 10^{-10}$ m, the planar density of Si ions is

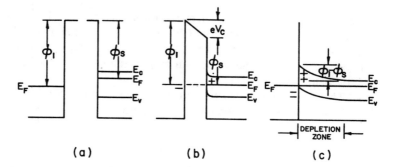

Figure 8-3   Metal and n-type semiconductor junction. (a) Prior to contact, (b) just prior to contact, and (c) after contact, in which the depletion zone and space-charge regions are indicated.

$\sim 4 \times 10^{18}$ m^{-2}.   (This neglects the low atomic packing factor, Section 3.12.3.) However, each Si atom contributes four valence electrons, so the planar density of valence electrons is $\sim 2 \times 10^{19}$ m^{-2}.   However, only a factor of about 10^{-7} of these electrons is available for conduction, giving an effective planar electron density of $\sim 2 \times 10^{12}$ m^{-2} (Section 6.3.1).

Such an effective electron density requires the transfer of the available electrons from $\sim 10^5$ planes in the semiconductors.   The volume of the semiconductor that contains the atomic planes from which the electrons have been transferred to the metal is called the *depletion zone* or *barrier layer*.

The depletion zone, having been stripped of its electrons, still contains the positively ionized donor states (Section 6.3.3).   These positively charged ions and the excess electrons, which were transferred to the metal, form an electric double layer that is much larger (by a factor of $\sim 10^3$) in width than the mean free path of an electron.   This extent of the depletion zone (with respect to the mean free path) makes it highly improbable that electrons will be found within it.   This ensures that the electrical resistivity of the depletion zone will approach that of an electrical insulator; hence this volume of the interface is termed the barrier layer.   A depletion zone of $\sim 10^5$ planes is normally $\sim 10^5 \times 5 \times 10^{-10} \sim 5 \times 10^{-5}$ m in width.

If the contact potential, V_c, is ~ 1 V, then the contact electric field across the depletion zone is $\bar{E}_c \sim 1 \div 5 \times 10^{-5} \sim 2 \times 10^4$ V m^{-1}.   This electric field affects the band structure of the depletion zone of the semiconductor, as shown in Fig. 8-3(c), by distorting it.   However, the band structure of the metal is left virtually unchanged by this because $\bar{E}_c$ for the semiconductor is smaller than that for the metal by a factor of $\sim 10^4$.

The distortion of the band structure of the semiconductor is explained in the following way.   The establishment of the depletion zone and its electric double layer, as noted previously, has the effect of a force excluding electrons.   The work required to balance this force must at least equal the potential energy of an electron in the depletion zone.   An electron entering the depletion zone has its potential energy increased as it traverses the electric double layer toward its surface. Thus, at the surface of the semiconductor, its PE is $\phi_c = eV_c$. This explains the upward distortion of the band structure of the semiconductor in the depletion zone. The contact field acts on the bands of the semiconductor in the same way as an equivalent, external electric field does.

The variation of this PE, $\phi(x)$, with respect to the potential $V(x)$ that is set up by $\bar{E}_c$ is approximated by starting with Poisson's equation:

$$\frac{d^2V(x)}{dx^2} = -\frac{\rho_x}{\varepsilon_0\varepsilon} \tag{8-47}$$

where ρ_x is the density of the charge that creates $\bar{E}_c$. The variable x is measured across the electric double layer beginning at the semiconductor surface to that portion of the semiconductor in which the field does not exist. So, if d is the width of the double layer, $0 \leqq x \leqq d$.

The potential $V(x) = -\phi(x)/e$. This is substituted into Eq. (8-47) to obtain

$$\frac{d^2\phi(x)}{dx^2} = \frac{e}{\varepsilon_0\varepsilon}\rho_x \tag{8-48}$$

The factor ρ_x is approximated by assuming that the donors belong to group V of the Periodic Table. If all the available electrons, originally in the depletion zone, now are at the surface of the metal, then $\rho_x \cong en_e$, and Eq. (8-48) becomes

$$\frac{d^2\phi(x)}{dx^2} \cong \frac{e^2n_e}{\varepsilon_0\varepsilon} \tag{8-49}$$

The integration of Eq. (8-49) results in

$$\phi(x) \cong \frac{e^2n_e}{2\varepsilon_0\varepsilon}(d^2 - x^2) \tag{8-50}$$

since $\phi(x) \cong 0$ at $x = d$.

The contact potential at the semiconductor surface is given by

$$\phi(x) = \phi_1 - \phi_s = \phi_c = eV_c, \qquad x = 0 \tag{8-51}$$

Thus, for the condition $x = 0$,

$$\phi_c = eV_c \cong \frac{e^2n_e}{2\varepsilon_0\varepsilon}d^2 \tag{8-52}$$

This is rearranged as

$$eV_c \cdot \frac{2\varepsilon_0\varepsilon}{e^2n_e} \cong d^2$$

The extent of the depletion zone is then found to be

$$d \cong \left(\frac{2\varepsilon_0\varepsilon V_c}{en_e}\right)^{1/2} \tag{8-53}$$

And, referring to the assumptions made for Eq. (8-49), n_e also approximates the dopant concentration.   Using values of $n_e \sim 10^{22}\,\text{m}^{-3}$, $\varepsilon \sim 10$, and $V_c \sim 1\,\text{V}$ in Eq. (8-53) gives $d \sim 10^{-5}\,\text{m}$, a distance that includes $\sim 10^5$ lattice planes.   This confirms the earlier approximation for the depletion zone.

If a voltage V is applied across the depletion zone such that the semiconductor is made to be positive with respect to the metal (reverse bias), Eq. (8-52) becomes

$$\phi_c + eV \cong \frac{e^2 n_e}{2\varepsilon_0 \varepsilon}\, d^2 \tag{8-54}$$

And Eq. (8-53) becomes

$$d \cong \left(\frac{2\varepsilon_0 \varepsilon (V_c + V)}{e n_e}\right)^{1/2} \tag{8-55}$$

Thus, the applied voltage increases the width of the depletion zone as well as the height of the barrier [Fig. 8-4(b)].

When the semiconductor is made to be negative (forward bias),

$$\phi_c - eV \cong \frac{e^2 n_e}{2\varepsilon_0 \varepsilon}\, d^2 \tag{8-56}$$

and

$$d \cong \left(\frac{2\varepsilon_0 \varepsilon (V_c - V)}{e n_e}\right)^{1/2} \tag{8-57}$$

Here the applied voltage decreases the width of the depletion zone and the barrier height [Fig. 8-4(a)].

The preceding discussions on the distortion of the bands of a semiconductor, in

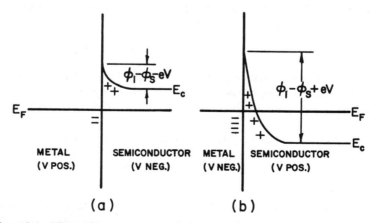

Figure 8-4 Effects of bias at a metal–semiconductor junction: (a) forward bias; (b) reverse bias.

the absence of an external electric field, are based on $\phi_S < \phi_1$. When the work function of the semiconductor is greater than that of the metal, electrons flow from the metal to the semiconductor. The surface of the semiconductor then becomes negatively charged, while that of the metal is positively charged. The parameter $\phi(x)$ decreases as the surface of the semiconductor is reached so that the bands are bent downward. Here E_c may be equal to or less than the Fermi level of the metal. The concentration of the carriers near the surface of the semiconductor increases. A potential barrier no longer exists. Here the volume of the semiconductor near its surface is an *accumulation region*.

8.7 METAL–SEMICONDUCTOR RECTIFICATION

Metal–semiconductor rectification depends on the formation of an internal barrier layer at the interface between the two component materials. If the semiconductor is *n* type, with $\phi_S < \phi_1$, electrons will diffuse from its conduction band into the metal and create positively ionized impurities. The result is a positive space-charge, or depletion, region in the semiconductor. The excess electrons in the metal induce a negative space charge. The double electric layers create an internal field in the interface region [Fig. 8-3(c)]. The space-charge region within the semiconductor must lie within the range of the decrease of the barrier potential difference, $\phi(x)$ [Eq. (8-50)]. At equilibrium, this potential difference results in the equalization of the Fermi levels [Eq. (8-41)], and the numbers of carriers in the barrier, or depletion, zone remain constant. The larger the difference between the work functions, the greater will be the distortion in the bands of the semiconductor [Eq. (8-52)]. This also causes E_c to become closer to the common E_F. Barriers of this kind are known as *Schottky barriers* (see Section 8.6).

At equilibrium, $E_{F1} = E_{FS}$ [Eq. (8-41)], and the diffusion of carriers across the barrier is the same in each direction. The result is two equal and opposite current densities. This may be expressed as

$$n_e = n_{e0} \quad \text{and} \quad j_e = j_{e0} \tag{8-58}$$

where n_{e0} and j_{e0} are, respectively, the equilibrium density and current density of the electrons. The application of an externally applied voltage, V, across the interface changes the bands of the semiconductor depending on its polarity [Eqs. (8-55) and (8-57)].

Let the voltage across the interface be such that

$$|V| < \frac{\phi_1 - \phi_S}{e} \tag{8-59}$$

This causes the energy barrier at the interface to change from that of the equilibrium case by an amount $\pm eV$, depending on the direction of the bias of the voltage. Using the Boltzmann statistics (Section 2.1), the effect of the applied voltage on the probability of finding an electron that can surmount the barrier is

$$P \propto \exp\left(-\frac{eV}{k_B T}\right) \tag{8-60}$$

So the density of electrons and the corresponding current density as a result of the influence of the external voltage are given, respectively, by the general forms

$$n_e \cong \pm n_{e0} \exp\left(-\frac{eV}{k_B T}\right) \tag{8-61a}$$

or

$$j_e \cong \pm j_e \exp\left(-\frac{eV}{k_B T}\right) \tag{8-61b}$$

An opposing flow of electrons, with a density of $\pm n_{e0}$, is always present. The sign of this countercurrent also depends on the polarity of the external voltage, so

$$n_e \cong \mp n_{e0} \tag{8-62}$$

The effects of the two flows of current are given by the general equation

$$n_e \cong \mp n_{e0} \pm n_{e0} \exp\left(-\frac{eV}{k_B T}\right)$$

or

$$n_e \cong \mp n_{e0}\left[1 \pm \exp\left(-\frac{eV}{k_B T}\right)\right] \tag{8-63a}$$

And the current density across the function is given by the general equation

$$j \cong \mp j_0\left[1 \pm \exp\left(-\frac{eV}{k_B T}\right)\right] \tag{8-63b}$$

Equations (8-63a) and (8-63b) require that the sign of the biased voltage be taken into account.

As an example, consider the effect of a forward-biased voltage. The factor V in Eq. (8-61a) is negative, so

$$n_e \cong n_{e0} \exp\left(\frac{eV}{k_B T}\right) \tag{8-64a}$$

The density of the counterflow of the electrons from the metal to the semiconductor is, from Eq. (8-62),

$$n_e \cong -n_{e0} \tag{8-64b}$$

Thus, the net electron density from the forward bias is

$$n_F \cong n_{e0} \exp\left(\frac{eV}{k_B T}\right) - n_{e0} \tag{8-65a}$$

or

$$n_F \cong n_{e0}\left[\exp\left(\frac{eV}{k_B T}\right) - 1\right] \qquad (8\text{-}65b)$$

And the corresponding current density is

$$j_F \cong j_0\left[\exp\left(\frac{eV}{k_B T}\right) - 1\right] \qquad (8\text{-}66)$$

Where a reverse-biased voltage is applied, the factor V in Eq. (8-61b) is positive and the sign of Eq. (8-62) is positive. These combine to give the net reverse electron density [Eq. (8-63a)] as

$$n_R \cong n_{e0}\left[1 - \exp\left(-\frac{eV}{k_B T}\right)\right] \qquad (8\text{-}67a)$$

and the corresponding reverse current density as

$$j_R \cong j_0\left[1 - \exp\left(-\frac{eV}{k_B T}\right)\right] \qquad (8\text{-}67b)$$

When a forward bias is applied, the height of the barrier is decreased [Eq. (8-51)], its width decreases [Eq. (8-57)], and the current density increases exponentially [Eq. (8-66)]. In contrast, the application of a reverse bias increases the barrier height [Eq. (8-51)], increases the width of the barrier [Eq. (8-55)], and results in a very small current density [Eq. (8-67b)] of $\sim j_0$. These behaviors are shown by the broken curve of Fig. 8-5. (Also see Fig. 8-4.)

It is thus seen that the direction of application of the external voltage determines the magnitude of the current density across a junction. This one-way flow permits the rectification of alternating current. The ratio of j_F/j_R is called the *rectification ratio*.

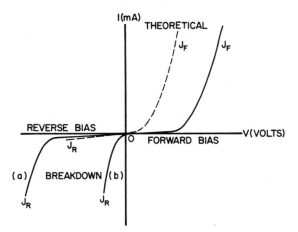

Figure 8-5 Schematic diagram of rectification at a p–n junction: (a) avalanche and (b) tunnel breakdown. The broken curve represents theoretical behavior. Note the threshold voltage for forward bias.

This ratio may vary from $\sim 10^{-1}$ to $\sim 10^5$, depending on the components of the junction.

Selenium oxide–normal metal diodes were discovered in 1883 and copper–cuprous oxide rectifiers were found in 1926. Both types were in use long before their mechanisms of operation were understood. Both types can supply relatively large dc currents for the operation of motors, relays, and battery chargers and are important in dc power supplies.

8.8 *P–N* JUNCTION RECTIFICATION

Semiconductor crystals can be made so that they consist of both *p*-type and *n*-type regions that are separated by a thin, interfacial (junction) layer. This junction usually is about 10^{-6} to 10^{-5} m wide. While, in actuality, concentration gradients of both donor and acceptor ions do exist across the boundary, which to a large extent may be controlled during manufacture, it is assumed for purposes of simplification that no such gradients exist. Thus, discontinuous concentration changes are postulated at the junction. The simple band models for the individual, bulk *p*- and *n*-type materials are as shown in Fig. 6-23. These band structures undergo a very rapid transition from that of one type to that of the other type of material at the junction. In some cases, the extent of the band transition must be kept small with respect to the mean free path of a carrier of average lifetime (see Section 8.9.1). The electrons and holes will diffuse to the opposite sides of the junction, in the same manner as described in the previous sections for electrons. The electrons diffuse into the *p*-type material and become minority carriers, creating positively charged donor ions in the *n*-type material by their absence. Holes diffuse into the *n*-type material and become minority carriers, creating negatively charged acceptor ions in the *p*-type material by their absence. The space-charge zone will attract all minority carriers within a volume that extends one mean free path of a carrier on either side of it. At equilibrium, two oppositely charged volumes build up, and a strong electric double layer and corresponding internal field are created in the carrier-depleted volume about the junction (Fig. 8-6). The depletion zone is described in greater detail in Sections 8.8.1 and 8.8.2.

In the absence of an applied electric field, at equilibrium, an electron diffusion will occur from the *n*- to the *p*-type that will equal that of the opposite direction. This is analogous to Eq. (8-58) and is denoted by

$$n_{-e} = n_{e0} = n_{+e} \qquad (8\text{-}68a)$$

where n_{-e} and n_{+e} are the densities of the minority carriers that have diffused from the *n* and *p* regions, respectively. A similar condition exists for the holes diffusing in the opposite direction from the *p*- to the *n*-type material giving, where $n_{h0} \neq n_{e0}$, and n_{+h} and n_{-h} are the densities of the minority carriers that have diffused from the *p* and *n* regions, respectively,

$$n_{+h} = n_{h0} = n_{-h} \qquad (8\text{-}68b)$$

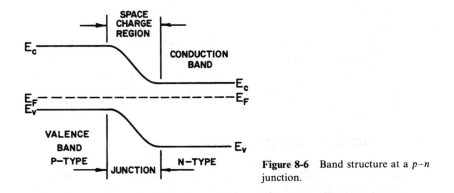

Figure 8-6 Band structure at a $p-n$ junction.

The application of a voltage across the junction gives, in the same way as shown for Eq. (8-61a), for the minority electrons,

$$n_{-e} = n_{e0} \exp\left(-\frac{eV}{k_B T}\right)$$ (8-69a)

and, similarly,

$$n_{+h} = n_{h0} \exp\left(-\frac{eV}{k_B T}\right)$$ (8-69b)

for the minority holes. The net density of the electron flow is

$$n_e = n_{-e} - n_{+e} = n_{e0}\left[\exp\left(-\frac{eV}{k_B T}\right) - 1\right]$$ (8-70a)

And that of the oppositely flowing holes is

$$n_h = n_{+h} - n_{-h} = n_{h0}\left[1 - \exp\left(-\frac{eV}{k_B T}\right)\right]$$ (8-70b)

The net density of carriers is $n_e - n_h$. This is given by the difference between Eqs. (8-70a) and (8-70b);

$$n = n_e - n_h = (n_{e0} - n_{h0})\left[\exp\left(-\frac{eV}{k_B T}\right) - 1\right]$$ (8-71)

Then the current density is obtained as

$$j = ne = e(n_{e0} - n_{h0})\left[\exp\left(-\frac{eV}{k_B T}\right) - 1\right]$$ (8-72a)

or

$$j = (j_{e0} - j_{h0})\left[\exp\left(-\frac{eV}{k_B T}\right) - 1\right]$$ (8-72b)

For forward bias,

$$j_F = (j_{e0} - j_{h0}) \left[\exp\left(\frac{eV}{k_B T}\right) - 1 \right] \tag{8-73}$$

where, as before, the current density increases exponentially with the voltage, j_{h0} being very small. In the case of reverse bias,

$$j_R = (j_{e0} - j_{h0}) \left[\exp\left(-\frac{eV}{k_B T}\right) - 1 \right] \tag{8-74}$$

The negative voltage causes the exponential term of Eq. (8-74) to become very small, and $j_R \cong j_{h0} - j_{e0}$, which also is very small. This behavior is essentially the same as that described for metal–semiconductor junctions (Section 8.7). This is shown schematically by the broken curve in Fig. 8-5. The ratio of j_F/j_R is very large, being of the order of about 10^8 to 10^9; this has been called the *rectification coefficient*.

The rectification capability of a *p–n* junction is apparent, provided that breakdown (Sections 8.9–8.9.3.2) is not permitted to take place. Characteristics of these kinds are the bases for transistors and other junction devices. It will be noted that the forward portion of the curve for a realistic junction device will show a threshold voltage beyond which the j_F becomes appreciable. A typical value for the threshold voltage is about 0.5 V.

8.8.1 Depletion Zone

As noted previously, the concentrations of the *n* and *p* dopants are considered to drop discontinuously to zero at the junction of the two materials. The electrons diffuse into the *p*-type material and the holes into the *n*-type material. In so doing, they leave behind ionized dopant ions of opposite sign on either side of the junction. The junction is virtually devoid of carriers and can act as an insulator. If the donors are from group V of the Periodic Table and the acceptors are from group III, and the concentration of each of the dopants is the same, the majority carrier densities will be the same in each portion. In addition, minority carriers of opposite sign will have diffused into each region.

The number of carriers in the depletion zone may be extimated by starting with the product of Eqs. (6-116) and (6-117) to obtain, for $m_e^* \cong m_h^* \cong m^*$,

$$n_{e0} \cdot n_{h0} \cong 4 \left(\frac{2\pi m^* k_B T}{h^2}\right)^3 \cdot \exp\left\{\frac{(E_F - E_c) + (E_v - E_F)}{k_B T}\right\} \tag{8-75a}$$

Considering the cubic factor to be small with respect to the exponential term, Eq. (8-75a) reduces to

$$n_{e0} \cdot n_{h0} \cong \text{constant} \exp\left[\frac{E_v - E_c}{k_B T}\right] \tag{8-75b}$$

The quantity $E_v - E_c$ is a constant for a given material under these conditions [Eq.

(6-119)], so

$$n_{e0} \cdot n_{h0} \cong \text{constant} = n_i^2 \tag{8-76}$$

in which n_i is the density of each type of intrinsic carrier; the constant n_i^2 is given by Eq. (8-75a). In a similar way,

$$n_{e0} \cdot n_{+h} \cong n_i^2 \tag{8-77a}$$

and

$$n_{h0} \cdot n_{-e} \cong n_i^2 \tag{8-77b}$$

Equations (8-77a) and (8-77b) are equated to give

$$n_{e0} \cdot n_{+h} \cong n_{h0} \cdot n_{-e} \cong n_i^2 \tag{8-78}$$

Now, using the same bases of approximations as those used in Section 8.6, $n_{e0} \cong n_{h0} \sim 10^{21}\,\text{m}^{-3}$ and $n_i \sim 10^{18}\,\text{m}^{-3}$, then, using Eq. (8-78), $n_{+h} \cong n_{-e} \sim 10^{15}\,\text{m}^{-3}$. Thus, it is seen that the numbers of carriers of opposite sign (minority carriers) is $\sim 10^{-6}$ times smaller than the majority carriers in either region.

The concentration differences between the two original unaffected volumes is the driving force for the diffusion of carriers of opposite sign into n- and p-type regions [Eq. (6-215)]. The resulting migrations of carriers ionizes the dopants and causes a positive charge in the n region and a negative charge in the p region. At equilibrium [Eqs. (8-68a) and (8-68b)], these unbalanced charges distort the bands of the n material in a downward direction and those of the p region in an upward direction with respect to the Fermi energy, shown schematically in Fig. 8-6. This bending of the bands is explained on the basis of the mechanisms discussed in Section 8.6. The buildup of the charges on either side of the junction of the p and n material creates a space-charge, or depletion, region (Sections 8.5, 8.5.1, and 8.6), a potential difference, V_c, and a contact potential barrier, ϕ_c, that serves to maintain the carrier distribution on either side of the junction.

If no barrier existed, the steep concentration gradients would ensure diffusion such that the densities of extrinsic carriers of both signs would ultimately be the same in both materials. The potential energy barrier, ϕ_c, required for a carrier to overcome the barrier is given by the general equation

$$\phi_c = k_B T \ln K \tag{8-79}$$

where K is the equilibrium constant of the reaction. If the reaction being considered is the equilibrium between the density of electrons in the n region and that of those in the p region, it may be expressed as

$$n_{e0} \rightleftarrows n_{-e} \tag{8-80}$$

This gives K directly from the Nernst distribution law as

$$K = \frac{n_{e0}}{n_{-e}} \tag{8-81}$$

In turn, Eq. (8-81) gives ϕ_c in terms of the carrier densities, since Eq. (8-81) converts Eq. (8-79) to

$$\phi_c = k_B T \ln \frac{n_{e0}}{n_{-e}} \tag{8-82}$$

Thus, the greater the disparity between the carrier concentrations of the two regions, the greater will be the value of ϕ_c. Using the data found earlier in this section, for room temperature ($T \sim 300$ K), $k_B \cong 8.63 \times 10^{-5}$ eV/K, and $K \cong 10^{21}/10^{15} \cong 10^6$, it is found that $\phi_c \cong 0.4$ eV. Where the holes are being considered, the same approach as that used to obtain Eq. (8-82) is taken.

8.8.2 Depletion Zone Width and Capacitance

Once the diffusion of the carriers has reached equilibrium, it may be considered that, at normal temperatures, all the dopants in the depletion zone are completely ionized (Section 6.3.3). As equilibrium is approached, an internal electric field is set up in a way analogous to that described in Section 8.5.1. The field grows in strength until equilibrium is reached. The resulting fields in depletion zones are sufficiently high as to ensure that virtually all the available carriers are swept out of the depletion regions in a way analogous to that described in Section 6.4.3, and they may be considered to approximate insulators. Thus, where N_{e0} and N_{h0} are the respective densities of the donor and acceptor dopants, and x_n and x_p are the boundaries of the n and p regions, the situation on either side of the junction is given, for each portion of the depletion zone, by

$$-N_{e0}ex_n = +N_{h0}ex_p \tag{8-83}$$

This is required for charge conservation to be preserved. This may be expressed in terms of the respective carrier densities ρ_{n0} and ρ_{h0} as

$$-\rho_{n0} = N_{n0}e_{n0} \tag{8-84a}$$

and by

$$\rho_{h0} = N_{h0}e_{h0} \tag{8-84b}$$

So

$$-\rho_{n0}x_n = \rho_{h0}x_p \tag{8-84c}$$

and

$$N_{e0}x_n = N_{h0}x_p \tag{8-85}$$

Using Eq. (8-84a), Poisson's equation [Eq. (8-47)] for the p region is

$$\frac{d^2V}{dx^2} = \frac{eN_{h0}}{\varepsilon_0\varepsilon} \tag{8-86}$$

Considering the center of the depletion zone as being a plane at which the discrete concentration changes take place, as assumed in Section 8.8, and using this plane as a reference, Eq. (8-86) is integrated twice from $-x_p$ to zero to obtain, for the p region,

$$V(-x_p) - V(0) = \frac{eN_{h0}x_p^2}{2\varepsilon_0\varepsilon} \tag{8-87}$$

And, similarly for the n region, doubly integrating from zero to x_n,

$$V(0) - V(x_n) = \frac{eN_{e0}x_n^2}{2\varepsilon_0\varepsilon} \tag{8-88}$$

The potential across the entire depletion zone is given by the sum of Eqs. (8-87) and (8-88). The contact potential is, after factoring the sum,

$$V_c = V(-x_p) - V(x_n) = \frac{e}{2\varepsilon_0\varepsilon}(N_{h0}x_p^2 + N_{e0}x_e^2) \tag{8-89}$$

Now using Eq. (8-85) to substitute for x_p^2 and x_n^2, Eq. (8-89) becomes

$$V_c = \frac{e}{2\varepsilon_0\varepsilon}\left(N_{h0}\frac{N_{e0}^2x_n^2}{N_{h0}^2} + N_{e0}\frac{N_{h0}^2x_p^2}{N_{e0}^2}\right) \tag{8-90a}$$

This reduces to

$$V_c = \frac{e}{2\varepsilon_0\varepsilon}\left(\frac{N_{e0}^2x_n^2}{N_{h0}} + \frac{N_{h0}^2x_p^2}{N_{e0}}\right) \tag{8-90b}$$

For charge conservation,

$$N_{e0} = N_{h0} \tag{8-91}$$

so it follows that

$$|n_{e0}e| = n_{h0}e \tag{8-92}$$

Now, using Eq. (8-76), as applied to the present case,

$$n_{e0}^2 \cong n_{h0}^2 \cong n_i^2 \tag{8-93}$$

For this to apply here, both x_n and x_p must be considered to have the same width as the total width of the junction, d. Thus,

$$x_n \cong x_p \cong d \tag{8-94}$$

Equations (8-93) and (8-94) simplify Eq. (8-90b) to read as

$$V_c \cong \frac{e}{2\varepsilon_0\varepsilon} \cdot n_i^2 d^2 \left(\frac{1}{n_{h0}} + \frac{1}{n_{e0}}\right) \tag{8-95}$$

This enables the width of the junction for moderately doped materials to be given by

$$d \cong \left(\frac{2\varepsilon_0\varepsilon V_c}{e} \cdot \frac{n_{h0} + n_{e0}}{n_i^2}\right)^{1/2} = \left(\frac{2\varepsilon_0\varepsilon V_c}{e} \cdot \frac{n_{h0} + n_{e0}}{n_{h0} \cdot n_{e0}}\right)^{1/2} \tag{8-96}$$

or, for $V_c = \phi_c/e$,

$$d \cong \left(\frac{2\varepsilon_0 \varepsilon \phi_c}{e^2} \cdot \frac{n_{h0} + n_{e0}}{n_{h0} \cdot n_{e0}} \right)^{1/2} \tag{8-96a}$$

It will be noted that when the dopant concentration in one region is considerably greater than that of the other, the width of the depletion region is governed by the concentration of the carriers of the opposite sign. For example, if $n_{e0} \gg n_{h0}$, Eq. (8-96a) may be approximated by

$$d \cong \left(\frac{2\varepsilon_0 \varepsilon \phi_c}{e^2} \cdot \frac{1}{n_{h0}} \right)^{1/2} \tag{8-97}$$

This finding is in agreement with Eq. (8-85). Equation (8-97) also clearly shows that the greater the concentration of dopant of a given sign, the smaller the portion of the depletion zone that may be ascribed to it.

The insulatorlike properties of the depletion zone permit the double layers to be considered as being the plates of a capacitor. On this basis, the charge on such a plate is $Q = e n_i d$. Thus, Eq. (8-95) may be written as

$$V_c \cong \frac{Q^2}{2\varepsilon_0 \varepsilon e} \left(\frac{n_{e0} + n_{h0}}{n_{e0} n_{h0}} \right)$$

and

$$Q \cong V_c^{1/2} \left(2\varepsilon_0 \varepsilon e \frac{n_{e0} n_{h0}}{n_{e0} + n_{h0}} \right)^{1/2}$$

So the capacitance, C, of the junction is, where V is a voltage that includes both V_c and the externally applied voltage,

$$C = \frac{\partial Q}{\partial V} = \left(\frac{\varepsilon_0 \varepsilon e}{2 V_c} \cdot \frac{n_{e0} n_{h0}}{n_{e0} + n_{h0}} \right)^{1/2} \tag{8-95a}$$

And, for the case in which $n_{e0} \gg n_{h0}$, the capacitance may be approximated in the same way as for Eq. (8-97) using

$$Q \cong V_c^{1/2} [2\varepsilon_0 \varepsilon e n_{h0}]^{1/2}$$

and the capacitance as

$$C \cong \left(\frac{\varepsilon_0 \varepsilon e n_{h0}}{2 V_c} \right)^{1/2} \tag{8-97a}$$

8.8.3 Effect of Temperature

An upper temperature limit exists beyond which rectification cannot take place at *p–n* junctions. This may be seen by means of Eq. (8-75a) by expressly including m_e, m_h, and Eq. (6-124). This gives the density of intrinsic carriers of either sign as

$$n_i = 2 \left(\frac{2\pi(m_e \cdot m_h)^{1/2} k_B T}{h^2} \right)^{3/2} \exp\left(\frac{-E_g}{2k_B T} \right) \tag{8-98}$$

The temperature at which $2k_B T = E_g$ reduces the exponential to a constant, and Eq. (8-98) becomes

$$n_i = \text{constant } (k_B T)^{3/2} = \text{constant } (E)^{3/2} \tag{8-99}$$

Under this condition, the temperature variation of the density of carriers is the same as that given for normal metals by Eq. (4-8). In this case, conduction much more closely corresponding to that of metals, rather than to that of semiconductors, is to be expected. $P-n$ junctions in silicon with $E_g \cong 1.1\,\text{eV}$ show rectification behavior up to about 150°C, while those in germanium, $E_g \cong 0.6\,\text{eV}$, show this up to close to 90°C, illustrating the controlling effect of E_g on the limiting temperature of rectification operation. The effect of Eq. (8-98) may be summarized by saying that the larger the value of E_g of a semiconductor containing a $p-n$ junction, the higher will be the temperatures at which it will retain its rectification properties. It will be recognized that this generalization must be qualified, because if $E_g \gg k_B T$, the material would be an electrical insulator rather than a semiconductor (Sections 4.5, 4.5.1, 4.6, and 4.7).

At temperatures below the limiting temperature, other more important factors and their effects on junctions must be taken into account. In the exhaustion range (Section 6.3.3), the densities of extrinsic carriers, being functions of dopant concentrations [Eqs. (8-84a) and (8-84b)] may be considered to be virtually independent of temperature. However, as shown by Eq. (8-98), the densities of intrinsic carriers show a steep increase with temperature. As the temperature increases, $n_i \gtrapprox n_{h0}$ and $n_i \gtrapprox n_{e0}$. Now, using Eq. (8-76), (8-77a), (8-77b), and (8-78),

$$n_{-e} = \frac{n_i^2}{n_{h0}} = \frac{n_{h0}^2}{n_{h0}} = n_{h0} \tag{8-100}$$

And, similarly, $n_{+h} = n_{e0}$. Then, since $n_{h0} = n_{e0}$ and $n_{-e} = n_{+h}$, Eq. (8-82) becomes, since $\ln(1) = 0$,

$$\phi_c = k_B T \ln \frac{n_{e0}}{n_{-e}} = 0 \tag{8-101a}$$

and

$$\phi_c = k_B T \ln \frac{n_{h0}}{n_{+h}} = 0 \tag{8-101b}$$

Thus, at temperatures below the upper limit given previously, the potential barrier at the $p-n$ junction vanishes and rectification can no longer take place; the effects of the junction vanish.

8.9 BREAKDOWN

It can be seen from Eqs. (8-73) and (8-74) and schematically in Fig. 8-5 that the ratio of j_R/j_F is very small, being of the order of $\sim 10^{-8}$ or less. This behavior continues with increasing reverse bias up to a certain point at which the resistance of the depletion zone rapidly decreases. The reverse current increases correspondingly when this takes place. This behavior is known as *breakdown.* When this occurs at reverse voltages of 1 V or less, it is caused by tunneling (Section 8.4) and is called *Zener breakdown* [Fig. 8-5(b)]. Where higher reverse voltages are required by breakdown, it results from the abrupt increase in carriers that occurs as a result of very rapid generation of carrier pairs. This mechanism is called *avalanche breakdown.* Breakdowns of these types restrict the applications of junction devices in many types of circuitry. However, some junction devices, sometimes known as *backward diodes,* are made to utilize this response. Breakdown can also result from thermal effects.

8.9.1 Zener or Tunnel Breakdown

When the dopant concentrations are high, of the order of $10^{24}\,\text{m}^{-3}$ or greater, and the concentration gradients are very steep, the width of the depletion zone is $\sim 10^{-7}\,\text{m}$ or smaller [Eqs. (8-96)], and large carrier densities, electrons and holes, are available. Here the carriers are degenerate [Eq. (2-76a)] and fill the available energy levels such that E_F is positioned so that it lies in the valence band of the *p*-type material and in the conduction band of the *n*-type material [Eqs. (6-163) and (6-164)]. This is shown schematically in Fig. 8-7.

These conditions are necessary for tunneling to take place. The high dopant concentrations result in the small depletion zone widths, as noted previously, that permit suitable penetration of the wave functions (Section 8.4). In addition, highly charged double layers are created with corresponding internal electric fields that are much greater than those described in previous sections, being about $10^8\,\text{V m}^{-1}$ or greater. Since the degenerate carriers of each type have energies close to E_F, and E_F lies inside both the valence and the conduction bands, carriers may transfer across the narrow junction with no change in energy. A carrier that approaches the depletion zone within a distance of one mean free path or less will be accelerated by the internal field and tunnel across the barrier.

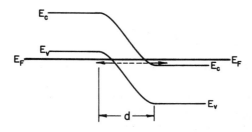

Figure 8-7 Direct tunneling across a junction between highly doped semiconductors. The junction width d is very small.

Under these conditions, a very small reverse bias can cause the appreciable current flow that is known as internal, tunnel, or Zener breakdown [Fig. 8-5(b)]. It is noted again, for emphasis, that tunneling can take place only where the dopant concentrations have steep gradients and are high enough to create the narrow depletion zone required by tunneling (Sections 8.4 and 8.8.2).

8.9.2 Avalanche Breakdown

The avalanche mechanism proceeds by a different mechanism than that involved in tunneling. Here, instead of tunneling *between* wells, the electrons are promoted *across* the gap, from the valence band to the conduction band, in increasing numbers. This does not require that the width of the depletion zone be small, since tunneling is not involved. Thus, the concentrations of dopants required for avalanching are smaller than those required for tunneling, the concentration gradients are not as steep, the widths of the depletion zones are greater, and the internal fields are much smaller.

The application of a suitable reverse voltage will cause a small fraction of the electrons to have energies greater than E_g. One such carrier may encounter an electron in the valence band and excite it up to the conduction band, leaving a hole or an ionized dopant behind. A considerable fraction of the energy of the initial electron is consumed in this process; two additional carriers, an electron and a hole, may be created. If it is assumed that the momentum of the initial electron is mv and its energy is $mv^2/2$, and that the masses of all three particles are the same, each particle is assumed to have a momentum of $mv/3$ and an energy of $(m/2)(v/3)^2$. The energy consumed in this reaction, $E(R)$, is the difference between the initial energy, $E(I)$, and the energy of the three particles after the reaction, $E(f)$. This gives

$$E(R) = E(I) - E(f) = \frac{mv^2}{2} - 3\left[\frac{m}{2}\left(\frac{v}{3}\right)^2\right] = \frac{mv^2}{3}$$

Thus,

$$E(R) = \frac{2E(I)}{3} \quad \text{and} \quad E(I) = \frac{3E(R)}{2} \tag{8-102}$$

If the collision is to result in pair production, the minimum energy must be such that it is greater than E_g. Thus, the minimum energy for an electron involved in the avalanche mechanism must be

$$E(I) \geqq \frac{3E_g}{2} \tag{8-103}$$

Thus, the voltage required to initiate avalanching will vary with the width of the energy gap.

The avalanche mechanism is based on the idea that a few such excited electrons can be responsible for the generation of progressively larger numbers of electron–hole pairs. The application of a suitable reverse voltage will create n_0 electrons per unit volume with energies sufficiently greater than E_g and an average mean free path λ. A fraction, f, of the n_0 electrons will be involved in pair production as they traverse λ. Thus, $n_0 f$ electrons and $n_0 f$ holes will be generated simultaneously. If it is assumed that the holes react in the same way as the electrons, since their ionization energies are similar, the holes also will create pairs. The effect of this is that $n_0 f^2$ electrons will have been energized. These contribute to this process by generating $n_0 f^3$ pairs. This process continues as long as the reverse voltage is applied and the result is that the density, n, of the avalanching particles is given by the geometric progression

$$n = n_0(1 + f + f^2 + f^3 + \cdots) = \frac{n_0}{1 - f} \tag{8-104}$$

It can be seen that as the reverse bias is increased, both n_0 and f will increase so that $n \gg n_0$. This increase in the carrier density results in the corresponding increase in the reverse current that is shown schematically in Fig. 8-5(a).

Compared to tunneling, avalanche breakdown occurs in materials with lower dopant concentrations and with wider junctions that permit higher reverse biases. The breakdown voltage in moderately doped materials depends on the depletion zones, where the concentration gradients of the dopants are not as sharp as those in tunnel diodes [Eqs. (8-96) and (8-97)]. These properties may be illustrated by approximating that Eq. (8-96) may be represented simply, for moderately doped materials, by

$$d \propto \left(\frac{V_c}{n_i^2}\right)^{1/2} = \frac{V_c^{1/2}}{n_i} \tag{8-105}$$

and by Eq. (8-97), for highly doped materials that contain dopant concentrations approaching those required for tunneling, as

$$d \propto \left(\frac{V_c}{n}\right)^{1/2} = \frac{V_c^{1/2}}{n^{1/2}} \tag{8-106}$$

where n represents the high dopant concentration and $n > n_i$. For a given V_c, the width of the depletion zone of a moderately doped material will change more slowly with dopant concentration than that of a highly doped material. And a given reverse bias, which would create a high internal field in the narrow depletion zone of a highly doped material and cause breakdown, would be insufficient to cause avalanching in a moderately doped material; higher reverse biases would cause avalanching before tunneling could take place. The voltage required for avalanching would depend on the value of E_g of the material [Eq. (8-103)]. This may be shown by comparing the electric fields required for breakdown in the following materials (in units of 10^7 V m^{-1}): Ge, 0.8; Si, 3.0; GaAs, 3.5; and SiO$_2$, 60.

8.9.3 Other Breakdown Mechanisms

In general, breakdown is to be avoided in all transistors except those designed for its employment. The sudden increase in the reverse current can have deleterious effects.

It should also be noted here that Ohm's law ($j = \sigma \bar{E}$) is obeyed only as long as the conductivity (or resistivity) of the material is not affected by the applied electric field. This is so because σ (or $1/\rho$) is the constant of proportionality that accounts for the linear relationship between the current density and the electric field. As implied in Eq. (6-24), the drift velocity of the nondegenerate electrons is small compared to their thermal velocity so that only the thermal velocity need be considered. The conductivity is unaffected by weak applied fields because $\langle \mathbf{v} \rangle$ is not a function of such applied fields.

As the applied field is increased, a value is reached beyond which the linear relationship no longer holds. Here the drift velocity becomes increasingly larger and can become appreciable with respect to $\langle \mathbf{v} \rangle$, so it may no longer be neglected. Thus, the conductivity no longer is independent of $\bar{E}$. This may best be examined by means of Eq. (6-109), where the mobility is the variable that clearly shows the effect of $\bar{E}$. The mobility is given by

$$\mu_e \propto \frac{\langle \lambda \rangle}{\langle \mathbf{v} \rangle} \tag{6-132}$$

At moderately high temperatures, $\langle \lambda \rangle$ approaches a small, essentially constant value [Eq. (5-202)]. So, using Eq. (6-133), $\mu_e \propto T^{-1/2}$. And, still treating these carriers classically, $\mu_e \propto E^{-1/2}$. Now, by means of Eq. (6-234), and recalling that $\tau(E)$ may be considered to be essentially constant for the present conditions, it may be approximated that $E \propto \bar{E}$. On this basis it may be approximated further that $\mu_e \propto \bar{E}^{-1/2}$. And, since $\sigma = ne\mu$, $\sigma \propto \bar{E}^{-1/2}$. Now, using this behavior of conductivity in Ohm's law, $j = \sigma\bar{E} \propto \bar{E}^{-1/2} \cdot \bar{E} = \bar{E}^{1/2}$; the linear relationship no longer holds in strong fields.

This parabolic behavior describes the variation of the current density with the electric field for strong fields. However, as the applied field is increased further, the drift velocity attains a limiting value that is independent of the field. The effect of this is that the current density is no longer a function of the field, but remains constant. It is said to reach a condition known as *drift-velocity saturation*. Further increases in the applied field can cause avalanching (Section 8.9.2), and a situation analogous to that described in Section 6.4.4 results if the equivalent thermal energy imparted to the electrons is not removed rapidly enough from the material.

8.9.3.1 Surface effects. Even if it is assumed that a perfect crystalline array exists within a semiconductor, a condition not possible according to the second law of thermodynamics, ions at or near its surface will be in disarray. This results because such ions have insufficient numbers of near neighbors to satisfy their bonding (Section 3.12.3). So, instead of the uniform band structure and carrier densities that exist within the crystal, the surface of the crystal will have an unsymmetrical band structure and a higher variation in its charge density at and near its surfaces. This asymmetry is large in semiconductors, as compared to that of metals, because of their comparatively low carrier densities. The resulting space-charge region distorts the band structure near the surface. In addition, the potential barrier at the surface (Fig. 4-5) adds to this distortion. The result of these factors is a surface depletion zone

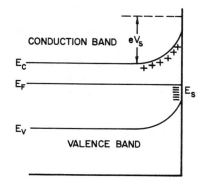

Figure 8-8 The space-charge region and depletion zone near the surface of an *n*-type semiconductor.

comparable to that in the semiconductor portion of Fig. 8-4(b). And in the nonideal case, in which crystalline imperfections are always present adjacent to the surface, the disruption of the periodicity of the lattice accentuates the band distortion and contributes to the depletion zone. These surface states are known as *Tamm states.*

This mechanism causes the states near the surfaces of *n*-type materials to become acceptors. Such states occupy levels in the band gap up to the Fermi level. These are filled with electrons from the valence band, and a narrow negatively charged, subsurface layer is set up. A barrier, thus, is created in much the same way as is described in Section 8.8.1. The volume of the barrier zone is swept free of electrons, and it becomes a surface-depletion zone. This may extend as much as $\sim 10^{-7}$ m into the volume of the crystal (Fig. 8-8).

Impurities chemisorbed on the surfaces of semiconductors also build up double layers that can add to the potential barrier in a way similar to that described in Section 8.1.1. (It must be realized that it is virtually impossible to keep the surfaces of real semiconductors perfectly clean.) The surface states created by contaminants lie within the gap. Their high, local concentrations are such that E_F lies near their average energy. Well-prepared surfaces usually contain less than $\sim 10^{14}$ m^{-2} impurity states. Contaminated materials can have surface impurity states in excess of $\sim 10^{17}$ m^{-2}. These states, also known as fast surface states, may fill or empty, depending on the direction of the bias, and result in undesirable, variable and/or unstable responses of semiconductor devices. The creation of surface barriers of the kinds described here may be responsible for breakdown.

The improper attachment of metallic contacts or leads to semiconductors can cause surface barriers and, consequently, rectification at their junctions (Section 8.7). This may be avoided by the use of metallurgical joints with shallow rather than sharp concentration gradients or by very high dopant concentrations in the volume in which the joint is made (Section 6.3.5).

8.9.3.2 Thermal effects. When the Joule heat created by reverse currents is such that it is not transferred out of the semiconductor device sufficiently rapidly, its temperature will rise. The electron density will increase, and breakdown can occur in devices in which this is not desired. The mechanism involved is that of avalanching (Section 8.9.2). Here the process is initiated by the excitation of the electrons as a result of the increased temperature, rather than by the reverse voltage.

A model similar to that outlined in Section 6.4.4 may be used to describe the thermal effects. In the case of interest here, thermal energy is the source of the excitation, rather than that imparted by an external electric field.

8.10 PHOTON-INDUCED MECHANISMS

The changes in the optical and electrical properties of semiconductors and of some insulators result from electromagnetic radiation. Here, instead of the thermal excitations noted in Chapter 6, the changes are induced by exposure to suitable radiation. Such incident photon energies, analogous to those given by Eq. (8-103), promote states within the band gap to the band edges or from band to band. The concepts presented earlier are extended and applied to other phenomena.

8.10.1 Carrier Generation, Recombination, and Trapping

The basic carrier mechanisms involved in electrical conduction in intrinsic and extrinsic semiconductors are outlined in the sections in Chapter 6. Additional means for carrier generation are by the tunneling and avalanche processes [Sections 8.9.1 and 8.9.2]. On the bases of the foregoing, it might be thought that carrier generation is primarily by a direct, single-step process [Fig. 8-9(a)]. This does not necessarily have to take place; multistep carrier generation is at least equally likely, depending on

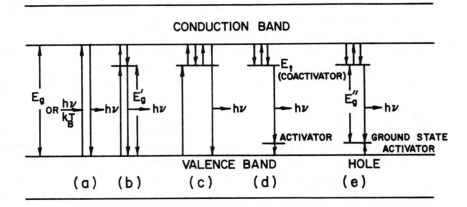

Figure 8-9 Schematic diagram of recombination processes: (a) direct recombination; (b) indirect recombination; (c) indirect recombination after multiple trapping by shallow traps; (d) phosphorescence at a luminescent center; (e) recombination at a luminescent center.

the material. Some small degree of multistep carrier generation is probable even in so-called intrinsic materials, because it is not possible to obtain them as perfect crystals or in the absolutely pure condition. Both of these factors contribute to multistep processes [see Fig. 8-9(b)].

Another indirect means of carrier promotion can occur by the *Auger effect*. This is based on the collision of two charged particles of the same sign. When one of these recombines with an oppositely charged particle, the released energy of the recombination is absorbed by the remaining particle. The increase in energy may promote it to the conduction band [Eq. (8-103)].

The indirect generation of carriers is initiated by the absorption of thermal or electromagnetic energy in the same ways as for direct generation. Here the intermediate states are provided by impurities and crystalline imperfections that occupy levels within the energy gap. These mechanisms are reversible, so phonons or photons are emitted when recombination occurs. The processes of generation and recombination may be considered to constitute a dynamic equilibrium for fixed conditions of temperature and/or applied energies. The generation of a carrier requires the absorption of a quantum of energy; its recombination is accompanied by the emission of that quantum of energy either as a phonon or photon.

An electron at a recombination center need not necessarily return to its original band; it may be excited back to the level from which it arrived. States that serve in this way are known as *traps*. This may be illustrated by noting that an electron in a trap near the conduction band may be cycled repeatedly between trap levels and the conduction band prior to its return to its original energy level [Fig. 8-9(c)], absorbing and emitting energy in the process. Since the emission of an electron by a trap may be considered as the absorption of a hole and the absorption of an electron as the emission of a hole, traps are considered to absorb and emit electrons and holes in the same way that they emit and absorb photons or phonons.

The degree to which a recombination center can act as a trap depends on the degree to which it is ionized. The greater the degree of its ionization, the deeper will be its potential well. The shallower the well (the lower the level of ionization), the closer it will be to a band edge and the greater the probability that it will act as a trap. The deeper the well, the closer it will be to the center of the gap and the smaller the probability that it will act as a trap. Traps are designated as being shallow or deep on this basis.

The probability that a carrier will escape from a trap varies as the Boltzmann factor, $\exp(-E_t/k_BT)$, in which E_t is the degree of ionization of the trap (the depth of its potential well). And, where $n(v)$ represents a function for the frequency of the particle in the trap, its rate of escape may be given by $n(v)\exp(-E_t/k_BT)$. [At ordinary temperatures, $n(v) \sim 10^8 \text{ s}^{-1}$.] Thus, where E_t is small, the trap state is most likely to provide a transient state for the absorbed carrier; where E_t is large, it is probable that the absorbed carrier will be considerably less available. This may be seen from the given rate of escape; the time that an electron spends in a trap will vary as $\exp(+E_t/k_BT)$.

8.10.1.1 Carrier lifetime. The density of carriers excited by electro-magnetic radiation may be much larger than that expected at thermal equilibrium. This results from trapping. The greater the probability of the capture of a carrier by a shallow trap, the slower will be the rate of decay and the longer the lifetime will be. The excess, nonequilibrium carriers can increase the number of nearly free carriers by a factor of 10^2, and these may exist as long as the irradiation continues. Each such carrier will continue to be excited for an average time τ. Thus, when the irradiation ceases, the recombination rate for electrons for the time τ is given as

$$R_e = -\frac{dn}{dt} \cong \frac{\Delta n}{\tau} \tag{8-107a}$$

where, using the prior notation,

$$n = n_e + n_{e0} \tag{8-107b}$$

and

$$\Delta n = n - n_{e0} = n_e \tag{8-107c}$$

Then, since n_{e0} is not a function of time,

$$R_e = -\frac{dn}{dt} = -\frac{dn_e}{dt} \cong \frac{n_e}{\tau} \tag{8-107d}$$

R_e is negative because it diminishes as n_e decreases. And rearrangement of Eq. (8-107d) gives

$$-\frac{dn_e}{n_e} \cong \frac{dt}{\tau} \tag{8-107e}$$

The integration of Eq. (8-107e) results in

$$\ln n_e \cong -t/\tau \tag{8-108}$$

And, using Eq. (8-107b) in Eq. (8-108), it becomes

$$\ln n_e = \ln(n - n_{e0}) = \ln \frac{n}{n_{e0}} \cong -\frac{t}{\tau} \tag{8-109a}$$

or, in exponential form, either as

$$\frac{n}{n_{e0}} \cong \exp\left(\frac{-t}{\tau}\right) \tag{8-109b}$$

or as

$$n \cong n_{e0} \exp\left(\frac{-t}{\tau}\right) \tag{8-109c}$$

Thus, the average lifetime is defined in the same way as is given for Eq. (6-12). The larger the value of τ, the longer will be the decay time. For $t = \tau$, recombination diminishes n by a factor of $1/2.7$. The parameter τ is also the decay time for the cessation of emissions after irradiation. And, for the condition where $t \gg \tau$, $n \to n_{e0}$, so $n_e \to 0$ and emission ceases. The same approach applied to the behavior of holes gives

$$n \cong n_{h0} \, \exp\!\left(\frac{-t}{\tau}\right) \tag{8-110}$$

in which n and n_{h0} now are their total and equilibrium densities, respectively, and τ is the average lifetime of a hole. When both holes and electrons are present, an approach similar to the one given previously yields

$$n_T \cong n_{T0} \, \exp\!\left(\frac{-t}{\tau_e + \tau_h}\right) \tag{8-111}$$

where n_T and n_{T0} are the total and equilibrium carrier densities, respectively, and τ_e and τ_h are their relaxation times.

The nearly free carriers generated by the radiation will diffuse in the semiconductor for times up to their average lifetime. The relationship between carrier mobility and diffusion coefficient are given by Eq. (6-221a). This equation may be reexpressed for a constant temperature as

$$\mu = \frac{De}{k_B T} = \text{constant} \cdot De \tag{8-112}$$

where μ and D are the mobility and diffusion coefficient, respectively, of the carrier under consideration. The distance, x, traveled by a carrier may be obtained from Fick's first law [Eq. (6-215)] as

$$x^2 \cong \alpha D \tau \tag{8-113}$$

where α is a constant of proportionality that depends on the concentration gradient of the radiation-generated carriers.

Recombination of a carrier will occur after time τ. This may take place directly, as shown in Fig. 8-9(a), or indirectly by impurity levels, as shown in Figs. 8-9(b), (c), and (d). A single quantum of energy will be emitted in direct recombination, and at least two quanta will be emitted by a multistep process. Where the quanta are sufficiently large, photons are emitted. This is called *radiative recombination*. And where phonons are given off, the process is considered as *nonradiative recombination*. In cases in which the energy gaps are relatively small, direct recombination is most prevalent. Where the gaps are relatively large, recombination generally is nonradiative. Gallium arsenide and gallium phosphide are important exceptions to this behavior. Where the radiation-generated carrier density is high, about 50% of the recombinations in these materials will be radiative. The radiation emitted by gallium phosphide largely results from indirect recombination similar to that indicated by Fig. 8-9(c).

Lifetimes are important factors in the quality control of semiconductor devices. They may be used to provide an index of impurity and imperfection concentrations.

The theoretical lifetime of an ideal semiconductor is of the order of seconds. Commercial materials have lifetimes that may vary from about 10^{-6} to 10^{-3} s and frequently are of the order of 10^{-5} s.

8.10.2 Photoluminescence

Thermally induced radiation, under equilibrium conditions, is discussed in Section 1.2 where the number of oscillators absorbing radiation is equal to those radiating quanta and dropping back to their original energies.   In the cases under consideration here, fewer electrons must be absorbing the radiation than those emitting it.   This means that the number of excited electrons that cause photon emission must be greater than that required by thermal equilibrium.   Such mechanisms thus involve nonequilibrium processes and, as such, may be considered as being types of inversions of carrier populations.

The spontaneous emission of photons by crystals is known as *luminescence*. This may be induced by exposure to radiation that ranges in wavelengths from those of gamma and x-rays and electrons to those with wavelengths in the radio, infrared, visible and ultraviolet portions of the spectrum.   The radiation emitted under these conditions is not a function of temperature; it is a function of the relation of the energy of the incident radiation with respect to that of the band gap.   Here, the fundamental relationship is given for the incident radiation by

$$\lambda_{max} \geqq \frac{1.24}{E_g} \tag{8-114}$$

Or, using Eq. (8-103) to take the energy of the reaction into account in Eq. (8-114), the emitted radiation is approximated by

$$\lambda_{max} > \frac{1.86}{E_g} \tag{8-115}$$

While some of the emitted radiation has the same wavelengths as the incident radiation [Eq. (8-114)], a large fraction of the emitted radiation is of longer wavelengths [Eq. (8-115)].   Some of the incident energy is absorbed as phonons. This difference in the wavelengths is *Stokes law of fluorescence*.   The overlapping of the distributions of the wavelengths of the incident and emitted wavelengths is called the *anti-Stokes luminescence* (or *anti-Stokes region*).   This results from the reaction of an incident photon with an ion that has been excited previously by a phonon.   The energy of the photon emitted by this ion, upon its return to its initial state, will be greater than that of the incident photon.   Its shorter wavelength accounts for the anti-Stokes region.

The efficiency of luminescence, or the degree at which energy is transferred, is defined as the ratio of the energy output (reradiation) to the energy input (irradiation). S. I. Vavilov showed that the efficiency of illumination increases linearly as a function of increasing λ up to a narrow maximum efficiency of ~ 0.8 and then very abruptly drops to zero at wavelengths at which the incident photons have insufficient energy to cause luminescence.

The decay times of such stimulated emissions divide these luminescent materials into two classes. Those that cease emission after about $\sim 10^{-9}$ to 10^{-6} s are classed as *fluorescent*; those that continue to emit photons for times greater than $\sim 10^{-6}$ s are classed as *phosphorescent*. Some of the latter may continue emissions for times up to the order of days. It will be noted that even the decay times of the fluorescent materials are very much greater than the 10^{-15} to 10^{-13} s times that are associated with the oscillation periods of photons. This may be illustrated by noting that the time period of an oscillation associated with $E_g = 1\,\text{eV}$ in Eq. (8-114) is 0.42×10^{-14} s.

8.10.2.1 Fluorescence.

Nearly perfect crystals of semiconductors or insulators with electron transitions such as those indicated by Fig. 8-9(a) show little or virtually no luminescence; their decay times are much too short. Appreciable fluorescence requires the presence of impurities and imperfections in the lattice. The impurities, known as *activators*, are the more important of these, even though they usually are present at concentrations of $<0.01\%$. The synthetic materials used for this purpose generally consist of three components: matrix materials such as the sulfides of Ca, Zn, and Cd, activators such as Mn, Ce, and Pb, and low-melting salts that act as solvents, such as $CaSiO_3$ and $Ca_3(PO)_4$.

The matrix and activator materials determine the spectrum of the output radiation and the efficiency of the energy transfer (Section 8.10.2). The activator and any impurity levels ($<0.0001\%$) lie within the energy gap [Fig. 8-9(d)]. The impurity content, other than the activators, usually is minimized since trapping is not required for fluorescences. Some impurities may inhibit fluorescence because they may promote nonradiative transitions.

It is more likely that an incident photon will be absorbed by an activator than by a state in the valence band. The resulting excited electron is promoted to the conduction band rather than to any of the comparatively few traps, since these fill quickly. It behaves in a nearly free way, within the conduction band, until it reacts with another activator, emits a photon, and returns to the activator level. The previously noted short decay times of fluorescent materials result from the short ionization times of the activators.

Some applications of luminescence include fluorescent lamps and television and oscilloscope tubes. Ultraviolet emissions also may be created by electrical discharges in low-pressure mercury vapor or argon lamps with fluorescent coatings on their inner surfaces. The illumination produced in this way is more efficient than that resulting from heated filaments.

8.10.2.2 Excitons.

Irradiation may generate particles of a type not discussed previously. When the energy of the incident photons approaches that of E_g, but is less than E_g, a type of electrically neutral particle consisting of both an electron and a hole may be created. Radiation at this level of energy causes an excited electron to remain paired with its hole, rather than promoting it to the conduction band; this excitation is insufficient for their complete separation. The separation between the electron and the hole is relatively large, being ~ 50 interionic spacings, because the dielectric screening between the two particles behaves in a way similar to that described by Eq. (6-170). Thus, even though the pair is localized at a given lattice site, the excitation extends over a volume of the crystal that includes many lattice sites. This permits the relatively high mobility of excitons because no net expenditure of energy is required for their motion. They have an average velocity, $\mathbf{v}$, of $\sim 10^5$ ms^{-1} at 300 K. Their kinetic energies may be described classically, in terms of their reduced mass, m_R, by $KE = \frac{1}{2}m_R\mathbf{v}^2 = \frac{3}{2}k_B T$.

An exciton may be considered to consist of an electron in "orbit" around its hole in a way analogous to a hydrogen atom. A treatment based on this analogy, similar to that used to obtain Eq. (6-171a), gives the allowed internal energies of an exciton as

$$E_{\text{exciton}} \cong -13.6\,\frac{m_R}{\varepsilon^2 n^2}\quad \text{eV} \tag{8-116}$$

where m_R is the reduced mass of the pair of particles, ε is the relative dielectric constant of the material, and n is an integer. The reduced mass is used in order that the pair of particles may be treated as a single particle. This factor is given by

$$m_R = \frac{m_e^* m_h^*}{m_e^* + m_h^*} \tag{8-117}$$

The energies given by Eq. (8-116) lie in the gap, just below E_c, in the approximate range

$$0.7E_g < E_{\text{exciton}} < 0.9E_g \tag{8-116a}$$

Excitons do not contribute directly to the photoconductivity, but they do play an important role in optical spectra in the visible and ultraviolet ranges in very thin crystals. The specimens must be very thin because exciton emissions are very readily absorbed by other ions in the crystal, a consequence of the exciton "size" and its overlap with many ions.

The mobility of excitons permits their diffusion through a crystal with relative ease. In so doing, they react with such scattering centers as lattice defects, impurities, and phonons. If they recombine so that the exciton pair returns to its ground state, a photon is emitted that contributes to the luminescence noted previously. If the pair separates or decomposes into its components and emits a phonon, its components may contribute to photoconductivity.

8.10.2.3 Phosphorescence.   In contrast to fluorescence, phosphorescent materials require shallow traps at levels just below the bottom edge of the conduction band. These traps, known as *coactivators*, may consist of any type of lattice imperfection, as well as impurity atoms.

An electron dropping from the conduction band to a trap level, or one that has been excited from an activator to a similar trap level, known as a coactivator, is sufficiently close to the bottom of the conduction band so that phonons may promote it back to the conduction band [Fig. 8-9(d)]. This is the same as saying that E_t (Section 8.10.1) is close to the energy of the average lattice vibration. As a consequence, electrons are readily excited to the conduction band and thus they may alternate between these sets of levels. Since the rate of escape from a trap varies as $\exp(-E_t/k_BT)$, the time spent by an electron in a trap will vary as $\exp(E_t/k_BT)$ and depend primarily on E_t. An electron alternating its behavior between diffusing in the lattice and being trapped thus represents a means of storage of electrons that are available for dropping back to activator sites and emitting radiation. The persistence of such radiation depends on the times that the electrons remain in such traps and the number of traps.

The spectrum of the radiation will be determined partly by the ionization levels of the coactivators. The longest wavelengths will depend primarily on E_g'' [Fig. 8-9(e)]. The shortest wavelengths are given by Eq. (8-114). Radiation with wavelengths between these are explained by interactions of excited ions with the lattice phonons. These excitations can also arise from interactions with excitons (Section 8.10.2.2) or by the acquisition of an electron from the conduction band. Depending on its position in the lattice at the instant of photon emission, the emitted photon may have an energy anywhere in the range between $h\nu_1$ and $h\nu_2$ [Fig. 8-10(a)], thus accounting for the spectrum between these limits. An electron transition, such as that at S_{E2}, starts from a higher initial energy than one from S_{E1}. This results from a higher energy of oscillation of the excited ion that arises from increments of energy absorbed from phonons. However, $h\nu_2$ may be less than $h\nu_1$. When this occurs, the energy difference results in the generation of a phonon accounting for this energy as $h(\nu_1 - \nu_2)$.

Radiative transitions may be suppressed by the presence of impurities that

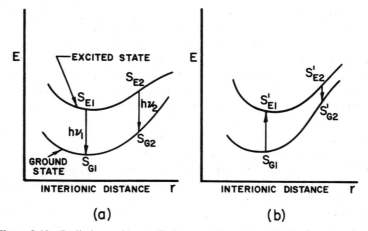

Figure 8-10 Radiative and nonradiative transitions: (a) radiation from transitions from the excited and ground states of an impurity ion; (b) nonradiative transitions from an excited impurity ion.

constitute very deep traps, or those that produce photons in the infrared spectrum, or those that generate only phonons. These impurities have been named "killers". After excitation to S'_{E1}, an ion may oscillate about S'_{E2} [Fig. 8-10(b)]. In dropping to S'_{G2}, either a photon in the infrared region or a phonon could be emitted. Either of these would then be followed by the emission of another phonon upon the return of the ion to state S_{G1}.

8.10.3 Photoconduction

Photoconduction is a consequence of increased carrier generation that results from electromagnetic irradiation in semiconductors and in some insulators. Those incident photons that obey Eqs. (8-114) and/or (8-115) can produce carriers in excess of those required by thermal equilibrium. (The electrical conductivity in the absence of irradiation is the *dark conductivity*.) Here E_g takes on a wider meaning than that used originally in Section 8.10.2; it can also include any energy increments between any donor states and the conduction band. This situation may be covered more generally by including all such energy differences as ΔE_i. Thus, the minimum photon energy for such carrier generation may be expressed as

$$hv = \Delta E_i \tag{8-117a}$$

and the maximum theoretical wavelength for photoconduction is, since $\lambda v = c$,

$$\lambda = \frac{ch}{\Delta E_i} \tag{8-117b}$$

As given, λ in Eq. (8-117b) represents the spectrum of radiation that can contribute to photoconduction by the promotion of electrons from any available impurity levels to the conduction band. The absorption spectrum of such a material will show a series of bands as a function of wavelength. The greatest absorption will occur for $\Delta E > E_g$ for the range of λ up to the photon energy equivalent of E_g, $\lambda(E_g)$. [This neglects, the almost complete absorption of photons with $\lambda \gg \lambda(E_g)$ that occurs at or very close to the surface.] At $\lambda \cong \lambda(E_g)$, the absorption decreases very steeply to that representing electron promotion from those impurity states requiring the next largest energy increment, such as those from a deep impurity level. In this way the absorption spectrum reveals the origin of the photoelectrons. The absorptions at the shortest λ are those requiring the greatest energy and originate from the valence band (the intrinsic components). Each successive absorption band indicates electron promotion requiring progressively smaller energies (the different extrinsic components) that are activated at successively longer wavelengths.

To observe this complete spectrum of wavelengths that contribute to photoconductivity, it is necessary to cool the specimens to cryogenic temperatures, well below the exhaustion range. This ensures that impurity states close to the bottom edge of the conduction band will not have been thermally depleted. If such materials were tested at room temperature, some absorptions would be absent and would indicate no activation by photons. They would appear to make no contribution to photoconductivity because thermal activation would have made such states unavailable for photon excitation. Those donor states that lie very close to the bottom edge of the conduction band may become evident only when illuminated by infrared radiation.

On this basis, it may be considered that, at normal temperatures, carrier generation may be ascribed to the direct process indicated in Fig. 8-9(a). The holes created have virtually no effect on the conductivity because of their rapid assimilation by recombination centers. Thus, when a crystal, with cross-sectional area A and length L (Fig. 8-11), is illuminated with suitable radiation, the rate of pair production will be at some value g per second and will also be representative of the rate of generation of electrons. The density of photon-generated electrons in the crystal per unit time is, where $\langle \tau \rangle$ is the average lifetime,

$$n_{pge} = \frac{g\langle \tau \rangle}{AL} \qquad (8\text{-}118)$$

And the photon-generated conductivity is, assuming that these electrons have the same mobility as those induced thermally,

$$\sigma_{pg} = n_{pge}e\mu_e = \frac{g\langle \tau \rangle}{AL}\, e\mu_e \qquad (8\text{-}119)$$

The carriers generated by thermal excitation are neglected in Eq. (8-119) because their density is small with respect to that given by Eq. (8-118). Now, by means of Ohm's law and the reciprocal of Eq. (8-118), the current in the crystal is

$$I_{pg} = \frac{V}{R} = \frac{VA}{\rho L} = \frac{VA}{L}\left(\frac{g\langle \lambda \rangle}{AL}\, e\mu_e\right) = \frac{g\langle \tau \rangle e\mu_e V}{L^2} \qquad (8\text{-}120)$$

The photon-induced current in a given specimen, per unit time of irradiation, is

$$I_{pg} = ge \qquad (8\text{-}121)$$

so the amplification is obtained from Eqs. (8-120) and (8-121) as

$$\gamma = \frac{g\langle \tau \rangle e\mu_e V}{L^2} \div ge = \frac{\langle \tau \rangle \mu_e V}{L^2} \qquad (8\text{-}122)$$

The effects of trapping and recombination have large influences on $\langle \tau \rangle$ and consequently on the other electrical properties in which it is an important factor. A

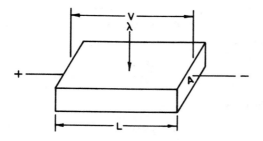

Figure 8-11 Schematic model of a photoconducting crystal with cross section A and a voltage V across its length L. The junctions at the leads are ohmic.

more useful measure may be obtained by the use of the actual lifetime of the carriers. This parameter may be obtained by starting with Eq. (8-109b) rewritten as

$$\frac{n_{pge}}{n_{pg0}} \cong \exp\left(\frac{-t}{\tau}\right) \tag{8-123}$$

where n_{pg0} is the density of photon-generated carriers when no trapping occurs and n_{pge} is the density when traps are present. Equation (8-123) may be rewritten and reexpressed by the first two terms of the series for e^x, $x = -t/\tau$, as

$$\frac{n_{pge}}{n_{pg0}} = \frac{1}{e^{t/\tau}} \cong \frac{1}{1 + t/\tau} \tag{8-124a}$$

Equation (8-124a) is inverted and rearranged as

$$\frac{n_{pg0}}{n_{pge}} - 1 \cong t/\tau \tag{8-124b}$$

or as

$$\frac{n_{pg0} - n_{pge}}{n_{pge}} \cong \frac{t}{\tau} \tag{8-124c}$$

And the lifetime is obtained as

$$\tau = \frac{t n_{pge}}{n_{pg0} - n_{pge}} \tag{8-124d}$$

The use of Eq. (8-124d) in Eq. (8-122) gives the amplification as

$$\gamma = \frac{\mu_e V}{L^2} \left(\frac{t n_{pge}}{(n_{pg0} - n_{pge})}\right) \tag{8-125}$$

The radiant energy, $E(\lambda)$, to which a semiconductor is exposed is increasingly absorbed as it penetrates below the surface of a semiconductor. The amount of this energy, $dE(\lambda)$, that is taken up by an element of volume of thickness dz, at a location z beneath its surface, varies as the number of photons, represented by $E(\lambda)$, traversing it and as its thickness. This is expressed, where α is the coefficient of proportionality, as

$$dE(\lambda) = -\alpha E(\lambda)\,dz \tag{8-126}$$

The coefficient α is the absorption coefficient because it is defined as the change in radiant energy per unit radiant energy per unit thickness. This may be seen from Eq. (8-126) in the form, for $\Delta z = 1$,

$$\alpha = -\frac{\Delta E(\lambda)}{E(\lambda)} \tag{8-126a}$$

The integration of Eq. (8-126) gives

$$E(\lambda) = E(\lambda)_0 \exp(-\alpha z) \tag{8-127}$$

The photons represented by $E(\lambda)$ consist of all of the photons ΔE_i described by Eq. (8-117a). The number of such photons that react to excite electrons in a unit volume of material is $\alpha E(\lambda)$ in unit time. This permits the rate of carrier generation to be given as

$$G_e = \alpha\beta E(\lambda) \qquad (8\text{-}128)$$

Here β is the average number of conduction electrons resulting from a reacting photon. As such, β constitutes a measure of the effectiveness of carrier excitation of photons. It has been called the *quantum yield*.

Beginning with the onset of illumination, the traps start to fill, and the density of the photon-generated carriers increases as demonstrated by a corresponding increase in the photoconductivity. These increases in the numbers of filled traps, and mobile carriers continue until G_e approaches the recombination rate, R_e [Eq. (8-107a)]. Here the traps approach maximum filling and the other factors increase at progressively slower rates until, at $G_e = R_e$, a dynamic equilibrium is reached in carrier production and a steady state is observed in the conductivity. Here, since the traps are not activated by holes, they have no effect while radiation continues. The steady state is described accordingly by means of Eqs. (8-107d) and (8-128) as

$$G_e = R_e = \alpha\beta E(\lambda) = \frac{\Delta n}{\tau} \qquad (8\text{-}129)$$

This permits an expression for the steady-state photoconductivity by providing the interrelationships for Δn as

$$\Delta n = G_e\tau = \alpha\beta E(\lambda)\tau \qquad (8\text{-}130)$$

Here $\Delta n \equiv n_{pge}$ of Eq. (8-119), so the photoconductivity is

$$\sigma_{pg} = n_{pge}e\mu_e = \alpha\beta E(\lambda)\tau_e e\mu_e \qquad (8\text{-}131)$$

Since the mobility is a function of τ_e [Eq. (6-7)] and τ_e is explicitly contained in Eq. (8-131), materials intended for use as photoconductors would be expected to have maximum values of τ_e.

The steady-state value of the conductivity will be maintained as long as the illumination continues. Once the illumination is turned off, the recombination will obey Eq. (8-123), so the decay in the photoconductivity is given by

$$\sigma_{pge} \cong \sigma_{pg0} \exp\left(\frac{-t}{\tau_e}\right) \qquad (8\text{-}132)$$

As noted in connection with Eqs. (6-12) and (8-109c), the larger that τ_e is, the longer the photoconduction will persist. Thus, if a sharp response to the cessation of illumination is desired, the maximum values for τ_e noted previously may not be desired. Trade-off must be made between conductivity and the persistence of the photoconductivity, or lag, to obtain the optimum value of τ_e for a given application.

The variation of G_e with $E(\lambda)$ [Eq. (8-128)] provides the basis for understanding the variable photoresistors noted in Section 6.3.5. In addition to the use of CdS and other crystals in the visible spectrum, PbS crystals are used in the infrared portion of the spectrum. The sensitivities of these materials is $\sim 10^5$ greater than that of the photocells noted in Section 8.1.1.

8.10.4 Population Inversion and Negative Absolute Temperature

The phenomena discussed in Sections 8.10.2.1 and 8.10.2.3 are examples of spontaneous photon emission. Stimulated emission occurs when radiation is emitted, such as those indicated from E_{e3} and E_{e2} to E_{e1} in Fig. 8-12(b). See Section 8.10.5.

The levels E_{e1} and E_{e2} are used as being illustrative of all transitions between lower and higher excited levels. Exposure to a given number of photons, N, will excite electrons from E_{e1} to E_{e2} (forward reaction) and from E_{e2} to E_{e1}, with photons being emitted during the reverse reaction. The difference between the number of particles participating in the forward reaction and that in the reverse reaction will vary as N, as the difference in the numbers of particles occupying each level $(n_2 - n_1)$ and as the probabilities of transitions of particles from each state to the other. Under equilibrium conditions, the probabilities, p^*, for transitions in either direction are the same. Thus, the difference in the occupancy of the two states (population difference) is given by

$$\Delta N(p^*) = N(n_2 - n_1)p^* \qquad (8\text{-}133)$$

Under these conditions, using Eq. (2-21), the probable number of particles in each state is given as

$$n_i \propto e^{-E_i/k_B T} \qquad (8\text{-}134)$$

And, since the constant of proportionality is the same for each, it will vanish as Eq. (8-134) is used here, and the relative occupancy of the levels is

$$\frac{n_2}{n_1} = \exp\left[\frac{-(E_2 - E_1)}{k_B T}\right] \qquad (8\text{-}135)$$

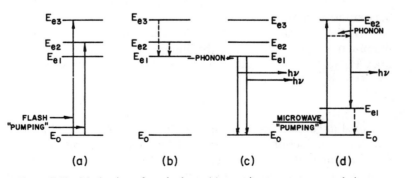

Figure 8-12 Mechanism of a ruby laser: (a) pumping to create a population inversion; (b) spontaneous, nonradiative transitions to a highly populated, metastable state; (c) stimulated emission resulting from the return to the ground state; (d) three-level maser.

Thus, n_2 must be less than n_1 because E_2 is designated as being greater than E_1 [Eq. (8-134)]. This being the case, the spontaneous emission is greater than any stimulated emission. This is the case under normal equilibrium conditions.

It follows, then, that for stimulated emission to occur a reversal of the normal relationship between n_1 and n_2 must be made to take place; n_2 must be greater than n_1. This nonequilibrium case is known as *population inversion*. The emissions resulting from such inversions are called *stimulated emission*.

It is helpful to understand these cases by reexpressing Eq. (8-135) in the following way. Equation (8-135) is rewritten in its logarithmic form as

$$\ln \frac{n_2}{n_1} = \frac{-(E_2 - E_1)}{k_B T} \qquad (8\text{-}136a)$$

Rearrangement results in

$$T = \frac{-(E_2 - E_1)}{k_B \, \ln(n_2/n_1)} \qquad (8\text{-}136b)$$

It will be seen that, for the equilibrium conditions given previously, T is positive because $\ln(n_2/n_1)$ is negative. And for the nonequilibrium case represented by a population inversion, T is negative because $\ln(n_2/n_1)$ is positive. How can this be the case? The concept of the temperature of a system is meaningful only when the system is at equilibrium. Departures of systems from equilibrium cannot be described by a single temperature because of the transient ranges in the distributions of the properties involved. If the population inversion for a given set of conditions (the system contains a finite number of levels and can absorb a finite amount of energy) is considered to be a kind of inverted equilibrium, the construct of a negative temperature may be considered to correspond to that "equilibrium." It turns out, as is shown later, that negative temperatures are much "hotter" than any positive temperatures because of the energies involved in the "equilibria" that they represent.

In such inverted populations, the levels E_i of Eq. (8-136b) no longer are discrete. They must conform to $E_i \pm \Delta E_i$ as given by Eq. (1-23) so that they agree with the principles of uncertainty and exclusion. Thus, the finite number of levels E_i must be considered as constituting small bands, each of width $\pm \Delta E_i$. This determines the line width of any stimulated radiation. Negative temperatures, thus, are strictly confined to quantized systems of this kind.

Where $T = 0\,\text{K}$, it follows from Eq. (8-136b), since the numerator must equal zero, that the energies must be equal and that the only occupied state is E_1. Thus, no population inversion is present and

$$E_i = E_{\min} = E_1 \qquad (8\text{-}137)$$

At the other equilibrium extremum, $E_{\max}$, where $T \to \infty$, again using Eq. (8-136b), $\ln(n_2/n_1)$ must approach zero. This condition requires that $n_2 \to n_1 = n$. Using the reciprocal of Eq. (8-134) for the population of each level, the total population is

$$\frac{2}{n} \exp\left(\frac{E_{\max}}{k_B T}\right) \cong \exp\left(\frac{E_1}{k_B T}\right) + \exp\left(\frac{E_2}{k_B T}\right) \qquad (8\text{-}138a)$$

so the maximum energy of the system at equilibrium is obtained by differentiating Eq. (8-138a) as

$$E_{max} \cong \frac{n}{2}(E_1 + E_2) \qquad (8\text{-}138b)$$

But, where a population inversion is present, the energy of the system must be greater than E_{max}. Since E_{max} is obtained by considering $T \to \infty$, and the equilibria of systems with population inversions are characterized by negative equilibrium temperatures, these negative temperatures must be greater than any infinite positive temperature. This also makes sense when it is considered that the higher energy of the electrons in the population inversion must flow to a lattice that may only be at positive temperatures.

8.10.5 Stimulated Emission

A comparatively new and increasingly useful type of emitted electromagnetic radiation is generated by the stimulation of population inversions. Devices that generate this type of radiation are called *masers* and *lasers* (microwave or light amplification by stimulated emission of radiation). Inorganic and organic crystals, liquids, and gases have been used for these purposes.

When a photon with energy hv, with a natural line width Δv, reacts with an excited ion whose energy above the ground state is "exactly" that of the incident photon, the reaction will stimulate the immediate emission of another photon, and the energy of the ion will drop back to a lower level. The emitted photon will have properties identical with that of the incident photon; it will have the same frequency, be in phase with, and have the same direction as the incident photon.

Under equilibrium conditions, states with lower energy normally are more highly populated than states at higher energies. Here more states are available for the absorption of energy than for energy emission. The Boltzmann equation describes this general situation as

$$N_e = N_0 \exp\left(\frac{-E_i}{k_B T}\right) \qquad (8\text{-}139)$$

in which N_e and N_0 are the numbers of excited and ground-state ions, respectively, and E_i is the energy of an excited state. When suitable excitation, or *pumping*, occurs (by electrical discharges in gases, by microwave oscillation, or by rapid optical flashing), the normal equilibrium conditions no longer exist, and N_e becomes greater than N_0. This is called a *population inversion* (Section 8.10.4). Lasers or masers are possible only when such an inversion is produced because more electrons must be emitting rather than absorbing energy. These must be replenished very rapidly to maintain the population inversion because the lifetime of an excited state is very short.

Lasers and masers require that the amplification produced by the stimulated emission be equal to the energy given up by the energy source. The excited ions are contained in a cavity resonator so that the resultant radiation originating from the excited ions forms a coherent standing wave in the cavity. Each additional photon, given up by an excited ion, adds to the intensity of the standing wave and amplifies it. The cavity for use in the microwave region consists of a container whose length, L, is given by $n\lambda = 2L$, where n is an integer and λ is the wavelength of the emitted photons. The length of the cavity is made to be much greater than its diameter. This gives a single mode of axial microwave oscillation for each L. Many axial modes of oscillation are possible with λ in the optical region because of the much shorter wavelengths. The shorter the cavity is, the fewer modes.

In an ideal cavity, one in which all the applied energy produces excited states that can generate coherent radiation, a stimulated ion reacts with a photon and spontaneously releases a phonon and a photon of $h\nu_e$ and then drops back to a lower state [Figs. 8-12(b) and (c)]. The simultaneously emitted phonon reacts with another excited ion. This produces another phonon and a photon of $h\nu_e$ and the process cascades. All the emitted photons have the same frequency, within the limits set by the uncertainty principle (or the same natural line width). This narrow frequency range is maintained because, as one ion returns to the ground state, the phonon emission, a nonradiative transition, permits another stimulated ion to relax into virtually the precise, equivalent, lattice vibrational state originally occupied by the first ion (Section 8.10.2.3). These relaxation times are of the order of 10^{-12} s. Thus, each emitting ion is at essentially the same phonon frequency as all other emitting ions at the time of its stimulation.

The energy levels of the stimulated ions must either be very narrow or the gaps between the levels must be wide or very distinct for this to occur. Situations other than these would be very inefficient, since only a few ions could react to the stimulating photons and phonons. This would result in low intensities and wide line widths. The natural line width decreases, under the preferred conditions, as the temperature decreases because the number of phonon modes decreases, and more ions may be stimulated for a given vibrational mode. See Sections 5.1.3 and following sections. And, as noted in Section 8.10.4, the lower limit of the stimulated radiation is determined by $\pm \Delta E_i$.

A common method of cavity formation is to place two mirrors, either dielectric or metallic, at the ends of the cavity in which the material is being stimulated. If the mirrors are planar, they must be as nearly parallel to each other as possible and be optically flat to avoid destructive interference. Spherical mirrors are less critical and more efficient. One either contains a small opening or is only partly reflecting. Part of the standing wave induced in the masing or lasing material leaves it by such means. The largest part of the radiation remains within the cavity. The emergent beam of radiation is of high intensity ($> 10^4$ W/cm^2). It also has a very small degree of dispersion.

Solids, liquids, and gases may be stimulated to emit coherent radiation. The ruby maser was one of the first of these materials; it also is used as a laser. Synthetic ruby consists of Al_2O_3 doped with paramagnetic Cr^{3+} ions. These are in substitutional solid solution on Al^{3+} sites in the HCP lattice. The Cr^{3+} ion has three states of ionization in a magnetic field. Its levels may split into three levels, while the O^{2-} and the Al^{3+} ions each have single ionized states that primarily are involved in the covalent bonding (Section 3.12.3). These ions also have low nuclear magnetic moments and, thus, do not affect the ionization levels of the Cr^{3+} ions. The high-purity Al_2O_3 and the very low concentration of Cr^{3+} ions assure that its levels will not be affected, or unnecessarily broadened, by the presence of other ions. A schematic diagram of the band structure of the Cr^{3+} ion is shown in Fig. 8-12(d)

The pumping photons required by lasers usually are provided by intense, short pulses of light in the green portion of the spectrum [Fig. 8-12(a)]. (The absorption of the green radiation is responsible for the color of rubies.) A spontaneous, non-radiative transition (phonon generation) [Fig. 8-12(b)] results in electron drops to level E_{e1}. Stimulated emission of the population inversion at E_{e1} then causes the electrons to drop back to the ground state [Fig. 8-12(c)]. The photons emitted by these transitions are reflected many times within the cavity. They are in phase with the standing wave in the cavity and thus account for the amplification by the laser. The high aspect ratio of the cavity minimizes any propagation of transverse radiation; other radiation not parallel to the cavity axis is lost and does not enter into the amplification process. The maser mechanism is the same as that for the laser. The only differences are that the pumping is by a microwave generator and the emission consists of amplified, coherent microwaves instead of light waves.

Many suitably doped compounds have been used for masers. The dopant almost invariably is a paramagnetic lanthanide or actinide ion capable of suitable degrees of ionization. In addition to Al_2O_3, other host crystals include CaF_2, SrF_2, BaF_2, $SrWO_4$, and $CaMoO_4$ doped with such ions as Pr^{3+}, Nd^{3+}, Sm^{2+}, Dy^{2+}, Ho^{3+}, Er^{3+}, Yb^{3+}, and U^{3+}.

$P-n$ junctions in semiconductors are the most efficient lasers. Coherent emission results from recombination adjacent to the line of the junction. Compounds such as GaAs, InAs, GaSb, InSb, and GaP are employed. These are doped to degeneracy on both sides of the junction so that the energy gaps correspond to the desired frequency of the emitted radiation. The junction must be very narrow and extremely straight for the generation of the standing wave. Forward-biased junctions may be pumped optically by an electron beam or by the injection of electrons or holes. These III−V compound junctions usually are operated below 77 K (liquid N_2 temperature).

Applications for lasers now vary widely and are employed in such diverse ways as melting, welding, eye and neurosurgery, and long-distance carrier waves for communications. Masers find wide use as amplifiers and carrier-wave generators for long-distance communications, especially when cryogenically cooled. Some of their more spectacular applications include radio astronomy and communications and telemetry in space exploration. Serious consideration has been given to their use in the transmission of solar energy from satellites.

8.11 BIBLIOGRAPHY

Cohen, M. M., *Introduction to the Quantum Theory of Semiconductors*, Gordon & Breach, New York, 1972.

Dekker, A. J., *Solid State Physics*, Prentice-Hall, Englewood Cliffs, N.J., 1957.

Pollock, D. D., *Physical Properties of Materials for Engineers*, Vols. I and II, CRC Press, Boca Raton, Fla., 1982.

Rosenberg, H. M., *The Solid State*, Oxford University Press, New York, 1978.

Solymar, L., and Walsh, D., *Lectures on the Electrical Properties of Materials*, Oxford University Press, New York, 1979.

Sproull, R. L., *Modern Physics*, 2nd ed., Wiley, New York, 1963.

Wolf, H. F., *Semiconductors*, Wiley-Interscience, New York, 1971.

8.12 PROBLEMS

8.1. Discuss the bonding differences in the physisorption and chemisorption mechanisms.

8.2. Show that $N_x \cong \mathbf{v}_x N(\mathbf{p})_x f(E)$, as given in Section 8.2.

8.3. Make a plot of Eq. (8-15) and Eq. (8-18). Show the potential barrier at the surface of a metal and its variation with distance away from the surface.

8.4. Make a sketch illustrating the basic differences between the Schottky effect and thermionic emission.

8.5. Derive Eq. (8-33) from Eq. (8-32).

8.6. Discuss the differences between field-emission microscopes and field-ion microscopes.

8.7. Explain why the Seebeck effect cannot be ascribed to the contact potential between two conductors.

8.8. Approximate the ratio of the thickness of a metal–metal double layer to the mean free path of an electron. (*Hint*: Refer to Problem 6.2.)

8.9. (a) Compare the metal–metal double layer of Problem 8.8 with the depletion zone in a metal–semiconductor junction.
 (b) Describe the depletion zone in terms of the mean free path of a nearly free electron.

8.10. Make graphs of the depletion zone widths, using arbitrary but constant values of V_c and other pertinent factors, to demonstrate the effects of forward and reverse biases in a semiconductor.

8.11. Graph Eqs. (8-66) and (8-67b) using an arbitrary value for j_0 and compare the results with those from Eqs. (8-73) and (8-74).

8.12. Approximate the densities of electrons on either side of a p–n junction if the potential barrier between them is 0.45 eV at room temperature.

8.13. Graph and explain the influence of Eq. (8-79) on the ratio of the densities of carriers in the n- and p-region on either side of a barrier at constant temperature.

8.14. Graph Eq. (8-96a), using a constant coefficient, to show the limit at which it approaches Eq. (8-97).

8.15. Approximate the breakdown energies for Ge, Si, and GaAs.

8.16. Use a simple band model to show the effect of trap energy on the probability of finding a carrier.

8.17. Plot the relative number of excess carriers for a relaxation time of 10^{-5} s.

8.18. Give an algebraic explanation for the anti-Stokes region.

8.19. Explain the energy limit in exciton production.

8.20. Approximate the average wavelength of an exciton in silicon at 300 K.

8.21. Explain why radiative and nonradiative transitions may take place within a given material.

8.22. Why does Eq. (8-117b) represent a spectrum of radiation suitable for photoconduction?

8.23. Determine if a maximum energy difference in the outer energy levels shown in Fig. 8-12 of $\sim 0.2\,\text{eV}$ and $E_{e1} - E_0 \sim 0.8\,\text{eV}$ are suitable conditions for a laser.

9 / BASIC SEMICONDUCTOR DEVICES

The devices discussed in the following sections are based on the properties of semiconductors and the ways in which the changes in these properties, as given in Chapters 6 and 8, lend themselves to practical applications. These have led to the development of many and varied devices. The earlier vacuum tubes have been almost entirely supplanted by solid-state junction devices. This result has been a revolution in solid-state circuitry that has been accompanied by a very high degree of miniaturization of electronic components. This includes packing densities of the order of 10^6 cm^{-2} active elements on a chip.

The devices presented here represent those basic properties most frequently used. If these are understood, the properties of new devices, as well as combinations and permutations of available devices, may also be understood and exploited. Thus, the diodes and transistors discussed here are presented in their most elementary form so that they may be perceived as building blocks. This approach includes graphs of electrical properties that are intended to illustrate functional relationships, not the engineering values of parameters.

9.1 SEMICONDUCTOR DIODE DEVICES

The devices discussed in the following sections have been included as being representative of some of the more commonly used diodes. The intent is to demonstrate the more important physical mechanisms involved, rather than to provide a catalogue of devices.

An important class of semiconductor diodes is based on single $p-n$ junctions. The readily controlled charge carriers and electric fields in the depletion zones in the neighborhood of junctions make these devices possible (Sections 8.8 and following sections). The variations in the behaviors of the carriers and the internal electric fields when influenced by voltages, (electric fields), radiation, and other external factors determine the properties, type, and applications of the devices. Since the junctions are small and the reactions take place at or near them, such devices can be, and are, made to be very small.

9.1.1 Semiconductor Diode Rectifiers

The fundamental properties of $p-n$ junctions are described in Sections 8.8 to 8.8.3. The polarity of the power supply is important in many electronic circuits. Power of the wrong sign can incapacitate many devices and circuits. Protection against this event may be obtained by the use of two oppositely oriented rectifiers in parallel between the supply and the circuit. Current of the wrong sign will be blocked by one of the rectifiers (negatively biased) and the circuit will be protected.

In a comparable way, excessive voltage (overvoltage) can be eliminated by means of two oppositely oriented rectifiers in series between the input and the circuit. These limit the input voltage to the desired value without entering into the circuit.

Two oppositely oriented rectifiers in parallel with each other, and connected in parallel with a suitable resistor across the terminals of the circuit, can provide protection against excessive current. In such an event, the rectifiers limit the excess current and allow only the desired amount of current to enter the circuit.

Rectifying diodes may be used as voltage regulators. This is based on their behavior when forward biased. Once ac current starts to flow, the voltage can be made to stay within relatively narrow limits. The initial portion of the solid curve in Fig. 8-5 can be made to be nearly zero for low voltages ($< \sim 0.5$ V) and to have relatively flat, steep slopes at higher voltages (> 0.8 V). This can be pictured as resulting from the fact that the resistance of the junction is very much greater than that of the bulk material (Sections 8.8.1 and 8.8.2). At higher voltages, the bulk resistance predominates and is virtually constant (at constant temperature) and results in nearly linear curves of current versus voltage with very steep slopes. In actuality, the negligible current at the low, initial voltages arises because the very low density of minority carriers, n_{-e}, is not a function of the dopant concentrations [Eq. (8-78)]. It is a consequence of the fact that n_{-e} is a function of E_g because E_F is eliminated from Eq. (8-75b). Carrier densities of these magnitudes continue to be trapped and virtually no current flows until all the traps are filled at a threshold voltage. Increased voltage increases the carrier density [Eq. (8-70a)], and the current increases accordingly.

Rectifying diodes can be used in suitable circuits to remove, or *clip*, the maximum or minimum voltages, or both, from ac supplies. Pairs of similarly oriented diodes, in appropriate circuits, may be used for half-wave rectification. Full-wave rectification may be obtained by the use of four rectifying diodes for dc power supplies.

9.1.2 Avalanche and Tunnel Diodes

Breakdown occurs in an avalanche diode at a given reverse voltage (Section 8.9.2). Diodes are manufactured to have specific breakdown voltages, within narrow limits. Very small increases beyond such voltages result in very large increases in current [Fig. 8-5(a)]. Diodes of this kind are available with power ratings from 250 mW to 50 W. The currents produced by these devices are limited by their power ratings; the Joule heat caused by excessive currents may cause burn-out.

One application of avalanche diodes is for the protection of meters. The diode is placed in the meter circuit so that, if excessive voltage is applied, virtually all the undesired current will be shunted through the diode rather than the meter circuit.

Since the avalanche breakdown voltage of a given diode is very nearly constant, it can be used as a voltage reference. Certain diodes used for this purpose have nearly equal avalanche and tunneling effects (Section 8.9.1) at about 5 V. These diodes are minimally affected by temperature because the tunneling and the avalanche effects have opposite temperature coefficients that approximately cancel each other. This characteristic also makes them useful for voltage stabilization in circuit components.

Very high doping is required for tunneling to take place (Section 8.9.1). This may be pictured as being responsible for the junction being in reverse, internal breakdown in the absence of an applied voltage. As a forward bias is applied, it begins to counteract and to diminish the internal breakdown, and correspondingly affects the tunneling carriers, until a maximum current is produced (Fig. 9-1). The internal breakdown then decreases with increasing forward bias up to about 0.3 to 0.4 V. This portion of the curve is called the *negative resistance region* because the current decreases with increasing voltage. Beyond this voltage range, the current–voltage curve is that of the usual *p–n* rectifying junction (Section 8.8).

The mechanism responsible for tunnel diodes is best understood by starting with the situation represented by Fig. 8-7. The high dopant concentrations broaden the valence and conduction bands so that the depletion zone is very narrow (Section 8.8.2), and the Fermi level lies within each of these bands. In addition, the density of minority carriers is, on a comparative basis, very small [Eq. (8-78)]. Under the

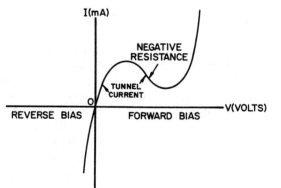

Figure 9-1 Current–voltage characteristics of a tunnel diode.

influence of a small forward bias, a component of the current will arise from electrons excited across the gap, by junction rectification, as described in Section 8.8. It will continue to increase with increasing voltage and add to the tunnel current. At first this component may be considered to be negligibly small with respect to that produced by direct tunneling (electrons excited through the barrier). As the voltage is increased, the direct tunneling continues as long as electrons in the conduction band can tunnel through the barrier to approximately equivalent levels in the valence band. As this proceeds, with increasing voltage, increasing numbers of states in the p-type material become available to the electrons and the current increases accordingly. The increasing voltage also displaces the two bands so that they tend to separate from one another in such a way that the overlap of E_v and E_c (Fig. 8-7) decreases. The maximum current occurs when the Fermi level of the p-type material becomes colinear with the bottom of the conduction band of the n-type material because the electrons then may be accommodated by any available p state. As the voltage is increased beyond this point, the increasing band separation increasingly diminishes the number of states accessible for direct electron tunneling, and the current decreases accordingly. A point is reached at which the top of the p-type valence band and the bottom of the n-type conduction band become colinear. Thus, as the voltage is increased, a situation is reached in which tunneling is no longer possible, and current can arise only from electrons promoted across the gap. Here those electrons that are being excited across the gap begin to make an appreciable contribution to the current. The current shows a minimum at this point. The increasing current observed thereafter is a result of junction rectification, as described in Section 8.8.

The response time in Zener, or tunnel, diodes is very short. Use has been made of this property to employ such devices in relatively simple circuits for oscillators that operate at frequencies up to the range of about 10^{11} Hz and for high-speed pulses of $\sim 10^{-10}$ s. Such very short response times also make them desirable for switching devices. In addition, they are used for microwave receivers because of their negligible breakdown voltages.

Tunnel diodes also are useful for converting the output of low-voltage, high-current devices to higher voltages. One such application has been their use in conjunction with thermoelectric generators.

9.1.3 Photoelectric Diodes

The electron–hole pairs created by photons near a $p–n$ junction that is close to the illuminated surface are accelerated by the large internal field in the space-charge region, causing a current to flow in a circuit (Sections 8.8.1 and 8.10.3). Any photo-excited minority carriers outside of this region can diffuse toward it, depending on their mean free paths and lifetimes (Section 8.10.1.1). Once they approach the depleted region, carriers whose sign is opposite to that of the region in which they are created are accelerated across the space-charge region to regions of the same sign. This increases the intensity of the electric field in the depletion zone. The effect is the same as increasing the forward-bias voltage. It takes place without requiring an

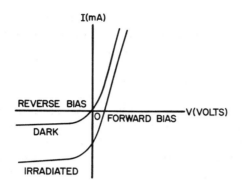

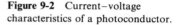

Figure 9-2 Current–voltage
characteristics of a photoconductor.

external voltage application. The phenomenon is known as the *photovoltaic effect.*
These devices convert electromagnetic energy into electrical energy; they are known as
solar cells when they are used to convert sunlight to electrical energy.

These devices may be used to amplify photon-induced current if a reverse bias
is applied; this must be large enough to cause avalanching to occur (Section 8.9.2).
The current generated in the absence of irradiation, the dark current, is negligibly
small. For a given bias, the density of the carriers, and therefore the current
generated, will vary directly as the intensity of the irradiation (Fig. 9-2). The resulting
current flow is from the positive terminal and has a negative sign. Thus, the power
($W = IV$) is negative, indicating that power generation, not utilization, is occurring.

Selenium cells are most effective for photographic purposes, since the wavelength at which their maximum response occurs is close to that of the human eye and
to that of sunlight. Radiation sources in the near-infrared and in the visible range are
detected more efficiently by Si cells. The theoretical efficiency of Si cells is about 11%
to 15%, with a possible maximum of about 20%. Silicon cells are used in preference to
Se cells because they are about 20 times more efficient. CdS cells are about half as
effective as Si cells, being about 7% efficient. However, their low cost compared to Si
makes them commercially attractive (see Section 8.10.3).

The photovoltaic mechanism is reversible under suitable conditions. A *p–n*
junction in a suitably doped material that permits direct recombination may produce
light. Under the application of a forward bias, the injected minority carriers, from the
external circuit, recombine and radiation is emitted. The recombination must take
place close to the surface of the material so that the emitted photons are not absorbed.
Gallium arsenide appears to be the best material for this (see Sections 9.1.4 and 9.1.5).

9.1.4 Light-Emitting Diodes

Semiconductor diodes that emit electromagnetic radiation in the visible portion of the
spectrum are called light-emitting diodes (Sections 8.10.2 and following sections).
The current–voltage curves for these diodes are similar to that shown in Fig. 8-5.
The threshold voltages for current flow, when forward biased, will be different for
different materials. For example, Ge, Si, and GaAs have threshold voltages of about

0.2, 0.6, and 1.0 V, respectively. Ge and Si emit only small amounts of infrared light, while GaAs is a source of much larger amounts of infrared radiation. GaAs–GaP materials with thresholds of between 1.4 and 1.8 V, depending on the dopant content, emit visible light in the range from red to amber (Section 9.1.3).

The application of a forward bias injects electrons from the *n*-type material into the conduction band of the *p*-type material. Recombination occurs when these electrons drop back to the valence band of the *p*-type material and a photon is emitted. However, the energy may be nonradiative and be given off as phonons. If significant amounts of trapping occur in the recombination process (Section 8.10.1), much of the energy may be absorbed as phonons (Sections 8.10.1.1 and 8.10.2.3). The most efficient photon-producing processes occur when direct recombination takes place. The energy of the emitted photon is determined by E_g. This is the equivalent of saying that E_g determines the wavelength of the emitted light [Eqs. (8-114) and (8-115)].

The radiative transitions that occur in GaP diodes are an exception to this behavior, being indirect transitions. In this case the recombination occurs between the two impurity levels provided by high concentrations of O and Zn dopants, in a way analogous to that indicated in Fig. 8-9(d). The resulting radiation is in the red range of the spectrum.

The efficiency of internal light generation is high in these devices, but comparatively little of this is emitted out of the surface of the devices. Much of the light is absorbed internally because of the long paths and low transparencies of the materials. In addition, some of the light that reaches the surface is reflected back by the surface and is absorbed.

Light-emitting diodes find considerable application in pilot lights and in digital read-out devices. These have very long lives ($\sim 2 \times 10^5$ h). In addition, their usefulness is increased by their compatibility with transistor logic circuitry.

9.1.5 Semiconductor Diode Lasers

Electroluminescent diodes give off light as a result of the promotion of injected electrons dropping from the conduction band to the valence band. The excited electrons emit a photon (spontaneous emission) in the drop back to the valence band. Diodes such as these are costly and are unable to produce high light intensities. Consequently, the limited applications already noted are made of these diodes (Section 9.1.4).

The best of the early semiconductor diode lasers were made from *p–n* junctions in GaAs. They were small, required about 2 V to operate, and were rugged and inexpensive. The high current densities required for their operation made it necessary that they be operated in short pulses ($\sim 10^{-6}$ s) at ordinary temperatures. The time between the pulses was about 10^{-3} s. This combination permitted adequate pumping without overheating. Continuous operation was possible only at cryogenic temperatures (77 K and lower).

The drift of electrons across a forward-biased junction brings them into the conduction band of the *p*-type region where they have energies of about E_g.

Electroluminescence occurs at low current densities. At higher current densities (2.5×10^4 A/cm^2), pair production increases, and stimulated emission and amplification (Section 8.10.5) occur. The optically flat ends, perpendicular to the junction, may be made to serve as partial mirrors, and a beam of coherent light is emitted from the *p*-type area adjacent to and parallel with the junction. Diodes such as these are called *homostructure lasers* because both parts of the diode are made from the same material, suitably doped GaAs.

The lasing region in GaAs diodes is wide because the mean free path of an injected electron may be of the order of several micrometers (μm). This is the reason for the high, critical, current density required for lasing in homostructure lasers. In addition, the comparatively wide lasing volume also results in the absorption of many photons, as noted in Section 9.1.4, rather than in photon amplification. Thus, the intensity of the emergent beam is diminished by the high internal losses.

These deficiencies are overcome by the use of what is known as a *double heterojunction*. One of these devices is shown schematically in Fig. 9-3. The central region is *p*-type GaAs. On either side of this are regions of *n*- and *p*-type Al$_x$Ga$_{1-x}$As. The Al$_x$Ga$_{1-x}$As has a larger value of E_g than does GaAs. The increase in E_g is directly proportional to the concentration of Al ions present. These layers are sandwiched between *n*- and *p*-type GaAs, respectively.

The larger values of E_g in the Al$_x$Ga$_{1-x}$As layers have very imporant functions in these heterojunctions. The junction between the *n*-type Al$_x$Ga$_{1-x}$As and *p*-type GaAs [Fig. 9-3(b)] results in a barrier for the holes in the *p*-type GaAs, but the conduction band is unchanged. When this junction is biased to inject electrons into the *p*-type GaAs, the holes are prevented from entering the *n*-type Al$_x$Ga$_{1-x}$As [Fig. 9-3(c)], the valence band remains unchanged, but the width of the conduction band in the Al$_x$Ga$_{1-x}$As is less than that of the *p*-type GaAs because of the larger value of E_g. This constitutes a potential barrier that prevents injected electrons in the conduction band of the *p*-type GaAs from entering that of the *p*-type Al$_x$Ga$_{1-x}$As. This barrier is unchanged by forward biasing. Thus, both types of carriers are contained within the central, *p*-type GaAs. Therefore, recombination and the photons that are generated are concentrated in a very small volume. The light is contained within the *p*-type GaAs layer because of the different refractive indexes of GaAs and AlGaAs.

The constraints that are inherent in the heterojunction prevent the electrons and the holes from diffusing out of the junction into a wider volume [curved arrows in Fig. 9-3(a)], and the abrupt changes in the indexes of refraction contain the photons within a small volume. When a forward bias is applied, electrons are reflected at the *p–p* junction and the holes are reflected at the *p–n* junction. Since both types of carriers are contained within the small volume, much lower current densities (1 to 3×10^3 A/cm^2) are required for lasing. The thin, active, *p*-type GaAs layer constitutes a much smaller volume than a homojunction for the confinement of the coherent, standing waves.

Even though the finished wafers containing these heterojunctions are small (about 0.5 mm $\times$ 0.1 mm $\times$ 75 to 125 μm thick), the heat dissipation necessitated by the required current densities is difficult to maintain. This is a result of the extremely low thermal conductivities of the materials involved (Sections 5.3 and Fig. 5-27). Devices

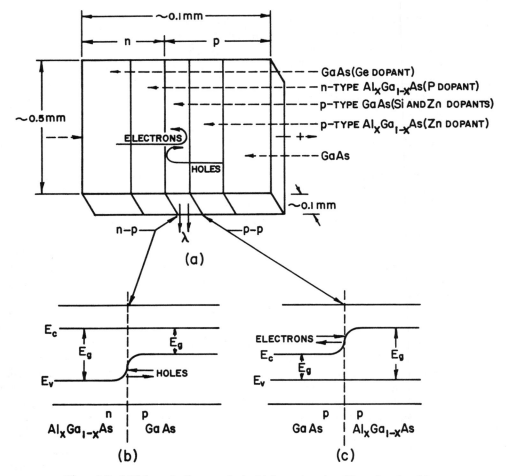

Figure 9-3   (a) Schematic diagram of a double heterojunction; (b) $n-p$ junction; (c) $p-p$
junction.

with smaller active zones, but with integral heat sinks, have been made. These have
operated continuously at temperatures as high as 100°C. Energy conversion
efficiencies of more than 10% have been reached.

9.1.6 Parametric Diodes

Almost all the devices discussed up to this point have been considered primarily under
conditions of forward bias. Reverse bias is given primary attention here because it
provides further insight into the nature of the depletion zone and explains the
operation of many useful devices. These include varactors, $p-i-n$ junctions, Schottky
diodes, and IMPATT diodes.

9.1.6.1 Varactors.   The properties of varactors are based on Eq. (8-95a). This may be rewritten by explicitly expressing the voltage, V, as the algebraic sum of the applied voltage V_A and V_c. The capacitance, then, is given as

$$C \cong \left(\frac{\varepsilon_0 \varepsilon e}{2(V_A + V_c)} \cdot \frac{n_{e0} n_{h0}}{n_{e0} + n_{h0}} \right)^{1/2} \tag{9-1}$$

In the absence of an applied voltage, C is constant because V_c is constant for given pairs of doped materials [Eq. (8-34)].

These devices are made so that the concentration of the n-type dopant is very much greater than that of the p-type material. The total width of the depletion zone approaches that in the p-type material in this case [Eqs. (8-85) and (8-97)]. Accordingly, the capacitance for this condition may be approximated from Eq. (8-97a) as being given by

$$C \cong \left(\frac{\varepsilon_0 \varepsilon e}{2(V_A + V_c)} \cdot n_{h0} \right)^{1/2} \tag{9-2}$$

Thus, $C \propto V_A^{-1/2}$, V_c being constant. It is for this reason that diodes of this kind are called *varactors* or *parametric diodes* [Fig. 9-4(a)]. Typical devices of this kind have depletion zones of 10^{-6} m and capacitances of $\sim 10^{-11}$ to $\sim 10^{-12}$ F.

Varactors are used for such applications as switching, tuning, and logic circuits. They are widely used in radio and television receivers for automatic frequency control. Another application is for parametric amplification in situations requiring stability and high speed.

In cases in which the depletion zones are small ($< 10^{-6}$ m) and the dopant concentrations vary nearly linearly from the center of the zone outward, $C \propto V^{-1/3}$. Devices with $dC/dV \propto V^{-4/3}$ are more useful for certain applications than those with $dC/dV \propto V^{-3/2}$, as derived from Eq. (9-2).

9.1.6.2 $P–i–n$ junctions.   The insertion of a thin layer of intrinsic semiconductor, designated by i, between the p- and n-type materials further enhances the capacitance-voltage effect. Devices of this kind are known as $p–i–n$ diodes.

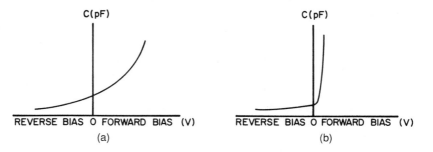

Figure 9-4 Electrical characteristics of (a) a varactor and (b) a $p–i–n$ device. (Modified from T. H. Beeforth and H. J. Goldsmid, *Physics of Solid State Devices*, pp. 48 and 52, Pion, London, 1970.)

The operation of a p–i–n device is similar to that of a p–n junction in that holes cross the i zone to the n-type material and electrons cross in the opposite way. The carrier densities in the i zone are negligibly small compared to those of the other regions (Section 8.8.1). Thus, the full extent of the resulting composite i region serves as a very wide depletion region, and a large increase in the charge densities of the carriers occurs on the p- and n-sides of the new interfaces. These factors cause a practically uniform electric field in the composite i region (Section 8.8.2). Since the intrinsic regions in p–i–n devices are similar to, but wider than, the space-charge regions in varactors, their capacitance response to external voltages is similar but smaller (Section 9.1.6.1). However, since the applied voltages result in much smaller internal fields, a much larger voltage is required for breakdown (Section 8.9.2). It is for this reason that p–i–n diodes are used where high voltages are involved [Fig. 9-4(b)]. The smaller capacitance of a p–i–n compared to that of a p–n device permits faster response times and output frequency can be increased.

The advantages of p–i–n devices are that they can operate with power supplies with smaller voltages and are stable, small, inexpensive, and rugged.

Some typical applications of p–i–n devices include radio-frequency applications and switching and limiting devices (Section 9.1.6.4). When used as photodetecting junction devices, p–i–n diodes are more sensitive than p–n junctions, because the higher fields that can be applied greatly increase pair production (see Section 9.1.3).

9.1.6.3 Schottky diodes. The Schottky diode consists of a semiconductor separated from a metal by a thin, oxide film (see Section 8.6). The oxide film must be kept very thin in order to permit the carriers to tunnel across it (Section 8.9.1). The very thin oxide films that are normally present on most semiconductor materials frequently are sufficient for this purpose (Section 8.9.3.1). The device is completed by the deposition of a metal layer on the oxide film. These are known as metal–insulator–semiconductor devices or as m–i–s junctions.

When the work function of the metal is larger than that of the semiconductor, electrons will flow from the semiconductor to the metal (Fig. 8-3). The excess negative charge builds up at the surface of the metal until it is sufficient to prevent further electron flow. At equilibrium, in the absence of an external field, the net flow of carriers is the same in each direction. Forward biasing decreases the barrier; reverse biasing increases the barrier; thus, ϕ varies with the applied voltage, V_A (Fig. 8-4), and the current of electrons from the semiconductor to the metal will, accordingly, increase or decrease. Where the dopant concentration is high, the depletion zone will be very narrow [Eq. (8-85)], and additional electrons may tunnel across the barrier (Section 8.9.1). Holes also may tunnel if the value of $(\phi_M - \phi_S) \cong E_g$ (Fig. 8-3).

The Schottky potential barrier, V_S, may be obtained from reexpressing Eq. (8-97) more generally because the barrier varies depending on the magnitude and sign of V_A. So, using $V = V_A + V_c$ and $\phi = (V_A + V_c)e = Ve$, Eq. (8-97) becomes, for $d \cong x_p$ [Eq. (8-94)],

$$d \cong \left(\frac{2\varepsilon_0 \varepsilon V}{e} \cdot \frac{1}{n_{h0}} \right)^{1/2} \tag{9-3}$$

and

$$V_S \cong \frac{en_{h0}d^2}{2\varepsilon_0\varepsilon} \tag{9-4}$$

The capacitance of the space-charge region may be obtained by again letting $n_{e0} \gg n_{h0}$, for this case, the same as Eq. (8-97a) was obtained from (8-95a). Under forward bias, Eq. (8-66) may be written more generally as

$$j = j_S \left[\exp\left(\frac{eV}{k_B T}\right) - 1 \right] \tag{9-5}$$

for the mechanisms involved here, where j_S is the Schottky current density.

The width of the barrier, in the absence of an external potential, may be obtained from Eq. (9-3) as

$$d \cong \left(\frac{2\varepsilon_0\varepsilon V_c}{en_{h0}}\right)^{1/2} \tag{9-6}$$

in which d is the total barrier width.

Other mechanisms contribute to j_S, depending on the materials. When the dopant concentration is high, d will be small. Then it is possible that additional contributions may be made to j_S, such as tunneling of holes and electrons, field-induced tunneling, and field emission [Sections 8.4 and 8.9.1]. The latter contribution occurs when high electric fields lower the barrier significantly more than normal fields and more high-energy (*hot*) electrons enter into the conduction process. It will also be noted that surface effects have not been taken into consideration in this analysis (Sections 8.1.1 and 8.9.3.1).

Schottky diodes are now made from Si or GaAs. They are used for pulse-shaping and switching and limiting functions because of the very short lifetimes of the carriers. The Schottky diode has an advantage over the ordinary diode because its current–voltage behavior in the reverse-bias mode is similar to that shown in Fig. 8-5(a) for general p–n junctions, but it gives much larger currents at saturation. The tunnel diode in the same mode gives smaller currents at relatively small voltages. This may be expressed for ordinary p–n Ge devices as

$$j_S \sim j_0 \times 10^6$$

In other words, the Schottky diode saturation current density is about 10^6 times larger than that of an ordinary Ge device. Another important characteristic is its short turn-off times that result from the previously noted very short carrier lifetimes. The mean free paths of the normal and hot electrons are very short.

9.1.6.4 IMPATT diodes.
An IMPATT diode (impact, avalanche, transit-time diode) consists of a dc, reverse-bias, p–n junction in which the reverse voltage is maintained at a level that is slightly less than that required for breakdown. It is placed in the circuit in such a way that a given time lapse occurs before the circuit reacts to avalanching.

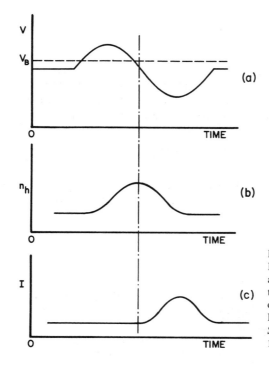

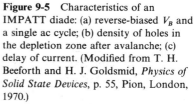

Figure 9-5 Characteristics of an IMPATT diode: (a) reverse-biased V_B and a single ac cycle; (b) density of holes in the depletion zone after avalanche; (c) delay of current. (Modified from T. H. Beeforth and H. J. Goldsmid, *Physics of Solid State Devices*, p. 55, Pion, London, 1970.)

When an alternating current is transmitted through the reverse-biased device [Fig. 9-5(a)], the density of electrons adjacent to the depletion zone will decay as a result of avalanching, and the density of holes will increase (Fig. 9-5(b)]. These pass out of the depletion zone and an additional lag occurs in the circuit because of the lower mobility of the holes [Eq. (6-134)]. The remaining current pulse is shown in Fig. 9-5(c). The wave shape is changed, and the resulting current is 180° out of phase with the original.

The use of these devices in tuned circuits results in continuous oscillation. *P–i–n* diodes are also used for this type of application (Section 9.1.6.2). The larger voltages possible with a *p–i–n* device and the high degree of carrier depletion upon reverse bias permit their use at higher power levels. Both types of devices are very useful in the generation of microwaves. They are competitive with klystron and magnetron tubes, and they have the additional advantage of compatibility with solid-state circuitry.

9.1.6.5 High-frequency pulse diodes. Diodes capable of alternating their properties as a result of the application of very short impulses at high frequencies play an important part in electronic circuitry. Such devices may be considered as high-speed switches that respond by alternating between forward and reverse properties when activated by rapid, alternating changes in voltage.

One way to visualize the operation of these diodes is to consider the $p-n$ diode as constituting a capacitor. The forward pulse charges it, and the reverse pulse discharges it.

The concentration gradients of the dopants again are considered to be very sharp at the junction of the p and n materials as described in Section 8.8. The arrival of a forward-biased pulse at the diode decreases the potential barrier at its junction in a way similar to that shown in Fig. 8-4(a). The electron density and the current density will increase exponentially and overshoot Eqs. (8-71) and (8-73), respectively, so the densities of the carriers of each sign will be very much greater than their equilibrium densities at their respective boundaries of the depletion zone. As the carriers diffuse across the boundary, they become minority carriers and then recombine. At the instant that the forward bias is applied, the current density will be very high because of the very high density of carriers. The carrier density decreases as the diffusion of the carriers progresses into material of opposite sign; the current diminishes accordingly. A dynamic equilibrium in the minority carrier densities then is reached, after a time equivalent to the minority carrier lifetime has elapsed, during which the carrier density and current density are at their equilibrium values expected from Eqs. (8-71) and (8-73).

Upon the arrival of the next pulse (reverse biased), the remaining, high, carrier concentrations at the boundary edges of opposite sign again cause an overshoot. Thus, a high current, opposite in sign to that created by the forward bias, is generated largely by the consumption of carriers that were present as a result of the forward pulse. This decays, in a time equivalent to that of carrier lifetime, to its equilibrium value.

In both the forward- and reverse-biased conditions, the high, precursor currents at the beginning of each pulse are limited by the resistance of the bulk materials of the diode. The time for the reaction to each pulse may be taken as varying as the minority relaxation time so that this factor controls the rate of switching. Any factor that diminishes this time increases the switching rate. If, as noted previously, the diode is considered as a capacitor, the rate of switching may be increased by decreasing the area, A, of the $p-n$ junction (the plates of the capacitor), since $C = A\varepsilon_0\varepsilon/d$. Another approach to this problem is to make the distance, d, between the contacts as short as possible. Typical switching rates of 10^9 s^{-1} and frequencies of 10^9 Hz are available.

High-frequency, pulse diodes are used in those electronic circuits that are based on very short signals at very high rates of reception. These include VHF electronic circuitry and computer applications.

9.1.7 Transistors

Transistors may be classified into two major kinds: bipolar-junction transistors (BJT) and field-effect transistors (FET). Both types are named for their operating mechanisms. BJTs are based on the contributions of both holes and electrons, thereby deriving their name. FETs are based on the flow of a single type of carrier, either electrons or holes (but not both), in which the resistance or the current of the device may be controlled by an electric field. FETs, thus, are unipolar devices.

The general configuration of a BJT consists of two regions of similarly doped material within a single crystal that are separated by a thin, oppositely doped region. This may be considered as consisting of two diodes that are in intimate contact with each other. These individual diodes have the same properties as described in Sections 8.8 and following sections and 9.1.1. However, their monolithic construction and their common, oppositely doped region make a very significant difference. In a reverse-biased, single diode, the minority carriers have their origin within the semiconductor or from the ohmic contact. However, in the BJT, the second diode (emitter) can be a large source of minority carriers when it is forward biased, because it constitutes a part of the first diode. The current collected by the first, reverse-biased diode (collector) is essentially not a function of the amount of the bias; it collects all minority carriers inside a region that extends about one carrier mean free path from the depletion zone, regardless of the origin of the carriers. Therefore, the number of carriers, and consequently the current collected by the first diode, can be controlled by the voltage across the second diode. The control of the voltage of the second diode and its consequent effects on the carriers are the basis for the application of these transistors.

The field-effect transistor (FET or JFET) is based on the $p-n$ junction. It may also have a three-component configuration ($n-p-n$ or $p-n-p$). The unaffected portion of the component that contains the depletion zone is called the *channel*. Ohmic connections to this component connect the channel to a source and drain. The outer materials in the three-component devices are connected to a single terminal known as a *gate*. When a reverse bias is applied between the channels and the two outer, oppositely doped regions (gates), a current flows in the channels, the two depleted zones widen and move into the adjacent central material, and the effective conduction cross section of the channels is reduced. Thus, varying the bias between the channels and the gates, or between the source and the drain, results in variations in the conduction by the channels. This is the same as changing the resistance between the source and drain. Virtually no current is drawn by the gates because these junctions are reverse biased. If the voltage along a channel (between source and drain) is small compared to that required for the effective reduction of a channel, the resistance of a channel will be independent of the drain current. When this is the case, the FET effectively acts as a linear resistor; its resistance may be varied with the source–drain voltage. However, when the voltage across a channel effectively reduces the width of that channel, the drain current is virtually independent of the voltage between the source and the drain. Thus, the FET becomes a current source in which the current is practically independent of the voltage between the source and the drain. Under these conditions, the current may be varied by the voltage between the gate and the source so that the FET behaves as a triode.

Simpler, single-junction FET devices may be made from a single $p-n$ junction. The source and the drain are connected to either side of one component of the junction. This half of the device contains the channel. The gate is connected to the other oppositely doped component of the device. The operation of this device is the same as that described previously, but contains one channel instead of two. This simple device will be used for purposes of explanation in Section 9.1.7.2.

The operation of BJT and FET devices is described in the following section. Both are readily adaptable to microminiaturization. As in the previous sections, the graphs are intended to show functional relationships rather than actual values because of the variations in the properties of devices made by different sources.

9.1.7.1 Bipolar junction devices.

BJTs are composite devices consisting of three alternately doped layers in a semiconductor single crystal. The layers may be $n-p-n$ or $p-n-p$, the central material in each case being very thin. Both configurations operate in the same way, but the polarities of the charges and potentials will be opposite in each case. The $n-p-n$ device is discussed here. In any event, the BJT may be considered to consist of two $p-n$ junctions in intimate proximity, as shown in Fig. 9-6(a).

When V_{eb} and V_{cb} [Fig. 9-6(b)] are made negative, both junctions are reverse biased; only the very small, reverse-biased currents will flow as in Fig. 8-5(a). This current is the sum of the two reverse currents. The band structure adjacent to the two junctions is shown schematically in Fig. 9-7(a). This reverse current must be very small because the density of carriers flowing between the emitter and the base is the algebraic sum $n_{c1} + n_{p1}$, and the density flowing from the base to the collector is $n_{c2} + n_{p2}$. Each component of these sums virtually cancels the other. When this condition is present, the transistor is off and it behaves as though the circuit were open. This is one mode of operation and is used for on–off and cutoff circuitry applications.

When a forward bias is imposed on the base [Fig. 9-7(b)], the collector–base ($p-n$) junction will again be reverse biased. Using the Boltzmann statistics, the density of electrons (minority carriers) adjacent to the depletion zone at the forward-biased $n-p$

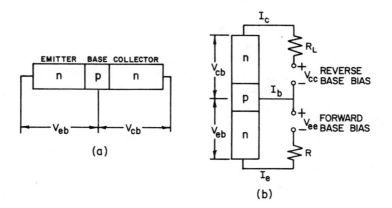

Figure 9-6 Bipolar junction transistor: (a) schematic diagram showing greatly magnified base; (b) circuit.

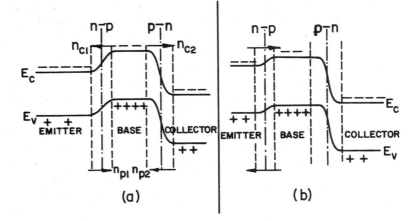

Figure 9-7 Band structure of a bipolar transistor: (a) reverse bias on both junctions; (b) forward bias on both junctions.

junction (emitter–base junction) is

$$n_{e(n,p)} = n_e \exp\!\left(\frac{eV_{eb}}{k_B T}\right)$$

And, in the same way, the density of minority carriers on the base side of the depletion zone at the p–n junction (base–collector junction) is

$$n_{e(p,n)} = n_e \exp\!\left(\frac{eV_{cb}}{k_B T}\right)$$

It can be seen from the difference in the heights of the barriers at the two junctions that $n_{e(n,p)} > n_{e(p,n)}$. And, since the base contains a large number of majority carriers, virtually no internal field is present in this region. Thus, virtually no opposition exists that would prevent the diffusion of the electrons across the base, to the collector, at a fairly uniform rate. This leads to a linear excess charge distribution across the base where $n_{e(n,p)} \to n_{e(p,n)} \to 0$ at the base–collector interface.

The emitter dopant concentration is made to be significantly higher than that of the base so that the depletion zone in the emitter is very small compared to in the base (Section 8.8.2) and the forward bias across the n–p junction results in virtually all carriers originating from the emitter entering the base. (This is also called carrier injection into the base.) This almost entirely precludes carrier injection from the base to the emitter. The carrier migration from the emitter to the collector is optimized, as previously noted, by keeping the base very thin. This minimizes recombination and allows almost all the carriers from the emitter to enter the collector. In other words, almost all the current will enter the collector. In actual devices, the fraction of the current entering the collector from the emitter, α, is at least 0.95 and may reach 0.98 to

0.99 +.   However, if the temperature becomes too high, the number of carriers becomes excessive and *thermal runaway* takes place (Section 8.9.3.2.

The very small amount of carrier recombination that does take place in the base is important since it causes a small flow of external current into the base. A small number of holes must enter via the base connection. This maintains the density of the majority carriers in the base at a level very close to that required by equilibrium. Therefore, this small current must equal the rate of recombination of the electrons and holes in the *p*-type base.

The current entering a device must equal the currents leaving it (Kirchhoff's current law: $\Sigma I = 0$). In the present case, this is expressed as

$$I_e = I_b + I_c \tag{9-7}$$

where the subscripts e, b, and c denote emitter, base, and collector, respectively. The external current entering the base that is needed to compensate for the carrier recombination in the base is, where α is defined, as previously noted, by $\alpha = I_c/I_e$,

$$I_b = (1 - \alpha)I_e \tag{9-8}$$

Since I_b is the input current and the output current is I_c, the ratio of these is the gain, γ. This is given by

$$\frac{I_c}{I_b} = \gamma \tag{9-9}$$

This gain may be increased by employing highly doped emitter material and very thin sectioned base zones. The gain is reasonably constant over a wide range of values of I_c, but it decreases at low currents because carrier recombination in the depletion region of the emitter–base junction can affect an appreciable fraction of the relatively low carrier densities.

The relationship between γ and α may be obtained by substituting Eq. (9-8) into Eq. (9-9) along with the definition $\alpha = I_c/I_e$:

$$\gamma = \frac{I_c}{I_b} = \frac{I_c}{(1 - \alpha)I_e} = \frac{\alpha}{1 - \alpha} \tag{9-10a}$$

or

$$\alpha = \frac{\gamma}{\gamma + 1} \tag{9-10b}$$

These relationships give insight into the amplification characteristics of BJT devices starting with Eq. (9-7) rearranged and then divided through by I_b to give

$$\frac{I_c}{I_b} = \frac{I_e}{I_b} - 1$$

And, by means of Eq. (9-9), this becomes

$$\gamma = \frac{I_e}{I_b} - 1, \qquad \gamma + 1 = \frac{I_e}{I_b}$$

Multiplying both sides of this equation by I_c gives, since $I_c/I_b = \gamma$,

$$(\gamma + 1)I_c = \frac{I_e}{I_b} I_c = \gamma I_e$$

This is expressed as

$$I_c = \frac{I_e \gamma}{\gamma + 1} \qquad (9\text{-}11)$$

Note was made previously of the very large fraction of the current entering the collector from the emitter. This is, using Eq. (9-10b) in Eq. (9-11),

$$I_c = \alpha I_e \qquad (9\text{-}12)$$

Since $\alpha \to 1$, $I_c \to I_e$, and, from Eq. (9-7), it follows that I_b is relatively small:

$$I_c \gg I_b \qquad (9\text{-}13)$$

Because $I_c/I_b = \gamma$ (Eq. (9-9) and $I_c \gg I_b$, the dc current amplification, γ, is much greater than unity. Usually, γ is made to range between 20 and 500, depending on the device. This is the basis for the use of BJTs as a means of amplification. (It should be noted that it is possible to have $\gamma \sim 10^4$, but such devices have undesirable stabilities and frequency responses. These result from such factors as the capacitance of the base, the configuration of the device, and the leads.)

A transistor operated in this way shows linear behavior of I_c versus I_b [Eq. (9-9) and Fig. 9-8(a)]. Since the density of carriers crossing from the emitter to the base varies as $\exp(eV_{eb}/k_BT)$, the base current, I_b, is also an exponential function of V_{eb}, as shown in Fig. 9-8(b). Many circuits make use of this exponential dependence of the base current. This is called an *active* operating mode and constitutes another manner of operation.

In this active mode, I_b is small (μA) and the approximation may be made that the emitter current will equal the current between the emitter and the collector, I_{ec}, and that this current will vary as the voltage between the emitter and collector, V_{ec}; $I_e \cong I_{ec} \propto V_{ec}$. Then, using Eq. (9-7),

$$I_c \cong V_{ec} - I_b \qquad (9\text{-}14)$$

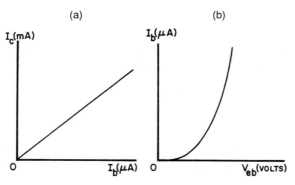

Figure 9-8 Electrical characteristics of a bipolar transistor: (a) base–collector current; (b) base current–voltage. (Modified from T. H. Beeforth and H. J. Goldsmid, *Physics of Solid State Devices*, p. 65, Pion, London, 1970.)

This behavior is plotted in Fig. 9-9. The initial linear response corresponds to that shown in Fig. 9-8(a) and by Eq. (9-9). But when V_{ec} causes a reverse bias at the collector junction, it has no further influence on I_c, which remains essentially constant. The variable factor I_b, then, becomes the controlling factor. This response is known as *saturation* and is indicated by the constant behavior of I_c versus V_{ec} in Fig. 9-9.

When V_{eb} is increased, the carrier density will increase, I_c will increase, and the voltage between the emitter and the collector will decrease [Fig. 9-6(b)]:

$$V_{ec} = V_{cc} - I_c R_L \qquad (9\text{-}15)$$

Continued increase in V_{eb} will cause the junction to reach a condition in which $V_{ec} = V_{eb}$ and the reverse bias vanishes. When this is the case, $V_{eb} = 0$, and the collector current is

$$I_c = \frac{V_{cc} - V_{eb}}{R_L}$$

However, since $V_{eb} \ll V_{cc}$, V_{eb} may be neglected, and it may be approximated that

$$I_c \cong \frac{V_{cc}}{R_L} \qquad (9\text{-}16)$$

Now, using Eq. (9-16) with Eq. (9-9), the base current is

$$I_b = \frac{I_c}{\gamma} = \frac{V_{cc}}{\gamma R_L} \qquad (9\text{-}16a)$$

An increase in I_b in excess of that given by Eq. (9-16a) causes the $n{-}p$ and $p{-}n$ junctions to become forward biased. This causes the injection of carriers into the base from the collector and results in a current, I_{cb}. Carriers also diffuse from the emitter to the collector, giving rise to I_{ec}. The net collector current is given by

$$I_c = I_{ec} - I_{cb}$$

I_c cannot be reversed even if additional carriers are injected into the base from the

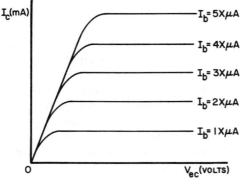

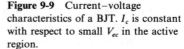

Figure 9-9 Current–voltage characteristics of a BJT. I_c is constant with respect to small V_{ec} in the active region.

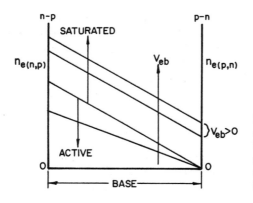

Figure 9-10 Influences of minority carrier concentrations as affected by V_{eb} in the active and saturated conditions in a BJT. (Modified from T. H. Beeforth and H. J. Goldsmid, *Physics of Solid State Devices*, p. 66, Pion, London, 1970.)

collector, since the collector is positive and is connected to the positive side of V_{cc}. The value of $n_{e(n,p)}$ will decrease linearly from the $n-p$ junction to the $p-n$ junction by diffusion. However, because of the positive potential ($V_{eb} > V_{ec}$), $n_{e(n,p)} \to n_{e(p,n)} > 0$. This means that $V_{eb} > V_{bc} > 0$. V_{ec}, as previously noted, is small and may be neglected in comparison with V_{cc}, so it may be omitted in Eq. (9-14). Thus, V_{eb} accounts for the uniform increments caused by I_b in the saturation mode (Fig. 9-9). This is in contrast to the active mode in which the minority carrier density in the base is $n_{e(n,p)} \to n_{e(p,n)} \to 0$ at the $p-n$ junction. The influence of V_{eb} on the operation of these devices is shown in Fig. 9-10.

Since the collector current is saturated for the condition given by Eq. (9-14), the effect of increasing V_{eb} is that of increasing I_b and, therefore, of increasing the minority carrier density within the base. The rate of diffusion of the carriers is essentially the same as in the unsaturated case; the main difference is that more carriers are present. This accounts for the shifts in the densities of carriers with increasing V_{eb}, the parallel slopes in the saturation region of Fig. 9-10, and the uniform increments of I_c induced by constant incremenlts of I_b in Fig. 9-9.

In summary, when $V_{eb} \leqslant 0$, $I_c = 0$. When $V_{eb} > 0$, but is insufficient to cause saturation, the active mode prevails and linear behavior takes place up to the limiting condition given by Eqs. (9-16) and (9-17). Beyond this limiting condition, the transistor is saturated, and I_c becomes constant versus V_{ec} and is a function of V_{eb} and I_b (Figs. 9-9 and 9-10).

The control of V_{eb} also enables the use of the transistor as a switch. BJTs applied for this purpose may be used in the megahertz to gigahertz range of frequencies.

9.1.7.2 Field-effect transistors.
Transistors based on the field effect (FET) are very important in electronic technology, particularly in integrated circuits. A depletion layer exists in the semiconductor material adjacent to a $p-n$ junction, in the absence of an applied voltage (Section 8.8). The operation of these devices depends on the way in which changes in the depleted zones affect device properties in response to variations in external voltages.

A single-junction FET, which now is used rarely, will be employed to explain the mechanisms involved. This is shown in Fig. 9-11. The mechanism is the same as that occurring at each of the junctions in the more common $n-p-n$ or $p-n-p$ devices.

When the n-type channel material is forward biased by V_{GS} (the gate–source potential), the p-type gate is negatively biased; the extent of the depletion zone will increase progressively and uniformly into the n-type material as a function of $V_{GS}^{1/2}$ [Eq. (9-3) and Fig. 9-11(a)]. The effect of the progressively diminishing channel width is to correspondingly decrease the density of the carriers that can flow along it. This will continue, as indicated by the broken lines in Fig. 9-11(a) for increasing voltages, until the depletion zone extends entirely across the channel. Complete occupation of the n-type material by the depleted zone occurs when the gate reverse bias reaches a limiting value of V_P; the channel now is *pinched off*, and the flow of current ceases. Using the relationship previously found for a Schottky barrier, Eq. (9-4) may be employed to express V_P:

$$V \cong \frac{eNd^2}{2\varepsilon_0\varepsilon} \rightarrow V_P \cong \frac{eNW^2}{2\varepsilon_0\varepsilon} \qquad (9\text{-}17)$$

In this case, N is the density of ionized donors of a single type that provide either n_{e0} or n_{h0}, as given by Eqs. (8-95) and (8-97), in the channel, and $d \rightarrow W$, the total width of the channel material. Since the channel may be in either p- or n-type material, the sign of N, accordingly, will be negative or positive, and V_P will be positive for channels in p-type materials and negative for channels in n-type materials.

Thus, in a way analogous to the behavior of the Schottky barrier, variations in the gate–source potential, V_{GS}, result in changes in the effective cross section of the channel. In turn, this changes the conductance of the channel. Another factor is introduced by this voltage that also affects the size of the channel. The current flowing along the channel (drain current, or I_{SD}) sets up another potential difference, V_{SD} (source–drain potential), which increases progressively along the channel from the

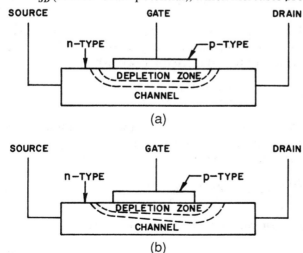

Figure 9-11 Schematic diagram of a JFET: (a) successive broken curves indicating increases in the depletion zone as a function of increasing V_{GS}; (b) asymmetry in depletion zone as a function of V_{SD} with V_{GS} held constant. (Modified from T. H. Beeforth and H. J. Goldsmid, *Physics of Solid State Devices*, p. 74, Pion, London, 1970.)

source to the drain. This is in addition to the V_{GS} potential already present. The result is that the depleted zone becomes asymmetric, again shown by broken lines for increasing voltages in Fig. 9-11(b), with its greatest asymmetrical extent where V_{SD} is largest. This, in effect, tapers the channel so that it becomes funnellike. The result is that the FET acts like a variable resistor, under these conditions, whose resistance varies as a function of V_{GS}. The voltage range in which this behavior occurs is known as the *ohmic region* [Fig. 9-12(a)].

In the operation of this device, V_{GS} may be insufficient to cause pinch-off by itself, but a given value may cause a sufficient potential with V_{SD} so that pinch-off may occur as a result of both potentials. The critical pinch-off value of V_{SD} is denoted by $V_{P(SD)}$. This condition is expressed by

$$V_{P(SD)} = V_{GS} - V_P \tag{9-18}$$

as indicated in Fig. 9-12(a). In an n-type channel, V_{GS} and V_P are negative and $V_{P(SD)}$ is positive, as shown in Fig. 9-12(a).

The behavior of the nondepleted portion, or the active, remaining region of the channel will be examined by a method that is based on that of Schockley (1952). The channel is treated as being asymmetric. The potential in the channel at any point x along it is given by V_{SC}, with $x = 0$ at the source end of the n-type channel material. If the potential across the gate–channel junction is V_{GC} and that across the source and the channel is V_{SC}, then

$$V_{GC} = V_{GS} - V_{SC} \tag{9-19}$$

Let the total width of the n-type channel material be W. The width of the depletion zone, based on Eqs. (9-3) and (9-17), is given for any point x along its length, L, as

$$W(x) = \left(\frac{2\varepsilon_0 \varepsilon V_{GC}}{eN}\right)^{1/2} \tag{9-20}$$

So the width of the channel is $W - W(x)$ at any point x. This enables the calculation of the potential drop along an element of length $dW(x)$, in the channel, by using Ohm's law based on the form $[W - W(x)] \, dV = I\rho \, dW(x)$:

$$dV_{SC} = \frac{I_{SD}\rho \, dW(x)}{W - [2\varepsilon_0\varepsilon V_{GC}/eN]^{1/2}} \tag{9-21}$$

in which I_{SD} is the current flowing between the source and the drain and ρ is the resistivity of the channel. Equation (9-21) may be rearranged, and substitution is made for V_{GC} using Eq. (9-19), so that it becomes

$$I_{SD}\rho \frac{dW(x)}{dV_{SC}} = W - \left[\frac{2\varepsilon_0\varepsilon(V_{SC} - V_{GS})}{eN}\right]^{1/2} \tag{9-22}$$

Equation (9-17) may be used to simplify Eq. (9-22) by noting that

$$\frac{2\varepsilon_0\varepsilon}{eN} \cong \frac{W^2}{V_P} \tag{9-23}$$

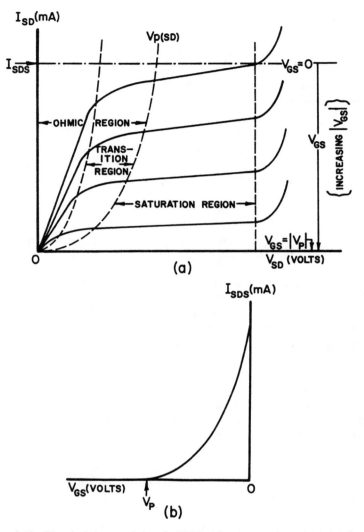

Figure 9-12   Electrical characteristics of a JFET with an *n*-type channel: (a) the effect of V_{GS} on the current–voltage (V_{SD}) behavior; (b) maximum saturation current (I_{SDS}) as a function of V_{GS}. (Modified from T. H. Beeforth and H. J. Goldsmid, *Physics of Solid State Devices*, p. 77, Pion, London, 1970.)

Equation (9-23) is substituted into Eq. (9-22) to obtain

$$I_{SD}\rho\,\frac{dW(x)}{dV_{SC}} = W - \left[\frac{W^2(V_{SC} - V_{GS})}{V_P}\right]^{1/2}$$

or

$$I_{SD}\rho\,dW(x) \cong W\left[\frac{1 - (V_{SC} - V_{GS})}{V_P}\right]^{1/2} dV_{SC} \qquad (9\text{-}24)$$

Equation (9-24) is integrated from $x = 0$ to $x = L$, the length of the channel, to give

$$I_{SD}\rho L \cong \left[V_{SC} + \frac{2}{3} V_P \left(\frac{V_{GS} - V_{SC}}{V_P} \right)^{3/2} \right]\Bigg|_0^{V_{SD}} \qquad (9\text{-}25)$$

The channel potential varies from $V_{SC} = 0$ at $x = 0$ to $V_{SC} = V_{SD}$ at $x = L$. The application of these limits results in

$$I_{SD}\rho L \cong W \left[V_{SD} + \frac{2}{3} V_P \left(\frac{V_{GS} - V_{SD}}{V_P} \right)^{3/2} - 0 - \frac{2}{3} V_P \left(\frac{V_{GS} - 0}{V_P} \right)^{3/2} \right]$$

or

$$I_{SD} \cong \frac{W}{\rho L} \left[V_{SD} + \frac{2}{3} V_P \left(\frac{V_{GS} - V_{SD}}{V_P} \right)^{3/2} - \frac{2}{3} V_P \left(\frac{V_{GS}}{V_P} \right)^{3/2} \right]$$

and, upon factoring,

$$I_{SD} \cong \frac{2W V_P}{3\rho L} \left[\frac{3V_{SD}}{2V_P} + \left(\frac{V_{GS} - V_{SD}}{V_P} \right)^{3/2} - \left(\frac{V_{GS}}{V_P} \right)^{3/2} \right] \qquad (9\text{-}26)$$

This analysis holds reasonably well for single, monolithic devices. However, in integrated circuitry, where planar diffusion or implantation techniques are used to make devices in the form of functional blocks rather than individual components, the channels are a result of dopant diffusion. The concentration profiles in any such diffusion processes are exponential, with negative slopes. Therefore, even though the channels may be small, the dopant ion concentrations are not uniform. An approximation for the behavior in this case [Eq. (9-27)] is given later.

When V_{SD} approaches V_P, very high internal, electric fields are produced because the channel becomes extremely narrow. A point is reached beyond which I_{SD} should become very small. This should simultaneously decrease V_{SD} and cause the channel to become wider. So, when V_{SD} becomes greater than V_P, the width of the channel approaches a nearly constant value and the device is said to be saturated. After saturation occurs, I_{SD} remains approximately constant [Fig. 9-12(a)], especially at higher values of $|V_{GS}|$. The distribution of the potentials within the channel, consequently, also remains nearly constant. The extent of the saturation region as a function of V_{SD} is limited by breakdown that occurs when the internal fields become sufficiently large (Section 8.9.2).

The saturation current may be calculated by means of Eq. (9-26) by setting $V_{SD} = V_P$. However, for the reasons previously given, this is not entirely satisfactory. I_{SD} may be more closely approximated from the relationship

$$I_{SD} \cong I_{SDS} \left(\frac{1 - V_{GS}}{V_P} \right)^2 \qquad (9\text{-}27)$$

in the region in which $V_{SD} \geqslant 2V_P$, where I_{SDS} is the maximum saturation current for $V_{GS} = 0$ [Figs. 9-12(a) and (b)].

The gate reacts only when the junction is in reverse bias. Thus, values of I_{GS} are in the 10^{-9}-A range. V_{GS} normally is less than 10 V, giving a resistance of about 10^9 Ω.

This high resistance vanishes if forward biasing occurs. Therefore, V_{GS} and V_{SD} must be opposite in sign.

I_{SD} primarily is a function of V_{GS} because I_{GS} is negligibly small, as noted previously. The amplification of an FET is found, from Eq. (9-27), as follows:

$$\gamma = \frac{dI_{SD}}{dV_{GS}} = \frac{2I_{SDS}}{V_P} \frac{1 - V_{GS}}{V_P} \tag{9-28}$$

(The amplification, γ, is actually the mutual conductance.) Solving Eq. (9-27) for $1 - V_{GS}/V_P$ and substituting this into Eq. (9-29) gives

$$\gamma = \frac{2I_{SDS}}{V_P} \left(\frac{I_{SD}}{I_{SDS}} \right)^{1/2} = \frac{2I_{SDS}^{1/2}}{V_P} I_{SD}^{1/2} \tag{9-29}$$

Thus, γ varies as $I_{SD}^{1/2}$, since I_{SDS} and V_P are constants.

In the saturation region, FETs may be used as sources of controlled, nearly constant currents. They may also be made to act as controllable resistors whose values may be made to vary in the ohmic region. As such, they are very versatile devices.

Compared to BJTs, FETs have lower-noise, high-input impedance (apparent ac resistance analogous to dc resistance.) In addition, their mechanism depends on the flow of majority carriers, and the current is affected primarily by the temperature variation of the carrier mobility (Section 6.3). BJTs are minority carrier devices in which the number of carriers able to cross a depletion zone without recombination varies exponentially with temperature (Section 9.1.7.1). This can lead to thermal runaway (Section 8.8.3). Thus, FETs are less sensitive to thermal effects. The gate current does tend to increase with temperature and steps must be taken to prevent it from decreasing the bias. The packing density of FETs is greater in integrated circuitry than that of BJTs. Also of prime commercial importance is the fact that FETs are less complicated to manufacture.

The FETs do not have as good a frequency response as BJTs, having useful gains up to frequencies of about 0.5 GHz as compared to 5 GHz or higher for BJTs. The reason for this is that the width of the base of a BJT may be made to be much smaller than that of the channel of a FET. In addition, the gains, γ in Eq. (9-9), are usually much larger than the γ [Eq. (9-29)] for a given frequency range. However, the FETs have greater linearity and stability when used in low-impedance amplifiers.

The FET has a smaller power capacity than the BJT. This also results from the comparative size of the small base in a BJT compared to the relatively large channel in an FET. The larger voltage drop across the channel limits the current to a greater degree than the base of a BJT, thus limiting the power capacity of the FET.

FETs are widely used in such applications as square-law amplifiers [Eq. (9-27)]. They are also used for square-law detection and analog multipliers. Their linearity and stability are responsible for their extensive use in RF amplifiers in radio receivers. Another application is for the amplification of signals from detectors of optical radiation.

9.1.7.3 Metal-oxide, Silicon FET (MOSFET).

The channel in the FET is within the volume of one of the doped semiconductor members of a p–n junction. However, it is possible to induce a channel adjacent to the surface of a semiconductor.

It was shown (Section 8.9.3.1) that the unsaturated bonds of the ions adjacent to the surface of a semiconductor act as traps for electrons. The net result is a thin layer of negative charge that forms at and just below the surface. This has the effect of repelling electrons in the conduction band and results in an adjacent volume of positively charged donor ions. And, in n-type material, this depleted zone is more like p-type than n-type material (see Fig. 8-8).

When the n-type dopant concentration is relatively low, the density of holes that is induced by the negative, surface charge will be larger than the density of electrons in the thin, subsurface volume. When this is the case, the surface material, for all practical purposes, may be considered to be converted to p-type. The close proximity of this p-type subsurface layer between the oxidized surface and the n-type interior material, or substrate, induces a depletion zone between the p- and n-type materials in the n-type interior material. In a similar fashion, p-type interior material will have an n-type surface layer (Sections 8.8.1 and 8.8.2). Behavior of this kind is called *surface inversion* [Eqs. (8-85) and (8-96a)].

Devices whose operation depends on such inversions are considered to act in the enhancement mode because of the relatively free flow of carriers from the source to the drain. In the case given, the holes readily enter into this flow. Where the holes are depleted by reversing the bias, the device is considered to act in the depletion mode. These modes are discussed later.

There is no theoretical difference between devices based on n- or p-type interior, or substrate, materials. However, since the surface states at the SiO_2–Si boundary are predisposed to induce a type surface depletion layer similar to that indicated in Fig. 8-8, an inversion layer is more readily attained in n-type substrates. Circuits have been designed that use devices with both types of substrates in the enhancement mode.

The density of the surface states has been shown to be very important in these devices. The early FETs were very ineffective, despite the theoretical prediction of their efficient operation. Schockley and Pearson (1948), working on the basis of Bardeen's prediction (1947), showed that surface states were present and were responsible for breakdown or leakage at lower voltages than for p–n junctions. In so doing, they discovered the BJT (Sections 9.1.7 and 9.1.7.1). The development of FETs was delayed until about 1953 when cleaner processing techniques were developed. It was not until integrated circuitry was made possible that FETs became competitive with BJTs (Section 9.1.7.2).

The surface charge of a semiconductor also may be varied when another charge is placed on a metallic electrode (gate) that is insulated from the semiconductor by the oxide film at the surface of the semiconductor. A negative bias of V_{GS} on the gate electrode decreases the induced positive surface states in an n-type semiconductor and thus diminishes the surface inversion layer. Conversely, a positive bias of V_{GS} on the gate electrode increases that inversion layer. Inversion layers may be induced in

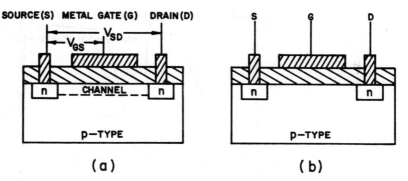

Figure 9-13 Schematic diagrams of MOSFETs: (a) induced *n*-type channel in a lightly doped *p*-type substrate; (b) heavily doped *p*-type substrate. The terminals are metallic in each case.

semiconductors with relatively high dopant concentrations by this external means. Devices employing either of these techniques act like FETs and are called MOSFETs.. They are also known as insulated-gate FETs (IGFETs) or metal-insulated semiconductor transistors (MISTs) (Fig. 9-13).

The operating mechanism of a MOSFET is based on the controlled variation of the inversion layer between the two *n*-type volumes shown in Fig. 9-13. In a relatively lightly doped substrate [(Fig. 9-12(a)], conduction occurs with $V_{GS} = 0$. Variations in V_{GS} change the shape and the extent of the inversion layer; increasingly reverse biased values of V_{GS} decrease the *p*-type inversion layer in *n*-type material until pinch-off occurs [Figs. 9-12(a) and (b)]. Positive bias, in the same way, increases the width of the inversion zone and I_{SD} increases.

The same mechanism takes place in highly doped substrates. Heavily doped MOSFETs cannot conduct until an inversion channel is created between the source and the drain because of the *p–n* junctions. In this case, the onset of current flow requires higher values of V_{GS} as the dopant concentration increases. This is shown in Fig. 9-14(a). Transistors of this kind are called enhancement-type MOSFETs.

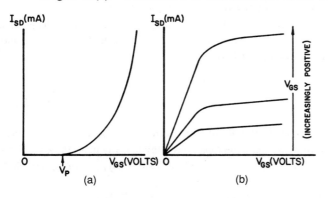

Figure 9-14 Electrical characteristics of enhancement-type MOSFETs.

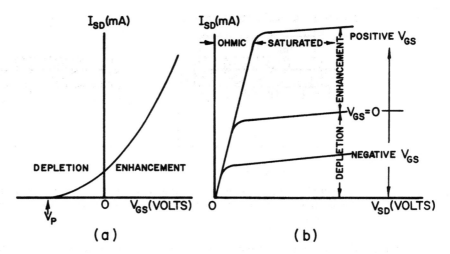

Figure 9-15 Influence of V_{GS} on the electrical characteristics indicating the depletion and enhancement modes. (Modified from T. H. Beeforth and H. J. Goldsmid, *Physics of Solid State Devices*, p. 79, Pion, London, 1970.)

As previously noted, MOSFETS are also constructed with either *n*- or *p*-type channels between the terminals. These may operate in either the depletion or the enhancement modes. They react in the same way as FETs. Their current–voltage characteristics are shown in Fig. 9-15. These are made in order to avoid reliance on the highly variable surface charges.

The similarities between Figs. 9-12, 9-14, and 9-15 are apparent. All of the curves show a V_P. In each case, the curves of I_{SD} versus V_{SD} show ohmic and saturation regions. These devices also follow a square-law amplification, different from Eq. (9-27), that varies linearly with V_{GS} at saturation.

MOSFETs may be used in virtually all the applications described for FETs, except that they have useful gains at frequencies up to about 30 GHz. They generate considerably more noise than FETs because of surface effects.

Some materials that have been used for these devices include CdS, CdSe, GaAs, Ge, InAs, InSb, PbS, PbTe, Si, and SnO_2. One of the most common types consists of an Si substrate and an SiO_2 insulating layer.

9.2 BIBLIOGRAPHY

BEEFORTH, T. H., and GOLDSMID, H. J., *Physics of Solid State Devices*, Pion, London, 1970.

BYLANDER, E. G., *Materials for Semiconductor Functions*, Hayden, Rochelle Park, N.J., 1971.

COHEN, M. M., *Introduction to the Quantum Theory of Semiconductors*, Gordon & Breach, New York, 1972.

DEKKER, A. J., *Solid State Physics*, Prentice-Hall, Englewood Cliffs, N.J., 1957.

DRISCOLL, F. F., and COUGHLIN, R. F., *Solid State Devices and Applications*, Prentice-Hall, Englewood Cliffs, N.J., 1975.

FELDMAN, J. M., *Physics and Circuit Properties of Transistors*, Wiley, New York, 1972.

KITTEL, C., *Introduction to Solid State Physics*, 3rd ed., Wiley, New York, 1966.

LECK, J. H., *Theory of Semiconductor Junction Devices*, Pergamon, Elmsford, N.Y., 1967.

SEITZ, F., *Modern Theory of Solids*, McGraw-Hill, New York, 1940.

SPROULL, R. L., *Modern Physics*, 2nd ed., Wiley, New York, 1963.

STRINGER, J., *An Introduction to the Electron Theory of Solids*, Pergamon, Elmsford, N.Y., 1967.

WILKES, P., *Solid State Theory in Metallurgy*, Cambridge University Press, New York, 1973.

WOLF, H. F., *Semiconductors*, Wiley-Interscience, New York, 1971.

9.3 PROBLEMS

9.1. Show how oppositely oriented rectifiers can be used to protect circuitry from current overload.

9.2. Explain the application of rectifying diodes for full-wave rectification.

9.3. Why are very high dopant concentrations required for tunnel diodes?

9.4. Describe the mechanism responsible for current minima in tunnel diodes.

9.5. Explain the basis for all photoconductive devices.

9.6. Why should the emittted irradiation from GaAs be greater than those from Ge or Si?

9.7. Describe the basic scheme for heterojunction lasers.

9.8. Does an upper limit exist in the doping of varactors?

9.9. Why should a p–i–n junction have a higher breakdown voltage than a p–n junction?

9.10. Compare the properties of p–i–n and Schottky junctions.

9.11. Discuss the role played by relaxation times on the operation of high-frequency diodes.

9.12. Explain the advantages of a BJT as compared to a single diode.

9.13. Discuss the differences in the operation of BJTs and FETs.

9.14. Why is it relatively easy to obtain very high values for the gain of a BJT?

9.15. Give the conditions under which BJTs may be used as sources of nearly constant current.

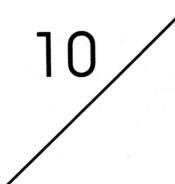

MAGNETISM

10.1 DIAMAGNETISM AND PARAMAGNETISM

There are no nonmagnetic materials. All matter is affected in one way or another by magnetic fields, however small their reactions may be. Materials have been classified on the bases of the magnitude, sign, and effect of temperature on their magnetic properties. This is the rationale of such classifications as diamagnetic, paramagnetic, ferromagnetic, ferrimagnetic and antiferromagnetic materials.

The magnetic properties of many materials result from combinations of two of these magnetic phenomena. For example, the properties of some materials can be shown to arise from combinations of diamagnetism and paramagnetism. The diamagnetic contribution arises from the motions of electrons in their "orbits", while the paramagnetic contribution comes from the alignment of ions that possess permanent magnetic moments. The larger of these two influences determines whether the material is diamagnetic or paramagnetic. In cases where the resultant effect is small with respect to the applied field, such materials often are practically but incorrectly considered as being nonmagnetic.

The concepts and relationships introduced here to explain diamagnetism and paramagnetism are also employed to explain other magnetic effects.

10.1.1 Basic Concepts and Units

The phenomena involved in magnetic effects may be presented in the *cgs system*, also called the emu, or electromagnetic, system, and the *mks*, or SI, *system*. The mks system does not provide the same simplicity for magnetic properties as it does for electrical properties. And, since the overwhelming preponderance of the data in the literature are given in the cgs system, this system is used here. However, the corresponding mks expressions for the basic relationships are provided in this section and are identified as SI.

10.1.1.1 Magnetic poles.   Consider a bar magnet. Each end of the bar contains regions at or near its ends that appear to be the sources of attraction or repulsion. Such regions are known as *magnetic poles*; one will be north seeking, the other south seeking. These poles are present in pairs. If the bar magnet is cut into two or more pieces, each piece will have its own set of poles. Until quite recently it appeared to be impossible for a single pole to exist. Experimental evidence now seems to indicate that isolated poles may exist. The *cgs* definition of a unit pole is given as a pole that induces a force of 1 dyne upon another identical pole when they are separated by a distance of 1 cm.

The attractive force, F, between two poles, p_1 and p_2, separated by a distance d is given by the general expression

$$F = K_1 \cdot \frac{p_1 p_2}{d^2} \quad \text{dynes} \tag{10-1}$$

where the constant of proportionality, K_1, is equated to unity in the cgs system. The constant K_1 in the mks system is $1/4\pi\mu_0$, where μ_0 is the permeability of vacuum (free space) and $\mu_0 = 4\pi \times 10^{-7}$ weber (Wb)$\cdot$A$^{-1}\cdot$m^{-1}. Thus, in the mks system,

$$F = \frac{1}{4\pi\mu_0} \cdot \frac{p_1 p_2}{d^2} \quad \text{newtons (N)} \tag{10-1 SI}$$

where 1 newton = 10^5 dynes. The force acting on the second pole is a result of the magnetic field around the first pole. This may be expressed by

$$F = K_2 p H \tag{10-2}$$

where p is the pole strength and H is the strength, or intensity, of the field. A unit field is defined in the cgs system as one that induces a force of 1 dyne upon a pole of unit strength. This also is known as an oersted (Oe). The constant $K_2 = 1$ in the cgs system, and its units are dynes. The constant $k_2 = 10^{-5}$ in the mks system, and its units are newtons.

A field intensity of 1 Oe induces a force of 1 dyne upon a unit pole. And a unit pole induces a force of 1 dyne at a distance 1 cm from a second unit pole. From this, the field intensity of a unit pole at a distance of 1 cm from the second pole must be 1 Oe. And, on the basis of Eqs. (10-1) and (10-2), for $K_1 = K_2 = 1$,

$$F = \frac{p_1 p_2}{d^2} = p_1 H$$

Then, in general terms, for the cgs system,

$$H = \frac{p}{d^2} \quad \text{Oe} \tag{10-3}$$

And, for the mks system,

$$H = \frac{1}{4\pi\mu_0} \cdot \frac{p}{d^2} \quad \text{ampere-turns-m}^{-1} \qquad (10\text{-}3 \text{ SI})$$

The field intensity, H, is described by the number of lines of force per unit area, so 1 Oe equals 1 line of force per cm^2. The unit of one line of force per cm^2 is equal to 1 maxwell·cm^{-2}. The latter is also called a gauss (G). A spherical surface of unit radius with its center at a unit pole has a surface of 4π cm^2. One oersted is defined as one line of force per cm^2, so 4π lines of force intersect this reference surface. In this way, a pole p may be considered to have $4\pi p$ lines of force emanating from it.

In the mks system, the pole strength is in webers. The unit of the weber (Wb) is (in emu) 1 dyne Oe^{-1}. It is related to other units as follows:

$$1 \text{ Wb/m}^2 = 1 \text{ N} \cdot \text{A}^{-1} \cdot \text{m}^{-1} = 1 \text{ tesla} = 10^4 \text{ G}$$

10.1.1.2 Magnetic moment. Consider a magnetic dipole, such as a bar magnet, where the poles are separated by a distance d, as shown in Fig. 10-1, placed in a uniform magnetic field H. The moment exerted by this couple is

$$m = pHd \sin \theta \qquad (10\text{-}4)$$

The effect of the magnetic field is to cause the dipole to rotate until it becomes parallel to the field. Here the forces induced by the field on the poles are equal and opposite and no further motion occurs. This reaction is characteristic of paramagnetic materials. The magnetic moment is obtained for $\theta = \pi/2$ and $H = 1$ Oe as

$$m = pd \qquad (10\text{-}5)$$

The units of m are in ergs·Oe^{-1} in the cgs system.

The magnet in any orientation other than that parallel to H will possess a potential energy

$$dE(\theta) = pH \, d(\sin \theta) \, d\theta \qquad (10\text{-}6)$$

This, after using Eq. (10-5) in Eq. (10-6) and integrating, becomes

$$E(\theta) = \int_0^\theta mH \sin \theta \, d\theta$$

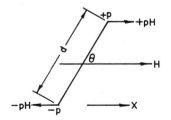

Figure 10-1 Magnetic dipole acted on by a uniform magnetic field.

The potential energy is taken as zero at $\theta = \pi/2$, so the potential energy is given by

$$E(\theta) = -mH \cos \theta \qquad (10\text{-}7)$$

When the magnet is parallel to the field, $\theta = 0$ and

$$E(\theta) = -mH \qquad (10\text{-}8)$$

It should be noted that m and H are actually vectors. Equation (10-8) may be used to obtain another expression for the force acting on the dipole as a result of a nonhomogeneous field by applying Eq. (3-1). This gives, using Eq. (10-5),

$$-\frac{\partial E(\theta)}{\partial x} = F = -(-m)\frac{\partial H}{\partial x} = pd\frac{\partial H}{\partial x} \qquad (10\text{-}9)$$

In a uniform field, $\partial H/\partial x$ is zero and a magnet parallel to the field will remain in that position. Where the field is not uniform, $\partial H/\partial x$ is nonzero and the magnet will rotate.

10.1.1.3 Magnetization.

When a material is placed in a magnetic field, a magnetic moment per unit volume, M, is induced in it. This is the *magnetization*, or the intensity of magnetization, and is given by

$$M = \frac{m}{V} \qquad (10\text{-}10)$$

where V is the volume. It also is known as the *volume unit susceptibility*.

This also may be expressed as the pole strength per unit area of the magnet. Equation (10-10) may be reexpressed using Eq. (10-6) as

$$M = \frac{pd}{V}$$

Then, dividing the numerator and denominator by d,

$$M = \frac{pd/d}{V/d} = \frac{p}{Ad/d} = \frac{p}{A} \qquad (10\text{-}11)$$

where A is the cross section of the magnet.

The units of M are $ergs \cdot Oe^{-1} \cdot cm^{-3}$. Here the electromagnetic unit, *emu*, is $ergs \cdot Oe^{-1}$, so M very frequently is given in terms of $emu \cdot cm^{-3}$. The corresponding unit of M in the *mks* system is $Wb \cdot m^{-2}$. ($1 \; Wb = 10^{8} \; emu$.)

Consider the situation in which a material of magnetization M is placed in an applied field H. As shown in Section 10.1.1.1, $4\pi M$ lines of force are associated with M. Thus, the magnetic induction is

$$B = H + 4\pi M \qquad (10\text{-}12)$$

because the lines of force of the applied field add to those already in the material. The quantities B, H, and M are parallel vectors, so they are treated here as scalar quantities. The units of B are, by custom, gauss, while those of H are oersteds. Both of these units are the same, being the maxwell $\cdot$ cm^{-2}. The unit of B in the mks system is the tesla (Wb $\cdot$ m^{-2}).

The vectors M and H also are proportional to each other. This is expressed as

$$M = \chi H \tag{10-13}$$

where the constant of proportionality, χ, is known as the *magnetic susceptibility*. The cgs units of χ are ergs $\cdot$ (Oe $\cdot$ cm^3)$^{-1}$/Oe or emu $\cdot$ (cm$^3 \cdot$ Oe)$^{-1}$. The mks units of χ are Wb $\cdot$ A$^{-1} \cdot$ m^{-1}.

As given by Eq. (10-10), M is the magnetization per unit volume. The susceptibility is expressed in other useful ways as indicated in Table 10-1.

The proportionality between B and H is expressed as

$$B = \mu H \tag{10-14}$$

where the constant of proportionality, μ, is the permeability of a substance. Equations (10-12) and (10-14) are equated to give

$$B = H + 4\pi M = \mu H \tag{10-15}$$

These equations may be used to obtain other useful relationships between these important properties. Reexpressing Eq. (10-15) gives

$$H(1 - \mu) = -4\pi M \tag{10-16}$$

Or, using Eq. (10-13),

$$1 - \mu = -4\pi \frac{M}{H} = -4\pi\chi$$

so that

$$\mu = 1 + 4\pi\chi \tag{10-17}$$

This relationship permits the classification of magnetic materials as shown in Table 10-2.

It should be noted that, for the mks system, Eq. (10-12) may be defined as

$$B = \mu_0 H + \mu_0 M \tag{10-18}$$

TABLE 10-1   EXPRESSIONS FOR MAGNETIC SUSCEPTIBILITY

Name	Symbol	Relation to χ	cgs units	Basis
Atomic	χ_A	χAW	emu (g-atom)$^{-1}$ Oe^{-1}	Atomic weight, AW
Mass	χ_m	χ/ρ	emu $\cdot$ g$^{-1} \cdot$ Oe^{-1}	Density, ρ
Molecular	χ_M	χM (mol)	emu $\cdot$ (g-mol)$^{-1}$ Oe^{-1}	Molecular weight, M (mol)

TABLE 10-2 CLASSIFICATION OF
MAGNETIC MATERIALS, Eq. (10-17)

Material	χ	μ
Diamagnetic	< 0	< 1
Free space (vacuum)	0	1
Paramagnetic	$0 < \chi < 1$	> 1
Antiferromagnetic	$0 < \chi < 1$	> 1
Ferromagnetic	$\gg 1$	$\gg 1$
Ferrimagnetic	$\gg 1$	$\gg 1$

This may be reexpressed as

$$M = \frac{B}{\mu_0} - H \qquad (10\text{-}19)$$

Upon dividing Eq. (10-19) through by H,

$$\frac{M}{H} = \frac{B}{H}\frac{1}{\mu_0} - 1$$

so that, by means of Eqs. (10-13) and (10-14), this may be given as

$$\chi = \frac{\mu}{\mu_0} - 1$$

Or

$$\mu_r = \frac{\mu}{\mu_0} = 1 + \chi \qquad (10\text{-}20)$$

where μ_r is the relative permeability.

The susceptibilities of some elements are given in Table 10-3.

10.1.2 Diamagnetism

Electrons in filled levels usually may be considered as having zero net spin and moment. The motion of a single electron in its "orbit" creates a magnetic moment because it is a moving charge. In the absence of a magnetic field, these orbital moments are balanced in atoms with closed shells, such as the noble gases (see Sections 3.9 to 3.11). The same is true for ionically bonded solids (Section 3.12.2) and for covalently bonded gases and solids, including organic compounds (Section 3.12.3). The magnetic moment of an ion in a solid is the vector sum of its orbital and spin motions. Diamagnetism is a result of the change in electron orbital motion in an external field. This causes magnetization opposite to that of the applied field, forces such materials away from high regions of field intensity, and is the reason for negative

TABLE   10-3 MASS   SUSCEPTIBILITIES   OF   SOME   ELEMENTS
(emu/g $\cdot$ Oe $\times 10^6$)

Element	χ_m	Element	χ_m	Element	χ_m
H	-1.97	Si	-0.13	Mo	$+0.04$
He	-0.47	P	-0.90	Pd	$+5.4$
Li	$+0.50$				
Be	-1.00	Ca	$+1.10$	Ag	-0.20
B	-0.69	Ti	$+1.25$	Cd	-0.18
C	-0.49	V	$+1.4$	Sn	-0.25
N	-0.80	Cr	$+3.08$	Sb	-0.87
O	$+106.2$	Mn	$+11.8$	La	$+1.04$
Na	$+0.51$	Cu	-0.86	Ta	$+0.93$
Mg	$+0.55$	Zn	-1.57	W	$+0.28$
Al	$+0.65$	Zr	-0.45	Pt	$+1.10$
		Nb	$+1.5$	Bi	-1.35

From J. K. Stanley, *Electrical and Magnetic Properties of Metals*, p. 212, American
Society for Metals, Metals Park, Ohio, 1963.

values for χ (Table 10-2). It should be noted that while almost all the noted classes of
materials would be expected to be diamagnetic, all show exceptions. Thus the closed-
shell configuration of electrons may be used only as a guide.

An electron in "orbit" in a Bohr-type atom (Section 1.5) may be considered to
constitute a current flowing in a resistanceless loop. The application of an external
field will induce a current and a magnetic field that is opposite to that of the external
field. The dielectric moment is a result of the current induced by the field. The fact
that all substances will show a diamagnetic response, however small, is the basis for
the earlier statement that there are no nonmagnetic materials, neglecting nuclear
magnetic responses.

Consider all the Z electrons, each with charge e, in a Bohr atom to be in an
"orbit" of average "radius" r that constitutes a resistanceless loop. The time required
for an electron to complete one lap around the loop is $2\pi/\omega_0$, ω_0 being its angular
velocity in the absence of an external field. The current I in the loop is obtained from
$I = \text{charge/time}$. Thus, for Z electrons, the current is

$$I = \frac{Ze}{2\pi/\omega_0} = \frac{Ze\omega_0}{2\pi} \tag{10-21}$$

The magnetic moment of an electron in this orbit is given by the area of the loop times
the current. Thus, in cgs units, the magnetic moment of Z electrons, perpendicular to
the plane of the loop, is

$$\mu_m = \frac{IA}{c} = \frac{1}{c} \cdot \frac{Ze\omega_0}{2\pi} \cdot \pi r^2 = \frac{Ze\omega_0 r^2}{2c} \tag{10-22}$$

where A is the area of the loop. The angular momentum (Section 1.5) is $\mathbf{p}(\theta) = mr^2\omega_0$, so Eq. (10-22) may be reexpressed as

$$\mu_m = Z\left(\frac{e}{2mc}\right)\mathbf{p}(\theta) \tag{10-23a}$$

And, where the momentum must be quantized, as shown in Section 1.5, since $\mathbf{p}(\theta) = nh/2\pi$, Eq. (10-23a) becomes

$$\mu_m = \frac{Z(e/2mc)nh}{2\pi} = \frac{Zneh}{4\pi mc} \tag{10-23b}$$

so the magnetic moment of a single electron in the lowest Bohr "orbit," $n = 1$, is $eh/4\pi mc$. It turns out, as will be shown later, that the spin magnetic moment is identical to the orbital magnetic moment derived here [Eqs. (10-86) and (10-90)]. This value is known as the *Bohr magneton*, μ_B, and has a value of 9.27×10^{-21} erg$\cdot$Oe^{-1}.

The change in the behavior of an electron in a filled shell may be shown by examining the forces acting on such an electron. In the absence of an external magnetic field, the force exerted by the electron is

$$F = m\omega_0^2 r \tag{10-24}$$

The application of a field normal to the plane of the orbit of the electron causes another force, F_n, to affect the electron. This force is vectorially opposite in direction to F of Eq. (10-24). Thus, the net force acting on the electron is

$$F = m\omega_0^2 r - F_n \tag{10-25}$$

Here, F_n, the Lorentz force previously employed in Sections 7.2.1, 7.2.2, and 7.2.3, is given by

$$F_n = \frac{e}{c}(\vec{v} \times \vec{H}) = \frac{e}{c}\,vH = \frac{e}{c}\,\omega r H \tag{10-26}$$

since $v = \omega r$. The net force acting on the electron is obtained by the substitution of Eq. (10-26) into Eq. (10-25) to obtain

$$F = m\omega_0^2 r - \frac{e}{c}\,\omega r H = m\omega^2 r \tag{10-27}$$

since the application of F_n has changed ω_0 to ω. Equation (10-27) is rearranged as

$$mr(\omega^2 - \omega_0^2) = -\frac{e}{c}\,\omega r H \tag{10-28}$$

The difference of the two squares can be factored and, noting that r vanishes, may be expressed as

$$\omega - \omega_0 = -\frac{eH}{mc}\cdot\frac{\omega}{\omega + \omega_0} \tag{10-29}$$

Since most external fields can induce only small changes in ω, it may be approximated that $\omega \cong \omega_0$. This substitution causes ω to vanish from the right side of Eq. (10-29), since $\omega + \omega_0 \cong 2\omega$, and

$$\omega - \omega_0 \cong -\frac{eH}{2mc} \cong \Delta\omega = \omega_L \qquad (10\text{-}30)$$

The effect of the applied magnetic field is to cause the electron to precess in a way analogous to that of a wobbling toy top (Fig. 10-2). This behavior of electrons is known as *Larmor precession* and is denoted by ω_L.

The diamagnetic magnetization and susceptibility, M and χ_D, may be obtained by starting with the differentiated form of Eq. (10-22). This is

$$\Delta\mu_m = \frac{Zer^2}{2c} \Delta\omega \qquad (10\text{-}31)$$

And, where $\Delta\omega$ is given by Eq. (10-30) and Eq. (10-22) is used after multiplying the numerator and denominator by ω,

$$\Delta\mu_m = -\frac{Zer^2}{2c} \cdot \frac{eH}{2mc} = -\frac{Zer^2\omega}{2c\omega} \cdot \frac{eH}{2mc} = -\frac{\mu_m}{\omega} \cdot \frac{eH}{4mc} \qquad (10\text{-}32)$$

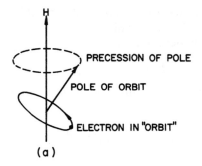

(a)

(b)

Figure 10-2 Larmor precession: (a) precession of pole of orbit with frequency ω_L; (b) mechanical and magnetic moments involved.

The second factor in Eq. (10-32) is just ω_L as given by Eq. (10-30). Thus Eq. (10-32) may be expressed as

$$\Delta\mu_m \cong \frac{\omega_L}{\omega}\, \mu_m \tag{10-33}$$

The magnetization of a solid of unit volume that contains N ions, where each ion contains multiple, filled orbits of "radii" r_i, is, based on Eq. (10-32):

$$M = N\,\Delta\mu_m = -N\, \frac{Ze^2}{4mc^2}\, H \sum_i r_i^2 \tag{10-34}$$

And the diamagnetic susceptibility is obtained by the use of Eq. (10-13) as

$$\chi_D = \frac{M}{H} = -N\, \frac{Ze^2}{4mc^2} \sum_i r_i^2 \tag{10-35}$$

in which the summations are made over all the "radii" of the closed electron shells of a single ion.

A more direct approximation for the magnetization may be obtained from Eq. (10-32) if it is assumed that the filled electron shells constitute spherical charge distributions of average "radius" r. The mean-square "radius" $\langle r^2 \rangle$ is used based on

$$\langle r^2 \rangle \cong \tfrac{2}{3}r^2 \tag{10-36}$$

This approximation gives

$$M = -N\, \frac{Ze^2 \langle r^2 \rangle H}{6mc^2} \tag{10-37}$$

and

$$\chi_D = -N\, \frac{Ze^2 \langle r^2 \rangle}{6mc^2} \tag{10-38}$$

As indicated by Eqs. (10-35) and (10-38), the diamagnetic susceptibilities are not functions of temperature.

It should be noted that all atoms have a diamagnetic component, especially in strong fields, and all substances show this effect even though their bonding results in the formation of completed electron shells. This component should be subtracted from such susceptibilities as paramagnetic or ferromagnetic for most accurate results. For practical purposes, this component, usually $\ll 10^{-5}$ emu$\cdot g^{-1}\cdot$Oe^{-1}, may be neglected.

10.1.2.1 Ionic and covalent materials.   The classical analysis presented in Section 10.1.2 is based on the assumption that the dipoles rotate freely in response to the external field. This accounts for its applicability to the noble gases and to some others with multionic molecules in which all the electron levels are filled and constitute closed shells. However, where the electrons are not paired and when the variations of angular momentum that occur along molecular axes in all multionic species are taken into account, an additional positive, temperature-independent term must be included in Eqs. (10-35) and (10-38). The same is true for real, anisotropic crystals. This contribution is known as *Van Vleck paramagnetism.* It is responsible for many of the exceptions noted at the beginning of this section. However, in most cases this contribution is smaller than that given by the equations noted, and such materials are diamagnetic.

Another case where paramagnetic effects can predominate over the diamagnetic effects occurs when the electrons undergo transitions to other states as a consequence of the applied field. In so doing, their magnetic moments change, and the magnetic susceptibility changes accordingly. Changes of this kind usually do not occur in cases in which strong bonds are formed by the ionic or covalent mechanisms (Sections 3.12.1 and 3.12.3). Here the constituent ions form closed-shell configurations and, despite the fact that some degree of bond hybridization is always present, they are usually diamagnetic. It must be noted that, even for compounds where the bonding is virtually all ionic, the susceptibility of a given species of ions is not constant. Such differences appear to vary inversely as the molecular weight of the salt.

Reasonable approximations of the susceptibilities of many organic molecules may be obtained from the algebraic sum of the susceptibilities of their constituents. This holds for polycrystalline or amorphous organic materials because of their random orientations. Anisotropic magnetic properties are to be expected of organic single crystals. Where trigonal hybridization occurs (Section 3.12.3), hexagonal rings can occur in graphite and in organic compounds, and the π electrons are relatively mobile. The application of a field normal to such rings causes the π electrons to travel around the rings, thus accounting for the relatively high magnetic susceptibilities of these materials.

10.1.2.2 Metals. The diamagnetic properties of metals are a consequence of the translational motion of the nearly free electrons (Section 3.12.4). In the presence of a magnetic field, the Lorentz force (Sections 7.2.1, 7.2.2, and 7.2.3.) has the effect of inducing the electrons to travel along helical paths in which the axis of the helix is parallel to the field. This motion is treated by projecting a turn of the helix onto a plane perpendicular to this axis. The resulting circle is used to evaluate the forces acting on the electron.

On this basis, the centrifical force is equated to the normal force as given by Eq. (10-26) to obtain

$$\frac{e\mathbf{v}H}{c} = \frac{m\mathbf{v}^2}{r} \tag{10-39}$$

The factor $\mathbf{v}^2$ vanishes from Eq. (10-39) when use is made of $\omega = \mathbf{v}/r$ to obtain

$$\omega = \frac{eH}{mc} \tag{10-40}$$

Here ω, the angular frequency of the electron in the spiral path, is called the *cyclotron frequency*. It will be noted that $\omega = |2\omega_L|$ of Eq. (10-30).

The quantum mechanic treatment is based on the energies affecting the electron. This energy is given by the sum

$$E_n = E_{xy} + E_z \tag{10-41}$$

Here E_{xy} is the energy component in the $x-y$ plane perpendicular to the field, and E_z is the component parallel to the field. The projected motion of the electron on the $x-y$ plane is treated as that of a simple harmonic oscillator using Eq. (5-138). This is included in Eq. (10-41) to obtain, after changing the variable from v to ω,

$$E_n = \left(n + \frac{1}{2}\right)hv = \left(n + \frac{1}{2}\right)h \cdot \frac{\omega}{2\pi} + E_z \tag{10-42}$$

The component E_z, being parallel to the field, does not affect the energy of oscillation, so Eq. (10-42) becomes, after using Eq. (10-40) for ω,

$$E_n = \frac{(n + \frac{1}{2})h}{2\pi} \cdot \frac{eH}{mc} \tag{10-43}$$

This equation contains a basic, inherent property of electrons called the Bohr magneton, μ_B. This quantity, derived later in Eq. (10-87), is given as

$$\mu_B = \frac{eh}{4\pi mc} = 9.27 \times 10^{-21} \text{ erg/Oe} \tag{10-44}$$

so that Eq. (10-42) may be written as

$$E_n \frac{2n + 1}{2} \cdot 2\mu_B H = (2n + 1)\mu_B H \tag{10-45}$$

Here n may equal zero.

The electrons excited by the magnetic field may now occupy states that are more widely separated than they would be in the quasicontinuum that they occupy in the absence of the field. These states, called *Landau levels*, are more highly degenerate than is the case for the original quasicontinuum because many electrons occupy a Landau level (Fig. 10-3). the gaps between the allowed Landau levels may be determined from Eqs. (1-5), (10-40), and (10-44) as

$$E_g = hv = \frac{h\omega}{2\pi} = \frac{h}{2\pi} \cdot \frac{eH}{mc} = 2\mu_B H = \Delta E \tag{10-46}$$

A verification of Eq. (10-46) may be obtained by examining the energy gap between Landau states. Let the coefficient of a given Landau state [Eq. (10-45)] be $2n + 1$. Then that of the next highest state may be expressed as $[2(n + 1) + 1]$. Thus, the energy difference between the adjacent states is $\pm 2\mu_B H$ and is the same as that given by Eq. (10-46).

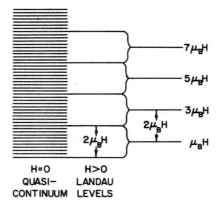

Figure 10-3 Schematic diagram of the change from a quasicontinuum to the degenerate Landau levels that results from the application of a magnetic field.

The highly degenerate Landau states do not violate the requirement of exclusion (Section 3.9). This may be shown by starting with Eq. (4-3). This is multiplied by a factor of 2 so that each type of spin is taken into account. This gives

$$N(\mathbf{p}) \, d\mathbf{p} = \frac{8\pi V \mathbf{p}^2 \, d\mathbf{p}}{h^3} \tag{10-47}$$

The density of such states for a given momentum $\mathbf{p}$ within the range $\mathbf{p}$ to $\mathbf{p} + d\mathbf{p}$ is obtained by dividing Eq. (10-47) through by $\mathbf{p}$ to obtain

$$\frac{N(\mathbf{p}) \, d\mathbf{p}}{\mathbf{p}} = \frac{8\pi V \mathbf{p} \, d\mathbf{p}}{h^3} \tag{10-48}$$

The range $d\mathbf{p}$ is reexpressed, in terms of its component in the $x-y$ plane, $d\mathbf{p}_{xy}$, and that parallel to the field, $d\mathbf{p}_z$, so Eq. (10-48) becomes

$$\frac{N(\mathbf{p}) \, d\mathbf{p}}{\mathbf{p}} = \frac{8\pi V \mathbf{p}_{xy} \, d\mathbf{p}_{xy} \, d\mathbf{p}_z}{h^3} \tag{10-49}$$

This expression is simplified by noting that $\mathbf{p}_{xy} \, d\mathbf{p}_{xy} = \frac{1}{2} d(\mathbf{p}_{xy}^2)$. And, using $E_{xy} = \mathbf{p}_{xy}^2 / 2m$, $m \, dE_{xy} = \frac{1}{2} d(\mathbf{p}_{xy}^2)$, so $\mathbf{p}_{xy} \, d\mathbf{p}_{xy} = m \, dE_{xy}$. And, using Eq. (10-46), $\mathbf{p}_{xy} \, d\mathbf{p}_{xy} \cong m \, \Delta E_{xy} = 2m\mu_B H$, this is substituted into Eq. (10-49) to obtain

$$\frac{N(\mathbf{p}) \, d\mathbf{p}}{\mathbf{p}} \cong \frac{16\pi V m \mu_B H \, d\mathbf{p}_z}{h^3} \tag{10-50}$$

Equation (10-44) is now used for μ_B in Eq. (10-50):

$$\frac{N(\mathbf{p}) \, d\mathbf{p}}{\mathbf{p}} \cong \frac{16\pi V m H \, d\mathbf{p}_z}{h^3} \cdot \frac{eh}{4\pi mc} = \frac{4VeH \, d\mathbf{p}_z}{ch^2} \tag{10-51}$$

The implications of Eq. (10-51) may be understood more readily by examining the factor $d\mathbf{p}_z$. This is done by expressing $\mathbf{p}_z$ in terms of the de Broglie equation as

$$\mathbf{p}_z = \frac{h}{\lambda_z} = \frac{h}{2\pi} \cdot \frac{2\pi}{\lambda_z} = \frac{h}{2\pi} \bar{\mathbf{k}}_z \tag{10-52}$$

The result being sought is obtained by differentiating Eq. (10-52) to obtain

$$d\mathbf{p}_z = \frac{h}{2\pi} d\bar{\mathbf{k}}_z \tag{10-53}$$

This enables the reexpression of Eq. (10-51) as

$$\frac{N(\mathbf{p})\, d\mathbf{p}}{\mathbf{p}} = \frac{4VeH}{ch^2} \cdot \frac{h}{2\pi} d\bar{\mathbf{k}}_z = \frac{2VeH}{\pi ch} d\bar{\mathbf{k}}_z \tag{10-54}$$

Thus, the density of states within a given Landau level is a function of the wave vectors of the electrons that are essentially parallel to the external field. The exclusion principle is observed as long as the excited electrons have slightly differing directions. In effect, this means that the Landau levels shown in Fig. 10-3 are, in actuality, very small ranges of states and not single states, as might be construed from the figure.

The contribution of the valence electrons to the dielectric properties of metals can be evaluated with reference to Figs. 4-3(c) and 10-3. Starting with Eq. (4-6), the density of electron states near $E = E_F$ is

$$N(E_F) = f \cdot N(E, 0) = \tfrac{1}{2} N(E, 0) \tag{10-55}$$

and the total number of electrons is approximated as

$$N \cong N(E, 0) \cdot E_F \tag{10-56}$$

Then the density of states is given by

$$N(E, 0) \cong \frac{N}{E_F} \tag{10-57}$$

Equation (10-57) is substituted into Eq. (10-55) to get

$$N(E_F) \cong \frac{N}{2E_F} \tag{10-58}$$

Given the large numbers of valence electrons, it may be assumed that equal numbers have opposite spins. In terms of spin, Eq. (10-58) is

$$N(E_F)_s \cong \frac{N}{4E_F} \tag{10-59}$$

Now, by means of Eq. (10-46), for $N(E_F)_s$ electrons,

$$\Delta E \cong -2N(E_F)_s \mu_B H \tag{10-60}$$

If it is further approximated that

$$M \cong -\Delta E \mu_B \tag{10-61}$$

Then, using Eq. (10-59) in Eq. (10-61),

$$M \cong -2N(E_F)_s \mu_B^2 H = -2 \frac{N}{4E_F} \mu_B^2 H = -N \frac{\mu_B^2 H}{2E_F} \tag{10-62}$$

or, using $E_F = k_B T_F$,

$$M \cong -N \frac{\mu_B^2 H}{2k_B T_F} \tag{10-62a}$$

This gives the susceptibility as

$$\chi_D = \frac{M}{H} = -N \frac{\mu_B^2}{2E_F} = -N \frac{\mu_B^2}{2k_B T_F} \tag{10-63}$$

Both M and χ_D are independent of temperature because E_F is virtually independent of temperature. It will also be understood that both M and χ_D are negative for diamagnetic materials (see Table 10-3).

As noted in Section 10.1.2 and by Eqs. (10-62) and (10-63), the diamagnetic effect in metals is independent of temperature. This effect in metals increases as the atomic number increases. This is the reason for the predominance of paramagnetism in such metals as gold and lead. A variation in the susceptibility of some metals occurs as a function of the intensity of the field and is a periodic function of $1/H$. This periodic behavior, known as the *de Haas–van Alphen effect*, is used to determine the Fermi surface.

10.1.3 Paramagnetism

In contrast to diamagnetic materials, paramagnetic materials are such that the magnetic components of the constituent ions are incompletely canceled; their ions have a net magnetic moment and act as magnetic dipoles. These dipoles are randomly oriented and cancel each other in the absence of an external magnetic field. The dipoles tend to align themselves parallel to an external magnetic field. The degree of alignment determines the susceptibility of the material. Thermal energy increases the vibrational activity of the ions, diminishes the alignment of the dipoles, and decreases the susceptibility. At a sufficiently high temperature, the alignment is completely destroyed and paramagnetism vanishes.

The theory of paramagnetism is also important because it provides a direct approach to ferromagnetism. Both phenomena are based on permanent magnetic dipoles.

10.1.3.1 Classical theory. Consider a unit volume of N freely moving, noninteracting ions, each with a dipole moment μ_m in a magnetic field H, as shown in Fig. 10-4. The magnetic moment of each ion has a potential energy that is given by Eq. (10-7) written, since the dipoles are oriented with respect to the direction of the

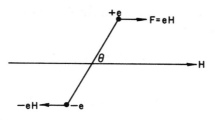

Figure 10-4   Sketch of a magnetic dipole.

field, as

$$E(\theta) = -\mu_m H \cos \theta \tag{10-64}$$

The magnetization of the substance is

$$M = N\mu_m \langle \cos \theta \rangle \tag{10-65}$$

where $\langle \cos \theta \rangle$ is the average of all N dipoles' angular orientations in relation to the field direction.

The method described by Eq. (2-23) is used to determine $\langle \cos \theta \rangle$ so that M and χ_p may be calculated. Here a reference sphere of unit radius is considered to be within the unit volume of N ions. The density of the dipole moments that intersect the spherical reference surface will be uniform over the entire surface because the freely moving, noninteracting dipoles are randomly oriented in the absence of a field. In a field, the density of dipoles no longer is random because the dipoles tend to align themselves parallel to the field. Here the number of the dipoles is measured by the density of their intersections with the reference surface area. This surface area, A, is thus a measure of the dipole density and, in turn, is measured by the solid angle that it subtends. The most probable solid angle is obtained from the preceding method, using $\beta = E(\theta)/k_B T$ for convenience, as

$$\langle \cos \theta \rangle = \frac{\displaystyle\int_0^\pi \exp(-\beta) \cos \theta \, dA}{\displaystyle\int_0^\pi \exp(-\beta) \, dA} \tag{10-66}$$

Equation (10-66) is a measure of the most probable density of the orientations of the dipoles as a result of exposure to a magnetic field. Use now is made of the following relationships:

$$x^2 + y^2 = 1 \tag{10-67a}$$

$$x = \cos \theta, \qquad dx = -\sin \theta \, d\theta \tag{10-67b}$$

$$y = \sin \theta, \qquad dy = \cos \theta \, d\theta \tag{10-67c}$$

and the surface area of the reference surface is given by

$$A = 2\pi \int y \left[1 + \left(\frac{dy}{dx} \right)^2 \right]^{1/2} dx \tag{10-67d}$$

The factor dA needed for Eq. (10-66) is obtained by substituting Eqs. (10-67b) and (10-67c) into the differential form of Eq. (10-67d):

$$dA = 2\pi \, \sin \, \theta [(dx)^2 + (dy)^2]^{1/2} \tag{10-67e}$$

or

$$dA = 2\pi \, \sin \, \theta [\sin^2 \, \theta + \cos^2 \, \theta]^{1/2} \, d\theta$$

from which, since $\sin^2 \theta + \cos^2 \theta = 1$,

$$dA = 2\pi \, \sin \, \theta \, d\theta \tag{10-68}$$

Equation (10-68) is substituted into Eq. (10-66) to obtain

$$\langle \cos \, \theta \rangle = \frac{\displaystyle\int_0^\pi \sin \, \theta \, \cos \, \theta \, \exp(-\beta) \, d\theta}{\displaystyle\int_0^\pi \sin \, \theta \, \exp(-\beta) \, d\theta} \tag{10-69}$$

since the factor 2π vanishes.

Equation (10-64) is now used to substitute for $E(\theta)$ in β of Eq. (10-69), the result being

$$\langle \cos \, \theta \rangle = \frac{\displaystyle\int_0^\pi \sin \, \theta \, \cos \, \theta \, \exp[(\mu_m H/k_B T) \cos \, \theta] \, d\theta}{\displaystyle\int_0^\pi \sin \, \theta \, \exp[(\mu_m H/k_B T) \cos \, \theta] \, d\theta} \tag{10-70}$$

Another convenient substitution may be made by letting

$$\frac{\mu_m H}{k_B T} \cos \, \theta = a \, \cos \, \theta = x \tag{10-71}$$

Then, differentiation and rearrangement of Eq. (10-71) in its simplified form gives

$$d\theta = -\frac{dx}{a \, \sin \, \theta} \tag{10-72}$$

Equations (10-71) and (10-72) are substituted into Eq. (10-70) to obtain

$$\langle \cos \, \theta \rangle = \frac{\displaystyle\int_0^\pi \sin \, \theta \, \cos \, \theta \, e^x \cdot [dx/(a \, \sin \, \theta)]}{\displaystyle\int_0^\pi \sin \, \theta \, e^x \cdot [dx/(a \, \sin \, \theta)]} = \frac{\displaystyle\int_0^\pi \cos \, \theta \, e^x \, dx}{\displaystyle\int_0^\pi e^x \, dx} \tag{10-73}$$

since $\sin \, \theta$, a, and the negative sign vanish. Now, to restore Eq. (10-73) so that it contains the form of Eq. (10-71), its numerator is multiplied by a/a. Thus,

$$\langle \cos \theta \rangle = \frac{(1/a) \int_0^\pi a \cos \theta \, e^x \, dx}{\int_0^\pi e^x \, dx} = \frac{1}{a} \frac{\int_{-a}^a xe^x \, dx}{\int_{-a}^a e^x \, dx} \tag{10-74}$$

where the new limits are obtained from Eq. (10-71). The numerator of Eq. (10-74) is of the type

$$\int xe^{bx} = \frac{e^{bx}}{b^2} (bx - 1) \tag{10-75a}$$

The factor $b = 1$ in Eq. (10-74), so Eq. (10-75a) reduces to

$$\int xe^x = e^x(x - 1) \tag{10-75b}$$

The application of Eq. (10-75b) to Eq. (10-74) and the integration of the denominator results in

$$\langle \cos \theta \rangle = \frac{1}{a} \frac{e^a(a - 1) - e^{-a}(-a - 1)}{e^a - e^{-a}} \tag{10-76}$$

Equation (10-76) reduces to

$$\langle \cos \theta \rangle = \frac{(e^a + e^{-a}) - (1/a)(e^a - e^{-a})}{e^a - e^{-a}} \tag{10-77}$$

Further simplification results from performing the indicated division:

$$\langle \cos \theta \rangle = \frac{e^a + e^{-a}}{e^a - e^{-a}} - \frac{1}{a} \tag{10-78a}$$

Or, noting that the exponential fraction in Eq. (10-78a) is coth a,

$$L(a) = \langle \cos \theta \rangle = \coth a - \frac{1}{a} \tag{10-78b}$$

Equation (10-78b) is the *Langevin function*. This function is shown in Fig. 10-5.

   An approximate solution to Eq. (10-78b) may be obtained for the conditions where H is small and/or T is large. This is the same as saying that $\mu_m H \ll k_B T$ or that $\mu_m H / k_B T = a \ll 1$. The approximation can be made, for such small values, that coth a may be given by the series

$$\coth a \cong \frac{1}{a} + \frac{a}{3} - \frac{a^3}{45} + \cdots \tag{10-79a}$$

Thus, Eq. (10-78b) is approximated, using the first two terms of this series, as

$$L(a) \cong \frac{1}{a} + \frac{a}{3} - \frac{1}{a} = \frac{a}{3} \tag{10-79b}$$

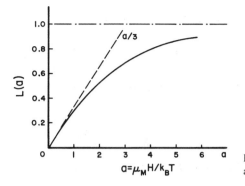

Figure 10-5 Langevin's function $L(a)$ as a function of the variable a.

Or, using Eq. (10-71),

$$L(a) \cong \frac{1}{3} \frac{\mu_m H}{k_B T} \tag{10-80}$$

Equation (10-80) is substituted into Eq. (10-65) to obtain

$$M \cong N\mu_m \cdot L(a) = N\mu_m \cdot \frac{\mu_m H}{3k_B T} \tag{10-81}$$

Since $N\mu_m$ gives the greatest magnetic moment possible for a unit volume of a given material, it is the equivalent of saying that *all* the dipoles are parallel to the external magnetic field. If this condition is designated as M_0, Eq. (10-81) may be reexpressed as

$$M = N\mu_m \cdot L(a) = M_0 L(a) \tag{10-82}$$

Used in this way, M_0 represents the limiting case of complete dipole alignment, or polarizability, or maximum magnetic saturation, and, because of thermal oscillation, $M/M_0 \cong L(a) < 1$. $L(a)$ defines the fraction of alignment actually attained. It can be seen (Fig. 10-5) that $L(a)$ approaches its maximum for $a > 5$. It may also be seen from Eq. (10-81) that the polarizability of a single magnetic dipole is $\mu_m^2/3k_B T$.

The paramagnetic susceptibility is obtained directly from Eq. (10-81) as

$$\chi_p = \frac{M}{H} = \frac{N\mu_m^2}{3k_B T} \tag{10-83}$$

Equation (10-83) may be expressed as shown in Table 10-1. It will be noted that, for a given value of H, χ_p is a function of T^{-1}. This behavior is known as the *Curie law* and is given as

$$\chi_p = \frac{C}{T}, \qquad C = \frac{N\mu_m^2}{3k_B} \tag{10-84}$$

where C is called the Curie constant (P. Curie, 1895). This is shown in Fig. 10-6.

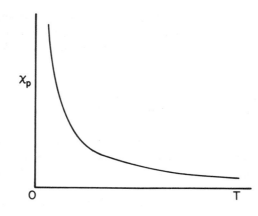

Figure 10-6 Schematic diagram of the paramagnetic susceptibility of a freely rotating, magnetic dipole as in the Langevin theory.

The unusually high value of χ_m of oxygen has been used as a basis for its analyses in flue gases. It is used here to give an approximation of μ_m using the datum from Table 10-3 in Eq. (10-83) and recalling that the molecular weight of O_2 is 32. This gives, for $T = 300$ K,

$$\mu_m \cong [6.02 \times 10^{23}(\text{atom}\cdot\text{mol}^{-1})]^{-1/2}[3 \times 32(\text{g}\cdot\text{mol}^{-1}) \times 1.38 \times 10^{-16}(\text{erg}\cdot\text{deg}^{-1})$$

$$\times 3 \times 10^2(\text{deg}) \times 1.06 \times 10^{-4}(\text{emu}\cdot\text{g}^{-1}\cdot\text{Oe}^{-1}]^{1/2}$$

$$\cong \left(\frac{42.1129 \times 10^{-17}}{6.02 \times 10^{23}}\right)^{1/2} = (6.998 \times 10^{-40})^{1/2}$$

$$\cong 2.65 \times 10^{-20}(\text{erg}\cdot\text{Oe}^{-1}\cdot\text{molecule}^{-1} \text{ of } O_2)$$

This magnetic moment of an oxygen molecule can be reexpressed in terms of Bohr magnetons simply by using $\mu_B \cong 0.93 \times 10^{-20}$ erg$\cdot$Oe^{-1} as the unit. This gives $2.65 \times 10^{-20} \div 0.93 \times 10^{-20} = 2.85\mu_B\cdot$molecule^{-1} of O_2. Now, having approximated the magnetic moment of an oxygen molecule, the value of the variable a in Eq. (10-71) may be approximated. Some of the experimental work on the analyses of oxygen in flue gases was done with permanent magnets with $H \cong 5 \times 10^3$ Oe. Using this value in Eq. (10-71) provides

$$a \cong \frac{\mu_m H}{k_B T} = \frac{2.65 \times 10^{-20}\,(\text{erg}\cdot\text{Oe}^{-1}\cdot\text{molecule}^{-1}) \times 5 \times 10^3\,(\text{Oe})}{1.38 \times 10^{-16}\,(\text{erg}\cdot\text{deg}^{-1}\cdot\text{molecule}^{-1}) \times 3 \times 10^2\,(\text{deg})}$$

$$\cong 3.2 \times 10^{-3}$$

Values of a of this order of magnitude are typical. This agrees with the assumption of very small values of a used to obtain Eq. (10-79a) and, consequently, Eq. (10-80).

The Langevin theory is based on the postulations that the ions or molecules are free to rotate in an applied field and are noninteracting, and that the resultant dipolar alignment is diminished by increasing temperature until it vanishes. Not all paramagnetic materials are composed of noninteracting, freely rotating components. Such materials have susceptibilities that are described by the Curie–Weiss law [Eq. (10-166)].

10.1.3.2 Quantum theory. The classical treatment of paramagnetic dipoles permits them to follow Eq. (10-64) without restriction. In effect, the classical particles are permitted to range within a continuous energy spectrum (Section 1.2). However, electron spin and orbital angular momentum are not permitted to behave in this way. They must assume discrete, quantized states. Thus, the quantum theory treats the behavior of the dipoles as being described by discrete states (Sections 3.9–3.11).

Since angular orbital momentum and spin are of importance in the understanding of paramagnetism and other magnetic phenomena, a brief review of some of their properties, discussed in Section 2.5 and the previously noted sections, is given here. The angular orbital momentum is designated by the second quantum number, l. Its variation, in integral steps, from 0 to $n - 1$ can be considered to describe the ellipticity of an electron "orbit". The angular orbital momentum is

$$\mathbf{p}(l) = \frac{h}{2\pi} \left[l(l + 1) \right]^{1/2} \tag{10.85}$$

Thus, using Eq. (10-23b), the magnetic orbital moment of an electron is

$$\mu_l = \frac{e}{2mc} \cdot \frac{h}{2\pi} \left[l(l + 1) \right]^{1/2} \tag{10-86}$$

This is another one of the ways of showing that μ_B is an inherent, natural unit and that the Bohr magneton is given by

$$\mu_B = \frac{eh}{4\pi mc} \tag{10-87}$$

Thus,

$$\mu_l = \mu_B \mathbf{p}(l) = \mu_B [l(l + 1)]^{1/2} \tag{10-88}$$

It was previously noted that the third quantum number is related to ϕ, Section 3.10. Thus, it is a measure of the angular momentum relative to the z axis. Where the applied field is parallel to this axis, this component of momentum is given by

$$\mathbf{p}(\phi) = \frac{h}{2\pi} m_l \tag{10-89}$$

where m_l can take on the $2l + 1$ values that range from $-l$ to l, including zero. These values are degenerate in the absence of a field because the z axis is undefined for this case. In the presence of a magnetic field, each value of l takes on a discrete orientation with respect to H, so its projection to the z axis also gives the discrete values of m_l (Fig. 10-7). This is the explanation of the spectral line splitting known as the *Zeeman effect*.

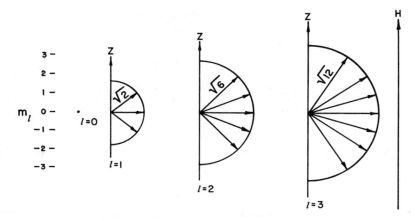

Figure 10-7   Allowed values of m_l arising from quantized orientations of angular momenta with respect to an external magnetic field.

Spin, designated by m_s, also had its origin in the explanation of spectra (Sections 2.5 and 3.9–3.11).   Spin can only have the values $\pm\frac{1}{2}$.   The intrinsic angular momentum due to an electron's spin about its own axis is

$$\mathbf{p}(s) = \frac{h}{2\pi} S = \frac{h}{2\pi} [s(s + 1)]^{1/2} \tag{10-90a}$$

Again using Eq. (10-23), the magnetic spin component of electron motion is

$$\mu_s = \frac{e}{2mc} \cdot \frac{h}{2\pi} [s(s + 1)]^{1/2} = \mu_B [s(s + 1)]^{1/2} \tag{10-90b}$$

It will be seen that the natural unit for the spin magnetic moment as given by Eq. (10-90b) is identical to that given for the orbital magnetic moment of Eqs. (10-86) and (10-87) and that these are the same as that given by Eq. (10-23b).   The allowed values of spin are shown in Fig. 10-8.

Intrinsic and angular orbital moments of electrons must be taken into account to determine their magnetic moments.   This is done by means of the *total quantum*

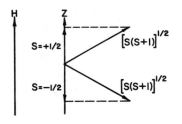

Figure 10-8 Projection of spin vectors parallel to an applied magnetic field.

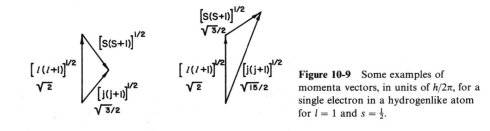

Figure 10-9 Some examples of momenta vectors, in units of $h/2\pi$, for a single electron in a hydrogenlike atom for $l = 1$ and $s = \frac{1}{2}$.

number j, also known as the *inner quantum number*. The total momentum is given as

$$\mathbf{p}(j) = \frac{h}{2\pi} \left[j(j + 1) \right]^{1/2} \tag{10-91}$$

in which $j = l \pm s$. Simple illustrations of this are given in Fig. 10-9.

Since the vectors l and s become oriented with respect to a magnetic field, so does their vector sum, j. These precess around the direction of the applied field, H. This mechanism, indicated in Fig. 10-10 for a single electron, is called *Russell–Saunders coupling*. Where strong mutual interactions occur among spin momentum vectors and/or between orbital momentum vectors, such as in the incomplete d bands of transition elements, more complex results are to be expected. The interactions between different s or l levels can be large with respect to s–l interactions. Under these conditions

$$S = \Sigma s \tag{10-92a}$$

$$L = \Sigma l \tag{10-92b}$$

and,

$$J = L + S \tag{10-92c}$$

It is important to note that S, L, and J can be calculated only for isolated atoms.

Two conditions, known as *Hund's rules*, determine the sums given by Eq. (10-92c). One states that spins are added to maximize S to the extent permitted by exclusion. The other is that orbital momenta are added to maximize L to the same

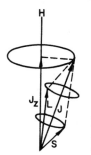

Figure 10-10 Russell–Saunders coupling showing the vector sum $J = L + S$.

extent permitted by the first rule. As written, Eq. (10-92c) accounts for the orbit and spin vectors in the same general direction and applies to cases in which parallel coupling occurs in bands that are more than half-filled. Where antiparallel coupling occurs and a band is less than half-filled,

$$J = L - S \qquad (10\text{-}92d)$$

Hund's rules are illustrated using $3d$ electrons. One electron has $S = \frac{1}{2}$; five electrons have $S = \frac{5}{2}$. The $3d$ band consists of two half-bands that can accommodate five electrons of opposite spin in each to make a total of ten electrons of both spins. Six electrons more than half-fill the d band, so $S = 5/2 - 1/2 = 4/2 = 2$. This illustrates the first rule. The value of L for six electrons is obtained by examining the two half-bands. For the filled half-band, $L = 2 + 1 + 0 - 1 - 2 = 0$. The unfilled half-band containing the sixth electron would have $L = 2$. So, for the six d electrons, $L = 0 + 2 = 2$. The total quantum number is $J = L + S = 2 + 4$. For the case of three d electrons, only one of the half-bands need be considered. Here, $L = 2 + 1 + 0 = 3$ and $S = 3/2$, so $J = 3 - 3/2 = 3/2$. This is an example of the second rule [Eq. (10-92d)]. In this way, Eqs. (10-92) provide relatively simple means for determining the total momenta of electrons.

10.1.3.3 Spectroscopic splitting factor.

Consider an electron in the Bohr atom (Section 1.5) with a principal quantum number $n = 1$. Its unit of angular momentum is $\mathbf{p}(\theta) = h/2\pi$. Its effect on the magnetic moment may be described using Eq. (10-23) as

$$\mu_l = \frac{e}{2mc} \cdot \mathbf{p}(\theta) = \frac{e}{2mc} \cdot \frac{h}{2\pi} \qquad (10\text{-}93a)$$

Its unit of spin momentum is $\frac{1}{2} \cdot h/2\pi$, since spin can be only $\pm 1/2$. Again using Eq. (10-23), its effect on the magnetic moment is

$$\mu_s = \frac{e}{mc} \cdot \mathbf{p}(s) = \frac{e}{mc} \cdot \frac{h}{4\pi} \qquad (10\text{-}93b)$$

It is seen that $\mathbf{p}(s) = \mathbf{p}(\theta)/2$. This enables Eqs. (10-93a) and (10-93b) to be expressed by a single equation, by including a factor g, where $\mathbf{p}$ is the total momentum, as

$$\mu = g \cdot \frac{e}{2mc} \cdot \mathbf{p} \qquad (10\text{-}93c)$$

When $g = 2$, $\mu = \mu_s$, and when $g = 1$, $\mu = \mu_l$. The factor g is the *Landé spectroscopic splitting factor*.

It was noted previously (Section 10.1.3.2) that spectral lines split under the influence of magnetic fields. This is known as the *Zeeman effect*, which describes a line associated with a given energy level as splitting into two or more lines in the presence of a field. This is a result of the L–S coupling described here and in Section 10.1.3.2 and shown in Fig. 10-11.

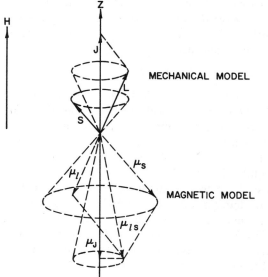

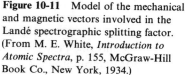

Figure 10-11 Model of the mechanical and magnetic vectors involved in the Landé spectrographic splitting factor. (From M. E. White, *Introduction to Atomic Spectra*, p. 155, McGraw-Hill Book Co., New York, 1934.)

The following notation is used in the figure and in the following analysis: μ_{ls} is the vector sum of $\mu_l + \mu_s$, $(S; J)$ is the cosine of the angle between vectors S and J (this must not be confused with the dot product), and $\hbar = h/2\pi$. It will be noted that the mechanical sums J and μ_J are shown as being collinear (for ease of visualization) and parallel to H. It will be noted that $\mu_s = 2\mu_B S$ because an electron has $S = \frac{1}{2}$ and $\mu_s = \mu_B \neq 1/2\mu_B$ [Eqs. (10-87) and (10-93b)]. The sum $\mu_l + \mu_s = \mu_{ls} = -\mu_B(l + 2s)$ is not collinear with J. All the mechanical and magnetic components precess around J. This causes the component $\mu_{ls} - \mu_J$ not to contribute the magnetic moment, but the component of $\mu_{ls} = \mu_J$ does give the magnetic moment.

On the basis of the foregoing,

$$\mu_s = 2S \frac{e\hbar}{2mc} \tag{10-94}$$

$$\mu_l = L \frac{e\hbar}{2mc} \tag{10-95}$$

$$\mu_{sj} = 2S \frac{e\hbar}{2mc} (S; J) \tag{10-96}$$

$$\mu_{lj} = L \frac{e\hbar}{2mc} (L; J) \tag{10-97}$$

and

$$\mu_j = \mu_{sj} + \mu_{lj} = \frac{e\hbar}{2mc} \cdot [2S(S; J) + L(L; J)] \tag{10-98}$$

The second factor of Eq. (10-98) is the vector sum of the spin and orbit contributions. As such, it is equal to Jg. Stated algebraically,

$$2S(S; J) + L(L; J) = Jg \tag{10-99}$$

The law of cosines,

$$a^2 = b^2 + c^2 - 2bc \cos A \tag{10-100}$$

is applied so that Eq. (10-99) is simplified by the following expressions derived by means of Eq. (10-100) for use in Eq. (10-99):

$$L(L; J) = \frac{J^2 + L^2 - S^2}{2J}$$

$$S(S; J) = \frac{S^2 + J^2 - L^2}{2J}$$

These substitutions into Eq. (10-99) give

$$2\left(\frac{S^2 + J^2 - L^2}{2J}\right) + \frac{J^2 + L^2 - S^2}{2J} = Jg \tag{10-101}$$

Equation (10-101) reduces to

$$g = 1 + \frac{J^2 + S^2 - L^2}{2J^2} \tag{10-102}$$

Equation (10-102) is further simplified by recalling that $S = [S(S + 1)]^{1/2}$, $L = L[(L + 1)]^{1/2}$, and $J = [J(J + 1)]^{1/2}$, and their substitution results in

$$g = 1 + \frac{J(J + 1) + S(S + 1) - L(L + 1)}{2J(J + 1)} \tag{10-103}$$

For spin alone, $S = J = \frac{1}{2}$, $L = 0$, and $g = 2$. For orbital motion only, $L = J = 1$, $S = 0$, and $g = 1$. Thus, the results obtained here agree with those obtained from Eq. (10-93c).

It should be noted the precession of J about an axis parallel to H (Fig. 10-11) is much more likely than the simplest state that is depicted in Fig. 10-10. Thus, J precesses around an axis parallel to the field, and its component parallel to the field, J_z, is allowed $2J + 1$ values. This fact is of greater utility in ferromagnetism than in paramagnetism.

The preceding analysis is consistent with Eq. (10-93c). This can be shown by dividing Eq. (10-98) by Eq. (10-99). This results in

$$\frac{\mu_j}{Jg} = \frac{(eh/2mc)[2S(S; J) + L(L; J)]}{[2S(S; J) + L(L; J)]} = \frac{eh}{2mc} \tag{10-104}$$

And, upon rearrangement of Eq. (10-104) and using Eq. (10-93c),

$$\mu_j = g\mu_B J \tag{10-105a}$$

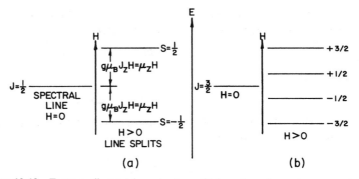

Figure 10-12 Zeeman effect: (a) for spin alone; (b) for spin and angular momentum.
Note that $2J + 1$ states are allowed.

or, more generally, as

$$\mu_z = g\mu_B J_z \tag{10-105b}$$

The splitting of spectral lines by a magnetic field into other sets of new lines (the Zeeman effect) is shown in Fig. 10-12. The energy difference between the new lines is

$$\Delta E = 2|\mu_J|H = 2|g\mu_B J_z|H, \quad \text{erg} \cdot \text{Oe}^{-1} \cdot \text{Oe} \tag{10-106}$$

The use of absolute values in Eq. (10-106) is required because J_z may have $2J + 1$ values centered about the original value. Half of these will be negative with respect to the original value. This also accounts for the factor 2 in Eq. (10-106).

The use of the spectroscopic splitting factor permits wider use of Eq. (10-30) for Larmor precession. This equation now may be written as

$$\omega_L = -g\,\frac{eH}{2mc} \tag{10-107}$$

And, since $g = 1$ for orbital momentum, Eq. (10-107) is the same as Eq. (10-30) for this case. The use of Eq. (10-87) permits the reexpression of Eq. (10-107) as

$$\omega_L = \frac{g\mu_B}{h}\,H \tag{10-108}$$

and the gyromagnetic ratio is given by

$$\gamma = \frac{g\mu_B}{h} \tag{10-109}$$

This enables spin to be taken into account as well as "orbital" motion. Paramagnetic resonance, or electron spin resonance, takes place when oscillating fields, with frequencies determined by Eq. (10-108), cause energy absorptions that induce electron transitions.

Except for diamagnetic materials, electron spin resonance occurs in all other materials. This includes those materials in which the spins interact through exchange forces, including ferro-, antiferro- and ferrimagnetic materials. The influence of this behavior is discussed in Section 10.5.3 for ferromagnetic materials.

10.1.3.4 Compounds of paramagnetic ions. Paramagnetic behavior is shown by compounds composed of ions whose electrons are unpaired, that is, unbalanced. Such electrons separate into two groups in the presence of a magnetic field, those with positive and those with negative magnetic moments, in a way analogous to that indicated in Fig. 10-12.

The fraction of ions with positive momenta is given by

$$\frac{N_+}{N} = \frac{N_+}{N_+ + N_-} = \frac{\exp x}{\exp x + \exp -x} \tag{10-110}$$

in which $x = \Delta E / k_B T = g\mu_B J H / k_B T$, where ΔE is obtained from Eq. (10-106), and similarly

$$\frac{N_-}{N} = \frac{\exp -x}{\exp x + \exp -x} \tag{10-111}$$

The difference between N_+ and N_- is a measure of the momentum imbalance. This is expressed as

$$N_+ - N_- = N \frac{e^x}{e^x + e^{-x}} - N \frac{e^{-x}}{e^x + e^{-x}}$$

and reduces to

$$N_+ - N_- = N \frac{e^x - e^{-x}}{e^x + e^{-x}} \tag{10-112}$$

The fraction in Eq. (10-112) is an expression for tanh x. Thus,

$$N_+ - N_- = N \tanh x \tag{10-113}$$

And, since each ion has a magnetic moment as given by Eq. (10-106), the magnetization of the solid is given by

$$M = (N_+ - N_-)\mu_z = N J g |\mu_B| \tanh x \tag{10-114}$$

For the same set of conditions as were employed to obtain Eqs. (10-79a) and (10-79b), it follows that $x \ll 1$. This simples the situation given by Eq. (10-114) because, for $x \ll 1$, tanh $x \cong x$. Thus,

$$M = (N_+ - N_-)\mu_z = N J g |\mu_B| x = N J g |\mu_B| \frac{g|\mu_B| J H}{k_B T} \tag{10-115}$$

And, for spin alone, where $J = \frac{1}{2}$ and $g = 2$, Eq. (10-115) reduces to

$$M = \frac{N \mu_B^2 H}{k_B T} \tag{10-116}$$

as in Fig. 10-13, and the susceptibility is

$$\chi_p = \frac{M}{H} = \frac{N \mu_B^2}{k_B T} \tag{10-117}$$

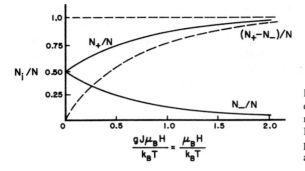

Figure 10-13 Ratios of spin momenta of electrons on ions ($g = 2$, $J = \frac{1}{2}$) with respect to an external magnetic field. Ions designated by N_+/N have spins parallel to the field and N_-/N are antiparallel.

It was found, as a result of the classical Langevin theory [Eq. (10-83] that

$$\chi_p = \frac{N\mu_m^2}{3k_BT} \tag{10-83}$$

Equating Eqs. (10-117) and (10-83) gives

$$\mu_B^2 = \frac{\mu_m^2}{3} \tag{10-118}$$

The relationship given by Eq. (10-118) can be explained in the following way. Let the effective number of Bohr magnetons be n_B. Then Eq. (10-118) may be reexpressed as

$$n_B^2\mu_B^2 = \mu_m^2 \tag{10-119}$$

It will be shown later [Eq. (10-127b)] that for spin alone, $n_B^2 = 3$, so

$$\frac{N\mu_m^2}{3k_BT} \equiv \frac{N\mu_B^2}{k_BT} \tag{10-120}$$

and Eq. (10-120) is another illustration of the Bohr correspondence principle.

10.1.3.5 Total quantum number.   The importance of J is that it is the vector sum of the orbital and spin quantum numbers [Eq. (10-92c)]. As such, all magnetic properties can be shown to be dependent on J. This will be demonstrated by showing M, χ_P and n_B to be functions of J.

Starting with the potential energy of an ion [Eq. (10-106)]

$$\text{PE} = -\mu_J H \tag{10-121}$$

where

$$\mu_J = g\mu_B J \tag{10-105a}$$

gives the magnetic energy of a single ion as $J \cdot g\mu_B H$. These relationships are used in the Boltzmann averaging method, for N ions in a unit volume, to obtain its probable magnetization, by summing over all discrete orientations of J, as

$$M \cong N \frac{\sum_{-J}^{J} J g \mu_B \exp(J \cdot g \mu_B H / k_B T)}{\sum_{-J}^{J} \exp(J \cdot g \mu_B H / k_B T)} \tag{10-122}$$

or as

$$M \cong N g \mu_B \frac{\sum_{-J}^{J} J \exp(J\beta)}{\sum_{-J}^{J} \exp(J\beta)} \tag{10-123}$$

where

$$\beta = \frac{g \mu_B H}{k_B T} \tag{10-123a}$$

Where $J \cdot \beta < 1$, the situation for weak fields and/or high temperatures, the series approximation $\exp J\beta \cong 1 + J\beta$ may be used. This enables Eq. (10-122) to be written as

$$M \cong N g \mu_B \frac{\sum_{-J}^{J} J(1 + J\beta)}{\sum_{-J}^{J}(1 + J\beta)} \tag{10-124}$$

The indicated summations are performed. In the case of the numerator, only terms of the form $(J - n_i)^2$ remain, since those involving J vanish. So

$$\sum_{-J}^{J} J(1 + J\beta) = 2 \sum_{-J}^{J} J^2 \beta = 2 \cdot \frac{J}{6}(J + 1)(2J + 1) \cdot \beta \tag{10-125}$$

In summing the denominator, all terms involving J vanish and the summation from $-J$ to J is just $(2J + 1) \cdot 1$. These results are substituted into Eq. (10-124) along with Eq. (10-123a) to obtain

$$M \cong \frac{N g^2 \mu_B^2 H}{k_B T} \cdot \frac{J(J + 1)(2J + 1)}{3(2J + 1)} \tag{10-126a}$$

$$M \cong \frac{N g^2 \mu_B^2 H}{k_B T} \cdot \frac{J(J + 1)}{3} \tag{10-126b}$$

It will be noted that when spin alone is considered, Eq. (10-126b) reduces to Eq. (10-116). And the susceptibility is

$$\chi_P = \frac{N g^2 \mu_B^2}{3 k_B T} \cdot J(J + 1) \tag{10-127}$$

Another important result may be obtained from Eq. (10-126b) by referring to Eq. (10-105b). From these equations, it is seen that

$$\mu_J^2 = g^2 J(J + 1) \mu_B^2 \tag{10-127a}$$

from which

$$n_B^2 = \frac{\mu_J^2}{\mu_B^2} = g^2 J(J + 1) \tag{10-127b}$$

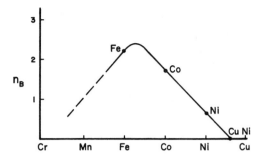

Figure 10-14 Effective number of Bohr magnetons of the first transition series as a function of atomic number.

and the effective number of Bohr magnetons is obtained as

$$n_B = g[J(J + 1)]^{1/2} \qquad (10\text{-}127c)$$

This is a very useful relationship. See Eq. (10-120). This sometimes is designated as n_{eff}.

The ferromagnetic transition elements are such that the "orbits" of their outer electrons are strongly coupled to the electric fields set up by the atoms in their cubic lattices. In the presence of a magnetic field, this very strong electron–lattice interaction makes it impossible for their orbital magnetic moments to become aligned with the field, a condition where the electrons are considered to be quenched. The spins are virtually unaffected because of the weak L–S coupling. The result of this is that the L component of J is zero [Eq. (10-92c)], so $J = L + S = 0 + S$ and the effective number of bohr magnetons is

$$n_B = g[S(S + 1)]^{1/2} \qquad (10\text{-}128)$$

This is shown in Fig. 10-14. It will also be remembered that the hybridized s–d electrons of these elements affects n_B (Section 7.1.18.2, 7.1.21, and 7.1.23). And, as noted in Section 10.1.3.2, since J cannot be calculated for ions in a solid, n_B cannot be calculated, but must be determined experimentally.

10.1.3.6 The Brillouin function. The analysis given in the previous section, for the lower range of β [Eq. (10-128)], is based on weak fields and/or high temperatures. The presentation given here is for the upper range of β, that is, for strong fields and/or low temperatures.

Equation (10-123) is given again, for convenience:

$$M = Ng\mu_B \frac{\sum_{-J}^{J} J \exp(J\beta)}{\sum_{-J}^{J} \exp(J\beta)}, \qquad \beta = \frac{g\mu_B H}{k_B T} \qquad (10\text{-}123)$$

It will be noted that the numerator of Eq. (10-123) is the derivative of its denominator, so it may be reexpressed as

$$M = Ng\mu_B \frac{d}{d\beta} \left[\ln \sum_{-J}^{J} e^{J\beta} \right] \qquad (10\text{-}129)$$

The summation in Eq. (10-129) is expanded as a series and factored to give

$$M = Ng\mu \frac{d}{d\beta} \ln\{e^{J\beta}[1 + e^{-\beta} + e^{-2\beta} + \cdots + e^{-2J\beta+2\beta} + e^{-2J\beta+\beta} + e^{-2J\beta}]\}$$

(10-130)

The geometric series is summed to obtain upon simplification

$$M = Ng\mu_B \frac{d}{d\beta}\left(\ln e^{J\beta} \cdot \frac{1 - \exp[(2J+1)\beta]}{1 - \exp(-\beta)}\right) = Ng\mu_B \frac{d}{d\beta}[\ln f(J\beta)] \quad (10\text{-}131)$$

The indicated multiplication is performed and the quantity in the brackets is

$$\ln f(J, \beta) = \ln \frac{e^{J\beta} - e^{-(J+1)\beta}}{1 - e^{-\beta}}$$

(10-132)

Equation (10-132) can be converted to a more useful form when its numerator and denominator are multiplied by $\exp(\beta/2)$ to obtain

$$\ln f(J, \beta) = \ln \frac{e^{(J+1/2)\beta} - e^{-(J+1/2)\beta}}{e^{\beta/2} - e^{-\beta/2}}$$

(10-133)

The numerator and denominator of Eq. (10-133) are of the form $2 \sinh x$, so it may be reexpressed as

$$\ln f(J, \beta) = \ln \frac{\sinh[(J+1/2)\beta]}{\sinh(\beta/2)}$$

(10-134)

where the factor 2 vanishes. Equation (10-134) is rewritten for convenience as

$$\ln f(J, \beta) = \ln \sinh\left[\left(J + \frac{1}{2}\right)\beta\right] - \ln \sinh \frac{\beta}{2}$$

(10-134a)

and the derivative is taken as

$$\frac{d}{d\beta}[\ln f(J, \beta)] = \frac{(J+1/2)\cosh[(J+1/2)\beta]}{\sinh[(J+1/2)\beta]} - \frac{1/2 \cosh(\beta/2)}{\sinh(\beta/2)}$$

(10-135)

or simplified as

$$\frac{d}{d\beta}[\ln f(J, \beta)] = \left(J + \frac{1}{2}\right)\coth\left[\left(J + \frac{1}{2}\right)\beta\right] - \tfrac{1}{2}\coth\frac{\beta}{2}$$

$$= \frac{2J+1}{2}\coth\frac{2J+1}{2}\beta - \tfrac{1}{2}\coth\frac{\beta}{2}$$

(10-136)

Now let

$$a' = J\beta$$

(10-137a)

$$\frac{d}{d\beta}[\ln f(J, \beta)] = \frac{2J+1}{2}\coth\frac{2J+1}{2J}a' - \tfrac{1}{2}\coth\frac{a'}{2J}$$

(10-137b)

Equation (10-137b) is multiplied by J/J to obtain

$$\frac{d}{d\beta}\left[\ln f(J, \beta)\right] = J\left[\frac{2J+1}{2J}\coth\left(\frac{2J+1}{2J}\,a'\right) - \frac{1}{2J}\coth\left(\frac{a'}{2J}\right)\right] \quad (10\text{-}138a)$$

Equation (10-138a) may be rewritten as

$$\frac{d}{d\beta}\left[\ln f(J, \beta)\right] = JB_J(a') \quad (10\text{-}138b)$$

where $B_J(a')$ is the Brillouin function.   Equation (10-129) now becomes

$$M = Ng\mu_B JB_J(a') = Ng\mu_B JB_J\left(\frac{Jg\mu_B H}{k_B T}\right) \quad (10\text{-}139)$$

It will be noted that $Ng\mu_B J$ is the maximum possible magnetization, or saturation, of the unit volume.   Thus, Eq. (10-139) may be written as

$$M = M_0 B_J(a') = M_0 B_J\left(\frac{Jg\mu_B H}{k_B T}\right) \quad (10\text{-}140)$$

where M_0 is the maximum saturation magnetization and the Brillouin function gives the fractions of saturation, M/M_0, expected for variations in H and T.

In the case in which J is very large, where all allowed values of J are present, the first term of Eq. (10-138a) reduces to $\coth a'$.   An approximation of the second term of Eq. (10-138a) may be obtained, for large J, from the first term of the series.

$$\coth a' = \frac{1}{a'} + \frac{a'}{3} - \frac{(a')^3}{45}\cdots \quad (10\text{-}141a)$$

This gives the second term of Eq. (10-138a) as $1/2J \cdot 2J/a' \cong 1/a'$.   Thus, under this condition, $B_J(a')$ approaches $L(a)$ [Eq. (10-78b)]:

$$B_J(a') \cong \coth a' - \frac{1}{a'} = L(a') \quad (10\text{-}141b)$$

Thus, Eq. (10-141b) also conforms to the Bohr correspondence principle.

For the case of strong fields and/or high temperatures, it may be approximated that the first term in Eq. (10-138a) reduces to $\coth a' \cong a'$ and that its second term is negligible.   This results in simplifying Eq. (10-139) as

$$M \cong Ng\mu_B J \cdot a' = Ng\mu_B J \cdot \frac{Jg\mu_B H}{k_B T} \quad (10\text{-}142a)$$

or

$$M \cong \frac{Ng^2\mu_B^2 J^2 H}{k_B T} \quad (10\text{-}142b)$$

And, where $J = S = \frac{1}{2}$, Eq. (10-142b) is reduced to Eq. (10-116).

10.1.3.7 Pauli paramagnetism.   The behavior of the "nearly free" valence electrons of normal metals is the basis for the paramagnetic properties of metals. In this case, χ_P is not affected by temperature, while that of other materials, noted in previous sections, is an inverse function of temperature. This behavior, first explained by Pauli, is known as *Pauli paramagnetism*.

The application of a magnetic field affects only those electrons with energies in the neighborhood of E_F. Using the approximation previously used in Section 5.1.9, the very small fraction of such electrons may be approximated by T/T_F. The application of this fraction to Eq. (10-117) gives

$$\chi_P \cong \frac{N\mu_B^2}{k_B T} \cdot \frac{T}{T_F} = \frac{N\mu_B^2}{k_B T_F} \tag{10-143}$$

a relationship independent of temperature for all practical purposes. In the absence of an external magnetic field, the electrons with opposite spins may be considered to balance each other as indicated by the broken curves of Fig. 10-15. When a magnetic field is applied, the bands split according to Eq. (10-106) and in a way analogous to that shown in Fig. 10-12(a). This is incorporated into the solid curves of Fig. 10-15. In effect, those electrons with spins parallel to the field are lowered in energy, with respect to the antiparallel electrons, by an amount equal to $2\mu_B H$. This spin imbalance is responsible for the behavior noted previously and is taken into account in the following way.

For a given unit volume, at ordinary temperatures, the number of spins parallel to the field is approximately given as

$$N_+ \cong \tfrac{1}{2} \int_0^{E_F} f \cdot N(E + \mu_B H)\, dE = \tfrac{1}{2} \int_0^{E_F} f \cdot N(E)\, dE + \tfrac{1}{2}\mu_B H N(E_F) \tag{10-144a}$$

and the number of antiparallel electrons is given similarly by

$$N_- \cong \tfrac{1}{2} \int_0^{E_F} f \cdot N(E)\, dE - \tfrac{1}{2}\mu_B H N(E_F) \tag{10-144b}$$

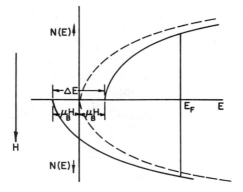

Figure 10-15 Effect of an external magnetic field on the densities of spin states of electrons in a normal metal.

The lower limit of Eq. (10-144a) should have been $-\mu_B H$ and that of Eq. (10-144b) should have been $\mu_B H$. The lower limit of zero, used in each of these equations for simplification, is permissible because the number of states between zero and $\pm \mu_B$ is negligibly small when compared to the total number of states.

The change in the number of states in each half-band is the difference between Eqs. (10-144a) and (10-144b). Or

$$\Delta N = (N_+ - N_-) = \mu_B H N(E_F) \tag{10-145}$$

The magnetic field has diminished the number of antiparallel electrons by an amount ΔN and has increased the number of parallel electrons by the same amount. Thus, the resulting magnetization for the unit volume is

$$M = 2\Delta N \mu_B = 2N(E_F)\mu_B^2 H \tag{10-146}$$

because $2\Delta N$ [Eq. (10-145)] is the spin imbalance for the unit volume under consideration.

The remaining task is to obtain a suitable expression for $N(E_F)$ in Eq. (10-146). This is accomplished by starting with Eq. (4-5) given in the range $E \cong E_F$ as

$$N(E_F) = \frac{2\pi}{h^3}(2m)^{3/2}E_F^{1/2} \tag{4-5}$$

since unit volume is being considered, and $V = 1$. And, from Eq. (4-5),

$$E_F^{1/2} = \frac{h}{2^{3/2}m^{1/2}}\left(\frac{3N}{\pi}\right)^{1/3} \tag{10-147}$$

The inclusion of Eq. (10-147) in Eq. (4-5) gives

$$N(E_F) = \frac{2\pi}{h^3}(2m)^{3/2} \cdot \frac{h}{2^{3/2}m^{1/2}}\left(\frac{3N}{\pi}\right)^{1/3}$$

and reduces to

$$N(E_F) = \frac{2\pi m}{h^2}\left(\frac{3N}{\pi}\right)^{1/3} \tag{10-148}$$

The substitution of Eq. (10-148) into Eq. (10-146) gives the result being sought:

$$M = 2 \cdot \frac{2\pi m}{h^2}\left(\frac{3N}{\pi}\right)^{1/3}\mu_B^2 H \tag{10-149}$$

The numerator and denominator of Eq. (10-149) are multiplied by E_F, using the square of Eq. (10-147) in the numerator, to obtain

$$M = \frac{4\pi m}{h^2}\left(\frac{3N}{\pi}\right)^{1/3} \cdot \mu_B^2 H \cdot \frac{(h^2/8m)(3N/\pi)^{2/3}H}{E_F} = \frac{3N\mu_B^2 H}{2E_F} \tag{10-150}$$

and, from this,

$$\chi_P = \frac{3N\mu_B^2}{2E_F} \tag{10-151}$$

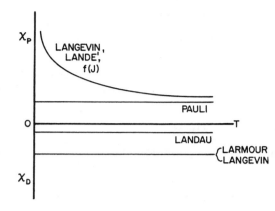

Figure 10-16 Schematic diagram of paramagnetic and diamagnetic susceptibilities as functions of temperature. The susceptibility scales on each side of zero have been expanded for clarity.

It will be noted that, while the Pauli paramagnetic susceptibility of metals is quite small, it is three times larger than the Landau diamagnetic susceptibility as given by Eq. (10-63). Corrections for the very small effects of temperature may be made by use of Eq. (2-84). This weak spin paramagnetism is smaller than might have been expected because only about 1% of the valence electrons are affected by the field (Sections 5.1.10.1 and 8.5.1).

The various types of diamagnetic and paramagnetic behaviors discussed thus far are plotted schematically in Fig. 10-16 for purposes of comparison.

10.1.3.8 Paramagnetic cooling. Temperatures very close to absolute zero have been obtained by the adiabatic magnetization and demagnetization of paramagnetic materials, usually salts. The magnetic moments of such materials may be considered to be virtually random in the absence of a magnetic field. This randomness is described by the entropy, S, as given by Boltzmann's fundamental equation, originally given as Eq. (2-17):

$$S = k_B \ln W \qquad (10\text{-}152)$$

in which W is the number of different ways N ions of a given specimen can have their electrons distributed throughout all the permitted $2J + 1$ states allowed to each ion (Sections 2.1, 10.1.3.2 and 10.1.3.3). Thus, the number of ways that the electrons of N ions are in these states is

$$W = (2J + 1)^N \qquad (10\text{-}153)$$

This permits Eq. (10-152) to be written as

$$S = Nk_B \ln(2J + 1) \qquad (10\text{-}154)$$

At saturation magnetization, the magnetic moments reach their approach to maximum parallel alignment with the field and the magnetic entropy is minimized. At cryogenic temperatures (usually that of liquid helium), the disordering effects of lattice vibrations on the magnetic dipoles may be approximated as being negligible, so the magnetic entropy may be taken as zero. Under these conditions, the change in entropy as a result of the application of the field may be approximated, by using Eq. (10-154), as

$$|\Delta S| \cong S - 0 = Nk_B \ln(2J + 1) \qquad (10\text{-}155)$$

The decrease in entropy corresponds to an ordered behavior that is representative of a lower temperature. This magnetic alignment therefore releases thermal energy. The quantity of thermal energy is given by

$$\Delta Q = T\Delta S \qquad (10\text{-}156)$$

This increment of heat must be removed through the lattice of the material to the coolant of the cryostat. When this heat is removed and thermal equilibrium is reached, the coolant is removed and the field is turned off under adiabatic conditions. The magnetic moments again assume their virtually random distribution and their original high magnetic entropy, a condition representative of a higher temperature. The only available thermal energy source for this is the lattice of the sample. The vibrational energy of the lattice is thus drained to the electrons by the required increase in their magnetic dipole entropy and the lattice cools. The changes in the entropy of the lattice are relatively small because of the relatively good crystalline alignment and long phonon wavelengths at cryogenic temperatures.

Repeated cycles of this kind are used to obtain temperatures of $\sim 10^{-3}\,\mathrm{K}$. It does not appear to be possible to achieve much lower temperatures by this method alone, since perfect order of the magnetic dipoles cannot be achieved as a result of dipole–dipole interactions, reactions with nuclear dipoles, and limitations arising from the fact that the complete randomness of Eq. (10-153) cannot be attained.

10.1.3.9 Paramagnetic materials. Materials consisting of ions with completed electron levels are expected to be diamagnetic because the orbital and spin components of their electrons are balanced. They have no net magnetic moment. Where the electron levels are incompletely filled, a magnetic moment will result from imbalance of electron spin and/or orbital components.

The metal ions in salts of transition metals are responsible for the magnetic properties of these materials. Here the effect is almost entirely a result of spin because the orbital effects appear to be quenched by the crystal fields [Eq. (10-128)] and $J = 0 + S$. This is verified by the experimental determinations of the Landé factors with values close to 2. Most of these materials show typical Curie law behavior [Eq. (10-84)] or a modification of the more general *Curie–Weiss* law [Eq. (10-166)], which is given as

$$\chi_P = \frac{C}{T - \theta} - \alpha \qquad (10\text{-}157)$$

in which θ is the Weiss constant and α is a constant, independent of temperature, that includes diamagnetic effects. The values of θ usually are small and increase slightly as a function of the magnetic ion concentration. It is of interest to note that even though the bases for the Curie law [Eqs. (10-83) and (10-117)] are that of freely moving, noninteracting ions, the law does hold, to a great extent, for such closely packed crystalline solids as carbonates, oxides, salts, and sulfates of the first transition group of elements in which the orbital components are quenched by the crystal fields.

The compounds of the rare earth elements are highly paramagnetic and, with the exceptions of samarium and europium, closely follow Eq. (10-157) in the form of Eq. (10-127). These compounds, unlike those of the transition elements, require that the orbital component be taken into account so that $J = L + S$. The reason for this is that the 4f electrons are very effectively screened by those in the outer 5s and 5p states. This screening counteracts the quenching effects of the crystal field, and the orbital components in the 4f levels remain unaffected.

The magnetic properties of normal metals are the resultant of three effects. The first of these is diamagnetic and arises from the closed shells of the ion cores (Section 10.1.2). Another diamagnetic component is a result of the motion of the nearly free valence electrons (Section 10.1.2.2). A third component is that of the Pauli paramagnetism (Section 10.1.3.7). The metal will be diamagnetic or paramagnetic depending on the relative magnitudes of the diamagnetic and paramagnetic components. When the sum of these is positive, weak paramagnetism results. Materials showing this behavior do not obey Eqs. (10-84) or (10-157) because Eqs. (10-38) and (10-63) are independent of temperature, and Eq. (10-151) is virtually independent of temperature. In a similar way, small quantities of ferromagnetic impurities present in either diamagnetic or paramagnetic materials can swamp out their inherent properties.

10.1.4 Molecular Field Theory

It was noted in Section 10.1.3.1 that the Curie law [Eq. (10-84)] is based on the postulations that the magnetic dipoles are free to rotate in a magnetic field, are noninteractive and that their alignment is diminished with increasing temperatures, by increasing amplitudes of ionic oscillations, until their array becomes random. However, many paramagnetic materials do not obey Eq. (10-84). P. Weiss (1907) assumed that in some cases the magnetic dipoles were not completely free to rotate in a field because of a spontaneous, cooperative interaction between the dipoles of which the material is composed. This interaction was considered by Weiss to arise from a "molecular" field, actually a mathematical construct that is the equivalent of an inherent internal field, which acts on and aligns the electron spins. These molecular fields were assumed to exist in small, spontaneously magnetized volumes, or *domains*, within the material. Each such domain consists of a volume in which all the magnetic dipoles are aligned in a given direction. The molecular field within a domain was considered to result from the magnetizations of the surrounding domains in an unspecified way. In present-day terminology, the molecular field would be known as an atomic or an ionic field because of the structures of metals and of the inorganic compounds to which the theory is applied.

The Weiss molecular field theory is based on the postulation that the forces tending to align the magnetic moment on an ion are proportional to the degree to which the ions within the volume that surround it are also aligned. This cooperative effect of the magnetic dipoles is considered to constitute the internal molecular field that is given by

$$H_m = \lambda M \qquad (10\text{-}158)$$

in which λ is the molecular field constant and H_m is the molecular field. This internal field can be very large, being $\sim 10^7$ Oe for iron, for example. This is a magnetic field intensity that is very much greater than any field produced experimentally up to the present time. The total field acting on a ferromagnetic material, then, is given by

$$H_T = H + \lambda M = H + H_m \qquad (10\text{-}159)$$

It was shown that

$$M = Ng\mu_B B_J(a') \qquad (10\text{-}139)$$

in which, from Eqs. (10-123a) and (10-137a),

$$a' = J\beta = \frac{Jg\mu_B H}{k_B T} \qquad (10\text{-}139a)$$

In ferromagnetic materials, the field intensity is given by H_T [Eq. (10-159)], rather than by H alone. Thus, for this case, a' in Eq. (10-142a) becomes

$$a' = \frac{Jg\mu_B (H + \lambda M)}{k_B T} \qquad (10\text{-}160)$$

Since ferromagnetic materials are magnetized spontaneously, $H = 0$, and Eq. (10-160) becomes

$$a' = \frac{Jg\mu_B \lambda M}{k_B T} \qquad (10\text{-}161)$$

and their magnetizations are given by rearranging Eq. (10-161) as

$$M = \frac{a' k_B T}{Jg\mu_B \lambda} \qquad (10\text{-}162)$$

The magnetization as a function of temperature may be obtained by the graphical simultaneous solution of Eqs. (10-139) and (10-162), as shown in Fig. 10-17. The degree of spontaneous magnetization increases as $T_c - T$ increases. Here T_c is the Curie temperature, that is, the temperature at and above which the ferromagnetism vanishes (the magnetic dipoles become random) and the material becomes paramagnetic. The relative magnetization, M_T/M_0, is normalized with respect to T_c in Fig. 10-18.

The paramagnetic properties of these materials can be described by starting with Eq. (10-126) for $T > T_c$. This gives the magnetization directly, because the molecular

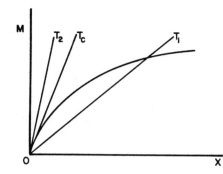

Figure 10-17 Method for the graphical solutions of the Brillouin function, Eqs. (10-139) and (10-162).

field is zero, as

$$M \cong \frac{Ng^2\mu_B^2 H}{3k_B T} \cdot J(J + 1), \qquad T > T_c \tag{10-126b}$$

Their ferromagnetic properties are given by the use of H_T, rather than by H. This results in

$$M \cong \frac{Ng^2\mu_B^2(H + \lambda M)}{3k_B T} \cdot J(J + 1), \qquad T < T_c \tag{10-163}$$

Equation (10-163) is of the form

$$M \cong \frac{C(H + \lambda M)}{T} \tag{10-164}$$

in which all the constants are included in the constant C:

$$C = \frac{Ng^2\mu_B^2 J(J + 1)}{3k_B} = \frac{Nn_B^2\mu_B^2}{3k_B} \tag{10-164a}$$

when Eq. (10-127b) is used.

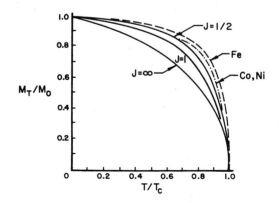

Figure 10-18 Relative saturation magnetization of Fe, Co, and Ni with respect to temperature, normalized with respect to the Curie temperatures. (From R. M. Bozorth, *Ferromagnetism*, p. 431, Van Nostrand Reinhold Co., New York, 1951.)

Equation (10-164) may be rearranged as

$$M(T - C\lambda) = CH \tag{10-164b}$$

Then, for ferromagnetic materials at $T > T_c$, in the paramagnetic region,

$$\chi_P = \frac{M}{H} = \frac{C}{T - C\lambda} = \frac{C}{T - T_c} \tag{10-165}$$

For paramagnetic materials, Eq. (10-165) is usually written as

$$\chi_P = \frac{C}{T - \theta} \tag{10-166}$$

Equations (10-165) and (10-166) are forms of the Curie–Weiss law. Where $\theta = 0$, Eq. (10-166) reduces to the Weiss law [Eq. (10-84)].

The Curie temperature may be obtained for ferromagnetic materials from Eqs. (10-164a) and (10-165) as

$$T_c = C\lambda = \frac{Ng^2\mu_B^2 J(J + 1)\lambda}{3k_B} = \frac{Nn_B^2\mu_B^2\lambda}{3k_B} \tag{10-167}$$

when treated in the same way as Eq. (10-164a). Negative values of T_c are indications of antiferromagnetic behavior in which the magnetic dipoles are in antiparallel alignments.

It can be seen (Fig. 10-18) that, as T approaches T_c, T/T_c approaches unity and the magnetization approaches zero. This is a result of the randomizing effects of increasing temperature. The increasing amplitudes of the lattice oscillations disrupt the good lattice arrays that are necessary for the effective, cooperative interactions that give rise to the molecular fields. This temperature-induced disorder is explained by the temperature variations of the Brillouin function [Eq. (10-138b)].

10.1.5 Variation of the Brillouin Function with Temperature

The temperature variation of the Brillouin function [Eq. (10-139)] is given, for convenience, using the more commonly used notation of $B_J(x)$ instead of $B_J(a')$ as in Eq. (10-139):

$$M = Ng\mu_B J B_J(x) \tag{10-139b}$$

in which

$$B_J(x) = \frac{2J + 1}{2J} \coth\left(\frac{2J + 1}{2J} x\right) - \frac{1}{2J} \coth\frac{x}{2J} \tag{10-168}$$

and where

$$x = a' = J\beta = J\frac{g\mu_B H}{k_B T} \quad \text{or} \quad T = \frac{g\mu_B H}{k_B x} \tag{10-169}$$

from Eqs. (10-137a) and (10-123a), respectively.

For maximum magnetization, $B_J(x)$ must approach unity in Eq. (10-139a) because the maximum magnetization is $Ng\mu_B J$. This can be shown by examining the hyperbolic functions in Eq. (10-168). Maximum magnetization means that J and x must be large so that, in this case,

$$\coth\left(\frac{2J+1}{2J}x\right) \cong \coth x \tag{10-170}$$

This approximation reduces Eq. (10-168) to

$$B_J(x) \cong \frac{2J+1}{2J}\coth x - \frac{1}{2J}\coth x \tag{10-171}$$

And this is further simplified by collecting terms to get

$$B_J(x) \cong \frac{2J+1-1}{2J}\coth x \cong \coth x \tag{10-172}$$

The limiting value of $B_J(x)$ now is obtained for very large x by noting that the coefficient is equal to unity and by using the exponential equivalent of $\coth x$ to obtain

$$B_J(x) \cong \frac{e^x + e^{-x}}{e^x - e^{-x}} \tag{10-173}$$

from which it is found that

$$\lim_{x\to\infty} B_J(x) = 1 \tag{10-174}$$

This is so because $e^{-x} \to 0$ for $x \to \infty$ and the resulting fraction $e^x/e^x \to 1$. And, in addition, since $x \to \infty$, $T \to 0\,\text{K}$ [Eq. (10-169)]. Thus, the magnetization is maximum at $T = 0\,\text{K}$, and the alignment of the dipoles is maximum.

The effect of increasing temperature is obtained by examining Eq. (10-169) for increasing T. This is the same as considering $x \to 0$. Here, Eq. (10-171) is evaluated by considering only the $\coth x$ terms in their exponential form and neglecting the coefficients. In this way,

$$B_J(x) \propto \coth = \frac{e^x + e^{-x}}{e^x - e^{-x}} \tag{10-175}$$

The limit of Eq. (10-175) cannot be taken in its present form because, for the case $x \to 0$, $B_J(x) \to 2/0$. This is avoided by the use of L'Hopital's theorem, where the differentiation of the numerator and denominator of Eq. (10-175) gives

$$B_J(x) = \frac{xe^x - xe^{-x}}{xe^x + xe^{-x}} = \frac{e^x - e^{-x}}{e^x + e^{-x}}$$

So that

$$\lim_{x\to 0} B_J(x) = \lim_{x\to 0}\frac{e^x - e^{-x}}{e^x + e^{-x}} = \frac{0}{1+1} = 0 \tag{10-176}$$

Thus, for $x \to 0$, T [Eq. (10-169)] must be large, and the alignment of the dipoles is random.

The essence of this analysis is that Eq. (10-167) is maximum at $T = 0$ K because of Eq. (10-174). And $B_J(x)$ diminishes and approaches zero for large T [Eq. (10-176)], as is shown in Fig. 10-18 for $J = \infty$. Expressed mathematically, since the maximum magnetization is, for $B_J(x) = 1$ [Eq. (10-174)],

$$M_0 = Ng\mu_B J \cdot B_J(x) = Ng\mu_B J \cdot 1 \qquad (10\text{-}177)$$

and, at any other temperature, the magnetization is

$$M_T = Ng\mu_B J B_J(x) \qquad (10\text{-}178)$$

so that the relative magnetization is given by the ratio of Eq. (10-178) to Eq. (10-177); and, since the coefficients $Ng\mu_B J$ vanish,

$$\frac{M_T}{M_0} = B_J(x) \qquad (10\text{-}178a)$$

In the case of ferromagnetic materials, in the absence of an external field, Eq. (10-169) becomes, using Eq. (10-158),

$$x = J \frac{g\mu_B \lambda M}{k_B T} \qquad (10\text{-}179)$$

The numerator and denominator of Eq. (10-179) are multiplied by M_0 to obtain

$$x = \frac{Jg\mu_B \lambda M}{k_B T} \cdot \frac{M_0}{M_0}$$

This is rearranged to get

$$\frac{M_T}{M_0} = \frac{k_B T}{Jg\mu_B \lambda M_0} \cdot x \qquad (10\text{-}180)$$

But, since spin alone need be considered, so that $J = \frac{1}{2}$ and $g = 2$, Eq. (10-180) reduces to

$$\frac{M_T}{M_0} = \frac{k_B T}{\mu_B \lambda M_0} \cdot x \qquad (10\text{-}180a)$$

The slope of the Brillouin function [Eq. (10-141b)] at $x \cong 0$ is found to be $\frac{1}{3}$ from Eqs. (10-141a) and (10-141b). When this is applied to the fraction in Eq. (10-180a), it is found that the slope is

$$\frac{k_B T_c}{\mu_B \lambda M_0} = \frac{1}{3} \qquad (10\text{-}181)$$

since $T = T_c$ for $B_J(x) = 0$ for ferromagnetism. And, by rearrangement,

$$T_c = \frac{\mu_B \lambda M_0}{3k_B} \qquad (10\text{-}182)$$

Equation (10-181) is rearranged again and multiplied through by T to obtain

$$\frac{k_B T}{\mu_B \lambda M_0} = \frac{T}{3T_c} \qquad (10\text{-}183)$$

Thus, Eq. (10-183) represents the slope of the line that gives the molecular field at any given temperature. This is applied to Eq. (10-180a) to find

$$\frac{M_T}{M_0} = \frac{T}{3T_c} x \qquad (10\text{-}183)$$

The plot of Eq. (10-183) depends on T, while that of Eq. (10-178a) does not depend on T. Thus, since $T/3T_c$ is the slope of Eq. (10-183), it will determine the intersection of the line of Eq. (10-183) with $B_J(x)$, as shown in Fig. 10-19. Each such nonzero point of intersection represents a graphical solution of the simultaneous equations given by Eqs. (10-178a) and (10-183) for a given temperature. This means that M_T/M_0 is normalized with respect to T/T_c. The ferromagnetic Curie temperature, T_c, is that temperature at which M_T/M_0 is zero. As a result of this, any ferromagnetic material, regardless of its values of M_0 and T_c, will have the same values of M_T/M_0 for any selected value of T/T_c. This is known as the *law of corresponding states*. It is inexact only to the extent of the temperature variations between materials that cause changes in the numbers of magnetic dipoles per unit volume. Such very small differences can be taken into account by using $3\alpha_L \cong \alpha_V$ (Section 5.2) to determine the volume change and the corresponding change in the number of dipoles in that volume.

The results given by Eq. (10-183) may be obtained in another way. Starting with M_T given by Eq. (10-162) and M_0 given by Eq. (10-177),

$$\frac{M_T}{M_0} = \frac{k_B T x}{J g \mu_B \lambda} \cdot \frac{1}{N g \mu_B J} = \frac{k_B T x}{N g^2 \mu_B^2 J^2 \lambda} \qquad (10\text{-}184)$$

An expression for λ is obtained from Eq. (10-167) as

$$\lambda = \frac{3 k_B T_c}{N g^2 J (J + 1) \mu_B^2} \qquad (10\text{-}185)$$

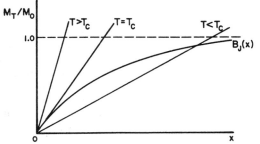

Figure 10-19 Method for graphical solutions for relative magnetizations based on the Brillouin function as given by Eq. (10-178a) with Eq. (10-183).

that is substituted into Eq. (10-184) to obtain

$$\frac{M_T}{M_0} = \frac{k_B Tx}{Ng^2\mu_B^2 J^2} \cdot \frac{Ng^2\mu_B^2 J(J+1)}{3k_B T_c} = \frac{(J+1)Tx}{3JT_c} \tag{10-186}$$

For ferromagnetic materials, $J = \frac{1}{2}$, so Eq. (10-186) reduces to Eq. (10-183). The agreement of Eq. (10-186), for $J = \frac{1}{2}$, is shown in Fig. 10-18. This is a result of the quenching of the orbital momenta by the crystal field. The case for $J = \infty$ is the classical case wherein the Brillouin function approaches the Langevin function [Eq. (10-141b)].

At temperatures considerably lower than T_c, the application of strong external fields adds little to the magnetization because of the large magnitude of the molecular field. This is shown by the intersection of Eq. (10-183) with the relatively flat portion of Eq. (10-178a) in Fig. 10-19.

The high values of M_T for $T < T_c$ in Eq. (10-178) result from the high values of $B_J(x)$, where x is given by Eq. (10-179). This accounts for the great difference in M in the ranges $T < T_c$ and $T \gg T_c$. This is the equivalent of saying that the spins (dipoles) are maximumly aligned in the lower temperature range and can attain only a very small fraction of this degree of alignment in the higher temperature range [Eq. (10-176)] when an external magnetic field is present. This great contrast in spin alignment is the basic difference between ferromagnetic and paramagnetic properties (see Section 10.2).

Above T_c, the ferromagnetic properties change to paramagnetic properties. These may be expressed most simply by the reciprocal of Eq. (10-127) as

$$\frac{1}{\chi_P} = \frac{3k_B T}{Ng^2\mu_B^2 J(J+1)}, \quad T > T_P \tag{10-187}$$

and $T = T_P$ for $1/\chi_P = 0$. The parameter T_P is the *paramagnetic Curie temperature.*

The linearity expressed by Eq. (10-187) does not hold in the relatively small range of temperatures just above T_c, as is shown in Fig. 10-20. This behavior is representative of those of iron, cobalt, nickel, and gadolinium. The temperature intercept of Eq. (10-187) is denoted by T_P. The deviation from linearity in the

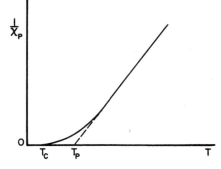

Figure 10-20 Schematic diagram of the reciprocal of the paramagnetic susceptibilities of ferromagnetic materials as a function of temperature. T_P is the paramagnetic Curie temperature (see Sections 7.1.21 and 7.1.21.1). (From A. H. Morrish, *Physical Principles of Magnetism*, p. 269, John Wiley & Sons, New York, 1965.)

temperature range from T_c to T_P is given as

$$\chi_P \propto (T - T_c)^{-4/3} \qquad (10\text{-}188)$$

instead of the variation of $(T - T_c)^{-1}$ given in Eq. (10-165). In the paramagnetic range, T_P is the same as θ in the Curie–Weiss equation [Eq. (10-166)]. The difference between T_P and T_c is usually ~ 30 K or less. This is an indication that the transition between the two magnetic states is gradual. In nickel and some of its alloys, this transition appears to take place over ranges of up to ~ 300 K [Sections 7.1.21 and 7.1.21.1] and can be shown to result from residual electron spin clusters in these ranges.

10.2 FERROMAGNETISM

Ferromagnetic materials, in contrast to diamagnetic and paramagnetic substances, can show strong responses to the application of an external magnetic field. The elements showing the most prominent ferromagnetic properties are Fe, Co, Ni, and Gd. Ferromagnetic properties are also shown by the rare earth elements at low temperatures. Many alloys and compounds of both classes of elements are also ferromagnetic.

Two important properties characterize ferromagnetic materials. The first is that the application of comparatively low applied magnetic fields may induce relatively large magnetizations in these materials. Second, they may show the presence of residual magnetization in the absence of external fields.

The Weiss molecular field theory (Section 10.1.4) is an attempt to explain these properties by means of domains and the strong, internal, molecular field. The magnetization within a domain is the vector sum of the parallel magnetic dipoles within the domain. The resultant of all such vectors in a given specimen gives its magnetization for the conditions to which the specimen is subjected.

The application of a colloidal suspension of magnetic particles to a polished surface of a ferromagnetic material outlines the domains. The high density of the magnetic lines of force attracts the particles to the domain boundaries. They form an outline of all the domains that intersect the surface. Domain boundary changes may be observed by the motion of the colloidal particles when magnetic fields are applied. These are known as *Bitter patterns*, after the inventor of this method (F. Bitter, 1931).

Ferromagnetism vanishes from ferromagnetic materials above T_c and they become paramagnetic [Eq. (10-165)]. Their paramagnetic behavior is given by Eqs. (10-127). It frequently is convenient to reexpress this equation as Eq. (10-187). Here the linear extrapolation of $T > T_c$ to $1/\chi_P \to 0$ gives the parameter T_P, as shown in Fig. 10-20 and Table 10-4.

It appears that T_∞ [Eq. (7-131)] is a more sensitive measure of the complete disappearance of ferromagnetism than is T_P [Eq. (10-187)]. The data given in Section 7.1.21 show that $T_\infty \cong 873$ K as compared to $T_P = 650$ K for nickel in Table 10-4. A similar situation is also present for the case of Alumel (90.4 Ni, 3.0 Mn, 2.4 Si, and

TABLE 10-4 CURIE TEMPERATURES AND EXTRAPOLATED VALUES OF FERROMAGNETIC MATERIALS

Element	T_c (K)[a]	T_P (K)[a]	T_∞ (K)[b]
Fe	1043	1093	1340[c]
Co	1403	1428	1540[d]
Ni	631	650[e]	670[f]
Gd	289	302.5	—

[a]Abstracted from A. H. Morrish, *Physical Principles of Magnetism*, p. 270, Wiley, New York, 1965.
[b]Estimated from heat capacity data.
[c]G. A. Moore and T. R. Shives, *Metals Handbook*, p. 206, American Society for Metals, Metals Park, Ohio, 1961, and R. L. Orr and J. Chapman, *Trans. AIME*, 229, 630, 1962.
[d]A. H. Morrish, *Physical Principles of Magnetism*, p. 272, Wiley, New York, 1965.
[e]See Sections 7.1.20, 7.1.21, and 7.1.21.1.
[f]J. Crangle, *Magnetic Properties of Solids*, p. 7, Edward Arnold, London, 1977.

4.2 Al). This alloy has nominal values of $T_c \cong 443$ K and $T_\infty \cong 773$ K (Section 7.1.21.1). Each of these materials shows that $T_\infty - T_c \gg T_P - T_c$. These large differences are considered to indicate the presence of spin clusters, vestiges of the former domains, at temperatures considerably above that at which small amounts of residual ferromagnetism are detectable by magnetic means.

The great contrast in the magnetic properties of a material in the ranges $T < T_c$ [Eq. (10-163)] and $T > T_\infty$ [Eq. (7-131)] is not due to such factors as μ_j [Eq. (10-105a)] nor to n_B [Eq. (10-127c)] since any changes in these parameters may be considered as being very small compared to those shown by M. The reason for this difference in M is a result of vastly greater extent of spin alignment in the range $T < T_c$ as compared to that for $T > T_\infty$ (see Sections 10.1.5 and 10.2.3). One way of indicating this behavior is given in Fig. 10-21.

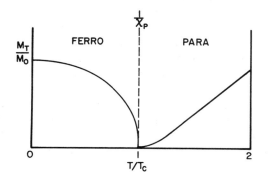

Figure 10-21 Schematic diagram of the relative ferromagnetic and reciprocal paramagnetic properties of ferromagnetic materials as functions of temperature, normalized with respect to the Curie temperature.

10.2.1 Exchange Integral

The molecular field theory (Section 10.1.4) is based on the postulation that the collective alignment of the magnetic dipoles (Sections 10.1.5 and 10.2) is a result of an unspecified, internal magnetic field [Eq. (10-158)]. Attempts to calculate this molecular field by methods such as the classical Lorentz internal field are too small by a factor of $\sim 10^{-3}$. Thus, the internal field cannot be ascribed to simple magnetic dipole interactions.

The modern explanation of the Weiss field is based on the concept of exchange as described in Section 3.12.2. At small separations between two atoms, the overlapping of their probability density functions is such that the valence electrons may no longer be considered to be associated with their original atoms. The electrons now must be thought of as being associated simultaneously with both atoms in a new, lower-energy state as given by Eq. (3-78) for the symmetric case (antiparallel spin). Here the exchange energy was denoted by E_{IS}; this factor will be given by E_x in the following sections. The relatively large decreases in electron energy that result from exchange are described in Section 3.12.3 and some are shown in Table 3-7.

The molecular field postulated by Weiss was "explained" by Heisenberg on the basis of exchange. Two electrons of spins S_i and S_j are postulated to have a potential energy, in units of $h/2\pi$, of

$$E_x = -2J_e S_i S_j \cos \phi_{ij} \qquad (10\text{-}189)$$

in which J_e is the exchange integral and ϕ_{ij} is the angle between their spin vectors. E_x is minimum when J_e is positive and the spins are parallel; $\cos \phi = 1$. E_x is maximum for positive values of J_e when the spins are antiparallel; $\cos \phi = -1$. E_x is also minimum when J_e is negative and the spins are antiparallel; $\cos \phi = -1$. Ferromagnetism arises from parallel spins, so J_e must be positive for this to take place. A rigorous treatment is yet to be given as to why this must be the case for ferromagnetic elements when it is well established that J_e must be negative for the much less complex hydrogen molecule. Negative values of J_e describe the situation for the antiparallel spin configurations of antiferromagnetic and ferrimagnetic materials. Values of $J_e \cong 0$ correspond to diamagnetism.

As can be seen from Eq. (10-189), the exchange energy and consequently the exchange forces arise from nearest-neighbor interactions. Each ion in the solid reacts with its near neighbors in pairs. The sum of all such reactions is the exchange energy. The corresponding summation of forces is the quantum mechanic equivalent of the molecular field. Despite the explanation given by the appropriate summation of Eq. (10-189), it still remains only as a reasonable postulation, since it has yet to explain the properties of Fe, Ni, and Co, to say nothing of the more complex rare earth elements. Despite these inadequacies, the concept of exchange forces has provided a very useful means to help to explain magnetic phenomena.

The concept of exchange, originally used by Heitler and London (1927) to explain covalent bonding (similar to that given in Section 3.12.3), was applied to ferromagnetism by Heisenberg (1928). It is essentially electrostatic in nature and provides a basis for the calculation of E_x of the correct magnitude.

According to the Pauli exclusion principle, only two electrons of opposite spin may occupy a given state. Changes in the relative directions of two interacting spins must result in changing their spatial charge distributions (Section 2.5). One way to picture this is to consider each state as an "orbit". The average "radius" of the "orbit" of an electron with parallel spin will be different from that of one with an antiparallel spin. The differing radial distances in each case cause the coulombic energies to be different. The exchange integral, then, is a measure of the way in which changes in spin orientations affect the exchange energy of the interacting electrons. When a significant overlap of the probability density functions takes place, the symmetrical wave functions require that the two spins occupying a given state must be opposite to each other. A reversal of one of the spins must change the orbit of the other electron. This constitutes a redistribution of the charges and, in effect, correlates the two spins in which their scalar product gives their energy of interaction. This energy is the spin contribution to the total Hamiltonian, H_x. Algebraically, this energy is given, for two z nearest neighbors with parallel spins, as

$$H_x = -2zJ_eS^2 = E_x \tag{10-190}$$

Using Eq. (10-178) and noting that $J \equiv S$ for ferromagnetism, the magnetic moment for one spin, in units of $h/2\pi$, is

$$M = g\mu_B S \tag{10-191}$$

Multiplying both sides of Eq. (10-191) by the molecular field, H_m, gives, since the molecular field and the exchange reaction are equivalent,

$$H_m M = H_m g\mu_B S \tag{10-192}$$

From which it follows that, since the product of H and M is in units of energy,

$$H_m g\mu_B S = k_B T \tag{10-193}$$

So Eqs. (10-190) and (10-192) may be equated. Thus,

$$E_x = -2zJ_eS^2 = -H_m g\mu_B S \tag{10-194}$$

For crystalline solids, Eq. (10-194) must be related to the volume of a unit cell, V_0. This is done by multiplying both sides of Eq. (10-191) by λ and then dividing them by V_0:

$$\frac{H_m}{V_0} = \lambda \frac{M}{V_0} = \lambda \frac{g\mu_B S}{V_0} \tag{10-195}$$

Equation (10-195) is substituted into Eq. (10-194) to give the exchange energy for a unit cell as

$$E_x = -2zJ_eS^2 = -\lambda \frac{g\mu_B S}{V_0} \cdot g\mu_B S = -\lambda \frac{g^2\mu_B^2 S^2}{V_0} \tag{10-196a}$$

Or, since S^2 and V_0 vanish, the exchange energy for a unit cell is

$$E_x = -J_e = -\frac{\lambda g^2\mu_B^2}{2} \tag{10-196b}$$

because $V_0 = f(z)$ and $1/V_0$ corresponds to z nearest neighbors. This results in a simple relationship between the exchange integral and the Weiss molecular field constant given by

$$2J_e = \lambda g^2 \mu_B^2 \tag{10-196c}$$

Equation (10-167) may be reexpressed for the ferromagnetic case, using $J \equiv S$, for the present case in which $N \equiv z$, as

$$T_c = \frac{zg^2 \mu_B^2 S(S+1)}{3k_B} \lambda \tag{10-197}$$

to obtain a relationship between T_c and λ. An expression for λ is found, from Eq. (10-197), to be

$$\lambda = \frac{3k_B T_c}{zg^2 \mu_B^2 S(S+1)} \tag{10-198}$$

This is substituted into Eq. (10-196b) to get

$$J_e = \lambda g^2 \mu_B^2 z/2 = \frac{3k_B T_c}{zg^2 \mu_B^2 S(S+1)} \cdot \frac{g^2 \mu_B^2}{2} = \frac{3k_B T_c}{2zS(S+1)} \tag{10-199}$$

For ferromagnetism, Eq. (10-199) may be reduced, for $S = \frac{1}{2}$, to

$$J_e = \frac{2k_B T_c}{z} \tag{10-199a}$$

This is the same as Eq. (7-131) when the latter is integrated between zero and T_c. It will be noted that Eq. (10-199a) gives results similar to the more rigorous Eq. (7-129). A comparison of these results is given in Table 10-5 in the more useful form of $J_e/k_B T_c = f(z)$. The slightly higher values given by Eq. (7-129) are considered to be more representative than those of Eq. (10-199a).

Another useful result of the exchange theory is that it provides a model with which to explain the magnetic behaviors of the transition elements. Positive values of J_e are used to explain ferromagnetic behavior, while negative values of J_e are considered to represent antiferromagnetic and ferrimagnetic behaviors. H. Bethe postulated that J_e as a function of r_a/r_d could be used to explain magnetic properties,

TABLE 10-5   CALCULATED VALUES FOR $J_e/k_B T_c$

Crystal Type	z	$J_e/k_B T_c$ Eq. (10-199a)	$J_e/k_B T_c$ Eq. (7-129)
Simple Cubic	6	0.33	0.41
BCC	8	0.25	0.29
FCC	12	0.17	0.18

where r_d is the average "radius" of the d shells and r_a is one-half of the distance between the nuclei of two atoms. The d electrons are of interest because their net spin alignment is considered to be the primary cause of magnetic properties.

When two atoms of the same transition species are brought sufficiently close together so that they begin to interact, r_a/r_d reaches a value at which their probability density functions begin to overlap. This results in a small, positive value for J_e [see Eq. (3-78)]. As the atoms are brought more closely together, r_a/r_d decreases, and J_e becomes increasingly larger until a maximum value of J_e is obtained. This occurs when the maximum degree of parallel spin is attained. As the atoms are brought more closely together, their spins must become increasingly more antiparallel (because of the exclusion principle) so that J_e diminishes, goes through zero, and becomes negative. The latter is the case for antiferromagnetism. Thus, ferromagnetism is present when the interionic conditions optimize the overlap of parallel d-level spins. This implies that J_e, and consequently ferromagnetism, is primarily a function of interionic distance. However, this does not necessarily lead to the requirement that ferromagnetic materials be crystalline, since amorphous CoAg alloys are ferromagnetic. This is indicated in Fig. 10-22. This curve gives a good representation of the relative values of the Curie temperatures of Gd, Ni, Co, and Fe. It also shows Mn as being antiferromagnetic (when it transforms from the paramagnetic to the antiferromagnetic state at ~ 100 K). While not shown in the figure, Cr, antiferromagnetic below ~ 310 K, is correctly placed by its value of r_a/r_d (Section 10.2.8).

J. C. Slater, using a similar approach, considered that r_a/r_d had to be approximately equal to, but not too much greater than 3.0 for ferromagnetic behavior. His findings are summarized, from *Physical Review*, Vol. 36, p. 57, 1930, as follows:

Metal	Cr	Mn	Fe	Co	Ni	Gd
r_a/r_d	2.60	2.94	3.26	3.64	3.94	3.1

The values of r_a/r_d for the elements of the first series of transition elements follow the same sequence as those of Bethe, but the value for Gd is considerably different from that of Bethe as shown in the figure. Curves of the type shown in Fig. 10-22 are

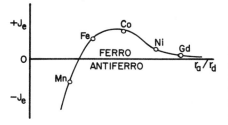

Figure 10-22 Schematic diagram of the exchange integral as a function of the ratio of half of the interionic distance to the "radius" of the d band for ferromagnetic elements compared to that of Mn.

known as Bethe–Slater curves. The sequence of relationships between r_a/r_d and J_e (Fig. 10-22) between J_e and λ [Eq. (10-196c)] and between λ and H_m [Eq. (10-158)] clearly indicates the connection of the Bethe–Slater ratios to fundamental magnetic properties.

While explanations such as those of Bethe and Slater are semiempirical, they are useful in the explanation of the properties of alloys of nonferromagnetic elements (see Sections 10.2.8 and 10.2.9).

A useful means for describing the relative magnetization at relatively low temperatures was given by F. Bloch (1930). This is based on the concept of exchange and the spin waves that logically follow from it. The concept of spin waves is based on the assumption that all the spins in a half-band are parallel at 0 K (Section 10.2.7). Consider the effect of a small increase in energy that would cause the reversal of only one of the spins. All the other unaffected ions have exactly the same probability that the spin of one of its electrons would be reversed. This being the case, the electron that underwent the original spin reversal would have a high probability of being associated with any one of the other ions. This must be so, because if it remained associated with its original ion, the exchange forces would have a high probability of causing it to flip back to its original direction of spin. But the original energy increment prevents this return. However, this energy can be absorbed readily if all the other electrons share in the single reversal of the original spin. Such sharing involves a much smaller energy increment when divided among all the other electrons involved, as compared to that associated with a single, direct reversal of one spin. This permits the "passage" of the electron successively to all nearest neighbors. The passage does not occur as a single spin reversal, but as a small, successive, angular difference in the spin vectors between nearest neighbors. In this way, the spin reversal of the original electron now may be regarded as a series of small angular differences in spin precession along its path. A line connecting the different spin vectors will have a wavelike shape. The passage of the electron during this gradual spin-reversal process can be dealt with by treating the reversal of the spin as a wave; hence the name *spin wave*.

Spin waves conform to the same conditions governing phonons. Quantized spin waves are known as *magnons*. Their behavior is similar to that of phonons (Sections 5.1.7 and 5.3.1). Magnons may be treated by means of the Bose–Einstein statistics because their number is variable and each changes the spin of the system by one unit, so $s = 1$ (the state of the system is unaffected by the interchange of opposite spins).

The spontaneous magnetization depends on each magnon. This effect is known as the Bloch $T^{3/2}$ law and is given by

$$\frac{\Delta M}{M_0} = \frac{0.1174}{N_0}\left(\frac{k_B T}{J_e}\right)^{3/2} \tag{10-200}$$

where ΔM is the magnetization change resulting from the spin waves, M_0 is the magnetization at 0 K, and N_0 is the number of ions in a unit cell in cubic lattices.

Equation (10-200) is most accurate for $T \le T/T_c = 0.3$. The similarity of magnons to phonons is responsible for their contributions to thermal conductivity in some antiferromagnetic materials and garnets, where the relaxation times are short.

Equation (10-200) may be expressed empirically as

$$\frac{M_T}{M_0} = 1 - \beta T^{3/2} \tag{10-201}$$

in which β is an experimentally determined constant. It reduces to Eq. (10-200) when

$$\beta = \frac{0.1174}{N_0}\left(\frac{k_B}{J_e}\right)^{3/2} \tag{10-201a}$$

Equation (10-201) may be used for $T > T/T_c = 0.3$ when the exponent of the temperature lies between $\frac{3}{2}$ and 2, depending on the material.

10.2.2 Exchange Energy

The alignment of the unbalanced spins gives rise to the magnetic properties of materials. This is treated by examining the electron populations of the bands involved. These are divided into half-bands, each of which contains electrons with spin of the same kind, but opposite to those in the other band. Diamagnetic and paramagnetic materials, in the absence of an external magnetic field, have equal numbers of electrons in each half-band. The spin imbalance responsible for ferromagnetism arises from unequal populations in the half-bands as indicated in Fig. 10-23.

The exchange energy arises from the coulombic energy involved in spin interactions, as noted in Section 10.2.1; it does not involve the interionic coulombic interactions that are independent of spin. The exclusion principle requires that only antiparallel spins may be accommodated in any given state. It therefore is unlikely that parallel spins would be closely associated. The variations in the exchange energy that take place while undergoing magnetic transitions affect the internal energy of the material and correspondingly change those properties, such as the heat capacity, the thermoelectric properties, the thermal expansion, and the thermal conductivity, that are based on the internal energy.

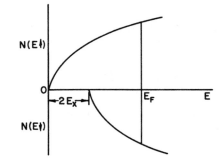

Figure 10-23 Schematic diagram of the densities of spin states of electrons in a ferromagnetic metal. The imbalance in the numbers of spin of opposite sign is responsible for ferromagnetism.

The properties of the exchange energy may be examined by starting with Eq. (10-158) expressed as an integral, and using Eq. (10-139) for parallel spin ($g = 2$, $J = \frac{1}{2}$ and $B_J(a') = 1$):

$$E_x = -\int_0^M H_m\, dM = -\int_0^M \lambda M\, dM = -\tfrac{1}{2}\lambda M^2 = -\tfrac{1}{2}\lambda N^2 \mu_B^2 \qquad (10\text{-}202)$$

E_x must be negative to conform to Eq. (10-189) for the case of parallel spin. It would be highly improbable for this to be the case if the internal energy were to be increased; the spontaneous alignment of spins could not occur if the potential energy were decreased (made more positive). This can be seen by examining the transition of just one electron from one spin to the other. One band loses an electron while the other gains an electron. The difference in the populations of the half-bands is two electrons. Thus, for the transfer of a single electron,

$$E_x = -2\lambda\mu_B^2 \qquad (10\text{-}203)$$

and E_x is the same as Eq. (10-202) for $N = 2$.

The decrease in E_x is a result of the transfer of an electron from one half-band to another of opposite spin. For this to be the case for spontaneous electron transfer,

$$E_x \leqq -2\lambda\mu_B^2 \qquad (10\text{-}204)$$

Equation (10-204) may be applied to an electron in a crystal lattice by the derivative of Eq. (3-25) so that the effects of changes in the direction of magnetization are given by

$$-E_x = -E_{\text{anis}} \leqq \Delta E \leqq \frac{2h^2\bar{k}}{8\pi^2 m}\Delta\bar{k} \leqq 2\lambda\mu_B^2 \qquad (10\text{-}205)$$

They become functions of the wave vectors. That wave vector providing the optimum lowering of E_x (greatest negative value of E_x) defines the crystallographic direction of easiest magnetization in the crystal, and E_{anis} is the anisotropy energy. This explains why the spontaneous magnetizations are different for different crystal directions as indicated in Fig. 10-24. The differences in the magnetization energies

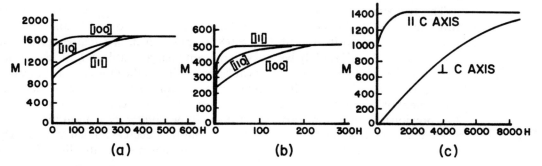

Figure 10-24   Magnetic anisotropy of single crystals of ferromagnetic elements: (a) iron, (b) nickel, and (c) cobalt. The units of M and H are gauss and oersteds, respectively. (From K. Honda and S. Kaya, *Sci. Rept. Tohokû University*, 15, 721, 1926.)

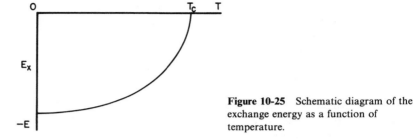

Figure 10-25 Schematic diagram of the exchange energy as a function of temperature.

between different crystallographic directions are called *anisotropy energies*. Their effects are apparent in Fig. 10-24. Also see Sections 10.2.4 and 10.2.5.

The general expression for the exchange energy as a function of temperature is obtained by using Eq. (10-139b) in Eq. (10-202) to obtain

$$E_x = -\tfrac{1}{2}\lambda M^2 = -\tfrac{1}{2}\lambda [Ng\mu_B J B_J(x)]^2 \qquad (10\text{-}206)$$

Here E_x is a function of $B_J(x)$. And, since $B_J(x) \to 1$ for $T \to 0$ and $B_J(x) \to 0$ for $T \to T_c$ (Section 10.1.5), E_x must show similar behavior, as indicated by Fig. 10-25.

10.2.3 Heat Capacity Changes

As heat is added to a ferromagnetic material, some of the thermal energy is absorbed in disrupting the aligned, parallel spins. This increases the internal energy by an amount equal to the diminished exchange energy. Since the heat capacity is a function of the internal energy, it also will be increased [Eq. (5-1)]. The increment contributed to the heat capacity of the solid is obtained from Eq. (10-202), since M is a function of T [Eq. (10-140) and Fig. 10-19], as

$$C_M \equiv \frac{\partial E_x}{\partial T} = -\frac{\partial}{\partial T}(1/2\lambda M^2) = -\lambda M \frac{dM}{dT} \qquad (10\text{-}207)$$

It will be noted that C_M is positive because dM/dT is negative.

The temperature variation of C_M may be derived from Fig. 10-25 using the relationship given by Eq. (10-207). The slope, $\partial E_x/\partial T$, is zero at 0 K and remains relatively small and positive until $T/T_c \cong 0.8$ is approached. At temperatures just below T_c, $C_M = \partial E_x/\partial T$ rapidly becomes significantly larger until, at T_c, it reaches its maximum value. A discontinuity occurs at T_c, according to the molecular field theory where C_M is predicted to drop discretely to zero. Accordingly, those properties that are dependent on the internal energy and therefore on the heat capacity, such as noted in the previous section, are also expected to show similar discontinuities. The experimental data do show discontinuities at T_c, but they decrease much more slowly than the precipitous drop predicted by this theory, as shown schematically in Fig. 10-26.

The experimental behavior of these properties is explained by the presence of spin clusters at $T > T_c$ because some spin alignment must exist to rationalize the

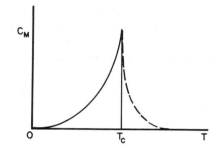

Figure 10-26 Schematic diagram of the Weiss theory of electron contribution to the heat capacity of a ferromagnetic material. The broken curve indicates the experimentally determined increment to the heat capacity that is not explained by the Weiss theory (see Sections 7.1.21 and 7.1.21.1).

observed behavior. Spin clusters are considered to consist of small, residual volumes that are the remnants of the original domains. Each of these volumes consists of small clusters of ions with aligned spins. The atoms on lattice sites surrounding the clusters have random spins. The aligned electron spins within each of the clusters set up local exchange fields in each cluster. The members and sizes of the clusters diminish progressively with $T > T_c$ and appear to vanish at temperatures considerably above T_c. The theory and another result of the influence of spin clusters are discussed and applied in Sections 7.1.21 and 7.1.21.1. [See Eq. (7-133).]

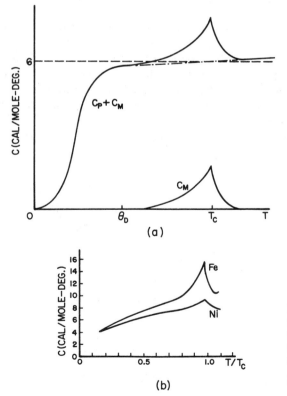

Figure 10-27 (a) Schematic diagram of the heat capacity of a ferromagnetic element showing the influence of the spin contribution. (b) Heat capacities of Fe and Ni. (From A. H. Morrish, *Physical Principles of Magnetism*, p. 272, John Wiley & Sons, New York, 1965.)

The continuous changes in the properties of spin clusters with increasing temperatures correspond to those in E_x and explain C_M. These are shown in Fig. 10-27(a) as an increment that is added to the heat capacity contributions of the lattice (Section 5.1.5) and of the electrons (Section 5.1.10.2). The magnitude and importance of the influence of C_M on the properties of some transition elements is shown in Fig. 10-27(b), where it normally would be expected at $C_V \cong C_P \cong 3R$ for $T \gtreqqless \theta_D$ [Eq. (5-7)]. The abnormal values for Fe and Ni for $T/T_c > 1$ are evidence of the energy required to dissipate the residual spin clusters.

10.2.4 Magnetic Domains

Domains may be observed by means of Bitter patterns (Section 10.2). A *domain* is a spontaneously magnetized volume in a crystal. Domain boundaries separate domains of different magnetic orientations. In the unmagnetized conditions, the orientations of the domains are random. This is the equivalent of saying that the magnetization of a single domain is given by a magnetic vector and that the magnetization of the crystal has the vector sum of all the domains within it equal to zero. The vector sums of the domains of magnetized materials are nonzero. Weiss postulated the presence of domains as a means for explaining the relative ease with which ferromagnetic materials are converted from the unmagnetized to the ferromagnetic conditions.

When a ferromagnetic specimen in the unmagnetized state is exposed to a magnetic field, those domains with vectors parallel to the field grow by the absorption of adjacent domains with less favorable orientations [Eq. (10-205)]. Some domains with vectors not quite as unfavorably oriented with respect to the applied field change their angular vectorial orientations to become parallel to the field by domain vector rotation. These may also absorb less favorably oriented domains. A domain will rotate from an easy to a hard direction of magnetization when suitably higher field intensities are applied. These mechanisms are indicated schematically in Fig. 10-28. These are directly related to E_{anis} [Eqs. (10-205)].

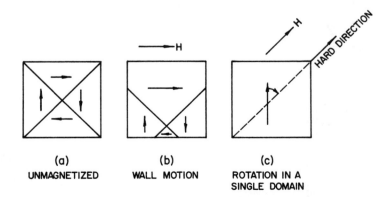

| (a) | (b) | (c) |
| UNMAGNETIZED | WALL MOTION | ROTATION IN A SINGLE DOMAIN |

Figure 10-28   Motion of domain walls as affected by an external magnetic field. (From A. H. Dekker, *Solid State Physics*, p. 476, Prentice-Hall, Englewood Cliffs, N.J., 1957.)

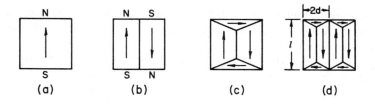

Figure 10-29   Progressively smaller domain sizes with correspondingly lower magnetic energies.

The motion of domain boundaries during the absorption of less favorably oriented domains occurs by small, discrete, irreversible increments; they are neither continuous nor reversible. The discontinuous domain motion may be detected by amplifying discrete sounds that emanate from a specimen during magnetization. This phenomenon is called the *Barkhausen effect*. Crystalline imperfections, fine precipitates, and inclusions can inhibit domain wall motion.

The magnetic energy of a crystal is greatest when the entire crystal consists of just one domain. This is a result of the large, external, magnetic field [Fig. 10-29(a)]. This energy is diminished by a factor of about $\frac{1}{2}$ when the crystal consists of two domains [Fig. 10-29(b)]. The domain size within the crystal may be diminished so that for n domains the energy is about $1/n$ that of the original, large domain [Eq. (10-222)]. A limit to the formation of such small domains occurs when the energy required for the formation a new boundary is greater than the lowering of the magnetic energy by the formation of a new domain.

The magnetic energy may also be diminished by the formation of domains that are so oriented that the magnetic flux can form a closed path within the material and no poles are formed either within the material or upon its surface [Fig. 10-29(c)]. These reduce the magnetic energy to zero. Such domains are called *domains of closure*.

The orientations of domains tend to be aligned in easy directions of magnetization [Fig. 10-29(d)]. Those of the domains of closure thus probably lie in harder directions of magnetization. Their anisotropy energy accordingly depends on the lattice type of the material.

10.2.5 Properties of Domains

Domain walls are the interfaces between domains of differing magnetic orientations. As such, they constitute a gradual change in the angular spin orientation from that of one domain to that of the other. The wall thus consists of a series of successively different spin orientations that is known as a *Bloch wall*. The exchange energy for two such adjacent spins is given by Eq. (10-189). And where the spin vectors have only slightly different angular orientations, it may be written as

$$E_x = -2J_e S^2 \cos \phi_{ij} \tag{10-208}$$

The small angular difference between adjacent spins minimizes (gives the maximum negative value of) E_x, for $J_e \cong 1$, because ϕ_{ij} is small and $\cos \phi_{ij}$ may be considered to approach unity.

Equation (10-208) may be reexpressed, using the first two terms of the series expression for $\cos \phi$,

$$\cos \phi_{ij} = 1 - \frac{\phi_{ij}^2}{2} + \frac{\phi_{ij}^4}{24} \cdots$$

as

$$E_x = -2J_e S^2 + J_e S^2 \phi_{ij} \tag{10-209a}$$

or

$$\gamma_x = J_e S^2 \phi_{ij}^2 \tag{10-209b}$$

since the first term in Eq. (10-209a) is a constant for both the wall and the domain and therefore need not be considered. Thus, for a wall whose width contains N ions, over which the angle ϕ between the spins of the two adjacent domains is shared equally by all the spins in the wall, the angle between adjacent spin orientations is

$$\phi_{ij} = \frac{\phi}{N} \tag{10-210}$$

Equation (10-210) is substituted into Eq. (10-209b) to obtain

$$\gamma_x = \frac{J_e S^2 \phi^2}{N^2} \tag{10-211}$$

The energy stored per unit area in each of the N planes in the boundary is

$$\gamma_x(a) = \frac{J_e S^2 \phi^2}{N^2 a^2} = \frac{J_e S^2 \phi}{N \delta a} \tag{10-212}$$

where a is the lattice parameter of a simple cubic lattice, and δ is the domain wall thickness and equals Na.

As the spins gradually rotate in small increments from easy to harder directions across the boundary, the anisotropy energy becomes increasingly greater. So, where K is the anisotropy constant per unit volume for the spin components in the plane of the wall, the anisotropy energy per unit area is

$$\gamma_{\text{anis}} = K \frac{Na^3}{a^2} = KNa = K\delta \tag{10-213}$$

where $Na = \delta$, the width of the wall. In the more general case, the anisotropy effects of spin components normal to the plane of the wall, K_2, are constant. Since all the spin components in the present case are parallel to the plane of the wall, $K_2 = 0$. (It should be noted that K as defined here frequently is denoted as K_1.)

The sum of Eqs. (10-212) and (10-213) gives the total energy stored in the boundary, by reexpressing $Na^2 = \delta a$, as

$$E_{\text{wall}} = \gamma_x(a) + \gamma_{\text{anis}} = \frac{J_e S^2 \phi^2}{\delta a} + K\delta \tag{10-214}$$

The minimum of wall energy is obtained from

$$\frac{\partial E_{\text{wall}}}{\partial \delta} = -\frac{J_e S^2 \phi^2}{\delta^2 a} + K \tag{10-215}$$

At the minimum, this derivative equals zero, so

$$K = \frac{J_e S^2 \phi^2}{\delta^2 a} \tag{10-216}$$

And, when Eq. (10-216) is substituted into Eq. (10-215), it is seen that $\gamma_{\text{anis}} = \gamma_x(a)$ for minimum energy. The minimum boundary thickness is obtained from Eq. (10-216) as

$$\delta = \left(\frac{J_e S^2 \phi^2}{Ka} \right)^{1/2} \tag{10-217}$$

The quantity on the right side of Eq. (10-217) is known as the domain wall thickness parameter. The substitution of Eq. (10-217) into Eq. (10-214) gives

$$E_{\text{wall}} = \left(\frac{J_e S^2 \phi^2 K}{a} \right)^{1/2} + \left(\frac{J_e S^2 \phi^2 K}{a} \right)^{1/2} = 2\left(\frac{J_e S^2 \phi^2 K}{a} \right)^{1/2} \tag{10-218}$$

A comparison of Eqs. (10-218) and (10-217) gives a convenient relationship between E_{wall} and δ as

$$E_{\text{wall}} = 2K\delta \tag{10-219}$$

Equation (10-218) may be made to be more convenient by the use of Eq. (10-199a). This results in

$$E_{\text{wall}} = 2\left(\frac{2k_B T_c}{z} \cdot \frac{S^2 \phi^2 K}{Ka} \right)^{1/2} \tag{10-220}$$

And, in a similar way, Eq. (10-217) becomes

$$\delta = \left(\frac{2k_B T_c}{z} \cdot \frac{S^2 \phi^2}{Ka} \right)^{1/2} \tag{10-221}$$

Some of the properties of domain walls, calculated by means of Eqs. (10-220) and (10-219), are given in Table 10-6 for equal increments of the 180° spin orientation and for $S = \frac{1}{2}$. Anisotropic effects are neglected. It will be noted that, while E_{wall} appears to be unexpectedly low, the energy density, E_{wall}/δ, is quite high.

The small values of ϕ_{ij} provide an insight into the relative ease of wall motion. Less than 1% of the energy is required for the propagation of small increments of spin orientation as one domain progressively absorbs another than that required for the

TABLE 10-6 SOME CALCULATED PROPERTIES OF DOMAIN WALLS

Parameter	Fe	Co	Ni
T_c (K)	1043	1403	631
K (erg cm$^{-3} \times 10^{-5}$)	4.2	4.1	0.5
a (Å)	2.48	2.51	2.48
z	8	12	12
E_{wall} (erg cm^{-2})	2.0	1.1	0.7
δ (Å)	240	1400	660
E_{wall}/δ (erg cm^{-3})	0.8×10^5	0.8×10^5	1×10^5
δ (atoms)	100	550	270
ϕ_{ij} (degree atom^{-1})	1.8	0.3	0.7

simultaneous reorientation of all spins to take place. This may be approximated by the use of Eq. (10-211) and the data in Table 10-6.

The domain spacing may be approximated, using Fig. 10-29(d), based on the observation of domains on the surface of the ferromagnetic crystal. The energy of the domain walls parallel to the direction of easy magnetization will vary as their length *l*. Their energy will also vary as the number of the domains, a factor that varies inversely as their width *d*. So the energy of the walls in the easy direction may be approximated as

$$E_w \cong E_{\text{wall}} \cdot \frac{l}{d} \qquad (10\text{-}222)$$

Now, assuming a crystal with a cubic lattice, the energy of the walls at the domains of closure will vary as $d/2$. This is given as

$$E_c = K \frac{d}{2} \qquad (10\text{-}223)$$

The domain boundary energy, per unit area of crystal surface, is given by the sum of Eqs. (10-222) and (10-223), or

$$E_{\text{dom}} = E_w + E_c \cong E_{\text{wall}} \cdot \frac{l}{d} + K \frac{d}{2} \qquad (10\text{-}224)$$

The minimum domain width per unit crystal surface is obtained from the differentiation of Eq. (10-224) and equating the result to zero:

$$\frac{\partial E_{\text{dom}}}{\partial d} \cong -E_{\text{wall}} \cdot \frac{l}{d^2} + \frac{K}{2} = 0 \qquad (10\text{-}225)$$

from which

$$d_{\text{min}} \cong \left(\frac{2l E_{\text{wall}}}{K} \right)^{1/2} \qquad (10\text{-}226)$$

When Eq. (10-226) is substituted into Eq. (10-224), the minimum domain energy, per unit crystal surface, is found to be

$$E_{\substack{\text{dom} \\ \text{(min)}}} \cong (2lKE_{\text{wall}})^{1/2} \qquad\qquad (10\text{-}227)$$

Using the values of E_{wall} and K for iron (Table 10-6) and $l = 1$ cm, it is found that $d_{\text{min}} \cong (2 \times 1 \times 2.0/4.2 \times 10^5)^{1/2} \cong 3 \times 10^{-2}$ cm, and $E_{\text{dom}} \cong (2 \times 1 \times 4.2 \times 10^5 \times 2.0)^{1/2} \cong 13 \times 10^2$ erg cm^{-2}.

The preceding calculations for the idealized domains pictured in Fig. 10-29(d) are seldom present. Many other shapes may be present depending on the purity and the crystal orientation of the plane of the surface being examined. The actual domain configurations are those that minimize the magnetic energy.

The grains of polycrystalline materials may also act similarly to single crystals. Their size, degree of crystalline perfection, and orientation affect the observed domain structure. Domains form within the grains oriented parallel to the easy magnetization direction. Where adjacent grains do not form domains of closure, free poles will be present. The grain size appears to determine the domain configuration in cases in which the number of free poles within the grains is small with respect to the number of their boundaries. In cases in which the thickness of the material is smaller than the average grain diameter, such as in thin sheet or foil, the thickness of the sheet appears to influence the domain structure. Other factors such as chemical homogeneity, inclusions, precipitates, internal stresses, and voids (in cast materials) can also adversely affect domain shapes and mobility. Acting as "pins," they can impede domain motion, an important factor in some permanent magnets (Section 10.5.1). Where the ferromagnetic phase is a microconstituent of a polyphase alloy; its size and distribution will affect the properties of the domains.

10.2.5.1 Bubble domains.

Single-crystal films of ferromagnetic or ferrimagnetic materials with thicknesses of $\sim 10^{-3}$ cm and whose easy directions of magnetization are perpendicular to their surfaces may form bubble domains. In the demagnetized condition, equal numbers of oppositely magnetized domains are present. These have convoluted shapes. They are transparent to red light and may also be observed by means of the Faraday effect. This makes use of the fact that the plane of polarization of the transmitted light is changed upon reflection by the direction of magnetization of the surface of the material. The domains are then studied with the use of a suitable microscope. Pulsed external magnetic fields parallel to the easy directions of magnetization (perpendicular to the surface of the film) cause the unfavorably oriented domains to become absorbed by the favorably oriented domains. The former are absorbed nonuniformly and break up into short segments after several pulses. These continue to be absorbed and, if the pulsed external (or bias) field is removed prior to their complete absorption, small, cylindrically shaped domains are observed. These bubble domains become very stable upon the application of a critical field.

The cylindrical domain shape is a result of the tendency to minimize the domain wall energy. This minimization of E_{dom} may be considered to be analogous to that shown for Eq. (10-225). The very high mobility of these domains results from the fact that the bubble is a single domain within a larger domain of opposite magnetic orientation. The spins composing the Bloch wall change from an orientation in one easy magnetic direction to the opposite orientation in the same easy direction. As a result, domain motion can occur by the application of fields of the order of 10^{-2} Oe. Thus, low fields parallel to the plane of the films cause the bubbles to flow across the films.

When the bubble density is high, they tend to array themselves so that they approach a uniform density. This occurs because the bubbles are magnetic dipoles of the same sign that repel each other at short distances. The result is a minimum distance between the bubbles that contributes toward their uniform density.

Garnet (antiferromagnetic) films, with thicknesses of $\sim 10^{-4}$ cm, have been widely investigated for their potential use in digital-based electronic devices. Such materials are of the general type $M_3Fe_5O_{12}$, in which M is a rare earth element, bismuth, or bismuth in combination with a rare earth. These are described in more detail in Section 10.4. They may have bubble densities of $\sim 10^5$ cm^{-2} with bubble diameters of $\sim 10^{-4}$ cm with velocities comparable to or greater than rates obtained from switching transistors. Such properties have been obtained using significantly smaller energies than those required by transistors.

The bubbles may be led through paths for use in electronic devices. One way of doing this is by means of the deposition of thin conducting loops, about the same size as the bubbles, onto the surfaces of the ferrimagnetic films. These are arrayed in grids. The sequential switching of the current from loop to loop changes the polarity of the loop and impels the bubble across the film. This is the *conductor-access method*. Another method, the *field-access method*, makes use of similarly sized, unsymmetric, magnetic configurations, deposited as thin films with T, Y, and bar shapes, to form the grids for the paths for the bubbles. The application of a periodically changing magnetic field induces correspondingly varying magnetic poles in the asymmetric deposited components (frequently Permalloy). An adjacent bubble is attracted to a pole of opposite magnetic sign and moves from one component to the next across the grid as the periodic field changes. Read-out rates greater than 10^5 bits s^{-1} have been attained by this technique.

Each of these methods lends itself to binary notation. The absence of a bubble on a site in either type of grid constitutes a zero; its presence constitutes a one.

Bubbles may be generated for use in each of these techniques. In the conductor-access method, bubbles on a U-shaped grid component may be made to split a bubble into two bubbles. In the field-access technique, a domain under a slightly asymmetric disc is split by the periodic field. In each case, energy from the splitting source is absorbed by the split bubbles so that they reach the minimum stable size. The split bubbles repel each other and move oppositely. By the same token, bubbles may be erased when the applied magnetic energy is sufficient to cause them to collapse to sizes smaller than the minimum stable size.

10.2.6 B–H Curves

Increasing magnetization can occur by means of domain wall motion, where more favorably oriented domains grow at the expense of those less favorably oriented with respect to the applied field. At considerably higher fields, the magnetization increases as a result of domain rotation (Figs. 10-28 and 10-30). The mechanisms indicated in these figures do not occur as sharp transitions from one to the other. In some cases, boundary motion and domain rotation may occur simultaneously in different portions of the crystal. In other cases, both mechanisms may be operative in the same part of the crystal.

Consider a demagnetized specimen, that is, one in which the domains are randomly oriented so that the net magnetization is zero, to which an external field is slowly applied. The domain boundaries move readily under the low, initial fields. And, if these fields are reversed, the boundary motion can be shown to be nearly reversible. As the field is increased, a large increase in M occurs for increasing H. Some of the units used in the following discussion were presented in Section 10.1.1 and are given again in Table 10-7 for convenience. The slope, M/H, increases very rapidly for small increases in H. The irreversible absorption of unfavorably oriented domains is the predominating mechanism during this stage. A point is reached, as H increases, at which M increases with diminishing values of M/H. This continues until the slope M/H begins to increase very slowly and nearly linearly, a point at which the specimen is said to be *saturated*. This last phase is largely a result of the rotation of the domain vectors so that their magnetizations are optimumly parallel to the external magnetic field, and is their intrinsic magnetization.

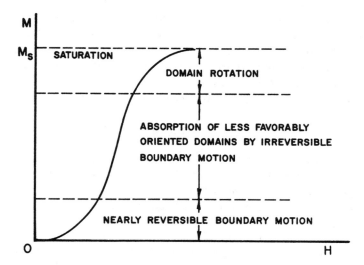

Figure 10-30 Responses of domains to external magnetic fields and their relationships to the magnetization diagram. (Modified from C. Kittel, *Introduction to Solid State Physics*, p. 489, John Wiley & Sons, New York, 1966.)

TABLE 10-7 COMMONLY USED MAGNETIC UNITS

Property	Symbol	cgs Units	Conversions	
			cgs	SI
Magnetic flux density	B	Gauss (G)	1 gauss $= 10^{-4}$ Wb·m^{-2}	1 Wb·m$^{-2} = 10^4$ G $= 1$ T
Magnetic field	H	Oersted (Oe)	1 Oe $= 79.6$ A·m^{-1}	1 Am$^{-1} = 12.6 \times 10^{-3}$ Oe
Magnetization	M	emu·cm^{-3}	1 emu·cm$^{-3} = 12.6$ Wb·m^{-2}	1 Wb·m$^{-2} = 796$ emu·cm^{-3}
Permeability (vacuum)			$\mu_{r,0} = 1$	$\mu_0 = 4\pi \times 10^{-7}$ Wb(A·m)$^{-1}$

1 emu $= 1$ erg·Oe^{-1}.
$B = H + 4\pi M = \mu H$ [Eqs. (10-12) and (10-14)].
$\mu = \mu_r \mu_0$ [Eq. (10-20)].

B–H diagrams, such as shown in Fig. 10-31, are widely used to indicate the magnetic behavior of materials. These are obtained in the same way as indicated for Fig. 10-30, but the direction of the applied field is reversed after the initial saturation to determine its effect on the magnetic properties in the first, second, and third quadrants. After saturation again is reached, in the third quadrant, the direction of the external magnetic field is changed back to its original direction and the curve is completed in the third, fourth, and first quadrants. The maximum magnetic flux density or saturation is given by B_s. B does not return to zero when $H = 0$. This residual flux density is indicated by B_r at $H = 0$. This is the spontaneous magnetization and is a consequence of the irreversible absorption of the less favorably oriented domains, in the absence of domain rotation; this is known as the *remanence*. The point at which $B = 0$ is the *coercive field* and is usually designated as H_c. It represents the magnetic field required to demagnetize the specimen. Anything that impedes domain wall motion increases H_c (Sections 10.2.4 and 10.2.5).

The irreversibility manifested as B_r is a consequence of the hysteresis that begins with the irreversible domain absorption. This spontaneous magnetization is responsible for the symmetrical hysteresis loops shown in the figure. The work required to go around the hysteresis loop is

$$W = \oint H\, dM \quad (\text{erg cm}^{-3}) \tag{10-228}$$

Most of this energy is dissipated largely as heat.

The characteristics of the hysteresis loop of a given material determine its applications. Where use as permanent magnets is intended, B_r and H_c should be large [Fig. 10-31(a)]. These ensure the retention of the desired high magnetic properties during use. If H_c were low, stray magnetic fields could diminish B_r. High, stable values of B_r are desirable for these applications. In other applications, such as transformer laminations, minimum hysteretic energy losses are desirable. This requirement dictates the use of materials with very small or negligible values of H_c and large values for B_r, as in Fig. 10-31(b). This combination of properties optimizes the magnetic properties because it minimizes the energy losses.

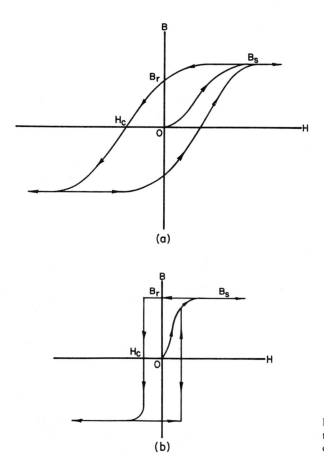

Figure 10-31 Some hysteresis loops: (a) typical; (b) rectangular. The *B* scale is compressed.

10.2.7 Band Structures of Ferromagnetic Elements

The band structures of transition elements and some of their alloys are discussed in Sections 7.1.8.2, 7.1.9.3, 7.1.10.2, and 7.1.20, where the hybridized nature of the *s* and *d* bands is considered in terms of the rigid-band model. Some of the effects of magnetic transitions of nickel and some of its alloys are given in Sections 7.1.21 and following sections. Their ferromagnetic properties are explained by considering electrons of opposite spin to occupy *d*-level half-bands, as discussed in Section 10.2.2 and sketched in Fig. 10-23. At temperatures below T_c, the half-bands are unequally occupied, causing a net spin imbalance, as shown in Table 10-8. These configurations have been postulated to explain such properties as heat capacity, Hall effect, and thermoelectricity, as well as ferromagnetism. At temperatures higher than T_c, the *d* half-bands are considered to be equally populated to account for their paramagnetic properties.

TABLE 10-8   BAND STRUCTURES OF IRON, COBALT AND NICKEL NEAR ROOM TEMPERATURE

| Element | Ground State | | Magnetic State | | | | | | | | |
	3d	4s	3d↑	3d↓	4s↑	4s↓	Holes 3d↑	Holes 3d↓	Spin Imbalance	$\mu_B \cdot$ion^{-1}	M_0 (emu·g^{-1} at 0 K)
Fe	6	2	4.8	2.6	0.3	0.3	0.2	2.4	2.2	2.22	221.9
Co	7	2	5.0	3.3	0.35	0.35	0	1.7	1.7	1.71	162.5
Ni	8	2	5.0	4.4	0.3	0.3	0	0.6	0.6	0.61	57.5

Based on R. M. Bozorth, *Ferromagnetism*, p. 438, Van Nostrand, 1951.

The s–d hybridization is responsible for the average, nonintegral values of half-band occupancy. It will be noted that the spin imbalance in each case corresponds closely to the experimentally determined number of Bohr magnetons per ion. This behavior is shown schematically in Fig. 10-32.

The number of unbalanced spins in BCC iron is shown as being 2.2. This is in close agreement with 2.22 μ_B ion^{-1}. The presence of holes in both half-bands also seems to confirm its alloying properties. This has been considered to be a result of the presence of electron configurations consisting primarily of $3d^8$ arrays. Configurations consisting of mixtures of $3d^7$ and $3d^9$ electron arrays have also been considered. This configuration could be more probable than that of the primary $3d^8$ because the hybridized $4s$ levels could interact with the d levels so as to minimize the energy of the $3d^7$–$3d^9$ configuration.

HCP cobalt, in its easy magnetic direction, shows a value of $\mu_B = 1.715$ that is in good agreement with the spin imbalance of 1.7 holes per ion. In a way similar to that noted for iron, this could be a result of different $3d^7$–$3d^9$ configurational mixtures. It

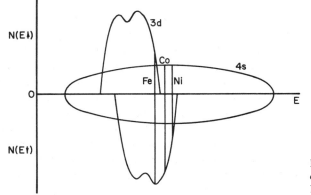

Figure 10-32 Schematic diagram of the densities of electron states of Fe, Co, and Ni.

may be probable that the Fermi level for cobalt should be shown at the same energy as that of the top of its spin-down half-band for this case, rather than as shown in Fig. 10-32.

Nickel has been shown to have 0.563 unbalanced spins at 0 K. This is in good agreement with the value of 0.57 shown in Fig. 7-18 and discussed in Sections 7.1.24 and 7.1.26. Here all the unbalanced spins are confined to just one half-band. It has been considered that this arises as a result of mixed $3d^8$, $3d^9$, and $3d^{10}$ configurations.

It should be noted again that, for the transition elements being considered here, the electron configuration of an ion in the crystalline state should not be considered as being fixed. Electron exchanges at the d level may occur between ion cores (Sections 3.12.2 and 10.2.1). These are known as *itinerant states*. This is the same as d-level hole transport from ion to ion. The averaging of such itinerant behavior over large numbers of ions in the crystalline state can account for the mixtures of $3d$ electron configurations postulated here. Such mixed configurations can help to account for the pronounced paramagnetic properties of some of the alloys of these metals. This is especially true for Cu–Ni alloys with high Cu contents.

10.2.8 Alloys of Iron, Cobalt, and Nickel

The magnetic properties of binary alloys of Fe, Co, and Ni appear to be a direct function of the amount of the alloying element present in solid solution in their lattices. This is shown, in terms of the simple, rigid-band model, for nickel-based alloys in Fig. 10-33.

An explanation of this behavior may be obtained by examining the changes in the band structures of the elemental metals that result from the presence of the alloys. The d band can contain up to ten electrons in an energy range of about 3 eV. This represents a high density of states. The s band can accept two electrons in a range of more than 7 eV; this is a much lower density of states than that of the d band. Since these elements can be described by the rigid-band model, the probability of finding an added electron in a band will vary directly with the density of states of that band. This results in a high probability that the s or p valence electrons of normal alloying elements, in substitutional solid solution in the lattice of the transition element, will be found in the d levels of the transition element because of its higher electron density. Ni–Cu alloys illustrate this mechanism. The substitution of a Ni ion by a Cu ion adds one electron to the system. The added electron will most probably occupy a d state (Sections 7.1.14.3 and 7.1.23 and following sections). This diminishes the number of unbalanced spins of the Ni ions and accordingly decreases the magnetization. As increasing numbers of Cu ions go into solution in the Ni lattice, the number of unbalanced Ni d-level spins decreases. A point is reached at which the electrons from the copper ions fill the d levels of the Ni ions to a maximum extent. This is expected to occur near 60 At% Cu–40 At% Ni. Magnetic measurements indicate that the critical filling of the d levels of Ni occurs between 53 and 62 At% Cu. This is in close agreement with the thermoelectric data (Section 7.1.24). At this composition, the spin imbalance is effectively zero, and the effective number of Bohr

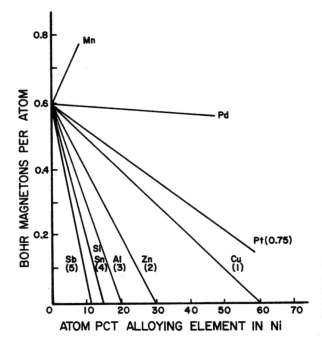

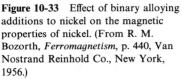

Figure 10-33 Effect of binary alloying additions to nickel on the magnetic properties of nickel. (From R. M. Bozorth, *Ferromagnetism*, p. 440, Van Nostrand Reinhold Co., New York, 1956.)

magnetons per atom of the alloy is zero, as shown in Fig. 10-33. The rate of decrease of the ferromagnetic properties of Ni would be expected to increase as the number of valence electrons contributed by normal metal solutes increases. This is verified by the data in Fig. 10-33, where the number of valence electrons contributed by each solute element is shown in parentheses. The compositions representing the transition to paramagnetic alloys decrease linearly as functions of increasing valence of the added normal metal.

On the other hand, Pd, which has an outer, hybridized s–d electronic configuration very close to that of Ni, shows almost no effect over a wide range of compositions, as might be expected. These two elements form a continuous series of solid solutions, but the change in the lattice parameter could account for the small decrease in the spontaneous magnetization as a result of small changes in r_a/r_d (Section 10.2.1). It will also be noted that Mn, which adds holes, has a positive ferromagnetic effect since its incompleted $3d$ levels add to the spin imbalance of the nickel ions.

Substitutional, iron-based, solid-solution alloys with normal metals show different behavior than that of alloys of nickel. The substitution of each normal metal ion in the iron lattice causes a decrease of about $2.2\mu_B$. It, therefore, appears that the effect of such ions is to diminish the number of Fe ions since electron interactions do not occur between the Fe and normal metal ions. Since the valence electrons from these solute ions do not affect the d levels of iron, the rigid-band model does not apply.   In contrast to this, Fe-based, alloys with other transition elements that have eight or more d electrons have properties that are readily explained by the rigid-band model.

The Co–Ni system forms a continuous series of ferromagnetic solid solutions. Alloys containing up to about 25 At% Ni show ordered, HCP structures at temperatures below about 422°C.   This has no effect on T_c as a function of composition, since it decreases continuously from 1121° to 361°C.   The effective number of magnetons of a binary "Co–Ni ion" is given by $n_B \cong 0.57\,C_{Ni} + 1.71\,(1 - C_{Ni})$, where C_{Ni} is the atom fraction of Ni ions.   The increase in the number of unbalanced spins with increasing amounts of Co (decreasing amounts of Ni) is an important basis for the application of the rigid-band model for the explanation of the properties of this alloy system.

The Fe–Co system is more complex than the alloys considered previously.   The Curie temperature is a complicated function of composition, and an ordered lattice structure occurs at the FeCo composition.   The composition close to $Fe_{0.7}Co_{0.3}$ has $n_B = 2.5$, the largest value for a binary alloy thus far observed.

The Fe–Ni system shows a greater variation in T_c as a function of composition than any of the systems discussed here.   Both Ni and Co increase the magnetization when in solution in Fe.   This behavior cannot be explained by the rigid-band model. It should be noted that the rigid-band model provides best results when the elements composing the binary alloys are next to each other in the Periodic Table, despite this exception.

These effects are shown for several binary alloy systems in Fig. 10-34.   Here the magnetization is plotted against the total number of electrons per atom of the alloy instead of atom percent.   This approach permits the presentation of a larger number of systems without considering the electron configurational details of the solutes.   As such, it constitutes a crude way of representing the rigid-band model.

Again, as discussed previously, the added elements either diminish or increase the spin imbalance of the d levels and, as a result, cause either a decrease or an increase in the magnetization of the resultant alloys.

Many other alloys of the transition elements show ferromagnetic behaviors.

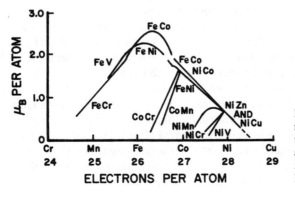

Figure 10-34   Effective number of Bohr magnetons per ion as a function of the total number of electrons per atom of alloy. (From R. M. Bozorth, *Ferromagnetism*, p. 440, Van Nostrand Reinhold Co., New York, 1956.)

These include alloys (actually intermetallic compounds) such as Fe_3Al, Au_4V, Au_4Mn, FeAl, and Cu_4Mn in which the magnetic behavior is a strong function of the degree of perfection of the two sublattices that constitute the ordered crystal structures of such compounds.   Some of these alloys, one component of which has no magnetic moment, may show that clusters of ions with "giant moments" are present.   This is the case for FeAl, where the concentration of the clusters decreases with increasing degrees of ordering.   It is considered that such behavior could be a consequence of segregation of the magnetic component.

Some additional binary alloy systems that contain ferromagnetic alloys include one transition element.   Some such systems are Mn–Al, Mn–As, Mn–Bi, Mn–N, Mn–P, Mn–Sb, Cr–Pt, Cr–S, Cr–Te, Co–Pt, and Fe–Pt.   While most of these systems do not contain a ferromagnetic component, they do include the elements Mn and Cr.   In these cases, it appears that the magnetic alloys within these alloy systems are such that the ratios r_a/r_d of the transition element result in positive values of J_e that are responsible for ferromagnetic properties.   This also holds for the Au_4V alloy noted earlier (see Section 10.2.1 and Fig. 10-22).   It has been shown that r_a/r_d is a function of J_e, that J_e is a function of λ [Eq. (10-196c)], and that the molecular field, in turn, is a function of λ [Eq. (10-158)].   Thus, the Bethe–Slater ratios are functions of the molecular field (Section 10.2.1). This accounts for the magnetic alloys in those systems noted previously that contain no ferromagnetic components.

10.2.9   Heusler-Type Alloys

Heusler (1898) first demonstrated that nonferrous alloys of Cu–Mn–Sn and Cu–Mn–Al showed spontaneous ferromagnetic behavior similar to that of nickel.   These alloys are close to the compositions given by Cu_2MnSn and Cu_2MnAl.   Comparable alloys, including such compositions as Cu_2MnIn and Cu_2MnGa, also have ferromagnetic properties.   Analogous compositions of silver-based alloys also show this behavior.

Alloys of this type are intermetallic compounds and all have a similar ordered lattice type.   The Mn and Sn or Mn and Al pairs of ions take alternate body-centered sites within a face-centered unit cell composed of eight unit cells of copper.   The ion pairs thus form a sublattice within that of the copper ions.

The properties of this class of alloys are highly dependent on the degree of perfection of the ordering of the superlattice.   Anything that disturbs this array diminishes the net magnetic moment and lowers the Curie temperature.   This is a result of the influence of the Bethe–Slater ratios that, in turn, control the sign of J_e (Section 10.2.1).   Such factors as compositional variations and stress can cause this. For example, disordered alloys are paramagnetic, and quenched alloys (stressed and largely disordered) show less desirable ferromagnetic properties than in the unstressed, ordered condition.   When such quenched alloys are aged (ordered and stress relieved) in the neighborhood of about 200°C for increasing times, their ferromagnetic properties increase with the time of the thermal treatment because their degrees of crystallographic perfection increase with the aging treatment.

The role of crystallographic order is to maintain the critical distance between the Mn ions. It ensures that the Mn ions are sufficiently far apart so that their d levels do not overlap. Here, the Bethe–Slater ratios are greater than that of elemental Mn and J_e is positive (Fig. 10-22). An empirical limit (at variance with Slater's calculation, Section 10.2.1) for compounds containing Mn, with direct exchange coupling among the Mn ions, appears to be a minimum distance of 2.8 Å between Mn atoms. MnAs and MnSb, with lattice spacings ($c/2$) of 2.84 and 2.89 Å, respectively, are ferromagnetic. A similar limit for Cr ions in Heusler-type alloys is 3.05 Å. Alloy combinations of this class must be such that the distances between the ions of the transition elements must be greater than such limits in order that the exchange integral can become positive and the exchange energy negative.

This model also predicts that both the exchange energy and the Curie temperature should reach maximum values at a given ratio of r_a/r_d. Empirical plots of the Curie temperatures of various binary and ternary Heusler-type alloys as a function of r_a/r_d show that such maxima occur at values of r_a/r_d, between 1.8 and 1.9. The exchange coupling increases sharply, as measured by T_c, from $r_a/r_d \cong 1.5$ up to a maximum that appears between 1.8 and 1.9. Thereafter, the exchange rapidly diminishes up to a value of about 2.1. It thus appears that the influence of the Bethe–Slater ratios have an effect on the ferromagnetic properties of alloys that is similar to that shown in Fig. 10-22 for the elemental metals.

Many other intermetallic compounds with Mn have been shown to have ferromagnetic properties. These include MnAl, MnAs, MnB, MnBi, Mn_4N, MnP, Mn_2Sb, MnSb, and Mn_4Sn. Other similar compounds involving chromium show weaker ferromagnetic behavior.

The discrepancies between the Bethe–Slater ratios for compounds, and those for the elemental transition metals involved in them, constitute a major basis for the criticisms of this theory. However, this approach does provide an insight into the ferromagnetic properties of nonferrous alloys that has not as yet been explained in any other way.

Ferromagnetic properties are almost never shown by ionically bonded compounds (Section 3.12.1). Only two such compounds are known: EuO ($T_c = 77$ K) and CrO_2 ($T_c = 400$ K). The latter has been given consideration for such applications as magnetic tapes. The absence of ferromagnetic properties in this class of compounds is considered to result from the high-strength bonding involved.

10.3 ANTIFERROMAGNETISM

Ferromagnetism has been shown to arise from a positive exchange interaction. The original explanation for the nonferromagnetic elements of the first transition series was based on the idea that a negative exchange reaction was responsible for their properties. It has since been shown that the properties of these elements may not necessarily be explained this simply (Section 10.3.1.4).

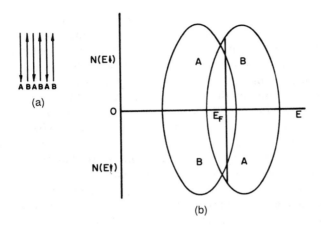

Figure 10-35 Electron spins in an antiferromagnetic material: (a) spin arrays on two sublattices; (b) densities of electron spin states in the d-level subbands.

The properties of virtually all antiferromagnetic compounds must arise from the effects of the relatively widely separated transition metal ions of which they are composed, since they have almost no free electrons. (Their electrical resistivities are higher than those of metals by a factor of at least 10^6.) The electron behaviors responsible for their magnetic properties thus are largely local and confined to the individual metal ions. Exchange mechanisms, therefore, must be considered to take place indirectly by way of intermediate nonmetal ions through a Heitler–London-like mechanism (Section 10.2.1), known as *superexchange*, because these ions are more widely separated than in the ferromagnetic case. This situation is more amenable to the application of the molecular field theory than is the case of ferromagnetic materials (Section 10.1.4), because of the localization of the magnetic ions. In the case of antiferromagnetism, J_e is negative and the spins of nearest neighbors are antiparallel (Sections 10.2.1 and 10.4).

Antiferromagnetism occurs in solids in which the magnetic moments are aligned in an antiparallel way so that their net effect may approach zero [Fig. 10-35(a)]. As the temperature increases, at relatively low temperatures, these materials show a small and continuous increase in magnetization. This continues up to a temperature, the Néel temperature, T_N, above which they become paramagnetic. The simplest model for this behavior assumes that two sets of antiparallel spins are present on at least two sublattices [Fig. 10-35(b)]. The spins on each of two sublattices are all parallel to each other, but are opposite in direction to the spins on the other sublattice. Spontaneous magnetization is not possible under such ideal conditions. As the temperature is increased, for $T < T_N$, the spins become slightly unbalanced and a small degree of spontaneous magnetization results. This small spontaneous magnetization increases with increasing temperatures up to T_N as the spin imbalance increases. At and above T_N, the thermally induced forces exceed the exchange forces and the magnetic moments become random, giving rise to paramagnetism.

10.3.1 Molecular Field Theory

The simplest case of a crystalline structure in which two sublattices may be pictured is that of the BCC lattice. This structure may be visualized as that of two identical, interpenetrating, simple-cubic sublattices where the A sublattice is composed of all the BCC corner positions, and the B sublattice contains all the remaining body-centered positions. An A atom, thus, always has an antiparallel B atom as its nearest neighbor and another parallel A atom as its next-nearest neighbor. In the same way, atoms on the B sublattice always have antiparallel A nearest neighbors and parallel B next-nearest neighbors. This means that the signs of the magnetizations of the sublattices are opposite to each other (see Section 10.4).

The molecular field affecting on A ion is proportional to and opposite to that on a B ion:

$$H_{mA} = -\lambda_{AA}M_A - \lambda_{AB}M_B \tag{10-229a}$$

in which λ_{AB} and λ_{AA} are the molecular field constants for nearest and next-nearest neighbors, respectively, and M_A and M_B are the respective magnetizations of the A and B sublattices. Since the signs of M_A and M_B are opposite, one of the terms in Eq. (10-229a) is positive. In a similar way, the molecular field affecting an ion on the B sublattice is

$$H_{mB} = -\lambda_{BB}M_B - \lambda_{BA}M_A \tag{10-229b}$$

It will be noted that, for a given species of ion occupying both sublattices, $\lambda_{AB} = \lambda_{BA}$, and $\lambda_{AA} = \lambda_{BB} = \lambda$. This approach neglects any interactions between second-nearest neighbors. It will be noted that J_e is negative, since nearest neighbors are antiparallel (Section 10.2.1), so λ is positive [Eq. (10-196a)]. Thus, when an external field H is present, the field on an ion on each sublattice, respectively, is

$$H_A = H - \lambda M_A - \lambda_{AB}M_B \tag{10-230a}$$

$$H_B = H - \lambda M_B - \lambda_{AB}M_A \tag{10-230b}$$

10.3.1.1 Temperatures above T_N.
The respective magnetizations of the sublattices, using Eq. (10-126b) in which $N/2$ equal the numbers of ions that are upon each one, are reexpressed as

$$M_A = \frac{N}{2} \cdot \frac{g^2\mu_B^2}{k_B T} \cdot \frac{J(J+1)}{3} = \frac{Ng^2\mu_B^2 S(S+1)}{6k_B T} H_A \tag{10-231a}$$

and

$$M_B = \frac{Ng^2\mu_B^2 S(S+1)}{6k_B T} H_B \tag{10-231b}$$

where J is replaced by S, since spin alone is being considered here. The magnetization is given by the sum of the magnetizations of both sublattices, given by Eqs. (10-231), as

$$M = M_A + M_B = \frac{Ng^2\mu_B^2S(S+1)}{6k_BT}(H_A + H_B) \tag{10-232}$$

The factor $(H_A + H_B)$ in Eq. (10-232) is obtained from the sum of Eqs. (10-230) as

$$H_A + H_B = 2H - (\lambda + \lambda_{AB})M_A - (\lambda + \lambda_{AB})M_B$$

$$H_A + H_B = 2H - (\lambda + \lambda_{AB})(M_A + M_B) = 2H - (\lambda + \lambda_{AB})M$$

since $M_A + M_B = M$. Thus, Eq. (10-132) becomes

$$M = \frac{Ng^2\mu_B^2S(S+1)}{6k_BT}[2H - (\lambda + \lambda_{AB})M] \tag{10-233}$$

Equation (10-233) may be used to obtain the paramagnetic susceptibility in the form of the Curie–Weiss relationship [Eq. (10-157)]. This is done by letting

$$C = \frac{Ng^2\mu_B^2S(S+1)}{3k_B} \tag{10-234}$$

It will be recognized that Eq. (10-234) is identical to Eq. (10-164a). The substitution of Eq. (10-234) into Eq. (10-233) simplifies it to read

$$M = \frac{C}{2T}[2H - (\lambda + \lambda_{AB})M] \tag{10-235a}$$

Or, expanding,

$$M = \frac{C}{T}H - \frac{C}{2T}(\lambda + \lambda_{AB})M$$

Then, upon rearrangement,

$$M\left[1 + \frac{C}{2T}(\lambda + \lambda_{AB})\right] = \frac{C}{T}H$$

This is solved for M to obtain

$$M = \frac{CH}{T[1 + C/2T(\lambda + \lambda_{AB})]} \tag{10-235b}$$

Equation (10-235b) reduces to

$$M = \frac{CH}{T + C/2(\lambda + \lambda_{AB})} \tag{10-235c}$$

So the paramagnetic susceptibility is

$$\chi_P = \frac{M}{H} = \frac{C}{T + C/2(\lambda + \lambda_{AB})} = \frac{C}{T + \theta} \tag{10-235d}$$

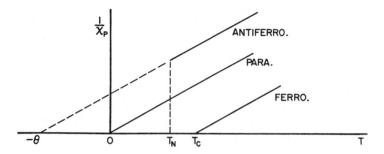

Figure 10-36   Schematic diagram of the relationships of the reciprocal paramagnetic susceptibility for different types of magnetic materials.

where, in the form of Eq. (10-166),

$$\theta = \frac{C}{2}(\lambda + \lambda_{AB}) \tag{10-236}$$

Thus, θ is inversely proportional to the molecular field constants.

The similarity of the relationship given by Eq. (10-235d) and that for the Curie–Weiss law for ferromagnets [Eq. (10-166)] will have been noted.   Here, the sign of θ is positive in Eq. (10-235d), while that of T_c is negative, but θ *is not equal to* T_N.   These differences are shown in Fig. 10-36 as variables affecting $1/\chi_P$.   Figure 10-37 also shows these differences for χ as a function T.

As shown in Fig. 10-36, $1/\chi_P$ is a linear function of T for $T \geqq T_N$ that extrapolates to zero at $T = -\theta$.   Thus, θ itself is negative. As such, antiferromagnetic materials

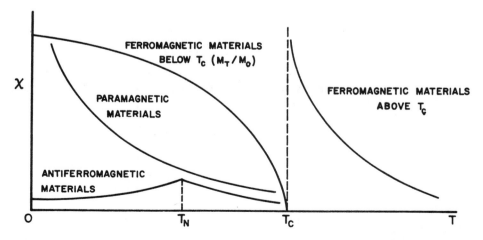

Figure 10-37   Comparative magnetic susceptibilities of different types of magnetic materials.

may be considered to obey the Curie–Weiss law [Eq. (10-166)] by reexpressing Eq. (10-235) as

$$\chi_P = \frac{C}{T - (-\theta)} \qquad (10\text{-}235e)$$

Since θ is intrinsically negative and is proportional to the molecular field constants [Eq. (10-236)], the molecular field is oriented opposite to the external field and tends to react against it. This is consistent with Eqs. (10-230) and, in essence, is the same as saying that J_e is negative.

10.3.1.2 Néel temperature.   The most direct and least complicated way of determining T_N is to begin at higher temperatures and to decrease them until $T = T_N$. The sublattices at T_N will then show their optimum spin imbalance, so the crystal will show its maximum spontaneous magnetization. So, for this condition, Eqs. (10-231a) and (10-231b) may be written in simplified form, using Eq. (10-234) for $H = 0$, as

$$M_A = \frac{C}{2} H_A = -\frac{C}{2T} (\lambda M_A + \lambda_{AB} M_B) \qquad (10\text{-}237a)$$

and

$$M_B = \frac{C}{2} H_B = -\frac{C}{2T} (\lambda M_B + \lambda_{AB} M_A) \qquad (10\text{-}237b)$$

These equations are expanded and rearranged as

$$M_A \left(1 + \frac{C}{2T} \lambda\right) + \frac{C}{2T} \lambda_{AB} M_B = 0 \qquad (10\text{-}238a)$$

and

$$M_B \left(1 + \frac{C}{2T} \lambda\right) + \frac{C}{2T} \lambda_{AB} M_A = 0 \qquad (10\text{-}238b)$$

The determinant of the coefficients of M_A and M_B of Eqs. (10-238) must be zero in order that meaningful values may be obtained for M_A and M_B. This determinant is

$$\begin{vmatrix} 1 + \dfrac{C}{2T} \lambda & \dfrac{C}{2T} \lambda_{AB} \\[2ex] \dfrac{C}{2T} \lambda_{AB} & 1 + \dfrac{C}{2T} \lambda \end{vmatrix} = 0$$

and, after taking the square root of the result, it reduces to

$$1 + \frac{C}{2T} \lambda = \frac{C}{2T} \lambda_{AB}$$

This may be rewritten as

$$T_N = \frac{C}{2}(\lambda_{AB} - \lambda)$$

So that the factor $1/2T$ vanishes, giving

$$2T + C\lambda = C\lambda_{AB}$$

so that

$$2T = C\lambda_{AB} - C\lambda = C(\lambda_{AB} - \lambda)$$

Then the Néel temperature is obtained as

$$T_N = \frac{C}{2}(\lambda_{AB} - \lambda) \qquad (10\text{-}239)$$

It is apparent from Eq. (10-239) that T_N will increase as the Weiss constant for the interaction of the two sublattices, λ_{AB}, increases and as those of the individual sublattices decrease. This may be rationalized by analogy with Eq. (10-196c). If λ_{AB} increases, so does J_e, and the exchange energy increases, too [Eq. (10-190)]. Such an increase in E_x affords greater resistance to the spin disarray that is induced by the increasing thermal energy. Hence, T_N is expected to increase accordingly. This effect would be augmented if the molecular field constants of the sublattices simultaneously decreased with increasing temperature.

The difference between θ and T_N, discussed in the previous section, may be emphasized by comparing Eq. (10-236) with Eq. (10-239). Thus,

$$\frac{\theta}{T_N} = \frac{C/2(\lambda + \lambda_{AB})}{C/2(\lambda_{AB} - \lambda)} = \frac{\lambda + \lambda_{AB}}{\lambda_{AB} - \lambda} \qquad (10\text{-}240)$$

As a consequence of Eq. (10-240), $\theta/T_N = 1$ when $\lambda = 0$, the case in which no magnetic interactions take place in either sublattice. And, where $\lambda \to \lambda_{AB}$, θ/T_N would be very large. This does not occur, since $\theta/T_N \gtrsim 3$ as shown in Table 10-9. Values significantly larger than this mean that next-nearest neighbors in the sublattices are not magnetically parallel, a condition requiring four sublattices for rationalization. Such a crystalline configuration can be derived from the construction of four simple cubic sublattices that can be evolved from a FCC crystal lattice. So, for the model presented here, $|\theta|$ must be more than T_N [Eq. (10-235e) and Fig. 10-36], and λ must be less than λ_{AB}.

10.3.1.3 Temperatures below T_N. Assuming that $M_A = M_B$ and that the easy direction of magnetization in each sublattice is parallel to that of the other, neglecting anisotropy effects, the properties of antiferromagnetic materials can be understood by examining the effects of an external magnetic field.

Where the external field is perpendicular to the molecular fields of the sublattices, it partially diminishes the tendency for parallel magnetizations by rotating

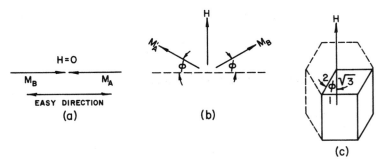

Figure 10-38 Antiferromagnetic materials: (a) magnetic vectors in the absence of a field; (b) magnetic vectors in a field perpendicular to the easy direction of magnetization; (c) application of a magnetic field perpendicular to a prism plane of a hexagonal lattice.

them by a small angle out of the easy direction. [E_x is diminished because J_e is slightly more positive than -1, and the spins are not quite parallel on each sublattice, Eq. (10-189).] The magnetization directions are now displaced by a small angle ϕ [Fig. 10-38(b)]. Thus, at equilibrium, neglecting any negligible changes in λ, and since $M_A = M_B$,

$$H = 2\lambda_{AB}M_A\phi \tag{10-241}$$

And, since $M = M_A + M_B$, the total magnetization is

$$M = 2M_A\phi \tag{10-242}$$

Then the susceptibility perpendicular to the easy direction is, from Eqs. (10-241) and (10-242),

$$\chi_\perp = \frac{M}{H} = \frac{2M_A\phi}{2\lambda_{AB}M_A\phi} = \frac{1}{\lambda_{AB}} \tag{10-243}$$

Thus, $\chi_\perp$ is a constant and is unaffected by temperature (Fig. 10-39).

At 0 K, $M_A + M_B = 0$ and $\chi = 0$ [Fig. 10-38(a)]. However, when an external field is applied parallel to the easy direction of magnetization and the temperature is increased, the magnetization of the sublattice parallel to the external field is increased and that antiparallel to the applied field is diminished. This difference between

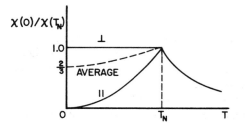

Figure 10-39 Relative magnetic susceptibility of antiferromagnetic materials as a function of direction and temperature.

magnetizations is the spontaneous magnetization. This increases continuously until, at T_N, $\chi_\| = \chi_\perp$.

The susceptibilities of materials with more complex lattice types or with polycrystalline or powdered structures may be described by a more general relationship. This is based on the idea that the complex crystal has an easy direction of magnetization or that the grains or the powders are randomly distributed. In the latter cases, the external field will be at some angle θ with the easy direction of a grain or of a single particle. Thus, the magnetic properties of the materials may be based relative to those of their easy directions. The subscripts $\|$ and $\perp$ denote properties parallel to and perpendicular to that of the easy directions. Thus, the susceptibility is given by

$$\chi = \frac{M}{H} = \frac{MH}{H^2} = \frac{M(H_\| + H_\perp)}{H^2} \tag{10-244}$$

and the two magnetic fields of interest are

$$H_\| = H \cos \phi \tag{10-245a}$$

and

$$H_\perp = H \sin \phi \tag{10-245b}$$

Equations (10-245) are substituted into Eq. (10-244) to obtain, after factoring H,

$$\chi = \frac{M(H \cos \phi + H \sin \phi)}{H^2} = \frac{M(\cos \phi + \sin \phi)}{H} \tag{10-246}$$

The susceptibility may also be expressed in terms of its component magnetizations as

$$\chi = \frac{M_\| \cos \phi + M_\perp \sin \phi}{H} \tag{10-247}$$

And the components of the susceptibilities are obtained, using Eq. (10-245a), as

$$\chi = \frac{M_\|}{H_\|} = \frac{M_\|}{H \cos \phi}, \qquad M_\| = \chi_\| H \cos \phi \tag{10-248a}$$

and, similarly,

$$\chi_\perp = \frac{M_\perp}{H_\perp} = \frac{M_\perp}{H \sin \phi}, \qquad M_\perp = \chi_\perp H \sin \phi \tag{10-248b}$$

Equations (10-248) are substituted into Eq. (10-247) to obtain, noting that H vanishes,

$$\chi = \chi_\| \cos^2 \phi + \chi_\perp \sin^2 \phi \tag{10-249}$$

Equation (10-249) may be applied to single crystals with complex crystal structures. However, for polycrystalline or powdered materials, the average values of the trigonometric components of Eq. (10-249) must be used. If it is assumed that the grains or particles have random orientations within the solids, the same method of

TABLE 10-9 PROPERTIES OF SOME ANTIFERROMAGNETIC CRYSTALS

Substance	Crystal Structure	T_N (K)	θ (K)	$\dfrac{\theta}{T_N}$	$\dfrac{\chi(0)}{\chi(T_N)}$
MnO	FCC	122	610	5.0	0.69
FeO	FCC	185	570	3.1	0.77
MnS	FCC	165	528	3.2	0.82
MnF_2	BCT	67	82	1.2	0.76
FeF_2	BCT	79	117	1.5	0.72
$FeCl_2$	Hexagonal layer	24	-48	-2.0	<0.2
$CoCl_2$	Hexagonal layer	25	38.1	1.5	—
Cr_2O_3	Rhombic	307	1070	3.5	0.76
Cr	BCC	310	—	—	—
Mn (α)	Complex	140	—	—	—

Abstracted from A. H. Morrish, *Physical Principles of Magnetism*, p. 463, John Wiley & Sons, Inc., New York, 1965, and C. Kittel, *Introduction to Solid State Physics*, 5th ed., p. 483, John Wiley & Sons, Inc., New York, 1976.

averaging used to obtain Eq. (10-36) is used to obtain $\langle \cos^2 \phi \rangle = \frac{1}{3}$ and $\langle \sin^2 \phi \rangle = \frac{2}{3}$. So, for randomly oriented materials, Eq. (10-249) becomes

$$\chi \cong \tfrac{1}{3}\chi_\parallel + \tfrac{2}{3}\chi_\perp \qquad (10\text{-}249\text{a})$$

Since $\chi_\perp$ is not a function of temperature [Eq. (10-243)] and $\chi_\parallel = 0$ at $T = 0\,\text{K}$, then $\chi(0) = \frac{2}{3}\chi_\perp$ at 0 K. And, at $T = T_N$, $\chi_\parallel = \chi_\perp$, so $\chi(T_N) = \chi_\perp$. These relationships give the ratio

$$\frac{\chi(0)}{\chi(T_N)} = \frac{\frac{2}{3}\chi_\perp}{\chi_\perp} = \frac{2}{3} \qquad (10\text{-}250)$$

Considering the simplifying assumptions that were made to obtain Eqs. (10-249) and (10-250), Eq. (10-250) compares well with the experimental data given in Table 10-9.

Equations (10-249) and (10-250) may be used to approximate the properties of antiferromagnetic materials with hexagonal lattice types. This is done by referring the susceptibilities to the rhombic unit cell. The assumptions are made that the easy direction is parallel to an edge of the base of the unit cell (parallel to the a axes) and that the magnetization in the c direction is small and may be neglected. If the external field is perpendicular to a prism plane and parallel to the base of the unit cell, $\sin \phi = \frac{1}{2}$ and $\cos \phi = \sqrt{3}/2$ [Fig. 10-38(c)]. This is so because H bisects one of the two equilateral triangles that constitute the equilateral parallelogram forming the base of the unit cell and $\phi = 30°$. These data substituted into Eq. (10-249) give

$$\chi \cong \tfrac{3}{4}\chi_\parallel + \tfrac{1}{4}\chi_\perp \qquad (10\text{-}249\text{b})$$

and, in the same way as for Eq. (10-250),

$$\frac{\chi(0)}{\chi(T_N)} \cong \frac{\frac{1}{4}\chi_\perp}{\chi_\perp} = \frac{1}{4} \qquad (10\text{-}250\text{a})$$

These results are in agreement with the data given for $FeCl_2$ in Table 10-9 and with other experimental data for $FeCO_3$ and CrSb. Such agreement could have been obtained for polycrystalline or for powdered materials only if the grains or particles were highly oriented with respect to the external field.

10.3.1.4 Metals. The metal ions are much more closely arrayed in their crystal lattices than are the compounds described in the preceding sections. In addition, metals have nearly free electrons in relatively wide bands, in contrast to the compounds that have virtually no free electrons and have filled valence bands. Thus, the individual, local magnetic moments postulated in Section 10.3 are absent. A further complication arises from the fact that magnetizations on the sublattices are complex and do not always oppose each other in the simple, essentially collinear way used in the previous descriptions. So, while some metals show relatively simple antiferromagnetic sublattices and distinct changes at T_N and obey the Curie–Weiss law for $T > T_N$, others behave in more complicated ways. The previously outlined theoretical approach is not applicable for most antiferromagnetic metals.

The behavior of α manganese (complex cubic structure) is essentially that which is to be expected, with $T_N \cong 130$ K. In contrast, chromium, with a much simpler BCC lattice, shows the expected antiferromagnetic behavior below $T_N \cong 310$ K. However, its mass susceptibility undergoes a large change in slope at that temperature. These elements do not follow the Curie–Weiss law for $T > T_N$. The value of $d\chi/dT$ for Cr remains positive for $T > T_N$.

Many intermetallic compounds show antiferromagnetic properties, a considerable number of which involve Mn or Cr. These are of the form MX or MX_2, in which M is either Mn or Cr and X is another metal. The spin configurations on the sublattices of these compounds are complex, and the magnetic orderings of many of these remain to be explained. Some binary Mn-rich alloys of Au, Cu, and Cr show no long-range order, but are antiferromagnetic. Many of these show no well-defined peaks at T_N; they do not obey the Curie–Weiss law.

Almost all the rare earth elements (the series from La to Lu) show antiferromagnetic properties. The elements Ce, Pr, Nd, Sm, and Eu are antiferromagnetic at 0 K and show a T_N that is followed by paramagnetic behavior. The elements Gd, Tb, Dy, Ho, Er, and Tm show ferromagnetic properties from 0 K to a T_c that is followed by antiferromagnetic states up to a T_N, where they change to paramagnetic behavior. With the exception of Gd, with a simple ferromagnetic structure, the other elements have different, complex magnetic structures in each magnetic regime. In addition to the complications noted previously for metal ions, these elements have unbalanced 4f levels deep within their electron "clouds". These 4f levels are highly shielded by the surrounding, completed electron levels. Thus shielded from a crystal field, they are unaffected by it, and their magnetic moments contain components arising from orbital motion as well as from spin. It is therefore not surprising that these elements show the ranges and kinds of magnetic properties noted previously.

10.4 FERRIMAGNETISM

Ferrimagnetism arises from an antiparallel array of magnetic dipoles similar to that described for antiferromagnetism, but, unlike the antiferromagnetic case, the dipoles are unbalanced [Fig. 10-40(a)]. The spin imbalance results in a spontaneous magnetism similar to that of ferromagnetism. These materials behave like ferromagnetic materials in that they are composed of domains, become saturated, and show hysteresis losses, but have virtually no free electrons. They also undergo a transition to the paramagnetic state at a critical temperature T_{FN}, *the ferrimagnetic Néel temperature.* (The notation T_{FN}, borrowed from A. H. Morrish, is used to distinguish it from T_c and T_N. It should, however, be noted that T_c is most commonly used for this transition.) They obey the Curie–Weiss law at temperatures above T_{FN}. These materials, double oxides of iron and those of another metal, are called ferrites, but they do not include the α solid solutions of iron that bear the same name.

Magnetite, or lodestone, is the oldest known ferrimagnetic material. It has a complex cubic lattice and the composition $FeO \cdot Fe_2O_3$. Other magnetic ferrites are formed by the substitution of Mn, Co, Ni, Cu, Mg, Zn, or Cd for the divalent Fe ion. With the exception of $CoO \cdot Fe_2O_3$, all the other ferrites are magnetically soft. Materials in which the trivalent ion is replaced by Mn, Co, Al, or Ga are known as *spinels* or as *ferrospinels*. These also have complex cubic lattices in which the divalent, nonmetallic ion usually is O, but may be S or Se. "Mixed" ferrites contain several divalent metallic ions each of which serves as a partial replacement for the divalent Fe ion in the "molecule." Two examples of such mixed ferrites are $Mn_{0.5}Zn_{0.5}O \cdot Fe_2O_3$ and $Ni_{0.3}Zn_{0.7}O \cdot Fe_2O_3$.

The cubic ferrites may be considered to have unit cells whose matrices consist of 32 close-packed O ions within which the other ions are located. In the "normal" cell, divalent ions occupy octahedral sites; these have 6 oxygen ions as nearest neighbors, and the 16 trivalent ions, having 4 nearest oxygen neighbors, are on tetrahedral sites [Fig. 10-40(b)]. The unit cell thus contains 56 ions or seven $MO \cdot Fe_2O_3$ "molecules". The "inverse" structure contains the divalent ions on octahedral sites, while the trivalent ions occupy equal numbers of tetrahedral and octahedral sites, [Figs. 10-40(b) and (c)].

Some ferrites, such as $BaO \cdot 6Fe_2O_3$, have complex hexagonal unit cells. The divalent Ba and O ions form a close-packed array, since they both are about the same "size" and are much larger than the Fe ions. The Fe ions occupy interstitial sites in this matrix. The magnetic properties arise only from the Fe ions, since the Ba and O ions have no magnetic moments. The unit cells of these compounds contain 64 ions (two "molecules") arranged within ten layers of four Ba or O ions. Four O ions are contained in eight of the layers, while two layers contain one Ba ion in each. The Fe ions are situated interstitially on 4 tetrahedral, 18 octahedral, and 2 hexahedral sites (with five O ion nearest neighbors), with magnetic moments perpendicular to the layers of oxygen that are normal to the c axis. The magnetic moments of the Fe ions are arranged so that 16 are arrayed antiparallel to the remaining 8 [Fig. 10-40(c)]. The easy direction of magnetization, thus, is parallel to the c axis.

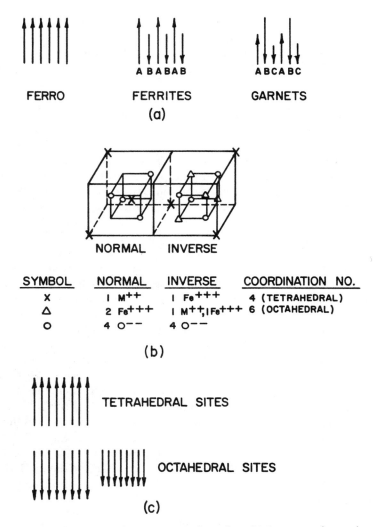

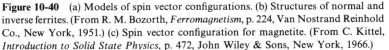

Figure 10-40   (a) Models of spin vector configurations. (b) Structures of normal and inverse ferrites. (From R. M. Bozorth, *Ferromagnetism*, p. 224, Van Nostrand Reinhold Co., New York, 1951.) (c) Spin vector configuration for magnetite. (From C. Kittel, *Introduction to Solid State Physics*, p. 472, John Wiley & Sons, New York, 1966.)

Garnets, materials with complex cubic structures, also have ferrimagnetic properties. Natural garnets have the composition $3MnO \cdot Al_2O_3 \cdot 3SiO_2$. Garnets may also be made in which divalent ions replace Mn ions and trivalent ions replace the Al ions. Other garnets have the composition $3M_2O_3 \cdot 5Fe_2O_3$. Here M may be one of the rare earth series of elements from Gd to Lu or Y. More complex, "mixed" garnets, such as $(EuY)_3 \cdot (GaFe)_5 \cdot O_{12}$ find applications in bubble-domain devices

(Section 10.2.5.1). With the exception of YIG (ytterbium-iron garnets), the other rare earth-containing garnets show magnetizations that decrease with temperature, go through zero, and then show net magnetizations of the opposite sign. The temperature at which the magnetization goes through zero, that is, the point at which the magnetizations of the sublattices exactly balance each other, is called the *compensation point*.

The unit cells of garnets contain 160 ions (4 "molecules") in a way analogous to that described for the cubic ferrites. The Fe ions are divided among 16 tetrahedral sites and 24 octahedral sites and show strong antiparallel behavior. The 24 M ions lie on octahedral sites between those of the Fe ions on similar sites. When rare earth ions occupy these sites, they are antiparallel to the net effect of the Fe ions. This accounts for their relatively weak magnetizations. YIG materials have much stronger magnetizations because the Y ion does not have a magnetic moment and, therefore, does not diminish the magnetic Fe imbalance.

As in the case of antiferromagnetics, these materials are insulators, and the magnetic ions are widely separated by nonmagnetic ions (ions with no net spin imbalance). Thus, if any significant exchange is to take place, the intermediate, nonmagnetic ions must participate indirectly to help to correlate the spins on the magnetic ions (Sections 10.2.1 and 10.3). This mechanism is known as *superexchange*. It may be illustrated by gradually bringing two magnetic ions, M^{2+}, close to an O^{2-} ion. At the start, the magnetic moments of the two M^{2+} ions are randomly oriented, while the oxygen ion, because of its closed inner shells, is nonmagnetic. The probability densities of the two outer electrons of the O^{2-} ion constitute two half-filled, collinear, looplike or dumbell-shaped "orbitals" that can accommodate two electrons of opposite spin. When an M^{2+} ion with its spin up is brought close enough to the O^{2-} ion, it will jointly occupy the "orbital" of the O^{2-} ion with spin down. The remaining "orbital" of the O^{2-} ion now contains an unbalanced spin-up electron. When a second M^{2+} ion is brought close to the O^{2-} ion, from a direction opposite to that of the first M^{2+} ion, the electron that it shares with the remaining O^{2-} "orbital" must have its spin down. If this electron did not have its spin down, its spin would have to change in accordance with the principle of exclusion. This sequence of events is simply illustrated as follows:

Original situation: M^{2+} $\uparrow O^{2-}\downarrow$ M^{2+}

First ion reaction: M^{2+} $\uparrow O^{2-}\downarrow\uparrow M^{2+}\downarrow$

Second ion reaction: $\uparrow M^{2+}\downarrow\uparrow O^{2-}\downarrow\uparrow M^{2+}\downarrow$

Thus, opposite spins on next-nearest M^{2+} ions are ensured through the intercession of the shared O^{2-} (Section 3.12.3). The energy of this type of bonding is minimized (the bond strength is maximized) when all three ions are collinear. Other nonlinear configurations are less energetically favorable.

This type of exchange may be most simply illustrated by an NaCl-type lattice that is composed of two interpenetrating FCC lattices. One of these contains the M^{2+} ions and the other contains the O^{2-} ions. The nearest neighbors of M^{2+} ions in the composite NaCl lattice are O^{2-} ions that are bonded by paired spins. Next nearest M^{2+} neighbors have opposite spins. This tendency for the concentration of opposite spins upon next-nearest magnetic neighbors, through the agency of non-magnetic nearest neighbors, driven by the consequent lowering of the bonding energy, constitutes the mechanism of superexchange. As noted for the case of antiferromagnetism, the localization of the magnetic moments on the M^+ ions makes it more capable of being explained by the molecular field theory than is the case for ferromagnetism.

When ions with no magnetic moment are added so that they substitute for some of the M^{2+} ions, the moments on the M^{2+} sublattice are decreased. This diminishes the amount of superexchange and results in decreased spontaneous magnetization and lowered T_{FN}.

10.4.1 Molecular Field Theory

The first explanation for the magnetic properties of ferrimagnets was given by L. Néel (1948). This approach considered that negative magnetic interactions took place between the ions on two cubic sublattices. An ion on a given A or B sublattice was treated as being affected by the molecular field of the other sublattice, $M_A \lambda_{AB}$ or $M_B \lambda_{BA}$, where $\lambda_{AB} = \lambda_{BA}$, as well as by that of its own molecular field, $M_A \lambda_A$ or $M_B \lambda_B$.

The molecular field acting on the A sublattice is

$$M_{mA} = \lambda_A M_A - \lambda_{AB} M_B \qquad (10\text{-}251a)$$

and that acting on the B sublattice is

$$H_{mB} = \lambda_B M_B - \lambda_{AB} M_A \qquad (10\text{-}251b)$$

Here the molecular field constant are positive, $\lambda_A \neq \lambda_B$, the negative signs arise from the assumption of antiparallel interactions, and only nearest neighbors are considered. Equations (10-251) are valid above and below T_{FN}.

When an external magnetic field H is applied, the magnetizations of the sublattices are

$$H_{mA} = H + \lambda_A M_A - \lambda_{AB} M_B \qquad (10\text{-}252a)$$

and

$$H_{mB} = H + \lambda_B M_B - \lambda_{AB} M_A \qquad (10\text{-}252b)$$

If it is assumed that the Curie law is followed in the paramagnetic region above T_{FN} [Eq. (10-84)], then the magnetizations may be treated classically in the form of

$$MT = \frac{N\mu_m^2 H_T}{3k_B} = CH_T \qquad (10\text{-}253)$$

where H_T is the total magnetic field, because of the localization of the magnetic ions. Thus,

$$M_A T = C(H + H_{mA}) \qquad (10\text{-}254a)$$

and

$$M_B T = C(H + H_{mB}) \tag{10-254b}$$

The paramagnetic susceptibility is obtained, by eliminating H_{mA}, H_{mB}, M_a, and M_b [Eqs. (10-251) and (10-254)], as

$$\chi_P = \frac{CT - \lambda_{AB}C^2\lambda_A\lambda_B(2 + \alpha + \beta)}{\lambda_{AB}CT(\alpha\lambda_A + \beta\lambda_B) + \lambda_{AB}^2C^2\lambda_A\lambda_B(\alpha\beta - 1)} \tag{10-255}$$

where

$$\alpha = \frac{\lambda_A}{\lambda_{AB}} \quad \text{and} \quad \beta = \frac{\lambda_B}{\lambda_{AB}}$$

It can be shown that Eq. (10-255) reduces to

$$\chi_P = \frac{C}{T + (C/\chi_0)} \tag{10-256}$$

for $T \gg T_{FN}$, where C/χ_0 is the linear extrapolated analog of θ in Fig. 10-36 and is treated in a way similar to that shown in Eq. (10-235a). The parameter χ_0 is given by

$$\frac{1}{\chi_0} = \lambda_{AB}(2\lambda_A\lambda_B - \alpha\lambda_A^2 - \beta\lambda_B^2) \tag{10-256a}$$

The value of $1/\chi_P$ is zero for $T = T_{FN}$, because χ_P is infinite at T_{FN}. Thus, the values of this parameter show a large curvature in the temperature range just above T_{FN}, after which the noted linear behavior is observed.

In the range of temperatures below T_{FN}, each sublattice is spontaneously magnetized oppositely to the other one, so

$$|M| = |M_A| - |M_B| \tag{10-257}$$

using Eq. (10-139), and since spin alone is considered here, the spin notation S is used instead of J; the magnetizations are

$$M_A = Ng\mu_B S \cdot B_S\left(\frac{Sg\mu_B H_{mA}}{k_B T}\right) \tag{10-258a}$$

$$M_B = Ng\mu_B S \cdot B_S\left(\frac{Sg\mu_B H_{mB}}{k_B T}\right) \tag{10-258b}$$

Absolute quantities are shown in Eq. (10-257), so the differences in directions of magnetizations of the sublattices are taken into account. This not only results in magnetization curves like that Fig. 10-41, but also in curves of many other shapes. Among these is one in which the net magnetization as a function of temperature is positive, goes through a compensation point, and then becomes negative. Other curves may show a minimum rather than a compensation point. Many other such curve shapes are possible, depending on the relative magnetizations [Eq. (10-257)].

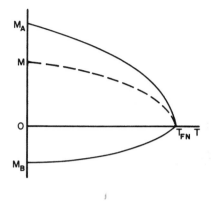

Figure 10-41 Magnetizations of sublattices A and B of a ferrimagnetic material showing the resultant magnetization M, Eq. (10-257). Note that the text uses the less common notation of T_{FN} for T_c for clarity.

10.5 MAGNETIC MATERIALS

Magnetic units are discussed in Section 10.1.1 and their applications to B–H curves are given in Section 10.2.6. A brief review of some of these is given here for convenience. H is the external magnetizing field in units of oersteds. B is the magnetic flux density, or induction, in gauss. The residual magnetization, or remanence, or residual induction, when H is absent, is denoted by B_r. The saturation magnetization is given by B_s. The coercive field, H_c, is a measure of the field required to demagnetize the material, that is, to reduce the residual magnetization to zero. The energy required by a full cycle of magnetization is given by the area within the B–H curve (Fig. 10-31).

The permeabilities of magnetically soft materials are high. They have relatively high values of B_r and B_s and very small values of H_c. Most engineering applications of these materials require that they absorb a minimum of energy in undergoing magnetic cycling. This requires that H_c be minimum for a maximum B_s, a condition that minimizes the area enclosed by the B–H curve. In contrast to this, it is desirable that magnetically hard materials have magnetic properties opposite to those of magnetically soft materials. Here, large values for both B_r and H_c and the resulting large-area B–H curves are required (Sections 10.2.4–10.2.6). This condition ensures that the magnetization will be large and "permanent".

The demagnetization curve, the portion of the B–H curve in the second quadrant of Fig. 10-31, may be used for the selection of magnetic materials. The values of B_r and H_c constitute the selection criteria. A more specific basis for material selection is given by the energy product (Fig. 10-42): the plot of the products of B and H for a series of points selected from the demagnetization curve given by the energy-product curve. Many engineers consider that the *energy product*, $(B_d \cdot H_d)_{max}$, is the preferred single measure for the selection of materials and/or comparison of hard magnetic materials.

Another measure, used primarily for the design of hard magnetic materials, is

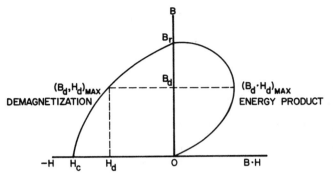

Figure 10-42 Schematic diagrams of the demagnetization and energy-product curves for a permanent magnet.

given by the *permanence coefficient*. This is the negative slope of B versus H:

$$\frac{B_d}{H_d} = \frac{A_g l_m}{A_m l_g} \qquad (10\text{-}259)$$

where B_d and H_d give the energy product for an operating point on the demagnetization curve, A_g and l_g are the respective cross-sectional area and length of the air gap, and A_m and l_m are the corresponding parameters of the magnet. This enables the optimum selection of materials for a given magnet configuration or the dimensions of the magnet to be made from a given material. Equation (10-259) must be regarded as an approximation because it neglects several important practical considerations.

10.5.1 Permanent Magnet Alloys

The early permanent magnets were made of high-carbon steels in the quenched and tempered condition. Later, steels containing tungsten, chromium, and cobalt were used (see Table 10-10). Some of these are given to show the effects of alloying constituents. The plain-carbon steels, now obsolete for magnetic purposes, have been included for purposes of comparison with other ferrous alloys. It will be noted that the values of B_r lie within a narrow range. However, the coercive forces and the maximum energy products increase considerably, indicating magnetic materials of increasing energy product and magnetic permanence as the alloy content increases.

Magnet steels, such as those shown in Table 10.10, should not be used at temperatures above 100°C because of the metallurgical changes that occur in these steels upon tempering. The resulting changes in their metallographic structure (martensite) are accompanied by decreases in magnetic properties. When such steels are maintained in the as-quenched condition, an aging effect occurs at room temperature. Some decrease occurs in their energy products. The largest portion of this change takes place immediately after quenching, but they approach a stable value in a few weeks. The magnetic properties of these steels are also adversely affected by mechanical shock. After repeated shocks, the properties approach some lower, approximately stable values. Thermal cycling and mechanical treatments, in excess of those anticipated in service, have been used to minimize such changes.

TABLE 10-10   TYPICAL PROPERTIES OF SOME MAGNET STEELS

Materials	Nominal Composition (weight percent)					B_r (G)	H_c (Oe)	$(B_d H_d)_{max}$ (millions)
	C	Mn	Cr	W	Co			
Carbon steel	0.65	0.85	—	—	—	10,000	42	0.18
Carbon steel	1.00	0.50	—	—	—	9,000	51	0.20
Chromium steel	1.00	0.50	3.50	—	—	9,500	66	0.29
Tungsten steel	0.70	0.50	0.50	6.00	—	9,500	74	0.33
Cobalt steel	0.70	0.35	2.50	8.25	17.00	9,500	170	0.65
Cobalt steel	0.80	0.55	5.75	3.75	36.00	9,750	240	0.93

Abstracted from "Magnetic Electrical and Other Special Purpose Materials", *Metals Handbook, Properties and Selection*, Vol. 1, 8th ed., T. Lyman, ed., American Society for Metals, Metals Park, Ohio, 1961, pp. 779–864.

The carbon-free alloys (Table 10-11) represent only a few of the commercial alloys. These are much less susceptible to aging (when properly treated) than are the steels. Remalloy, which was the first of this class of alloys to be produced, is representative of the Fe–Co–Mo alloys. These are dispersion hardened and can be forged and rolled. The magnetic properties of this class of alloys are intermediate between those of the alloy steels and of the Alnico alloys.

Two other members of this intermediate group, Cunife and Cunico, are precipitation hardening and can be worked mechanically. This permits their fabrication into sheets, thin tapes, and wires. The best properties of Cunife are obtained after successive cold reductions that total about 95% reduction in area, during which intermediate and final thermal treatments are used. Such treatments can approximately double the energy product of Cunife. The increases in B_r, H_c, and the energy product are anisotropic in wire, sheet, and tape. Here the maximum properties are parallel to the rolling direction. The optimum properties of Cunico are developed by a process consisting of cold working followed by reheating, rapid quenching from the annealing temperature, and subsequent aging treatment. The strains induced by the precipitates in both alloy types make domain boundary motion more difficult and result in increased values of H_c (Sections 10.2.4 and 10.2.5). The precipitates are considered to be single domains that act to increase the H_c.

The Remalloy alloys (Table 10-11) also show variations in magnetic properties as a result of thermal treatments that control the size and distribution of the precipitates. They are more machinable and less sensitive to warping or cracking than are the cobalt steels shown in Table 10-10.

The Alnico group of alloys is brittle and cannot be worked. These alloys are available in the sintered or cast form. The mechanical properties of the sintered alloys are much higher than those of the cast alloys because of their smaller grain size.

The most widely used permanent magnet alloys belong to the Alnico class. These alloys are composed primarily of Fe, Ni, Al, Co, and Cu to which Ti and other elements have been added. These alloys are known by other trade names in Europe and Japan.

TABLE 10-11 NOMINAL COMPOSITIONS AND PROPERTIES OF SOME SELECTED PERMANENT MAGNET MATERIALS

Material	Composition					T_c (°C)	B_r (G × 10⁻³)	H_c (Oe)	$(B_dH_d)_{max}$ (Oe × 10⁻⁶)
	Al	Ni	Co	Cu					
Cast Alnico alloys									
Alnico 1	12	21	4	3	Fe	780	7.10	440	1.4
Alnico 2	10	19	13	3	Fe	810	7.25	580	1.6
Alnico 3	12	25	—	3	Fe	760	7.00	470	1.4
Alnico 4	12	27	5	—	Fe	800	5.35	730	1.3
Alnico 5	8.5	14.5	24	3	Fe	900	12.50	620	5.25
Alnico 5–7	8	16	24	3	2Ti Fe	900	13.20	730	7.4
Alnico 7	8	18	24	4	5Ti Fe	840	8.57	1050	3.7
Alnico 8	8	14	38	3	8Ti Fe	860	8.30	1600	5.0
Alnico 9	7	15	35	4	5Ti Fe	—	10.50	1450	8.5
Alnico 12	6	18	24	—	8Ti Fe	—	6.00	950	1.7
Sintered Alnico alloys									
Alnico 2	10	17	12.5	6	Fe	810	6.70	525	1.5
Alnico 4	12	28	5	—	Fe	800	5.20	700	1.2
Alnico 5	8.5	14.5	24	3	Fe	900	10.40	600	3.60
Alnico 6	8	16	24	3	2Ti Fe	860	8.80	760	2.75
Alnico 8	7	15	35	4	5Ti Fe	860	7.60	1550	4.5
Lodex 30	9.9 Fe, 5.5 Co, 77 Pb, 8.6 Sb					980	4.00	1250	1.6
Lodex 33	21.9 Fe, 12.3 Co, 59.2 Pb, 6.6 Sb					980	8.00	865	3.2
Lodex 42	19.2 Fe, 10.8 Co, 63 Pb, 7 Sb					980	8.30	845	1.4
Cunife	20 Fe, 20 Ni, 60 Cu					410	5.40	550	1.5
Cunico	29 Co, 21 Ni, 50 Cu					860	3.40	680	0.8
Vicalloy I	39 Fe, 51 Co, 10 V					855	12.90	240	0.9
Vicalloy II	35 Fe, 52 Co, 13 V					855	9.05	415	2.3
Remalloy 1	17 Mo, 12 Co, 68 Fe					900	9.70	250	1.0
Remalloy 2	20 Mo, 12 Co, 68 Fe					900	8.55	340	1.2
Platinum cobalt	76.7 Pt, 23.3 Co					480	6.45	4450	9.2
Cobalt–rare earth I Co₅Sm						725	9.80	9000	21.0

Abstracted from *Metals Handbook, Desk Edition*, H. E. Boyer & T. L. Gall, Eds., American Society for Metals, pp. 20-8 and 20-9, Metals Park, Ohio, 1985.

The magnetic properties of Alnico-type alloys in the as-cast or as-sintered conditions are poor, but are greatly improved by thermal treatments. These consist of homogenization treatments at 1250°C, then cooling at a rate of about $1°C\ s^{-1}$, followed by aging at 600°C. Alnico alloys 5, 6, and 7 are cooled from the homogenization temperature in a magnetic field of about 1000 to 3000 Oe and then are reheated and aged in the absence of a field. Alloys treated magnetically have highly anisotropic properties; their optimum magnetic properties are in the direction of the magnetic field applied during cooling, while those transverse to this field are inferior. Alnico alloys 8 and 9 are given the magnetic treatment isothermally at 815°C for times up to 10 minutes. Alnico alloys 1 to 4 are given the above thermal treatment without the application of a magnetic field and, thus, have isotropic magnetic properties. Magnetic anisotropy is taken into account in the design of magnets for specific applications.

The nature of the precipitated phase, produced by the thermal treatment, is considered to be responsible for the magnetic properties of this class of alloys. The homogenization treatment results in an α (BCC) solid solution. Rodlike, ordered, α' precipitates (CsCl structure), high in Fe and Co, form during cooling. These highly magnetic particles precipitate preferentially parallel to the $\langle 100 \rangle$ directions of the weakly magnetic, Ni- and Al-rich α matrix. Optimum precipitation conditions occur at the beginning of precipitation when the α' particles have had sufficient time to precipitate, but not enough time to grow or to coalesce excessively. The aging process also increases the Fe and Co contents of the α' at the expense of the α constituent. This increases the H_c, $(B_d H_d)_{max}$, and the spontaneous magnetization of the α' and greatly diminishes the corresponding properties of the α matrix. It is thought that the magnetic hardness of these alloys arises from the shape anisotropy of the α' rodlets because alloying with Ti increases their length and the values of H_c (Section 10.2.5). Cooling in the presence of strong magnetic fields enhances the preferential precipitation mode of the rods so that their easy direction of magnetization is parallel to the direction of the external field. This increases B_r and H_c. Preferred orientations, induced by controlled directional cooling, in as-cast structures also enhance the directionality of the α' precipitates.

The high coercive forces of the Alnico alloys result from the presence of the precipitated phase. The single-domain nature of the highly dispersed, fine precipitates with high values of H_c are primarily responsible for the alloys' high energy products and high H_c. The precipitated particles are small enough so that they consist of a single domain and, having no domain walls, demagnetization must occur by the simultaneous randomization of all the spins of all the particles. This requires high fields, as evidenced by the H_c of the alloys.

Alnico 2 frequently is used in preference to alloys 1, 3, and 4 because of its higher energy product. Alnico 3 is used frequently for magnets with small cross sections. Alnico 4 is used, because of its high H_c, where demagnetizing conditions may be present. Alnico 5 is the most commonly used of this class of alloys because of its high energy product.

Unlike the magnet steels, the Alnico alloys are virtually unaffected by mechanical shock or aging at ordinary temperatures. Fluctuations in ambient temperature have negligible effects on their properties. However, large variations in temperature and inadvertent exposure to high, external magnetic fields can result in unstable magnetic properties. Compensation for such exposures may be made by subjecting these materials to magnetic fields of greater strength than those expected in a given application.

As is the case for Alnico alloys, the Lodex alloys derive their high values of H_c from the shape of their fine, magnetic, FeCo particles. The Pb content controls the packing fraction of the FeCo and serves as their matrix. They are very useful for small parts, despite their smaller values of $(B_dH_d)_{max}$, because they are fabricated by pressing without sintering. This powder technique permits easy manipulation of such powder parameters as particle aspect ratio, size, and packing fraction, which are extremely difficult to control in sintered or cast materials. It also enables the production of magnets with very close dimensional tolerances and so eliminates any necessity for grinding.

The Vicalloys may be considered to consist of an Fe–Co base to which V has been added. The Fe–Co γ phase is FCC and is not ferromagnetic. This phase decomposes into a ferromagnetic, brittle, disordered α phase, with a BCC lattice, and into an ordered α' phase, with a CsCl lattice, in the composition range near FeCo. Small additions of V lower this transformation temperature, greatly increase the ductility, and increase T_c and B_r. Vicalloy 1 is homogenized at 1200°C, quenched, and aged at 600°C to precipitate the α' in retained γ. As-cast Vicalloy 2 is hot worked (in the γ region) followed by cold working to about 90% reduction to transform any retained γ to α. The treatment is completed by aging at 600°C, during which some of the α phase transforms to α'. It then has its easy direction of magnetization parallel to the direction of cold working. The mechanism responsible for the magnetic properties of the Vicalloy alloys is not, as yet, understood. Their values of H_c are thought to result from the properties of fine precipitates.

CoPt is of equiatomic composition with an FCC structure above 825°C. It transforms into a long-range ordered structure of the CuAu–I lattice configuration. This structure is tetragonal, with a c/a ratio of approximately 0.97, in which the atoms of Co may be considered to occupy the corner positions of a face-centered cube and the Pt atoms occupy the face sites. The planes parallel to the c axis, thus, consist of alternate layers of each atom. Its most desirable magnetic properties are obtained when partial crystalline order exists. This is obtained by forming a disordered solid solution at 1000°C, cooling at a controlled rate to room temperature, followed by aging at 600°C. This results in very small, single-domain, ordered regions in a disordered CoPt solid solution. Each of these phases is ferromagnetic, but the high values of H_c and $(B_dH_d)_{max}$, which justify the cost of this material, are primarily caused by the small, single-domain regions.

The cobalt–rare earth alloys are more cost effective than CoPt. They provide much stronger magnetic properties than does CoPt at considerably lower cost. These also belong to a class of alloys of compositions that include RTr, RTr_2, RTr_5, and R_2Tr_{17}, in which R is yttrium or a rare earth element and Tr is a transition element, usually of the first transition series. The alloy Co_5Sm appears to be the most promising of these. These materials have hexagonal ($CaCu_5$) lattice structures, with easy magnetization directions parallel to their c axes. They have the highest values of anisotropy constants of all magnetic materials (Section 10.2.5). Since the powders of which they are composed are too large for them to be single domains, they must contain domain walls. Their high values of H_c indicate that resistance to domain wall motion must be responsible for their magnetic properties (Sections 10.2.4 and 10.2.5).

10.5.2 Barium Ferrites

The bases for the properties of these "ceramic" magnetic materials are given in Sections 10.4 and 10.4.1. The barium ferrites are the most widely used of this class of materials. They compete with the Alnico alloys because of their high values of H_c, despite their lower values of B_r, and their comparatively low cost. Another factor considered in their application is that their electrical resistivities are much larger than those of the Alnico alloys, being greater by a factor of at least 10^6.

The properties of ferrites (Table 10-12) are considerably different from those of the Alnico alloys. The curve shapes indicated by these data and applied in Eq. (10-259) indicate that the shapes of ferrite magnets must be considerably different from those made with Alnico materials.

Ferrites are widely used in radios, magnetic recording (drums, tapes, and discs), and television tube focusing magnets. Electrical measuring devices, such as ammeters and voltmeters, make use of ferrite magnets. They are also used widely in small dc motors, torque drives, and other mechanisms in situations in which small, strong, permanent magnets are required because of space limitations.

Two other ferrites, which do not include BaO, also find considerable use. These are $SrO \cdot 6Fe_2O_3$ and $PbO \cdot 6Fe_2O_3$. These are treated in the same way as for $BaO \cdot 6Fe_2O_3$.

10.5.3 Magnetically Soft Alloys

The most desirable soft-magnet materials are those with minimum hysteretic energy losses (Sections 10.2.6 and 10.5). This means that such materials must have very low values for H_c and high values for B_r such that their energy products are very small.

TABLE 10-12 NOMINAL COMPOSITIONS AND PROPERTIES OF SOME SELECTED FERRITES[a]

Sintered Ferrites[b]	Composition	T_c (°C)	B_r (G × 10⁻³)	H_c (Oe)	$(B_d H_d)_{max}$ (Oe × 10⁻⁶)
Ferrite 1	$BaO \cdot 6Fe_2O_3$	450	2.20	1800	1.0
Ferrite 2	$BaO \cdot 6Fe_2O_3$	450	3.80	2200	3.4
Ferrite 3	$BaO \cdot 6Fe_2O_3$	450	3.20	3000	2.5
Ferrite 4	$SrO \cdot 6Fe_2O_3$	460	4.00	2200	3.7
Ferrite 5	$SrO \cdot 6Fe_2O_3$	460	3.55	3150	3.0
Ferrite 6	$BaO \cdot 6Fe_2O_3$	—	3.60	2900	3.1
Ferrite 7	$SrO \cdot 6Fe_2O_3$	—	3.43	3330	2.9

Abstracted from *Metals Handbook—Desk Edition*, H. E. Boyer and T. L. Gall, eds., pp. 20-8 and 20-9, American Society for Metals, Metals Park, Ohio, 1985, and B. D. Cullity, *Introduction to Magnetic Materials*, p. 558, Addison-Wesley Publishing Co., Reading, Mass., 1972.
[a]Anisotropic properties after thermal treatment.
[b]Magnetic Materials Producers Association designations.

The electrical resistivity of pure iron is too low for its application to the alternating-current circuits that constitute the largest application of soft magnetic materials. The low resistivity permits excessive eddy-current losses in alternating fields. The elements silicon and/or aluminum are added to form substitutional solid solutions with increased electrical resistivities that minimize this difficulty. In so doing, they decrease B_r and B_s. However, these alloys can have deleterious effects because alloy contents in excess of about 4.5 Wt% silicon or 8 Wt% aluminum induce metallurgical factors that cause embrittlement. Both of these alloying elements have desirable effects in that they limit the alpha to gamma allotropic transformation to narrow loops that confine the transformation to a range of about 2 Wt% beyond the iron terminus. Thus, most commercial alloys of these kinds have BCC lattices. Thus, large grain sizes can be produced in these materials without undergoing alpha to gamma transformation. This has the effect of decreasing H_c, since the smaller percentage of grain boundaries offers much less resistance to domain wall motion.

Small amounts of impurity elements such as carbon, sulfur, nitrogen, and oxygen have deleterious effects in alloys because they can distort the crystal lattice, form precipitates and/or inclusions, and adversely affect domain formation and motion. Higher concentrations of some of these elements can result in the precipitation of particles within the grains of iron and cause similar effects. Because of this, the carbon content of iron-based alloys is maintained as low as possible; concentrations of about 0.005 Wt% are preferred, with a maximum content of 0.01 Wt% C.

The effects of magnetic anisotropy are discussed in Sections 10.2.1, 10.2.4, and 10.2.5. Special thermal and mechanical treatments are given to iron–silicon alloys so that crystal orientation of the grains is oriented such that the cube edges of the unit cell are aligned parallel to the plane of the sheet. This gives an easy direction of magnetization perpendicular to the plane of the sheet [Fig. 10-24(a)]. These materials also have large grain sizes to keep the percentage of grain-boundary material small.

Iron–silicon alloys have lower values of H_c than does ingot iron. Alloys are available that vary in composition from 0.5 to 4.5 Wt% Si. Almost all these alloys are grain oriented. The values of B_r of oriented materials are about twice that of ingot iron. The essentially binary iron–silicon alloys, in the absence of impurity elements, show little or no aging. This minimizes increases in the energy product with time and temperature because of the absence of precipitation.

Nickel has the effect of increasing the permeability of Fe–Ni alloys, with a maximum at about 80 Wt% Ni. Their energy products have maxima in the neighborhood of about 50 Wt%. Most of the available alloys have nickel contents between 40 and 60 Wt%. Alloys with contents of about 50% Ni have comparatively high values of B_r and low values of H_c.

Iron–cobalt alloys have values of B_r and H_c higher than those of ingot iron. Alloys are available that contain up to about 65 Wt% Co. The maximum value of B_s occurs in the alloy containing 34.5 Wt% Co. Fe–Co alloys with Co in excess of about

30 Wt% Co are brittle. Suitably processed alloys with small additions of chromium or vanadium minimize this tendency.

The Fe–Si alloys are used to form components made from laminations as a means of minimizing eddy-current losses. Such losses decrease with increasing silicon content because of the corresponding increase in electrical resistivity; their permeabilities increase and their values of H_c decrease. Some typical applications include pole pieces for ac and dc electric motors, cores for transformers, stators for rotating machines, communications equipment, high-efficiency distribution and power transformers, and large generators.

The alloys noted in Tables 10-13 and 10-14 are considered to be typical of the large numbers of commercially available, soft magnetic alloys. Those containing higher amounts of alloying elements generally possess high permeabilities with low hysteretic losses. They are widely used in such components as audio transformers, coils, relays, magnetic amplifiers, high-frequency coils, magnetic shields, and dc electromagnets.

Ferrous alloys containing 50 to 90 Wt% Ni are also available in the form of oriented sheets. They have high permeabilities and low H_c. Some of these alloys are available with rectangular hysteresis loops [Fig. 10-31(b)]. Other Fe–Co–Ni alloys with compositions within the ranges of about 15 to 25 Wt% Fe, 15 to 70 Wt% Co, and 10 to 70 Wt% Ni have similar properties. These alloys are useful because the narrow

TABLE 10-13 TYPICAL PROPERTIES OF SOFT MAGNETIC ALLOYS

Trade Name	Nominal Composition (wt%)	Maximum Permeability	Remanence B_r (G)	Coercive Force H_c (Oe)
Thermenol	16 Al, 3.5 Mo, Fe	60,000	2,070	0.018
16 Alfenol	16 Al, Fe	80,000	4,000	0.044
Sinimax	43 Ni, 3 Si, Fe	50,000	5,500	0.06
Monimax	48 Ni, 3 Mo, Fe	60,000	8,900	0.06
Supermalloy	79 Ni, 5 Mo, Fe	300,000 min	4,000 min	0.006
Mumetal	77 Ni, 5 Cu, 1.5 Cr, Fe	100,000	2,300	0.30
Supermendur	49 Co, 2 V, Fe	70,000	21,400	0.23
—	35 Co, 1 Cr, Fe	10,000	11,000	0.63
Ingot Iron	—	5,000	7,700	1.00
—	0.5 Si–Fe	3,000	—	0.90
—	1.75 Si–Fe	5,000	—	0.80
—	3.0 Si–Fe	8,000	—	0.70
—	Oriented 3.0 Si–Fe	50,000	12,000	0.09
—	Oriented 50% Ni–Fe	150,000	14,500	0.09
—	50% Ni–Fe	100,000	9,000	0.05

Abstracted from "Magnetic, Electrical and Other Special-Purpose Materials," *Metals Handbook, Properties and Selection*, Vol. 1, 8th ed., T. Lyman, ed., American Society for Metals, Metals Park, Ohio, 1961, pp. 779–864.

TABLE 10-14 TYPICAL ALLOYS OF IRON AND NICKEL AND COBALT FOR SOFT-MAGNET APPLICATIONS

Nominal Composition (Wt%)	Maximum Permeability	Remanence, B_r (G)	Coercive Force, H_c (Oe)	Approximate Curie Temperature (°C)
45 Ni, Fe	30,000	8,000	0.20	440
47 to 50 Ni, Fe	50,000	8,000	0.07	535
50 Ni, Fe	70,000	8,000	0.05	535
Oriented 50 Ni, Fe	50,000–100,000	14,500	0.10–0.20	535
48 Ni, 3 Mo, Fe	35,000	—	0.10	—
79 Ni, 4 Mo, Fe	100,000	5,000	0.05	421
79 Ni, 5 Mo, Fe	800,000	5,000	0.005	—
77 Ni, 5 Cu, 1.5 Cr, Fe	100,000	3,000	0.05	—
50 Co, Fe	5,000	14,000	2.00	—
49 Co, 2 V, Fe	4,500	14,000	2.00	526
35 Co, 1–2 others, Fe	10,000	13,000	1.00	—

Abstracted from "Magnetic, Electrical and Other Special-Purpose Materials," *Metals Handbook, Properties and Selection*, Vol. 1, 8th ed., T. Lyman, ed., American Society for Metals, Metals Park, Ohio, 1961, pp. 779–864.

rectangular loops provide nearly constant, high permeabilities. Thus, virtually constant, high values of B are obtained for small values of H that are maintained over wide ranges of H. These alloys are used in such applications as audio transformers, coils, relays, amplifier coils, magnetic shields, high-flux-density motors, and transformers and electromagnets.

10.5.4 Magnetically Soft Ferrites

The theory, compositions, and structures of ferrites are described in Sections 10.4 and 10.4.1. Many of these materials have relatively low values of B_s and T_c. Their very low coercive forces consequently result in very narrow hysteresis loops. They are thus very useful at high frequencies because their electrical resistivities are at least of the order of 10^6 greater than those of the high-permeability metallic alloys. This combination of properties minimizes eddy-current losses and eliminates the necessity for laminations. The hysteresis losses, which increase as a function of frequency, are much smaller for the ferrites than for the soft magnetic alloys. In addition, ferrites have high permeabilities, so the volume of a suitable ferrite required for a given magnetic effect will be considerably smaller than that required of a soft magnetic alloy. On the other hand, the saturation magnetizations of ferrites are smaller than those of the soft magnetic alloys. The properties of ferrites have revolutionized high- and ultrahigh-frequency electronics and high-power, high-field uses such as power transformers, generators, and motors.

Some ferrites are available that have rectangular hysteresis loops [Fig. 10-31(b)]. These have high saturation values that are induced by relatively small fields. While undergoing magnetization, a magnetization front exists such that the material is magnetized in opposite directions on either side of the front. The material is magnetized in one direction at saturation magnetization. These are widely used as computer memory elements.

Still other ferrites are used as components of electronic filters, microwave devices, and magnetic switches. The application of ferrites in memory and switching devices is made in the form of very thin films (about 10^{-5} cm) and tapes. This is done to prevent the presence of more than one domain across the thickness of the film. The reversal of magnetizations must thus result from domain rotation. This eliminates domain wall motion and minimizes the time for imprinting bits of information on the film.

As would be expected from their bonding and structures, ferrites are hard and brittle. Most bulk ferrites are compacted and sintered in, or very close to, their final configurations because of the difficulty of machining these ceramiclike materials. Any machining must be done by grinding.

The powder techniques that are used to fabricate these materials make the presence of voids unavoidable in these pressed and sintered materials. Porosities in the range of about 5% to 25% are to be expected. This varies with the powder particle sizes, compacting pressure, and the time and temperature of sintering. Variations in these parameters can affect the magnetic properties of the ferrites. Voids have undesirable effects because they impede domain wall motion by acting as pins. Increasing the final grain size by increasing sintering temperatures and times diminishes this effect; porosity at grain boundaries offers less resistance to domain boundary motion than do voids within the grains.

The compositions of most commercial ferrites are usually considered to be proprietary by their producers. These are mixed ferrites that are sold under various trade names. They are associated with given sets of magnetic properties.

Two general types of ferrites are commercially available. The $Ni_{0.3}Zn_{0.7}O \cdot Fe_2O_3$ ($H_c < 0.1$ Oe, $B_s \sim 320$ G, $\mu_0 \sim 100$) are used for VHF applications. Here losses are minimized when domain wall motion is suppressed. Accordingly, the grain sizes of the sintered materials are kept as small as possible and the porosity is kept high. The $Mn_{0.5}Zn_{0.5}O \cdot Fe_2O_3$ ferrites ($H_c \sim 0.1$ Oe, $B_s \sim 200$ G, $\mu_0 \sim 2000$) are used at lower frequencies. Here μ_0 is the initial permeabilities or the initial slope of the curve of B versus H.

Each of these ferrites contains Zn, an element that has the effect of decreasing the exchange forces between the sublattices (Section 10.4). This decreases the spontaneous magnetization and T_{FN}. In addition, the range of frequencies over which μ_0 remains essentially constant as a function of frequency (at its dc value) is also decreased as a result of decreased superexchange. This comes about as a result of the electron spin resonance that is induced by oscillating fields, as described at the end of Section 10.1.3.3. It is a result of the resonant response of the net magnetic moment of an ion, most frequently that arising from spin, with the frequency of the applied field.

The frequency, given by Eq. (10-108), is a direct function of the applied field. The magnitude of this field depends directly on the strength of the superexchange forces that are responsible for the easy direction of magnetization within each domain. Sufficiently high external fields eliminate the domains, and the spins then precess about a direction parallel to the direction of the field. Thus, the applied field must also be large enough to overcome the crystalline magnetic anisotropy. The initial dc frequency response of μ_0 of a ferrite thus remains unchanged up to frequencies of field oscillations that are less than that of ω_L (the resonant frequency); the electrons precess parallel to the applied field, out of their original axes, and energy is absorbed [Eq. (10-106]. This results in large decreases in magnetization, B, and consequently greatly reduces μ_0. These resonant frequencies range from ~ 10 to $\sim 100 \, MHz$, depending on the composition of the ferrite. The frequency range over which μ_0 is constant means that B will be constant, so these ferrites are desirable materials for applications requiring magnetic cores with constant values of B.

A major application of garnets is given in Section 10.2.5.1.

10.6 BIBLIOGRAPHY

ANDERSON, J. C., *Magnetism and Magnetic Materials*, Chapman and Hall, New York, 1968.

BATES, L. F., *Modern Magnetism*, Cambridge University Press, New York, 1961.

BERKOWITZ, A. E., and KNELLES, E., *Magnetism and Metallurgy*, Vol. 1, Academic Press, New York, 1969.

BOZORTH, R. M., *Ferromagnetism*, Van Nostrand Reinhold, New York, 1951.

CHIKAZUMI, S., *Physics of Paramagnetism*, Wiley, New York, 1966.

CRANGLE, J., *Magnetic Properties of Solids*, Edward Arnold, London, 1977.

CULLITY, B. D., *Introduction to Magnetic Materials*, Addison-Wesley, Reading, Mass., 1972.

DEKKER, A. J., *Solid State Physics*, Prentice-Hall, New York, 1957.

KITTEL, C., *Introduction to Solid State Physics*, Wiley, New York, 1976.

MARTIN, D. H., *Magnetism in Solids*, London Iliffe Books, London, 1967.

MORRISH, A. H., *Physical Principles of Magnetism*, Wiley, New York, 1965.

STONER, E. C., *Magnetism*, Methuen, London, 1936.

10.7 PROBLEMS

10.1. Convert the mass susceptibilities of Al, Cr, Zn, and Ag to atomic susceptibilities.

10.2. Why must filled electron shells be considered to have zero net spin and magnetic moment?

10.3. Why can a single electron in an otherwise empty band be considered to constitute a resistanceless current?

10.4. Use Eq. (10-38) to approximate the dielectric susceptibility of a BCC metal with a lattice parameter of $\sim 3 \, \text{Å}$ and $\langle r \rangle \sim 10^{-8} \, cm$ at $300 \, K$ and an atomic number of 11. Compare the results with the data in Table 10-3 and explain any differences.

10.5. Compare the results obtained from Eq. (10-151) for $E_F \cong 3.1$ eV with those obtained in Problem 10.4.

10.6. Estimate the molecular field constant for iron using $T_c = 770°C$ and $M_0 \sim 1.71$ emu·g^{-1}.

10.7. Approximate the Brillouin parameter for iron for the case in which the magnetization is 222 and 218 emu·g^{-1} at zero and 300 K, respectively. Explain the meaning of this result.

10.8. Show that the Brillouin function approaches the classical Langevin function at the appropriate limit.

10.9. Why does $J = \infty$ (Fig. 9-2) represent the classical case?

10.10. Show that the slope of the Brillouin function is $\frac{1}{3}$ for $x \to 0$.

10.11. Given that the lattice parameter and density of iron are approximately 2.87 Å and 7.9 g·cm^{-3}, respectively, calculate (a) its maximum magnetization in units of emu·cm^{-3} and (b) in units of emu·g^{-1}. (c) If the experimental values are about 1720 emu·cm^3 and 220 emu·g^{-1}, determine the experimental value of the gyromagnetic ratio.

10.12. Why should $T_\infty > T_P > T_c$?

10.13. Why is the exchange energy negative for ferromagnetic materials?

10.14. Evaluate the exchange integral for a cobalt atom if $T_c = 1131°C$.

10.15. If the relative magnetization of iron is 0.98 at 300 K, estimate the value of the empirical Bloch coefficient.

10.16. Explain the shape of the heat capacity curve of a ferromagnetic material.

10.17. Derive an expression for the change in entropy of a ferromagnetic material for $T \ll T_c$ in terms of the fraction of unbalanced spin.

10.18. Explain the necessity for the original postulation of the presence of domains in ferromagnetic materials.

10.19. Why do domains oriented favorably with respect to an external field absorb those less favorably oriented?

10.20. Why is the magnetic energy of a crystal consisting of a single domain greater than that of the same crystal consisting of many domains?

10.21. Why are the magnetic orientations of domains of closure in harder magnetization directions than those with which they are associated?

10.22. Explain the role of the Bloch wall in magnetization processes.

10.23. Why does the anisotropy energy change across a domain wall?

10.24. Explain why metallurgical microstructures affect domain properties.

10.25. Why do $B–H$ curves reflect domain properties?

10.26. (a) Calculate the values of H_m and λ for iron at 0 K using the data in Table 10-8.
(b) Verify the calculation for λ using Eq. (10-180a).

10.27. Explain the relationship of ionic spin imbalance to the ionic magnetization of ferromagnetic elements.

10.28. Explain the basis of the effects of alloying elements on the magnetic properties of ferromagnetic elements in terms of their band structures.

10.29. Give a possible reason for the importance of the Bethe–Slater ratios in explaining the magnetic properties of Heusler-type alloys.

10.30. Compare the basic mechanisms of antiferromagnetic materials with those responsible for ferromagnetism.

10.31. Explain the difference between the antiferromagnetic parameters θ and T_N.

10.32. Contrast the electronic structures of ferromagnetic rare earth elements with those of the ferromagnetic transition elements.

10.33. What is the utility of considering the lattices of antiferromagnetic materials to be composed of two or more sublattices?

10.34. Compare the parameter θ of antiferromagnetic materials with that of C/χ_0 for ferrimagnetic materials.

10.35. What approach can be used in the design of antiferromagnetic materials with tailored magnetizations as functions of temperature?

10.36. Contrast the desirable properties of magnetically soft and hard materials and explain their differences.

10.37. Explain the effect of the microstructural properties of magnetic precipitates on their magnetic properties.

10.38. Discuss some of the advantages and disadvantages of magnetic materials prepared by powder techniques.

10.39. Explain why the shapes of magnets made of ferrites must be different from those made of Alnico alloys.

10.40. Discuss the advantages and disadvantages of Fe–Si alloys used for soft magnets.

APPENDIX

SOME USEFUL CONSTANTS AND CONVERSIONS

Quantity	Symbol	cgs Units	mks (SI) Units
Energy			
1 calorie	cal	4.19×10^7 ergs	1 erg $= 2.39 \times 10^{-8}$ cal
1 erg	erg	10^{-7} J	1 J $= 10^7$ erg
1 electron volt	eV	1.602×10^{-12} erg	1 erg $= 6.25 \times 10^{11}$ eV
		23.05 kcal $\cdot$ mol^{-1}	5.50 kJ $\cdot$ mol^{-1}
Gases			
Avogadro's number	N_A	6.02×10^{23} mol^{-1}	6.02×10^{23} mol^{-1}
Gas constant	R	1.987 cal(mol $\cdot$ K)$^{-1}$	8.31 J (mol $\cdot$ K)$^{-1}$
Boltzmann's constant	k_B	1.38×10^{-16} erg $\cdot$ K^{-1}	1.38×10^{-23} J $\cdot$ K^{-1}
Atomic			
Planck's constant	h	6.63×10^{-27} erg $\cdot$ s	6.63×10^{-34} J $\cdot$ s
		4.14×10^{-15} eV $\cdot$ s	
Rydberg	Ry	13.60 eV	1.097×10^7 m
Bohr radius	r_0	0.529×10^{-8} cm	0.529×10^{-10} m
Electron charge	e	4.80×10^{-10} esu	1.60×10^{-19} C

SOME USEFUL CONSTANTS AND CONVERSIONS

Quantity	Symbol	cgs Units	mks (SI) Units
Temperature			
Practical scale	t	0°C	273.15 K
Thermodynamic scale	T	$+273.15$°C	0 K
Electrical			
Resistivity	ρ	Ω-cm	Ω-m
Conductivity	σ	$(\Omega\text{-cm})^{-1}$	$(\Omega\text{-m})^{-1}$
Magnetic			
Magnetic field intensity	H	oersted (Oe)	$79.6\ \mathrm{A}\cdot\mathrm{m}^{-1}$
Magnetization intensity	M	emu cm^{-3}	$12.57\ 10^{-4}\ \mathrm{Wb}\cdot\mathrm{m}^{-2}$
Magnetic induction intensity	B	gauss (G)	$10^{4}\ \mathrm{Wb}\cdot\mathrm{m}^{-2}$
			$1\ \mathrm{Wb}\cdot\mathrm{m}^{-2} = 10^{4}\ \mathrm{G}$
			$1\ \mathrm{A}\cdot\mathrm{m}^{-1} = 12.57\ \mathrm{Oe}$
Bohr magneton	μ_B	$9.27 \times 10^{-21}\ \mathrm{erg}\cdot\mathrm{Oe}^{-1}$	
General			
Angstrom	Å	10^{-8} cm	10^{-10} m
Micrometer	μm	10^{-4} cm	10^{-6} m
Velocity of light (in vacuum)	c	$2.998 \times 10^{10}\ \mathrm{cm}\cdot\mathrm{s}^{-1}$	$2.998 \times 10^{8}\ \mathrm{m}\cdot\mathrm{s}^{-1}$
Dielectric constant of vacuum	ε_0	1 (dimensionless)	$8.85 \times 10^{-12}\ \mathrm{F}\cdot\mathrm{m}^{-1}$
Wavelength associated with 1 eV	—	1.24×10^{-4} cm	1.24×10^{-6} m
Frequency associated with 1 eV	—	$2.42 \times 10^{14}\ \mathrm{s}^{-1}$	$2.42 \times 10^{14}\ \mathrm{s}^{-1}$
Boltzmann's constant per electron charge	k_B/e	$8.63 \times 10^{-5}\ \mathrm{V\ K}^{-1}$	

INDEX

627